AF412827

HANDBOOK
OF DIFFERENTIAL EQUATIONS

ORDINARY DIFFERENTIAL EQUATIONS

VOLUME II

HANDBOOK OF DIFFERENTIAL EQUATIONS

ORDINARY DIFFERENTIAL EQUATIONS

VOLUME II

Edited by

A. CAÑADA

Department of Mathematical Analysis, Faculty of Sciences,
University of Granada, Granada, Spain

P. DRÁBEK

Department of Mathematics, Faculty of Applied Sciences,
University of West Bohemia, Pilsen, Czech Republic

A. FONDA

Department of Mathematical Sciences, Faculty of Sciences,
University of Trieste, Trieste, Italy

2005

ELSEVIER

NORTH
HOLLAND

Amsterdam • Boston • Heidelberg • London • New York • Oxford •
Paris • San Diego • San Francisco • Singapore • Sydney • Tokyo

ELSEVIER B.V.
Radarweg 29
P.O. Box 211, 1000 AE Amsterdam
The Netherlands

ELSEVIER Inc.
525 B Street, Suite 1900
San Diego, CA 92101-4495
USA

ELSEVIER Ltd
The Boulevard, Langford Lane
Kidlington, Oxford OX5 1GB
UK

ELSEVIER Ltd
84 Theobalds Road
London WC1X 8RR
UK

First edition 2005

Library of Congress Cataloging in Publication Data: A catalog record is available from the Library of Congress.

British Library Cataloguing in Publication Data: A catalogue record is available from the British Library.

ISBN 0 444 52027 9
Set ISBN: 0 444 51742 1

♾ The paper used in this publication meets the requirements of ANSI/NISO Z39.48-1992 (Permanence of Paper).
Printed in The Netherlands.

Preface

This handbook is the second volume in a series devoted to self contained and up-to-date surveys in the theory of ordinary differential equations, written by leading researchers in the area. All contributors have made an additional effort to achieve readability for mathematicians and scientists from other related fields, in order to make the chapters of the volume accessible to a wide audience. These ideas faithfully reflect the spirit of this multivolume and the editors hope that it will become very useful for research, learning and teaching. We express our deepest gratitude to all contributors to this volume for their clearly written and elegant articles.

This volume consists of six chapters covering a variety of problems in ordinary differential equations. Both, pure mathematical research and real word applications are reflected pretty well by the contributions to this volume. They are presented in alphabetical order according to the name of the first author. The paper by Barbu and Lefter is dedicated to the discussion of the first order necessary and sufficient conditions of optimality in control problems governed by ordinary differential systems. The authors provide a complete analysis of the Pontriaghin maximum principle and dynamic programming equation. The paper by Bartsch and Szulkin is a survey on the most recent advances in the search of periodic and homoclinic solutions for Hamiltonian systems by the use of variational methods. After developing some basic principles of critical point theory, the authors consider a variety of situations where periodic solutions appear, and they show how to detect homoclinic solutions, including the so-called "multibump" solutions, as well. The contribution of Cârjă and Vrabie deals with differential equations on closed sets. After some preliminaries on Brezis–Browder ordering principle and Clarke's tangent cone, the authors concentrate on problems of viability and problems of invariance. Moreover, the case of Carathéodory solutions and differential inclusions are considered. The paper by Hirsch and Smith is dedicated to the theory of monotone dynamical systems which occur in many biological, chemical, physical and economic models. The authors give a unified presentation and a broad range of the applicability of this theory like differential equations with delay, second order quasilinear parabolic problems, etc. The paper by López-Gómez analyzes the dynamics of the positive solutions of a general class of planar periodic systems, including those of Lotka–Volterra type and a more general class of models simulating symbiotic interactions within global competitive environments. The mathematical analysis is focused on the study of coexistence states and the problem of ascertaining the structure, multiplicity and stability of these coexistence states in purely symbiotic and competitive environments. Finally, the paper by Ntouyas is a survey on nonlocal initial and boundary value problems. Here, some old and new results are established and the author shows how the nonlocal initial or boundary conditions generalize the classical ones, having many applications in physics and other areas of applied mathematics.

We thank again the Editors at Elsevier for efficient collaboration.

List of Contributors

Barbu, V., *"Al.I. Cuza" University, Iaşi, Romania, and "Octav Mayer" Institute of Mathematics, Romanian Academy, Iaşi, Romania* (Ch. 1)

Bartsch, T., *Universität Giessen, Giessen, Germany* (Ch. 2)

Cârjă, O., *"Al. I. Cuza" University, Iaşi, Romania* (Ch. 3)

Hirsch, M.W., *University of California, Berkeley, CA* (Ch. 4)

Lefter, C., *"Al.I. Cuza" University, Iaşi, Romania, and "Octav Mayer" Institute of Mathematics, Romanian Academy, Iaşi, Romania* (Ch. 1)

López-Gómez, J., *Universidad Complutense de Madrid, Madrid, Spain* (Ch. 5)

Ntouyas, S.K., *University of Ioannina, Ioannina, Greece* (Ch. 6)

Smith, H., *Arizona State University, Tempe, AZ* (Ch. 4)

Szulkin, A., *Stockholm University, Stockholm, Sweden* (Ch. 2)

Vrabie, I.I., *"Al. I. Cuza" University, Iaşi, Romania, and "Octav Mayer" Institute of Mathematics, Romanian Academy, Iaşi, Romania* (Ch. 3)

Contents

Preface — v
List of Contributors — vii
Contents of Volume 1 — xi

1. Optimal control of ordinary differential equations — 1
   V. Barbu and C. Lefter
2. Hamiltonian systems: periodic and homoclinic solutions by variational methods — 77
   T. Bartsch and A. Szulkin
3. Differential equations on closed sets — 147
   O. Cârjă and I.I. Vrabie
4. Monotone dynamical systems — 239
   M.W. Hirsch and H. Smith
5. Planar periodic systems of population dynamics — 359
   J. López-Gómez
6. Nonlocal initial and boundary value problems: a survey — 461
   S.K. Ntouyas

Author index — 559
Subject index — 565

Contents of Volume 1

Preface   v
List of Contributors   vii

1. A survey of recent results for initial and boundary value problems singular in the
    dependent variable   1
    R.P. Agarwal and D. O'Regan
2. The lower and upper solutions method for boundary value problems   69
    C. De Coster and P. Habets
3. Half-linear differential equations   161
    O. Došlý
4. Radial solutions of quasilinear elliptic differential equations   359
    J. Jacobsen and K. Schmitt
5. Integrability of polynomial differential systems   437
    J. Llibre
6. Global results for the forced pendulum equation   533
    J. Mawhin
7. Ważewski method and Conley index   591
    R. Srzednicki

Author Index   685
Subject Index   693

CHAPTER 1

Optimal Control of Ordinary Differential Equations

Viorel Barbu and Cătălin Lefter

University "Al.I. Cuza", Iaşi, Romania, and
Institute of Mathematics "Octav Mayer", Romanian Academy, Romania

Contents

1. Introduction 3
   1.1. The calculus of variations 4
   1.2. General form of optimal control problems 9
2. Preliminaries 10
   2.1. Elements of convex analysis 10
   2.2. Ekeland's variational principle 15
   2.3. Elements of differential geometry and exponential representation of flows 16
3. The Pontriaghin maximum principle 29
   3.1. The main theorem 29
   3.2. Proof of the maximum principle 32
   3.3. Convex optimal control problems 38
   3.4. Examples 47
   3.5. Reachable sets and optimal control problems 55
   3.6. Geometric form of Pontriaghin maximum principle 57
   3.7. Free time optimal control problems 59
4. The dynamic programming equation 62
   4.1. Optimal feedback controllers and smooth solutions to Hamilton–Jacobi equation 62
   4.2. Linear quadratic control problems 65
   4.3. Viscosity solutions 68
   4.4. On the relation between the two approaches in optimal control theory 72
References 74

HANDBOOK OF DIFFERENTIAL EQUATIONS
Ordinary Differential Equations, volume 2
Edited by A. Cañada, P. Drábek and A. Fonda

1

1. Introduction

The theory of control of differential equations has developed in several directions in close relation with the practical applications of the theory. Its evolution has shown that its methods and tools are drawn from a large spectrum of mathematical branches such as ordinary differential equations, real analysis, calculus of variations, mechanics, geometry. Without being exhaustive we just mention, as subbranches of the control theory, the controllability, the stabilizability, the observability, the optimization of differential systems and of stochastic equations or optimal control. For an introduction to these fields, and not only, see [22,33,38], as well as [2,25] and [26] for a geometric point of view.

The purpose of this work is to discuss the first order necessary and sufficient conditions of optimality in control problems governed by ordinary differential systems. We do not treat the optimal control of partial differential equations although all basic questions of the finite dimensional theory (existence of optimal control, maximum principle, dynamic programming) remain valid but the treatment requires more sophisticated methods because of the infinite dimensional nature of the problems (see [4,27,38]).

In Section 1 we present some aspects and ideas in the classical Calculus of variations that lead later, in the fifties, to the modern theory of optimal control for differential equations.

Section 2 presents some preliminary material. It contains elements of convex analysis and the generalized differential calculus for locally Lipschitz functionals, introduced by F.H. Clarke [10]. This will be needed for the proof of the maximum principle of Pontriaghin, under general hypotheses, in Section 3.1. We then discuss the exponential representation of flows, introduced by A. Agrachev and R. Gamkrelidze in order to give a geometric formulation to the maximum principle that we will describe in Sections 3.5, 3.6.

Section 3 is concerned with the Pontriaghin maximum principle for general Bolza problems. There are several proofs of this famous classical result and here, following F.H. Clarke's ideas (see [11]), we have adapted the simplest one relying on Ekeland's variational principle. Though the maximum principle given here is not in its most general form, it is however sufficiently general to cover most of significant applications. Some examples are treated in detail in Section 3.4. Since geometric control theory became in last years an important branch of mathematics (for an introduction to the theory see [2,26]), it is useful and interesting to give a geometric formulation of optimal control problems and, consequently, a geometric form of the maximum principle. Free time optimal problems are also considered as a special case.

In the last section we present the dynamic programming method in optimal control problems based on the partial differential equation of dynamic programming, or Bellman equation (see [7]). The central result of this chapter says that the value function is a viscosity solution to Bellman equation and that, if a classical solution exists, then an optimal control, in feedback form, is obtained. Applications to linear quadratic problems are given. We discuss also the relationship between the maximum principle and the Bellman equation and we will see in fact that the dynamic programming equation is the Hamilton–Jacobi equation for the Hamiltonian system given by the maximum principle.

1.1. *The calculus of variations*

In this section we point out the fundamental lines of development in the Calculus of variations. We will not impose rigorous assumptions on the functions entering the described problems, they will be as regular as needed. The main purpose is just to emphasize some fundamental ideas that will be reencountered, in a metamorphosed form, in the theory of optimal control for differential equations. For a rigorous presentation of the theory a large literature may be cited, however we restrict for instance to [8,24] and to a very nice survey of extremal problems in mathematics, including the problems of Calculus of variations, in [34].

Let M be an n-dimensional manifold, and M_0, M_1 be subsets (usually submanifolds) of M. $L : \mathbb{R} \times TM \to \mathbb{R}$ is the Lagrangean function, TM being the tangent bundle of M. The generic problem of the classical Calculus of variations consists in finding a curve, y^*, which minimizes a certain integral

$$J(y) = \int_{t_0}^{t_1} L\big(t, y(t), y'(t)\big)\, dt \tag{1.1}$$

in the space of curves

$$Y = \big\{ y : [t_0, t_1] \to M; \; y(t_j) \in M_j, \; j = 1, 2, \; y \text{ continuous and piecewise } C^1 \big\}.$$

The motivation for studying such problems comes from both geometry and classical mechanics.

EXAMPLES. *1. The brachistocrone.* The classical brachistocrone problem proposed by Johann Bernoulli in 1682, asks to find the curve, in a vertical plane, on which a material point, moving without friction under the action of its weight, is reaching the lower end of the curve in minimum time. More precisely, if the curve is joining two points $y(t_0) = y_0$, $y(t_1) = y_1$, then the time necessary for the material point to reach y_1 from y_0 is

$$T = \int_{t_0}^{t_1} \big(2g|y(t) - y_0|\big)^{-1/2} \sqrt{1 + \big(y'(x)\big)^2}\, dt.$$

The curve with this property is a cycloid.

2. The minimal surface of revolution. One is searching for the curve $y : [t_0, t_1] \to \mathbb{R}$, $y(t_0) = y_0$, $y(t_1) = y_1$, which generates the surface of revolution of least area. The functional to be minimized is

$$J(y) = 2\pi \int_{t_0}^{t_1} y(x) \sqrt{1 + \big(y'(t)\big)^2}\, dt.$$

The solution is the catenary.

3. Lagrangean mechanics. A mechanical system with a finite number of degrees of freedom is mathematically modelled by a manifold M and a Lagrangean function

$L : \mathbb{R} \times TM \to \mathbb{R}$ (see [3]). The manifold M is the configuration space of the mechanical system. The points $y \in M$ are generalized coordinates and the $y' \in TM$ are generalized speeds. The principle of least action of Maupertuis–d'Alembert–Lagrange states that the trajectories of the mechanical system are *extremal* for the functional J defined in (1.1). Consider the case of a system of N material points in the 3 dimensional space, moving under the action of mutual attraction forces. In this case the configuration space is $(\mathbb{R}^3)^N$, while the Lagrangean is

$$L = T - U \tag{1.2}$$

where T is the kinetic energy

$$T = \sum_{i=1}^{N} \frac{1}{2} m_i |x_i'|^2$$

and U is the potential energy

$$U(x_1, \ldots, x_N) = \sum_{i=1}^{N} \frac{k m_i m_j}{|x_i - x_j|},$$

k is an universal constant.

To make things clear we consider the simplest problem in the Calculus of variations when $M = \mathbb{R}^n$ and y_0, y_1 are fixed.

We consider the space of variations $\mathcal{Y} = \{ h : [t_0, t_1] \to \mathbb{R}^n ; \ h(t_0) = h(t_1) = 0, h \in C^1 \}$; if y^* is a minimum of J in Y, then the first variation

$$\delta J(y^*)h := \frac{\mathrm{d}}{\mathrm{d}s} J(y^* + sh) \bigg|_{s=0} = 0. \tag{1.3}$$

A curve that satisfies (1.3) is called *extremal* and this is only a *necessary condition* for a curve to realize the infimum of J. One easily computes

$$\delta J(y)h = \int_{t_0}^{t_1} L_y\big(t, y^*(t), (y^*)'(t)\big) \cdot h + L_{y'}\big(t, y^*(t), (y^*)'(t)\big) \cdot h' \, \mathrm{d}t$$

where L_y, $L_{y'}$ are the gradients of L with respect to y and y', respectively. If y^* is C^2, an integration by parts in the previous formula gives

$$L_y\big(t, y^*(t), (y^*)'(t)\big) - \frac{\mathrm{d}}{\mathrm{d}t} L_{y'}\big(t, y^*(t), (y^*)'(t)\big) = 0 \tag{1.4}$$

which are the *Euler–Lagrange* equations. It is a system of n differential equations of second order.

It may be proved that if the Hessian matrix $(L_{y'y'}) > 0$, then the regularity of L is inherited by the extremals, for instance if $L \in C^2$ then the extremals are C^2 and thus satisfy the Euler–Lagrange system. The proof of this fact is based on the first of the *Weierstrass–Erdmann necessary conditions* which state that, along each extremal, $L_{y'}$ and the Hamiltonian defined below in (1.6) are continuous.

Another necessary condition for the extremal y^* to realize the infimum of J is that $(L_{y'y'}) \geqslant 0$ along y^*. This is the *Legendre necessary condition*.

Suppose from now on that $(L_{y'y'})$ is a nondegenerate matrix at any point (t, y, y'). We set

$$p = L_{y'}(t, y, y'). \tag{1.5}$$

Since $(L_{y'y'})$ is nondegenerate, formula (1.5) defines a change of coordinates $(t, y, y') \to (t, y, p)$. From the geometric point of view it maps TM locally onto T^*M, the cotangent bundle. In mechanics p is called the generalized momentum of the system and in most applications its significance is of adjoint (or dual) variable. We consider the *Hamiltonian*

$$H(t, y, p) = (p, y') - L(t, y, y'). \tag{1.6}$$

If, moreover, L is convex in y' then $H = L^*$, the *Legendre transform* of L:

$$H(t, y, p) = \sup_{y'}\{(p, y') - L(t, y, y')\}.$$

For example, if L is given by (1.2) then $H = T + U$ and it is just the total energy of the system. If we compute the differential of H along an extremal, taking into account the Euler–Lagrange equations, we obtain

$$dH = -L_t\, dt - \frac{d}{dt}L_{y'}\, dy + y'\, dp.$$

Thus, through these transformations we obtain the *Hamiltonian equations*

$$\begin{cases} y' = \dfrac{\partial H}{\partial p}(t, y, p), \\[2mm] p' = -\dfrac{\partial H}{\partial y}(t, y, p). \end{cases} \tag{1.7}$$

Solutions of the Hamiltonian system are in fact extremals corresponding to the Lagrangean $\widetilde{L}(t, (y, p), (y', p')) = p \cdot y' - H(t, y, p)$ in T^*M. The projections on M are extremals for J. Roughly speaking, solving the Euler–Lagrange system is equivalent to solving the Hamiltonian system of $2n$ differential equations of first order. From the mechanics point of view these transforms give rise to the Hamiltonian mechanics which study the mechanical phenomena in the phase space T^*M while in mathematics this is the start point for the symplectic geometry (see for example [3,28]).

Consider now the more general case of end points lying on two submanifolds M_0, M_1. It may be shown that the first variation of J in y computed in an admissible variation h (assume that also t_0, t_1 are free) is

$$\delta J(y)h = \int_{t_0}^{t_1} L_y(t, y, y') \cdot h + L_{y'}(t, y, y') \cdot h' \, dt + \{p\delta y - H\delta t\}\big|_{t_0}^{t_1} \tag{1.8}$$

where

$$H\delta t\big|_{t_0}^{t_1} = H\big(t_1, y(t_1), p(t_1)\big)\delta t_1 - H\big(t_0, y(t_0), p(t_0)\big)\delta t_0,$$

$$p\delta y\big|_{t_0}^{t_1} = p(t_1)\delta y_1 - p(t_2)\delta y_0.$$

It turns out that $y^* \in C^2$ is extremal for J if y^* satisfies the Euler–Lagrange equations (1.4) and in addition

$$\{p\delta y - H\delta t\}\big|_{t_0}^{t_1} = 0. \tag{1.9}$$

These are *transversality conditions*. In case t_0, t_1 are fixed, these become

$$p(t_0) \perp M_0, \qquad p(t_1) \perp M_1.$$

Since $(L_{y'y'})$ is supposed to be nondegenerate, the Euler–Lagrange equations form a second order nondegenerate system of equations and this implies that the family of extremals starting at moment t_0 from a given point of $y_0 \in M$ cover a whole neighborhood V of (t_0, y_0) (we just vary the value of $y'(t_0)$ in the associated Cauchy problem and use some result on the differentiability of the solution with respect to the initial data, coupled with the inverse function theorem). We consider now the function $S : V \to \mathbb{R}$ defined by

$$S(t, y) = \int_{t_0}^{t} L\big(s, x(s), x'(s)\big) \, ds$$

where the integral is computed along the extremal $x(s)$ joining the points (t_0, y_0) and (t, y). It may be proved that S satisfies the first order nonlinear partial differential equation

$$S_t + H(t, y, S_y) = 0. \tag{1.10}$$

This is the *Hamilton–Jacobi* equation. This is strongly related to the Hamiltonian system (1.7) which is the system of characteristics associated to the partial differential equation (1.10) (see [16]).

A partial differential equation is usually a more complicated mathematical object than an ordinary differential system. Solving a first order partial differential system reduces to solving the corresponding characteristic system. This is the method of characteristics (see [16]).

However, this duality may be successfully used in a series of concrete situations to integrate the Hamiltonian systems appearing in mechanics or in the calculus of variations. This result, belonging to Hamilton and Jacobi, states that if a general solution for the Hamilton–Jacobi equation (1.10) is known, then the Hamiltonian system may be integrated (see [3,16,24]). More precisely, we assume that a general solution of (1.10) is $S = S(t, y_1, \ldots, y_n, \alpha_1, \ldots, \alpha_n)$ such that the matrix $\left(\frac{\partial^2 S}{\partial y_i \partial \alpha_j}\right)$ is nondegenerated. Then $\frac{\partial S}{\partial \alpha_j}$ are prime integrals and a general solution of the Hamiltonian system (1.7) is given by the $2n$ system of implicit equations:

$$\beta_i = \frac{\partial S}{\partial \alpha_i}, \qquad p_i = \frac{\partial S}{\partial y_i}.$$

In fact S is a generating function for the symplectic transform $(y_i, p_i) \to (\beta_i, \alpha_i)$ and in the new coordinates the system (1.7) has a simple form for which the Hamiltonian function is $\equiv 0$. A last remark is that a general solution to equation (1.10) may be found if variables of H are separated (see [3,24]).

We considered previously first order necessary conditions. Suppose that $(L_{y'y'}) > 0$. Let us take now the second variation

$$\delta^2 J(y)h := \frac{d^2}{ds^2} J(y + sh)\Big|_{s=0}.$$

This is a quadratic form denoted by

$$Q^y(h) = \int_{t_0}^{t_1} \Omega^y(t, h, h') \, dt$$

where the new Lagrangean

$$\Omega^y(t, h, h') = \big(L_{yy}(t, y, y')h, h\big) + 2\big(L_{yy'}(t, y, y')h, h'\big) \\ + \big(L_{y'y'}(t, y, y')h', h'\big).$$

Here $(\cdot, \cdot)$ denotes the scalar product in $\mathbb{R}^n$ and we assumed that the matrix $(L_{yy'})$ is symmetric (for $n = 1$ this is trivial, in higher dimensions the hypothesis simplifies computations but may be omitted). Clearly, if y^* realizes a global minimum of J, then the quadratic form $Q(y^*) \geqslant 0$. The positivity of Q is related to the notion of *conjugate point*. A point $\bar{t}$ is conjugate to t_0 along the extremal y^* if there exists a non trivial solution $h : [t_0, \bar{t}] \to \mathbb{R}^n$, $h(t_0) = h(\bar{t}) = 0$ of the second Euler equation:

$$\Omega_h^{y^*} - \frac{d}{dt}\Omega_{h'}^{y^*} = 0.$$

The *Jacobi necessary condition* states that if y^* realizes the infimum of J then the open interval (t_0, t_1) does not contain conjugate points to t_0. If y^* is just an extremal and the closed interval $[t_0, t_1]$ does not contain conjugate points to t_0, then y^* is a local weak minimum of J (in C^1 topology).

1.2. *General form of optimal control problems*

We consider the controlled differential equation

$$y'(t) = f\big(t, y(t), u(t)\big), \quad t \in [0, T]. \tag{1.11}$$

The input function $u : [0, T] \to \mathbb{R}^m$ is called *controller* or *control* and $y : [0, T] \to \mathbb{R}^n$ is the *state of the system*. We will assume that $u \in \mathcal{U}$ where $\mathcal{U}$ is the set of measurable, locally integrable functions which satisfy the *control constraints*:

$$u(t) \in U(t) \quad \text{a.e. } t \in [0, T] \tag{1.12}$$

where $U(t) \subset \mathbb{R}^n$ are given closed subsets. The differential system (1.11) is called the *state system*. We also consider a Lagrangean L and the *cost functional*

$$J(y, u) = \int_0^T L\big(t, y(t), u(t)\big)\, dt + g\big(y(0), y(T)\big). \tag{1.13}$$

A pair (y, u) is said to be *admissible pair* if it satisfies (1.11), (1.12) and $J(y, u) < +\infty$. The *optimal control problem* we consider is

$$\min\big\{J(y, u);\, \big(y(0), y(T)\big) \in C, (y, u) \text{ verifies } (1.11)\big\} \tag{1.14}$$

Here $C \subset \mathbb{R}^n \times \mathbb{R}^n$ is a given closed set.

A controller u^* for which the minimum in (1.14) is attained is called *optimal controller*. The corresponding states y^* are called *optimal states* while (y^*, u^*) will be referred as *optimal pairs*. By solution to (1.11) we mean an absolutely continuous function $y : [0, T] \to \mathbb{R}$ (i.e., $y \in AC([0, T]; \mathbb{R}^n)$) which satisfies almost everywhere the system (1.11). In the special case $f(t, y, u) \equiv u$, problem (1.14) reduces to the classical problem of calculus of variations that was discussed in Section 1.1. For different sets C we obtain different types of control problems. For example, if C contains one element, that is the initial and final states are given, we obtain a *Lagrange problem*. If the initial state of the system is given and the final one is free, $C = \{y_0\} \times \mathbb{R}$, one obtains a *Bolza problem*. A Bolza problem with the Lagrangean $L \equiv 0$ becomes a *Mayer problem*.

An optimal controller u^* is said to be a *bang-bang controller* if $u^* \in \partial U(t)$ a.e. $t \in (0, T)$ where ∂U stands for the topological boundary of U.

It should be said that the control constraints (1.12) as well as end point constraints $(y(0), y(T)) \in C$ can be implicitly incorporated into the cost functional J by redefining L and g as

$$\tilde{L}(t, y, u) = \begin{cases} L(t, u) & \text{if } u \in U(t), \\ +\infty & \text{otherwise,} \end{cases}$$

$$\tilde{g}(y_1, y_2) = \begin{cases} g(y_1, y_2) & \text{if } (y_1, y_2) \in C, \\ +\infty & \text{otherwise.} \end{cases}$$

Moreover, integral (isoperimetric) constraints of the form

$$\int_0^T h_i\big(t, y(t), u(t)\big)\, dt \leqslant \alpha_i, \quad i = 1, \ldots, l,$$

$$\int_0^T h_i\big(t, y(t), u(t)\big)\, dt = \alpha_i, \quad i = l+1, \ldots, m$$

can be implicitly inserted into problem (1.14) by redefining new state variables $\{z_1 \ldots, z_m\}$ and extending the state system (1.11) to

$$\begin{cases} y'(t) = f\big(t, y(t), u(t)\big), & t \in (0, T), \\ z'(t) = h\big(t, y(t), u(t)\big), \\ z(0) = 0, \quad z_i(T) \leqslant \alpha_i \text{ for } i = 1, \ldots, l, \quad z_i(T) = \alpha_i \text{ for } i = l+1, \ldots, m \end{cases}$$

where $h = \{h_i\}_{i=1}^m$. For the new state variable $X = (y, z)$ we have the end point constraints

$$\big(X(0), X(T)\big) \in K,$$

where

$$K = \Big\{\big((y_0, 0, \ldots, 0), (y_1, z)\big) \in \mathbb{R}^{n+m} \times \mathbb{R}^{n+m}, \ (y_0, y_1) \in C,$$

$$z_i \leqslant \alpha_i, i = 1, \ldots, l, \ z_i = \alpha_i, i = l+1, \ldots, m\Big\}.$$

2. Preliminaries

2.1. *Elements of convex analysis*

Here we shall briefly recall some basic results pertaining convex analysis and generalized gradients we are going to use in the formulation and in proof of the maximum principle.

Let X be a real Banach space with the norm $\|\cdot\|$ and dual X^*. Denote by $(\cdot, \cdot)$ the pairing between X and X^*.

The function $f : X \to \overline{\mathbb{R}} =]-\infty, +\infty]$ is said to be *convex* if

$$f\big(\lambda x + (1 - \lambda)y\big) \leqslant \lambda f(x) + (1 - \lambda)f(y), \quad 0 \leqslant \lambda \leqslant 1, \ x, y \in X. \tag{2.1}$$

The set $D(f) = \{x \in X;\ f(x) < \infty\}$ is called the *effective domain* of f and

$$E(f) = \big\{(x, \lambda) \in X \times \mathbb{R};\ f(x) \leqslant \lambda\big\} \tag{2.2}$$

is called the *epigraph* of f. The function f is said to be *lower semicontinuous* (l.s.c.) if

$$\liminf_{x \to x_0} f(x) \geqslant f(x_0).$$

The function f is said to be *proper* if $f \not\equiv +\infty$.

It is easily seen that a convex function is l.s.c. if and only if it is weakly lower semicontinuous. Indeed, f is l.s.c. if and only if every level set $\{x \in X;\ f(x) \leqslant \lambda\}$ is closed. Moreover, the level sets are also convex, by the convexity of f; the conclusion follows by the coincidence of convex closed sets and weakly closed sets.

Note also that, by Weierstrass theorem, if X is a reflexive Banach space and if f is convex, l.s.c. and $\lim_{\|x\| \to \infty} f(x) = +\infty$, then f attains its infimum on X.

We note without proof (see, e.g., [9,6]) the following result:

PROPOSITION 2.1. *Let $f : X \to \overline{\mathbb{R}}$ be a l.s.c. convex function. Then f is bounded from below by an affine function and f is continuous on* int $D(f)$.

Given a l.s.c. convex function $f : X \to \overline{\mathbb{R}}$, the mapping $\partial f : X \to X^*$ defined by

$$\partial f(x) = \left\{ w \in X^*;\ f(x) \leqslant f(u) + (w, x - u),\ \forall u \in X \right\} \tag{2.3}$$

is called the *subdifferential* of f. An element of $\partial f(x)$ is called *subgradient* of f at x.

The mapping ∂f is generally multivalued. The set

$$D(\partial f) = \left\{ x;\ \partial f(x) \neq \phi \right\}$$

is the domain of ∂f. It is easily seen that x_0 is a minimum point for f on X if and only if $0 \in \partial f(x_0)$.

We note also, without proof, some fundamental properties of ∂f (see, e.g., [6,9,31]).

PROPOSITION 2.2. *Let $f : X \to \overline{\mathbb{R}}$ be convex and l.s.c. Then* int $D(f) \subset D(\partial f)$.

Let C be a closed convex set and let $\mathbb{I}_C(x)$ be the indicator function of C, i.e.,

$$\mathbb{I}_C(x) = \begin{cases} 0, & x \in C, \\ +\infty, & x \notin C. \end{cases}$$

Clearly, $\mathbb{I}_C(x)$ is convex and l.s.c. Moreover, we have $D(\partial \mathbb{I}_C(x)) = C$ and

$$\partial \mathbb{I}_C(x) = \left\{ w \in X^*;\ (w, x - u) \geqslant 0,\ \forall u \in C \right\}. \tag{2.4}$$

$\partial \mathbb{I}_C(x)$ is precisely the *normal cone* to C at x, denoted $N_C(x)$.

If $F : X \to Y$ is a given function, X, Y Banach spaces, we set

$$F'(x, y) = \lim_{\lambda \to 0} \frac{F(x + \lambda y) - F(x)}{\lambda}$$

called the *directional derivative* of F in direction y.

By definition F is Gâteaux differentiable in x if $\exists DF(x) \in L(X, Y)$ such that

$$F'(x, v) = DF(x)v, \quad \forall v \in X.$$

In this case, DF is the Gâteaux derivative (differential) at x.

If $f : X \to \mathbb{R}$ is convex and Gâteaux differentiable in x, then it is subdifferentiable at x and $\partial f(x) = \nabla f(x)$.

In general, we have

PROPOSITION 2.3. *Let $f : X \to \overline{\mathbb{R}}$ be convex, l.s.c. and proper. Then, for each $x_0 \in D(\partial f)$*

$$\partial f(x_0) = \left\{ w \in X^*; \ f'(x_0, u) \geqslant (w, u), \ \forall w \in X \right\}. \tag{2.5}$$

If f is continuous at x_0, then

$$f'(x_0, u) = \sup\{(w, u); \ w \in \partial f(x_0)\}, \quad \forall u \in X. \tag{2.6}$$

Given $f : X \to \overline{\mathbb{R}}$, the function $f^* : X^* \to \overline{\mathbb{R}}$

$$f^*(p) = \sup\{(p, x) - f(x); \ x \in X\}$$

is called the *conjugate* of f, or the *Legendre transform* of f.

PROPOSITION 2.4. *Let $f : X \to \overline{\mathbb{R}}$ be convex, proper, l.s.c. Then the following conditions are equivalent*:
1. $x^* \in \partial f(x)$,
2. $f(x) + f^*(x^*) = (x^*, x)$,
3. $x \in \partial f^*(x^*)$.

In particular, $\partial f^* = (\partial f)^{-1}$ and $f = f^{**}$. In general, $\partial(f + g) \supset \partial f + \partial g$ and the inclusion is strict. We have, however,

PROPOSITION 2.5 (Rockafellar). *Let f and g be l.s.c. and convex on D. Assume that $D(f) \cap int \, D(g) \neq \phi$. Then*

$$\partial(f + g) = \partial f + \partial g. \tag{2.7}$$

We shall assume now that $X = H$ is a Hilbert space. Let $f : H \to \overline{\mathbb{R}}$ be convex, proper and l.s.c. Then ∂f is maximal monotone. In other words,

$$(y_1 - y_2, x_1 - x_2) \geqslant 0, \quad \forall(x_i, y_i) \in \partial f, \ i = 1, 2 \tag{2.8}$$

and

$$R(I + \lambda \partial f) = H, \quad \forall \lambda > 0. \tag{2.9}$$

$R(I + \lambda \partial f)$ is the range of $I + \lambda \partial f$.

The mapping

$$(\partial f)_\lambda = \lambda^{-1}\left(I - (I + \lambda \partial f)^{-1}\right), \quad \lambda > 0 \tag{2.10}$$

is called the *Yosida approximation* of f.

Denote by $f_\lambda : H \to \mathbb{R}$ the function

$$f_\lambda(x) = \inf\left\{ \frac{|x - y|^2}{2\lambda} + f(y); \ y \in H \right\}, \quad \lambda > 0$$

which is called the *regularization* of f (see [29]).

PROPOSITION 2.6 (Brezis [9]). *Let $f : H \to \overline{\mathbb{R}}$ be convex and l.s.c. Then f_λ is Fréchet differentiable on H, $\partial f_\lambda = \{\nabla f_\lambda\}$ and*

$$f_\lambda(x) = \frac{\lambda}{2}|\partial f_\lambda(x)|^2 + f\big((I + \lambda \partial f(x))^{-1}\big), \tag{2.11}$$

$$\lim_{\lambda \to 0} f_\lambda(x) = f(x), \quad \forall x \in H. \tag{2.12}$$

Consider the function $\mathbb{I}_g : L^p(\Omega) \to \overline{\mathbb{R}}$ defined by

$$\mathbb{I}_g(y) = \begin{cases} \displaystyle\int_\Omega g(x, y(x))\mathrm{d}x & \text{if } g(\cdot, y(\cdot)) \in L^1(\Omega), \\ +\infty & \text{otherwise} \end{cases} \tag{2.13}$$

where $g : \Omega \times \mathbb{R}^m \to \mathbb{R}$ is a function satisfying (Ω is a measurable subset of $\mathbb{R}^n$)
1. $g(x, \cdot) : \mathbb{R}^m \to \overline{\mathbb{R}}$ is convex and l.s.c. for a.e. $x \in \omega$.
2. g is $\mathcal{L} \times \mathcal{B}$ measurable, i.e. g is measurable with respect to the σ-algebra of subsets of $\Omega \times \mathbb{R}^m$ generated by products of Lebesgue sets in Ω and Borelian sets in $\mathbb{R}^m$.
3. $g(x, y) \geqslant (\alpha(x), y) + \beta(x)$, a.e. $x \in \Omega$, $y \in \mathbb{R}^m$, where

$$\alpha \in L^q(\Omega), \quad \beta \in L^1(\Omega), \quad \frac{1}{p} + \frac{1}{q} = 1.$$

4. $\exists y_0 \in L^p(\Omega)$ such that $\mathbb{I}_g(y_0) < +\infty$.

PROPOSITION 2.7. *Let $1 \leqslant p < \infty$. Then $\mathbb{I}_g$ is convex, l.s.c. and $\not\equiv +\infty$. Moreover,*

$$\partial \mathbb{I}_g(y) = \big\{w \in L^q(\Omega); \ w(x) \in \partial g(x, y(x)) \text{ a.e. } x \in \Omega \big\}. \tag{2.14}$$

EXAMPLE 2.1. Let

$$C = \big\{y \in L^p(\Omega); \ a \leqslant y(x) \leqslant b, \text{ a.e. } x \in \Omega \big\}.$$

Then

$$g(y) = \begin{cases} 0 & \text{if } a \leqslant y \leqslant b, \\ +\infty & \text{otherwise} \end{cases}$$

and so by (2.14)

$$
\begin{aligned}
N_C(y) &= \left\{ w \in L^q(\Omega); \ w(x) \in N_{[a,b]}\big(y(x)\big) \text{ a.e. } x \in \Omega \right\} \\
&= \big\{ w \in L^q(\Omega); \ w(x) = 0 \text{ if } a < y(x) < b, \\
&\qquad\qquad w(x) \leqslant 0 \text{ if } y(x) = a, \\
&\qquad\qquad w(x) \geqslant 0 \text{ if } y(x) = b \big\}.
\end{aligned}
\tag{2.15}
$$

The case $\mathbb{I}_g : L^\infty(\Omega) \to \overline{\mathbb{R}}$ is more delicate since in this case $\partial \mathbb{I}_g(y)$ takes values in a measure space on Ω (see [32]).

Generalized gradients   Let X be a Banach space of norm $\|\cdot\|$ and dual X^*. The function $f : X \to \mathbb{R}$ is said to be *locally Lipschitz continuous* if for any bounded subset M of X there exists a constant L_M such that

$$
\|f(x) - f(y)\| \leqslant L_M \|x - y\|, \quad \forall x, y \in M.
$$

The *directional derivative* of f in x is defined by

$$
f^0(x, v) = \limsup_{\substack{y \to x \\ \lambda \to 0}} \lambda^{-1}\big(f(y + \lambda v) - f(y)\big).
\tag{2.16}
$$

The function f^0 is finite, positively homogeneous in v and subadditive. Then, by the Hahn–Banach theorem, $\exists \eta \in X^*$ such that

$$
(\eta, v) \leqslant f^0(x, v), \quad \forall v \in X.
$$

By definition, the *generalized gradient* of f in x, denoted $\partial f(x)$, is the set

$$
\partial f(x) = \left\{ \eta \in X^*; \ (\eta, v) \leqslant f^0(x, v), \ \forall v \in X \right\}.
\tag{2.17}
$$

PROPOSITION 2.8 (See [10,12]). *For each $x \in X$, $\partial f(x)$ is a convex and w^*-compact subset of X^*. Moreover,*

$$
f^0(x, v) = \sup\left\{ (\eta, v); \ \eta \in \partial f(x) \right\}, \quad \forall v \in X
$$

and the map $\partial f : X \to 2^{X^*}$ is weakly star upper semicontinuous, i.e. if $x_n \to x$ and $\eta_n \to \eta$ weakly star in X^*, then $\eta \in \partial f(x)$.

If f is locally Lipschitz and Gâteaux differentiable, then $\partial f = Df$. Moreover, if f is convex and locally Lipschitz, then ∂f is precisely the subdifferential of f.

Given a closed subset C of X, denote by d_C the distance function

$$
d_C(x) = \inf\left\{ \|x - y\|; \ y \in C \right\}, \quad \forall x \in X.
$$

We can see that d_C is Lipschitzian

$$\left| d_C(x) - d_C(y) \right| \leqslant \|x - y\|, \quad \forall x, y \in X.$$

Let $x \in C$. The element $v \in X$ is said to be tangent to C in x if

$$d_C^0(x, v) = 0.$$

The set of all tangent elements v is denoted $T_C(x)$ (the *tangent cone* to C at x). The *normal cone* $N_C(x)$ to C at x is by definition

$$N_C(x) = \left\{ \eta \in X^*; \ (\eta, v) \leqslant 0, \ \forall v \in T_C(x) \right\}.$$

PROPOSITION 2.9. *The vector $h \in X$ is tangent to C in x if and only if $\forall \{x_n\} \subset C$ convergent to x and each $\{\lambda_n\} \to 0$, there is $\{h_n\} \to h$ such that*

$$x_n + \lambda_n h_n \in C, \quad \forall n.$$

PROPOSITION 2.10. *If f, g are locally Lipschitzian, then*

$$\partial(f + g)(x) \subset \partial f(x) + \partial g(x), \quad \forall x \in X.$$

If C is a closed subset of X and if f attains its minimum on C in x, then

$$0 \in \partial f(x) + N_C(x).$$

We refer to the book [12] for further properties of generalized gradients.

2.2. Ekeland's variational principle

Here we shall briefly recall, without proof, an important result known in literature as *Ekeland variational principle* [21].

THEOREM 2.1. *Let X be a complete metric space and $F : X \to \mathbb{R}$ be a l.s.c. function, $\not\equiv +\infty$ and bounded from below. Let $\varepsilon > 0$ and $x \in X$ be such that*

$$F(x) \leqslant \inf\{F(y); \ y \in X\} + \varepsilon. \tag{2.18}$$

Then there exists $x_\varepsilon \in X$ such that

$$F(x_\varepsilon) \leqslant F(x), \tag{2.19}$$

$$d(x_\varepsilon, x) \leqslant \sqrt{\varepsilon}, \tag{2.20}$$

$$F(x_\varepsilon) < F(y) + \sqrt{\varepsilon}\, d(x_\varepsilon, y), \quad \forall y \neq x_\varepsilon. \tag{2.21}$$

Roughly speaking, Theorem 2.1 says that x_ε is a minimum point of the function

$$y \to F(y) + \sqrt{\varepsilon}\, d(x_\varepsilon, y).$$

COROLLARY 2.1. *let X be a Banach space and $F : X \to \mathbb{R}$ be Gâteaux differentiable and bounded from below. Then $\forall \varepsilon > 0$, $\exists x_\varepsilon \in X$ such that*

$$F(x) \leqslant \inf\{F(y);\ y \in X\} + \varepsilon, \tag{2.22}$$

$$\left|\nabla F(x_\varepsilon)\right| \leqslant \sqrt{\varepsilon}. \tag{2.23}$$

One may thus construct a minimizing sequence of almost critical points.

2.3. *Elements of differential geometry and exponential representation of flows*

In what follows we present some basic facts concerning the operator calculus introduced by A. Agrachev and R. Gamkrelidze (see [1,2,23]) called exponential representation of flows or chronological calculus. This is a very elegant tool that allows to replace nonlinear objects such as manifolds, tangent vector fields, flows, diffeomorphisms with linear ones which will be functionals and operators on the algebra $C^\infty(M)$ of real infinitely differentiable functions on M. At the end of the section a variation of parameters formula will be given; this formula will show to be very useful in proving the geometric form of Pontriaghin maximum principle. We follow essentially the description in [2].

Differential equations on manifolds   In what follows M is a smooth n-dimensional manifold, $TM = \bigcup_{y \in M} T_y M$ is the tangent bundle.

We consider the Cauchy problem for the nonautonomous ordinary differential equation:

$$\begin{cases} y' = f^t(y) := f(t, y), \\ y(0) = y_0 \end{cases} \tag{2.24}$$

where f^t is a *nonautonomous vector field* on M, that is $f^t(y) \in T_y M$ for any $y \in M$, $t \in \mathbb{R}$. In the case $M = \mathbb{R}^n$ or a subdomain of $\mathbb{R}^n$ we have the following classical theorem of Carathéodory (see [15, Chapter 2, Theorem 1.1]):

THEOREM 2.2. *If f is measurable in t for each fixed y and continuous in y for every fixed t and there exists a L^1 function m_0 such that in a neighborhood of $(0, y_0)$*

$$\left|f(t, y)\right| \leqslant m_0(t),$$

then problem (2.24) has a local solution in the extended sense (see Section 1.2).

If for any fixed t, $f_i(t, \cdot)$ is C^1 and for any $(\bar{t}, \bar{y})$ there exists an L^1 function m_1 and neighborhood of $(\bar{t}, \bar{y})$ such that for any (t, y) in this neighborhood

$$\left|\frac{\partial f_i}{\partial y_j}(t, y)\right| \leqslant m_1(t),$$

then the solution is unique. Moreover, under this assumption the solution is C^1 with respect to the initial data.

In order to solve equation (2.24) in the case of a general manifold M, we represent it in local coordinates. Let $\varphi : \mathcal{N}(y_0) \subset M \to \mathcal{N}(x_0) \subset \mathbb{R}^n$, a local chart. In these coordinates the vector field f^t is represented as:

$$(\varphi_* f^t)(x) = \sum_i^n \tilde{f}_i(t, x) \frac{\partial}{\partial x_i} = \tilde{f}^t(x).$$

Here φ_* is the tangent map, or differential of φ. Solving problem (2.24) is equivalent to solving the following Cauchy problem in $\mathbb{R}^n$:

$$\begin{cases} x' = \tilde{f}(t, x), \\ x(0) = x_0. \end{cases} \tag{2.25}$$

In order to insure existence and uniqueness of a local solution, we will assume that $\tilde{f}$ satisfies the hypotheses of Theorem 2.2 which are in fact hypothesis on f since they do not depend on the choice of the local chart. Under these hypothesis, by the theorem of Carathéodory, problem (2.25) has a unique local solution $x(t, x_0)$ which is absolutely continuous with respect to t and C^1 with respect to the initial data x_0 and satisfies the equation almost everywhere. The solution of (2.24) is $y(t, y_0) = \varphi^{-1}(x(t, x_0))$ and one may prove that this is independent of the local chart. The solution of the Cauchy problem (2.24) is defined on a maximal interval that we will suppose to be $\mathbb{R}$ for all initial data. Such vector fields that determine global flows are called *complete*. This always happens if the manifold M is compact.

If we denote by F^t the *flow* defined by the equation (2.24): $F^t(y_0) = y(t, y_0)$, then $F^t \in \text{Diff}(M)$ the set of diffeomorphisms of the manifold M and equation (2.24) may be written

$$\begin{cases} \dfrac{\mathrm{d}}{\mathrm{d}t} F^t(y) = f^t \circ F^t(y), \quad y \in M, \\ F^0 = \text{Id}. \end{cases} \tag{2.26}$$

HYPOTHESES. We will suppose from now on that M is a C^∞ manifold and $\text{Diff}(M)$ denotes the set of C^∞ diffeomorphisms of M. Moreover, we will suppose that the nonautonomous vector field f^t is complete and in any local chart $\tilde{f}(t, x)$ is measurable with respect to t for any fixed x and C^∞ with respect to x for every fixed t and there exist locally integrable functions $m_k(t)$ such that locally

$$\left| D_x^k \tilde{f}(t, x) \right| \leqslant m_k(t).$$

These hypotheses insure that the Cauchy problem (2.24) has unique solution depending C^∞ on the initial data.

Exponential representation of flows We describe in the sequel how the chronological exponential is defined and we will see that topological and differential structures are translated in the new language into the weak convergence of functionals and operators.

Points are represented as algebra homomorphisms from $C^\infty(M)$ to $\mathbb{R}$. If $y \in M$ then it defines an algebra homomorphism $\hat{y} : C^\infty(M) \to \mathbb{R}$, $\hat{y}(\alpha) = \alpha(y)$. One may prove that for any algebra homomorphism $\psi : C^\infty(M) \to \mathbb{R}$, there exists an unique $y \in M$ such that $\psi = \hat{y}$ (see [2]).

Diffeomorphisms of the manifold M are represented as automorphisms of the algebra $C^\infty(M)$. More precisely, if $F \in \mathrm{Diff}(M)$ we define $\widehat{F} : C^\infty(M) \to C^\infty(M)$ as $\widehat{F}(\alpha) = \alpha \circ F$. More generally, if $F : M \to N$ is a smooth map between two manifolds, then it defines an algebra homomorphism $\widehat{F} : C^\infty(N) \to C^\infty(M)$ as $\widehat{F}(\beta) = \beta \circ F$ with $\beta \in C^\infty(N)$. Observe that if $F, G \in \mathrm{Diff}(M)$ then $\widehat{F \circ G} = \widehat{G} \circ \widehat{F}$.

Tangent vectors. Let $f \in T_y M$. Then, as is well known f may be seen either as tangent vector in y to a curve passing through y or as directional derivative, or Lie derivative, of functions in the point y in the direction f. For the first point of view one considers a smooth curve $y(t)$, $y(0) = y$, $y'(0) = v$. The second point of view is to consider the Lie derivative $L_f \alpha = \frac{\mathrm{d}}{\mathrm{d}t} \alpha(y(t))|_{t=0}$. Through the representation described above, we may construct $\hat{f} : C^\infty(M) \to \mathbb{R}$, $\hat{f}(\alpha) := \frac{\mathrm{d}}{\mathrm{d}t}[\hat{y}(t)(\alpha)]|_{t=0} = L_f \alpha$. Obviously, $\hat{f}$ is a linear functional on $C^\infty(M)$ and satisfies the Leibnitz rule

$$\hat{f}(\alpha\beta) = \alpha(y)\hat{f}(\beta) + \hat{f}(\alpha)\beta(y). \tag{2.27}$$

Any linear functional on $C^\infty(M)$ satisfying (2.27) corresponds in this way to a tangent vector.

Vector fields. Let $\mathrm{Vec}(M)$ be the set of smooth vector fields on M and let $f \in \mathrm{Vec}(M)$. Then f defines a linear operator $\hat{f} : C^\infty(M) \to C^\infty(M)$, $\hat{f}(\alpha)(y) = \widehat{f(y)}(\alpha)$. This operator satisfies the Leibnitz rule

$$\hat{f}(\alpha\beta) = \alpha\hat{f}(\beta) + \hat{f}(\alpha)\beta. \tag{2.28}$$

Any linear functional of $C^\infty(M)$ satisfying (2.28) is called *derivation* and corresponds to a unique vector field.

We study now the behaviour of tangent vectors and vector fields under the action of diffeomorphisms.

Let $F \in \mathrm{Diff}(M)$ and $g \in T_y M$ such that $g = \frac{\mathrm{d}}{\mathrm{d}t} y(t)|_{t=0}$. Then $F_* g \in T_{F(y)} M$ and is defined as $F_* g = \frac{\mathrm{d}}{\mathrm{d}t} F(y(t))|_{t=0}$. So, if $\alpha \in C^\infty(M)$, then

$$\widehat{F_* g}(\alpha) = \frac{\mathrm{d}}{\mathrm{d}t} \widehat{F(y(t))}(\alpha)\bigg|_{t=0} = \frac{\mathrm{d}}{\mathrm{d}t} \alpha\big(F(y(t))\big)\bigg|_{t=0} = \hat{g}(\alpha \circ F) = \hat{g} \circ \widehat{F}(\alpha).$$

So,

$$\widehat{F_* g} = \hat{g} \circ \widehat{F}. \tag{2.29}$$

In the same way, if $g \in \text{Vec}(M)$, since $\widehat{g(y)} = \hat{y} \circ \hat{g}$, $\widehat{F_*g(F(y))} = \widehat{F(y)} \circ \widehat{F_*g} = \hat{y} \circ \widehat{F} \circ \widehat{F_*g}$. On the other hand $\widehat{F_*g(F(y))} = \widehat{F_*(g(y))} = \hat{y} \circ \hat{g} \circ \widehat{F}$. As y is arbitrary, $\widehat{F} \circ \widehat{F_*g} = \hat{g} \circ \widehat{F}$ so

$$\widehat{F_*g} = \widehat{F}^{-1} \circ \hat{g} \circ \widehat{F} = \text{Ad}\,\widehat{F}^{-1}\hat{g}. \tag{2.30}$$

REMARK 2.1. The notation Ad comes from the theory of Lie groups where it stands for the adjoint representation of the group in the space of linear operators of the associated Lie algebra. In our situation the group of diffeomorphisms of the manifold stands for the Lie group and the associated Lie algebra is the algebra of vector fields. Through the described representation one obtains the group of automorphisms of $C^\infty(M)$ and the associated Lie algebra is the algebra of derivations of $C^\infty(M)$ (see, e.g., [3,28]).

The equation (2.24) becomes, through the described representation:

$$\begin{cases} \dfrac{d}{dt}\widehat{y(t)} = \widehat{y(t)} \circ \hat{f}^t, \\[2mm] \hat{y}(0) = \hat{y}_0 \end{cases} \tag{2.31}$$

so the flow defined by the equation satisfies

$$\begin{cases} \dfrac{d}{dt}\widehat{F^t} = \widehat{F^t} \circ \hat{f}^t, \\[2mm] F^0 = \text{Id}. \end{cases} \tag{2.32}$$

The flow $\widehat{F^t}$ is called *the right chronological exponential* and, in accordance with the linear case, is denoted by

$$\widehat{F^t} = \overrightarrow{\exp} \int_0^t \hat{f}^s \, ds. \tag{2.33}$$

In order to simplify notations, we will omit from now on the hat $\widehat{}$ unless confusion is possible and, usually, when we refer to diffeomorphisms and vector fields we mean their representations.

We observe however that at this point equations (2.31), (2.32) are not completely rigorous since we have not yet defined a topology in the corresponding spaces of functionals or operators on $C^\infty(M)$.

Topology We consider on $C^\infty(M)$ the topology of uniform convergence on compacta of all derivatives. More precisely, if $M = \Omega \subset \mathbb{R}^n$, for $\alpha \in C^\infty(M)$, $K \Subset M$ and $k = (k_1, \ldots, k_n)$, $k_i \geqslant 0$, we define the seminorms:

$$\|\alpha\|_{s,K} = \sup\{|D^k\alpha(y)|; \ |k| = k_1 + \cdots + k_n \leqslant s, y \in K\}.$$

This family of seminorms determines a topology on $C^\infty(M)$ which becomes a Fréchet space (locally convex topological linear space with a complete metric topology given by a translation invariant metric). In this topology $\alpha_m \to \alpha$ iff $\|\alpha_m - \alpha\|_{s,K} \to 0$ for all $s \geq 0$ and $K \Subset M$.

In the case of a general manifold, we choose a locally finite covering of M with charts $(V_i, \varphi_i)_{i \in I}$, $\varphi : V_i \to \mathcal{O}_i \subset \mathbb{R}^n$ diffeomorphisms and let $\{\alpha_i\}_{i \in I}$ be a partition of unity subordinated to this covering. We define the family of seminorms

$$\|\alpha\|_{s,K} = \sup\left\{ D^k\big[(\alpha_i \alpha) \circ \varphi^{-1}\big](y) \mid |k| \leq s, \varphi^{-1}(y) \in K, i \in I \right\}.$$

This family of seminorms depends on the choice of the atlas but the topology defined on $C^\infty(M)$ is independent of this choice. One could also proceed by using the Whitney theorem and considering M as a submanifold of some Euclidean space.

Once we have defined the topology on $C^\infty(M)$ we consider the space of linear continuous operators $L(C^\infty(M))$. The spaces $\mathrm{Diff}(M)$ and $\mathrm{Vec}(M)$, through the representation are linear subspaces. Indeed, one may easily verify that for $f \in \mathrm{Vec}(M)$ and $F \in \mathrm{Diff}(M)$

$$\big\|\hat{f}\alpha\big\|_{s,K} \leq C_1 \|\alpha\|_{s+1,K}, \qquad \big\|\widehat{F}\alpha\big\|_{s,K} \leq C_2 \|\alpha\|_{s,K}$$

where the constants $C_1 = C_1(s, K, f)$, $C_2 = C_2(s, K, F)$. We thus define a family of seminorms on $\mathrm{Vec}(M)$, respectively $\mathrm{Diff}(M)$:

$$\|f\|_{s,K} = \sup\left\{ \big\|\hat{f}\alpha\big\|_{s+1,K} \mid \|\alpha\|_{s,K} = 1 \right\},$$

$$\|F\|_{s,K} = \sup\left\{ \big\|\widehat{F}\alpha\big\|_{s,K} \mid \|\alpha\|_{s,K} = 1 \right\},$$

which define locally convex topologies. On these spaces we also may consider the weak topology induced from $C^\infty(M)$: $F_n \to F$ iff $F_n \alpha \to F\alpha$ for all $\alpha \in C^\infty(M)$ (the same for a sequence of vector fields).

Differentiability and integrability of families of functions or operators   First of all we define these properties on $C^\infty(M)$ which is a Fréchet space. In general, let X be a Fréchet space whose topology is defined by the family $\{p_k\}_{k \in \mathbf{N}}$ of seminorms. The metric on X is defined by

$$d(x, y) = \sum_{k \in \mathbf{N}} \frac{1}{2^k} \frac{p_k(x - y)}{1 + p_k(x - y)}.$$

Let $h : J \subset \mathbb{R} \to X$. The function h is differentiable in t_0 if there exists in X the limit

$$\lim_{t \to t_0} \frac{h(t) - h(t_0)}{t - t_0}.$$

The function h is Lipschitz continuous if $p_k \circ h$ is Lipschitz for all p_k. Differentiability and Lipschitz continuity may also be defined using the metric structure of X.

The function h is bounded if $p_k \circ h$ is bounded for all p_k.

For measurability and integrability we adapt the plan of development for the *Bochner integral* (see, e.g., [35]). A function h is called a step function if it may be represented as

$$h = \sum_{n \in \mathbf{N}} x_n \chi_{J_n}$$

where χ_{J_n} is the characteristic function of a measurable subset $J_n \subset J$. We call such a representation of h a σ-representation and it is obvious that this is not unique. We say that the function h is *strongly measurable* if h is the limit a.e. of a sequence of step functions. The function h is *weakly measurable* if $x^* \circ h$ is measurable for all $x^* \in X^*$. One may prove that if X is separable the two notions of measurability coincide (see Pettis theorem in [35] in the case X is a Banach space). If h is a step function then h is *integrable* if

$$\sum_n \mu(J_n) p_k(x_n) \leqslant \infty$$

for all the seminorms p_k. The integral of h is then defined as

$$\int_J h(t)\, dt = \sum_n \mu(J_n) x_n$$

and it may be shown that it is independent of the order of summation and of the σ-representation of h.

If h is a measurable function we say that it is *integrable* if there exists a sequence of integrable step functions $\{h_n\}_{n \in \mathbf{N}}$ such that for all k

$$\lim_{n \to +\infty} \int_J p_k\big(h(t) - h_n(t)\big)\, dt = 0.$$

In this case one may show that there exists

$$\lim_{n \to +\infty} \int_J h_n(t)\, dt$$

and this is independent of the sequence $\{h_n\}_{n \in \mathbf{N}}$ with the given properties. The limit is denoted by

$$\int_J h(t)\, dt$$

and is the integral of h on J.

For a family P^t, $t \in J \subset \mathbb{R}$ of linear continuous operators or linear continuous functionals on $C^\infty(M)$ the above notions (continuity, differentiability, boundedness, measurability, integrability) will be considered in the weak sense, that is the function $t \to P^t$ has one of these properties if $P^t \circ \alpha$ has the corresponding property for all $\alpha \in C^\infty(M)$. We will not discuss here the relation between the strong and weak properties.

At this point we see that the operator equation (2.32) makes sense and it can be easily proved that it has a unique solution. We point out the Leibnitz rule:

$$\frac{\mathrm{d}}{\mathrm{d}t} P^t \circ Q^t \Big|_{t=t_0} = \frac{\mathrm{d}}{\mathrm{d}t} P^t \Big|_{t=t_0} \circ Q^{t_0} + P^{t_0} \circ \frac{\mathrm{d}}{\mathrm{d}t} Q^t \Big|_{t=t_0}$$

for two functions $t \to P^t$, $t \to Q^t$ differentiable at t_0.

Consider now the flow F^t defined by (2.24) and $G^t = (F^t)^{-1}$. If we differentiate the identity $F^t \circ G^t = I$ we obtain

$$F^t \circ f^t \circ G^t + F^t \circ \frac{\mathrm{d}}{\mathrm{d}t} G^t = 0$$

and thus

$$\begin{cases} \dfrac{\mathrm{d}}{\mathrm{d}t} G^t = -f^t \circ G^t, \\ G^0 = \mathrm{Id}. \end{cases} \tag{2.34}$$

We define thus the *left chronological exponential*:

$$G^t = \overleftarrow{\exp} \int_0^t -f^t \, \mathrm{d}t.$$

Further properties and extensions   We have seen that $\widehat{F_* g} = \mathrm{Ad}\, \widehat{F}^{-1} \hat{g}$ for $F \in \mathrm{Diff}(M)$, $g \in \mathrm{Vec}(M)$. We compute now the differential $\frac{\mathrm{d}}{\mathrm{d}t}\big|_{t=0} \mathrm{Ad}(\widehat{F^t})$ for a flow F^t on M such that

$$\begin{cases} \dfrac{\mathrm{d}}{\mathrm{d}t} F^t \Big|_{t=0} = f \in \mathrm{Vec}\, M, \\ F^0 = \mathrm{Id}. \end{cases}$$

We have

$$\frac{\mathrm{d}}{\mathrm{d}t}\big|_{t=0} \big(\mathrm{Ad}\, F^t\big) g = f \circ g - g \circ f = [f, g] =: (\mathrm{ad}\, f) g.$$

In the particular case

$$F^t = \overrightarrow{\exp} \int_0^t f^s \, \mathrm{d}s$$

we obtain:

$$\begin{cases} \dfrac{\mathrm{d}}{\mathrm{d}t} (\mathrm{Ad}\, F^t) g = \big(\mathrm{Ad}\, F^t\big)\, \mathrm{ad}\, f^t g, \\ \mathrm{Ad}\, F^0 = \mathrm{Id} \end{cases}$$

so we may write, formally:

$$\mathrm{Ad}\left(\overrightarrow{\exp}\int_0^t f^s\,\mathrm{d}s\right)=\overrightarrow{\exp}\left(\int_0^t \mathrm{ad}\,f^s\,\mathrm{d}s\right).$$

Let now $F\in\mathrm{Diff}(M)$ and g^t a nonautonomous vector field. Then

$$F\circ\overrightarrow{\exp}\int_0^t g^s\,\mathrm{d}s\circ F^{-1}=\overrightarrow{\exp}\int_0^t (\mathrm{Ad}\,F g^s)\,\mathrm{d}s. \tag{2.35}$$

Indeed, the both sides of the equality verify the same Cauchy problem for the operator equation

$$\begin{cases}\dfrac{\mathrm{d}}{\mathrm{d}t}q^t=q^t\circ\left(\mathrm{Ad}\,F\,g^t\right),\\[2mm] q^0=\mathrm{Id}\end{cases}$$

and thus, by uniqueness of the solution, they coincide.

Now if we take again $G^t=(F^t)^{-1}$ and if we differentiate the identity $G^t\circ F^t=\mathrm{Id}$ we obtain that $\frac{\mathrm{d}}{\mathrm{d}t}G^t\circ F^t=-G^t\circ F^t\circ f^t$ and thus

$$\frac{\mathrm{d}}{\mathrm{d}t}G^t=-G^t\circ\left(\mathrm{Ad}\,F^t\right)f^t.$$

This gives the relationship between left and right chronological exponentials:

$$\overleftarrow{\exp}\int_0^t f^s\,\mathrm{d}s=\overrightarrow{\exp}\int_0^t \left(\mathrm{Ad}\,F^s\right)f^s\,\mathrm{d}s. \tag{2.36}$$

If $F\in\mathrm{Diff}(M)$, as we have seen, it defines an algebra automorphism of $C^\infty(M):\widehat{F}\alpha=\alpha\circ F=F^*\alpha$, where F^* is the pull back of C^∞ differential forms defined by F. This suggests the fact that $\widehat{F}$ may be extended, as algebra automorphism to the graded algebra $\Lambda(M)=\bigoplus\Lambda^k(M)$ of differential forms. If $\omega\in\Lambda^k(M)$ then we define

$$\widehat{F}\omega:=F^*\omega.$$

It is well known that F^* commutes with exterior differential:

$$F^*\circ d=d\circ F^*$$

and for $\omega_i\in\Lambda^{k_i}(M)$,

$$F^*(\omega_1\wedge\omega_2)=F^*(\omega_1)\wedge F^*(\omega_2).$$

So $\widehat{F}$ is an algebra automorphism for $\Lambda(M)$. We now consider a vector field:

$$\frac{\mathrm{d}}{\mathrm{d}t}F^t|_{t=0}=f,\quad F^0=\mathrm{Id}.$$

The action on $\Lambda^k(M)$ is the Lie derivative of differential forms:

$$\frac{\mathrm{d}}{\mathrm{d}t}\widehat{F^t}\omega\bigg|_{t=0} = \frac{\mathrm{d}}{\mathrm{d}t}(F^t)^*\omega\bigg|_{t=0} = L_f\omega.$$

We may thus understand the action of a vector field $f \in \mathrm{Vec}(M)$ as the Lie derivative of differential forms:

$$\hat{f} = L_f.$$

The chronological exponential may be thus written

$$F^t = \overrightarrow{\exp}\int_0^t L_{f^s}\,\mathrm{d}s.$$

We point out two fundamental properties of the Lie derivative:
Since $\widehat{F^t}\circ d = d \circ \widehat{F^t}$ one obtains that

$$\hat{f}\circ d = d \circ \hat{f} \quad \text{(equivalently } L_f \circ d = d \circ L_f\text{)}.$$

Denote by i_f the interior product of a differential form ω with a vector field f: $i_f\omega(f_1,\ldots,f_k) = \omega(f, f_1,\ldots, f_k)$, for $\omega \in \Lambda^k(M)$, $f_i \in \mathrm{Vec}\,M$. Then the classical *Cartan's formula* reads:

$$\hat{f} = d \circ i_f + i_f \circ d. \tag{2.37}$$

Variation of parameters formula	Consider the Cauchy problem for the linear differential equation in $\mathbb{R}^n$:

$$\begin{cases} y' = Ay + b(t), \\ y(0) = y_0. \end{cases}$$

The solution of the homogeneous equation ($b \equiv 0$) is $y(t) = \mathrm{e}^{At}y_0$. For the nonhomogeneous equation a solution may be found by the *variation of constants* or *variation of parameters method*. This consists in searching a solution of the form $y(t) = \mathrm{e}^{At}c(t)$ and an equation for $c(t)$ is obtained: $c'(t) = A(t)b(t)$. The solution is given by the *variation of constants formula*

$$y(t) = \mathrm{e}^{At}y_0 + \int_0^t \mathrm{e}^{A(t-s)}b(s)\,\mathrm{d}s. \tag{2.38}$$

We consider now the nonlinear differential equation

$$y' = f^t(y)$$

which generates the flow $F^t = \overrightarrow{\exp} \int_0^t f^s \, ds$. We consider also the perturbed equation

$$y' = f^t(y) + g^t(y)$$

which generates the flow $H^t = \overrightarrow{\exp} \int_0^t f^s + g^s \, ds$, depending on the perturbation g^t. We want to find an expression for this dependence. For this purpose one proceeds as in the linear case and search H^t in the form

$$H^t = G^t \circ F^t$$

where the flow G^t has to be deduced. Differentiating this equality we find

$$\frac{d}{dt} H^t = G^t \circ F^t \circ \left(f^t + g^t \right)$$

$$= \frac{d}{dt} G^t \circ F^t + G^t \circ \frac{d}{dt} F^t = \frac{d}{dt} G^t \circ F^t + G^t \circ F^t \circ f^t.$$

So,

$$\begin{cases} \dfrac{d}{dt} G^t = G^t \circ F^t \circ f^t \circ \left(F^t \right)^{-1} = G^t \circ \operatorname{Ad} F^t g^t, \\ G^0 = \operatorname{Id}. \end{cases}$$

We thus obtained

$$G^t = \overrightarrow{\exp} \int_0^t \operatorname{Ad} F^s g^s \, ds$$

and the first form of variations formula

$$H^t = \overrightarrow{\exp} \int_0^t \operatorname{Ad} F^s g^s \, ds \circ F^t. \tag{2.39}$$

We also obtain

$$H^t = F^t \circ \operatorname{Ad}\left(F^t \right)^{-1} \left(\overrightarrow{\exp} \int_0^t \operatorname{Ad} F^s g^s \, ds \right)$$

and by (2.35)

$$H^t = F^t \circ \overrightarrow{\exp} \int_0^t \operatorname{Ad}\left[\left(F^t \right)^{-1} \circ F^s \right] g^s \, ds = F^t \circ \overrightarrow{\exp} \int_0^t \left(F_s^t \right)_* g^s \, ds \tag{2.40}$$

where $F_s^t = \overrightarrow{\exp} \int_s^t f^\tau \, d\tau$.

The second form of variations formula may be thus written

$$\overrightarrow{\exp}\int_0^t f^s + g^s\,ds = \overrightarrow{\exp}\int_0^t f^s\,ds \circ \overrightarrow{\exp}\int_0^t \overrightarrow{\exp}\int_t^s \mathrm{ad}\, f^\tau\,d\tau\, g^s\,ds.$$

In the case when f, g are autonomous vector fields the formula becomes (compare with (2.38)):

$$e^{t(f+g)} = \overrightarrow{\exp}\int_0^t \mathrm{Ad}\, e^{sf} g\,ds \circ e^{tf} = e^{tf} \circ \overrightarrow{\exp}\int_0^t \mathrm{Ad}\, e^{(s-t)f} g\,ds.$$

Elements of symplectic geometry. Hamiltonian formalism

DEFINITION 2.1. A symplectic structure on a (necessarily odd dimensional) manifold N is a nondegenerate closed differential 2-form. A manifold with a symplectic structure ω is called a symplectic manifold

Let M be a manifold and $T^*M = \bigcup_{y \in M} T_q^*M$ be the cotangent bundle. If $(x_1, \ldots, x_n)$ are local coordinates on M then if $p \in T_y^*M$, $p = \sum_{i=1}^n p_i\,dx_i$, $(p_1, \ldots, p_n, x_1, \ldots, x_n)$ define the canonical local coordinates on T^*M. Define

$$\omega = \sum_{i=1}^n dp_i \wedge dx_i. \tag{2.41}$$

To see that the definition is independent of the local coordinates let $\pi : T^*M \to M$ be the canonical projection and the canonical 1-form on T^*M:

$$\omega_{1\xi}(w) = \xi \circ \pi_*(w), \quad \text{for } w \in T_\xi(T^*M).$$

If $w \in T(T^*M)$ then

$$w = \sum_{i=1}^n \xi_i \frac{\partial}{\partial p_i} + v_i \frac{\partial}{\partial x_i}.$$

Since

$$\pi_*\left(\frac{\partial}{\partial p_i}\right) = 0 \quad \text{and} \quad \pi_*\left(\frac{\partial}{\partial x_i}\right) = \frac{\partial}{\partial x_i}$$

one finds that

$$\omega_{1\xi}(w) = \sum_{i=1}^n p_i v_i$$

so

$$\omega_1 = \sum_{i=1}^{n} p_i \, \mathrm{d}x_i$$

and

$$\omega = \mathrm{d}\omega_1.$$

It is easy to see now that ω is a symplectic structure on T^*M which becomes a symplectic manifold.

Let now (N, ω) be a general symplectic manifold. Functions in $C^\infty(N)$ are called *Hamiltonians*. Let H be such a Hamiltonian. Then there exists a unique vector field on N denoted $\vec{H}$ such that

$$-i_{\vec{H}}\,\omega = \omega(\cdot, \vec{H}) = \mathrm{d}H.$$

$\vec{H}$ is called the *Hamiltonian vector field* of H and the corresponding flow is the *Hamiltonian flow*. The *Hamiltonian equation* is

$$\frac{\mathrm{d}}{\mathrm{d}t}\xi(t) = \vec{H}\left(\xi(t)\right) \tag{2.42}$$

and the Hamiltonian flow is

$$\Xi^t = \overrightarrow{\exp} \int_0^t \vec{H} \, \mathrm{d}s.$$

The *Poisson bracket* of the Hamiltonians α, β is defined as

$$\{\alpha, \beta\} = L_{\vec{\alpha}}\beta = \mathrm{d}\beta(\vec{\alpha}) = \omega(\vec{\alpha}, \vec{\beta}) = -\{\beta, \alpha\}.$$

One may prove that $(C^\infty(N), \{\cdot, \cdot\})$ is a Lie algebra and the map $H \to \vec{H}$ is a Lie algebra homomorphism from $C^\infty(N)$ to $\mathrm{Vec}(N)$. Bilinearity and antisymmetry are immediate. Jacobi identity as well as the fact that $\overrightarrow{\{\alpha, \beta\}} = [\vec{\alpha}, \vec{\beta}\,]$ are easy to prove if in local coordinates ω has the canonic form (2.41). We conclude since, by *Darboux theorem* (see [3]), there exists indeed a symplectic atlas on N such that ω in local coordinates is in canonical form. In these coordinates

$$\vec{H} = \sum_{i=1}^{n} \frac{\partial H}{\partial p_i}\frac{\partial}{\partial x_i} - \frac{\partial H}{\partial x_i}\frac{\partial}{\partial p_i}$$

and

$$\{\alpha, \beta\} = \sum_{i=1}^{n} \frac{\partial \alpha}{\partial p_i}\frac{\partial \beta}{\partial x_i} - \frac{\partial \alpha}{\partial x_i}\frac{\partial \beta}{\partial p_i}.$$

Moreover, the Hamiltonian system (2.42) is written in the form (1.7).

Now, if $F \in \mathrm{Diff}(N)$ preserves the symplectic structure, that is $F^*\omega = \omega$, then $\mathrm{Ad}\, F \vec{H} = \overrightarrow{FH}$. Indeed,

$$\omega\big(\cdot, \overrightarrow{FH}\big) = \mathrm{d}(FH) = \mathrm{d}(H \circ F) = H_* \circ F_* = \mathrm{d}H \circ F_*$$
$$= \omega\big(F_* \cdot, \vec{H}\big) = F^*\omega\big(\cdot, (F_*)^{-1}\vec{H}\big) = \omega\big(\cdot, \mathrm{Ad}\, F \vec{H}\big).$$

Prime integrals of Hamiltonian systems are those which commute with the Hamiltonian. Indeed, consider the equation (2.42). Then $\alpha \in C^\infty(M)$ is a prime integral iff $e^{t\vec{H}}\alpha = \mathrm{const}$ which is equivalent, by differentiation, to $\vec{H}\alpha = 0$ or $\{H, \alpha\} = 0$.

We consider again the case of the symplectic manifold T^*M. Given a nonautonomous vector field $f^t \in \mathrm{Vec}(M)$ we consider the Hamiltonian

$$\big(f^t\big)^{\#}(\xi) = \big(\xi, f^t\big).$$

In canonical symplectic coordinates (p_i, x_i), if $f^t = \sum_{i=1}^{n} f_i(t, x)\frac{\partial}{\partial x_i}$ then $(f^t)^{\#}(\xi) = \sum_{i=1}^{n} p_i f_i(t, x)$. Using the canonical coordinates it is easy to see that for $f, g \in \mathrm{Vec}(M)$

$$\big\{f^{\#}, g^{\#}\big\} = [f, g]^{\#}.$$

For a nonautonomous vector field f^t the Hamiltonian vector field on T^*M defined by $\overrightarrow{(f^t)^{\#}}$ is named the *Hamiltonian lift*. It satisfies

$$\pi_*\overrightarrow{\big(f^t\big)^{\#}} = f^t.$$

We want to establish now the relation between the flows determined by f^t, respectively by $\overrightarrow{(f^t)^{\#}}$. Let $F_\tau^t = \overrightarrow{\exp}\int_\tau^t f^s\, \mathrm{d}s$. Then $(F_\tau^t)^* \in \mathrm{Diff}(T^*M)$. Let

$$g^\tau = \frac{\mathrm{d}}{\mathrm{d}t}\bigg|_{t=\tau} \big(F_\tau^t\big)^*$$

PROPOSITION 2.11. $g^t = -\overrightarrow{(f^t)^{\#}}$ *and*

$$\big(F_\tau^t\big)^* = \overrightarrow{\exp}\int_t^\tau \overrightarrow{\big(f^s\big)^{\#}}\, \mathrm{d}s \tag{2.43}$$

PROOF. First of all, since, in the exponential representation on T^*M, $(F_\tau^{t+\varepsilon})^* = (F_t^{t+\varepsilon})^* \circ (F_\tau^t)^*$, it follows by differentiating with respect to ε in 0 that

$$\frac{\mathrm{d}}{\mathrm{d}t}\big(F_\tau^t\big)^* = g^t \circ \big(F_\tau^t\big)^*$$

so

$$\left(F_\tau^t\right)^* = \overleftarrow{\exp} \int_\tau^t g^s \, ds.$$

Since

$$\pi \circ \left(F_\tau^t\right)^* = \left(F_\tau^t\right)^{-1} \circ \pi$$

it follows by differentiation that

$$\pi_* g^t = -f^t. \tag{2.44}$$

On the other hand, the flow $(F_\tau^t)^*$ preserves the 1-form ω_1 and thus the symplectic form ω. By Cartan's formula (2.3)

$$0 = L_{g^t}\omega_1 = i_{g^t}\omega + d\omega_1(g^t)$$

which implies that

$$g^t = \overrightarrow{\omega_1(g^t)}$$

which means that the field g^t is Hamiltonian. By (2.44) and observing that g^t is linear homogeneous on the tangent space to the fibers, the first conclusion follows. Equality (2.43) is now immediate taking into account the relationship between the left and right chronological exponential. $\qquad\square$

3. The Pontriaghin maximum principle

The Pontriaghin maximum principle, developed by L. Pontriaghin and his collaborators (see [30]), is a set of first order *necessary conditions* of optimality in optimal control problems. This is expressed in terms of the dual linearized state system and reduces the problem to solving a two point boundary problem for a differential system, which is in fact a Hamiltonian system, composed of the state equation and the dual equation.

3.1. *The main theorem*

Throughout this section the following conditions will be assumed on the optimal control problem (1.14):

(i) The functions

$$L : [0, T] \times \mathbb{R}^n \times \mathbb{R}^m \to \mathbb{R}; \quad f : [0, T] \times \mathbb{R}^n \times \mathbb{R}^m \to \mathbb{R}^n$$

and L_y, f_y are measurable in t, continuous in (y, u) and

$$\left\| f_y(t, y, u) \right\| + \left\| L_y(t, y, u) \right\| \leqslant \alpha(t, u), \quad \forall t \in [0, T], \ \forall u \in U(t).$$

  (ii) For each $t \in [0, T]$, $U(t) \subset \mathbb{R}^m$ is closed and for each closed set $D \subset \mathbb{R}^m$ the set $\{t; \ D \cap U(t) \neq \emptyset\}$ is measurable.

  (iii) $g \in C^1(\mathbb{R}^n \times \mathbb{R}^n)$ and C is a closed convex subset of $\mathbb{R}^n \times \mathbb{R}^n$.

Here we have denoted by the same symbol $\| \cdot \|$ the norm in $L(\mathbb{R}^n, \mathbb{R}^n)$ and in $\mathbb{R}^n$.

Theorem 3.1 below is the celebrated Pontryagin's maximum principle for problem (1.14).

THEOREM 3.1. *Let (y^*, u^*) be an optimal pair in problem (1.14) such that $\alpha(t, u^*) \in L^1(0, T)$. Then there exists $p \in AC([0, T]; \mathbb{R}^n)$ and a constant λ which is equal to 0 or 1 such that*

$$\left\| p(t) \right\| + |\lambda| \neq 0, \quad \forall t \in [0, T]$$

and

$$p' = -f_y^*(t, y^*, u^*)p + \lambda L_y(t, y^*, u^*), \quad a.e. \ t \in [0, T], \tag{3.1}$$

$$\{p(0), -p(T)\} \in \lambda \nabla g\big(y^*(0), y^*(T)\big) + N_C\big(y^*(0), y^*(T)\big), \tag{3.2}$$

$$\big(p(t), f\big(t, y^*(t), u^*(t)\big)\big) - \lambda L\big(t, y^*(t), u^*(t)\big)$$
$$= \max_{u \in U(t)} \left\{ \big(p(t), f\big(t, y^*(t), u\big)\big) - \lambda L\big(t, y^*(t), u\big) \right\} \quad a.e. \ t \in [0, T]. \tag{3.3}$$

If $\lambda = 1$, the problem is called *normal* otherwise it is *abnormal*. If $C = \{y_0\} \times \mathbb{R}^n$ and $g(y_1, y_2) \equiv g_1(y_2)$ this is the *Bolza problem* with initial condition $y(0) = y_0$ and (3.2) reduces to

$$p(T) = -\lambda \nabla g_1\big(y(T)\big). \tag{3.4}$$

If $C = \{(y_1, y_2); \ y_1 = y_2\}$ and $g \equiv 0$, this is the periodic optimal control problem and (3.2) has the form

$$p(0) = p(T). \tag{3.5}$$

The end point boundary conditions (3.2) are also called *transversality conditions*.

An elementary approach to the maximum principle   In order to understand better how the optimality system (3.1)–(3.3) appears, as well as the proof we give in the next section, we discuss first the special case where $L(t, \cdot, \cdot)$, $f(t, \cdot, \cdot)$ are smooth (of class C^1 for instance), $g(y_1, y_2) = g_2(y_2) \in C^1$, $C = \{y_0\} \times \mathbb{R}^n$ and $U(t) \equiv U$ is a closed convex subset of $\mathbb{R}^m$. We shall assume also $f_y(t, y^*, u^*) \in L^\infty(0, T, \mathbb{R}^{2n})$, $f_u(t, y^*, u^*) \in L^\infty(0, T, \mathbb{R}^{2n})$, where

(y^*, u^*) is an optimal pair. By optimality we have for $\lambda \in (0, 1)$ and $\forall v \in L^2(0, T; \mathbb{R}^m)$, $u^*(t) + \lambda v(t) \in U$ *a.e. $t \in (0, T)$*

$$\int_0^T L(t, y^*, u^*) \, dt + g_2(y^*(T))$$

$$\leqslant \int_0^T L\big(t, y^{u^*+\lambda v}, u^* + \lambda v\big) \, dt + g_2\big(y^{u^*+\lambda v}(T)\big).$$

Here y^v is the solution of the Cauchy problem $y' = f(t, y, v)$, $y(0) = y_0$. Noticing that

$$z(t) = \lim_{\lambda \to 0} \frac{1}{\lambda}\big(y^{u^*+\lambda v}(t) - y^*(t)\big)$$

is the solution to the system in variations

$$\begin{cases} z' = f_y(t, y^*, u^*)z + f_u(t, y^*, u^*)v, & t \in (0, T), \\ z(0) = 0, \end{cases} \tag{3.6}$$

we find that

$$\int_0^T \big(L_y(t, y^*, u^*), z(t)\big) + \big(L_u(t, y^*, u^*), v(t)\big) \, dt + \big(\nabla g_2(y^*(T)), z(T)\big) \geqslant 0. \tag{3.7}$$

Next we define p the solution to the Cauchy problem

$$\begin{cases} p' = -f_y^*(t, y^*, u^*)p + L_y(t, y^*, u^*), & \text{a.e. } t \in (0, T), \\ p(T) = -\nabla g_2(y^*(T)). \end{cases}$$

Multiplying the latter by z and integrating on $(0, T)$ we find by (3.6), (3.7) that

$$\int_0^T \big(L_u(t, y^*, u^*) - f_u(t, y^*, u^*), v(t)\big) \, dt \geqslant 0.$$

Equivalently,

$$\int_0^T \big(L_u(t, y^*, u^*) - f_u(t, y^*, u^*), w(t) - u^*(t)\big) \, dt \geqslant 0$$

for all $w \in L^2(0, T; \mathbb{R}^m)$, $w(t) \in U$ a.e. $t \in (0, T)$. This means (see (2.4)) that

$$f_u\big(t, y^*(t), u^*(t)\big) - L_u\big(t, y^*(t), u^*(t)\big) \in N_U\big(u^*(t)\big) \quad \text{a.e. } t \in (0, T).$$

If f is linear in u and $u \to L(t, y, u)$ is convex the latter is equivalent with (3.3).

3.2. *Proof of the maximum principle*

The main idea of proof given here is essentially due to F. Clarke [11]. We may assume that

$$\|f(t, y^*, u)\| + \|L_y(t, y, u)\| + \|f_y(t, y, u)\| \leqslant \beta(t) \tag{3.8}$$

for a.e. $t \in [0, T]$ and all $(y, u) \in \mathbb{R}^n \times U(t)$ where $\beta \in L^1[0, T]$. This clearly implies that

$$\begin{aligned}
\|f(t, x, u) - f(t, y, u)\| &\leqslant \beta(t)\|x - y\|, \\
\|f(t, y, u)\| &\leqslant \beta(t)(\|y - y^*\| + 1)
\end{aligned} \tag{3.9}$$

and so (standard existence theory) the state system (1.11) has a unique absolutely continuous solution $y \in \mathrm{AC}([0, T]; \mathbb{R}^n)$ for each function $u(t) \in U(t)$, $\forall t \in (0, T)$. This can be achieved by replacing $U(t)$ by

$$\begin{aligned}
U_n(t) = \{u \in U(t); \; &\|f(t, y^*, u) - f(t, y^*, u^*)\| \leqslant n, \\
&\|f_y(t, y, u) - f_y(t, y, u^*)\| \leqslant n \; \forall y \in \mathbb{R}^n\}.
\end{aligned}$$

Indeed, if (y^*, u^*) is optimal then clearly it is optimal in problem (1.14) with control constraints $u(t) \in U_n(t)$ a.e. $t \in (0, T)$ and so if the maximum principle (3.1)–(3.3) is true in this case, there are $\{p_n, \lambda_n\}$ satisfying the conditions of Theorem 3.1 where U is replaced by $U_n(t)$.

By (3.1) and by assumption (ii) we see that

$$\|p_n'(t)\| \leqslant \alpha_0(t, u^*(t)), \quad \text{a.e. } t \in (0, T).$$

If $\{|p_n(0)|\}$ is bounded this implies that $\{p_n\}$ is compact in $C([0, T]; \mathbb{R}^n)$ and so on a subsequence

$$p_n \to p \quad \text{in } C([0, T]; \mathbb{R}^n)$$

where p satisfies Eqs. (3.1)–(3.3). Otherwise, we set $q_n = p_n(t)/\|p_n(0)\|$ and conclude as above that

$$p_n \to p \quad \text{in } C([0, T]; \mathbb{R}^n)$$

where p satisfies Eqs. (3.1)–(3.3) with $\lambda = 0$.

Now we come back to the proof of the maximum principle. We set

$$X = \{u : [0, T] \to \mathbb{R}^m \text{ measurable}; \; u(t) \in U(t), \text{ a.e. } t \in [0, T]\}. \tag{3.10}$$

It is easily seen that this is a complete metric space with the Ekeland distance

$$d(u, v) = m\{t \in [0, T]; \; u(t) \neq v(t)\}. \tag{3.11}$$

Let $Y = X \times \mathbb{R}^n$ endowed with the standard product metric and consider the function

$$\varphi : Y \to \mathbb{R}, \quad \varphi(u, y_0) = \int_0^T L\big(t, y^u(t), u(t)\big)\, dt + g\big(y_0, y^u(T)\big) \tag{3.12}$$

where

$$\begin{cases} (y^u)' = f(t, y^u, u) & \text{a.e. in } (0, T), \\ y^u(0) = y_0. \end{cases}$$

We define the function

$$\varphi_\varepsilon(u, y_0) = \Big[\big((\varphi(u, y_0) - \varphi(u^*, y^*(0)) + \varepsilon)^+\big)^2 + d_C^2\big(y_0, y^u(T)\big)\Big]^{1/2} \tag{3.13}$$

for all $(u, y_0) \in Y$. Here d_C is the distance to C.

By Ekeland's principle, Theorem 2.1, for each $\varepsilon > 0$ there exists $(u_\varepsilon, y_0^\varepsilon) \in Y$ such that $\varphi_\varepsilon(u_\varepsilon, y_0^\varepsilon) \leq \varepsilon$,

$$d(u_\varepsilon, u^*) + \big\| y_0^\varepsilon - y^*(0) \big\| \leq \sqrt{\varepsilon} \tag{3.14}$$

and

$$\varphi_\varepsilon(u_\varepsilon, y_0^\varepsilon) \leq \varphi_\varepsilon(u, y_0) + \sqrt{\varepsilon}\big(d(u_\varepsilon, u) + \| y_0^\varepsilon - y_0\|\big), \quad \forall (u, y_0) \in Y. \tag{3.15}$$

We take $y_0^\rho = y_0^\varepsilon + \rho z_0$ and set $y_\varepsilon = y^{u_\varepsilon}$, $y_\varepsilon(0) = y_0^\varepsilon$ with $\| z_0 \| = 1$. In (3.15) we take the "spike" admissible control

$$u_\rho(t) = \begin{cases} u(t), & t \in [t_0 - \rho, t_0], \\ u_\varepsilon(t), & t \in [0, T] \setminus (t_0 - \rho, t_0) \end{cases}$$

where $t_0 \in [0, T]$ is arbitrary but fixed and $u \in X$. Then we have

$$y_\rho(t) = y^{u_\rho}(t) = y_\varepsilon(t) + \rho z_\varepsilon(t) + \mathrm{o}(\rho)$$

where $z_\varepsilon \in BV([0, T]; \mathbb{R}^n)$ is the solution to the equation in variations

$$\begin{cases} z_\varepsilon' = f_y(t, y_\varepsilon, u_\varepsilon) z_\varepsilon & \text{a.e. } t \in (0, t_0) \cup (t_0, T), \\ z_\varepsilon(0) = z_0, \\ z_\varepsilon^+(t_0) - z_\varepsilon^-(t_0) = f\big(t_0, y_\varepsilon(t_0), u(t_0)\big) - f\big(t_0, y_\varepsilon(t_0), u_\varepsilon(t_0)\big). \end{cases} \tag{3.16}$$

(Here $BV([0, T]; \mathbb{R}^n)$ is the space of functions of bounded variation on $[0, T]$.)

By (3.15) we have

$$-2\sqrt{\varepsilon} \leq \frac{\varphi_\varepsilon(u_\rho, y_0^\rho) - \varphi_\varepsilon(u_\varepsilon, y_0^\varepsilon)}{\rho}$$

$$
= \frac{1}{\varphi_\varepsilon(u_\rho, y_0^\rho) + \varphi_\varepsilon(u_\varepsilon, y_0^\varepsilon)} \left[\frac{((\varphi(u_\rho, y_0^\rho) - \varphi(u^*, y^*(0)) + \varepsilon)^+)^2}{\rho} \right.
$$

$$
- \frac{((\varphi(u_\varepsilon, y_0^\varepsilon) - \varphi(u^*, y^*(0)) + \varepsilon)^+)^2}{\rho}
$$

$$
\left. + \frac{d_C^2(y_0^\rho, y_\rho(T)) - d_C^2(y_0^\varepsilon, y_\varepsilon(T))}{\rho} \right].
$$

Letting ρ tend to zero, we obtain after some calculation that

$$
-2\sqrt{\varepsilon} \leqslant \frac{(\varphi(u_\varepsilon, y_0^\rho) - \varphi_\varepsilon(u^*, y^*(0)) + \varepsilon)^+}{\varphi_\varepsilon(u_\varepsilon, y_0^\varepsilon)} \left[\int_0^T L_y(t, y_\varepsilon, u_\varepsilon) z_\varepsilon \, dt \right.
$$

$$
+ \left(\nabla g(y_0^\varepsilon, y_\varepsilon(T)), (z_0, z_\varepsilon(T)) \right) + L(t_0, y_\varepsilon(t_0), u(t_0))
$$

$$
\left. - L(t_0, y_\varepsilon(t_0), u_\varepsilon(t_0)) \right]
$$

$$
+ \frac{(\nabla d_C(y_0^\varepsilon, y_\varepsilon(T)), (z_0, z_\varepsilon(T)))}{\varphi_\varepsilon(u_\varepsilon, y_0^\varepsilon)} d_C(y_0^\varepsilon, y_\varepsilon(T)).
$$

Here t_0 is a Lebesgue point for $L(t, y_\varepsilon(t), u_\varepsilon(t))$ and $L(t, y_\varepsilon(t), u(t))$.

We remind that s is a *Lebesgue point* for a function $v : I \subset \mathbb{R} \to \mathbb{R}$ if

$$
\lim_{t \to s} \frac{1}{t-s} \int_s^t |v(\tau) - v(s)| \, d\tau = 0 \tag{3.17}
$$

and if $v \in L^1_{\text{loc}}$ then almost all points $s \in I$ are Lebesgue points of v. We have

$$
\begin{cases} |\nabla d_C(y_0^\varepsilon, y_\varepsilon(T))| = 1 & \text{if } (y_0^\varepsilon, y_\varepsilon(T)) \notin C, \\ d_C(y_0^\varepsilon, y_\varepsilon(T)) |\nabla d_C(y_0^\varepsilon, y_\varepsilon(T))| = 0 & \text{if } (y_0^\varepsilon, y_\varepsilon(T)) \in C. \end{cases}
$$

We set

$$
\begin{cases} \lambda_\varepsilon = \dfrac{(\varphi(u_\varepsilon, y_0^\varepsilon) - \varphi(u^*, y^*(0)) + \varepsilon)^+}{\varphi_\varepsilon(u_\varepsilon, y_0^\varepsilon)}, \\[2ex] \mu_\varepsilon = \dfrac{d_C(y_0^\varepsilon, y_\varepsilon(T))}{\varphi_\varepsilon(u_\varepsilon, y_0^\varepsilon)} \nabla d_C(y_0^\varepsilon, y_\varepsilon(T)). \end{cases}
$$

This yields

$$
\lambda_\varepsilon \left[\int_0^T L_y(t, y_\varepsilon, u_\varepsilon) z_\varepsilon \, dt + \left(\nabla g(y_0^\varepsilon, y_\varepsilon(T)), (z_0, z_\varepsilon(T)) \right) + L(t_0, y_\varepsilon(t_0), u(t_0)) \right.
$$

$$
\left. - L(t_0, y_\varepsilon(t_0), u_\varepsilon(t_0)) \right] + (\mu_\varepsilon, (z_0, z_\varepsilon(T))) \geqslant -2\sqrt{\varepsilon}, \tag{3.18}
$$

where $\lambda_\varepsilon > 0$ and

$$\lambda_\varepsilon^2 + \|\mu_\varepsilon\|^2 = 1, \quad \forall \varepsilon > 0. \tag{3.19}$$

Let $p_\varepsilon \in AC([0, T]; \mathbb{R}^n)$ be the solution to the backward differential system

$$\begin{cases} p_\varepsilon'(t) = -f_y^*(t, y_\varepsilon, u_\varepsilon)p_\varepsilon + \lambda_\varepsilon L_y(t, y_\varepsilon, u_\varepsilon), & t \in [0, T], \\ p_\varepsilon(T) = -P\big(\lambda_\varepsilon \nabla g(y_0^\varepsilon, y_\varepsilon(T)) + \mu_\varepsilon\big) \end{cases} \tag{3.20}$$

where P is the projection $P(y_1, y_2) = y_2$.

Substituting into (3.18), we get (via (3.16))

$$\lambda_\varepsilon \int_0^T L_y(t, y_\varepsilon, u_\varepsilon)z_\varepsilon \, dt$$

$$= -\int_0^T \big(p_\varepsilon(t), z_\varepsilon'(t) - f_y(t, y_\varepsilon, u_\varepsilon)z_\varepsilon\big) \, dt$$

$$- \big(p_\varepsilon(t_0), f\big(t_0, y_\varepsilon(t_0), u(t_0)\big) - f\big(t_0, y_\varepsilon(t_0), u_\varepsilon(t_0)\big)\big)$$

$$- \big(p_\varepsilon(0), z_0\big) - \big(P\big(\lambda_\varepsilon \nabla g(y_0^\varepsilon, y_\varepsilon(T)) + \mu_\varepsilon\big), z_\varepsilon(T)\big).$$

Substituting into (3.18) we obtain

$$-2\sqrt{\varepsilon} \leqslant \lambda_\varepsilon \big(L\big(t_0, y_\varepsilon(t_0), u(t_0)\big) - L\big(t_0, y_\varepsilon(t_0), u_\varepsilon(t_0)\big)\big)$$

$$- \big(p_\varepsilon(t_0), f\big(t_0, y_\varepsilon(t_0), u(t_0)\big) - f\big(t_0, y_\varepsilon(t_0), u_\varepsilon(t_0)\big)\big)$$

$$+ \lambda_\varepsilon \big(\nabla g(y_0^\varepsilon, y_\varepsilon(T)), (z_0, z_\varepsilon(T))\big) + \big(\mu_\varepsilon, (z_0, z_\varepsilon(T))\big)$$

$$- \big(p_\varepsilon(0), z_0\big) - \big(P\big(\lambda_\varepsilon \nabla g(y_0^\varepsilon, y_\varepsilon(T)) + \mu_\varepsilon\big), z_\varepsilon(T)\big) \tag{3.21}$$

for all $z_0 \in \mathbb{R}^n$ and $u(t_0) \in U(T_0)$. This yields

$$-2\sqrt{\varepsilon} \leqslant \lambda_\varepsilon \big(L\big(t_0, y_\varepsilon(t_0), u(t_0)\big) - L\big(t_0, y_\varepsilon(t_0), u_\varepsilon(t_0)\big)\big)$$

$$- \big(p_\varepsilon(t_0), f\big(t_0, y_\varepsilon(t_0), u(t_0)\big) - f\big(t_0, y_\varepsilon(t_0), u_\varepsilon(t_0)\big)\big)$$

$$- \big((p_\varepsilon(0), -(I - P)\big(\lambda_\varepsilon \nabla g(y_0^\varepsilon, y_\varepsilon(T)) + \mu_\varepsilon\big)), (z_0, z_\varepsilon(T))\big).$$

For $z_0 = 0$ we get

$$2\sqrt{\varepsilon} - \lambda_\varepsilon L\big(t_0, y_\varepsilon(t_0), u_\varepsilon(t_0)\big) + \big(f\big(t_0, y_\varepsilon(t_0), u_\varepsilon(t_0)\big), p_\varepsilon(t_0)\big)$$

$$\geqslant \max_{u \in U(t_0)} \big\{-\lambda_\varepsilon L\big(t_0, y_\varepsilon(t_0), u\big) + \big(f\big(t_0, y_\varepsilon(t_0), u\big), p_\varepsilon(t_0)\big)\big\} \tag{3.22}$$

and for $u = u_\varepsilon$ we see by (3.21) that

$$\big(p_\varepsilon(0), -p_\varepsilon(T)\big) = \lambda_\varepsilon \nabla g\big(y_\varepsilon(0), y_\varepsilon(T)\big) + \mu_\varepsilon + \eta_\varepsilon \tag{3.23}$$

where $\|\eta_\varepsilon\| \leqslant 2\sqrt{\varepsilon}$.

Then, on a subsequence, again denoted ε, we have

$$
\begin{aligned}
&\lambda_\varepsilon \to \lambda, \\
&\mu_\varepsilon \to \mu, \\
&p_\varepsilon \to p \qquad \text{in } C\big([0, T]; \mathbb{R}^n\big) \quad \text{(by Arzela theorem)}, \\
&y_0^\varepsilon \to y^*(0) \quad \text{in } C\big([0, T]; \mathbb{R}^n\big), \\
&m\big\{t \in [0, T]; \ u_\varepsilon(t) \neq u^*(t)\big\} \to 0.
\end{aligned}
\tag{3.24}
$$

Recalling that by (3.9)

$$
\big\| f(t, y_\varepsilon, u_\varepsilon) \big\| \leqslant \beta(t)\big(\|y_\varepsilon - y^*\| + 1\big)
$$

we infer that $\{y_\varepsilon\}$ is compact in $C([0, T]; \mathbb{R}^n)$ and $\{y_\varepsilon'\}$ is weakly compact in $L^1(0, T; \mathbb{R}^n)$. We have

$$
y_\varepsilon(t) = y_\varepsilon(0) + \int_0^t f\big(s, y_\varepsilon(s), u_\varepsilon(s)\big)\, ds
\tag{3.25}
$$

and

$$
\begin{aligned}
\big\| f\big(t, & y_\varepsilon(t), u_\varepsilon(t)\big) - f\big(t, y^*(t), u^*(t)\big)\big\| \\
&\leqslant \big\| f\big(t, y_\varepsilon(t), u_\varepsilon(t)\big) - f\big(t, y^*(t), u_\varepsilon(t)\big)\big\| \\
&\quad + \big\| f\big(t, y^*(t), u_\varepsilon(t)\big) - f\big(t, y^*(t), u^*(t)\big)\big\| \\
&\leqslant \beta(t)\big\| y_\varepsilon(t) - y^*(t)\big\| + \big\| f\big(t, y^*(t), u_\varepsilon(t)\big) - f\big(t, y^*(t), u^*(t)\big)\big\|.
\end{aligned}
\tag{3.26}
$$

On the other hand, since f is continuous in (y, u) we have

$$
\big\| f\big(t, y^*(t), u_\varepsilon(t)\big) - f\big(t, y^*(t), u^*(t)\big)\big\| \to 0 \quad \text{a.e. in } (0, T).
$$

Since

$$
\big\| f\big(t, y^*(t), u_\varepsilon(t)\big)\big\| \leqslant \beta(t) \qquad \text{a.e. } t \in [0, T]
$$

we infer by (3.26) and the Lebesgue dominated convergence theorem that

$$
f(t, y_\varepsilon, u_\varepsilon) \to f(t, y^*, u^*) \quad \text{in } L^1(0, T).
$$

Thus, by (3.25) we conclude that

$$
y_\varepsilon \to y^* \quad \text{in } C\big([0, T]; \mathbb{R}^n\big).
$$

By (3.23) we see that

$$
\big(p(0), -p(T)\big) = \lambda \nabla g\big(y^*(0), y^*(T)\big) + \mu
\tag{3.27}
$$

where $\mu \in N_C(y^*(0), y^*(T))$.

Indeed, since C is convex, d_C^2 is convex and we have that

$$\left(\mu_\varepsilon, \left(y_\varepsilon(0), y_\varepsilon(T)\right) - w\right) \geqslant 2\left(d_C^2\left(y_\varepsilon(0), y_\varepsilon(T)\right) - d_C^2(w)\right), \quad \forall w \in C$$

and passing to the limit we obtain

$$\left(\mu, \left(y^*(0), y^*(T)\right) - w\right) \geqslant 0, \quad \forall w \in C.$$

Since

$$f_y^*\left(t, y_\varepsilon(t), u_\varepsilon(t)\right) \to f_y^*\left(t, y^*(t), u^*(t)\right) \quad \text{a.e. } t \in (0, T),$$
$$L_y\left(t, y_\varepsilon(t), u_\varepsilon(t)\right) \to L_y\left(t, y^*(t), u^*(t)\right) \quad \text{a.e. } t \in (0, T)$$

by (3.8) and (3.20) it follows that

$$p' = -f_y^*(t, y^*, u^*)p + \lambda L_y(t, y^*, u^*) \quad \text{a.e. } t \in (0, T) \tag{3.28}$$

and by (3.22)

$$-\lambda L\left(t_0, y^*(t_0), u^*(t_0)\right) + \left(f\left(t_0, y^*(t_0), u^*(t_0)\right), p(t_0)\right)$$
$$= \max_{u \in U(t_0)} \left\{-\lambda L\left(t_0, y^*(t_0), u\right) + \left(f\left(t_0, y^*(t_0), u\right), p(t_0)\right)\right\} \quad \text{a.e. } t_0 \in [0, T].$$
$$\tag{3.29}$$

Recall also that $\lambda \geqslant 0$ and

$$|\lambda| + \|\mu\| = 1. \tag{3.30}$$

If $\lambda > 0$, then replacing p by p/λ we get (3.1)–(3.3) with $\lambda = 1$.

If $\lambda = 0$, then by (3.30) we see that $\mu \neq 0$ and so

$$\|p(0)\| + \|p(T)\| \neq 0.$$

Clearly, this implies that

$$\|p(t)\| \neq 0 \quad \forall t \in [0, T].$$

The proof of the theorem is complete.

REMARK 3.1. The problem (1.14) with state constraints

$$y(t) \in K \subset \mathbb{R}^m \quad \forall t \in [0, T] \tag{3.31}$$

where K is a closed convex subset of $\mathbb{R}^n$ can be treated similarly. In this case we take

$$\varphi_\varepsilon(u, y_0) = \Big[\big((\varphi(u, y_0) - \varphi(u^*, y^*(0)) + \varepsilon)^+\big)^2 + d_C^2(y_0, y^u(T))$$

$$+ \int_0^T d_K^2(y(t))\, dt \Big]^{1/2}.$$

We propose to the reader to obtain a optimality theorem of the type of Theorem 3.1 in the present situation.

REMARK 3.2. In Theorem 3.1 the condition that C is convex can be removed. (See [4] for a proof in this general case.)

3.3. *Convex optimal control problems*

We shall study here the problem (P):

$$\min\left\{ \int_0^T L\big(t, y(t), u(t)\big)\, dt + g\big(y(0), y(T)\big); \ \big(y(0), y(T)\big) \in C \right\} \tag{3.32}$$

subject to

$$y'(t) = A(t)y(t) + B(t)u(t) + f_0(t), \quad u \in L^1\big(0, T; \mathbb{R}^m\big)$$

$$u(t) \in U(t) \quad \text{a.e. } t \in [0, T]$$

where L is convex in (y, u), measurable in t, $g \in C^1(\mathbb{R}^n \times \mathbb{R}^n)$ is convex, C is closed and convex, $A(t) \in L(\mathbb{R}^n \times \mathbb{R}^n)$, $B(t) \in L(\mathbb{R}^m, \mathbb{R}^n)$, $A(\cdot)$, $B(\cdot)$, $f_0(\cdot)$ are integrable and $U(t) \subset \mathbb{R}^m$ is closed and convex.

We have by Theorem 3.1 the following sharpening of the maximum principle in this case:

COROLLARY 3.1. *If (y^*, u^*) is optimal and $u^* \in L^1(0, T)$, then there exists $p \in AC([0, T]; \mathbb{R}^n)$, $\lambda \in \{0, 1\}$ such that $\lambda + |p(t)| \neq 0$ and*

$$\begin{cases} p' = -A^*(t)p + \lambda L_y\big(t, y^*(t), u^*(t)\big) \quad \text{a.e. } t \in (0, T), \\ \big(p(0), -p(T)\big) - \lambda \nabla g\big(y^*(0), y^*(T)\big) \in N_C\big(y^*(0), y^*(T)\big), \end{cases} \tag{3.33}$$

$$B^* p(t) \in \lambda L_u\big(t, y^*(t), u^*(t)\big) + N_{U(t)}\big(u^*(t)\big) \quad \text{a.e. } t \in (0, T) \tag{3.34}$$

where $\partial L = (\partial_y L, \partial_u L)$ is the subdifferential of L.

In fact, (3.34) comes from the maximum principle

$$\lambda L\big(t, y^*(t), u^*(t)\big) - \big(f\big(t, y^*(t), u^*(t)\big), p(t)\big)$$
$$= \max_{u \in U(t)} \big\{\lambda L\big(t, y^*(t), u\big) - \big(f\big(t, y^*(t), u\big), p(t)\big)\big\} \tag{3.35}$$

where $f(t, y, u) \equiv A(t)y + B(t)u + f_0(t)$ and, because $\lambda L(t, y^*, \cdot)$ is convex, the maximum in (3.35) is attained for u^* satisfying (3.34).

COROLLARY 3.2. *Assume that $C = C_1 \times C_2$ and there exists (y, u) admissible such that $y(T) \in \operatorname{int} C_2$ or $y(0) \in \operatorname{int} C_1$. Then the problem is normal.*

PROOF. Assume that $\lambda = 0$ and $y(T) \in \operatorname{int} C_2$. Then $|p(t)| \neq 0$ and

$$\begin{cases} p' = -A^*(t)p & \text{a.e. } t \in [0, T], \\ p(0) \in N_{C_1}\big(y^*(0)\big), \quad -p(T) \in N_{C_2}\big(y^*(T)\big), \end{cases} \tag{3.36}$$

$$B^*(t)p(t) \in N_{U(t)}\big(u^*(t)\big), \quad \text{a.e. } t \in [0, T]. \tag{3.37}$$

We have

$$\begin{cases} y' = A(t)y + B(t) + f_0(t)u & \text{a.e. } t \in [0, T], \\ y(T) \in \operatorname{int} C_2. \end{cases} \tag{3.38}$$

This yields

$$(y^* - y)' = A(t)(y - y^*) + B(t)(u - u^*) \quad \text{a.e. } t \in [0, T]$$

and

$$-\big(p(T), y^*(T) - \rho w - y(T)\big) \geq 0, \quad \forall |w| = 1.$$

Then using (3.37) we get

$$\rho\big|p(T)\big| = -\big(p(T), y^*(T) - y(T)\big)$$
$$= -\big(p(0), y^*(0) - y_0\big) + \int_0^T \big(u - u^*, B^*(t)p(t)\big)\, dt \leq 0$$

(because $p(0) \in N_{C_1}(y^*(0))$). Hence $p(T) = 0$, contradiction. The corollary is proved. The case $y(0) \in \operatorname{int} C_1$ can be treated similarly. $\qquad\square$

COROLLARY 3.3. *Under assumption of Corollary 3.2, the system (3.33)–(3.34) is also sufficient for optimality.*

PROOF. As seen earlier, $\lambda = 1$. Let (y, u) be an arbitrary pair such that

$$y' = A(t)y + B(t)u + f_0(t) \quad \text{a.e. } t \in [0, T], \ u(t) \in U(t),$$

$$(y(0), y(t)) \in C_1 \times C_2.$$

We have (by convexity of C)

$$L(t, y^*, u^*) \leqslant L(t, y, u) + \big(L_y(t, y^*, u^*), y^* - y\big) + \big(L_u(t, y^*, u^*), u^* - u\big),$$

$$g\big(y^*(0), y^*(T)\big) \leqslant g\big(y(0), y(T)\big) + \big(y^*(0) - y(0), y^*(T) - y(T)\big)$$

$$\times \nabla g\big(y^*(0), y^*(T)\big).$$

Integrating on $[0, T]$ and using (3.33) and (3.34), we see that

$$\int_0^T L(t, y^*, u^*) \, dt + g\big(y^*(0), y^*(T)\big) \leqslant \int_0^T L(t, y, u) \, dt + g\big(y(0), y(T)\big)$$

as claimed. $\qquad\qquad\qquad\qquad\qquad\qquad\qquad\qquad\qquad\qquad\qquad\qquad\qquad\quad\square$

We shall prove now a maximum principle for the *convex control problem of Bolza* under more general conditions on L. We shall assume that

1. $L(t, \cdot, \cdot) : \mathbb{R}^n \times \mathbb{R}^n \to \mathbb{R}$ is convex, continuous and the Hamiltonian function

$$H(t, x, p) = \sup\big\{(p, u) - L(t, x, u) \colon u \in U(t)\big\}$$

  is finite and summable in t for each $(x, p) \in \mathbb{R}^n \times \mathbb{R}^m$. $U(t) \subset \mathbb{R}^m$ is closed and convex for each t. Moreover, assumption (ii) in Section 3.1 holds.
2. $g \in C(\mathbb{R}^n \times \mathbb{R}^n)$ is convex and $C = C_1 \times C_2$ where C_1, C_2 are convex and closed.
3. $f(t, y, u) = A(t)y + B(t)u + f_0(t)$ where $A(t) \in L^1(0, T; \mathbb{R}^n \times \mathbb{R}^n)$, $B(t) \in L^1(0, T; \mathbb{R}^m \times \mathbb{R}^n)$, $f_0 \in L^1(0, T)$.
4. There is (y, u) admissible such that either $y(0) \in \operatorname{int} C_1$ or $y(T) \in \operatorname{int} C_2$.

THEOREM 3.2. *Assume that* $(y^*, u^*) \in C([0, T]; \mathbb{R}^n) \times L^2([0, T]; \mathbb{R}^n)$ *is optimal. Then there exists* $p \in AC([0, T]; \mathbb{R}^n)$ *such that*

$$p' + A^*(t)p \in \partial_u L\big(t, y^*(t), u^*(t)\big) \quad \text{a.e. } t \in (0, T), \tag{3.39}$$

$$\big(p(0), -p(T)\big) \in \partial g\big(y^*(0), y^*(T)\big) + N_C\big(y^*(0), y^*(T)\big), \tag{3.40}$$

$$B^*p \in \partial_u L\big(t, y^*(t), u^*(t)\big) + N_{U(t)}\big(u^*(t)\big) \quad \text{a.e. } t \in (0, T). \tag{3.41}$$

Moreover, conditions (3.39)–(3.41) *are also sufficient for optimality.*

PROOF. For $\lambda > 0$ consider the problem

$$\min\bigg\{\int_0^T L_\lambda(t, y, u) \, dt + g_\lambda\big(y(0), y_\lambda(T)\big) + \frac{1}{2}\int_0^T \big|u(t) - u^*(t)\big|^2 \, dt$$

$$+ \frac{1}{2}|y(0) - y^*(0)|^2 + \frac{1}{2\lambda}\left(d_{C_1}^2\left(y_\lambda(0)\right) + d_{C_2}^2\left(y_\lambda(T)\right)\right), \ u(t) \in U(t) \right\} \quad (\mathrm{P}_\lambda)$$

subject to

$$y' = A(t)y + B(t)u + f_0(t) \quad \text{a.e. } t \in (0, T).$$

Problem (P_λ) has a unique solution (y_λ, u_λ).

LEMMA 3.1. *For $\lambda \to 0$ we have*

$$y_\lambda \to y^* \quad \text{in } C\left([0, T]; \mathbb{R}^n\right), \tag{3.42}$$

$$u_\lambda \to u^* \quad \text{in } L^2\left([0, T]; \mathbb{R}^n\right). \tag{3.43}$$

PROOF. Recall that (see Section 2.1)

$$L_\lambda(t, y, u) = \inf_{(z,v)} \left\{ \frac{|y - z|^2}{2\lambda} + \frac{|u - v|^2}{2\lambda} + L(t, z, v) \right\},$$

$$g_\lambda(y_1, y_2) = \inf_{(z_1, z_2)} \left\{ \frac{|y_1 - z_1|^2}{2\lambda} + \frac{|y_2 - z_2|^2}{2\lambda} + g(z_1, z_2) \right\}$$

and d_{C_1} (d_{C_2}) is the distance to C_1 (and C_2 respectively). We have

$$\int_0^T L_\lambda\left(t, y_\lambda(t), u_\lambda(t)\right) dt + g_\lambda\left(y_\lambda(0), y_\lambda(T)\right) + \frac{1}{2}\int_0^T \left|u_\lambda(t) - u^*(t)\right|^2 dt$$

$$+ \frac{1}{2}\left|y_\lambda(0) - y^*(0)\right|^2 + \frac{1}{2\lambda}\left(d_{C_1}^2\left(y_\lambda(0)\right) + d_{C_2}^2\left(y_\lambda(T)\right)\right)$$

$$\leqslant \int_0^T L\left(t, y^*(t), u^*(t)\right) dt + g\left(y^*(0), y^*(T)\right) \tag{3.44}$$

because $L_\lambda \leqslant L$, $g_\lambda \leqslant g$.

Let for $\lambda \to 0$,

$$u_\lambda \to \bar{u} \quad \text{weakly in } L^2\left(0, T; \mathbb{R}^m\right),$$

$$y_\lambda \to \bar{y} \quad \text{strongly in } C\left(0, T; \mathbb{R}^n\right).$$

Then by the Fatou lemma

$$\liminf_{\lambda \to 0} \int_0^T L_\lambda\left(t, y_\lambda(t), u_\lambda(t)\right) dt \geqslant \int_0^T L\left(t, \bar{y}(t), \bar{u}(t)\right) dt$$

and by the lower semicontinuity of g

$$\liminf_{\lambda \to 0} g_\lambda\left(y_\lambda(0), y_\lambda(T)\right) \geqslant g\left(\bar{y}(0), \bar{u}(T)\right).$$

Then by (3.44) we deduce that

$$y_\lambda(0) \to y^*(0), \quad u_\lambda \to u^* \quad \text{in } L^2(0, T; \mathbb{R}^m). \qquad \square$$

Now the maximum principle in (P_λ) yields $((\mathrm{P}_\lambda)$ is smooth and so Corollary 3.2 applies)

$$\begin{cases} p_\lambda' + A^*(t)p_\lambda = \nabla_y L_\lambda(t, y_\lambda, u_\lambda) \quad \text{a.e. } t \in (0, T), \\ (p_\lambda(0), -p_\lambda(T)) = \nabla g_\lambda(y_\lambda(0), y_\lambda(T)) + (y_\lambda(0) - y^*(0), 0) \\ \qquad\qquad\qquad + \dfrac{1}{\lambda}(\nabla d_{C_1}^2(y_\lambda(0)), \nabla d_{C_2}^2(y_\lambda(T))), \\ B^* p_\lambda = \partial_u L_\lambda(y_\lambda, u_\lambda) + u_\lambda - u^* + N_{U(t)}(u_\lambda(t)) \quad \text{a.e. } t \in (0, T). \end{cases} \qquad (3.45)$$

We shall prove now that $\{p_\lambda(t)\}$ is bounded in $\mathbb{R}^n$. We shall use the same argument as in the proof of Corollary 3.2. Indeed, we have

$$(p_\lambda(T), y_\lambda(T) - y(0) - \rho w) \geqslant (\nabla_2 g_\lambda(y_\lambda(0), y_\lambda(T)), y_\lambda(T) - y(0)) \geqslant M$$

because

$$\frac{1}{2}\nabla d_{C_2}^2(y) = \frac{1}{\lambda}\big(I - (I + \lambda \partial I_{C_2})^{-1}\big)(y).$$

Hence

$$\rho|p_\lambda(T)| \leqslant (p_\lambda(T), y_\lambda(T) - y(0)).$$

On the other hand, by (3.45), we have

$$-(p_\lambda(0), y_\lambda(0) - y(0)) + (p_\lambda(T), y_\lambda(T) - y(T))$$
$$= \int_0^T (\nabla_y L_\lambda(t, y_\lambda, u_\lambda), y_\lambda - y)\, dt + \int_0^T (B(u_\lambda - u), p_\lambda)\, dt$$

where

$$y' = Ay + Bu + f_0.$$

Then again by (3.45) we see that

$$(p_\lambda(T), y_\lambda(T) - y(T))$$
$$\geqslant (p_\lambda(0), y_\lambda(0) - y(0)) + \frac{1}{2\lambda}(d_{C_1}^2(y_\lambda(0)) - d_{C_2}^2(y(0))).$$

Substituting the latter in the previous inequalities we see that $\{p_\lambda(T)\}$ is bounded in $\mathbb{R}^n$.

On the other hand, the Hamiltonian function $H(t, y, p)$ is concave in y and convex in p. By assumption 1, it follows that for each $y_0 \in \mathbb{R}^n$ there is a neighborhood $\mathcal{V}(y_0)$ of y_0 and $\alpha \in L^1(0, T)$ such that

$$-H(t, y, 0) \leqslant \alpha(t), \quad \forall y \in \mathcal{V}(y_0), \ t \in (0, T).$$

Indeed, we may choose $\mathcal{V}(y_0)$ a simplex generated by $\{y_1, \dots, y_{N+1}\}$. Then

$$-H(t, y, 0) \leqslant -\sum_i \lambda_i H(t, y_i, 0) =: \alpha(t).$$

By the inequality

$$H(t, y, 0) - H(t, y, \rho w) \leqslant \rho(v, w), \quad \forall v \in \partial_p H(t, y, 0)$$

it follows that

$$\sup\{\|v\|; \ v \in \partial_p H(t, y, 0)\} \leqslant \beta(t), \quad \text{a.e. } t \in (0, T) \tag{3.46}$$

where $\beta \in L^1(0, T)$. We have

$$L_\lambda\big(t, y^*(t) + \rho w, v_0(t)\big) \leqslant L\big(t, y^*(t) + \rho w, v_0(t)\big) \leqslant \alpha(t)$$

for all $|w| = 1$ and $v_0(t) \in \partial_p H(t, y^*(t) + \rho w, 0)$. Because

$$L(t, y, v_0) = \sup\{(v_0, p) - H(t, y, p)\}$$

and

$$L(t, y, v_0) + H(t, y, 0) = 0.$$

We have, by (3.45) and the convexity of $L_\lambda(t, y, \cdot)$, that

$$\begin{aligned}
\big(p_\lambda' + A^* p_\lambda, y_\lambda - y^* - \rho w\big) + \big(B^* p_\lambda + u_\lambda - u^*, u_\lambda - v_0\big) \\
\geqslant L_\lambda(t, y_\lambda, u_\lambda) - L_\lambda(t, y^* + \rho w, v_0) \quad \text{a.e. } t \in (0, T).
\end{aligned}$$

Hence

$$\begin{aligned}
\rho\big|p_\lambda' + A^* p_\lambda\big| \leqslant \alpha(t) - L_\lambda(t, y_\lambda, u_\lambda) + \big(p_\lambda' + A^* p_\lambda, y_\lambda - y\big) \\
+ \big(B^* p_\lambda, u_\lambda - v_0\big) + (u_\lambda - u^*, u_\lambda - v_0).
\end{aligned} \tag{3.47}$$

This yields (L_λ is bounded from below by an affine function)

$$\rho \int_0^T |p_\lambda' + A^* p_\lambda|\, dt \leq C + \left(p_\lambda(T), y_\lambda(T) - y^*(T)\right)$$
$$- \left(p_\lambda(0), y_\lambda(0) - y^*(0)\right) + C\left|p_\lambda(t)\right|\left|u_\lambda(t) - v_0(t)\right|$$
$$+ |u_\lambda - u^*|^2 + |u_\lambda - u^*||u_\lambda - v_0| \quad \text{a.e. } t \in (0, T).$$

Since $\int_0^T |u_\lambda|^2\, dt$ and $|p_\lambda(0)|$, $|p_\lambda(T)|$ are bounded it follows by Gronwall's lemma that

$$|p_\lambda(t)| \leq C, \quad \forall t \in [0, T],$$
$$\int_0^T |p_\lambda'(t) + A^* p_\lambda(t)|\, dt \leq C.$$

As a matter of fact, by (3.47) we see that $\{p_\lambda'\}$ is weakly compact in $L^1(0, T; \mathbb{R}^n)$ (the Dunford–Pettis theorem). Then on a subsequence, again denoted λ, we have

$$\begin{aligned}
p_\lambda(t) &\to p(t) \quad \text{uniformly on } [0, T], \\
p_\lambda' &\to p' \qquad \text{weakly in } L^1(0, T; \mathbb{R}^n).
\end{aligned} \tag{3.48}$$

Moreover, letting $\lambda \to 0$ into (see the second equation in (3.45))

$$\left(p_\lambda(T), y_\lambda(0) - \xi\right) - \left(p_\lambda(T), y_\lambda(T) - \eta\right)$$
$$\geq g_\lambda\left(y_\lambda(0), y_\lambda(T)\right) - g_\lambda(\xi, \eta) + \frac{1}{\lambda}\left(d_{C_1}^2\left(y_\lambda(0)\right) + d_{C_2}^2\left(y_\lambda(T)\right)\right)$$
$$+ \left(y_\lambda(0) - y^*(0), y_\lambda(0) - \xi\right), \quad \forall(\xi, \eta) \in C$$

we see that

$$\left(p(0), y_\lambda(0) - \xi\right) - \left(p(T), y^*(T) - \eta\right) \geq g\left(y^*(0), y^*(T)\right) - g(\xi, \eta),$$
$$\forall(\xi, \eta) \in C.$$

Then letting $\lambda \to 0$ we see that the transversality condition (3.40) holds.

Next, we let $\lambda \to 0$ into the inequality (see (3.45))

$$\int_0^T \left(p_\lambda' + A^* p_\lambda, y_\lambda - y\right) dt + \int_0^T \left(B^* p_\lambda, u_\lambda - u\right) dt$$
$$\geq \int_0^T L_\lambda\left(t, y_\lambda, u_\lambda\right) dt + \frac{1}{2}\int_0^T |u_\lambda - u^*|^2\, dt - \int_0^T L_\lambda(t, y, u)\, dt$$
$$- \frac{1}{2}\int_0^T |u - u^*|^2\, dt.$$

By weak lower semicontinuity of the convex integrand we know that (see Lemma 3.1)

$$\liminf_{\lambda \to 0} \int_0^T L_\lambda(t, y_\lambda, u_\lambda)\, dt \geqslant \int_0^T L(t, y^*, u^*)\, dt.$$

Hence

$$\int_0^T \left(p' + A^* p, y^* - y \right) dt + \int_0^T \left(B^* p, u^* - u \right) dt$$

$$\geqslant \int_0^T L(t, y^*, u^*)\, dt - \int_0^T L(t, y, u)\, dt,$$

$$\forall (y, u) \in L^\infty(0, T; \mathbb{R}^n) \times L^2(0, T; \mathbb{R}^m).$$

The latter implies that

$$p'(t) + A^*(t) p(t) \in \partial_p L\big(t, y^*(t), u^*(t)\big) \qquad \text{a.e. } t \in (0, T),$$
$$B^*(t) p(t) \in \partial_u L\big(t, y^*(t), u^*(t)\big) + N_{U(t)}\big(u^*(t)\big) \quad \text{a.e. } t \in (0, T)$$

as claimed.

The sufficiency of conditions (3.39)–(3.41) for optimality is immediate. It relies on the obvious inequalities

$$L(t, y^*, u^*) \leqslant L(t, y, u) + \big(L_y(t, y^*, u^*), y^* - y \big)$$
$$+ \big(L_u(t, y^*, u^*) + \eta, u^* - u \big) \quad \text{a.e. } t \in (0, T), \ \eta \in N_{U(t)}\big(u^*(t)\big)$$

where (y, u) is any pair of functions such that $u(t) \in U(t)$. If we take (y, u) such that $y(0) \in C_1$ and $y(T) \in C_2$

$$y' = A(t) y + B(t) u + f_0(t)$$

and we integrate on $(0, T)$ we get by (3.40)–(3.41)

$$\int_0^T L(t, y^*, u^*)\, dt \leqslant \int_0^T L(t, y, u)\, dt + \big(p(T), y^*(T) - y(T) \big)$$

$$- \big(p(0), y^*(0) - y(0) \big)$$

$$\leqslant \int_0^T L(t, y, u)\, dt - g\big(y^*(0), y^*(T)\big) + g\big(y(0), y(T)\big).$$

Hence (y^*, u^*) is optimal in problem (P_λ). $\qquad\qquad\square$

The dual problem   Define the functions

$$M(t, q, w) = \sup\{(q, v) + (w, y) - L(t, y, v); \ v \in U(t)\},$$
$$m(p_1, p_2) = \sup\{(p_1, x_1) - (p_2, x_2) - g(x_1, x_2); \ x_1 \in C_1, \ x_2 \in C_2\}. \tag{3.49}$$

The problem

$$\min\left\{\int_0^T M\big(t, B^*(t)p, w(t)\big)\, dt + \int_0^T \big(f_0(t), p(t)\big)\, dt \right.$$
$$\left. + m\big(p(0), p(T)\big); \ p' + A^*(t)p = w \ \text{a.e. in } (0, T)\right\} \tag{P*}$$

is called the dual of (P). We have

THEOREM 3.3. *Under assumptions of Theorem 3.2 the pair* (y^*, u^*) *is optimal in problem* (P) *if and only if* (P*) *has a solution* (p^*, w^*). *We have*

$$\inf P + \inf P^* = 0. \tag{3.50}$$

The proof is immediate. It relies on the conjugacy relationship between L and M.

It turns out that in certain situations the dual problem (P*) is simpler than the primal problem (P). In particular, control constraints $u(t) \in U(t)$ disappear in the dual problem. Let us illustrate this on the following simple example:

$$\min\left\{\frac{1}{2}\int_0^T \|y(t)\|^2\, dt; \ y'(t) = A(t)y(t) + u(t), \ y(0) = y_0,\right.$$
$$\left. \|y(T)\| \leqslant 1, \|u(t)\| \leqslant \rho \ \text{a.e. } t \in (0, T)\right\}$$

where $\|\cdot\|$ is the Euclidean norm.

In this case

$$M(t, q, w) = \frac{1}{2}\|w\|^2 + \rho\|q\| \quad \forall (q, w) \in \mathbb{R}^n \times \mathbb{R}^n,$$
$$m(p_1, p_2) = (p_1, y_0) - \|p_2\| \quad \forall (p_1, p_2) \in \mathbb{R}^n \times \mathbb{R}^n$$

and so the dual control problem (P*) is

$$\min\left\{\int_0^T \left(\rho\|p(t)\| + \frac{1}{2}\|w(t)\|^2\right) dt + \big(p(0), y_0\big) + \|p(T)\|; \right.$$
$$\left. p' + A^*(t)p = w \ \text{a.e. in } (0, T)\right\}.$$

Comments The results presented above are essentially due to R.T. Rockafellar (see [32]). In infinite dimensional spaces such results were established in [6].

3.4. *Examples*

3.4.1. *The optimal control of the prey–predator system* We shall treat here some specific problems from different areas of interest. Consider the Volterra prey–predator system

$$\begin{cases} x'(t) = x(t)\big(\lambda_1 - \mu_1 u(t) y(t)\big), & t \in [0, T], \\ y'(t) = y(t)\big(-\lambda_2 + \mu_2 u(t) x(t)\big), & t \in [0, T], \\ x(0) = x_0, \quad y(0) = y_0, & x_0, y_0 > 0 \end{cases} \tag{3.51}$$

where x is the prey, y the predator and $0 \leqslant u(t) \leqslant 1$ is the segregation rate; λ_i, μ_i are positive constants.

Consider the optimal control problem (see [4,37]):

$$\min\big\{-\big(x(T) + y(T)\big); \ 0 \leqslant u \leqslant 1\big\}. \tag{3.52}$$

This is a Bolza optimal control problem where

$$L \equiv 0, \quad g(x, y) = -(x + y),$$

$$U = \{u \in \mathbb{R}; \ 0 \leqslant u \leqslant 1\},$$

$$f(x, y, u) = \begin{pmatrix} x(\lambda_1 - \mu_1 u y) \\ y(-\lambda_2 + \mu_2 u x) \end{pmatrix}.$$

The maximum principle (see Theorem 3.1) yields

$$\begin{cases} p_1' = -(\lambda_1 - \mu_1 u^* y^*) p_1 - \mu_2 u^* y^* p_2, & t \in [0, T], \\ p_2' = \mu_1 u^* x^* p_1 - (-\lambda_2 + \mu_2 u^* x^*) p_2, & t \in [0, T], \\ p_1(T) = 1, \quad p_2(T) = 1, \end{cases} \tag{3.53}$$

$$u^*(t) = \operatorname*{argmax}_{u \in U}\big\{x^*(t)\big(\lambda_1 - \mu_1 u y^*(t)\big) p_1(t) + y^*(t)\big(-\lambda_2 + \mu_2 u x^*(t)\big) p_2(t)\big\}$$

$$\text{a.e. } t \in [0, T]. \tag{3.54}$$

Equivalently

$$u^*(t) = \begin{cases} 0 & \text{if } \mu_2 p_2(t) - \mu_1 p_1(t) < 0, \\ 1 & \text{if } \mu_2 p_2(t) - \mu_1 p_1(t) > 0. \end{cases} \tag{3.55}$$

(Since $x_0, y_0 > 0$, we have $x^*(t) > 0$, $y^*(t) > 0$.)

We shall discuss the form of the optimal control according to the sign of $\mu_2 - \mu_1$. We note first that always u^* is a *bang–bang controller* because the set of zeros of the function $\mu_2 p_2 - \mu_1 p_1$ consists of a finite number of points.

1. $\mu_2 - \mu_1 < 0$.

   In this case

$$(\mu_2 p_2 - \mu_1 p_1)(T) = \mu_2 - \mu_1 < 0. \tag{3.56}$$

Hence

$$\mu_2 p_2 - \mu_1 p_1 < 0$$

in a maximal interval $[T - \delta, T]$. On this interval we have by (3.56) that $u^*(t) = 0$ and so

$$\begin{cases} p_1' = -\lambda_1 p, & t \in (T - \delta, T), \\ p_2' = \lambda_2 p_2, & t \in (T - \delta, T), \\ p_1(T) = p_2(T) = 1. \end{cases}$$

Hence

$$p_1(t) = e^{\lambda_1(T-t)}, \quad p_2(t) = e^{\lambda_2(t-T)} \quad \text{on } [T - \delta, T].$$

Since the function $\mu_2 e^{\lambda_2(t-T)} - \mu_1 e^{-\lambda_1(T-t)}$ is increasing, it follows that $(T - \delta, T) = (0, T)$. In other words, in this case $u^*(t) = 0$, $\forall t \in (0, T)$.

2. $\mu_2 - \mu_1 > 0$.

   Then

$$\mu_2 p_2(t) - \mu_1 p_1(t) > 0 \quad \text{for } t \in [T - \varepsilon, T]$$

and

$$u^*(t) = 0 \quad \text{for } t \in [T - \varepsilon, T].$$

(We may assume that $t \in [T - \varepsilon, T]$ is maximal with this property.) Let us prove that $t_1 = T - \varepsilon$ is a switching point for u^*, i.e.,

$$\mu_2 p_2(t) - \mu_1 p_1(t) < 0 \quad \text{for } 0 \leqslant t \leqslant T - \varepsilon.$$

We have on $(T - \varepsilon, T)$

$$\begin{cases} p_1' = -p_1(\lambda_1 - \mu_1 y^*) - \mu_2 y^* p_2, & t \in (T - \varepsilon, T), \\ p_2' = -p_2(\mu_2 x^* - \lambda_2) + \mu_1 x^* p_1 \end{cases} \tag{3.57}$$

i.e.,

$$\begin{cases} p_1' = -\lambda_1 p_1 + y^*(\mu_1 p_1 - \mu_2 p_2), & t \in (t_1, T), \\ p_2' = \lambda_2 p_2 + x^*(\mu_1 p_1 - \mu_2 p_2), & t \in (t_1, T). \end{cases}$$

Hence

$$p_1(t) \geqslant e^{\lambda_1(T-t)} \geqslant 1, \quad p_1(t_1) = \frac{\mu_2}{\mu_1} p_2(t_1) \geqslant \frac{\mu_2}{\mu_1} > 0.$$

Note that $\mu_1 p_1(t) - \mu_2 p_2(t) = \varphi(t)$ satisfies the equation

$$\varphi'(t) = u^* \varphi(t)(\mu_2 x^* - \mu_1 y^*) - \lambda_2 \mu_2 p_2 - \lambda_1 \mu_1 p_1, \quad t \in (0, t_1).$$

Hence

$$\varphi(t) = C e^{\int_t^{t_1} u^*(\mu_2 x^* - \mu_1 y^*)\, ds}$$
$$+ \int_t^{t_1} e^{\int_t^s u^*(\mu_2 x^* - \mu_1 y^*)\, dr} (\lambda_2 \mu_2 p_2 + \lambda_1 \mu_1 p_1)\, ds \quad \text{for } t \in (0, t_1).$$

Since $\varphi(t_1) = 0$ we see that $C = 0$ and

$$\lambda_2 \mu_2 p_2(t_1) + \lambda_1 \mu_1 p_1(t_1) \geqslant \lambda_2 \mu_2 + \lambda_1 \mu_1 > 0$$

we conclude that $\varphi(t) > 0$ in a left neighborhood of t_1. Hence

$$u^*(t) = 0 \quad \text{for } t \in (t_1 - \varepsilon, t_1) = (t_2, t_1).$$

But as seen above, $\mu_2 p_2 - \mu_1 p_1$ is increasing on (t_2, t_1) and so

$$\mu_2 p_2(t) - \mu_1 p_1(t) < 0$$

for all $t < t_1$. Hence $t_1 = 0$ and so

$$u^*(t) = \begin{cases} 1 & \text{for } t_1 < t \leqslant T, \\ 0 & \text{for } 0 \leqslant t < t_1 \end{cases} \tag{3.58}$$

where the switching point t_1 can be computed from the equation

$$\mu_2 p_2(t) - \mu_1 p_1(t) = 0.$$

3. $\mu_2 - \mu_1 = 0$.
   In this case it follows that $u^*(t) = 0$ for all $t \in (0, T)$.

3.4.2. *Periodic solutions to Hamiltonian systems*   Consider the Hamiltonian system

$$\begin{cases} y'(t) = \partial_p H\big(y(t), p(t)\big) + f_1(t), & t \in (0, T), \\ p'(t) = -\partial_y H\big(y(t), p(t)\big) + f_2(t), & t \in (0, T) \end{cases} \tag{3.59}$$

with periodic conditions

$$y(0) = y(T), \quad p(0) = p(T). \tag{3.60}$$

Assumptions:

1. $H : \mathbb{R}^n \times \mathbb{R}^n \to \mathbb{R}$ is convex, of class C^1 and

$$\gamma_1 |y|^2 + \gamma_2 |p|^2 + C_1 \leqslant H(y, p) < \frac{\pi}{T} \left(|y|^2 + |p|^2 \right) + C_2 \quad \forall y, p \in \mathbb{R}$$

where $\gamma_1, \gamma_2 > 0$.

2. $f_1, f_2 \in L^2(0, T; \mathbb{R}^n)$.

A special case of (3.59) is $f_1 = 0$, $H(y, p) = g(y) + \frac{1}{2}|p|^2$. In this case, (3.59) reduces to

$$y' = p, \quad p' = -g'(y) + f_2(t)$$

i.e.,

$$y'' + g'(y) = f_2(t), \quad y(0) = y(t), \quad y'(0) = y'(T).$$

Following Clarke and Ekeland [13], we may reduce problem (3.59), (3.60) to the optimal control problem

$$\min \left\{ \int_0^T \left(G\big(v(t) - f_2(t), u(t) - f_1(t)\big) - \big(y(t), v(t)\big) \right) dt \right\} \tag{3.61}$$

subject to

$$\begin{aligned} y'(t) &= u(t), \quad z'(t) = -v(t) \quad \text{a.e. } t \in (0, T), \\ y(0) &= y(T), \quad z(0) = z(T) \end{aligned} \tag{3.62}$$

where $G = H^*$ is the conjugate of H, i.e.,

$$G(q_1, q_2) = \sup_{(y,p)} \big\{ (y, p_1) + (p, q_2) - H(y, p) \big\}.$$

If (y^*, z^*, u^*, v^*) is optimal the maximum principle yields (see Theorem 3.1)

$$\begin{cases} q_1'(t) = -\lambda v^*(t) & \text{a.e. } t \in (0, T), \\ q_2'(t) = 0 & \text{a.e. } t \in (0, T), \\ q_1(0) = q_1(T), \end{cases} \tag{3.63}$$

$$\begin{aligned} \big\{ u^*(t), v^*(t) \big\} = \operatorname*{argmax}_{\{u,v\}} \big\{ &(u, q_1(t)) - (v, q_2(t)) \\ &- \lambda G\big(v - f_2(t), u - f_1(t)\big) + \lambda\big(y^*(t), v\big) \big\}. \end{aligned} \tag{3.64}$$

Clearly, $\lambda = 1$ because otherwise (i.e. if $\lambda = 0$), $q_1 \equiv q_2 \equiv 0$ which contradicts (3.64). By (3.62)–(3.64) we have $q_2 \equiv C_2$, $q_1 - z^* \equiv C_1$ and by (3.64) we see that

$$\{y^*(t) - q_2(t), q_1(t)\} \in \partial G\big(v^*(t) - f_2(t), u^*(t) - f_1(t)\big) \quad \text{a.e. } t \in [0, T].$$

Since $(\partial G)^{-1} = \partial H$ we get (see Proposition 2.4)

$$\{v^*(t) - f_2(t), u^*(t) - f_1(t)\} \in \partial H\big(y^*(t) - C_2, z^*(t) + C_1\big).$$

We set

$$\bar{y} = y^*(t) - C_2, \quad p = z^*(t) + C_1. \tag{3.65}$$

We have (see (3.63))

$$\begin{cases} \bar{y}' \in \partial_p H(\bar{y}, p) + f_1(t) & \text{a.e. } t \in (0, T), \\ p' \in -\partial_y H(\bar{y}, p) + f_2(t) & \text{a.e. } t \in (0, T), \\ \bar{y}(0) = \bar{y}(T), \quad p(0) = p(T) \end{cases}$$

i.e., $(\bar{y}, p)$ is a solution of (3.59). We note also that, by assumption 1, we have

$$G(q_1, q_2) \leqslant \sup_{\{y, p\}} \left\{(y, p_1) + (p, q_2) - \gamma_1 |y|^2 - \gamma_2 |p|^2\right\} - C_1 < \infty,$$

$$\forall (q_1, q_2) \in \mathbb{R}^n \times \mathbb{R}^n$$

i.e., G is continuous on $\mathbb{R}^n \times \mathbb{R}^n$.

To conclude the proof, it remains to show that problem (3.61) has at least one solution. Let $\{y_n, z_n, u_n, v_n\}$ be such that

$$y_n' = u_n, \quad z_n' = -v_n, \quad y_n(0) = y_n(T), \quad z(0) = z_n(T),$$

$$d \leqslant \int_0^T G(v_n - f_2, u_n - f_1)\, dt - \int_0^T (y_n, v_n)\, dt \leqslant d + \frac{1}{n}. \tag{3.66}$$

where d is the infimum in (3.61).

By definition of G and by assumption 1, we have

$$G(v_n - f_2, u_n - f_1) - (y_n, u_n)$$

$$\geqslant (y, v_n - f_2) + (p, u_n - f_1) - (y_n, v_n) - \omega\big(|y|^2 + |p|^2\big)$$

$$\forall (y, p) \in \mathbb{R}^n \times \mathbb{R}^n$$

where $\omega < \frac{\pi}{T}$. This yields

$$G(v_n - f_2, u_n - f_1) - (y_n, u_n) \geqslant \frac{1}{4\omega}\big((v_n - f_2)^2 + (u_n - f_1)^2\big) - (y_n, v_n).$$

$$\tag{3.67}$$

On the other hand, the periodic solution to $y' = u$ is expressed as

$$y(t) = \sum_{m \neq 0} \frac{u^m e^{i\mu_m t}}{i\mu_m}, \quad \mu_m = \frac{2m\pi}{T}.$$

Hence

$$y_n(t) = \sum_{m \neq 0} \frac{u_n^m e^{i\mu_m t}}{i\mu_m}, \quad u_n^m = \frac{1}{T} \int_0^T u_n(t) e^{-i\mu_m t}\, dt.$$

This yields

$$\int_0^T (y_n, v_n)\, dt = \sum_{m \neq 0} \frac{u_n^m v_n^m}{i\mu_m}, \quad v_n^m = \frac{1}{T} \int_0^T v_n e^{-i\mu_m t}\, dt.$$

Hence by the Parseval formula

$$\int_0^T (y_n, v_n)\, dt \leqslant \frac{T}{4\pi} \int_0^T \left(u_n^2 + v_n^2 \right) dt.$$

Then, by (3.67) and by assumption 1, we have

$$\int_0^T \left(G(v_n - f_2, u_n - f_1) - (y_n, u_n) \right) dt \geqslant \alpha \int_0^T \left(|u_n|^2 + |v_n|^2 \right) dt + C.$$

Hence the sequences $\{u_n\}$, $\{v_n\}$ are bounded in $L^2(0, T; \mathbb{R}^n)$.

On a subsequence we have

$$\begin{aligned}
u_n &\to u^* \quad \text{weakly in } L^2(0, T; \mathbb{R}^n), \\
v_n &\to v^* \quad \text{weakly in } L^2(0, T; \mathbb{R}^n).
\end{aligned} \tag{3.68}$$

We set

$$\bar{y}_n(t) = y_n(t) - \int_0^T y_n(t)\, dt,$$

$$\bar{z}_n(t) = z_n(t) - \int_0^T z_n(t)\, dt.$$

Then

$$\int_0^T \bar{y}_n\, dt = \int_0^T \bar{z}_n\, dt = 0, \quad \bar{y}_n' = u_n, \quad \bar{y}_n' = -v_n$$

and $\{\bar{y}_n\}$, $\{\bar{z}_n\}$ are bounded in $H^1(0, T; \mathbb{R}^n)$. Hence

$$
\begin{aligned}
\bar{y}_n \to y^* \quad &\text{strongly in } C([0, T]; \mathbb{R}^n) \\
&\text{weakly in } H^1(0, T; \mathbb{R}^n), \\
\bar{z}_n \to z^* \quad &\text{strongly in } C([0, T]; \mathbb{R}^n) \\
&\text{weakly in } H^1(0, T; \mathbb{R}^n)
\end{aligned}
\tag{3.69}
$$

where $(y^*)' = u^*$, $(z^*)' = -v^*$, a.e. $t \in [0, T]$.

Clearly, $\{y^*, z^*, u^*, v^*\}$ is optimal in (3.61). This follows by (3.66) using (3.68), (3.69) and the fact that the convex integrand

$$
(u, v) \to \int_0^T G(v - f_2, u - f_1) \, \mathrm{d}t
$$

is weakly lower semicontinuous in $L^2(0, T; \mathbb{R}^n) \times L^2(0, T; \mathbb{R}^n)$. We have proved therefore

THEOREM 3.4. *Under assumptions 1 and 2 there exists a solution* $(y, p) \in H^1(0, T; \mathbb{R}^n) \times H^1(0, T; \mathbb{R}^n)$ *to the periodic problem* (3.59), (3.60).

3.4.3. *An application to resonant systems*   Consider the problem

$$
\min\left\{ \int_0^T u(t) \, \mathrm{d}t; \ u \in U \right\}
$$

where

$$
U = \{u : (0, 1) \to \mathbb{R}, \ \text{measurable}, \ 0 \leqslant u \leqslant B, \ \text{meas}[t: u(t) > 0] > 0,
$$

$$
y'' + uy = 0, \ y'(0) = y'(1) = 0 \text{ has at least one nontrivial solution}\}.
$$

THEOREM 3.5. *Assume that* $B > \pi^2$. *Then every optimal pair* (y^*, u^*) *satisfies*

$$
u^*(t) = \begin{cases} B, & 0 \leqslant t \leqslant t_1, \\ 0, & t_1 < t < t_2, \\ B, & t_2 \leqslant t < 1. \end{cases}
\tag{3.70}
$$

PROOF. The state system is

$$
y' = z, \quad z' = -uy, \quad z(0) = z(1) = 0.
$$

By the maximum principle, Theorem 3.1, there exists $\lambda = 0, 1$ and (p_1, p_2) such that

$$
\begin{aligned}
p_1' = up_2, \quad p_2' &= -p_1 \quad \text{a.e. } t \in [0, 1], \\
p_2'(0) = 0, \quad p_2'(1) &= 0,
\end{aligned}
\tag{3.71}
$$

$$
u^*(t) = \underset{u \in U}{\operatorname{argmax}}\{-uy^*(t)p_2(t) - \lambda u\} \quad \text{a.e. } t \in [0, T].
\tag{3.72}
$$

We note that $p_2 = p$ is the solution to

$$\begin{cases} p'' + u^* p = 0 & \text{a.e. } t \in [0, T], \\ p(0) = p(1) = 0. \end{cases}$$

Hence p and y^* are both solutions to $y'' + u^* y = 0$ and therefore $p = Cy^*$ on $(0, 1)$. If $\lambda = 0$, by (3.72) we see that

$$u^*(t) = \operatorname*{argmax}_{u \in U} \left\{ -C\big(y^*(t)\big)^2 u \right\}.$$

Hence, $u^*(t) = 0$ if $C > 0$ and $u^*(t) = B$ if $C < 0$. However $u^* \neq 0$ and $u^*(t) \neq B$ because we know that $u = \pi^2$ is admissible in our problem and $B > \pi^2$. Hence $\lambda \neq 0$ and so $\lambda = 1$. Then by (3.72) it follows that

$$-\big(y^*(t)p(t) + 1\big) \in N_U\big(u^*(t)\big) \subset N_{[0, B]}\big(u^*(t)\big) \quad \text{a.e. } t \in [0, 1].$$

Hence (see (2.15))

$$u^*(t) = \begin{cases} 0 & \text{if } y^*(t)p(t) + 1 > 0, \\ B & \text{if } y^*(t)p(t) + 1 < 0. \end{cases}$$

Equivalently

$$u^*(t) = \begin{cases} 0 & \text{if } C(y^*(t))^2 + 1 > 0, \\ B & \text{if } C(y^*(t))^2 + 1 < 0. \end{cases} \tag{3.73}$$

(This follows by an elementary argument as in [36].)

On the other hand, if u^* is optimal, then $|(y^*)'(t)| \neq 0$, $\forall t \in (0, 1)$. One can compute exactly (t_1, t_2) and find

$$v = \int_0^T u^*(t)\, dt.$$

From here, we may conclude that for any $u \in L^\infty(0, T)$ with $0 \leqslant u \leqslant B$, if

$$\int_0^T u(t)\, dt < v,$$

then the equation

$$y'' + uy = 0, \quad y'(0) = y'(1) = 0$$

has only the trivial solution $y = 0$.

This result is relevant in the theory of nonlinear resonant problems of the form

$$y'' + f(y) = 0, \quad y'(0) = y'(1) = 0. \qquad \square$$

3.5. *Reachable sets and optimal control problems*

In what follows we consider the controlled equation (1.11) supposed for simplicity to be autonomous:

$$y' = f^u(y) := f(y, u)$$

with the initial data

$$y(0) = y_0.$$

The hypotheses on f and the controller set $\mathcal{U}$ are such that for any $u \in \mathcal{U}$ the nonautonomous vector field $f^{u(t)}$ satisfies the hypotheses in Section 2.3. For example we will suppose that the controller set $\mathcal{U} = \{u : [0, T] \to U : u \in L^\infty(0, T)\}$ and $U \subset \mathbb{R}^m$ is closed. The vector field f is assumed to be C^∞ in y for any fixed $u \in U$ and f, $\frac{\partial f}{\partial y}$ are continuous in (y, u).

The *reachable set* (or *attainable set*) at moment t is defined as follows:

$$\mathcal{R}_{y_0}(t) = \left\{ y_0 \circ \overrightarrow{\exp} \int_0^t f^{u(\tau)} \, d\tau \;\middle|\; u \in \mathcal{U} \right\}.$$

We denote by $y^u(t) = \overrightarrow{\exp} \int_0^t f^{u(\tau)} \, d\tau$ and the flow corresponding to the equation $F_s^t = \overrightarrow{\exp} \int_s^t f^{u(\tau)} \, d\tau$.

We prove in this section that optimal control problems reduce to the study of *reachable sets*.

THEOREM 3.6. *If* $y^{u^*}(T) \in \partial\mathcal{R}_{y_0}(T)$, *the boundary of* $\mathcal{R}_{y_0}(T)$, *then, for all* $0 < s < T$, $y^{u^*}(s) \in \partial\mathcal{R}_{y_0}(s)$.

PROOF. It is clear that $F_s^T(\mathcal{R}_{y_0}(s)) \subset \mathcal{R}_{y_0}(T)$ and, since F_s^T is a diffeomorphism, $F_s^T(\text{int}\,\mathcal{R}_{y_0}(s)) \subset \text{int}\,\mathcal{R}_{y_0}(T)$ and the conclusion is immediate. $\qquad\square$

We will still call such trajectories *optimal* and the corresponding controller *optimal control*.

Optimal control problems   We consider the following controlled equation

$$\begin{cases} y' = f^u(y), \\ y(0) = y_0. \end{cases} \tag{3.74}$$

The optimal control problem is to find the admissible strategy $u : [0, T] \to U$ such that the pair (u, y^u) minimizes a cost functional J. Depending on the form of the cost functional J we distinguished in Section 1.2 three types of optimal control problems: Lagrange, Mayer and Bolza.

We consider first of all the case of the *Lagrange problem* with fixed end points y_0, y_1 and cost functional $J(y, u) = \int_0^T L(y(t), u(t)) \, dt$, with the Lagrangean L satisfying similar hypotheses with f. We introduce the new variable j and consider the new system:

$$\begin{cases} j' = L(t, y, u), \\ y' = f^u(y) \end{cases}$$
(3.75)

with initial data

$$y(0) = y_0, \quad j(0) = 0.$$

We consider the set of controllers $u \in \mathcal{U}$ such that $y^u(T) = y_1$. Then it is clear that, if (u^*, y^{u^*}) is an optimal pair then

$$\left(j(T), y^{u^*}(T) \right) \in \partial \mathcal{R}_{(0, y_0)}(T)$$

where $R_{(0, y_0)}$ refers to the reachable set for system (3.75). This reduction has some inconvenient in the sense that it does not distinguish between trajectories realizing the minimum of J and those realizing the maximum. That is why one considers the extended set of controllers $\widetilde{\mathcal{U}} = \{(u, v) \mid u \in \mathcal{U}, v \in [0, +\infty)\}$ and the problem

$$\begin{cases} j' = L(t, y, u) + v, \\ y' = f^u(y), \\ y(0) = q_0, \quad j(0) = 0. \end{cases}$$
(3.76)

It is now clear that if u^* is optimal for the Lagrange problem then $(\tilde{u}, 0)$ is optimal in (3.76).

Since *Mayer problem* is just a particular case of a *Bolza problem*, we describe the reduction for the latter. In this case the cost functional is $J(y, u) = g(y(T)) + \int_0^T L(t, y(t), u(t)) \, dt$ and the functions g, L satisfy similar hypotheses to f. We introduce the new variable j, the extended set of controllers $\widetilde{U}$ and consider the new system:

$$\begin{cases} j' = L(t, y, u) + \displaystyle\sum_{i=1}^{n} \frac{\partial l}{\partial y_i}(y) f_i(y, u) + v, \\ y' = f^u(y), \\ q(0) = q_0, \quad j(0) = 0. \end{cases}$$
(3.77)

As in the case of the Lagrange problem, if u^* is optimal for the Bolza problem, then $(u^*, 0)$ is optimal in (3.77).

We have thus seen that the study of optimal control problems reduces to the study of reachable sets, more precisely to the study of those trajectories whose final points belong to the boundary of the reachable set at that moment.

3.6. *Geometric form of Pontriaghin maximum principle*

In this section we present a coordinate free version of the maximum principle (see also Section 3.1). The proof we give here follows the ideas in [2].

For $u \in \mathcal{U}$ we consider the Hamiltonian $H^u := (f^u)^\#$.

THEOREM 3.7. *Suppose that for the admissible control* $u^*(t), t \in [0, T]$, $y^{u^*}(T) \in \partial \mathcal{R}_{y_0}(T)$. *Then there exists a non-zero Lipschitz curve in the cotangent bundle* $\xi(t) \in T^*_{y^{u^*}(t)} M$ *solution of the Hamiltonian equation*

$$\frac{\mathrm{d}}{\mathrm{d}t}\xi(t) = \overrightarrow{H}^{u^*(t)}(\xi(t))$$

and the following maximality condition is satisfied

$$\overrightarrow{H}^{u^*(t)}(\xi(t)) = \max_{u \in U} \overrightarrow{H}^{u(t)}(\xi(t)).$$

PROOF. The idea of the proof is to consider as curve $\xi(t)$ a curve of covectors "orthogonal" at each point to the boundary of the reachable set: $\partial \mathcal{R}_{y_0}(t)$.

Step 1. Let $\mathbf{T}: \mathbb{R}^N \to \mathbb{R}^n$ be Lipschitz continuous, $\mathbf{T}(0) = 0$ and differentiable at 0. Denote by $\mathbf{T}_0 = D\mathbf{T}(0)$ and by

$$\mathbb{R}^N_+ = \left\{(x_1, \ldots, x_N) \in \mathbb{R}^N \mid x_i \geqslant 0\right\}.$$

We prove that if $\mathbf{T}_0(\mathbb{R}^N_+) = \mathbb{R}^n$, then for any neighborhood $\mathcal{V}$ of 0 in $\mathbb{R}^N$, $0 \in \mathrm{int}\,\mathbf{T}(\mathcal{V} \cap \mathbb{R}^N_+)$.

By the convexity of $\mathbb{R}^N_+$ it turns out that $\mathbf{T}_0|_{\mathrm{int}\,\mathbb{R}^N_+}$ is surjective and let $y \in \mathrm{int}\,\mathbb{R}^N_+$ with $\mathbf{T}_0(y) = 0$ and $\delta > 0$ such that $y + B_\delta \subset \mathrm{int}\,\mathbb{R}^N_+$.

It is clear that one may find an n-dimensional linear subspace $X \subset \mathbb{R}^N$ such that $\mathbf{T}_0(X) = \mathbb{R}^n$ so $\mathbf{T}_0|_X$ is invertible. Denote by $D = B_\delta \cap X$ and for $\varepsilon > 0$ the continuous functions $\mathbf{T}_\varepsilon : D \to \mathbb{R}^n$:

$$\mathbf{T}_\varepsilon(v) = \frac{1}{\varepsilon}\mathbf{T}\big(\varepsilon(y + v)\big).$$

One may easily check that $\mathbf{T}_\varepsilon \to \mathbf{T}_0$ uniformly on $\overline{D}$ and since $0 \in \mathrm{int}\,\mathbf{T}_0(D)$ it turns out, using an usual argument based on degree theory, that $0 \in \mathrm{int}\,\mathbf{T}_\varepsilon(D)$ for $\varepsilon > 0$ small enough or, equivalently, $0 \in \mathrm{int}\,\mathbf{T}(\varepsilon(y + B_\delta))$ which concludes the Step 1.

Step 2. We consider an admissible control $u(t)$ and compute the end point of the trajectory using the second form of variations of parameters formula (2.40):

$$y^u(T) = y_0 \circ \overrightarrow{\exp} \int_0^T f^{u^*(t)} + \big(f^{u(t)} - f^{u^*(t)}\big)\,\mathrm{d}t$$

$$= y^{u^*}(T) \circ \overrightarrow{\exp} \int_0^T (F_t^T)_*\big(f^{u(t)} - f^{u^*(t)}\big)\,\mathrm{d}t. \tag{3.78}$$

Denote by

$$g^{t,u} = \left(F_t^T\right)_* \left(f^u - f^{u^*(t)}\right)$$

and let $\mathcal{L}$ be the set of Lebesgue points of u^* (see (3.17)). We suppose that the convex cone $\mathcal{W}$ generated by $\{g^{t,u}(y^{u^*}(T)) \mid t \in \mathcal{L}, u \in U\}$ is the whole tangent space $T_{y^{u^*}(T)}M$. We prove that, in this case, necessarily $y_1 := y^{u^*}(T) \in \operatorname{int} \mathcal{R}_{y_0}(T)$.

Indeed, one may find a finite set of points $0 < t_1 < \cdots < t_N < T$ and $u_1, \ldots, u_N \in U$ such that $\mathcal{W}$ is generated by the finite set $\{g^{t_i, u_i} \mid i = 1, \ldots, N\}$. For $x = (x_1, \ldots, x_N) \in \mathbb{R}_+^N$ we consider the strategy:

$$u^x(t) = \begin{cases} u_i, & t \in [t_i, t_i + x_i], \\ u^*(t), & t \in [0, T] \setminus \bigcup_{i=1}^{N} [t_i, t_i + x_i]. \end{cases}$$

Variation of parameters formula (3.78) gives:

$$y^{u^x}(T) = y_0 \circ \overrightarrow{\exp} \int_0^T f^{u^x(t)} \, dt$$

$$= y^{u^*}(T) \circ \overrightarrow{\exp} \int_{t_1}^{t_1 + x_1} g^{t, u_1} \, dt \circ \cdots \circ \overrightarrow{\exp} \int_{t_N}^{t_N + x_N} g^{t, u_n} \, dt. \tag{3.79}$$

If one considers the map:

$$\mathbf{T}(x_1, \ldots, x_N) = y^{u^x}(T), \quad x_1, \ldots, x_N \in \mathbb{R}$$

then it is easy to verify that $\mathbf{T}$ is Lipschitz continuous, $\mathbf{T}(0) = y_1$ and

$$\left. \frac{\partial \mathbf{T}}{\partial x_i} \right|_{(0,\ldots,0)} = g^{t_i, u_i}(y_1).$$

The hypotheses of Step 1 are verified and, consequently,

$$y_1 \in \operatorname{int} \mathbf{T}\left(\mathcal{V} \cap \mathbb{R}_+^N\right)$$

for any neighborhood $\mathcal{V}$ of 0 in $\mathbb{R}^N$ and, since u^x is an admissible strategy we find that

$$y_1 \in \operatorname{int} \mathcal{R}_{y_0}(T).$$

Step 3. Suppose now that $y^{u^*}(T) \in \partial \mathcal{R}_{y_0}(T)$. Since $0 \in \partial \mathcal{W}$, there exists a support hyperplane defined by $\xi(T) \in T^*_{y^{u^*}(T)}M, \xi(T) \neq 0$ such that

$$\xi(T)\left(g^{T,u}\left(y^{u^*}(T)\right)\right) \leqslant 0 \quad \text{a.e. } t \in [0, T], u \in U.$$

This means exactly that

$$\left[(F_t^T)^* \xi(T)\right]\left(f^{u^*}(y^{u^*}(t))\right) \geqslant \left[(F_t^T)^* \xi(T)\right]\left(f^u(y^u(t))\right).$$

Denote by

$$\xi(t) = (F_t^T)^* \xi(T).$$

The maximality condition is satisfied and, by (2.43)

$$\frac{\mathrm{d}}{\mathrm{d}t}\xi(t) = \overrightarrow{H}^{u^*(t)}(\xi(t))$$

which concludes the proof. $\qquad\square$

It is now easy to recover the maximum principle for optimal control problems, as is expressed in Section 3.1 by using the transformation described in Section 3.5 and the geometric form of maximum principle, Theorem 3.7.

3.7. *Free time optimal control problems*

The free time optimal control problem

$$\min_{(u,T)}\left\{\int_0^T L(t, y(t), u(t))\,\mathrm{d}t + g(y(0), y(T)); \ (y(0), y(T)) \in C\right\} \tag{3.80}$$

subject to

$$y' = f(y, u) \quad \text{a.e. } t \in (0, T)$$

can be reduced to a fixed time optimal control problem of the form (1.14) by substitution

$$t = \int_0^s w^2(\tau)\,\mathrm{d}\tau, \quad 0 \leqslant s \leqslant 1,$$
$$z(s) = y(t(s)), \qquad v(s) = u(t(s)).$$

This yields

$$\min_{(v,w)}\left\{\int_0^1 L(t(s), z(s), v(s))w^2(s)\,\mathrm{d}s + g(z(0), z(1)); \ (z(0), z(1)) \in C\right\}$$

subject to

$$z'(s) = f(z(s), v(s))w^2(s) \quad \text{a.e. } s \in (0, 1).$$

In this case, by Theorem 3.1 we find that the maximum principle has the form

$$
\begin{aligned}
& p'(t) = -f_y(y^*, u^*)p + \lambda L_y(y^*, u^*), \\
& (p(0), -p(T)) \in \lambda \nabla g(y^*(0), y^*(T)) + N_C(y^*(0), y^*(T)), \\
& (p(t), f(t, y^*(t), u^*(t))) - \lambda L(t, y^*(t), u^*(t)) \\
& \quad = \max_{(u \in U(t))} \{(p(t), f(y^*(t), u^*)) - \lambda L(y^*(t), u)\}, \\
& (p(t), f(t, y^*(t), u^*(t))) - \lambda L(t, y^*(t), u^*(t)) = 0 \quad \text{a.e. } t \in [0, T]
\end{aligned}
\tag{3.81}
$$

where $\lambda = 0, 1$ and $\|p(t)\| + \lambda \neq 0, \forall t \in [0, T]$.

In the special case, $L \equiv 1$, $g \equiv 0$, (3.80) reduces to the optimal time problem

$$
\min_{(T, u)} \{T; \ (y(0), y(T)) \in C, \ u \in U(t), \ \text{a.e. } t \in (0, T)\}.
\tag{3.82}
$$

We leave to the reader to deduce the maximum principle (from (3.81)) in this case. We note, however, that if $g = 0$, $C = \{y_0\} \times \{y_1\}$, then it reads as

$$
\begin{aligned}
& p'(t) = -f_y(t, y^*, u^*)p && \text{a.e. in } (0, T), \\
& (p(t), f(t, y^*(t), u^*(t))) - \lambda = 0 && \text{a.e. } t \in (0, T), \\
& (p(t), f(t, y^*(t), u^*(t))) = \max_{u \in U(t)} (p(t), f(t, y^*(t), u))
\end{aligned}
$$

where $\lambda \in \{0, 1\}$ and $|p(t)| + \lambda \neq 0, \forall t \in [0, T]$. Hence $|p(t)| \neq 0, \forall t \in [0, T]$.

From the geometric point of view, in the case of Lagrange or Bolza problems considered in Section 3.5, if the final time T is free the conclusion is that for the equivalent problems (3.75), respectively (3.77), the optimal pair $((u^*, 0), (j^*, y^{u^*}))$ satisfy, for $\varepsilon > 0$ small enough

$$
(j^*(T), y^{u^*}(T)) \in \partial \left(\bigcup_{|T-t|<\varepsilon} \mathcal{R}_{(0, y_0)}(t) \right).
\tag{3.83}
$$

So, in general we obtain the following maximum principle for free time problems:

THEOREM 3.8. *Suppose that for the admissible control $u^*(t)$, $t \in [0, T]$,*

$$
y^{u^*}(T) \in \partial \left(\bigcup_{|T-t|<\varepsilon} \mathcal{R}_{y_0}(t) \right)
$$

Then there exists a non-zero Lipschitz curve in the cotangent bundle $\xi(t) \in T^*_{y^{u^*}(t)}$ solution of the Hamiltonian equation

$$
\frac{d}{dt} \xi(t) = \overrightarrow{H}^{u^*(t)}(\xi(t))
$$

and the following maximality condition is satisfied

$$H^{u^*(t)}\big(\xi(t)\big) = \max_{u \in U} H^{u(t)}\big(\xi(t)\big). \tag{3.84}$$

Moreover, the following condition holds:

$$H^{u^*(t)}\big(\xi(t)\big) = 0 \quad a.e.\ t \in [0, T]. \tag{3.85}$$

PROOF. As before, the case of free time may be reduced to the fixed time situation by considering reparameterizations of the trajectories of the initial system and introduce a supplementary control related to the time rescaling in the way described below. First of all, one considers a reparameterization (a new time scale):

$$t = \gamma(s), \quad \gamma' > 0.$$

Then the solution $y^u(s) = y^u(\gamma(s))$ satisfies

$$\frac{\mathrm{d}}{\mathrm{d}s} y^u(s) = \gamma'(s) f^{u(\gamma(s))}(y).$$

The modified system is

$$\begin{cases} y' = \varphi f^u(y), \quad u \in U, |\varphi - 1| < \dfrac{\varepsilon}{T} < 1, \\ y(0) = y_0. \end{cases}$$

Admissible controls are measurable, bounded functions of the form $v(t) = (\varphi(t), u(t))$, $|\varphi - 1| \leqslant \frac{\varepsilon}{T}$ and the corresponding solution is denoted by $y^v(t)$. If we denote by $v^*(t) = (1, u^*(t))$ then it is clear that, since $y^{u^*}(T) \in \partial(\bigcup_{|T-t|<\varepsilon} \mathcal{R}_{y_0}(t))$ it follows that $y^{v^*}(T) \in \partial \mathcal{R}_{y_0}(T)$ and, at this point, the maximum principle may be applied. The Hamiltonian is

$$H^v(\xi) = \varphi H^u(\xi), \quad v = (\varphi, u).$$

The Hamiltonian system for H^{v^*} is the same as for H^{u^*}. The maximality condition becomes

$$H_{v^*}\big(\xi(t)\big) = H^{u^*(t)}\big(\xi(t)\big) = \max_{|\varphi-1|<\varepsilon T, u \in U} \varphi H^u\big(\xi(t)\big)$$

from which conditions (3.84), (3.85) follow. $\qquad\square$

Transversality conditions If in the optimal control problem of Lagrange we consider the initial and final states belonging to some submanifolds $y_0 \in M_0$, $y_1 \in M_1$, then supplementary transversality conditions appear, that is, the adjoint flow should satisfy (compare with (3.2))

$$\xi(0) \text{ orthogonal to } M_0,$$

$$\xi(T) \text{ orthogonal to } M_1.$$

4. The dynamic programming equation

The optimal controller u^* determined on $[0, T]$ via the optimality system (maximum principle) is also referred as *open loop optimal controller*. An alternative way is to look for an optimal controller expressed in feedback form. Such a controller is called *closed loop optimal controller* and it is determined from a Hamilton–Jacobi equation.

4.1. *Optimal feedback controllers and smooth solutions to Hamilton–Jacobi equation*

Consider the optimal control problem

$$\min\left\{\int_0^T L\big(t, y(t), u(t)\big)\, dt + \ell\big(y(T)\big); \ y'(t) = f\big(t, y(t), u(t)\big),\right.$$

$$\left. y(0) = y_0, \ u(t) \in U(t), \ \text{a.e. } t \in [0, T]\right\}. \tag{4.1}$$

This is a special case of problem (1.14) where $g(y_1, y_2) = l(y_2)$ and $C = \{y_0\} \times \mathbb{R}^n$. The control $u : [0, T] \to \mathbb{R}^m$ is of course measurable. We shall assume as usually that

$$L : [0, T] \times \mathbb{R} \times \mathbb{R}^m \to \mathbb{R}, \quad f : [0, T] \times \mathbb{R}^n \times \mathbb{R}^m \to \mathbb{R}, \quad \ell : \mathbb{R}^n \to \mathbb{R}$$

satisfy the assumption

$$\big\| L_y(t, y, u) \big\| + \big\| f_y(t, y, u) \big\| \leqslant \alpha(t, u), \quad \forall u \in U(t), \ t \in [0, T]$$

and ℓ, L, f are measurable in t, continuous in u while $U(t)$ is closed and $t \to U(t)$ measurable.

A function

$$V : [0, T] \times \mathbb{R}^n \to \mathbb{R}^m$$

Borel measurable is said to be a *feedback controller* for the system $y' = f(t, y, u)$ if for each $(t_0, y_0) \in [0, T] \times \mathbb{R}^n$ the Cauchy problem

$$\begin{cases} y'(t) = f\big(t, y(t), V\big(t, y(t)\big)\big), & \text{a.e. } t \in (t_0, T), \\ y(t_0) = y_0 \end{cases} \tag{4.2}$$

has at least one absolutely continuous solution $y \in C([0, T]; \mathbb{R}^n)$. The system (4.2) is called a *closed loop system*.

A control $u(t) = V(t, y(t))$ where y is the solution to (4.2) is called *feedback controller*. In other words, a *feedback controller* (or *control*) is an input function (control) which is expressed as a function of time and of the state of the system in the present time.

If $u(t) = V(t, y(t))$ is optimal for problem (4.1), then this feedback control is called *optimal feedback control*.

The representation of an optimal control in the feedback form (i.e. as a feedback controller) is called the *synthesis problem* of optimal control. The function V is also called *synthesis function* for the corresponding optimal control problem.

The synthesis problem is closely related to a first order partial differential equation. This is the *Hamilton–Jacobi–Bellman* or the *dynamic programming equation* associated to problem (4.1) and has the following form:

$$\begin{cases} \varphi_t(t, x) - \sup_{u \in U(t)} \left\{ -\left(\varphi_x(t, x), f(t, x, u)\right) - L(t, x, u) \right\} = 0 \\ \qquad\qquad\qquad\qquad \forall x \in \mathbb{R}^n, \ t \in [0, T], \\ \varphi(T, x) = \ell(x) \quad \forall x \in \mathbb{R}^n \end{cases} \tag{4.3}$$

where

$$\varphi_t = \frac{\partial \varphi}{\partial t}, \quad \varphi_x = \frac{\partial \varphi}{\partial x}.$$

If we denote by $H : [0, T] \times \mathbb{R}^n \times \mathbb{R}^m \to \mathbb{R}$ the Hamiltonian function

$$H(t, x, p) = \sup\left\{ -\left(p, f(t, x, u)\right) - L(t, x, u); \ u \in U(t) \right\} \tag{4.4}$$

then we may rewrite (4.3) as (compare with Eq. (1.10)):

$$\begin{cases} \varphi_t(t, x) - H\left(t, x, \varphi_x(t, x)\right) = 0, & t \in (0, T), x \in \mathbb{R}^n, \\ \varphi(T, x) = \ell(x), & x \in \mathbb{R}^n. \end{cases} \tag{4.5}$$

In order to make clear the relationship between this equation and problem (4.1) we introduce the function

$$\Phi(t, x) = \operatorname*{argsup}_{u \in U(t)} \left\{ -\left(\varphi_x(t, x), f(t, x, u)\right) - L(t, x, u) \right\}. \tag{4.6}$$

Let $\psi : [0, T] \times \mathbb{R}^n \to \mathbb{R}$ be the optimal value function associated with the optimal control problem, i.e.,

$$\psi(t, x) = \inf_u \left\{ \int_t^T L\left(x, y(s), u(s)\right) ds + \ell\left(y(T)\right); \right.$$

$$\left. y'(s) = f\left(s, y(s), u(s)\right), \ s \in (t, T), \ y(t) = x \right\}. \tag{4.7}$$

THEOREM 4.1. *Let* $\varphi \in C^1([0, T] \times \mathbb{R}^n)$ *be a solution to Hamilton–Jacobi equation, with the Cauchy condition* $\varphi(T) = \ell(x)$. *Assume that* Φ *is a feedback controller. Then*

$$\varphi(t, x) = \psi(t, x), \quad \forall(t, x) \in (0, T) \times \mathbb{R}^n$$

and Φ *is an optimal feedback controller.*

PROOF. Let (t, x) be fixed and let y^t be the solution to

$$\left(y^t\right)'(s) = f\left(s, y^t, \Phi\left(s, y^t(s)\right)\right), \quad s \in (t, T), \qquad y^t(t) = x. \tag{4.8}$$

We have

$$\frac{d}{ds}\varphi\left(s, y^t(s)\right) = \varphi_s\left(s, y^t(s)\right) + \left(\varphi_y\left(s, y^t(s)\right), f\left(s, y^t(s), \Phi\left(s, y^t(s)\right)\right)\right)$$

$$\text{a.e. } s \in (t, T), \tag{4.9}$$

Then by (4.6) we see that

$$\frac{d}{ds}\varphi\left(s, y^t(s)\right) = \varphi_s\left(s, y^t(s)\right) - H\left(s, y^t(s), \varphi_y\left(s, y^t(s)\right)\right)$$

$$- L\left(s, y^t(s), \Phi\left(s, y^t(s)\right)\right) \quad \text{a.e. } s \in (t, T).$$

By (4.5) we get

$$\frac{d}{ds}\varphi\left(s, y^t(s)\right) = -L\left(s, y^t(s), \Phi\left(s, y^t(s)\right)\right) \quad \text{a.e. } s \in (t, T).$$

Hence

$$\varphi(t, x) = \int_t^T L\left(s, y^t(s), \Phi\left(s, y^t(s)\right)\right) ds + \ell\left(y^t(T)\right) \geqslant \psi(t, x). \tag{4.10}$$

On the other hand, if (y, v) is any admissible pair into problem (4.1), i.e.,

$$v(s) \in U(s) \quad \text{a.e. } s \in (t, T), \quad L\left(s, y(s), v(s)\right) \in L^1(t, T)$$

and

$$\begin{cases} y'(s) = f\left(s, y(s), v(s)\right) & \text{a.e. } s \in (t, T), \\ y(t) = x \end{cases}$$

we have

$$\frac{d}{ds}\varphi\left(s, y(s)\right) = \varphi_s\left(s, y(s)\right) + \left(\varphi_x\left(s, y(s)\right), f\left(s, y(s), v(s)\right)\right)$$

$$\geqslant \varphi_s\left(s, y(s)\right) - H\left(s, y(s), \varphi_x\left(s, y(s)\right)\right) - L\left(s, y(s), v(s)\right)$$

$$\text{a.e. } s \in (t, T).$$

Hence

$$\frac{d}{ds}\varphi\left(s, y(s)\right) \geqslant -L\left(s, y(s), v(s)\right) \quad \text{a.e. } s \in (t, T)$$

and so

$$\varphi(t, x) \leqslant \int_t^T L\big(s, y(s), v(s)\big)\, \mathrm{d}s + \ell\big(y(T)\big). \tag{4.11}$$

Since (y, v) was arbitrary we conclude by (4.10) and (4.11) that $\varphi = \psi$ as claimed. Moreover, by (4.10) it follows that (for $t = 0$) $u = \Phi(t, y(t))$ is an optimal feedback controller. This completes the proof. $\qquad\square$

In this way, if the Hamilton–Jacobi equation has a C^1 solution then the synthesis problem reduces to this equation. Let us consider, in the next section, a simple special case.

4.2. *Linear quadratic control problems*

Let $U(t) \equiv \mathbb{R}^m$ and let

$$\begin{cases} f(t, y, u) \equiv b(t) + A(t)x + B(t)u & \forall x \in \mathbb{R}^n, u \in \mathbb{R}^m, t \in (0, T), \\ L(t, y, u) \equiv \dfrac{1}{2}\big(Q(t)x, x\big) + \dfrac{1}{2}|u|^2, \\ \ell(x) = \dfrac{1}{2}(P_0 x, x) \end{cases} \tag{4.12}$$

where

$$A(t) \in L^\infty\big(0, T; L(\mathbb{R}^n, \mathbb{R}^n)\big), \quad B(t) \in L^\infty\big(0, T; L(\mathbb{R}^m, \mathbb{R}^n)\big),$$
$$P_0 \in L\big(\mathbb{R}^n, \mathbb{R}^n\big), \quad b \in L^\infty\big(0, T; \mathbb{R}^n\big), \quad Q \in L^\infty\big(0, T; L(\mathbb{R}^n, \mathbb{R}^n)\big).$$

Here we have denoted by $(\cdot, \cdot)$ the scalar product and by $|\cdot|$ the Euclidean norm in $\mathbb{R}^n$ and $\mathbb{R}^m$, respectively. We shall assume further that P_0, $Q(t)$ are symmetric positive matrices, i.e.,

$$(P_0 x, x) \geqslant 0 \qquad \forall x \in \mathbb{R}^n, \tag{4.13}$$

$$\big(Q(t)x, x\big) \geqslant 0 \quad \forall x \in \mathbb{R}^n,\ t \in [0, T]. \tag{4.14}$$

In this case the Hamiltonian function is (see (4.4))

$$\begin{aligned} H(t, x, p) &= \sup_{u \in \mathbb{R}^m} \left\{ -\big(p, A(t)x + B(t)u + b\big) - \frac{1}{2}|u|^2 - \frac{1}{2}\big(Q(t)x, x\big) \right\} \\ &= -\frac{1}{2}\big(Q(t)x, x\big) - \big(p, b(t)\big) - \big(p, A(t)x\big) \\ &\quad + \sup_{u \in \mathbb{R}^m} \left\{ -\big(B^*(t)p, u\big) - \frac{1}{2}|u|^2 \right\} \\ &= -\frac{1}{2}\big(Q(t)x, x\big) + \frac{1}{2}\big|B^*(t)p\big|^2 - \big(p, b(t)\big) - \big(A^*(t)p, x\big) \end{aligned}$$

and so the Hamilton–Jacobi equation (4.5) becomes

$$\begin{cases} \varphi_t(t,x) - \dfrac{1}{2}\big|B^*(t)\varphi_x(t,x)\big|^2 + \big(\varphi_x(t,x),b(t)\big) \\[2mm] \qquad + \big(A(t)x,\varphi_x(t,x)\big) + \dfrac{1}{2}\big(Q(t)x,x\big) = 0, \quad t\in(0,T),\ x\in\mathbb{R}^n, \\[2mm] \varphi(T,x) = \dfrac{1}{2}(P_0 x,x). \end{cases} \qquad (4.15)$$

Moreover, by (4.6) we see that

$$\Phi(t,x) = -B^*(t)\varphi_x(t,x) \quad \forall t\in(0,T),\ x\in\mathbb{R}^n. \qquad (4.16)$$

The form of Eq. (4.15) suggests that φ should be a quadratic function in x. Indeed, if we look for φ under the form

$$\varphi(t,x) = \dfrac{1}{2}\big(P(t)x,x\big) + \big(r(t),x\big) \quad \forall t\in(0,T),\ x\in\mathbb{R}^n$$

we obtain by (4.15) (note that $\varphi_x = P(t)x + r(t)$)

$$\dfrac{1}{2}\big(P'(t)x,x\big) + \big(r(t),x\big) - \dfrac{1}{2}\big|B^*(t)P(t)x + B^*(t)r(t)\big|^2$$

$$\qquad + \big(A(t)x, P(t)x + r(t)\big) + \big(P(t)x + r(t), b(t)\big) + \dfrac{1}{2}\big(Q(t)x,x\big) = 0,$$

$$P(T) = P_0, \quad r(T) = 0.$$

Differentiating with respect to x, we get

$$P'(t)x + r'(t) - P(t)B(t)B^*(t)P(t)x + A^*(t)P(t)x + P(t)A(t)x$$

$$\qquad + A^*(t)r - P(t)B(t)B^*(t)r(t) + P(t)b(t) + Q(t)x = 0.$$

Hence $P(t)$ must be a solution to the Riccati equation

$$\begin{cases} P'(t) + A^*(t)P(t) + P(t)A(t) - P(t)B(t)B^*(t)P(t) + Q(t) = 0, \\ \quad t\in(0,T), \\ P(T) = P_0 \end{cases} \qquad (4.17)$$

while r satisfies the linear equation

$$\begin{cases} r'(t) + A^*(t)r(t) - P(t)B(t)B^*(t)r(t) + P(t)b(t) = 0 \\ \quad \text{a.e. } t\in(0,T), \\ r(T) = 0. \end{cases} \qquad (4.18)$$

The optimal feedback control is expressed as (see (4.16))

$$u(t) = \Phi(t,y) = -B^*(t)P(t)x - B^*(t)r(t), \quad t\in(0,T). \qquad (4.19)$$

In this way the synthesis problem for the linear quadratic optimal control problem (4.1), (4.12) reduces to the Riccati equation (4.17).

It turns out that under our assumptions the Cauchy problem (4.17) has a unique solution, globally defined (see [4,38])

$$P \in W^{1,\infty}\big([0,T]; \mathbb{R}^n\big), \quad P(t) = P^*(t), \quad P(t) \geqslant 0 \quad \forall t \in [0,T]$$

and this fact proves that a feedback operator for this problem exists.

REMARK 4.1. The hypothesis $Q \geqslant 0$ is essential for existence of optimal pairs. If Q is not positive, then it is possible that an optimal pair does not exist and the dynamic programming equation (4.3) does not have a solution defined for $t \in [0,T]$. More precisely, if we still search for a solution as a quadratic form, then the Riccati equation (4.17) may not have a globally defined solution. This is deeply related to the existence of *conjugate points*, almost as in the classical calculus of variations. For more details on conjugate points see [26,2].

EXAMPLE. Let us consider the problem

$$\min\left\{ \int_0^T \big(y^2 + u^2\big)\,dt;\ y' = y + u,\ y(0) = y_0 \right\}. \tag{4.20}$$

In this case the optimal feedback control is given by

$$u(t) = -P(t)y$$

where P is the solution of the equation

$$\begin{cases} P'(t) + 2P(t) - P^2(t) + 2 = 0, & t \in (0,T), \\ P(T) = 0. \end{cases}$$

We get

$$P(t) = 2\sqrt{3}\,C\,\frac{e^{-2t/\sqrt{3}}}{1 - C e^{-2t/\sqrt{3}}} + 1 + \sqrt{3}$$

where C is to be determined from the condition $P(T) = 0$.

The above discussion remains true for infinite dimensional linear quadratic problems of the form (4.1), (4.2) where $\mathbb{R}^n$ and $\mathbb{R}^m$ are replaced by the Hilbert spaces H and U and

$$B(t) \in L^\infty\big(0,T; L(U,H)\big), \quad P_0 \in L(H,H)$$

and

$$A(t) : H \to H$$

is a family of linear closed operators which generate an evolution operator in H. In this case, however the Riccati equation (4.17) should be understood in a weak sense.

4.3. *Viscosity solutions*

Coming back to Eq. (4.5) it must be said, however, that in general it has not a classical solution. Thus a new concept of solution has been introduced by M.G. Crandall and P.L. Lions in [20] (see also [17]).

Consider the equation

$$\varphi_t(t, x) - H(t, x, \varphi_x) = 0, \quad (t, x) \in Q = [0, T] \times \Omega. \tag{4.21}$$

The function $\varphi \in C(Q)$ is said to be a viscosity solution to Eq. (4.1) if for each $\chi \in C^1(\overline{Q})$ the following conditions hold:

1. If $\varphi - \chi$ has local maximum at $(t_0, x_0) \in Q$ then

$$\chi_t(t_0, x_0) - H\big(t_0, x_0, \chi_x(t_0, x_0)\big) \geq 0.$$

2. If $\varphi - \chi$ has a local minimum at (t_0, x_0) then

$$\chi_t(t_0, x_0) - H\big(t_0, x_0, \chi_x(t_0, x_0)\big) \leq 0.$$

We have

THEOREM 4.2. *The optimal value function ψ defined by (4.7) is a viscosity solution to Hamilton–Jacobi equation (4.21).*

PROOF. Let $(t_0, x_0) \in Q = (0, T) \times \mathbb{R}^n$ and let $\chi \in C^1([0, T] \times \mathbb{R}^n)$ such that $\psi - \chi$ has a local maximum at (t_0, x_0), i.e.,

$$\psi(t_0, x_0) - \chi(t_0, x_0) \geq \psi(t, x) - \chi(t, x), \quad \forall(t, x) \in V \tag{4.22}$$

where V is a neighborhood of (t_0, x_0).

Let (y, u) be an admissible pair on $[t_0, T]$, such that $y(t_0) = x_0$ and u is continuous on $[t_0, T]$. We have

$$\begin{cases} y'(s) = f\big(s, y(s), u(s)\big) & \text{a.e. } s \in (t_0, T), u(s) \in U(s), \\ y(t_0) = x_0. \end{cases} \tag{4.23}$$

In (4.22) we take $t \geq t_0$ and $x = y(t)$ and we have

$$\chi(t_0, x_0) - \chi\big(t, y(t)\big) \leq \psi(t_0, x_0) - \psi\big(t, y(t)\big)$$

$$\leq \int_{t_0}^{t} L\big(\tau, y(\tau), u(\tau)\big) \, d\tau, \quad t \geq t_0. \tag{4.24}$$

Indeed for each $0 \leqslant t \leqslant s \leqslant T$ we have

$$\psi(t, x) = \inf\left\{\int_t^s L\big(\tau, y(\tau, t, x), u(\tau)\big)\, d\tau + \psi\big(s, y(s, t, x, u)\big); \ u(\tau) \in U\right\}.$$
(4.25)

Here (s, t, x, u) is the solution to system (4.23) with the initial point (t, x). This equality is the so called dynamic programming principle and follows by a direct calculation. Now if we divide (4.24) by $t - t_0$ and let $t \to t_0$ we get

$$-\chi_t(t_0, x_0) - \big(\chi_x(t_0, x_0), f(t_0, x_0, u(t_0))\big) \leqslant L(t_0, x_0, u(t_0)).$$

Since $u(t_0)$ is arbitrary in $U(t_0)$ we obtain

$$-\chi_t(t_0, x_0) + H\big(t_0, x_0, \chi_x(t_0, x_0)\big) \leqslant 0$$

as claimed.

Assume now that $\psi - \chi$ has a local minimum at (t_0, x_0). We have

$$\chi(t_0, x_0) - \chi(t, x) \geqslant \psi(t_0, x_0) - \psi(t, x) \quad \forall (t, x) \in V$$

where V is a neighborhood of (t_0, x_0). Let (y, u) be an admissible pair such that $y(t_0) = x_0$. For $\varepsilon > 0$ sufficiently small we have

$$\psi\big(t_0 + \varepsilon, y(t_0 + \varepsilon)\big) - \psi(t_0, x_0) \geqslant \chi\big(t_0 + \varepsilon, y(t_0 + \varepsilon)\big) - \chi(t_0, x_0)$$
$$= \int_{t_0}^{t_0 + \varepsilon} \frac{d}{dt} \chi\big(t, y(t)\big)\, dt$$
$$= \int_{t_0}^{t_0 + \varepsilon} \chi_t\big(t, y(t)\big) + \big(\chi_y\big(t, y(t)\big), f\big(t, y(t), u(t)\big)\big)\, dt.$$
(4.26)

If

$$\chi_t(t_0, x_0) - H\big(t_0, x_0, \chi_x(t_0, x_0)\big) \geqslant \lambda > 0$$

then we have

$$\chi_t(t_0, x_0) + \big(\chi_x(t_0, x_0), f(t_0, x_0, u)\big) + L\big(t, y(t), u(t)\big) \geqslant \lambda, \quad \forall u \in U(t_0).$$

Hence

$$\chi_t\big(t, y(t)\big) + \big(\chi_y\big(t, y(t)\big), f\big(t, y(t), u(t)\big)\big) + L\big(t, y(t), u(t)\big) \geqslant \frac{\lambda}{2}$$

for $t_0 \leqslant t \leqslant t_0 + \varepsilon$, ε sufficiently small.

Then by (4.26) we see that

$$\psi\big(t_0 + \varepsilon, y(t_0 + \varepsilon)\big) - \psi(t_0, x_0) \geqslant \frac{\varepsilon\lambda}{2} - \int_{t_0}^{t_0+\varepsilon} L\big(t, y(t), u(t)\big)\, dt.$$

The latter contradicts the principle of dynamic programming (4.25). This completes the proof.

In particular, Theorem 4.2 implies the existence of a viscous solution for the Hamilton–Jacobi equation (4.5) with final Cauchy condition.

The uniqueness of such a viscosity solution is more delicate and follows by some sharp arguments developed by Crandall, Lions and Ishii (see [17–20]).   □

Let us conclude this section with a representation result for the viscosity solution to Hamilton–Jacobi equation in the case of a convex Hamiltonian, i.e. (see [4,5])

$$f(t, x, u) \equiv Ax + Bu, \quad U \text{ convex},$$
$$L(t, x, u) \equiv L(x, u), \quad L \text{ convex}.$$

Denote by M the conjugate function

$$M(q, w) = \sup\big\{(q, v) + (w, y) - L(y, v); \ v \in U, \ y \in \mathbb{R}^n\big\}.$$

Then, as seen earlier, we have

$$\psi(t, x) = -\inf\bigg\{\int_t^T M\big(B^* p, w\big)\, ds + \ell^*\big(-p(T)\big) + \big(p(t), x\big);$$

$$p' = -A^* p + w \text{ a.e. in } (t, T)\bigg\}$$

$$= -\inf_q\bigg\{\ell^*(-q) + \inf_w\bigg\{\int_t^T M\big(B^* p, w\big)\, ds + \big(p(t), x\big);$$

$$p' = -A^* p + w, \ p(T) = q\bigg\}\bigg\}$$

$$= -\inf_q\big\{\ell^*(-q) + \theta(t, q)\big\}$$

where

$$\theta(t, q) = \inf\bigg\{\int_t^T M\big(B^* p, w\big)\, ds + \big(p(t), x\big); \ p' = -A^* p + w, \ p(T) = q\bigg\}$$

$$= \inf\bigg\{\int_t^{T-t} M\big(B^* z, v\big)\, ds + \big(z(T - t), x\big); \ z' = A^* z + v, \ z(0) = q\bigg\}$$

$$= \inf_w\bigg\{\int_t^T M\big(B^* z, w\big)\, ds + \big(z(T), x\big); \ z' = A^* z + w, \ z(t) = q\bigg\}.$$

In other words, θ is the solution to Hamilton–Jacobi equation

$$\begin{cases} \theta_t(t,q) + \left(A^*q, \theta_q(t,q)\right) - \sup_w\left\{-(\theta_q(t,q), Bw) - M(B^*q, w); w \in \mathbb{R}^n\right\} = 0, \\ \theta(T,q) = (x,q). \end{cases} \tag{4.27}$$

We have proved therefore

THEOREM 4.3. *The viscosity solution to Hamilton–Jacobi equation*

$$\begin{cases} \varphi_t(t,x) + \left(Ax, \varphi_x(t,x)\right) - \sup_{u \in U}\left\{(-\varphi_x, Bu) - L(x,u)\right\} = 0, \\ \varphi(T,x) = \ell(x) \end{cases} \tag{4.28}$$

is given by

$$\varphi(t,x) = -\inf_q\left\{\ell^*(-q) + \theta(t,q)\right\} \tag{4.29}$$

where θ is solution to (4.27).

Let us consider the special case

$$L(x,u) \equiv h(u) + g(x), \quad A \equiv 0, \quad B \equiv I.$$

Then Eq. (4.28) reduces to (h^* is the conjugate of h)

$$\begin{cases} \varphi_t(t,x) - h^*\left(-\varphi_x(t,x)\right) + g(x) = 0, \\ \varphi(T,x) = \ell(x) \end{cases} \tag{4.30}$$

and so Theorem 4.3 yields

$$\varphi(t,x) = -\inf_q\left\{\ell^*(-q) + \theta(t,q)\right\} \tag{4.31}$$

where

$$\begin{cases} \theta_t(t,q) - g^*\left(-\theta_q(t,q)\right) + h^*\left(-B^*q\right) = 0, \\ \theta(T,q) = (x,q) \quad \forall t \in [0,T], \ q \in \mathbb{R}^n. \end{cases} \tag{4.32}$$

For $g = 0$ the solution φ to the problem

$$\begin{cases} \varphi_t(t,x) - h^*\left(-\varphi(t,x)\right) = 0 \\ \varphi(T,x) = \ell(x) \end{cases} \tag{4.33}$$

                    V. Barbu and C. Lefter

is therefore given by (Lax–Hopf formula)

$$\varphi(t, x) = -\inf_q \{\ell^*(-q) + (x, q) + (T - t)h^*(-q)\}$$

$$= \inf_y \left\{ \ell(y) + (T - t)h\left(\frac{y - x}{T - t}\right) \right\}, \quad t \in (0, T). \tag{4.34}$$

As is easily seen, this formula remains true for a non convex function ℓ.

EXAMPLE. The equation

$$\begin{cases} \varphi_t - c\|\varphi_x\| = 0, & x \in \mathbb{R}^n, t \in (0, T), \\ \varphi(0, x) = \varphi_0(x) \end{cases} \tag{4.35}$$

is called the *eikonal equation* and models the flame propagation ($\varphi(t, x)$ is the characteristic equation of the burnt region Ω_t at moment t).

For $\psi = -\varphi$ we get

$$\begin{cases} \psi_t + c\|\psi_x\| = 0, \\ \psi(0, x) = \ell(x), \quad \ell = -\varphi_0 \end{cases}$$

and by (4.34) we get

$$\psi(t, x) = \inf_y \left\{ \ell(y) + th\left(\frac{y - x}{t}\right) \right\}$$

$$= \inf_y \{ \ell(y); \|y - x\| \leqslant ct \}$$

$$= \inf_y \{ -\varphi_0(y); \|y - x\| \leqslant ct \}$$

where

$$h(u) = \begin{cases} 0, & |u| \leqslant c, \\ +\infty, & |u| > c. \end{cases}$$

Finally

$$\varphi(t, x) = \sup \{ \varphi_0(y); \|x - y\| \leqslant ct \}.$$

4.4. *On the relation between the two approaches in optimal control theory*

We make now more precise the relationship between the maximum principle and the dynamic programming equation.

We consider the optimal problem (4.1). As we have seen in Section 3.1, if u^* is an optimal control, then there exist $\lambda \in \{0, 1\}$ and $p(t)$ (not both 0), solution to the following system:

$$\begin{cases} y' = f(t, y^*, u^*) = \dfrac{\partial}{\partial p} H^{u^*}(t, y^*, p), \\[2mm] p' = -\big(p, f_y(t, y^*, u^*)\big) + \lambda L_y(t, y^*, u^*) = -\dfrac{\partial}{\partial y} H^{u^*}(t, y^*, p), \\[2mm] p(T) = -\nabla l\big(y^*(T)\big) \end{cases} \qquad (4.36)$$

where $H^u(t, y, p) = (p, f(t, y, u)) - \lambda L$ and

$$H^{u^*(t)}\big(t, y^*(t), p(t)\big) = \max_{u \in U} H^u\big(t, y * (t), p(t)\big).$$

In Sections 4.1, 4.3, we derived a partial differential equation satisfied in the viscosity sense by the value function ψ:

$$\begin{cases} \psi_t - H(t, x, \psi_x) = 0, \\ \psi(T, x) = l(x) \end{cases} \qquad (4.37)$$

and $H(t, x, p) = \max_u H^u(t, x, -p)$. In fact, if we denote by $S = -\psi$, then

$$S_t + H^{u^*}(t, x, S_x) = 0$$

and this is just the Hamilton–Jacobi equation (see Section 1.1) corresponding to the Hamiltonian system (4.36) that appears in the maximum principle of Pontriaghin.

The next theorem states for this control problem the maximum principle by making use of the Bellman equation. The hypotheses are more restrictive than in Theorem 3.1. Anyhow, this makes more precise the relationship between the two branches of control theory and it is in a certain sense the analogous of the Jacobi theorem in calculus of variations. For a general setting of this problem we refer to [14] (see also [22]).

THEOREM 4.4. *Suppose that u^* is an optimal control and the value function ψ is C^2. Then the problem is normal and if we denote by*

$$p = -\psi_x\big(t, y^*(t)\big)$$

then p satisfies (4.36) with $\lambda = 1$.

PROOF. As seen in Theorem 4.2, ψ is a viscosity solution of (4.37) and since it is C^1, it is a classical solution. Thus

$$\frac{\mathrm{d}}{\mathrm{d}t} p(t) = -\psi_{tx}\big(t, y^*(t)\big) - \psi_{xx}\big(t, y^*(t)\big) f(t, y^*, u^*). \qquad (4.38)$$

By the other hand, by the dynamic programming principle

$$\frac{\mathrm{d}}{\mathrm{d}t}\psi\big(t, y^*(t)\big) = -L\big(t, y^*(t), u^*(t)\big)$$
$$= \psi_t\big(t, y^*(t)\big) + \big(\psi_x\big(t, y^*(t)\big), f\big(t, y^*(t), u^*(t)\big)\big).$$

This, combined with Bellman equation tells us that $(y^*(t), u^*(t))$ realizes the infimum of $(x, u) \to \psi_t(t, x) + (\psi_x(t, x), f(t, x, u)) + L(t, x, u)$. This implies that the derivative with respect to x computed in $(t, y^*(t), u^*(t))$ is 0 that is

$$\psi_{xt}\big(t, y^*(t)\big) + \psi_{xx}\big(t, y^*(t)\big) f\big(t, y^*(t), u^*(t)\big)$$
$$+ \big(\psi_x\big(t, y^*(t)\big), f_y\big(t, y^*(t), u^*(t)\big)\big) + L_y\big(t, y^*(t), u^*(t)\big) = 0$$

which combined with (4.38) concludes the proof. $\qquad\square$

References

[1] A.A. Agrachev and R.V. Gamkrelidze, *Exponential representation of flows and a chronological enumeration*, Mat. Sb. (N.S.) **107(149)** (1978), 467–532, 639.

[2] A.A. Agrachev and Y.L. Sachkov, *Control Theory from the Geometric Viewpoint*, Encyclopaedia Math. Sci., Vol. 87, Springer-Verlag, Berlin (2004).

[3] V.I. Arnold, *Mathematical Methods of Classical Mechanics*, Grad. Texts in Math., Vol. 60, second edition, Springer-Verlag, New York (1989). Translated from Russian by K. Vogtmann and A. Weinstein.

[4] V. Barbu, *Mathematical Methods in Optimization of Differential Systems*, Math. Appl., Vol. 310, Kluwer Academic, Dordrecht (1994). Translated and revised from the 1989 Romanian original.

[5] V. Barbu and G. Da Prato, *Hamilton–Jacobi Equations in Hilbert Spaces*, Pitman Res. Notes Math. Ser., Vol. 86, Pitman, Boston, MA (1983).

[6] V. Barbu and Th. Precupanu, *Convexity and Optimization in Banach Spaces*, Math. Appl. (East European Series), Vol. 10, Romanian edition, D. Reidel, Dordrecht (1986).

[7] R. Bellman, *Dynamic Programming*, Princeton Univ. Press, Princeton, NJ (1957).

[8] G.A. Bliss, *Calculus of Variations*, Carus Math. Monogr., Vol. 1, sixth edition, Math. Assoc. America, Washington; The Open Court Publishing Company, La Salle, IL (1971).

[9] H. Brezis, *Opérateurs maximaux monotones et semi-groupes de contractions dans les espaces de Hilbert*, North-Holland Math. Stud., Vol. 5, North-Holland, Amsterdam (1973).

[10] F.H. Clarke, *Generalized gradients and applications*, Trans. Amer. Math. Soc. **205** (1975), 247–262.

[11] F.H. Clarke, *The maximum principle under minimal hypotheses*, SIAM J. Control Optim. **14** (1976), 1078–1091.

[12] F.H. Clarke, *Optimization and Nonsmooth Analysis*, Canadian Math. Soc. Ser. Monogr. Adv. Texts, Wiley-Interscience, New York (1983).

[13] F.H. Clarke and I. Ekeland, *Nonlinear oscillations and boundary value problems for Hamiltonian systems*. Arch. Rational Mech. Anal. **78** (1982), 315–333.

[14] F.H. Clarke and R.B. Vinter, *The relationship between the maximum principle and dynamic programming*, SIAM J. Control Optim. **25** (1987), 1291–1311.

[15] E.A. Coddington and N. Levinson, *Theory of Ordinary Differential Equations*, McGraw-Hill, New York (1955).

[16] R. Courant and D. Hilbert, *Methods of Mathematical Physics*, Vol. II: *Partial Differential Equations* (Vol. II by R. Courant), Wiley-Interscience, New York (1962).

[17] M.G. Crandall, L.C. Evans and P.-L. Lions, *Some properties of viscosity solutions of Hamilton–Jacobi equations*, Trans. Amer. Math. Soc. **282** (1984), 487–502.

[18] M.G. Crandall, H. Ishii and P.-L. Lions, *Uniqueness of viscosity solutions of Hamilton–Jacobi equations revisited*, J. Math. Soc. Japan **39** (1987), 581–596.

[19] M.G. Crandall, H. Ishii and P.-L. Lions, *User's guide to viscosity solutions of second order partial differential equations*, Bull. Amer. Math. Soc. (N.S.) **27** (1992), 1–67.

[20] M.G. Crandall and P.-L. Lions, *Viscosity solutions of Hamilton–Jacobi equations*, Trans. Amer. Math. Soc. **277** (1983), 1–42.

[21] I. Ekeland, *On the variational principle*, J. Math. Anal. Appl. **47** (1974), 324–353.

[22] W.H. Fleming and R.W. Rishel, *Deterministic and Stochastic Optimal Control*, Appl. Math. (N.Y.), Vol. 1, Springer-Verlag, New York (1975).

[23] R. Gamkrelidze, *Exponential representation of solutions of ordinary differential equations*, Proceedings of Equadiff IV, Prague 1977, Lecture Notes in Math., Vol. 703, Springer-Verlag (1977), 118–129.

[24] I.M. Gelfand and S.V. Fomin, *Calculus of Variations*, translated from Russian and edited by R.A. Silverman, reprint of the 1963 original, Dover, Mineola, NY (2000).

[25] A. Isidori, *Nonlinear Control Systems*, Comm. Control Engrg. Ser., third edition, Springer-Verlag, Berlin (1995).

[26] V. Jurdjevic, *Geometric Control Theory*, Cambridge Stud. Adv. Math., Vol. 52, Cambridge Univ. Press, Cambridge (1997).

[27] J.-L. Lions, *Optimal Control of Systems Governed by Partial Differential Equations*, translated from French by S.K. Mitter, Grundlehren Math. Wiss., Band 170, Springer-Verlag, New York (1971).

[28] J.E. Marsden and T.S. Ratiu, *Introduction to Mechanics and Symmetry*, Texts Appl. Math., Vol. 17, second edition, Springer-Verlag, New York (1999). A basic exposition of classical mechanical systems.

[29] J. Moreau, *Fonctionnelles convexes, Séminaire sur les Équations aux Dérivées Partielles (1966–1967). II*, Collège de France, Paris (1967).

[30] L. Pontriaghin, V. Boltianski, R. Gamkrelidze and E. Mishchenko, *Théorie mathématique des processus optimaux*, Mir, Moscow (1974). Traduit du russe par D. Embarek.

[31] R.T. Rockafellar, *Convex Analysis*, Princeton Math. Ser., Vol. 28, Princeton Univ. Press, Princeton, NJ (1970).

[32] R.T. Rockafellar, *Integral functionals, normal integrands and measurable selections*, Nonlinear Operators and the Calculus of Variations (Summer School, Univ. Libre Bruxelles, Brussels, 1975), Lecture Notes in Math., Vol. 543, Springer-Verlag, Berlin (1976), 157–207.

[33] E.D. Sontag, *Mathematical Control Theory*, Texts Appl. Math., Vol. 6, second edition, Springer-Verlag, New York (1998). Deterministic finite-dimensional systems.

[34] V.M. Tikhomirov, *Stories about Maxima and Minima*, translated from Russian by A. Shenitzer, Math. World, Vol. 1, Amer. Math. Soc., Providence, RI; Math. Assoc. America, Washington, DC (1990).

[35] I.I. Vrabie, C_0-*Semigroups and Applications*, North-Holland Math. Stud., Vol. 191, North-Holland, Amsterdam (2003).

[36] H.Z. Wang and Y. Li, *Neumann boundary value problems for second-order ordinary differential equations across resonance*, SIAM J. Control Optim. **33** (1995), 1312–1325.

[37] S. Yosida, *An optimal control problem of the prey-predator system*, Funkcial. Ekvac. **25** (1982), 283–293.

[38] J. Zabczyk, *Mathematical Control Theory: An Introduction*, Systems Control Found. Appl., Birkhäuser, Boston, MA (1992).

CHAPTER 2

Hamiltonian Systems: Periodic and Homoclinic Solutions by Variational Methods

Thomas Bartsch

Mathematisches Institut, Universität Giessen, Arndtstr. 2, 35392 Giessen, Germany

Andrzej Szulkin

Department of Mathematics, Stockholm University, 106 91 Stockholm, Sweden

Contents

0. Introduction 79
1. Critical point theory 81
   1.1. Basic critical point theory 81
   1.2. Critical point theory for strongly indefinite functionals 90
2. Periodic solutions 98
   2.1. Variational setting for periodic solutions 98
   2.2. Periodic solutions near equilibria 106
   2.3. Fixed energy problem 110
   2.4. Superlinear systems 115
   2.5. Asymptotically linear systems 119
   2.6. Spatially symmetric Hamiltonian systems 121
3. Homoclinic solutions 125
   3.1. Variational setting for homoclinic solutions 125
   3.2. Existence of homoclinics 130
   3.3. Multiple homoclinic solutions 135
   3.4. Multibump solutions and relation to the Bernoulli shift 138
References 142

HANDBOOK OF DIFFERENTIAL EQUATIONS
Ordinary Differential Equations, volume 2
Edited by A. Cañada, P. Drábek and A. Fonda
© 2005 Elsevier B.V. All rights reserved

0. Introduction

The complex dynamical behavior of Hamiltonian systems has attracted mathematicians and physicists ever since Newton wrote down the differential equations describing planetary motions and derived Kepler's ellipses as solutions. Hamiltonian systems can be investigated from different points of view and using a large variety of analytical and geometric tools. The variational treatment of Hamiltonian systems goes back to Poincaré who investigated periodic solutions of conservative systems with two degrees of freedom using a version of the least action principle. It took however a long time to turn this principle into a useful tool for finding periodic solutions of a general Hamiltonian system

$$
\begin{cases}
\dot{p} = -H_q(p, q, t), \\
\dot{q} = H_p(p, q, t)
\end{cases}
\tag{HS}
$$

as critical points of the Hamiltonian action functional

$$
\Phi(p, q) = \int_0^{2\pi} p \cdot \dot{q}\, dt - \int_0^{2\pi} H(p, q, t)\, dt
$$

defined on a suitable space of 2π-periodic functions $(p, q) : \mathbb{R} \to \mathbb{R}^N \times \mathbb{R}^N$. The reason is that this functional is unbounded from below and from above so that the classical methods from the calculus of variations do not apply. Even worse, the quadratic form

$$
(p, q) \mapsto \int_0^{2\pi} p \cdot \dot{q}\, dt
$$

has infinite-dimensional positive and negative eigenspaces. Therefore Φ is said to be strongly indefinite. For strongly indefinite functionals refined variational methods like Morse theory or Lusternik–Schnirelmann theory still do not apply. These were originally developed for the closely related problem of finding geodesics and extended to many other ordinary and partial differential equations, in particular to the second order Hamiltonian system

$$
\ddot{q} = -V_q(q, t)
$$

where the associated Lagrangian functional

$$
J(q) = \frac{1}{2} \int_0^{2\pi} \dot{q}^2\, dt - \int_0^{2\pi} V(q, t)\, dt
$$

is not strongly indefinite.

A major breakthrough was the pioneering paper [78] of Rabinowitz from 1978 who obtained for the first time periodic solutions of the first order system (HS) by the above mentioned variational principle. Some general critical point theory for indefinite functionals was subsequently developed in the 1979 paper [19] by Benci and Rabinowitz. Since then the number of papers on variational methods for strongly indefinite functionals and on applications to Hamiltonian systems has been growing enormously. These methods are not restricted to periodic solutions but can also be used to find heteroclinic or homoclinic orbits and to prove complex dynamics. In fact, they can even be applied to infinite-dimensional Hamiltonian systems and strongly indefinite partial differential equations having a variational structure.

The goal of this chapter is to present an introduction to variational methods for strongly indefinite functionals like Φ and its applications to the Hamiltonian system (HS). The chapter is divided into three sections. Section 1 is concerned with critical point theory, Section 2 with periodic solutions, and Section 3 with homoclinic solutions of (HS). We give proofs or sketches of proofs for selected basic theorems and refer to the literature for more advanced results. No effort is being made to be as general as possible. Neither did we try to write a comprehensive survey on (HS). The recent survey [81] of Rabinowitz in Volume 1A of the Handbook of Dynamical Systems facilitated our task considerably. We chose our topics somewhat complementary to those treated in [81] and concentrated on the first order system (HS), though a certain overlap cannot and should not be avoided. As a consequence we do not discuss second order systems nor do we discuss convex Hamiltonian systems where one can work with the dual action functional which is not strongly indefinite. One more topic which we have not included—though it has recently attracted attention of many researchers—is the problem of finding heteroclinic solutions by variational methods. These and many more topics are being treated in a number of well written monographs dealing with variational methods for Hamiltonian systems, in particular [1,5,33,52,69,73,80]. Further references can be found in these books and in Rabinowitz' survey [81]. Naturally, the choice of the topics is also influenced by our own research experience.

Restricting ourselves to variational methods we do not touch upon the dynamical systems approach to Hamiltonian systems which includes perturbation theory, normal forms, stability, KAM theory, etc. An introduction to these topics can be found for instance in the textbooks [47,75]. Also we do not enter the realm of symplectic topology and Floer homology dealing with Hamiltonian systems on symplectic manifolds. Here we refer the reader to the monograph [74] and the references therein.

We conclude this introduction with a more detailed description of the contents. In Section 1 we consider pertinent results in critical point theory. Particular emphasis is put on a rather simple and direct approach to strongly indefinite functionals.

Section 2 is concerned with periodic solutions of (HS). We present a unified approach, via a finite-dimensional reduction in order to show the existence of one solution, and via a Galerkin-type method in order to find more solutions. Subsection 2.2 concerns the existence of periodic solutions near equilibria (Lyapunov-type results) and in Subsection 2.3 the fixed energy problem is considered (finding solutions of a priori unknown period which lie on a prescribed energy surface). The remaining subsections consider the existence and the number of periodic solutions under different growth conditions on the Hamiltonian and for spatially symmetric Hamiltonians.

Section 3 deals with homoclinic solutions for (HS) with time-periodic Hamiltonian. Here we present a few basic existence and multiplicity results and discuss a relation to the

Bernoulli shift and complicated dynamics. The proofs are more sketchy than in Section 2 because we did not want to enter too much into technicalities which are more complex than in the periodic case. Moreover, the subject of this section is still rapidly developing and has not been systematized in the same way as the periodic solution problem.

1. Critical point theory

1.1. *Basic critical point theory*

Let E be a real Hilbert space with an inner product $\langle .,. \rangle$ and Φ a functional in $C^1(E, \mathbb{R})$. Via the Riesz representation theorem we shall identify the Fréchet derivative $\Phi'(x) \in E^*$ with a corresponding element of E, and we shall write $\langle \Phi'(x), y \rangle$ rather than $\Phi'(x)y$. Our goal here is to discuss those methods of critical point theory which will be useful in our applications to Hamiltonian systems. In particular, although most of the results presented here can be easily extended to real Banach spaces, we do not carry out such extension as it will not be needed for our purposes.

Recall that $\{x_j\}$ is said to be a *Palais–Smale sequence* (a (PS)-*sequence* in short) if $\Phi(x_j)$ is bounded and $\Phi'(x_j) \to 0$. The functional Φ satisfies the *Palais–Smale condition* (the (PS)-*condition*) if each (PS)-sequence possesses a convergent subsequence. If $\Phi(x_j) \to c$ and $\Phi'(x_j) \to 0$, we shall sometimes refer to $\{x_j\}$ as a (PS)$_c$-sequence. Φ satisfies the Palais–Smale condition at the level c (the (PS)$_c$-condition) if every (PS)$_c$-sequence has a convergent subsequence.

We shall frequently use the following notation:

$$\Phi^c := \{x \in E \colon \Phi(x) \leqslant c\},$$

$$K := \{x \in E \colon \Phi'(x) = 0\}, \qquad K_c := \{x \in K \colon \Phi(x) = c\}.$$

One of the basic technical tools in critical point theory is the deformation lemma. Below we state a version of it, called the quantitative deformation lemma. It is due to Willem [101], see also [23, Theorem I.3.4] and [102, Lemma 2.3].

A continuous mapping $\eta : A \times [0, 1] \to E$, where $A \subset E$, is said to be a *a deformation of A in E* if $\eta(x, 0) = x$ for all $x \in A$. Denote the distance from x to the set B by $d(x, B)$.

LEMMA 1.1. *Suppose $\Phi \in C^1(E, \mathbb{R})$ and let $c \in \mathbb{R}$, $\bar{\varepsilon}, \delta > 0$ and a set $N \subset E$ be given. If*

$$\left\| \Phi'(x) \right\| \geqslant \delta \quad \textit{whenever} \quad d(x, E \setminus N) \leqslant \delta \quad \textit{and} \quad \left| \Phi(x) - c \right| \leqslant \bar{\varepsilon}, \tag{1.1}$$

then there exists an $\varepsilon \in (0, \bar{\varepsilon})$, depending only on $\bar{\varepsilon}$ and δ, and a deformation $\eta : E \times [0, 1] \to E$ such that:
  (i) *$\eta(x, t) = x$ whenever $|\Phi(x) - c| \geqslant \bar{\varepsilon}$;*
  (ii) *$\eta(\Phi^{c+\varepsilon} \setminus N, 1) \subset \Phi^{c-\varepsilon}$ and $\eta(\Phi^{c+\varepsilon}, 1) \subset \Phi^{c-\varepsilon} \cup N$;*
  (iii) *The mapping $t \mapsto \Phi(\eta(x, t))$ is nonincreasing for each $x \in E$.*

PROOF. Since the argument is well known, we omit some details. A complete proof may be found, e.g., in [23, Theorem I.3.4] or [102, Lemma 2.3].

A mapping $V : E \setminus K \to E$ is said to be a *pseudo-gradient vector field* for Φ if V is locally Lipschitz continuous and satisfies

$$\|V(x)\| \leqslant 2\|\Phi'(x)\|, \qquad \langle \Phi'(x), V(x) \rangle \geqslant \|\Phi'(x)\|^2 \tag{1.2}$$

for each $x \in E \setminus K$. It is well known and not difficult to prove that any $\Phi \in C^1(E, \mathbb{R})$ has a pseudo-gradient vector field; see, e.g., [23,80,102].

Let $\chi : E \to [0, 1]$ be a locally Lipschitz continuous function such that

$$\chi(x) = \begin{cases} 0 & \text{if } |\Phi(x) - c| \geqslant \bar{\varepsilon} \text{ or } d(x, E \setminus N) \geqslant \delta, \\ 1 & \text{if } |\Phi(x) - c| \leqslant \bar{\varepsilon}/2 \text{ and } d(x, E \setminus N) \leqslant \delta/2 \end{cases}$$

and consider the Cauchy problem

$$\frac{d\eta}{dt} = -\frac{1}{2}\delta \chi\big(\eta(x, t)\big) \frac{V(\eta(x, t))}{\|V(\eta(x, t))\|}, \qquad \eta(x, 0) = x.$$

Since the vector field above is locally Lipschitz continuous and bounded, $\eta(x, t)$ is uniquely determined and continuous for each $(x, t) \in E \times \mathbb{R}$. It is now easy to see that (i) is satisfied. Moreover,

$$\frac{d}{dt}\Phi\big(\eta(x, t)\big) = \left\langle \Phi'\big(\eta(x, t)\big), \frac{d\eta}{dt} \right\rangle \leqslant -\frac{\delta}{4}\chi\big(\eta(x, t)\big)\big\|\Phi'\big(\eta(x, t)\big)\big\| \tag{1.3}$$

according to (1.2). Hence also (iii) holds.

Let $x \in \Phi^{c+\varepsilon} \setminus N$ and $0 < \varepsilon \leqslant \bar{\varepsilon}/2$. In order to establish the first part of (ii) we must show that $\Phi(\eta(x, 1)) \leqslant c - \varepsilon$. Since

$$\|\eta(x, t) - x\| \leqslant \int_0^t \left\| \frac{d\eta}{ds} \right\| ds \leqslant \frac{1}{2}\delta t,$$

$d(\eta(x, t), E \setminus N) \leqslant \delta/2$ whenever $0 \leqslant t \leqslant 1$. We may assume $\Phi(\eta(x, 1)) \geqslant c - \bar{\varepsilon}/2$ (otherwise we are done). Then, according to (1.3) and the definition of χ,

$$\Phi\big(\eta(x, 1)\big) = \Phi(x) + \int_0^1 \frac{d}{dt}\Phi\big(\eta(x, t)\big)\,dt \leqslant \Phi(x) - \frac{\delta}{4}\int_0^1 \big\|\Phi'\big(\eta(x, t)\big)\big\|\,dt$$

$$\leqslant c + \varepsilon - \frac{\delta^2}{4}.$$

Hence $\Phi(\eta(x, 1)) \leqslant c - \varepsilon$ if we choose $\varepsilon \leqslant \min\{\bar{\varepsilon}/2, \delta^2/8\}$.

In order to prove the second part of (ii) it remains to observe that if $x \in \Phi^{c+\varepsilon}$ and $\eta(x, 1) \notin N$, then $d(\eta(x, t), E \setminus N) \leqslant \delta/2$ and therefore again $\Phi(\eta(x, 1)) \leqslant c - \varepsilon$. $\qquad\square$

We emphasize that the constant ε is independent of the functional Φ and the space E as long as Φ satisfies (1.1). We shall make repeated use of this fact.

It is easy to see that if Φ satisfies (PS) and N is a neighbourhood of K_c, then there exist $\bar{\varepsilon}, \delta > 0$ such that (1.1) holds.

Next we introduce the concept of local linking, due to Li and Liu [60]. Let $\Phi \in C^1(E, \mathbb{R})$ and denote the ball of radius r and center at the origin by B_r. The corresponding sphere will be denoted by S_r. The function Φ is said to satisfy the local linking condition at 0 if there exists a subspace $F_0 \subset E$ and $\alpha, \rho > 0$ such that F_0 and $F_0^{\perp}$ have positive dimension,

$$\Phi \leqslant 0 \quad \text{on } F_0 \cap \bar{B}_\rho, \qquad \Phi \leqslant -\alpha \quad \text{on } F_0 \cap S_\rho \tag{1.4}$$

and

$$\Phi \geqslant 0 \quad \text{on } F_0^{\perp} \cap \bar{B}_\rho, \qquad \Phi \geqslant \alpha \quad \text{on } F_0^{\perp} \cap S_\rho. \tag{1.5}$$

We shall denote the inner product of x and y in $\mathbb{R}^m$ by $x \cdot y$ and we set $|x| := (x \cdot x)^{1/2}$. For a symmetric matrix B we denote the Morse index of the quadratic form corresponding to B by $M^-(B)$.

THEOREM 1.2. *Suppose* $\Phi \in C^1(\mathbb{R}^m, \mathbb{R})$ *satisfies the local linking conditions* (1.4) *and* (1.5) *for some* $F_0 \subset \mathbb{R}^m$. *Then* Φ *has a critical point* $\bar{x}$ *with* $|\Phi(\bar{x})| \geqslant \alpha$ *in each of the following two cases:*
  (i) *There exists* $R > 0$ *such that* $\Phi < 0$ *in* $\mathbb{R}^m \setminus B_R$;
  (ii) $\Phi(x) = \frac{1}{2}Bx \cdot x + \psi(x)$, *where* $\psi'(x) = o(|x|)$ *as* $|x| \to \infty$, B *is a symmetric invertible matrix and* $M^-(B) > \dim F_0$.

PROOF. We first consider case (ii) which is more difficult. If there exists a critical point $\bar{x}$ with $\Phi(\bar{x}) \leqslant -\alpha$, we are done. If there is no such point, then there exists a pseudogradient vector field V whose domain contains $\Phi^{-\alpha}$, and since $|\Phi'(x)|$ is bounded away from 0 as $|x|$ is large, $|\Phi'(x)| \geqslant \delta$ (where $\delta > 0$) whenever $x \in \Phi^{-\alpha}$. Hence the Cauchy problem

$$\frac{d\gamma}{dt} = -V\big(\gamma(x,t)\big), \qquad \gamma(x,0) = x$$

has a solution for all $x \in \Phi^{-\alpha}$, $t \geqslant 0$ and

$$\Phi\big(\gamma(x,t)\big) = \Phi(x) + \int_0^t \frac{d}{ds}\Phi\big(\gamma(x,s)\big)\, ds \leqslant -\alpha - \int_0^t \big|\Phi'\big(\gamma(x,s)\big)\big|^2\, ds$$

$$\leqslant -\delta^2 t. \tag{1.6}$$

Choose $R > 0$ such that $|\psi'(x)| \leqslant \frac{1}{2}\lambda_0 |x|$ for all $|x| \geqslant R$, where $\lambda_0 := \{\inf |\lambda_j|: \lambda_j$ is an eigenvalue of $B\}$. Let $\mathbb{R}^m = F^+ \oplus F^-$, with $F^{\pm}$ respectively being the positive and the negative space of B. For $x \in \mathbb{R}^m$ write $x = x^+ + x^-$, $x^{\pm} \in F^{\pm}$. By (1.6) and the form of Φ, $|\gamma(x,T)^-| \geqslant R$ for any $x \in F_0 \cap S_\rho$ provided T is large enough.

Let S^n and D^{n+1} be the unit sphere and the unit closed ball in $\mathbb{R}^{n+1}$. Recall that a space X is called l-connected if any mapping from S^n to X, $0 \leqslant n \leqslant l$, can be extended to a mapping from D^{n+1} to X (cf. [84, Section 1.8]). We want to show that the set $\Phi^{-\alpha} \cap$

$\{x \in \mathbb{R}^m : |x^-| \geqslant R\}$ is $(k-1)$-connected if $k < M^-(B)$, possibly after choosing a larger R. This will imply in particular that any homeomorphic image of S^{k-1} contained in this set is contractible there. Let $r(x,t) := (1-t)x^+ + x^-, 0 \leqslant t \leqslant 1$. Then r is a strong deformation retraction of $\Phi^{-\alpha} \cap \{x \in \mathbb{R}^m : |x^-| \geqslant R\}$ onto $F^- \setminus B_R$. To see this, we only need to verify that $r(x,t) \in \Phi^{-\alpha}$ for all x, t. Suppose first $Bx^+ \cdot x^+ \leqslant -\frac{1}{2} Bx^- \cdot x^-$. Then $|x^+| \leqslant C|x^-|$ for some C; thus $\psi((1-t)x^+ + x^-) = o(|x^-|^2)$ as $|x^-| \to \infty$ and

$$\Phi(r(x,t)) = \frac{1}{2}(1-t)^2 Bx^+ \cdot x^+ + \frac{1}{2} Bx^- \cdot x^- + \psi((1-t)x^+ + x^-)$$

$$\leqslant \frac{1}{4} Bx^- \cdot x^- + \psi((1-t)x^+ + x^-) \leqslant -\alpha$$

if R is large enough. Let now $Bx^+ \cdot x^+ \geqslant -\frac{1}{2} Bx^- \cdot x^-$; then $|x^-| \leqslant D|x^+|$ and

$$\frac{\mathrm{d}}{\mathrm{d}t}\Phi(r(x,t)) = -\Phi'(r(x,t)) \cdot x^+$$

$$= -Bx^+ \cdot x^+ - \psi'((1-t)x^+ + x^-) \cdot x^+ \leqslant 0,$$

again provided R is large enough. Hence in this case $\Phi(r(x,t)) \leqslant \Phi(x) \leqslant -\alpha$. Since $F^- \setminus B_R$ is homeomorphic to $S^{l-1} \times [R, \infty)$, where $l := M^-(B)$, $F^- \setminus B_R$ is $(k-1)$-connected for any $k < l$. It follows that so is the set $\Phi^{-\alpha} \cap \{x \in \mathbb{R}^m : |x^-| \geqslant R\}$.

The set $\{\gamma(x,T) : x \in F_0 \cap S_\rho\}$ is contained in $\Phi^{-\alpha} \cap \{x \in \mathbb{R}^m : |x^-| \geqslant R\}$ and homeomorphic to S^{k-1}, $k < M^-(B)$. Hence it can be contracted to a point x^* in $\Phi^{-\alpha} \cap \{x \in \mathbb{R}^m : |x^-| \geqslant R\}$. Denote this contraction by γ_0, let $D_0 := F_0 \cap \bar{B}_\rho$, $D := D_0 \times [0,1]$ and define a mapping $f : \partial D \to \mathbb{R}^m$ by setting

$$f(x_0, s) := \begin{cases} x_0, & s = 0, \ x_0 \in D_0, \\ \gamma(x_0, 2sT), & 0 \leqslant s \leqslant \frac{1}{2}, \ x_0 \in \partial D_0 = F_0 \cap S_\rho, \\ \gamma_0(\gamma(x_0, T), 2s - 1), & \frac{1}{2} \leqslant s \leqslant 1, \ x_0 \in \partial D_0, \\ x^*, & s = 1, \ x_0 \in D_0. \end{cases} \tag{1.7}$$

It is clear from the construction that $\Phi(f(x_0, s)) \leqslant 0$ whenever $(x_0, s) \in \partial D$ and < 0 if $x_0 \in \partial D$. Let

$$\Gamma := \{g \in C(D, \mathbb{R}^m) : g|_{\partial D} = f\} \tag{1.8}$$

and

$$c := \inf_{g \in \Gamma} \max_{(x_0, s) \in D} \Phi(g(x_0, s)). \tag{1.9}$$

We shall show that $f(\partial D)$ links $F_0^\perp \cap S_\rho$ in the sense that if $g \in \Gamma$, then $g(x_0, s) \in F_0^\perp \cap S_\rho$ for some $(x_0, s) \in D$. Assuming this, we see that the maximum in (1.9) is always $\geqslant \alpha$, and hence $c \geqslant \alpha$. We claim c is a critical value. Otherwise $K_c = \emptyset$, so (1.1) holds for g with $N = \emptyset$, $\bar{\varepsilon} \in (0, c)$ because Φ satisfies the (PS)-condition as a consequence of (ii). Let $\varepsilon < \bar{\varepsilon}$ and η be as in Lemma 1.1. Let $g \in \Gamma$ be such that $g(x_0, s) \in \Phi^{c+\varepsilon}$ for all $(x_0, s) \in D$. Since $\Phi(f(x_0, s)) \leqslant 0$ if $(x_0, s) \in \partial D$ and $\eta(x, 1) = x$ for $x \in \Phi^0$, the mapping $(x_0, s) \mapsto \eta(g(x_0, s), 1)$ is in Γ. But this is impossible according to the definition of c because $\Phi(\eta(g(x_0, s), 1)) \leqslant c - \varepsilon$ for all $(x_0, s) \in D$.

It remains to show that $f(\partial D)$ links $F_0^\perp \cap S_\rho$. Write $x = x_0 + x_0^\perp \in F_0 \oplus F_0^\perp$, $D_0^\perp = F_0^\perp \cap \overline{B}_\rho$. For $g \in \Gamma$ we consider the map

$$G : \left(D_0 \oplus D_0^\perp\right) \times [0, 1] \to \mathbb{R}^m, \quad G(x, s) = x_0^\perp - g(x_0, s).$$

If there is no linking, then $G(x, s) \neq 0$ for some $g \in \Gamma$ and all $x_0 \in D_0$, $x_0^\perp \in \partial D_0^\perp$, $0 \leqslant s \leqslant 1$. For $x_0 \in \partial D_0$ we have $g(x_0, s) = f(x_0, s)$, hence $\Phi(g(x_0, s)) < 0$ and $G(x, s) \neq 0$ (because $\Phi(x_0^\perp) \geqslant 0$). It follows that $G(x, s) \neq 0$ when $x \in \partial(D_0 \oplus D_0^\perp)$ and G is an admissible homotopy for Brouwer's degree. Hence

$$\deg\left(G(., 0), D_0 \oplus D_0^\perp, 0\right) = \deg\left(G(., 1), D_0 \oplus D_0^\perp, 0\right).$$

Since $G(x, 0) = x_0^\perp - g(x_0, 0) = x_0^\perp - f(x_0, 0) = x_0^\perp - x_0$, the degree on the left-hand side above is $(-1)^{\dim F_0}$. On the other hand, $f(x_0, 1) = x^*$, where x^* is a point outside $D_0 \oplus D_0^\perp$; hence $G(x, 1) \neq 0$ for any $x \in D_0 \oplus D_0^\perp$ and the degree is 0. This contradiction completes the proof of (ii).

In case (i) the argument is similar but simpler. Suppose there is no critical point $\bar{x}$ with $\Phi(\bar{x}) \leqslant -\alpha$. Since $\Phi \leqslant 0$ in $\mathbb{R}^m \setminus B_R$, $|\gamma(x_0, T)| \geqslant R$ for some $T > 0$ and all $x_0 \in F_0 \cap S_\rho$. It is obvious that the set $\{\gamma(x_0, T) : x_0 \in F_0 \cap S_\rho\}$ (which is homeomorphic to a sphere of dimension $\leqslant m - 2$) can be contracted to a point in $\mathbb{R}^m \setminus B_R$. Now we can proceed as above. Note only that Φ satisfies the (PS)$_c$-condition because any (PS)$_c$-sequence lies eventually in B_R. $\qquad\square$

We shall need the following extension of Theorem 1.2:

THEOREM 1.3. *Suppose* $\Phi \in C^1(\mathbb{R}^m, \mathbb{R})$ *satisfies the local linking conditions* (1.4) *and* (1.5) *for some* $F_0 \subset \mathbb{R}^m$. *If there exist subspaces* $\widetilde{F} \supset F \supset F_0$, $\widetilde{F} \neq F$, *and* $R > 0$ *such that* $\Phi < 0$ *on* $\widetilde{F} \setminus B_R$ *and* $\Phi|_F$ *has no critical point* $x \in \Phi^{-\alpha}$, *then* $\Phi'(\bar{x}) = 0$ *for some* $\bar{x}$ *with* $\alpha \leqslant \Phi(\bar{x}) \leqslant \max_{x \in \widetilde{F} \cap \overline{B}_{R+1}} \Phi(x)$.

PROOF. This time we obtain γ by solving the Cauchy problem

$$\frac{d\gamma}{dt} = -\chi\big(\gamma(x, t)\big) V\big(\gamma(x, t)\big), \quad \gamma(x, 0) = x,$$

where $\chi \in C^\infty(\mathbb{R}^m, [0, 1])$ is such that $\chi = 1$ on B_R, $\chi = 0$ on $\mathbb{R}^m \setminus B_{R+1}$ and $V : F \cap \Phi^{-\alpha} \to F$ is a pseudogradient vector field for $\Phi|_F$. Now we proceed as in the proof of case (i) above and obtain T such that $\gamma(x_0, T) \in F \cap (B_{R+1} \setminus B_R)$ whenever $x_0 \in F_0 \cap S_\rho$.

This set can be contracted to a point in $\widetilde{F} \cap (\overline{B}_{R+1} \setminus B_R)$, hence we obtain a map $f \in C(\partial D, \widetilde{F} \cap \overline{B}_{R+1})$ as in (1.7). Since there exists $g \in \Gamma \cap C(D, \widetilde{F} \cap \overline{B}_{R+1})$ where Γ is as in (1.8) it follows that

$$c \leqslant \max_{(x_0, s) \in D} \Phi\big(g(x_0, s)\big) \leqslant \max_{x \in \widetilde{F} \cap \overline{B}_{R+1}} \Phi(x). \qquad \square$$

An infinite-dimensional version of the linking theorems (in a setting which corresponds to Theorems 1.2 and 1.3) may be found in [62]. However, we shall only make use of the finite-dimensional versions stated above.

If the functional Φ is invariant with respect to a representation of some symmetry group, then Φ usually has multiple critical points. In order to exploit such symmetries, we introduce index theories.

Let E be a Hilbert space and

$$\Sigma := \{A = C \cap O \subset E : C \text{ is closed, } O \text{ is open and } -A = A\}. \qquad (1.10)$$

Intersections of an open and a closed set (of a topological space) are called *locally closed*. Thus Σ consists of the locally closed symmetric subsets of E. Let $A \in \Sigma$, $A \neq \emptyset$. The *genus* of A, denoted $\gamma(A)$, is the smallest integer k such that there exists an odd mapping $f \in C(A, \mathbb{R}^k \setminus \{0\})$. If such a mapping does not exist for any k, then $\gamma(A) := +\infty$. Finally, $\gamma(\emptyset) = 0$. Equivalently, $\gamma(A) = 1$ if $A \neq \emptyset$ and if there exists an odd map $A \to \{+1, -1\}$; $\gamma(A) \leqslant k$ if A can be covered by k subsets $A_1, \ldots, A_k \in \Sigma$ such that $\gamma(A_j) \leqslant 1$.

PROPOSITION 1.4. *The two definitions of genus given above are equivalent for $A \in \Sigma$.*

PROOF. If $f : A \to \mathbb{R}^k \setminus \{0\}$ is as in the first definition, then the sets $A_j := \{x \in A : f_j(x) \neq 0\}$, $j = 1, \ldots, k$, cover A, are open in A and $-A_j = A_j$, hence $A_j \in \Sigma$. The map $f_j / |f_j| : A_j \to \{+1, -1\}$ shows that $\gamma(A_j) \leqslant 1$.

Suppose $\gamma(A) \leqslant k$ in the sense of the second definition. Since $A \cap A_j \in \Sigma$, we may assume $A_j \subset A$, $A = C \cap O$, $A_j = C_j \cap O_j$ and $C_j \subset C$, $O_j \subset O$, where C, C_j, O, O_j are as in the definition of Σ. If $f_j : A_j \to \{+1, -1\}$ is odd, we may extend it to a continuous map $f_j : O_j \to \mathbb{R}$. This is a consequence of Tietze's theorem because A_j is a closed subset of O_j. Replacing $f(x)$ by $\frac{1}{2}(f(x) - f(-x))$ we may assume that the extension is also odd. Let $\pi_j : A \to [0, 1]$, $j = 1, \ldots, k$, be a partition of unity subordinated to the covering $O_1, \ldots, O_k$ of A. Replacing $\pi_j(x)$ by $\frac{1}{2}(\pi_j(x) + \pi_j(-x))$ we may assume that all π_j are even. Now the map

$$f : A \to \mathbb{R}^k, \quad f(x) = \big(\pi_1(x) f_1(x), \ldots, \pi_k(x) f_k(x)\big)$$

is well defined, continuous, odd, and satisfies $f(A) \subset \mathbb{R}^k \setminus \{0\}$. $\qquad \square$

The above definitions of genus do not need to coincide for arbitrary subsets $A = -A$ which are not locally closed.

PROPOSITION 1.5. *Let $A, B \in \Sigma$.*

   (i) *If there exists an odd mapping $g \in C(A, B)$, then $\gamma(A) \leqslant \gamma(B)$.*

   (ii) *$\gamma(A \cup B) \leqslant \gamma(A) + \gamma(B)$.*

   (iii) *There exists an open neighbourhood $N \in \Sigma$ of A such that $\gamma(A) = \gamma(N)$.*

   (iv) *If A is compact and $0 \notin A$, then $\gamma(A) < \infty$.*

   (v) *If $U \in \Sigma$ is an open bounded neighbourhood of $0 \in \mathbb{R}^l$, then $\gamma(\partial U) = l$. In particular, $\gamma(S^{l-1}) = l$, where S^{l-1} is the unit sphere in $\mathbb{R}^l$.*

   (vi) *If X is a subspace of codimension m in E and $\gamma(A) > m$, then $A \cap X \neq \emptyset$.*

   (vii) *If $0 \notin A$ and $i(A) \geqslant 2$, then A is an infinite set.*

A proof of this classical result may be found, e.g., in [80,85] if Σ contains only closed sets. This restriction is however not needed; see Proposition 1.7 below.

Let G be a compact topological group. A *representation* T of G in a Hilbert space E is a family $\{T_g\}_{g \in G}$ of bounded linear operators $T_g : E \to E$ such that $T_e = id$ (where e is the unit element of G and id the identity mapping), $T_{g_1 g_2} = T_{g_1} T_{g_2}$ and the mapping $(g, x) \mapsto T_g x$ is continuous. T is *an isometric* representation if each T_g is an isometry. A set $A \subset E$ is called *T-invariant* if $T_g A = A$ for all $g \in G$. When there is no risk of ambiguity we shall say A is G-invariant or simply invariant. The set

$$\mathcal{O}(x) := \{T_g x \colon g \in G\}$$

will be called the *orbit* of x and

$$E^G := \{x \in E \colon T_g x = x \text{ for all } g \in G\}$$

the set of *fixed points* of the representation T. Obviously, E^G is a closed subspace of E and $\mathcal{O}(x) = \{x\}$ if and only if $x \in E^G$.

   Let

$$\Sigma := \{A \subset E \colon A \text{ is locally closed and } T_g A = A \text{ for all } g \in G\}. \tag{1.11}$$

Note that the definition (1.11) of Σ coincides with (1.10) if $G = \mathbb{Z}/2 \equiv \{1, -1\}$ and $T_{\pm 1} x = \pm x$. A mapping $f : E \to \mathbb{R}$ is said to be *T-invariant* (or simply invariant) if $f(T_g x) = x$ for all $g \in G$ and $x \in E$. If T and S are two (possibly different) representations of G in E and F, then a mapping $f : E \to F$ is *equivariant* with respect to T and S (or equivariant) if $f(T_g x) = S_g f(x)$ for all $g \in G$, $x \in E$. Finally, if $f : E \to F$, we set

$$f_G(x) := \int_G S_{g^{-1}} f(T_g x) \, dg, \tag{1.12}$$

where the integration is performed with respect to the normalized Haar measure. It is easy to see that f_G is equivariant. As a special case, for $G = \mathbb{Z}/2$ acting via the antipodal map on E and F we have $f_G(x) = \frac{1}{2}(f(x) - f(-x))$, so f_G is odd. If G acts trivially on F (i.e., $S_{\pm 1} x = x$) we obtain $f_G(x) = \frac{1}{2}(f(x) + f(-x))$, so f_G is even.

If $\Phi \in C^1(E, \mathbb{R})$ is invariant with respect to an isometric representation T of G, then it is easy to see that $\Phi'(T_g x) = T_g \Phi'(x)$ for all $x \in E$, $g \in G$. Hence x is a critical point of Φ if and only if so are all $y \in \mathcal{O}(x)$. The set $\mathcal{O}(x)$ will be called a *critical orbit* of Φ.

In what follows we restrict our attention to isometric representations of $G = \mathbb{Z}/p$, where $p \geqslant 2$ is a prime, and $G = S^1 = \mathbb{R}/2\pi\mathbb{Z}$. If $G = S^1$, we do not distinguish between $\theta \in \mathbb{R}$ and the corresponding element of G, and we may also identify this element with $e^{i\theta}$. The same applies for $G = \mathbb{Z}/p \subset S^1$, where we identify the elements of G with roots of unity, again represented as $e^{i\theta}$.

Next we define an index $i : \Sigma \to \mathbb{N}_0 \cup \{\infty\}$ for $G = S^1$ and $G = \mathbb{Z}/p$, p a prime number. In the case $p = 2$ we recover the genus. For $A \in \Sigma$, $A \neq \emptyset$, we define $i(A) = 1$ if there exists a continuous map $f : A \to G \subset \mathbb{C} \setminus \{0\}$ such that $f(T_\theta x) = e^{in\theta} f(x)$ for some $n \in \mathbb{N}$ and all x, θ ($n/p \notin \mathbb{N}$ if $G = \mathbb{Z}/p$). And $i(A) \leqslant k$ if A can be covered by k sets $A_1, \ldots, A_k \in \Sigma$ such that $i(A_j) \leqslant 1$. If such a covering does not exist for any k, then $i(A) := +\infty$. Finally, we set $i(\emptyset) := 0$. We have a version of Proposition 1.4 for $G = S^1$.

PROPOSITION 1.6. *If $G = S^1$, then $i(A)$ is the smallest integer k for which there exists a mapping $f \in C(A, \mathbb{C}^k \setminus \{0\})$ such that $f(T_\theta x) = e^{in\theta} f(x)$ for some $n \in \mathbb{N}$ and all x, θ.*

The proof is similar to that of Proposition 1.4. Note only that (1.12) needs to be used and if $f_j(T_\theta x) = e^{in_j\theta} f_j(x)$, then $f(x) = (f_1(x)^{n/n_1}, \ldots, f_k(x)^{n/n_k})$ where n is the least common multiple of $n_1, \ldots, n_k$. The corresponding version for $G = \mathbb{Z}/p$, p an odd prime, requires spaces lying between $\mathbb{C}^k \setminus \{0\}$ and $\mathbb{C}^{k+1} \setminus \{0\}$; see [10, Proposition 2.9].

The above definition is due to Benci [17,18] in the case $G = S^1$ and to Krasnosel'skii [56] for $G = \mathbb{Z}/p$. Benci used in fact mappings $f \in C(A, \mathbb{C}^k \setminus \{0\})$ as in Proposition 1.6. Let us also remark that a different, cohomological index, has been introduced by Fadell and Rabinowitz [36] for $G = \mathbb{Z}/2$ and $G = S^1$, and by Bartsch [10, Example 4.5] for $G = \mathbb{Z}/p$. While the geometrical indexes of Krasnosel'skii and Benci are much more elementary, the cohomological indexes have some additional properties (which will not be needed here).

Since we only consider isometric representations, it is easy to see that the orthogonal complement $\widetilde{E} := (E^G)^\perp$ is invariant. In order to formulate the properties of the index for $G = S^1$ and $G = \mathbb{Z}/p$ we set

$$d_G := 1 + \dim G = \begin{cases} 1 & \text{for } G = \mathbb{Z}/p, \\ 2 & \text{for } G = S^1. \end{cases}$$

PROPOSITION 1.7. *Suppose $G = S^1$ or $G = \mathbb{Z}/p$, where p is a prime, and let $A, B \in \Sigma$.*
  (i) *If there exists an equivariant mapping $g \in C(A, B)$, then $i(A) \leqslant i(B)$.*
  (ii) *$i(A \cup B) \leqslant i(A) + i(B)$.*
  (iii) *There exists an open neighbourhood $N \in \Sigma$ of A such that $i(A) = i(N)$.*
  (iv) *If A is compact and $A \cap E^G = \emptyset$, then $i(A) < \infty$.*
  (v) *If U is an open bounded invariant neighbourhood of 0 in a finite-dimensional invariant subspace X of $\widetilde{E}$, then $i(\partial U) = \frac{1}{d_G} \dim X$.*
  (vi) *If X is an invariant subspace of $\widetilde{E}$ with finite codimension and if $i(A) > \frac{1}{d_G} \operatorname{codim}_{\widetilde{E}} X$, then $A \cap (E^G \oplus X) \neq \emptyset$.*

(vii) *If $A \cap E^G \neq \emptyset$, then $i(A) = +\infty$. If $A \cap E^G = \emptyset$ and $i(A) \geqslant 2$, then A contains infinitely many orbits.*

PROOF. (i) Let $i(B) = k < \infty$ (otherwise there is nothing to prove) and $B_1, \ldots, B_k$ be a covering of B as in the definition of the index $i(B)$. Then $g^{-1}(B_1), \ldots, g^{-1}(B_k)$ is a covering of A as in the definition of $i(A)$, hence $i(A) \leqslant k$.

(ii)–(iv) are obvious.

(v) It follows easily from Proposition 1.4 or 1.6 that $i(\partial U) \leqslant \frac{1}{d_G} \dim X$ if $G = \mathbb{Z}/2$ or $G = S^1$, respectively. In the $\mathbb{Z}/p$-case for $p \geqslant 3$ we may identify X with $\mathbb{C}^l$ and take the covering $A_j := \{z \in \mathbb{C}^l : z_j \neq 0, \ p \arg(z_j) \neq 0 \bmod 2\pi\}$, $B_j := \{z \in \mathbb{C}^l : z_j \neq 0, \ p \arg(z_j) \neq \pi \bmod 2\pi\}$, $j = 1, \ldots, k$, of $\mathbb{C}^l \setminus \{0\}$ in order to see that $i(\mathbb{C}^l \setminus \{0\}) \leqslant 2l = \dim X$.

The reverse inequality is a consequence of the Borsuk–Ulam theorem. A proof for $G = S^1$ may be found in [73, Theorem 5.4], and for $G = \mathbb{Z}/p$ in [9].

(vi) Let Y be the orthogonal complement of X in $\widetilde{E}$. Then Y is invariant and $\dim Y = \operatorname{codim}_{\widetilde{E}} X$. Suppose $A \cap (E^G \oplus X) = \emptyset$ and let $f(x) = P_Y x$ where P_Y denotes the orthogonal projector onto Y. Then $f : A \to Y \setminus \{0\}$. If $G = \mathbb{Z}/2$ or $G = S^1$ this implies $i(A) \leqslant \frac{1}{d_G} \dim Y$ by Propositions 1.4, 1.6, respectively. If $G = \mathbb{Z}/p$, $p \geqslant 3$, we identify Y with $\mathbb{C}^m$ and write $f = (f_1, \ldots, f_m)$. It follows from the Peter–Weyl theorem (see [73, Theorem 5.1], where the case $G = S^1$ is considered) that $f_j(T_\theta x) = e^{in_j\theta} f(x)$ ($n_j \neq 0 \bmod p$). Let $g_j(x) := f_j(x)^{n/n_j}$, where n is the least common multiple of $n_1, \ldots, n_k$. Then $g : A \to \mathbb{C}^m \setminus \{0\}$ and $g(T_\theta x) = e^{in\theta} g(x)$, so $i(A) \leqslant i(\mathbb{C}^m \setminus \{0\}) \leqslant 2m = \operatorname{codim}_{\widetilde{E}} X$, a contradiction.

(vii) Suppose $A \cap E^G \neq \emptyset$ and there exists a covering $A_1, \ldots, A_k$ of A as in the definition. Then $A_j \cap E^G \neq \emptyset$ for some j. For each $x \in A_j \cap E^G$ we have $f(T_\theta x) = f(x)$. So if $f_j(T_\theta x) = e^{in\theta} f_j(x)$, with n as before, then $f_j(x) = 0$. Thus there is no mapping $f_j : A_j \to G \subset \mathbb{C} \setminus \{0\}$ as required in the definition of index, hence $i(A) = +\infty$. If $A \cap E^G = \emptyset$ and A consists of k orbits $\mathcal{O}(x_1), \ldots, \mathcal{O}(x_k)$, then we let $n_j \geqslant 1$ be the largest integer such that $2\pi/n_j \in G$ and $T_{2\pi/n_j} x_j = x_j$. If $G = \mathbb{Z}/p$ then all $n_j = 1$. We define $f : A \to G$ by setting $f(T_\theta x_j) = e^{in\theta}$, where n is the least common multiple of $n_1, \ldots, n_k$. $\qquad\square$

It is easy to prove an equivariant version of the deformation Lemma 1.1 for invariant functionals $\Phi : E \to \mathbb{R}$. One simply observes that if V is a pseudo-gradient vector field for Φ then

$$V_G(x) := \int_G T_g^{-1} V(T_g x) \, \mathrm{d}g \tag{1.13}$$

is an equivariant pseudo-gradient vector field for Φ. Integrating V_G as in the proof of Lemma 1.1 yields an equivariant deformation η.

THEOREM 1.8. *Suppose $f \in C^1(S^{n-1}, \mathbb{R})$ is invariant with respect to a representation of $\mathbb{Z}/p$ in $\mathbb{R}^n$ without nontrivial fixed points. Then f has at least n $\mathbb{Z}/p$-orbits of critical points.*

THEOREM 1.9. *Suppose $f \in C^1(S^{2n-1}, \mathbb{R})$ is invariant with respect to a representation of S^1 in $\mathbb{R}^{2n}$ without nontrivial fixed points. Then f has at least n S^1-orbits of critical points.*

PROOF (outline). The proofs of Theorems 1.8 and 1.9 are standard. Set $M = S^{d_G \cdot n - 1}$ and suppose f has finitely many critical orbits $\mathcal{O}_j = \mathcal{O}(x_j)$, $j = 1, \ldots, k$. We may assume that the critical values $c_j = f(x_j)$ are ordered: $c_1 \leqslant \cdots \leqslant c_k$. Then using the properties of the index and an equivariant deformation lemma for functionals defined on manifolds one sees that $i(f^{c_j}) \leqslant j$. The result follows from $i(M) = n$.

We present a simpler proof which works if f' is locally Lipschitz continuous. Let η be the negative gradient flow of f on M and consider the sets

$$A_j := \{x \in M : \eta(x, t) \to \mathcal{O}_j \text{ as } t \to \infty\}$$

and

$$B_j := \bigcup_{i=1}^{j} A_i \quad \text{for } j = 0, \ldots, k.$$

Then $B_0 = \emptyset$, $B_k = M$, and it is not difficult to see that B_{j-1} is an open subset of B_j. Consequently all A_j are locally closed. Using the flow η one constructs an equivariant map $f_j : A_j \to \mathcal{O}_j$. This implies $i(A_j) \leqslant i(\mathcal{O}_j) = 1$, and therefore $n = i(M) = i(B_k) \leqslant k$. This proof does not need the equivariant deformation lemma, and it produces directly a covering of M as in the definition of the index. $\qquad\square$

REMARK 1.10. One can define index theories satisfying properties 1.7(i)–(iv), (vii) for arbitrary compact Lie groups G. However, properties 1.7(v), (vi) which are important for applications and computations, cannot be extended in general, except for a very restricted class of groups. This has been investigated in detail in [10]. In certain applications the representation of G in E is of a special form which allows to obtain similar results as above. In order to formulate this, call a finite-dimensional representation space $V \cong \mathbb{R}^n$ of the compact Lie group G admissible if every equivariant map $\overline{\mathcal{O}} \to V^{k-1}$, $\mathcal{O} \subset V^k$ a bounded open and invariant neighbourhood of 0 in V^k, has a zero on $\partial\mathcal{O}$. Clearly the antipodal action of $\mathbb{Z}/2$ on $\mathbb{R}$ is admissible as are the nontrivial representations of $\mathbb{Z}/p$ or S^1 in $\mathbb{R}^2$. Let $E = \bigoplus_{j=1}^{\infty} E_j$ be the Hilbert space sum of the finite-dimensional Hilbert spaces E_j such that each E_j is isomorphic to V as a representation space of G. For instance, $E = L^2(S^1, V)$ with the representation of G given by $(T_g x)(t) = T_g(x(t))$ has this property. The same is true for subspaces like $H^1(S^1, V)$ or $H^{1/2}(S^1, V)$. For an invariant, locally closed set $A \subset E$ let $i(A) = 1$ if $A \neq \emptyset$ and there exists a continuous equivariant map $A \to SV = \{v \in V : \|v\| = 1\}$. And let $i(A) \leqslant k$ if A can be covered by $A_1, \ldots, A_k \in \Sigma$ with $i(A) \leqslant 1$. Proposition 1.7 can be extended to this index theory. See [6,16] for applications to Hamiltonian systems.

1.2. *Critical point theory for strongly indefinite functionals*

As will be explained in Section 2.1, functionals naturally corresponding to Hamiltonian systems are strongly indefinite. This means that they are of the form $\Phi(z) = \frac{1}{2}\langle Lz, z\rangle -$

$\psi(z)$ where $L: E \to E$ is a selfadjoint Fredholm operator with negative and positive eigenspace both infinite-dimensional, and the same is true for the Hessian $\Phi''(z)$ of a critical point z of Φ. In order to study such functionals it will be convenient to use a variant of the Palais–Smale condition that allows a reduction to the finite-dimensional case and leads to simpler proofs. We shall also present two useful critical point theorems which apply when the Palais–Smale condition does not hold. These will be needed for the existence of homoclinic solutions.

First we introduce certain sequences of finite-dimensional subspaces and replace the Palais–Smale condition by another one which is adapted to these sequences.

Let $\{E_n\}_{n \geqslant 1}$ be a sequence of finite-dimensional subspaces such that $E_n \subset E_{n+1}$ for all n and

$$E = \overline{\bigcup_{n=1}^{\infty} E_n}.$$

Let $P_n: E \to E_n$ denote the orthogonal projection. Then $\{x_j\}$ is called a (PS)*-*sequence* for Φ (with respect to $\{E_n\}$) if $\Phi(x_j)$ is bounded, each $x_j \in E_{n_j}$ for some n_j, $n_j \to \infty$ and $P_{n_j} \Phi'(x_j) \to 0$ as $j \to \infty$. Φ is said to satisfy the (PS)*-*condition* if each (PS)*-sequence has a convergent subsequence. It is easy to see that K_c is compact for each c if (PS)* holds. Indeed, let $x_j \in K_c$, then we can find $n_j \geqslant j$ such that $\|y_j - x_j\| \leqslant 1/j$, $\Phi(y_j) \to c$ and $P_{n_j} \Phi'(y_j) \to 0$, where $y_j := P_{n_j} x_j$. Hence $\{y_j\}$, and therefore also $\{x_j\}$, has a convergent subsequence. We shall repeatedly use the notation

$$\Phi_n := \Phi|_{E_n} \quad \text{and} \quad A_n := A \cap E_n.$$

Observe that $\Phi_n'(x) = P_n \Phi'(x)$ for all $x \in E_n$.

The condition (PS)* (in a slightly different form) has been introduced independently by Bahri and Berestycki [7,8], and Li and Liu [60].

LEMMA 1.11. *If Φ satisfies* (PS)* *and N is a neighbourhood of K_c, then there exist $\bar{\varepsilon}, \delta > 0$ and $n_0 \geqslant 1$ such that $\|\Phi_n'(x)\| \geqslant \delta$ whenever $d(x, E \setminus N) \leqslant \delta$, $|\Phi(x) - c| \leqslant \bar{\varepsilon}$ and $n \geqslant n_0$.*

PROOF. If the conclusion is false, then we find a sequence $\{x_j\}$ such that $x_j \in E_{n_j}$ for some $n_j \geqslant j$, $d(x_j, E \setminus N) \to 0$, $\Phi_{n_j}(x_j) \to c$ and $\Phi_{n_j}'(x_j) \to 0$. Hence $\{x_j\}$ is a (PS)*-sequence. Passing to a subsequence, $x_j \to \bar{x} \in K_c$. However, since K_c is compact, the sequence $\{x_j\}$ is bounded away from K_c and therefore $\bar{x} \notin K_c$, a contradiction. $\qquad\square$

Next we introduce the notion of limit index in order to deal with symmetric functionals. As in Section 1.1 we consider the groups $G = \mathbb{Z}/p$, where p is a prime, or $G = S^1$ and their isometric representations in E. The group $\mathbb{Z}/2$ always acts via the antipodal map (i.e., $T_{\pm 1} x = \pm x$) so that obviously $E^G = \{0\}$. The reason for going beyond the usual index is that we need to distinguish between certain infinite-dimensional sets having $i(A) = \infty$; in particular, we need to compare different spheres of infinite dimension and codimension.

Let $\{E_n\}$ be a sequence of subspaces as above and suppose in addition that each E_n is G-invariant and $E^G \subset E_n$ for some n. Let $\{d_n\}$ be a sequence of integers and

$$\mathcal{E} := \{E_n, d_n\}_{n=1}^{\infty}.$$

The limit index of $A \in \Sigma$ with respect to $\mathcal{E}$, $i_{\mathcal{E}}(A)$, is defined by

$$i_{\mathcal{E}}(A) := \limsup_{n \to \infty} \big(i(A_n) - d_n\big).$$

Clearly $i_{\mathcal{E}}(A) = \infty$ if $A \cap E^G \neq \emptyset$. The limit index, in a somewhat different form, has been introduced by Y.Q. Li [63], see also [92]. A special case is the limit genus, $\gamma_{\mathcal{E}}(A)$. We note that $i_{\mathcal{E}}(A)$ can take the values $+\infty$ or $-\infty$ and if $E_n = E$, $d_n = 0$ for all n, then $i_{\mathcal{E}}(A) = i(A)$, and similarly for the genus.

REMARK 1.12. The limit index is patterned on the notion of limit relative category introduced by Fournier et al. in [42]. Recall that if Y is a closed subset of X, then a closed set $A \subset X$ is said to be of category k in X relative to Y, denoted $\mathrm{cat}_{X,Y}(A) = k$, if k is the least integer such that there exist closed sets $A_0, \ldots, A_k \subset X$, $A_0 \supset Y$, which cover A, all A_j, $1 \leqslant j \leqslant k$, are contractible in X and there exists a deformation $h : A_0 \times [0,1] \to X$ with $h(A_0, 1) \subset Y$ and $h(Y, t) \subset Y$ for all $t \in [0,1]$. If $Y = \emptyset$ (and $A_0 = \emptyset$), then $\mathrm{cat}_X(A) = \mathrm{cat}_{X,\emptyset}(A)$ is the usual Lusternik–Schnirelman category of A in X. For $X \subset E$ and using the above notation for subsets of E, the limit relative category $\mathrm{cat}_{X,Y}^{\infty}(A)$ is by definition equal to $\limsup_{n \to \infty} \mathrm{cat}_{X_n, Y_n}(A_n)$. Note that unlike for the limit index, the limit category is necessarily a nonnegative integer. Note also that if D is the unit closed ball and S its boundary in an infinite-dimensional Hilbert space, then $\mathrm{cat}_{D,S}(D) = \mathrm{cat}_S(S) = 0$ while $\mathrm{cat}_{D,S}^{\infty}(D) = \mathrm{cat}_S^{\infty}(S) = 1$.

Below we formulate some properties of $i_{\mathcal{E}}$ which automatically hold for $\gamma_{\mathcal{E}}$. As before $\widetilde{E}$ is the orthogonal complement of E^G. It follows from the invariance of E_n that the dimension of $\widetilde{E}_n = E_n \cap \widetilde{E}$ is even except when $G = \mathbb{Z}/2$. Recall the notation $d_G = 1 + \dim G$.

PROPOSITION 1.13. *Let $A, B \in \Sigma$.*

    (i) *If for almost all n there exists an equivariant mapping $g_n \in C(A_n, B_n)$, then $i_{\mathcal{E}}(A) \leqslant i_{\mathcal{E}}(B)$.*

    (ii) *$i_{\mathcal{E}}(A \cup B) \leqslant i_{\mathcal{E}}(A) + i(B)$ if $i_{\mathcal{E}}(A) \neq -\infty$.*

    (iii) *Let $l \in \mathbb{Z}$, $R > 0$. If Y is an invariant subspace of $\widetilde{E}$ such that $\dim Y_n = (d_n + l)d_G$ for almost all n, then $i_{\mathcal{E}}(Y \cap S_R) = l$.*

    (iv) *Let $m \in \mathbb{Z}$. If X is an invariant subspace of $\widetilde{E}$ such that $\mathrm{codim}_{\widetilde{E}_n} X_n = (d_n + m)d_G$ for almost all n and if $i_{\mathcal{E}}(A) > m$, then $A \cap (E^G \oplus X) \neq \emptyset$.*

PROOF. (i) It follows from (i) of Proposition 1.7 that $i(A_n) - d_n \leqslant i(B_n) - d_n$. So passing to the limit as $n \to \infty$ we obtain the conclusion.

    (ii) $i(A_n \cup B_n) - d_n \leqslant i(A_n) - d_n + i(B_n) \leqslant (i(A_n) - d_n) + i(B)$. Now we can pass to the limit again.

(iii) This follows from (v) of Proposition 1.7.

(iv) There exists a number n such that $E^G \subset E_n$, $\operatorname{codim}_{\widetilde{E}_n} X_n = (d_n + m)d_G$ and $i(A_n) > d_n + m$. So by (vi) of Proposition 1.7, $\emptyset \neq A_n \cap (E^G \oplus X_n) \subset A \cap (E^G \oplus X)$. $\qquad \square$

Recall that if x is a critical point of an invariant functional Φ, then so are all $y \in \mathcal{O}(x)$. We have the following results concerning the existence of critical orbits.

THEOREM 1.14. *Suppose that* $\Phi \in C^1(E, \mathbb{R})$ *is* G-*invariant, satisfies* (PS)* *and* $\Phi(0) = 0$. *Moreover, suppose there exist numbers* $\rho > 0$, $\alpha < \beta < 0$, *integers* $m < l$, *and invariant subspaces* $X, Y \subset \widetilde{E}$ *such that:*

(i) $E^G \subset E_n$ *for almost all* n;

(ii) $\operatorname{codim}_{\widetilde{E}_n} X_n = (d_n + m)d_G$ *and* $\dim Y_n = (d_n + l)d_G$ *for almost all* n;

(iii) $\Phi|_{Y \cap S_\rho} \leqslant \beta$;

(iv) $\Phi|_{E^G \oplus X} \geqslant \alpha$ *and* $\Phi|_{E^G} \geqslant 0$.

Then Φ *has at least* $l - m$ *distinct critical orbits* $\mathcal{O}(x_j)$ *such that* $\mathcal{O}(x_j) \cap E^G = \emptyset$. *The corresponding critical values can be characterized as*

$$c_j = \inf_{i_{\mathcal{E}}(A) \geqslant j} \sup_{x \in A} \Phi(x), \quad m + 1 \leqslant j \leqslant l,$$

and are contained in the interval $[\alpha, \beta]$.

PROOF. It is clear that $\{A \in \Sigma : i(A) \geqslant j + 1\} \subset \{A \in \Sigma : i(A) \geqslant j\}$, hence $c_{m+1} \leqslant c_{m+2} \leqslant \cdots \leqslant c_l$. According to (iii) of Proposition 1.13, $i_{\mathcal{E}}(Y \cap S_\rho) \geqslant l$, hence by (iii), $c_l \leqslant \beta$. Suppose $i_{\mathcal{E}}(A) \geqslant m + 1$. Then $A \cap (E^G \oplus X) \neq \emptyset$ by (iv) of Proposition 1.13 and it follows from (iv) that $c_{m+1} \geqslant \alpha$. Moreover, (iv) implies $K_{c_j} \cap E^G = \emptyset$.

Suppose $c := c_j = \cdots = c_{j+p}$ for some $p \geqslant 0$. The proof will be complete if we can show that $i(K_c) \geqslant p + 1$ (because either all c_j are distinct and $K_{c_j} \neq \emptyset$, or $i(K_{c_j}) \geqslant 2$ for some j and K_{c_j} contains infinitely many orbits according to (vii) of Proposition 1.7). By (iii) of Proposition 1.7 there exists a neighbourhood $N \in \Sigma$ such that $i(N) = i(K_c)$, and for this N we may find $\bar\varepsilon, \delta > 0$ and $n_0 \geqslant 1$ such that the conclusion of Lemma 1.11 holds. It follows from Lemma 1.1 that we can find an $\varepsilon > 0$ such that for each $n \geqslant n_0$ there exists a deformation $\eta_n : E_n \times [0, 1] \to E_n$ with $\eta_n(\Phi_n^{c+\varepsilon}, 1) \subset \Phi_n^{c-\varepsilon} \cup N_n$. Moreover, using (1.13) we may assume that $\eta_n(., t)$ is equivariant for each t. So by (i) and (ii) of Proposition 1.13 and the definition of c,

$$j + p \leqslant i_{\mathcal{E}}(\Phi^{c+\varepsilon}) \leqslant i_{\mathcal{E}}(\Phi^{c-\varepsilon} \cup N) \leqslant i_{\mathcal{E}}(\Phi^{c-\varepsilon}) + i(N) < j + i(N). \tag{1.14}$$

Hence $i(K_c) = i(N) > p$. $\qquad \square$

Applying Theorem 1.14 to $-\Phi$ we immediately obtain the following result which will be more convenient in our applications:

COROLLARY 1.15. *Suppose that* $\Phi \in C^1(E, \mathbb{R})$ *is* G-*invariant, satisfies* (PS)* *and* $\Phi(0) = 0$. *Moreover, suppose there exist numbers* $\rho > 0$, $0 < \alpha < \beta$, *integers* $m < l$, *and invariant subspaces* $X, Y \subset \widetilde{E}$ *such that:*

(i) $E^G \subset E_n$ *for almost all* n;

(ii) $\operatorname{codim}_{\widetilde{E}_n} X_n = (d_n + m)d_G$ *and* $\dim Y_n = (d_n + l)d_G$ *for almost all* n;

(iii) $\Phi|_{Y \cap S_\rho} \geqslant \alpha$;

(iv) $\Phi|_{E^G \oplus X} \leqslant \beta$ *and* $\Phi|_{E^G} \leqslant 0$.

Then Φ *has at least* $l - m$ *distinct critical orbits* $\mathcal{O}(x_j)$ *such that* $\mathcal{O}(x_j) \cap E^G = \emptyset$. *The corresponding critical values can be characterized as*

$$c_j = \sup_{i_{\mathcal{E}}(A) \geqslant j} \inf_{x \in A} \Phi(x), \quad m + 1 \leqslant j \leqslant l,$$

and are contained in the interval $[\alpha, \beta]$.

COROLLARY 1.16. *If the hypotheses of Theorem 1.14 or Corollary 1.15 are satisfied with* $l \in \mathbb{Z}$ *fixed and* $m \in \mathbb{Z}$ *arbitrarily small, then* Φ *has infinitely many geometrically distinct critical orbits* $\mathcal{O}(x_j)$ *such that* $\mathcal{O}(x_j) \cap E^G = \emptyset$. *Moreover,* $c_j \to -\infty$ *in Theorem 1.14 and* $c_j \to \infty$ *in Corollary 1.15 as* $j \to -\infty$.

PROOF. It suffices to consider the case of Theorem 1.14. The value c_j is defined for all $j \leqslant l$, $j \in \mathbb{Z}$ and since the sequence $\{c_j\}$ is nondecreasing, either $c_j \to -\infty$ and we are done, or $c_j \to c \in \mathbb{R}$ as $j \to -\infty$. In the second case K_c is nonempty and compact according to (PS)*. Let $N \in \Sigma$ be a neighbourhood of K_c such that $i(N) = i(K_c) < \infty$ and let $\varepsilon > 0$ be as in Lemma 1.1. Since $c + \varepsilon \geqslant c_{j_0}$ for some j_0 and $c - \varepsilon < c_j$ for all $j \leqslant l$, we have (cf. (1.14))

$$j_0 \leqslant i_{\mathcal{E}}\left(\Phi^{c+\varepsilon}\right) \leqslant i_{\mathcal{E}}\left(\Phi^{c-\varepsilon}\right) + i(N) = -\infty,$$

a contradiction. $\qquad\square$

REMARK 1.17. Proposition 1.13, Theorem 1.14 and Corollaries 1.15 and 1.16 are valid if $G = \mathbb{Z}/2$ and $T_{\pm 1}x = \pm x$ (i.e., Φ is even). For this G, $i(A)$ is just the genus $\gamma(A)$. If $G = \mathbb{Z}/p$ and $p \geqslant 3$, then $l - m$ is necessarily an even integer.

We now state a critical point theorem which needs tools from algebraic topology.

THEOREM 1.18. *Let* M *be a compact differentiable manifold and* $\Phi : E \times M \to \mathbb{R}$ *a* C^1*-functional defined on the product of the Hilbert space* E *and* M. *Suppose* Φ *satisfies* (PS)*, *there exist numbers* $\rho > 0$, $\alpha < \beta \leqslant \gamma$ *and subspaces* W, Y, *where* $E = W \oplus Y$, $W_n \subset W$, $Y_n \subset Y$, $\dim W_n \geqslant 1$, *such that:*

(i) $\Phi|_{(W \cap S_\rho) \times M} \leqslant \alpha$;

(ii) $\Phi|_{Y \times M} \geqslant \beta$;

(iii) $\Phi|_{(W \cap \overline{B}_\rho) \times M} \leqslant \gamma$.

Then Φ *possesses at least* $\operatorname{cupl}(M) + 1$ *critical points.*

Here $\operatorname{cupl}(M)$ denotes the cuplength of M with respect to singular cohomology theory with coefficients in an arbitrary field. For a proof we refer to Fournier et al. [42]. The argument there uses the limit relative category (see Remark 1.12) and is in the spirit of Theorem 1.14. In particular, the numbers c_j are defined by minimaxing over sets $A \supset D$ with $\mathrm{cat}^{\infty}_{C,D}(A) \geqslant j$, where $(C, D) := (E \times M, (W \cap S_\rho) \times M)$. An important role is played by the inequality $\mathrm{cat}^{\infty}_{C,D}((W \cap \overline{B}_\rho) \times M) \geqslant \mathrm{cupl}(M) + 1$. A related result can be found in [90].

For applications to homoclinic solutions one has to deal with functionals where neither the (PS)- nor the (PS)*-condition holds. We present two abstract critical point theorems which are helpful in this case. The proofs involve again a reduction to a finite-dimensional situation.

THEOREM 1.19. *Let E be a separable Hilbert space with the orthogonal decomposition $E = E^+ \oplus E^-$, $z = z^+ + z^-$, and suppose $\Phi \in C^1(E, \mathbb{R})$ satisfies the hypotheses:*
  (i) $\Phi(z) = \frac{1}{2}(\|z^+\|^2 - \|z^-\|^2) - \psi(z)$ *where $\psi \in C^1(E, \mathbb{R})$ is bounded below, weakly sequentially lower semicontinuous with $\psi' : E \to E$ weakly sequentially continuous;*
  (ii) $\Phi(0) = 0$ *and there are constants $\kappa, \rho > 0$ such that $\Phi(z) > \kappa$ for every $z \in S_\rho \cap E^+$;*
  (iii) *there exists $e \in E^+$ with $\|e\| = 1$, and $R > \rho$ such that $\Phi(z) \leqslant 0$ for $z \in \partial M$ where $M = \{z = z^- + \zeta e : z^- \in X^-, \|z\| \leqslant R, \zeta \geqslant 0\}$.*
Then there exists a sequence $\{z_j\}$ in E such that $\Phi'(z_j) \to 0$ and $\Phi(z_j) \to c$ for some $c \in [\kappa, m]$, where $m := \sup \Phi(\overline{M})$.

The theorem is due to Kryszewski and Szulkin [58]. Some compactness is hidden in condition (i) where the weak topology is used. In the applications the concentration-compactness method, see [65], can sometimes be used in order to obtain an actual critical point. Of course, if the Palais–Smale condition holds then there exists a critical point at the level c.

PROOF (outline). Let $P^{\pm} : E \to E^{\pm}$ be the orthogonal projections. We choose a Hilbert basis $\{e_k\}_{k \in \mathbb{N}}$ of E^- and define the norm

$$|||u||| := \max \left\{ \|P^+ u\|, \sum_{k=1}^{\infty} \frac{|\langle u, e_k \rangle|}{2^k} \right\}.$$

The topology induced on E by this norm will be denoted by τ. On subsets $\{u \in E : \|P^- u\| \leqslant R\}$ this topology coincides with the weak $\times$ strong product topology $(E^-, w) \times (E^+, \| \cdot \|)$ on E. In particular, for a $\| \cdot \|$-bounded sequence $\{u_j\}$ in E^- we have $u_j \rightharpoonup u$ if and only if $u_j \to u$ with respect to $||| \cdot |||$. Given a finite-dimensional subspace $F \subset E^+$, $\| \cdot \|$-bounded subsets of $E^- \oplus F$ are $||| \cdot |||$-precompact.

We prove the theorem arguing indirectly. Suppose there exists $\alpha > 0$ with $\|\Phi'(u)\| \geqslant \alpha$ for all $u \in \Phi^m_\kappa := \{u \in E : \kappa \leqslant \Phi(u) \leqslant m\}$. Then we construct a deformation $h : I \times \Phi^m \to \Phi^m$, $I = [0, 1]$, with the properties:
  (h_1) $h : I \times \Phi^m \to \Phi^m$ is continuous with respect to the $\| \cdot \|$-topology on Φ^m, and with respect to the τ-topology;

(h_2) $h(0, u) = u$ for all $u \in \Phi^m$;

(h_3) $\Phi(h(t, u)) \leqslant \Phi(u)$ for all $t \in I$, $u \in \Phi_\kappa^m$;

(h_4) each $(t, u) \in I \times \Phi^m$ has a τ-open neighbourhood W such that the set $\{v - h(s, v): (s, v) \in W\}$ is contained in a finite-dimensional subspace of E;

(h_5) $h(1, \Phi^m) \subset \Phi^\kappa$.

This leads to a contradiction as follows. Since M is τ-compact, by (h_1) and (h_4) there exists a finite-dimensional subspace $F \subset E$ containing the set $\{v - h(s, v): (s, v) \in I \times M\}$, hence $h(I \times (M \cap F)) \subset F$. Since

$$h(I \times \partial M) \subset F \cap \Phi^\kappa \subset F \setminus \left(S_\rho \cap E^+\right)$$

a standard argument using the Brouwer degree yields $h(1, M \cap F) \cap S_\rho \cap E^+ \neq \emptyset$. (The sets $F \cap \partial M$ and $F \cap S_\rho \cap E^+$ link in F.) Now condition (ii) of the theorem implies $h(1, M \cap F) \not\subset \Phi^\kappa$, contradicting ($h_5$).

It remains to construct a deformation h as above. For each $u \in \Phi_\kappa^m$ we choose a pseudo-gradient vector $w(u) \in E$, that is $\|w(u)\| \leqslant 2$ and $\langle \Phi'(u), w(u) \rangle > \|\Phi'(u)\|$ (this definition differs somewhat from (1.2)). By condition (i) of the theorem there exists a τ-open neighbourhood $N(u)$ of u in E such that $\langle \Phi'(v), w(u) \rangle > \|\Phi'(u)\|$ for all $v \in N(u) \cap \Phi_\kappa^m$. If $\Phi(u) < \kappa$ we set $N(u) = \{v \in E: \Phi(v) < \kappa\}$. As a consequence of condition (i) of the theorem this set is τ-open. Let $(\pi_j)_{j \in J}$ be a τ-Lipschitz continuous partition of unity of Φ^m subordinated to the covering $N(u)$, $u \in \Phi^m$. This exists because the τ-topology is metric. Clearly, the $\pi_j : E \to [0, 1]$ are also $\|\cdot\|$-Lipschitz continuous. For each $j \in J$ there exists $u_j \in \Phi^m$ with $\operatorname{supp} \pi_j \subset N(u_j)$. We set $w_j = w(u_j)$ and define the vector field

$$f : \Phi^m \to E, \quad f(u) := \frac{m - \kappa}{\alpha} \sum_{j \in J} \pi_j(u) w_j .$$

This vector field is locally Lipschitz continuous and τ-locally Lipschitz continuous. It is also τ-locally finite-dimensional. Thus we may integrate it and obtain a flow $\eta : [0, \infty) \times \Phi^m \to \Phi^m$. It is easy to see that the restriction of η to $[0, 1] \times \Phi^m$ satisfies the properties (h_1)–(h_5). $\square$

REMARK 1.20. A sequence $\{z_j\}$ is called a *Cerami sequence* if $\Phi(z_j)$ is bounded and $(1 + \|z_j\|)\Phi'(z_j) \to 0$. This definition has been introduced by Cerami in [22]. Note in particular that if $\{z_j\}$ is as above, then $\langle \Phi'(z_j), z_j \rangle \to 0$ which does not need to be the case for an (a priori unbounded) (PS)-sequence. It has been shown in [59] that under the hypotheses of Theorem 1.19 a stronger conclusion holds: there exists a Cerami sequence $\{z_j\}$ such that $\Phi(z_j) \to c \in [\kappa, m]$.

The next result of this section deals with $\mathbb{Z}/p$-invariant functionals $\Phi \in C^1(E, \mathbb{R})$. As a substitute for the (PS)- or (PS)*-condition we introduce the concept of (PS)-attractor. Given an interval $I \subset \mathbb{R}$, we call a set $\mathcal{A} \subset E$ a (PS)$_I$-attractor if for any (PS)$_c$-sequence $\{z_j\}$ with $c \in I$, and any $\varepsilon, \delta > 0$ one has $z_j \in U_\varepsilon(\mathcal{A} \cap \Phi_{c-\delta}^{c+\delta})$ provided j is large enough. Here $U_\varepsilon(F)$ denotes the ε-neighbourhood of F in E.

THEOREM 1.21. *Let E be a separable Hilbert space with an isometric representation of the group $G = \mathbb{Z}/p$, where p is a prime, such that $E^{\mathbb{Z}/p} = \{0\}$. Let $E = E^+ \oplus E^-$, $z = z^+ + z^-$, be an orthogonal decomposition and $E^\pm$ be $\mathbb{Z}/p$-invariant. Let $\Phi \in C^1(E, \mathbb{R})$ be a $\mathbb{Z}/p$-invariant functional satisfying the following conditions:*

(i) *$\Phi(z) = \frac{1}{2}(\|z^+\|^2 - \|z^-\|^2) - \psi(z)$ where $\psi \in C^1(E, \mathbb{R})$ is bounded below, weakly sequentially lower semicontinuous with $\psi' : E \to E$ weakly sequentially continuous;*

(ii) *$\Phi(0) = 0$ and there exist $\kappa, \rho > 0$ such that $\Phi(z) > \kappa$ for every $z \in S_\rho \cap E^+$;*

(iii) *there exists a strictly increasing sequence of finite-dimensional $\mathbb{Z}/p$-invariant subspaces $F_n \subset E^+$ such that $\sup \Phi(E_n) < \infty$ where $E_n := E^- \oplus F_n$, and an increasing sequence of real numbers $R_n > 0$ with $\sup \Phi(E_n \setminus B_{R_n}) < \inf \Phi(B_\rho)$;*

(iv) *for any compact interval $I \subset (0, \infty)$ there exists a $(PS)_I$-attractor $\mathcal{A}$ such that $\inf\{\|z^+ - w^+\| : z, w \in \mathcal{A}, z^+ \neq w^+\} > 0$.*

Then Φ has an unbounded sequence $\{c_j\}$ of positive critical values.

PROOF (outline). Let τ be the topology on E introduced in the proof of Theorem 1.19. For $c \in \mathbb{R}$ we consider the set $\mathcal{M}(c)$ of maps $g : \Phi^c \to E$ satisfying:

(P_1) g is τ-continuous and equivariant;

(P_2) $g(\Phi^a) \subset \Phi^a$ for all $a \geq \inf \Phi(B_\rho) - 1$ where ρ is from condition (ii);

(P_3) each $u \in \Phi^c$ has a τ-open neigbourhood $W \subset E$ such that the set $(\mathrm{id} - g)(W \cap \Phi^c)$ is contained in a finite-dimensional linear subspace of E.

Let i be the $\mathbb{Z}/p$-index from Section 1.1 and set

$$i_0(c) := \min_{g \in \mathcal{M}(c)} i\big(g(\Phi^c) \cap S_\rho \cap E^+\big) \in \mathbb{N}_0 \cup \{\infty\}.$$

Clearly i_0 is nondecreasing and $i_0(c) = 0$ for $c \leq \kappa$ where κ is from (ii). i_0 is a kind of *pseudoindex* in the sense of Benci's paper [18]. Now we define the values

$$c_k := \inf\big\{c > 0 : i_0(c) \geq k\big\}.$$

One can show that $i_0(c)$ is finite for every $c \in \mathbb{R}$ and can only change at a critical level of Φ. In order to see the latter, given an interval $[c, d]$ without critical values one needs to construct maps $g \in \mathcal{M}(d)$ with $g(\Phi^d) \subset \Phi^c$. Such a map can be obtained as time-1-map of a deformation as in the proof of Theorem 1.19. Of course one has to make sure that the deformation is equivariant which is the case if the vector field is equivariant. This can be easily achieved, see (1.13). Given a finite-dimensional subspace $F_n \subset E^+$ from condition (iii) one next proves that $i_0(c) \geq \dim F_n$ for any $c \geq \Phi(E^- \oplus F_n)$. This is a consequence of the properties of the index stated in Proposition 1.7. No extension of the index to infinite dimensions is needed.

Details of the proof, of a slightly more general result in fact, can be found in [12] for $p = 2$ and in [13] for p an odd prime. $\qquad\square$

2. Periodic solutions

2.1. *Variational setting for periodic solutions*

In this section we reformulate the problem of existence of 2π-periodic solutions of the Hamiltonian system

$$\dot{z} = JH_z(z,t) \tag{2.1}$$

in terms of the existence of critical points of a suitable functional and we collect some basic facts about this functional. When looking for periodic solutions of (2.1) we shall always assume that the Hamiltonian $H = H(z,t)$ satisfies the following conditions:

(H$_1$) $H \in C(\mathbb{R}^{2N} \times \mathbb{R}, \mathbb{R})$, $H_z \in C(\mathbb{R}^{2N} \times \mathbb{R}, \mathbb{R}^{2N})$ and $H(0,t) \equiv 0$;

(H$_2$) H is 2π-periodic in the t-variable;

(H$_3$) $|H_z(z,t)| \leqslant c(1 + |z|^{s-1})$ for some $c > 0$ and $s \in (2, \infty)$.

We note that it causes no loss of generality to assume $H(0,t) \equiv 0$. Occasionally we shall need two additional conditions:

(H$_4$) $H_{zz} \in C(\mathbb{R}^{2N} \times \mathbb{R}, \mathbb{R}^{4N^2})$;

(H$_5$) $|H_{zz}(z,t)| \leqslant d(1 + |z|^{s-2})$ for some $d > 0$ and $s \in (2, \infty)$.

Clearly, (H$_5$) implies (H$_3$).

Let $E := H^{1/2}(S^1, \mathbb{R}^{2N})$ be the Sobolev space of 2π-periodic $\mathbb{R}^{2N}$-valued functions

$$z(t) = a_0 + \sum_{k=1}^{\infty}(a_k \cos kt + b_k \sin kt), \quad a_0, a_k, b_k \in \mathbb{R}^{2N} \tag{2.2}$$

such that $\sum_{k=1}^{\infty} k(|a_k|^2 + |b_k|^2) < \infty$. Then E is a Hilbert space with an inner product

$$\langle z, w \rangle := 2\pi a_0 \cdot a_0' + \pi \sum_{k=1}^{\infty} k\big(a_k \cdot a_k' + b_k \cdot b_k'\big), \tag{2.3}$$

where a_k', b_k' are the Fourier coefficients of w. It is well known that the Sobolev embedding $E \hookrightarrow L^q(S^1, \mathbb{R}^{2N})$ is compact for any $q \in [1, \infty)$ (see, e.g., [2]) but $z \in E$ does not imply z is bounded. There is a natural action of $\mathbb{R}$ on $L^q(S^1, \mathbb{R}^{2N})$ and E given by time translation:

$$(T_\theta z)(t) := z(t + \theta) \quad \text{for } \theta, t \in \mathbb{R}.$$

Since the functions z are 2π-periodic in t, T induces an isometric representation of $G = S^1 \equiv \mathbb{R}/2\pi\mathbb{Z}$. In the notation of Section 1.1, we have $\mathcal{O}(z_1) = \mathcal{O}(z_2)$ if and only if $z_2(t) = z_1(t + \theta)$ for some θ and all $t \in \mathbb{R}$. Let

$$\Phi(z) := \frac{1}{2} \int_0^{2\pi} (-J\dot{z} \cdot z)\, dt - \int_0^{2\pi} H(z,t)\, dt$$

and

$$\psi(z) := \int_0^{2\pi} H(z,t)\,dt.$$

PROPOSITION 2.1. *If H satisfies* (H_1)–(H_3), *then* $\Phi \in C^1(E, \mathbb{R})$ *and* $\Phi'(z) = 0$ *if and only if z is a 2π-periodic solution of* (2.1). *Moreover, ψ' is completely continuous in the sense that $\psi'(z_j) \to \psi'(z)$ whenever $z_j \rightharpoonup z$. If, in addition, H satisfies* (H_4) *and* (H_5), *then* $\Phi \in C^2(E, \mathbb{R})$ *and $\psi''(z)$ is a compact linear operator for each z.*

PROOF. We only outline the argument. The details may be found, e.g., in [80, Appendix B] or [102, Appendix A and Lemma 2.16]. Although the results in [102] concern elliptic partial differential equations, the proofs are easy to adapt to our situation.

Let $s' = s/(s-1)$ be the conjugate exponent. By (H_3),

$$H_z : L^s\big(S^1, \mathbb{R}^{2N}\big) \to L^{s'}\big(S^1, \mathbb{R}^{2N}\big)$$

is a continuous mapping, and using this one shows that $\psi \in C^1(E, \mathbb{R})$ and

$$\langle \psi'(z), w \rangle = \int_0^{2\pi} H_z(z,t) \cdot w\,dt.$$

Moreover, it follows by the compact embedding of E into $L^s(S^1, \mathbb{R}^{2N})$ that ψ' is completely continuous.

Since $-J\dot{z} \cdot w = \dot{z} \cdot Jw$, the bilinear form $(z, w) \mapsto \int_0^{2\pi}(-J\dot{z} \cdot w)\,dt$ is (formally) self-adjoint. According to (2.2) and (2.3),

$$\int_0^{2\pi}(-J\dot{z} \cdot w)\,dt = \pi \sum_{k=1}^{\infty} k\big(-Jb_k \cdot a_k' + Ja_k \cdot b_k'\big),$$

hence this form is continuous in E and the quadratic form $z \mapsto \int_0^{2\pi}(-J\dot{z} \cdot z)\,dt$ is of class C^1. Now it is easy to see that $\Phi'(z) = 0$ if and only if z is a 2π-periodic solution of (2.1). Moreover, by elementary regularity theory, $z \in C^1(S^1, \mathbb{R}^{2N})$.

If (H_4) and (H_5) are satisfied, then, referring to the arguments in [80], [102] again, we see that $\psi \in C^2(E, \mathbb{R})$ and

$$\langle \psi''(z)w, y \rangle = \int_0^{2\pi} H_{zz}(z,t)w \cdot y\,dt.$$

Since ψ' is completely continuous, $\psi''(z)$ is a compact linear operator. $\quad\square$

Note that complete continuity of ψ' implies weak continuity of ψ (i.e., $\psi(z_j) \to \psi(z)$ whenever $z_j \rightharpoonup z$).

REMARK 2.2. If system (2.1) is *autonomous*, i.e., $H = H(z)$, then $\Phi(T_\theta z) = \Phi(z)$ for all $\theta \in \mathbb{R}$. Thus Φ is T-invariant. Two 2π-periodic solutions z_1, z_2 of an autonomous system are *geometrically distinct* if and only if $\mathcal{O}(z_1) \neq \mathcal{O}(z_2)$. When $H = H(z)$, we shall write $H'(z)$ instead of $H_z(z)$.

Let $z(t) = a_k \cos kt \pm Ja_k \sin kt$. Then

$$\int_0^{2\pi} (-J\dot{z} \cdot z)\, dt = \pm 2\pi k |a_k|^2 = \pm \|z\|^2.$$

It follows that E has the orthogonal decomposition $E = E^+ \oplus E^0 \oplus E^-$, where

$$E^0 = \left\{ z \in E \colon z \equiv a_0 \in \mathbb{R}^{2N} \right\},$$

$$E^{\pm} = \left\{ z \in E \colon z(t) = \sum_{k=1}^{\infty} a_k \cos kt \pm Ja_k \sin kt, \ a_k \in \mathbb{R}^{2N} \right\}$$

and if $z = z^0 + z^+ + z^-$, then

$$\int_0^{2\pi} (-J\dot{z} \cdot z)\, dt = \left\| z^+ \right\|^2 - \left\| z^- \right\|^2.$$

Hence

$$\Phi(z) = \frac{1}{2} \int_0^{2\pi} (-J\dot{z} \cdot z)\, dt - \int_0^{2\pi} H(z, t)\, dt = \frac{1}{2} \left\| z^+ \right\|^2 - \frac{1}{2} \left\| z^- \right\|^2 - \psi(z). \tag{2.4}$$

REMARK 2.3. Φ is called the action functional and the fact that z is a 2π-periodic solution of (2.1) if and only if $\Phi'(z) = 0$ is the least action (or the Euler–Maupertuis) principle. However, although the solutions z are critical points (or extremals) of Φ, they can never be minima (or maxima). Indeed, let $a \in \mathbb{R}^{2N}$ and $z_j = a \cos jt \pm Ja \sin jt$. Then $\Phi(z_j) \to \pm\infty$, so Φ is unbounded below and above. Moreover, using (2.4) and the weak continuity of ψ, it is easy to see that Φ has neither local maxima nor minima.

The first to develop a variational method for finding periodic solutions of a Hamiltonian system as critical points of the action functional was Rabinowitz [78]. Among other he has shown that a star-shaped compact energy surface necessarily carries a closed Hamiltonian orbit (see Section 2.3 for a discussion of this problem).

LEMMA 2.4. *Suppose* (H_1)–(H_3) *are satisfied.*
    (i) *If* $H(z, t) = \frac{1}{2}A_0(t)z \cdot z + G_0(z, t)$ *and* $(G_0)_z(z, t) = o(|z|)$ *uniformly in* t *as* $z \to 0$, *then* $\psi_0'(z) = o(\|z\|)$ *and* $\psi_0(z) = o(\|z\|^2)$ *as* $z \to 0$, *where* $\psi_0(z) := \int_0^{2\pi} G_0(z, t)\, dt.$

(ii) *If $H(z,t) = \frac{1}{2}A_\infty(t)z \cdot z + G_\infty(z,t)$ and $(G_\infty)_z(z,t) = \mathrm{o}(|z|)$ uniformly in t as $|z| \to \infty$, then $\psi'_\infty(z) = \mathrm{o}(\|z\|)$ and $\psi_\infty(z) = \mathrm{o}(\|z\|^2)$ as $\|z\| \to \infty$, where $\psi_\infty(z) := \int_0^{2\pi} G_\infty(z,t)\,dt$.*

PROOF. Since $H(0,t) = 0$, $\psi_0(0) = \psi_\infty(0) = 0$. It follows from (H$_3$) that for each $\varepsilon > 0$ there is a $c_1(\varepsilon) > 0$ such that $|(G_0)_z(z,t)| \leqslant \varepsilon|z| + c_1(\varepsilon)|z|^{s-1}$. Hence by the Sobolev inequality,

$$\left|\langle \psi'_0(z), w\rangle\right| \leqslant \int_0^{2\pi} \left(\varepsilon|z| + c_1(\varepsilon)|z|^{s-1}\right)|w|\,dt \leqslant \left(\varepsilon\|z\| + c_2(\varepsilon)\|z\|^{s-1}\right)\|w\|.$$

Taking the supremum over $\|w\| \leqslant 1$ and letting $z \to 0$ we see that $\psi'_0(z) = \mathrm{o}(\|z\|)$ as $z \to 0$. Since

$$\psi_0(z) = \int_0^1 \frac{d}{ds}\psi_0(sz)\,ds = \int_0^1 \langle \psi'_0(sz), z\rangle\,ds,$$

$\psi_0(z) = \mathrm{o}(\|z\|^2)$.

Similarly, for each $\varepsilon > 0$ there is a $c_3(\varepsilon)$ such that $|(G_\infty)_z(z,t)| \leqslant \varepsilon|z| + c_3(\varepsilon)$. So

$$\left|\langle \psi'_\infty(z), w\rangle\right| \leqslant \left(\varepsilon\|z\| + c_4(\varepsilon)\right)\|w\|$$

and $\psi'_\infty(z) = \mathrm{o}(\|z\|)$ as $\|z\| \to \infty$. Since

$$|\psi_\infty(z)| = \left|\int_0^1 \langle \psi'_\infty(sz), z\rangle\,ds\right| \leqslant \frac{1}{2}\varepsilon\|z\|^2 + c_4(\varepsilon)\|z\|,$$

$\psi_\infty(z) = \mathrm{o}(\|z\|^2)$ as $\|z\| \to \infty$. $\qquad\square$

Let now

$$\widetilde{E}_n := \left\{z \in E\colon z(t) = \sum_{k=1}^n (a_k \cos kt + b_k \sin kt)\right\} \quad \text{and} \quad E_n := E^0 \oplus \widetilde{E}_n. \tag{2.5}$$

Then $E_n \subset E_{n+1}$ for all n,

$$E = \overline{\bigcup_{n=1}^\infty E_n}$$

and

$$E_n^0 = E^0, \quad \widetilde{E}_n^\pm = E_n^\pm = E^\pm \cap E_n. \tag{2.6}$$

Moreover, $E^{S^1} = E^0$ (so $E^{S^1} \subset E_n$ for all n) and each subspace E_n is S^1-invariant (or more precisely, T-invariant). When using limit index theories we shall have $\mathcal{E} = \{E_n, d_n\}$, where

$$d_n \cdot d_G = 2N(1+n) = \dim E^0 + \frac{1}{2} \dim \widetilde{E}_n = \dim E^0 + \dim \widetilde{E}_n^+.$$

The orthogonal projection $E \to E_n$ will be denoted by P_n.

PROPOSITION 2.5. *Suppose* (H_1)–(H_3) *are satisfied and let* $\{z_j\}$ *be a sequence such that* $z_j \in E_{n_j}$ *for some* $n_j, n_j \to \infty$ *and* $P_{n_j} \Phi'(z_j) \to 0$ *as* $j \to \infty$. *Then* $\{z_j\}$ *has a convergent subsequence in each of the following two cases:*
  (i) $H(z,t) = \frac{1}{2} A_\infty(t)z \cdot z + G_\infty(z,t)$, *where* $(G_\infty)_z(z,t) = \mathrm{o}(|z|)$ *as* $|z| \to \infty$ *and* $z = 0$ *is the only* 2π-*periodic solution of the linear system*

$$\dot{z} = J A_\infty(t)z.$$

  (ii) $\Phi(z_j)$ *is bounded above and there exist* $\mu > \max\{2, s-1\}$ *and* $R > 0$ *such that*

$$0 < \mu H(z,t) \leqslant z \cdot H_z(z,t) \quad \text{for all } |z| \geqslant R.$$

So in particular, (PS)* *holds if* H *satisfies one of the conditions above.*

It follows upon integration that (ii) implies

$$H(z,t) \geqslant a_1 |z|^\mu - a_2 \tag{2.7}$$

for some $a_1, a_2 > 0$. Hence H grows superquadratically and H_z superlinearly as $|z| \to \infty$. Note also that $\mu \leqslant s$ according to (H_3).

PROOF. Let $z_j \in E_{n_j}$, $n_j \to \infty$ and $P_{n_j} \Phi'(z_j) \to 0$ as $j \to \infty$. Since $P_{n_j} \Phi'(z_j) = z_j^+ - z_j^- - P_{n_j} \psi'(z_j) \to 0$ and ψ' is completely continuous, it follows that if $\{z_j\}$ is bounded, then $z_j \to z$ after passing to a subsequence. Moreover, $\Phi'(z) = 0$. Hence it remains to show that $\{z_j\}$ must be bounded.

Suppose (i) is satisfied and let

$$\langle B_\infty z, w \rangle := \int_0^{2\pi} A_\infty(t)z \cdot w \, \mathrm{d}t, \qquad L_\infty z := z^+ - z^- - B_\infty z.$$

Then $B_\infty : E \to E$ is a compact linear operator (cf. Proposition 2.1). Since $L_\infty z = 0$ if and only if $\dot{z} = J A_\infty(t)z$, L_∞ is invertible and it follows that $(P_n L_\infty |_{E_n})^{-1}$ is uniformly bounded for large n. Hence

$$(P_{n_j} L_\infty)^{-1} P_{n_j} \Phi'(z_j) = z_j - (P_{n_j} L_\infty)^{-1} P_{n_j} \psi'_\infty(z_j) \to 0$$

and $\{z_j\}$ is bounded because $\psi'_\infty(z) = \mathrm{o}(\|z\|)$ as $\|z\| \to \infty$.

Suppose now (ii) holds. Below $c_1, c_2, \ldots$ will denote different constants whose exact values are insignificant. Since $z_j \in E_{n_j}$,

$$
\begin{aligned}
c_1\|z_j\| + c_2 &\geqslant \Phi(z_j) - \frac{1}{2}\langle P_{n_j}\Phi'(z_j), z_j\rangle \\
&= \int_0^{2\pi}\left(\frac{1}{2}z_j \cdot H_z(z_j, t) - H(z_j, t)\right)\mathrm{d}t \\
&\geqslant \left(\frac{\mu}{2} - 1\right)\int_0^{2\pi} H(z_j, t)\,\mathrm{d}t \\
&\geqslant c_3\|z_j\|_\mu^\mu - c_4,
\end{aligned}
\tag{2.8}
$$

where the last inequality follows from (2.7). Since E^0 is finite dimensional, $\|z_j^0\| \leqslant c_5\|z_j\|_\mu$ and by (2.8),

$$
\|z_j^0\| \leqslant c_6\|z_j\|^{1/\mu} + c_7.
\tag{2.9}
$$

By the Hölder and Sobolev inequalities,

$$
\begin{aligned}
\|z_j^+\|^2 &= \langle\Phi'(z_j), z_j^+\rangle + \int_0^{2\pi} H_z(z_j, t)\cdot z_j^+\,\mathrm{d}t \\
&\leqslant c_8\|z_j^+\| + c_9\int_0^{2\pi}|z_j|^{s-1}|z_j^+|\,\mathrm{d}t \\
&\leqslant c_8\|z_j^+\| + c_{10}\|z_j\|_\mu^{s-1}\|z_j^+\|
\end{aligned}
\tag{2.10}
$$

(here we have used that $\mu > s - 1$) and a similar inequality holds for z_j^-. Hence

$$
\|z_j^\pm\| \leqslant c_8 + c_{10}\|z_j\|_\mu^{s-1}
$$

and by (2.8),

$$
\|z_j^\pm\| \leqslant c_{11} + c_{12}\|z_j\|^{(s-1)/\mu}.
$$

This and (2.9) combined imply $\{z_j\}$ is bounded. $\qquad\square$

REMARK 2.6. (a) If $H(z, t) = \frac{1}{2}A(t)z \cdot z + G(z, t)$, where A is a symmetric $2N \times 2N$ matrix with periodic entries and G satisfies the superlinearity condition of Proposition 2.5, it is easy to see by an elementary computation that so does H, possibly with a smaller $\mu > \max\{2, s - 1\}$ and a larger R.

(b) In some problems the growth restriction (H$_3$) may be removed and the condition $\mu > \max\{2, s - 1\}$ can be replaced by $\mu > 2$. For this purpose one introduces a modified

Hamiltonian H_K such that $H_K(z,t) = H(z,t)$ for $|z| \leqslant K$ and $H_K(z,t) = C|z|^s$ for $|z| \geqslant K+1$ and some convenient C, s. Then the modified functional satisfies (PS)*, one can apply a suitable variational method to obtain one or more solutions which are uniformly bounded independently of K. Hence these solutions satisfy the original equation for K large. We shall comment on that when appropriate.

Let now H satisfy (H$_1$)–(H$_5$). For each fixed z we have $\langle \Phi''(z)w, w \rangle = \|w^+\|^2 - \|w^-\|^2 - \langle \psi''(z)w, w \rangle$, and since $\psi''(z)$ is a compact linear operator, it is easy to see that the quadratic forms $\pm\Phi''(z)$ have infinite Morse index. However, it is possible to define a certain relative index which will always be finite. Let A be a symmetric $2N \times 2N$ constant matrix and

$$\langle Lz, w \rangle := \int_0^{2\pi} (-J\dot{z} - Az) \cdot w \, dt. \tag{2.11}$$

It follows from (2.2) and (2.3) that

$$\langle Lz, z \rangle = -2\pi A a_0 \cdot a_0 + \pi \sum_{k=1}^{\infty} k\left(\left(-Jb_k - \frac{1}{k}Aa_k\right) \cdot a_k + \left(Ja_k - \frac{1}{k}Ab_k\right) \cdot b_k\right). \tag{2.12}$$

The restriction of this quadratic form to a subspace corresponding to a fixed $k \geqslant 1$ is represented by the $(4N \times 4N)$-matrix $\pi k T_k(A)$, where

$$T_k(A) := \begin{pmatrix} -\dfrac{1}{k}A & -J \\ J & -\dfrac{1}{k}A \end{pmatrix}.$$

Let $M^+(\cdot)$ and $M^-(\cdot)$ respectively denote the number of positive and negative eigenvalues of a symmetric matrix (counted with their multiplicities) and let $M^0(\cdot)$ be the dimension of the nullspace of this matrix. Then $M^0(T_k(A)) = 0$ and $M^\pm(T_k(A)) = 2N$ for all k large enough. Indeed, a simple computation shows that the matrix

$$\begin{pmatrix} 0 & -J \\ J & 0 \end{pmatrix}$$

has the eigenvalues ± 1, each of multiplicity $2N$, so by a simple perturbation argument, $M^\pm(T_k(A)) = 2N$ for almost all k (cf. [3, Section 12], [4, Section 2]). Therefore the following numbers are well defined and finite:

$$i^-(A) := M^+(A) - N + \sum_{k=1}^{\infty} \left(M^-\big(T_k(A)\big) - 2N\right), \tag{2.13}$$

$$i^+(A) := M^-(A) - N + \sum_{k=1}^{\infty} \left(M^+\left(T_k(A)\right) - 2N \right) \tag{2.14}$$

and

$$i^0(A) := M^0(A) + \sum_{k=1}^{\infty} M^0\left(T_k(A)\right). \tag{2.15}$$

Clearly, $i^{\pm}(A)$, $i^0(A)$ are finite and $i^-(A) + i^+(A) + i^0(A) = 0$. The quantity $i^-(A)$ is a relative Morse index in the sense that it provides a measure for the difference between the negative parts of the quadratic forms $z \mapsto \int_0^{2\pi} (-J\dot{z} - Az) \cdot z \, dt$ and $z \mapsto \int_0^{2\pi} (-J\dot{z} \cdot z) \, dt$.

It is easy to see that $i^0(A) = \dim N(L)$ is the number of linearly independent 2π-periodic solutions of the linear system

$$\dot{z} = JAz. \tag{2.16}$$

Hence in particular $i^0(A) \leqslant 2N$. Moreover, $i^0(A) = 0$ if and only if $\sigma(JA) \cap i\mathbb{Z} = \emptyset$. Indeed, it follows from (2.12) (or by substituting (2.2) into (2.16)) that (2.16) has a nontrivial 2π-periodic solution if and only if either A is singular (so $0 \in \sigma(JA)$) or

$$-kJb_k = Aa_k \quad \text{and} \quad kJa_k = Ab_k \tag{2.17}$$

for some $(a_k, b_k) \neq (0, 0)$ and $k \geqslant 1$. (2.17) is equivalent to

$$JA(a_k - ib_k) = ik(a_k - ib_k).$$

Hence $\pm ik \in \sigma(JA)$ and (2.16) has a nontrivial 2π-periodic solution if and only if $\sigma(JA) \cap i\mathbb{Z} \neq \emptyset$. We also see that P_n commutes with L, hence if $E = E^+(L) \oplus E^0(L) \oplus E^-(L)$ is the orthogonal decomposition corresponding to the positive, zero and negative part of the spectrum of L and $E_n^{\pm}(L) := E^{\pm}(L) \cap E_n$, $E_n^0(L) := E^0(L) \cap E_n$, then $E_n = E_n^+(L) \oplus E_n^0(L) \oplus E_n^-(L)$ is an orthogonal decomposition into the positive, zero and negative part of $L_n := P_n L|_{E_n}$. Note that $E^0(L) = N(L)$ and $E_n^0(L) = E^0(L)$ for almost all n.

In one of our applications we shall need a slight extension of Proposition 2.5.

COROLLARY 2.7. *Suppose H is as in (a) of Remark 2.6 and A is a constant matrix. If $\{z_j\}$ is a sequence such that $\Phi(z_j)$ is bounded above, $z_j = w_j^+ + w_j^0 + w_j^- \in E_{m_j}^+(L) \oplus E^0(L) \oplus E_{n_j}^-(L)$, $m_j, n_j \to \infty$ and $(P_{m_j}^+ + P^0 + P_{n_j}^-)\Phi'(z_j) \to 0$ as $j \to \infty$ ($P_n^{\pm}$ and P^0 denote the orthogonal projections $E \to E_n^{\pm}(L)$, $E \to E^0(L)$), then $\{z_j\}$ has a convergent subsequence.*

The proof follows by inspection of the argument of (ii) in Proposition 2.5. Note in particular that (2.8) holds with $G(z,t) = H(z,t) - \frac{1}{2}Az \cdot z$ replacing H and $E^0(L)$ replacing E^0. Moreover, since

$$\langle \pm L_n z, z \rangle \geqslant \varepsilon \|z\|^2 \quad \text{for some } \varepsilon > 0 \text{ and all } z \in E_n^{\pm}(L) \tag{2.18}$$

(ε independent of n), also (2.10) can be easily adapted.

REMARK 2.8. (a) If A is a nonconstant matrix with 2π-periodic entries, the definitions of $i^{\pm}(A)$ and $i^0(A)$ no longer make sense. Therefore we need some other quantities to measure the size of the positive and the negative part of L. Assume for simplicity that the operator L corresponding to A is invertible and let $M_n^-(L)$ be the Morse index of the quadratic form $z \mapsto \langle Lz, z \rangle$ restricted to E_n. Let

$$j^-(A) := \lim_{n \to \infty} \left(M_n^-(L) - (1 + 2n)N \right)$$

and $j^+(A) := j^-(-A)$. It can be shown that $j^{\pm}(A)$ are well defined and finite, and $j^{\pm}(A) = i^{\pm}(A)$ whenever A is a constant matrix [57, Sections 7 and 5]. See also the references below.

(b) Morse-type indices for Hamiltonian systems have been introduced by Amann and Zehnder [3,4] and Benci [18]. In [3,4] and [18] computational formulas for these indices are also discussed. Our definitions of $i^{\pm}(A)$, $i^0(A)$ follow Li and Liu [61] (more precisely, the indices $i^{\pm}(A)$ as defined here differ from those in [61] by N). The number $j^-(A)$ equals the Conley–Zehnder (or Maslov) index of the fundamental solution $\gamma : [0, 2\pi] \to Sp(2N)$ of the equation $\dot{z}(t) = JA(t)$. Here $Sp(2N)$ denotes the group of symplectic $2N \times 2N$-matrices. Recall that a matrix C is symplectic if $C^t J C = J$. See the books of Abbondandolo [1], Chang [23, Section IV.1] and in particular Long [68,69] for a comprehensive discussion of the Conley–Zehnder index.

2.2. *Periodic solutions near equilibria*

The first existence and multiplicity results for periodic solutions of (2.1) are concerned with solutions near an equilibrium. The classical results of Lyapunov [72], Weinstein [98], and Moser [76] have been very influential for the development of the theory and can be proved using the basic variational methods from Section 1.1.

We consider the autonomous Hamiltonian system

$$\dot{z} = JH'(z) \tag{2.19}$$

where the Hamiltonian $H : \mathbb{R}^{2N} \to \mathbb{R}$ is of class C^2. Since the vector field H' is of class C^1, each initial value problem has a unique solution $z = z(t)$ defined on some maximal interval I. Furthermore,

$$\frac{\mathrm{d}}{\mathrm{d}t} H\big(z(t)\big) = H'\big(z(t)\big) \cdot \dot{z}(t) = -J\dot{z}(t) \cdot \dot{z}(t) = 0, \tag{2.20}$$

hence $H(z(t))$ is constant for all $t \in I$. Throughout this section we assume that H has 0 as critical point. We first consider the case where 0 is nondegenerate. The constant function $z \equiv 0$ is then an isolated, hyperbolic stationary solution of (2.19). It is well known that periodic orbits near 0 can only exist if $JH''(0)$ has purely imaginary eigenvalues. The

Lyapunov center theorem states that if $JH''(0)$ has a pair of purely imaginary eigenvalues $\pm i\omega$ which are simple, and if no integer multiples $\pm ik\omega$ are eigenvalues of $JH''(0)$ then (2.19) has a one-parameter family of periodic solutions emanating from the equilibrium point. More precisely, let $E(\pm i\omega) \subset \mathbb{R}^{2N}$ be the two-dimensional eigenspace associated to $\pm i\omega$ and let $\sigma \in \{\pm 2\}$ be the signature of the quadratic form $Q(z) = \frac{1}{2}H''(0)z \cdot z$ on $E(\pm i\omega)$. Then for each $\varepsilon > 0$ small enough there exists a periodic solution of (2.19) on the energy surface $H = H(0) + \sigma\varepsilon^2$ with period converging to $2\pi/\omega$ as $\varepsilon \to 0$.

If $\pm i\omega$ is not simple, or if integer multiples are eigenvalues then there may be no periodic solutions near 0 as elementary examples show; see [76,27]. In order to formulate a sufficient condition we assume that all eigenvalues of $JH''(0)$ which are of the form $ik\omega$, $k \in \mathbb{Z}$, are semisimple, i.e., their geometric and algebraic multiplicities are equal. Let $E_\omega \subset \mathbb{R}^{2N}$ be the generalized eigenspace of $JH''(0)$ corresponding to the eigenvalues of the form $\pm ik\omega$, $k \in \mathbb{N}$. Let $Q : \mathbb{R}^{2N} \to \mathbb{R}$, $Q(z) = \frac{1}{2}H''(0)z \cdot z$, be the quadratic part of H at 0 and let $\sigma = \sigma(\omega) \in \mathbb{Z}$ be the signature of the quadratic form $Q|E_\omega$ on E_ω. Observe that σ is automatically an even integer.

THEOREM 2.9. *If $\sigma \neq 0$ then one of the following statements hold.*
  (i) *There exists a sequence of nonconstant T_k-periodic solutions z_k of (2.19) which lie on the energy surface $H = H(0)$ with $z_k \to 0$ and $T_k \to 2\pi/\omega$ as $k \to \infty$. The period T_k is not necessarily minimal.*
  (ii) *For $\varepsilon > 0$ small enough there are at least $|\sigma/2|$ nonconstant periodic solutions z_j^ε, $j = 1, \ldots, |\sigma/2|$, of (2.19) on the energy surface $H = H(0) + \sigma\varepsilon^2$ with (not necessarily minimal) period T_j^ε. These solutions converge towards 0 as $\varepsilon \to 0$. Moreover, $T_j^\varepsilon \to 2\pi/\omega$ as $\varepsilon \to 0$.*

This theorem is due to Bartsch [11]. It generalizes the Weinstein–Moser theorem [98, 76] which corresponds to the case where $Q|E_\omega$ is positive or negative definite, hence $|\sigma| = \dim E_\omega$. Observe that the energy surfaces $H = c$ are not necessarily compact for c close to $H(0)$.

PROOF (outline). We may assume $H(0) = 0$. The τ-periodic solutions of (2.19) correspond to 2π-periodic solutions of

$$\dot{z} = \frac{\tau}{2\pi} JH'(z). \tag{2.21}$$

These in turn correspond to critical points of the action functional

$$\mathcal{A}(z) = \int_0^{2\pi} J\dot{z}(t) \cdot z(t)\,dt$$

restricted to the surface $\{z \in E \colon \psi(z) = 2\pi H(0) + \lambda\}$ where $E = H^{1/2}(S^1, \mathbb{R}^{2N}))$ and $\psi(z) = \int_0^{2\pi} H(z(t))\,dt$ are as in Section 2.1. The period appears as Lagrange multiplier in this approach. After performing a Lyapunov–Schmidt reduction of the equation

$$\mathcal{A}'(z) = \frac{\tau}{2\pi}\psi'(z)$$

near $\tau = 2\pi/\omega$ and $z \equiv 0$, one is left with the problem of finding critical points of a function

$$\mathcal{A}_0(v) = \mathcal{A}\big(v + \bar{w}(v)\big)$$

constrained to the level set $\{v \in V : \psi_0(v) = \lambda\}$. Here V is the kernel of the linearization

$$E \ni z \mapsto \mathcal{A}''(0)z - \frac{\tau}{2\pi}\psi''(0)z \in E,$$

$\bar{w} : V \supset U \to V^\perp \subset E$ is defined on a neighborhood U of 0 in V, and $\psi_0(v) := \psi(v + \bar{w}(v))$. Thus $V \cong E_\omega$ and one checks that $\mathcal{A}_0$ and ψ_0 are of class C^1 and that $\psi_0''(0)$ exists. In fact:

$$\langle \psi_0''(0)v, w \rangle = \int_0^{2\pi} H''(0)v(t) \cdot w(t)\, dt,$$

hence we can apply the Morse lemma to ψ_0 near 0. After a change of coordinates ψ_0 looks (in the sense of the Morse lemma) near 0 like the nondegenerate quadratic form

$$q : V \to \mathbb{R}, \quad v \mapsto \frac{1}{2}H''(0)v \cdot v.$$

Therefore the level surfaces $\psi_0^{-1}(\lambda)$ look locally like the level surfaces $q^{-1}(\lambda)$. If q is positive definite, hence $\sigma = \dim V = \dim E_\omega$ (which is just the situation of the Weinstein–Moser theorem), one can conclude the proof easily upon observing that the functionals $\mathcal{A}$, ψ and, hence $\mathcal{A}_0$ and ψ_0 are invariant under the representation of $S^1 = \mathbb{R}/2\pi\mathbb{Z}$ in E induced by the time shifts. Moreover, $\psi_0^{-1}(\lambda) \cong q^{-1}(\lambda)$ is diffeomorphic to the unit sphere SV of V for $\lambda > 0$ small. By Theorem 1.9 any C^1-functional $SV \to \mathbb{R}$ which is invariant under the action of S^1 has at least $\frac{1}{2}\dim V = \frac{1}{2}\dim E_\omega$ S^1-orbits of critical points. If q is negative definite then $\psi_0^{-1}(\lambda)$ is diffeomorphic to the unit sphere SV of V for $\lambda < 0$ close to 0, and one obtains $|\sigma/2|$ critical orbits on these levels.

This elementary argument from S^1-equivariant critical point theory does not work if q is indefinite. Instead one looks at the flow φ_λ on $\Sigma_\lambda := \psi_0^{-1}(\lambda)$ which is essentially induced by the negative gradient of $\mathcal{A}_0|\Sigma_\lambda$. Since $\mathcal{A}_0$ and ψ_0 are only of class C^1 the gradient vector field is of class C^0, so it may not be integrable and has to be replaced by a pseudogradient vector field which leaves Σ_λ invariant for all λ. Next one observes that the hypersurfaces Σ_λ undergo a surgery as λ passes $H(0) = 0$. If $2n^+$ (respectively $2n^-$) is the maximal dimension of a subspace of V on which q is positive (respectively negative) definite then Σ_ε is obtained from $\Sigma_{-\varepsilon}$ upon replacing a handle of type $B^{2n^+} \times S^{2n^--1}$ by $S^{2n^+} \times B^{2n^-}$. It is this change in the topology of Σ_λ near 0 which forces the existence of stationary orbits of φ near the origin. In order to analyze the influence of this surgery on the flow φ_λ one has to use methods from equivariant Conley index theory and Borel cohomology. The difference $|n^+ - n^-| = |\sigma|/2$ is a lower bound for the number of stationary S^1-orbits of φ_λ on Σ_λ if $\lambda > 0$ is small and $\sigma \cdot (\lambda - H(0)) > 0$. $\square$

It is also possible to parameterize the nontrivial periodic orbits near an equilibrium by their period. The following result is due to Fadell and Rabinowitz [36].

THEOREM 2.10. *If $\sigma \neq 0$ then one of the following statements hold.*
  (i) *There exists a sequence of nonconstant periodic orbits $z_k \to 0$ of (2.19) with (not necessarily minimal) period $T = 2\pi/\omega$.*
  (ii) *There exist integers $k, l \geqslant 0$ with $k + l \geqslant |\sigma|/2$, and there exists $\varepsilon > 0$ such that for each $\tau \in (T - \varepsilon, T)$ (2.19) has at least k periodic orbits z_j^τ, $j = 1, \ldots, k$ with (not necessarily minimal) period τ. And for each $\tau \in (T, T + \varepsilon)$ (2.19) has at least l periodic orbits z_j^τ, $j = k + 1, \ldots, k + l$ with (not necessarily minimal) period τ. Moreover, $z_j^\tau \to 0$ as $\tau \to T = 2\pi/\omega$.*

The proof uses a cohomological index theory. The integers k, l (and thus the direction of the bifurcating solutions with the period as parameter) are not determined by $H''(0)$—unlike case (ii) in Theorem 2.9.

Now we consider the case of a degenerate equilibrium. Suppose first that 0 is an isolated critical point of H, so there are no stationary orbits of (2.19) in a neighbourhood of the origin. Let $i\omega$ be an eigenvalue of $JH''(0)$ and let $F_\omega \subset \mathbb{R}^{2N}$ be the generalized eigenspace of $JH''(0)$ corresponding to $\pm i\omega$. Thus $F_\omega \subset E_\omega$ does not contain generalized eigenvectors of $JH''(0)$ corresponding to multiples $\pm ik\omega$ with $|k| \geqslant 2$. Let $\sigma_1 = \sigma_1(\omega)$ be the signature of the quadratic form $Q|F_\omega$. Since we allow $H''(0)$ to have a nontrivial kernel we also need the critical groups $C^q(H, 0) = \check{H}^q(H^0, H^0 \setminus \{0\})$ associated to $0 \in \mathbb{R}^{2N}$ as a critical point of the Hamiltonian H. Here $\check{H}^*$ denotes the Čech (or Alexander–Spanier) cohomology with coefficients in an arbitrary field.

THEOREM 2.11. *If $\sigma_1 \neq 0$ and $C^q(H, 0) \neq 0$ for some $q \in \mathbb{Z}$, then there exists a sequence z_k of nonconstant periodic orbits of (2.19) with (not necessarily minimal) period T_n such that $\|z_n\|_{L^\infty} \to 0$ and $T_n \to 2\pi/\omega$.*

The result has been proved by Szulkin in [91] using Morse theoretic methods. It is unknown whether the solutions obtained in Theorems 2.9–2.11 lie on connected branches of periodic solutions. Continua of periodic solutions however do exist under stronger hypotheses when degree theoretic methods apply. We state one such result in this direction.

THEOREM 2.12. *If $\sigma_1 \neq 0$ and the local degree $\deg(\nabla H, 0)$ of ∇H at the isolated critical point 0 is nontrivial, then there exists a connected branch of periodic solutions of (2.19) near 0.*

For a proof see the paper [27] by Dancer and Rybicki. They work in the space $W^1(S^1, \mathbb{R}^{2N})$ and apply a degree for S^1-gradient maps to the bifurcation equations associated to $\dot{y} = \lambda H'(y)$. 2π-periodic solutions $y(s)$ of this equation correspond to $2\pi/\lambda$-periodic solutions of (2.19). The degree theory allows a classical Rabinowitz type argument yielding a global continuum of solutions in $\mathbb{R} \times W^1(S^1, \mathbb{R}^{2N})$ that bifurcates from $(\lambda_0, 0)$ with $\lambda_0 = \omega$.

Observe that $\sigma = \sum_{k=1}^{\infty} \sigma_k$ where $\sigma_k = \sigma_k(\omega) = \sigma_1(k\omega)$ is the signature of $Q|F_k$, F_k the generalized eigenspace of $JH''(0)$ corresponding to $\pm ik\omega$. Also observe (see [23, Theorem II.3.2]) that the local degree can be expressed in terms of the critical groups as

$$\deg(H', 0) = \sum_{q=0}^{\infty} (-1)^q \dim C^q(H, 0).$$

Thus the hypotheses of Theorem 2.11 are weaker than those of Theorem 2.12. Correspondingly, the conclusion is also weaker.

Since the above results require only local conditions on the Hamiltonian near a stationary point they immediately generalize to Hamiltonian systems on a symplectic manifold (W, Ω). The last result that we state in this section deals with periodic orbits near a manifold M of equilibria. This result can in general not be reduced to the special case $W = \mathbb{R}^{2N}$ with the standard symplectic structure $\Omega = \sum_{k=1}^{N} dp_i \wedge dq_i$ because the manifold M need not lie in a symplectic neighbourhood chart. We therefore state it in the general setting.

THEOREM 2.13. *Let (W, Ω) be a symplectic manifold and let $H : W \to \mathbb{R}$ be a smooth Hamiltonian. Suppose there exists a compact symplectic submanifold $M \subset H^{-1}(c) \subset W$ which is a Bott-nondegenerate manifold of minima of H. Then there exists a sequence of nonconstant periodic trajectories of the Hamiltonian flow associated to H which converge to M.*

The result is due to Ginzburg and Kerman [45]. It clearly applies to (2.19) where $W = \mathbb{R}^{2N}$ and Ω is as above. Compared with the Weinstein–Moser theorem where M is a point, Theorem 2.13 does not yield periodic orbits on all energy surfaces close to M, and neither does it yield a multiplicity result. We refer to [45] and the references therein for further results on periodic orbits of Hamiltonian flows near manifolds of equilibria.

2.3. *Fixed energy problem*

Let $H \in C^2(\mathbb{R}^{2N}, \mathbb{R})$ and suppose $D := \{z \in \mathbb{R}^{2N} : H(z) \leqslant 1\}$ is a compact subset of $\mathbb{R}^{2N}$ such that $H'(z) \neq 0$ for all $z \in S := H^{-1}(1)$. Then S is a compact hypersurface of class C^2 and we may assume without loss of generality that 0 is in the interior of D. We consider the autonomous Hamiltonian system (2.19). If $z(t_0) \in S$ then $z(t) \in S$ for all t because $H(z(t))$ is constant along solutions of (2.19) (see (2.20)). Since S is compact, $z(t)$ exists for all $t \in \mathbb{R}$.

We will be interested in the existence of *closed Hamiltonian orbits* on S, i.e., the sets $\text{Orb}(z) := \{z(t) : t \in \mathbb{R}\}$, where $z = z(t)$ is a periodic solution of (2.19) with $z(t) \in S$. Here we use the notation $\text{Orb}(z)$ for closed orbits in order to distinguish them from S^1-orbits $\mathcal{O}(z)$ defined in Section 2.1. If $\widetilde{H} \in C^2(\mathbb{R}^{2N}, \mathbb{R})$ is another Hamiltonian such that $S = \widetilde{H}^{-1}(c)$ for some c and $\widetilde{H}'(z) \neq 0$ on S, then $H'(z)$ and $\widetilde{H}'(z)$ are parallel and nowhere zero on S. It follows that two solutions z and $\tilde{z}$ of the corresponding Hamiltonian systems are equivalent up to reparameterization if $z(t_0) = \tilde{z}(\tilde{t}_0)$ for some $t_0, \tilde{t}_0 \in \mathbb{R}$. In particular, the orbits $\mathrm{Orb}(z)$ and $\mathrm{Orb}(\tilde{z})$ coincide (see [80] for a detailed argument). Consequently, closed orbits depend only on S and not on the particular choice of a Hamiltonian having the properties given above. It is also possible to define closed orbits without referring to any Hamiltonian: given a compact surface S of class C^2, one may look for periodic solutions of the system $\dot{z} = JN(z)$, where $N(z)$ is the unit outer normal to S at z.

Throughout this section we assume that S satisfies the following condition:

($\mathcal{S}$) S is a compact hypersurface of class C^2 in $\mathbb{R}^{2N}$, S bounds a starshaped neighbourhood of the origin and all $z \in S$ are transversal to S.

It follows from ($\mathcal{S}$) that for each $z \in \mathbb{R}^{2N} \setminus \{0\}$ there exists a unique $\alpha(z) > 0$ such that $z/\alpha(z) \in S$. Let $\alpha(0) := 0$ and

$$H(z) := \alpha(z)^4. \tag{2.22}$$

Clearly $\alpha(sz) = s^4 \alpha(z)$ for all $s \geqslant 0$, hence H is positively homogeneous of degree 4. Moreover, $S = H^{-1}(1)$, $H \in C^2(\mathbb{R}^{2N}, \mathbb{R})$ and (by Euler's identities) $H'(z) \cdot z = 4H(z) \neq 0$ whenever $z \neq 0$. In particular, $H'(z) \neq 0$ on S.

Suppose z is a periodic solution of (2.19) with the Hamiltonian H given by (2.22). If $\mathrm{Orb}(z) \subset H^{-1}(\lambda)$ then $\hat{z}(t) := \lambda^{-1/4} z(t/\sqrt{\lambda})$ is a periodic solution of (2.19) on $H^{-1}(1)$. If z has minimal period 2π then $\hat{z}$ has minimal period $T := 2\pi\sqrt{\lambda}$. On the other hand, given a periodic solution z of (2.19) on $H^{-1}(1)$ with minimal period T then $\tilde{z}(t) := (T/2\pi)^{1/2} z(Tt/2\pi)$ is a periodic solution of (2.19) on $H^{-1}((T/2\pi)^2)$ with minimal period 2π. One easily checks that $\hat{\tilde{z}} = z$ and $\tilde{\hat{z}} = z$. Given two periodic solutions z_1, z_2 of (2.19) on $H^{-1}(1)$ with minimal period T and having the same orbit $\mathrm{Orb}(z_1) = \mathrm{Orb}(z_2) \subset H^{-1}(1)$ then there exists $\theta \in \mathbb{R}$ with $z_2(t) = z_1(t + \theta)$ for all t. The corresponding solutions $\tilde{z}_1, \tilde{z}_2$ with minimal period 2π then satisfy $\tilde{z}_2(t) = \tilde{z}_1(t + \tilde{\theta})$ for some $\tilde{\theta} \in \mathbb{R}$, hence they are not geometrically distinct in the sense of Remark 2.2. If z_1, z_2 have different orbits $\mathrm{Orb}(z_1), \mathrm{Orb}(z_2) \subset H^{-1}(1)$ then $\tilde{z}_1, \tilde{z}_2$ are geometrically distinct.

We summarize the above considerations in the following

THEOREM 2.14. *Let S be a hypersurface satisfying ($\mathcal{S}$) and let H be defined by (2.22). A periodic solution $z(t)$ of (2.19) on $H^{-1}(\lambda)$ yields a periodic solution $\lambda^{-1/4} z(t/\sqrt{\lambda})$ on $S = H^{-1}(1)$. Moreover, there is a one-to-one correspondence between closed orbits on S and geometrically distinct periodic solutions of (2.19) with minimal period 2π.*

We emphasize the importance of the assumption on the minimality of the period. If z is a solution of (2.19) with minimal period T and $\mathrm{Orb}(z) \subset S$, then $z(t)$ covers $\mathrm{Orb}(z)$ k times as t goes from 0 to kT. A corresponding solution $\tilde{z}_k(t) := (kT/2\pi)^{1/2} z(kTt/2\pi)$ has minimal period $2\pi/k$, hence $\tilde{z}_k$ and $\tilde{z}_m$ are geometrically distinct if $k \neq m$, yet $\mathrm{Orb}(\hat{\tilde{z}}_k) = \mathrm{Orb}(\hat{\tilde{z}}_m) = \mathrm{Orb}(z)$.

Now we state the first main result of this section. It is due to Rabinowitz [78] and, if S bounds a convex neighbourhood of the origin to Weinstein [99].

THEOREM 2.15. *Let S be a hypersurface satisfying ($\mathcal{S}$). Then S contains a closed Hamiltonian orbit.*

PROOF. By Theorem 2.14 it suffices to show that (2.19) with H given by (2.22) has a 2π-periodic solution $z \neq 0$. Let r be the largest and R the smallest number such that

$$r \leqslant |z| \leqslant R \quad \text{for all } z \in S. \tag{2.23}$$

Then

$$\frac{|z|^4}{R^4} \leqslant H(z) \leqslant \frac{|z|^4}{r^4}, \quad \text{for all } z \in \mathbb{R}^{2N}. \tag{2.24}$$

The functional

$$\Phi(z) = \frac{1}{2} \int_0^{2\pi} (-J\dot{z} \cdot z)\,\mathrm{d}t - \int_0^{2\pi} H(z)\,\mathrm{d}t = \frac{1}{2}\|z^+\|^2 - \frac{1}{2}\|z^-\|^2 - \psi(z)$$

is S^1-invariant. Moreover, $H'(z) \cdot z = 4H(z) > 0$, hence Φ satisfies (PS)* according to Proposition 2.5. Let $\widetilde{E}_n$, E_n be given by (2.5) and let $2d_n = 2N(1+n)$, $Y = E^+$ and $X = E_1^+ \oplus E^-$ (cf. (2.6)). Then $E^{S^1} = E^0 \subset E_n$ for all n, $\operatorname{codim}_{\widetilde{E}_n} X_n = 2(d_n - 2N)$ and $\dim Y_n = 2(d_n - N)$. Hence (i) and (ii) of Corollary 1.15 are satisfied, with $l = -N$ and $m = -2N$. By (2.24) and Lemma 2.4, $\Phi|_{Y \cap S_\rho} \geqslant \alpha$ for some $\alpha, \rho > 0$. Since

$$\Phi(z) \leqslant \frac{1}{2}\|z^+\|^2 - \frac{1}{2}\|z^-\|^2 - \frac{1}{R^4} \int_0^{2\pi} |z|^4\,\mathrm{d}t \tag{2.25}$$

and $\dim E_1^+ < \infty$, $\Phi(z) \to -\infty$ as $\|z\| \to \infty$, $z \in X$. Finally, $\Phi|_{E^0} \leqslant 0$ because $H \geqslant 0$. It follows that also (iii) and (iv) of Corollary 1.15 hold. Hence (2.19) has at least N geometrically distinct 2π-periodic solutions $z \neq 0$, and by Theorem 2.14, the hypersurface S carries a closed Hamiltonian orbit. $\qquad\square$

It is not necessary to exploit the S^1-symmetry in order to show the existence of one 2π-periodic solution $z \neq 0$. However, the argument presented here will be needed below.

REMARK 2.16. Since the above proof gives no information on the minimal period of the N geometrically distinct 2π-periodic solutions, we do not know whether they correspond to distinct closed orbits on S. System (2.19) has in fact infinitely geometrically distinct 2π-periodic solutions. Indeed, we may replace $X = E_1^+ \oplus E^1$ by $X = E_r^+ \oplus E^-$ for any positive integer r and use Corollary 1.16 (see also Theorem 2.19). On the other hand, if $z = (p_1, \ldots, p_N, q_1, \ldots, q_N)$ and

$$S = \left\{ z \in \mathbb{R}^{2N} : \sum_{j=1}^{N} \frac{1}{2}\alpha_j \left(p_j^2 + q_j^2\right) = 1 \right\},$$

where $\alpha_1, \ldots, \alpha_N$ are rationally independent positive numbers, then it is easy to see that S has exactly N distinct closed orbits.

In an answer to a conjecture of Weinstein, Viterbo [96] has generalized Theorem 2.15 to all compact hypersurfaces admitting a so-called contact structure. Subsequently his proof has been simplified (and a more general result obtained) by Hofer and Zehnder [51]. Struwe [86], building upon the work of Hofer and Zehnder, proved that given an interval $[a, b]$ of regular values of H such that the hypersurfaces $S_c := H^{-1}(c) \subset \mathbb{R}^{2N}$, $c \in [a, b]$, are compact, the set $\{c \in [a, b]:\ S_c$ carries a closed Hamiltonian orbit$\}$ has full measure $b - a$. It has been shown by counterexamples of Ginzburg [43] and Herman [49] that in general a compact hypersurface may not have any closed Hamiltonian orbit (see also [44] and the references there).

In view of Remark 2.16 it is natural to ask whether each S satisfying (S) must necessarily have N distinct closed Hamiltonian orbits. We shall show that this is indeed the case under an additional geometric condition.

Denote the tangent hyperplane to S at w by $T_w(S)$, suppose S satisfies (S) and let ρ be the largest number such that

$$T_w(S) \cap \{z \in \mathbb{R}^{2N}:\ |z| < \rho\} = \emptyset \quad \text{for all } w \in S. \tag{2.26}$$

Then ρ is the minimum of the distances from $T_w(S)$ to the origin over all $w \in S$. It follows from (S) that ρ is well defined; moreover, if r is as in (2.23) and S bounds a convex set, then $\rho = r$.

THEOREM 2.17. *Let S be a hypersurface satisfying (S) and suppose $R^2 < 2\rho^2$, where R, ρ are as in (2.23), (2.26). Then S contains at least N distinct closed Hamiltonian orbits.*

PROOF. It follows from the proof of Theorem 2.15 that (2.19) (with H given by (2.22)) has at least N geometrically distinct 2π-periodic solutions. According to Theorem 2.14 it suffices to show that these solutions have minimal period 2π.

Recall from the proof of Theorem 2.15 that $l = -N$ and $m = -2N$, so invoking Corollary 1.15 we have

$$c_j = \sup_{i_\mathcal{E}(A) \geqslant j} \inf_{z \in A} \Phi(z), \quad -2N + 1 \leqslant j \leqslant -N.$$

Since $\operatorname{codim}_{\widetilde{E}_n} X_n = 2(d_n - 2N)$, it follows from (iv) of Proposition 1.13 that if $i_\mathcal{E}(A) \geqslant -2N + 1$, then $A \cap (E^G \oplus X) = A \cap (E_1^+ \oplus E^0 \oplus E^-) \neq \emptyset$. Hence

$$c_j \leqslant \sup\{\Phi(z):\ z \in E_1^+ \oplus E^0 \oplus E^-\}. \tag{2.27}$$

Let $z \in E_1^+ \oplus E^0 \oplus E^-$. Using (2.25), the fact that $\|z^+\| = \|z^+\|_2$ for $z^+ \in E_1^+$ and the Hölder inequality, we obtain

$$\Phi(z) \leqslant \frac{1}{2}\|z^+\|^2 - \frac{1}{2}\|z^-\|^2 - \frac{1}{R^4}\|z\|_4^4 \leqslant \frac{1}{2}\|z^+\|_2^2 - \frac{1}{R^4}\|z\|_4^4$$

$$\leqslant \frac{1}{2}\|z\|_2^2 - \frac{1}{R^4}\|z\|_4^4 \leqslant \sqrt{\frac{\pi}{2}}\|z\|_4^2 - \frac{1}{R^4}\|z\|_4^4 \leqslant \frac{\pi R^4}{8}.$$

This and (2.27) imply

$$c_j \leqslant \frac{\pi R^4}{8}. \tag{2.28}$$

By the definition of ρ and the homogeneity of H,

$$\rho \left| H'(z) \right| \leqslant z \cdot H'(z) = 4H(z) = 4 = 4H(z)^{3/4}$$

whenever $z \in S$. By the homogeneity again,

$$\rho \left| H'(z) \right| \leqslant 4H(z)^{3/4} \quad \text{for all } z \in \mathbb{R}^{2N}. \tag{2.29}$$

Let now $z = z(t)$ be a 2π-periodic solution of (2.19). Since $H(z(t))$ is constant,

$$\begin{aligned}
\Phi(z) &= \frac{1}{2} \int_0^{2\pi} (-J\dot{z} \cdot z)\, dt - \int_0^{2\pi} H(z)\, dt \\
&= \int_0^{2\pi} \left(\frac{1}{2} H'(z) \cdot z - H(z) \right) dt = \int_0^{2\pi} H(z)\, dt = 2\pi H(z).
\end{aligned} \tag{2.30}$$

Suppose z has minimal period $2\pi/m$ and write $z = \bar{z} + \tilde{z}$, $\bar{z} \in E^0$, $\tilde{z} \in E^+ \oplus E^-$. By Wirtinger's inequality,

$$\|\tilde{z}\|_2 \leqslant \frac{1}{m} \|\dot{\tilde{z}}\|_2,$$

and it follows using (2.30), (2.29) that

$$\begin{aligned}
2\pi H(z) = \Phi(z) &= \frac{1}{2} \int_0^{2\pi} (-J\dot{\tilde{z}} \cdot z)\, dt - \int_0^{2\pi} H(z)\, dt \\
&\leqslant \frac{1}{2} \|\dot{\tilde{z}}\|_2 \|\tilde{z}\|_2 - 2\pi H(z) \leqslant \frac{1}{2m} \|\dot{z}\|_2^2 - 2\pi H(z) \\
&= \frac{1}{2m} \int_0^{2\pi} \left| H'(z) \right|^2 dt - 2\pi H(z) \leqslant \frac{8}{m\rho^2} \int_0^{2\pi} H(z)^{3/2}\, dt - 2\pi H(z) \\
&= 2\pi \left(\frac{8H(z)^{3/2}}{m\rho^2} - H(z) \right).
\end{aligned}$$

Hence

$$H(z) \geqslant \frac{m^2 \rho^4}{16}$$

and

$$\Phi(z) = 2\pi H(z) \geqslant \frac{\pi m^2 \rho^4}{8}.$$

Since $R^2 < 2\rho^2$,

$$\Phi(z) > \frac{\pi m^2 R^4}{32}.$$

On the other hand, if the solution z corresponds to c_j, $-2N + 1 \leqslant j \leqslant -N$, then $\Phi(z) \leqslant \pi R^4/8$ according to (2.28). Hence $m = 1$ and z has minimal period 2π. $\qquad\square$

Theorem 2.17 is due to Ekeland and Lasry [35] (see also [33]) in the case of S bounding a compact strictly convex region and $R^2 < 2r^2$, and by Berestycki et al. [20] in the more general case considered here. We would also like to mention a result by Girardi and Matzeu [46] showing that if S satisfies (S), then the condition $R^2 < 2\rho^2$ may be replaced by $R^2 < \sqrt{3}\rho r$ in Theorem 2.17.

It has been a longstanding conjecture (see, e.g., [33, p. 235]) that if S bounds a compact strictly convex set, then the minimal number of distinct closed Hamiltonian orbits such S must carry is N. Ekeland and Lassoued [34] and Szulkin [89] have shown that S carries at least 2 such orbits if its Gaussian curvature is positive everywhere. In a recent work Liu, Long and Zhu [67] have shown that if in addition S is symmetric about the origin, the number of such orbits is at least N, and for general (possibly nonsymmetric) S as above, Long and Zhu [71] have shown the existence of at least $[\frac{N}{2}] + 1$ closed orbits ($[a]$ denotes the integer part of a). They also make a new conjecture that $[\frac{N}{2}] + 1$ (and not N) is the lower bound for the number of closed Hamiltonian orbits. See also Long's book [69] for a detailed discussion.

In the case $N = 2$, Hofer, Wysocki and Zehnder [50] proved that if S bounds a strictly convex set then there are either two or infinitely many closed Hamiltonian orbits on S. In [32] Ekeland has shown that a generic S bounding a compact convex set and having positive Gaussian curvature carries infinitely many closed Hamiltonian orbits. This result has been partially generalized by Viterbo [97] to hypersurfaces satisfying a condition similar to (S). The question of the existence of infinitely many closed orbits is extensively discussed in [33,69] where many additional references may be found.

2.4. *Superlinear systems*

Throughout this section we assume that H satisfies (H_1)–(H_3), $H(z, t) = \frac{1}{2}Az \cdot z + G(z, t)$, where A is a symmetric $2N \times 2N$ matrix, $G_z(z, t) = o(z)$ uniformly in t as $z \to 0$, and there exist $\mu > \max\{2, s - 1\}$ and $R > 0$ such that

$$0 < \mu G(z, t) \leqslant z \cdot G_z(z, t) \quad \text{for all } |z| \geqslant R. \tag{2.31}$$

Recall from (2.7) that the last condition implies G (and hence H) is superquadratic and H_z superlinear.

THEOREM 2.18. *Suppose H satisfies the hypotheses given above and $\sigma(JA) \cap i\mathbb{Z} = \emptyset$. Then the system (2.1) has a 2π-periodic solution $z \neq 0$.*

PROOF. Let

$$\Phi(z) = \frac{1}{2} \int_0^{2\pi} (-J\dot{z} - Az) \cdot z \, dt - \int_0^{2\pi} G(z,t) \, dt$$

and let E_n be given by (2.5). Denote the linear operator corresponding to the quadratic part of Φ by L (cf. (2.11)). Since $\sigma(JA) \cap i\mathbb{Z} = \emptyset$, L is invertible, $E = E^+(L) \oplus E^-(L)$ and $E_n = E_n^+(L) \oplus E_n^-(L)$ (see the discussion and notation preceding Corollary 2.7). It follows using (2.18) and Lemma 2.4 that there exist $\alpha, \rho > 0$ such that

$$\Phi \leqslant 0 \quad \text{on } E_n^-(L) \cap \overline{B}_\rho, \qquad \Phi \leqslant -\alpha \quad \text{on } E_n^-(L) \cap S_\rho \tag{2.32}$$

and

$$\Phi \geqslant 0 \quad \text{on } E_n^+(L) \cap \overline{B}_\rho, \qquad \Phi \geqslant \alpha \quad \text{on } E_n^+(L) \cap S_\rho \tag{2.33}$$

for all n. If Φ has a critical point $z \in \Phi^{-\alpha}$, then $z \neq 0$, so z is a solution of (2.1) we were looking for. Suppose no such z exists. We shall complete the proof by showing that in this case $\Phi'(z) = 0$ for some z with $\Phi(z) \geqslant \alpha$.

We claim that $\Phi_{mn} := \Phi|_{E_m^+(L) \oplus E_n^-(L)}$ has no critical point $z \in \Phi_{mn}^{-\alpha}$ whenever $m, n \geqslant n_0$ and n_0 is large enough. Indeed, otherwise there is a sequence $\{z_j\} \subset \Phi^{-\alpha}$ such that $z_j \in E_{m_j}^+(L) \oplus E_{n_j}^-(L)$, $m_j, n_j \to \infty$ and $\Phi'_{m_j n_j}(z_j) = 0$. By Corollary 2.7, $z_j \to z$ after passing to a subsequence, so $\Phi(z) \leqslant -\alpha$ and $\Phi'(z) = 0$, a contradiction. Hence we may choose n_0 so that $\Phi_{n_0 n}$ has no critical point in $\Phi^{-\alpha}$ for any $n \geqslant n_0$. Let $z = w^+ + w^- \in E_{n_0+1}^+(L) \oplus E_n^-(L)$. Then

$$\Phi(z) = \frac{1}{2}\langle Lw^+, w^+ \rangle + \frac{1}{2}\langle Lw^-, w^- \rangle - \int_0^{2\pi} G(z,t) \, dt, \tag{2.34}$$

and since $G(z,t) \geqslant a_1|z|^\mu - a_2$ according to (2.7), it follows that $\Phi(z) \leqslant 0$ whenever $|z| \geqslant R$. Moreover, since n_0 is fixed, R does not depend on n. If $n \geqslant n_0 + 1$, then by Corollary 1.3 (with E_n corresponding to $\mathbb{R}^m$, $F_0 = E_n^-(L)$, $F = E_{n_0}^+(L) \oplus E_n^-(L)$ and $\widetilde{F} = E_{n_0+1}^+ \oplus E_n^-(L)$) there exists $z_n \in E_n$ such that $\Phi'_n(z_n) = 0$ and $\alpha \leqslant \Phi(z_n) \leqslant \sup_{\overline{B}_{R+1}} \Phi$. Applying Proposition 2.5 to the sequence $\{z_n\}$ we obtain a critical point z with $\Phi(z) \geqslant \alpha$. $\square$

Next we prove that the autonomous system

$$\dot{z} = JH'(z) = J\big(Az + G'(z)\big) \tag{2.35}$$

with superquadratic Hamiltonian has infinitely many geometrically distinct T-periodic solutions for any $T > 0$. Since there is a one-to-one correspondence between T-periodic solutions for the system $\dot{z} = JH'(z)$ and 2π-periodic solutions for $\dot{z} = \lambda JH'(z)$, where $\lambda = T/2\pi$ (this can be easily seen by substituting $\tau = t/\lambda$), we may assume without loss of generality that $T = 2\pi$.

THEOREM 2.19. *Suppose $H(z) = \frac{1}{2}Az \cdot z + G(z)$ satisfies* (H$_1$), (H$_3$), $H(z) \geq 0$ *for all* $z \in \mathbb{R}^{2N}$, *G satisfies* (2.31) *and* $G'(z) \to 0$ *as* $z \to 0$. *Then the system* (2.35) *has a sequence* $\{z_j\}$ *of nonconstant 2π-periodic solutions such that* $\|z_j\|_\infty \to \infty$.

If one can show that for each $T > 0$ the system (2.35) has a nonconstant T-periodic solution $z^T \neq 0$, then the number of geometrically distinct nonconstant 2π-periodic solutions is in fact infinite. Indeed, let $z_k = z^{2\pi/k}$, then z_k and z_l may coincide for some $k \neq l$, yet the sequence $\{z_k\}$ will contain infinitely many distinct elements. However, the result stated above shows much more: the solutions z_j have amplitude which goes to infinity with j.

PROOF OF THEOREM 2.19. We verify the hypotheses of Corollary 1.15. By Proposition 2.5, Φ satisfies (PS)*, and obviously, $E^{S^1} = E^0 \subset E_n$. Let $2d_n = 2N(1+n)$, $X = (E_r^+(L) \oplus E^0(L) \oplus E^-(L)) \cap \widetilde{E}$, where r is a positive integer, and $Y = E^+(L) \cap \widetilde{E}$ (we use the notation of the preceding proof). Employing (2.14) and recalling that $M^+(T_k(A)) = 2N$ for large k, we have for n, r large enough ($n \geq r$),

$$\dim Y_n = \sum_{k=1}^{n} M^+(T_k(A)) = 2nN + \sum_{k=1}^{n} (M^+(T_k(A)) - 2N)$$

$$= 2d_n + i^+(A) - M^-(A) - N =: 2(d_n + l) \tag{2.36}$$

and

$$\operatorname{codim}_{\widetilde{E}_n} X_n = \sum_{k=r+1}^{n} M^+(T_k(A)) = 2Nn - 2Nr = 2d_n - 2N(r+1)$$

$$=: 2(d_n + m).$$

It follows from (2.33) that $\Phi|_{Y \cap S_\rho} \geq \alpha$ and from (2.34) with $z = w^+ + w^0 + w^- \in E_r^+(L) \oplus E^0(L) \oplus E^-(L)$ that $\Phi(z) \to -\infty$ whenever $\|z\| \to \infty$, $z \in E^0 \oplus X$. Hence $\Phi|_{E^0 \oplus X} \leq \beta$ for some β. Moreover, $\Phi|_{E^0} \leq 0$ because $H \geq 0$. We have verified the hypotheses of Corollary 1.15 for all r large enough. Since $m \to -\infty$ as $r \to \infty$, we conclude from Corollary 1.16 that (2.35) has a sequence $\{z_j\}$ of nonconstant 2π-periodic solutions such that $\Phi(z_j) \to \infty$.

It remains to show that $\|z_j\|_\infty \to \infty$. By (H$_3$),

$$c_j = \Phi(z_j) = \Phi(z_j) - \frac{1}{2}\langle \Phi'(z_j), z_j \rangle = \int_0^{2\pi} \left(\frac{1}{2} z_j \cdot H'(z_j) - H(z_j) \right) dt$$

$$\leq \tilde{c} \int_0^{2\pi} (1 + |z|^s) \, dt \leq 2\pi \tilde{c} (1 + \|z\|_\infty^s), \tag{2.37}$$

and the conclusion follows because $c_j \to \infty$. $\qquad\square$

The assumption $H \geqslant 0$ is not necessary. Below we show how the proof of Theorem 2.19 can be modified in order to remove it.

COROLLARY 2.20. *The conclusion of Theorem 2.19 remains valid without the condition $H \geqslant 0$.*

PROOF. Since G satisfies (2.31), so does H according to Remark 2.6; hence H is bounded below. Let

$$\tilde{\alpha} := \max_{z \in E^0} \Phi(z)$$

and let $r_0 < r < n$ be positive integers. Let $X = (E_r^+(L) \oplus E^0(L) \oplus E^-(L)) \cap \widetilde{E}$ as before and $Y = E_{r_0}(L)^\perp \cap E^+(L) \cap \widetilde{E}$. Then $\dim Y_n = 2(d_n + l)$ and we still have $m < l$ if $r - r_0$ is large enough. Define $S = Y \cap \{z \in E : \|z\|_s = 1\}$ and note that S is radially homeomorphic to the unit sphere in Y. We claim that $\Phi|_{Y \cap S} \geqslant \alpha > \tilde{\alpha}$ for all large r_0. Assuming this for the moment, we find r_0 such that the condition above is satisfied. Since $\Phi|_{E^0} \leqslant \tilde{\alpha}$, we can easily see by modifying the argument of Theorem 1.14 that the conclusion of Corollary 1.15 (and hence also of Corollary 1.16) holds.

It remains to prove the claim. Arguing by contradiction, we find $r_j \to \infty$ and $z_j \in E_+^{r_j}(L) \cap \widetilde{E}$ such that $\|z_j\|_s = 1$ and $\Phi(z_j) \leqslant \alpha$. Hence

$$\alpha \geqslant \Phi(z_j) = \frac{1}{2}\langle Lz_j, z_j \rangle - \int_0^{2\pi} G(z_j)\, dt$$

$$\geqslant \varepsilon \|z_j\|^2 - \tilde{c}(\|z_j\|_s^s + 1) = \varepsilon \|z_j\|^2 - 2\tilde{c},$$

so $\{z_j\}$ is bounded in E. Passing to a subsequence, $z_j \rightharpoonup z$ in E and $z_j \to z$ in $L^s(S^1, \mathbb{R}^{2N})$. It follows that $\|z\|_s = 1$; in particular, $z \neq 0$. On the other hand, $z_j \in E_{r_j}(L)^\perp \cap E^+(L)$ implies $z_j \rightharpoonup 0$, a contradiction. A somewhat different argument will be given in the proof of Theorem 2.25. $\qquad\square$

The first result on the existence of a nontrivial periodic solution of (2.1) is due to Rabinowitz [78]. Theorem 2.18 may be found in [62]. The result contained there is in fact more general: the case $\sigma(JA) \cap i\mathbb{Z} \neq \emptyset$ is allowed if G has constant sign for small $|z|$. Also some Hamiltonians not satisfying the requirement $\mu > s - 1$ are allowed; for this purpose a truncation argument indicated in Remark 2.6 is employed. Other extensions of Theorem 2.18 are due to Felmer [38] and Long and Xu [70]. Corollary 2.19 is due to Rabinowitz [79] (in [79] no growth restriction (H_3) is needed; again, this is achieved by truncation).

An interesting question concerning the autonomous system (2.35) is whether one can find solutions with prescribed minimal period. Results in this direction, mainly for convex Hamiltonians, can be found in Ekeland's book [33, Section IV.5] and in Long's book [69, Chapter 13].

2.5. *Asymptotically linear systems*

In this section we assume that in addition to (H_1)–(H_3) H satisfies the conditions

$$H(z, t) = \frac{1}{2} A_0 z \cdot z + G_0(z, t),$$

$$\text{where } (G_0)_z(z, t) = o(z) \text{ uniformly in } t \text{ as } z \to 0 \tag{2.38}$$

and

$$H(z, t) = \frac{1}{2} A_\infty z \cdot z + G_\infty(z, t),$$

$$\text{where } (G_\infty)_z(z, t) = o(|z|) \text{ uniformly in } t \text{ as } |z| \to \infty. \tag{2.39}$$

Here A_0, A_∞ are $2N \times 2N$ constant matrices. We assume for simplicity that the system (2.1) is *nonresonant* at the origin and at infinity, that is, $\sigma(J A_0) \cap i\mathbb{Z} = \sigma(J A_\infty) \cap i\mathbb{Z} = \emptyset$. This terminology is justified by the fact that the systems $\dot{z} = J A_0 z$ and $\dot{z} = J A_\infty z$ have no other 2π-periodic solutions than $z = 0$.

As a first result in this section we give a sufficient condition for the existence of a non-trivial 2π-periodic solution of (2.1).

THEOREM 2.21. *Suppose H satisfies (H_1)–(H_3), (2.38), (2.39) and $\sigma(J A_0) \cap i\mathbb{Z} = \sigma(J A_\infty) \cap i\mathbb{Z} = \emptyset$. If $i^-(A_0) \neq i^-(A_\infty)$, then the system (2.1) has a 2π-periodic solution $z \neq 0$.*

PROOF. Suppose $i^-(A_0) < i^-(A_\infty)$. The same argument applied to $-\Phi$ will give the conclusion for $i^-(A_0) > i^-(A_\infty)$. Let L_0 and L_∞ be given by (2.11), with respectively $A = A_0$ and $A = A_\infty$. As in the proof of Theorem 2.18, we see that (2.32) and (2.33) are satisfied, with L_0 replacing L. Suppose Φ has no other critical points than 0. Then $\Phi_n = \Phi|_{E_n}$ has no critical points with $|\Phi_n(z)| \geq \alpha$ provided $n \geq n_0$ and n_0 is large enough. For otherwise we find $z_j \in E_{n_j}$ such that $n_j \to \infty$ and $\Phi'_{n_j}(z_j) = 0$. According to Proposition 2.5, $z_j \to z$ after passing to a subsequence, hence z is a critical point and $|\Phi(z)| \geq \alpha$ which is impossible. Fix $n \geq n_0$. Now we invoke Theorem 1.2. It follows from Lemma 2.4 that $\Phi'_n(z) = P_n L_\infty|_{E_n}(z) + o(\|z\|)$ as $\|z\| \to \infty$. If n is large enough, then (with $2d_n = 2N(1 + n)$ as in the proof of Theorem 2.19)

$$\dim E_n^-(L_0) = M^+(A_0) + \sum_{k=1}^{n} M^-\left(T_k(A_0)\right) = 2d_n + i^-(A_0) - N,$$

and similarly,

$$M^-(P_n L_\infty|_{E_n}) = \dim E_n^-(L_\infty) = 2d_n + i^-(A_\infty) - N.$$

Since $i^-(A_0) < i^-(A_\infty)$, Theorem 1.2 with $F_0 = E_n^-(L_0)$ and $B = P_n L_\infty|_{E_n}$ implies that Φ_n has a critical point z such that $|\Phi_n(z)| \geq \alpha$. This contradiction completes the proof. $\square$

As in the preceding section, we now turn our attention to the autonomous case.

THEOREM 2.22. *Suppose* $H = H(z)$ *satisfies* (H_1), (H_3), (2.38), (2.39) *and* $\sigma(JA_0) \cap i\mathbb{Z} = \sigma(JA_\infty) \cap i\mathbb{Z} = \emptyset$. *If* $H \geqslant 0$ *and* $i^-(A_0) < i^-(A_\infty)$, *then the system* (2.1) *has at least* $\frac{1}{2}(i^-(A_\infty) - i^-(A_0))$ *geometrically distinct nonconstant* 2π-*periodic solutions.*

PROOF. We verify the assumptions of Corollary 1.15. Φ satisfies (PS)*, $E^{S^1} = E_0 \subset E_n$ and $\Phi|_{E^0} \leqslant 0$. Let $X = E^-(L_\infty) \cap \widetilde{E}$ and $Y = E^+(L_0) \cap \widetilde{E}$. Since $H \geqslant 0$ and $\sigma(JA_0) \cap i\mathbb{Z} = \sigma(JA_\infty) \cap i\mathbb{Z} = \emptyset$, A_0 and A_∞ are positive definite and $M^+(A_0) = M^+(A_\infty) = 2N$. Therefore (cf. (2.36))

$$\dim Y_n = 2d_n + i^+(A_0) - N =: 2(d_n + l)$$

and

$$\operatorname{codim}_{\widetilde{E}_n} X_n = \sum_{k=1}^{n} M^+\big(T_k(A_\infty)\big) = 2d_n + i^+(A_\infty) - N =: 2(d_n + m).$$

Since $i^+(A_0) = -i^-(A_0)$ and $i^+(A_\infty) = -i^-(A_\infty)$, $\frac{1}{2}(i^-(A_\infty) - i^-(A_0)) = l - m$. Finally, $\Phi|_{Y \cap S_\rho} \geqslant \alpha > 0$ according to (2.33) and since $E^-(L_\infty) = E^0 \oplus X$, it is easy to see that $\Phi|_{E^0 \oplus X} \leqslant \beta$ for an appropriate $\beta > \alpha$. Corollary 1.15 yields at least $l - m$ geometrically distinct 2π-periodic solutions with $\Phi \geqslant \alpha$. Since $\Phi > 0$ these solutions cannot be constant. $\qquad\square$

In the next theorem we drop the hypothesis $H \geqslant 0$ and require H to be even in z.

THEOREM 2.23. *Suppose* $H = H(z)$ *satisfies* (H_1), (H_3), (2.38), (2.39) *and* $\sigma(JA_0) \cap i\mathbb{Z} = \sigma(JA_\infty) \cap i\mathbb{Z} = \emptyset$. *If* $H(-z) = H(z)$ *for all* $z \in \mathbb{R}^{2N}$, *then the system* (2.1) *has at least* $\frac{1}{2}|i^-(A_\infty) - i^-(A_0)|$ *geometrically distinct* 2π-*periodic solutions* $z \neq 0$.

PROOF. We only sketch the argument. Since Φ is an even functional, in view of Remark 1.17 we may apply Theorem 1.14 if $i^-(A_0) > i^-(A_\infty)$ and Corollary 1.15 if $i^-(A_0) < i^-(A_\infty)$ in order to get $|i^-(A_\infty) - i^-(A_0)|$ pairs of nonzero 2π-periodic solutions (here we use genus instead of index and disregard the S^1-symmetry). For a critical value c the set K_c consists of critical S^1-orbits, some of them may correspond to constant solutions, the other ones are homeomorphic to S^1. So if K_c contains a nonconstant solution, then $\gamma(K_c) \geqslant 2$. On the other hand, if $\gamma(K_c) > 2$, it is easy to see that K_c contains infinitely many geometrically distinct critical orbits. Hence the number of nonzero geometrically distinct critical orbits is at least $\frac{1}{2}|i^-(A_\infty) - i^-(A_0)|$. $\qquad\square$

REMARK 2.24. (a) The argument of Theorem 2.23 does not guarantee the existence of nonconstant solutions.

(b) If $H = H(z, t)$ is even in z and satisfies the other assumptions of the above theorem, then the same argument asserts the existence of at least $|i^-(A_\infty) - i^-(A_0)|$ pairs of nontrivial 2π-periodic solutions (see [4] or [92]). Hamiltonian systems with spatial symmetries will be further discussed in Section 2.6 below.

(c) As we have mentioned in Section 2.4, there is a one-to-one correspondence between 2π-periodic solutions of the system $\dot{z} = JH'(z)$ and T-periodic solutions of $\dot{z} = \lambda JH'(z)$, where $\lambda = 2\pi/T$. Hence, in view of the results of that section, there exist nonconstant T-periodic solutions of any period T whenever H is autonomous and superquadratic. Here the situation is different. If $H \geqslant 0$, then $i^-(\lambda A_0) = i^-(\lambda A_\infty) = N$ for all small $\lambda > 0$. Thus Theorem 2.22 gives no nonconstant solutions of small period T. This is not surprising, for if H' is Lipschitz continuous with Lipschitz constant M, then each nonconstant periodic solution must have period $T \geqslant 2\pi/M$ according to Theorem 4.3 in [20]. On the other hand, it is easy to give examples where $i^-(\lambda A_\infty) - i^-(\lambda A_0) \to \infty$ as $\lambda \to \infty$. So the number of geometrically distinct T-periodic solutions will go to infinity with T. However, there may be no solutions of arbitrarily large *minimal* period (see [92], Remark 6.3 for an example).

There is an extensive literature concerning the existence of one or two nontrivial solutions of (2.1) in the framework of Theorem 2.21. Usually the argument is based on an infinite-dimensional Morse theory and it is possible to weaken the nonresonance conditions at zero and infinity. Also, it is not necessary to have constant matrices A_0 and A_∞. The first results related to Theorem 2.21 may be found in Amann and Zehnder [3,4]. For other results and more references, see, e.g., Abbondandolo [1], Chang [23], Guo [48], Izydorek [53], Kryszewski and Szulkin [57], Li and Liu [61], Szulkin and Zou [93]. Theorem 2.22 is due to Amann and Zehnder [4] and Benci [18]. It has been extended by Degiovanni and Olian Fannio [28], see also [92]. While the proof in [28] uses a cohomological index theory (like the one in [36]) and a variant of Benci's pseudoindex [18], the argument in [92] is based on a relative limit index (which is a generalization of the limit index i_ε). Another extension, using Conley index theory, has been carried out by Izydorek [54]. Some other aspects of the problem (an estimate of the number of T-periodic solutions in terms of the so-called twist number) are discussed in Abbondandolo [1]. However, in all results related to Theorem 2.22 we know of, the assumptions on H are rather restrictive. This is briefly discussed in Remark 6.4 of [92]. Theorem 2.23 is due to Benci [18]. For results about solutions of the autonomous equation (2.1) with prescribed minimal period we refer again to Long's book [69, Section 13.3].

2.6. *Spatially symmetric Hamiltonian systems*

In this section we consider the non-autonomous Hamiltonian system (2.1) when H is invariant with respect to certain group representations in $\mathbb{R}^{2N}$. More precisely, we consider two different kinds of symmetries:
- A compact Lie group G acts on $\mathbb{R}^{2N}$ via an orthogonal and symplectic representation; the standard example is the antipodal action of $\mathbb{Z}/2$ (i.e., H is even in z).
- The infinite group $\mathbb{Z}^k$ acts on $\mathbb{R}^{2N}$ via space translations; the standard example is $\mathbb{Z}^{2N}$ leading to a Hamiltonian system on the torus $T^{2N} := \mathbb{R}^{2N}/\mathbb{Z}^{2N}$. Another example is $\mathbb{Z}^N$ acting via translation of the q-variables which leads to a Hamiltonian system on the cotangent space T^*T^N of the N-dimensional torus.

In the first case we may think of G as a closed subgroup of $O(2N) \cap Sp(2N)$. We shall always assume that $H : \mathbb{R}^{2N} \times \mathbb{R} \to \mathbb{R}$ satisfies the hypotheses (H_1)–(H_3) from Section 2.1.

First we treat the compact group case and require:

(S) The compact group G acts on $\mathbb{R}^{2N}$ via an orthogonal and symplectic representation T such that the action is fixed point free on $\mathbb{R}^{2N} \setminus \{0\}$ (i.e., $(\mathbb{R}^{2N})^G = \{0\}$). H is invariant with respect to T: $H(T_g z, t) = H(z, t)$ for all $g \in G$, $z \in \mathbb{R}^{2N}$, $t \in \mathbb{R}$.

By an orthogonal and symplectic representation we mean that the matrix of T_g is in $O(2N) \cap Sp(2N)$ for all $g \in G$. Clearly, if (S) holds and $z(t)$ is a periodic solution of (2.1) then so is $T_g z(t)$ for every $g \in G$. Thus one has to count G-orbits of periodic solutions and not just periodic solutions.

THEOREM 2.25. *Suppose* (H_1)–(H_3) *and* (S) *hold for G of prime order. If H is superquadratic in the sense of* (2.31) *then the system* (2.1) *has a sequence of 2π-periodic solutions z_j such that $\|z_j\|_\infty \to \infty$.*

If H is invariant with respect to an orthogonal symplectic representation of a more general compact Lie group G then one can apply Theorem 2.25 provided there exists a subgroup $G_1 \subset G$ of prime order having 0 as the only fixed point. This may or may not be the case. It is always the case for $G = S^1$ or more generally, for $G = (S^1)^k$ a torus, acting without nontrivial fixed points. A general existence result in this direction works for *admissible* group actions; see Remark 1.10.

PROOF. We want to apply Corollary 1.16 to the usual action functional

$$\Phi(z) = \frac{1}{2} \int_0^{2\pi} (-J\dot{z} \cdot z)\, dt - \int_0^{2\pi} H(z, t)\, dt.$$

Although the proof of Corollary 2.20 could be used here with minor changes, we provide a slightly different argument as we have mentioned earlier. By Proposition 2.5 the (PS)*-condition holds.

Recall the spaces $E_n \subset E$ from (2.5). We choose $k_0 \in \mathbb{N}$ and set $Y := E_{k_0}^\perp \cap E^+$. Now we claim that for k_0 large enough, there exists $\rho, \alpha > 0$ so that Φ satisfies condition (iii) from Corollary 1.15, that is,

$$\Phi(z) \geqslant \alpha \quad \text{for } z \in Y \text{ with } \|z\| = \rho. \tag{2.40}$$

In order to see this we first observe that

$$\|z\| \geqslant \sqrt{k_0} \|z\|_2 \quad \text{holds for } z \in Y,$$

and that there exists $c_1 > 0$ with

$$|H(z, t)| \leqslant c_1 \big(|z|^s + 1\big)$$

by (H_3). Using the continuous embedding $E \hookrightarrow L^{2s-2}$ we obtain

$$\|z\|_s^s \leqslant \|z\|_2 \cdot \|z\|_{2s-2}^{s-1} \leqslant \frac{c_2}{\sqrt{k_0}} \|z\| \cdot \|z\|^{s-1} \quad \text{for } z \in Y.$$

This implies

$$\Phi(z) \geqslant \frac{1}{2}\|z\|^2 - c_1\left(\|z\|_s^s - 2\pi\right) \geqslant \frac{1}{2}\|z\|^2 - \frac{c_1 c_2}{\sqrt{k_0}}\|z\|^s - 2c_1\pi$$

for every $z \in Y$. Setting $\rho = \left(\frac{\sqrt{k_0}}{sc_1c_2}\right)^{1/(s-2)}$ we therefore have for $z \in Y$ with $\|z\| = \rho$:

$$\Phi(z) \geqslant \left(\frac{1}{2} - \frac{1}{s}\right)\left(\frac{\sqrt{k_0}}{sc_1c_2}\right)^{2/(s-2)} - 2c_1\pi > 0$$

provided k_0 is large. Thus we may fix $k_0 \in \mathbb{N}$ so that (2.40) holds.

Next we define $X_k := E^- + E_k$ for $k \in \mathbb{N}$. Then $\sup \Phi(X_k) < \infty$ because $\Phi(z) \to -\infty$ for $z \in X_k$ with $\|z\| \to \infty$ as we have seen earlier. In order to apply Corollary 1.16 with $X = X_k$ it remains to check the dimension condition from Corollary 1.16. Recall that $d_G = 1$ for $G = \mathbb{Z}/p$. Setting $d_n = 2N(1+n)$ we have

$$\dim Y_n = 2N(n - k_0) = d_n + l$$

with $l = -2N(k_0 + 1)$ and

$$\operatorname{codim}_{E_n}(X_k \cap E_n) = 2N(n - k) = d_n + m(k)$$

with $m(k) = -2Nk - 2N$ ($E_n = \widetilde{E}_n$ here because $E^G = \{0\}$). Clearly $l - m(k) \to \infty$ as $k \to \infty$. Now the theorem follows from Corollary 1.16 and (2.37). $\qquad\square$

Comparing Theorem 2.25 with Theorems 2.18 and 2.19 we see that from the variational point of view the spatial symmetry condition (S) has the same effect as the S^1-symmetry of the autonomous problem. This is also true for asymptotically linear Hamiltonian systems— as we already observed in Remark 2.24(b). We state one multiplicity result in this setting.

THEOREM 2.26. *Suppose H satisfies (H_1)–(H_3) and is asymptotically quadratic in the sense of (2.38) and (2.39). Suppose moreover that $\sigma(JA_0) \cap i\mathbb{Z} = \sigma(JA_\infty) \cap i\mathbb{Z} = \emptyset$, and let $i^-(A_0)$ and $i^-(A_\infty)$ be the Morse indices defined in (2.13). If (S) holds for G of prime order or for $G \cong S^1$ then the system (2.1) has at least $\frac{1}{d_G}|i^-(A_\infty) - i^-(A_0)|$ G-orbits of nontrivial 2π-periodic solutions.*

PROOF. The result follows from Theorem 1.14 or Corollary 1.15; cf. also the proof of Theorem 2.23. $\qquad\square$

Theorem 2.26 is also true for $G = (\mathbb{Z}/p)^k$ a p-torus and $d_G = 1$, or $G = (S^1)^k$ a torus and $d_G = 2$, or if G acts freely on $\mathbb{R}^{2N} \setminus \{0\}$ and $d_G = 1 + \dim(G)$. G acting freely means that $T_g z = z$ for some $z \neq 0$ implies that g is the identity. Such actions exist only for a very restricted class of Lie groups. It does not help much if a subgroup G_1 of G acts freely (or without nontrivial fixed points if G_1 is a torus or p-torus), because a G-orbit consists of several G_1-orbits. So the multiple G_1-orbits of periodic solutions may correspond to just one G-orbit of periodic solutions.

There are various extensions of Theorem 2.26 when the linearized equations at 0 or at ∞ have nontrivial 2π-periodic solutions, mostly for even Hamiltonians; see for instance [54] and the references therein.

Now we consider the spatially periodic case. The classical result is due to Conley and Zehnder [24] and deals with the case of $\mathbb{Z}^{2N}$, that is H is periodic in all variables. Then periodic solutions appear in $\mathbb{Z}^{2N}$-orbits.

THEOREM 2.27. *Suppose $H \in C^1(\mathbb{R}^{2N} \times \mathbb{R})$ is 2π-periodic in all variables. Then (2.1) has at least $2N + 1$ distinct $\mathbb{Z}^{2N}$-orbits of 2π-periodic solutions.*

PROOF. We consider the decomposition $E = E^+ \oplus E^0 \oplus E^-$ from Section 2.1 and observe that

$$\Phi(z + 2\pi k) = \Phi(z) \quad \text{for every } z \in E, \ k \in \mathbb{Z}^{2N}.$$

Setting $M = T^{2N} = E^0/2\pi\mathbb{Z}^{2N}$ we obtain an induced C^1-functional

$$\Psi : \left(E^+ \oplus E^-\right) \times M \to \mathbb{R}, \quad \Psi\left(z^+, z^-, z^0 + 2\pi\mathbb{Z}^{2N}\right) = \Phi\left(z^+ + z^- + z^0\right).$$

Critical points of Ψ correspond to $\mathbb{Z}^{2N}$-orbits of 2π-periodic solutions of (2.1). The conclusion follows from Theorem 1.18 and the fact that $\mathrm{cupl}(T^{2N}) = 2N + 1$. More precisely, we let $W = E^-$ and $Y = E^+$. Since H is bounded, β as in (ii) of Theorem 1.18 exists and taking ρ large enough, we also find $\alpha < \beta$ and γ. Finally, it is easy to see that (PS)*-sequences are bounded (cf. Proposition 2.5), consequently, the (PS)*-condition is satisfied. $\qquad\square$

Using Theorem 1.18 one can also treat more general periodic symmetries, for instance when $H(p, q, t)$ is invariant under $\mathbb{Z}^N$ acting on the q-variables by translations. Then one needs some condition on the behavior of $H(p, q, t)$ as $|p| \to \infty$. Results in this direction have been obtained by a number of authors, see [23,37,42,40,66,90].

If the periodic solutions are non-degenerate then Conley and Zehnder [24] used Morse theoretic arguments to prove:

THEOREM 2.28. *Suppose $H \in C^2(\mathbb{R}^{2N} \times \mathbb{R})$ is 2π-periodic in all variables and all 2π-periodic solutions of (2.1) are non-degenerate. Then (2.1) has at least 2^{2N} distinct $\mathbb{Z}^{2N}$-orbits of 2π-periodic solutions.*

The Morse theoretic arguments involve in particular the Conley–Zehnder index; see Section 2.1, in particular Remark 2.8. Theorems 2.27 and 2.28 are special cases of the

Arnold conjecture. This states that a Hamiltonian flow on a compact symplectic manifold M has at least $\mathrm{cat}(M)$ periodic solutions. If all periodic solutions are non-degenerate then it has at least $\sum_{i=0}^{\dim M} \dim H_i(M)$ critical points where $H_i(M)$ denotes the ith homology group of M with coefficients in an arbitrary field. Theorems 2.27 and 2.28 correspond to the case $M = T^{2N}$ where $\mathrm{cat}(M) = 2N + 1$ and $\dim H_i(M) = \binom{2N}{i}$, so that $\sum_{i=0}^{\dim M} \dim H_i(M) = 2^{2N}$. The interested reader can find results and many references concerning the Arnold conjecture in the book [52] and in the paper [41].

We conclude this section with a theorem on Hamiltonian systems where the Hamiltonian is both even and spatially periodic in the z-variables:

$$H(z + 2\pi k, t + 2\pi) = H(z, t) = H(-z, t) \quad \text{for all } z \in \mathbb{R}^{2N}, \ k \in \mathbb{Z}^{2N}, \ t \in \mathbb{R}.$$
$$(2.41)$$

It follows that $H_z(z, t) = 0$ for all $z \in (\pi\mathbb{Z})^{2N}$, hence modulo the $\mathbb{Z}^{2N}$-action (2.1) has at least 2^{2N} stationary solutions $z(t) \equiv c_i$ with $c_i \in \{0, \pi\}$, $i = 1, \dots, 2N$. Thus the Arnold conjecture is trivially satisfied if (2.41) holds. A natural question to ask is whether there exist 2π-periodic solutions in addition to the 2^{2N} trivial equilibria.

THEOREM 2.29. *Suppose $H \in C^2(\mathbb{R}^{2N} \times \mathbb{R})$ satisfies (2.41) and suppose all 2π-periodic solutions of (2.1) are non-degenerate. For $z \in \pi\mathbb{Z}^{2N}$ let $A_z(t) := H_{zz}(z, t)$ and let $j^-(A_z) \in \mathbb{Z}$ be the Conley–Zehnder index. Then (2.1) has at least $k = \max |j^-(A_z)| - N$ pairs $\pm z_1, \dots, \pm z_k$ of 2π-periodic solutions which lie on different $\mathbb{Z}^{2N}$-orbits.*

Theorem 2.29 has been proved in [14], an extension to Hamiltonian systems on $T^* T^N$, where H is even in z and periodic in the q-variables can be found in [15].

3. Homoclinic solutions

3.1. *Variational setting for homoclinic solutions*

Up to now we have been concerned with periodic solutions of Hamiltonian systems. In this part we turn our attention to homoclinic solutions of the system

$$\dot{z} = J H_z(z, t). \tag{3.1}$$

We assume H satisfies the following hypotheses:
- $(\widetilde{H}_1)$ $H \in C(\mathbb{R}^{2N} \times \mathbb{R}, \mathbb{R})$, $H_z \in C(\mathbb{R}^{2N} \times \mathbb{R}, \mathbb{R}^{2N})$ and $H(0, t) \equiv 0$;
- $(\widetilde{H}_2)$ H is 1-periodic in the t-variable;
- $(\widetilde{H}_3)$ $|H_z(z, t)| \leqslant c(1 + |z|^{s-1})$ for some $c > 0$ and $s \in (2, \infty)$;
- $(\widetilde{H}_4)$ $H(z, t) = \frac{1}{2} Az \cdot z + G(z, t)$, where A is a constant symmetric $2N \times 2N$-matrix, $\sigma(JA) \cap i\mathbb{R} = \emptyset$ and $G_z(z, t)/|z| \to 0$ uniformly in t as $z \to 0$.

Note that the hypotheses $(\widetilde{H}_1)$–$(\widetilde{H}_3)$ are the same as (H_1)–(H_3) in Section 2.1 except that the period is normalized to 1 and not 2π (which is slightly more convenient here).

Let z_0 be a 1-periodic solution of (3.1). A solution z is said to be *homoclinic* (or *doubly asymptotic*) to z_0 if $z \not\equiv z_0$ and $|z(t) - z_0(t)| \to 0$ as $|t| \to \infty$. It has been shown by Coti Zelati, Ekeland and Séré in [25] that if H satisfies $(\widetilde{H}_1)$–$(\widetilde{H}_3)$, then in many cases a symplectic change of variables will reduce the problem of finding homoclinics to z_0 to that of finding solutions homoclinic to 0 for the system (3.1), with a new Hamiltonian H satisfying $(\widetilde{H}_1)$–$(\widetilde{H}_4)$. Therefore in what follows we only consider solutions which are homoclinic to 0 (i.e., $z \not\equiv 0$ and $z(t) \to 0$ as $|t| \to \infty$), or *homoclinic solutions* for short. Recall that if $(\widetilde{H}_4)$ holds, then 0 is called a *hyperbolic point*.

Let $E := H^{1/2}(\mathbb{R}, \mathbb{R}^{2N})$ be the Sobolev space of functions $z \in L^2(\mathbb{R}, \mathbb{R}^{2N})$ such that

$$\int_{\mathbb{R}} \left(1 + \xi^2\right)^{1/2} |\hat{z}(\xi)|^2 \, d\xi < \infty,$$

where $\hat{z}$ is the Fourier transform of z. E is a Hilbert space with an inner product

$$(z, w) := \int_{\mathbb{R}} \left(1 + \xi^2\right)^{1/2} \hat{z}(\xi) \cdot \overline{\hat{w}(\xi)} \, d\xi.$$

The Sobolev embedding $E \hookrightarrow L^q(\mathbb{R}, \mathbb{R}^{2N})$ is continuous for any $q \in [2, \infty)$ (see, e.g., [2] or Section 10 in [87]) but not compact. Indeed, let $z_j(t) := z(t - j)$, where $z \not\equiv 0$; then $z_j \rightharpoonup 0$ in E as $j \to \infty$ but $z_j \not\to 0$ in L^q. However, the embedding $E \hookrightarrow L^q_{\mathrm{loc}}(\mathbb{R}, \mathbb{R}^{2N})$ is compact.

Now we introduce a more convenient inner product. It follows from $(\widetilde{H}_4)$ that $-\mathrm{i}\xi J - A$ is invertible and $(-\mathrm{i}\xi J - A)^{-1}$ is uniformly bounded with respect to $\xi \in \mathbb{R}$. Using this fact, the equality $-J\dot{z} \cdot w = \dot{z} \cdot Jw$ and Plancherel's formula it can be shown that the mapping $L : E \to E$ given by

$$(Lz, w) = \int_{\mathbb{R}} (-J\dot{z} - Az) \cdot w \, dt$$

is bounded, selfadjoint and invertible (see Section 10 in [87] for the details). It is also shown in [87] that the spectrum $\sigma(-J(d/dt) - A)$ is unbounded below and above in $H^1(\mathbb{R}, \mathbb{R}^{2N})$; hence $E = E^+ \oplus E^-$, where $E^\pm$ are L-invariant infinite-dimensional spaces such that the quadratic form (Lz, z) is positive definite on E^+ and negative definite on E^-. Therefore we can define a new equivalent inner product in E by setting

$$\langle z, w \rangle := \left(Lz^+, w^+\right) - \left(Lz^-, w^-\right),$$

where $z^\pm, w^\pm \in E^\pm$. If $\| \cdot \|$ denotes the corresponding norm, we have

$$\int_{\mathbb{R}} (-J\dot{z} - Az) \cdot z \, dt = (Lz, z) = \left\| z^+ \right\|^2 - \left\| z^- \right\|^2. \tag{3.2}$$

Let

$$\Phi(z) := \frac{1}{2} \int_{\mathbb{R}} (-J\dot{z} - Az) \cdot z \, dt - \int_{\mathbb{R}} G(z, t) \, dt$$

and

$$\psi(z) := \int_{\mathbb{R}} G(z, t)\, dt.$$

Then

$$\Phi(z) = \frac{1}{2}\|z^+\|^2 - \frac{1}{2}\|z^-\|^2 - \psi(z). \tag{3.3}$$

PROPOSITION 3.1. *If H satisfies $(\widetilde{H}_1)$–$(\widetilde{H}_4)$, then $\Phi \in C^1(E, \mathbb{R})$ and z is a homoclinic solution of (3.1) if and only if $z \neq 0$ and $\Phi'(z) = 0$. Moreover, ψ' and Φ' are weakly sequentially continuous.*

PROOF. We outline the argument. By $(\widetilde{H}_3)$ and $(\widetilde{H}_4)$, $|G_z(z, t)| \leqslant c(|z| + |z|^{s-1})$ for some constant $c > 0$. Hence $\psi \in C^1(E, \mathbb{R})$ and

$$\langle \psi'(z), w \rangle = \int_{\mathbb{R}} G_z(z, t) \cdot w\, dt$$

according to Lemma 3.10 in [102] (although in [102] E is the Sobolev space $H^1(\mathbb{R}^N)$, an inspection of the proof shows that the argument remains valid in our case). Having this, it is easy to see from (3.2) and (3.3) that $\Phi \in C^1(E, \mathbb{R})$ and $\Phi'(z) = 0$ if and only if $z \in E$ and z is a weak solution of (3.1). Since $z \in L^q(\mathbb{R}, \mathbb{R}^{2N})$ for all $q \in [2, \infty)$, $G_z(z(.), .) \in L^2(\mathbb{R}, \mathbb{R}^{2N})$; hence $z \in H^1(\mathbb{R}, \mathbb{R}^{2N})$. In particular, z is continuous by the Sobolev embedding theorem, and consequently, $z \in C^1(\mathbb{R}, \mathbb{R}^{2N})$, i.e., z is a classical solution of (3.1).

It is well known that if $z \in H^1(\mathbb{R}, \mathbb{R}^{2N})$, then $z(t) \to 0$ as $|t| \to \infty$. For the reader's convenience we include a proof. Let $t \leqslant u \leqslant t + 1$. Given $\varepsilon > 0$, there is R such that if $|t| \geqslant R$, then $\|z\|_{L^2((t,t+1),\mathbb{R}^{2N})} < \varepsilon$ and $\|\dot{z}\|_{L^2((t,t+1),\mathbb{R}^{2N})} < \varepsilon$. Hence $|z(u)| < \varepsilon$ for some $u \in [t, t+1]$ and, using Hölder's inequality,

$$|z(t)| \leqslant |z(u)| + |z(u) - z(t)| \leqslant \varepsilon + \int_t^u |\dot{z}(\tau)|\, d\tau$$

$$\leqslant \varepsilon + \|\dot{z}\|_{L^2((t,t+1),\mathbb{R}^{2N})} < 2\varepsilon.$$

Finally, weak sequential continuity follows from the compact embedding $E \hookrightarrow L^q_{\mathrm{loc}}$, $2 \leqslant q < \infty$. Indeed, if $z_j \rightharpoonup z$ in E, then $z_j \to z$ in $L^2_{\mathrm{loc}}(\mathbb{R}, \mathbb{R}^{2N}) \cap L^s_{\mathrm{loc}}(\mathbb{R}, \mathbb{R}^{2N})$, and invoking the argument of Lemma 3.10 in [102] again, we see that $\langle \psi'(z_j), w \rangle \to \langle \psi'(z), w \rangle$ for all $w \in E$, i.e., $\psi'(z_j) \rightharpoonup \psi'(z)$. Clearly, $\Phi'(z_j) \rightharpoonup \Phi'(z)$ as well. $\qquad\square$

It will be important in what follows that the functional Φ is invariant with respect to the representation of the group $\mathbb{Z}$ of integers given by

$$(T_a z)(t) := z(t + a), \quad a \in \mathbb{Z} \tag{3.4}$$

(this is an immediate consequence of the periodicity of G). Moreover, since the linear operator L is $\mathbb{Z}$-invariant, so are the subspaces $E^{\pm}$.

It follows from the $\mathbb{Z}$-invariance of Φ that Φ' is $\mathbb{Z}$-equivariant; hence if $z = z(t)$ is a homoclinic, so are all $T_a z$, $a \in \mathbb{Z}$. Therefore Φ cannot satisfy the Palais–Smale condition at any critical level $c \neq 0$. Setting $\mathcal{O}(z) := \{T_a z : a \in \mathbb{Z}\}$ (cf. Section 1.1), we call two homoclinic solutions z_1, z_2 *geometrically distinct* if $\mathcal{O}(z_1) \neq \mathcal{O}(z_2)$.

LEMMA 3.2. *If H satisfies $(\widetilde{H}_1)$–$(\widetilde{H}_4)$, then $\psi'(z) = \mathrm{o}(\|z\|)$ and $\psi(z) = \mathrm{o}(\|z\|^2)$ as $z \to 0$.*

We omit the argument which is exactly the same as in Lemma 2.4. Next we turn our attention to Palais–Smale sequences.

PROPOSITION 3.3. *If H satisfies $(\widetilde{H}_1)$–$(\widetilde{H}_4)$ and there are $\mu > \max\{2, s - 1\}$ and $\delta > 0$ such that*

$$\delta |z|^{\mu} \leqslant \mu G(z, t) \leqslant z \cdot G_z(z, t) \quad \text{for all } z, t, \tag{3.5}$$

then each (PS)-*sequence $\{z_j\}$ for Φ is bounded. Moreover, if $\Phi(z_j) \to c$, then $c \geqslant 0$, and if $c = 0$, then $z_j \to 0$.*

PROOF. As in (2.8), we have

$$c_1 \|z_j\| + c_2 \geqslant \Phi(z_j) - \frac{1}{2}\langle \Phi'(z_j), z_j \rangle \geqslant \left(\frac{\mu}{2} - 1\right) \int_{\mathbb{R}} G(z_j, t)\, dt \geqslant c_3 \|z_j\|_{\mu}^{\mu}. \tag{3.6}$$

Since for each $\varepsilon > 0$ there is $C(\varepsilon)$ such that $|G_z(z, t)| \leqslant \varepsilon |z| + C(\varepsilon)|z|^{s-1}$ (by $(\widetilde{H}_3)$ and $(\widetilde{H}_4)$), we see as in (2.10) that

$$\|z_j^{\pm}\|^2 \leqslant \alpha_j \|z_j^{\pm}\| + c_4 \varepsilon \|z_j\| \|z_j^{\pm}\| + c_5(\varepsilon) \|z_j\|_{\mu}^{s-1} \|z_j^{\pm}\|, \tag{3.7}$$

where $\alpha_j := \|\Phi'(z_j)\| \to 0$. Hence choosing ε small enough, we have $\|z_j\| \leqslant c_6 + c_7 \|z_j\|_{\mu}^{s-1}$, and taking (3.6) into account,

$$\|z_j\| \leqslant c_8 + c_9 \|z_j\|^{(s-1)/\mu}.$$

It follows that $\{z_j\}$ is bounded.

Now we obtain from (3.6) that if $\Phi(z_j) \to c$, then

$$c \geqslant c_3 \limsup_{j \to \infty} \|z_j\|^{\mu},$$

so $c \geqslant 0$. If $c = 0$, then $z_j \to 0$ in L^{μ} and passing to a subsequence, $z_j \rightharpoonup z$ in E. Hence $z = 0$. Letting $j \to \infty$ in (3.7) we see that $z_j \to 0$ also in E. $\qquad\square$

Condition (3.5) is rather restrictive. Other conditions (always including the inequality $0 < \mu G(z,t) \leqslant z \cdot G_z(z,t)$ for all $z \neq 0$ and some $\mu > 2$) which imply boundedness of (PS)-sequences may be found, e.g., in [6] and [31].

We shall need the following result which is a special case of P.L. Lions' vanishing lemma (see, e.g., [65] or [102]):

LEMMA 3.4. *Let $r > 0$ be given and let $\{z_j\}$ be a bounded sequence in E. If*

$$\lim_{j \to \infty} \sup_{a \in \mathbb{R}} \int_{a-r}^{a+r} |z_j|^2 \, dt = 0, \tag{3.8}$$

then $z_j \to 0$ in $L^q(\mathbb{R}, \mathbb{R}^{2N})$ for all $q \in (2, \infty)$.

PROOF. In [65,102] the case of $E = H^1(\mathbb{R}^N)$ has been considered. Below we adapt the argument of [102, Lemma 1.21] to our situation.

Let $q \in (2, 4)$. By Hölder's inequality,

$$\|z_j\|^q_{L^q((a-r,a+r),\mathbb{R}^{2N})} \leqslant \|z_j\|^{q-2}_{L^2((a-r,a+r),\mathbb{R}^{2N})} \|z_j\|^2_{L^p((a-r,a+r),\mathbb{R}^{2N})},$$

where p satisfies $(q-2)/2 + 2/p = 1$. Hence

$$\|z_j\|^q_{L^q((a-r,a+r),\mathbb{R}^{2N})} \leqslant \sup_{b \in \mathbb{R}} \left(\|z_j\|^{q-2}_{L^2((b-r,b+r),\mathbb{R}^{2N})} \right) \|z_j\|^2_{L^p((a-r,a+r),\mathbb{R}^{2N})}$$

$$\leqslant C \sup_{b \in \mathbb{R}} \left(\|z_j\|^{q-2}_{L^2((b-r,b+r),\mathbb{R}^{2N})} \right) \|z_j\|^2_{H^{1/2}((a-r,a+r),\mathbb{R}^{2N})}. \tag{3.9}$$

Here we may use the norm in $H^{1/2}((a-r, a+r), \mathbb{R}^{2N})$ given by

$$\|z\|^2_{H^{1/2}((a-r,a+r),\mathbb{R}^{2N})} = \|z\|^2_{L^2((a-r,a+r),\mathbb{R}^{2N})} + \int_{a-r}^{a+r} \int_{a-r}^{a+r} \frac{|z(t) - z(s)|^2}{(t-s)^2} \, ds \, dt$$

(see [2, Theorem 7.48]). Covering $\mathbb{R}$ by intervals $(a_n - r, a_n + r)$, $n \in \mathbb{Z}$, in such a way that each $t \in \mathbb{R}$ is contained in at most 2 of them and taking the sum with respect to n in (3.9), we obtain

$$\|z_j\|^q_q \leqslant 2C \sup_{a \in \mathbb{R}} \left(\|z_j\|^{q-2}_{L^2((a-r,a+r),\mathbb{R}^{2N})} \right) |z_j|^2_E,$$

where

$$|z_j|^2_E := \|z_j\|^2_2 + \int_{\mathbb{R}} \int_{\mathbb{R}} \frac{|z(t) - z(s)|^2}{(t-s)^2} \, ds \, dt.$$

According to Theorem 7.12 in [64], the norms $|\cdot|_E$ and $\|\cdot\|$ are equivalent. Since $\{z_j\}$ is bounded in E, it follows that $z_j \to 0$ in L^q.

If $q \geqslant 4$, we can choose $q_0 \in (2,4)$ and $p > q$. Then by Hölder's inequality, $\|z_j\|_q^q \leqslant \|z_j\|_{q_0}^{(1-\lambda)q}\|z_j\|_p^{\lambda q}$, where $(1-\lambda)q/q_0 + \lambda q/p = 1$, and the conclusion follows because $\|z_j\|_{q_0} \to 0$ and $\|z_j\|_p$ is bounded. $\square$

PROPOSITION 3.5. *Suppose* $(\widetilde{H}_1)$–$(\widetilde{H}_4)$ *are satisfied and* $\{z_j\}$ *is a bounded* (PS)-*sequence such that* $\Phi(z_j) \to c > 0$. *Then* (3.1) *has a homoclinic solution.*

PROOF. Suppose first $\{z_j\}$ is vanishing in the sense that (3.8) is satisfied. It is clear that (3.7) holds for $\mu = s$. Since $\alpha_j = \|\Phi'(z_j)\| \to 0$ and $z_j \to 0$ in L^s according to Lemma 3.4, it follows from (3.7) with ε appropriately small that $z_j \to 0$ in E; hence $\Phi(z_j) \to 0$. This contradiction shows that $\{z_j\}$ cannot be vanishing. Therefore there exist $\delta > 0$ and a_j such that, up to a subsequence,

$$\int_{a_j-r}^{a_j+r} |z_j|^2 \, dt \geqslant \delta \tag{3.10}$$

for almost all j. Choosing a larger r if necessary we may assume $a_j \in \mathbb{Z}$. Let $\tilde{z}_j(t) := z_j(a_j + t)$. It follows from the $\mathbb{Z}$-invariance of Φ that $\{\tilde{z}_j\}$ is a bounded (PS)-sequence and $\Phi(\tilde{z}_j) \to c$. Hence $\tilde{z}_j \rightharpoonup \tilde{z}$ in E and $\tilde{z}_j \to \tilde{z}$ in L^2_{loc} after passing to a subsequence. Moreover, since

$$\int_{-r}^{r} |\tilde{z}_j|^2 \, dt = \int_{a_j-r}^{a_j+r} |z_j|^2 \, dt, \tag{3.11}$$

$\tilde{z} \neq 0$. According to Proposition 3.1, Φ' is weakly sequentially continuous. Thus $\Phi'(\tilde{z}) = 0$ and the conclusion follows. $\square$

3.2. *Existence of homoclinics*

Our first result in this section asserts that if G_z is superlinear, then (3.1) has at least 1 homoclinic.

THEOREM 3.6. *Suppose* H *satisfies* $(\widetilde{H}_1)$–$(\widetilde{H}_4)$ *and* G *satisfies* (3.5) *with* $\mu > \max\{2, s-1\}$. *Then* (3.1) *has a homoclinic solution.*

PROOF. According to (3.3), the functional Φ corresponding to (3.1) has the form required in Theorem 1.19. Moreover, $G \geqslant 0$ and therefore $\psi \geqslant 0$. Let $z_j \rightharpoonup z$. Then $z_j \to z$ in $L^2_{\text{loc}}(\mathbb{R}, \mathbb{R}^{2N})$ and $z_j \to z$ a.e. in $\mathbb{R}$ after passing to a subsequence, so it follows from Fatou's lemma that ψ is weakly sequentially lower semicontinuous. Moreover, ψ' is weakly sequentially continuous according to Proposition 3.1. Hence (i) of Theorem 1.19 holds, and so does (ii) because $\Phi(z) = \frac{1}{2}\|z\|^2 + o(\|z\|^2)$ whenever $z \to 0$, $z \in E^+$.

We shall verify (iii). Let $z_0 \in E^+$, $\|z_0\| = 1$. Since there exists a continuous projection from the closure of $\mathbb{R}z_0 \oplus E^-$ in $L^\mu(\mathbb{R}, \mathbb{R}^{2N})$ to $\mathbb{R}z_0$ and $G(z, t) \geq \delta\mu^{-1}|z|^\mu$,

$$\Phi\left(z^- + \zeta z_0\right) \leq \frac{\zeta^2}{2} - \frac{1}{2}\left\|z^-\right\|^2 - \delta\mu^{-1}\left\|z^- + \zeta z_0\right\|_\mu^\mu$$

$$\leq \frac{\zeta^2}{2} - \frac{1}{2}\left\|z^-\right\|^2 - \delta_0 \zeta^\mu \|z_0\|_\mu^\mu$$

for some $\delta_0 > 0$, and it follows that $\Phi(z^- + \zeta z_0) \leq 0$ whenever $\|z^- + \zeta z_0\|$ is large enough. Obviously, $\Phi(z^-) \leq 0$ for all $z^- \in E^-$.

We have shown that the hypotheses of Theorem 1.19 are satisfied. Hence there exists a (PS)-sequence $\{z_j\}$ such that $\Phi(z_j) \to c > 0$ and it remains to invoke Propositions 3.3 and 3.5. $\qquad\square$

Next we turn our attention to asymptotically linear systems. Suppose $\sigma(JA) \cap i\mathbb{R} = \emptyset$, let λ_1 be the smallest positive and λ_{-1} the largest negative λ such that $\sigma(J(A + \lambda I)) \cap i\mathbb{R} \neq \emptyset$ and set $\lambda_0 := \min\{\lambda_1, -\lambda_{-1}\}$. Then

$$\lambda_1 = \inf\left\{\|z\|^2 : z \in E^+, \|z\|_2 = 1\right\},$$

$$\lambda_{-1} = -\inf\left\{\|z\|^2 : z \in E^-, \|z\|_2 = 1\right\}, \tag{3.12}$$

$$\|z\|^2 \geq \lambda_0\|z\|^2 \quad \text{for all } z \in E$$

(see Section 10 of [87] for a detailed argument). We shall need the following two additional assumptions on G:

($\widetilde{\mathrm{H}}_5$) $G(z, t) = \frac{1}{2}A_\infty(t)z \cdot z + F(z, t)$, where $A_\infty(t)z \cdot z \geq \lambda|z|^2$ for some $\lambda > \lambda_1$ and $F_z(z, t)/|z| \to 0$ uniformly in t as $|z| \to \infty$;

($\widetilde{\mathrm{H}}_6$) $G(z, t) \geq 0$ and $\frac{1}{2}G_z(z, t) \cdot z - G(z, t) \geq \alpha(|z|)$, where $\alpha(0) = 0$ and $\alpha(|z|)$ is positive and bounded away from 0 whenever z is bounded away from 0.

A simple example of a function G which satisfies ($\widetilde{\mathrm{H}}_5$) and ($\widetilde{\mathrm{H}}_6$) is given by $G(z, t) = a(t)B(|z|)$, where a is 1-periodic, $a(t) \geq a_0 > 0$ for all t, $B \in C^1(\mathbb{R}, \mathbb{R})$, $B(0) = B'(0) = 0$, $B'(s)/s$ is strictly increasing, tends to 0 as $s \to 0$ and to $\lambda > \lambda_1/a_0$ as $s \to \infty$. That ($\widetilde{\mathrm{H}}_6$) holds follows from the identity

$$\frac{1}{2}B'(s)s - B(s) = \int_0^s \left(\frac{B'(s)}{s} - \frac{B'(\sigma)}{\sigma}\right)\sigma \, d\sigma.$$

LEMMA 3.7. *If H satisfies ($\widetilde{\mathrm{H}}_1$)–($\widetilde{\mathrm{H}}_6$), then each Cerami sequence $\{z_j\}$ (see Remark 1.20 for a definition) is bounded.*

PROOF. Suppose $\{z_j\}$ is unbounded and let $w_j := z_j/\|z_j\|$. We may assume taking a subsequence that $w_j \rightharpoonup w$. We shall obtain a contradiction by showing that $\{w_j\}$ is neither vanishing (in the sense that (3.8) holds) nor nonvanishing.

Assume first $\{w_j\}$ is nonvanishing. As in the proof of Proposition 3.5, we find $a_j \in \mathbb{Z}$ such that, passing to a subsequence, $\widetilde{w}_j(t) := w_j(a_j + t)$ satisfy

$$\int_{-r}^{r} |\widetilde{w}_j|^2 \, dt \geq \delta > 0$$

for j large. Passing to a subsequence once more, $\widetilde{w}_j \rightharpoonup \widetilde{w}$ in E and $\widetilde{w}_j \to \widetilde{w}$ in L^2_{loc} and a.e. in $\mathbb{R}$. In particular, $\widetilde{w} \neq 0$. Since $\|\Phi'(z_j)\| = \|\Phi'(\widetilde{z}_j)\|$, it follows that $\Phi'(\widetilde{z}_j)/\|\widetilde{z}_j\| \to 0$ and therefore

$$\|\widetilde{z}_j\|^{-1}\langle \Phi'(\widetilde{z}_j), v \rangle = \langle \widetilde{w}_j^+, v \rangle - \langle \widetilde{w}_j^-, v \rangle - \int_{\mathbb{R}} A_\infty(t)\widetilde{w}_j \cdot v \, dt$$

$$- \int_{\mathbb{R}} \frac{F_z(\widetilde{z}_j, t) \cdot v}{|\widetilde{z}_j|} |\widetilde{w}_j| \, dt \to 0$$

for all $v \in C_0^\infty(\mathbb{R}, \mathbb{R}^{2N})$. Since $|F_z(z, t)| \leq a|z|$ for some $a > 0$, $F_z(z, t) = \mathrm{o}(|z|)$ as $|z| \to \infty$ and $\operatorname{supp} v$ is bounded, we see by the dominated convergence theorem that the last integral on the right-hand side above tends to 0. Consequently, letting $j \to \infty$, we obtain

$$\dot{\widetilde{w}} = J\big(A + A_\infty(t)\big)\widetilde{w}$$

which contradicts the fact that the operator $-J(d/dt) - (A + A_\infty(t))$ has no eigenvalues in $L^2(\mathbb{R}, \mathbb{R}^{2N})$ [31, Proposition 2.2].

Suppose now $\{w_j\}$ is vanishing. Then

$$\|z_j\|^{-1}\langle \Phi'(z_j), w_j^+ \rangle = \|w_j^+\|^2 - \int_{\mathbb{R}} \frac{G_z(z_j, t) \cdot w_j^+}{|z_j|} |w_j| \, dt \to 0$$

and

$$\|z_j\|^{-1}\langle \Phi'(z_j), w_j^- \rangle = -\|w_j^-\|^2 - \int_{\mathbb{R}} \frac{G_z(z_j, t) \cdot w_j^-}{|z_j|} |w_j| \, dt \to 0.$$

Since $\|w_j\| = 1$,

$$\int_{\mathbb{R}} \frac{G_z(z_j, t) \cdot (w_j^+ - w_j^-)}{|z_j|} |w_j| \, dt \to 1.$$

Let

$$I_j := \big\{ t \in \mathbb{R} : |z_j(t)| \leq \varepsilon \big\},$$

where $\varepsilon > 0$ has been chosen so that $|G_z(z,t)| \leqslant \frac{1}{2}\lambda_0|z|$ whenever $|z| \leqslant \varepsilon$ (such ε exists according to $(\widetilde{H}_4)$). Since w_j^+ and w_j^- are orthogonal in L^2, it follows from (3.12) that

$$\int_{I_j} \frac{G_z(z_j,t) \cdot (w_j^+ - w_j^-)}{|z_j|} \, |w_j| \, dt \leqslant \frac{1}{2}\lambda_0 \|w_j\|_2^2 \leqslant \frac{1}{2}$$

and therefore, since $|G_z(z,t)| \leqslant a_0|z|$ for some $a_0 > 0$,

$$\frac{1}{4} \leqslant \int_{\mathbb{R}\setminus I_j} \frac{G_z(z_j,t) \cdot (w_j^+ - w_j^-)}{|z_j|} \, |w_j| \, dt \leqslant 2a_0 \int_{\mathbb{R}\setminus I_j} |w_j|^2 \, dt$$

$$\leqslant a_0 \operatorname{meas}(\mathbb{R} \setminus I_j)^{(p-2)/p} \|w_j\|_p^{2/p}$$

for almost all j ($p > 2$ arbitrary but fixed). As $\{w_j\}$ is vanishing, $w_j \to 0$ in $L^p(\mathbb{R}, \mathbb{R}^{2N})$ according to Lemma 3.4 and consequently, $\operatorname{meas}(\mathbb{R} \setminus I_j) \to \infty$. Let $\alpha_0 := \inf_{|z|>\varepsilon} \alpha(|z|)$. Then $\alpha_0 > 0$ and $(\widetilde{H}_6)$ implies

$$\Phi(z_j) - \frac{1}{2}\langle \Phi'(z_j), z_j \rangle = \int_{\mathbb{R}} \left(\frac{1}{2} G_z(z_j,t) - G(z_j,t) \right) dt$$

$$\geqslant \int_{\mathbb{R}\setminus I_j} \left(\frac{1}{2} G_z(z_j,t) - G(z_j,t) \right) dt \geqslant \int_{\mathbb{R}\setminus I_j} \alpha_0 \, dt \to \infty.$$

However, $\Phi(z_j)$ is bounded and since $\{z_j\}$ is a Cerami sequence, $\langle \Phi'(z_j), z_j \rangle \to 0$. Therefore the left-hand side above must be bounded, a contradiction. $\qquad\square$

The idea of showing boundedness of $\{z_j\}$ by excluding both vanishing and nonvanishing of $\{w_j\}$ goes back to Jeanjean [55].

THEOREM 3.8. *Suppose H satisfies $(\widetilde{H}_1)$–$(\widetilde{H}_6)$. Then (3.1) has a homoclinic solution.*

PROOF. We shall use Theorem 1.19 again, this time together with Remark 1.20. Clearly, (i) and (ii) still hold. So if we can show that also (iii) is satisfied, then in view of Remark 1.20 there exists a Cerami sequence $\{z_j\}$ with $\Phi(z_j) \to c > 0$. By Lemma 3.7, $\{z_j\}$ is bounded, hence it is a (PS)-sequence as well and we can invoke Propositions 3.3 and 3.5 in the same way as before.

It remains to verify (iii) of Theorem 1.19. According to (3.12) and since $\lambda > \lambda_1$ there exists $z_0 \in E^+$, $\|z_0\| = 1$, such that

$$1 = \|z_0\|^2 < \lambda\|z_0\|_2^2. \tag{3.13}$$

Since $\Phi|_{E^-} \leqslant 0$, it suffices to show that

$$\Phi(z^- + \zeta z_0) = \frac{\zeta^2}{2} - \frac{1}{2}\|z^-\|^2 - \int_{\mathbb{R}} G(z^- + \zeta z_0, t) \, dt \leqslant 0 \tag{3.14}$$

whenever $\|z^- + \zeta z_0\|$ is large enough. Assuming the contrary, we find z_j^- and ζ_j such that $z_j = z_j^- + \zeta_j z_0$ satisfies $\|z_j\| \to \infty$ and the reverse inequality holds in (3.14). Setting $w_j := z_j/\|z_j\| = (z_j^- + \zeta_j z_0)/\|z_j\| = w_j^- + \eta_j z_0$, we obtain

$$\frac{\eta_j^2}{2} - \frac{1}{2}\|w_j^-\|^2 - \int_{\mathbb{R}} \frac{G(z_j, t)}{\|z_j\|^2}\, dt \geqslant 0,$$

and since $G \geqslant 0$,

$$\frac{\eta_j^2}{2} - \frac{1}{2}\|w_j^-\|^2 - \int_I \frac{G(z_j, t)}{\|z_j\|^2}\, dt \geqslant 0, \tag{3.15}$$

where I is a bounded interval (to be specified). Passing to a subsequence, $\eta_j \to \eta \in [0, 1]$, $w_j^- \rightharpoonup w^-$ in E and $w_j^- \to w^-$ in L^2_{loc} and a.e. in $\mathbb{R}$. It follows from (3.15) that $\eta_j \geqslant \|w_j^-\|$, hence $\eta > 0$ because $\eta_j^2 + \|w_j^-\|^2 = 1$. In view of (3.13) and since z_0 and w^- are orthogonal in L^2,

$$\eta^2 - \|w^-\|^2 - \int_{\mathbb{R}} A_\infty(t)w \cdot w\, dt \leqslant \eta^2 - \|w^-\|^2 - \lambda\|w\|_2^2$$

$$= \eta^2\big(1 - \lambda\|z_0\|_2^2\big) - \|w^-\|^2 - \lambda\|w^-\|_2^2$$

$$< 0,$$

hence there exists a bounded interval I such that

$$\eta^2 - \|w^-\|^2 - \int_I A_\infty(t)w \cdot w\, dt < 0. \tag{3.16}$$

On the other hand,

$$\int_I \frac{G(z_j, t)}{\|z_j\|^2}\, dt = \frac{1}{2}\int_I A_\infty(t)w_j \cdot w_j\, dt + \int_I \frac{F(z_j, t)}{|z_j|^2}\,|w_j|^2\, dt,$$

and since I is bounded, $|F(z, t)| \leqslant a|z|^2$ for some $a > 0$ and $F(z, t) = \mathrm{o}(|z|^2)$ as $|z| \to \infty$, it follows from the dominated convergence theorem that the second integral on the right-hand side above tends to 0. Consequently, passing to the limit in (3.15) we obtain

$$\eta^2 - \|w^-\|^2 - \int_I A_\infty(t)w \cdot w\, dt \geqslant 0,$$

a contradiction to (3.16). $\qquad\qquad\qquad\qquad\qquad\qquad\qquad\qquad\qquad\qquad\qquad\square$

While several results concerning the existence of homoclinic solutions for second order systems (e.g., of Newtonian or Lagrangian type) may be found in the literature, much less seems to be known about (3.1) under conditions similar to $(\widetilde{H}_1)$–$(\widetilde{H}_4)$. The first paper to use modern variational methods for finding homoclinic solutions seems to be [25] of Coti Zelati, Ekeland and Séré. It has been shown there that if H satisfies $(\widetilde{H}_1)$–$(\widetilde{H}_4)$, $G = G(z, t)$ is superquadratic in an appropriate sense and convex in z, then (3.1) has at least 2 geometrically distinct homoclinics. If $G = G(z)$, there is at least 1 homoclinic. The convexity of G was used in order to reformulate the problem in terms of a dual functional which is better behaved than Φ (see also a comment in Section 3.4). A result comparable to our Theorem 3.6, but stronger in the sense that there is no growth restriction on G, is due to Tanaka [95]. His proof is rather different from ours: he shows using a linking argument and fine estimates that there is a sequence of $2\pi k_j$-periodic solutions z_j of (3.1) which tend to a homoclinic as $j \to \infty$. A somewhat different result has been obtained by Ding and Willem [31]. Their function G is also superquadratic but they allow the matrix A to be time-dependent (and 1-periodic) and moreover, they allow 0 to be the left endpoint of a gap of the spectrum of $-J(\mathrm{d}/\mathrm{d}t) - A(t)$ (more precisely, $\sigma(-J(\mathrm{d}/\mathrm{d}t) - A(t)) \cap (0, \alpha) = \emptyset$ for some $\alpha > 0$). See also Xu [103], where the superlinearity condition has been weakened with the aid of a truncation argument. Theorem 3.8 is due to Szulkin and Zou [94]; however, the argument presented here is simpler.

Finally we would like to mention that if A is time-dependent and $|A(t)| \to \infty$ in an appropriate sense as $|t| \to \infty$, then it can be shown that in many cases Φ satisfies (PS)* and methods similar to those developed in Sections 2.4–2.6 become available, see, e.g., Ding [29]. Results concerning bifurcation of homoclinics may be found in Stuart [87] and Secchi and Stuart [82].

3.3. *Multiple homoclinic solutions*

In Section 2.4 we have seen that in the autonomous case if H is superquadratic, then the Hamiltonian system has infinitely many periodic solutions whose amplitude tends to infinity. The proof relied in a crucial way on the S^1-invariance of the corresponding functional. For homoclinics the situation is very different. Let $z = (p, q) \in \mathbb{R}^2$ and

$$H(z) = \frac{1}{2}p^2 - \frac{1}{2}q^2 + \frac{1}{4}q^4. \tag{3.17}$$

The corresponding Hamiltonian system reduces to a second order equation $-\ddot{q} = V'(q)$, where the potential $V(q) = \frac{1}{4}q^4 - \frac{1}{2}q^2$. It is easy to see that there exists a homoclinic z_0 and

$$S := \left\{ \pm z_0(t - a) \colon a \in \mathbb{R} \right\} \tag{3.18}$$

is the set of all homoclinics. So although S consists of infinitely (in fact uncountably) many geometrically distinct $\mathbb{Z}$-orbits (in the sense of Section 3.1), it contains only two homoclinics which are really distinct, the reason for this being that Φ is invariant with respect to the representation (3.4) of $\mathbb{R}$ rather than $\mathbb{Z}$. In particular, there are no homoclinics of large amplitude.

Below we shall assume that, in addition to the $\mathbb{Z}$-invariance, Φ is also invariant with respect to a representation of $\mathbb{Z}/p$ (p a prime) in $\mathbb{R}^{2N}$ and (S) of Section 2.6 is satisfied. We also recall from Section 2.6 that if H is even in z, then (S) holds with $\mathcal{G} = \mathbb{Z}/2$ (here we denote groups by $\mathcal{G}$ in order to distinguish them from functions $G = G(z, t)$). Our aim is to show that there are infinitely many geometrically distinct homoclinics provided H is superquadratic. Note that since $\mathbb{Z}/p$ is finite, (3.1) has infinitely many homoclinics which are geometrically distinct when both the representations of $\mathbb{Z}$ and $\mathbb{Z}/p$ are taken into account if and only if it has infinitely many geometrically distinct homoclinics with respect to the representation of $\mathbb{Z}$ only.

THEOREM 3.9. *Suppose H and G satisfy* $(\widetilde{H}_1)$–$(\widetilde{H}_4)$ *and (3.5) with* $\mu > \max\{2, s - 1\}$, *there exist $c_0, \varepsilon_0 > 0$ such that*

$$\left| G_z(z + w, t) - G_z(z, t) \right| \leqslant c_0 |w| \left(1 + |z|^{s-1} \right) \quad \text{whenever} \quad |w| \leqslant \varepsilon_0 \qquad (3.19)$$

and (S) of Section 2.6 holds for $\mathcal{G} = \mathbb{Z}/p$, p a prime. Then (3.1) has infinitely many geometrically distinct homoclinic solutions.

Clearly, (3.19) is satisfied if H_{zz} is continuous and $|H_{zz}(z, t)| \leqslant c(1 + |z|^{s-1})$ for some $c > 0$.

According to our comments in Section 2.6, if H is invariant with respect to an orthogonal and symplectic representation of a group $\mathcal{G}$, if $\mathbb{Z}/p \subset \mathcal{G}$ and $(\mathbb{R}^{2N})^{\mathbb{Z}/p} = \{0\}$, then the conclusion of the above theorem remains valid. However, if the system is autonomous or $\mathcal{G}$ is infinite, then already the existence of one homoclinic (which follows from Theorem 3.6) implies that there are infinitely many homoclinics which are geometrically distinct in the $\mathbb{Z} \times \mathbb{Z}/p$-sense but not in the sense of a representation of the larger group.

An important step in the proof of Theorem 3.9 is the following:

PROPOSITION 3.10. *Suppose $(\widetilde{H}_1)$–$(\widetilde{H}_4)$, (3.5) and (3.19) are satisfied and let $\{z_j\}$ be a $(PS)_c$-sequence with $c > 0$. Then there exist (not necessarily distinct) homoclinics $w_1, \ldots, w_k$ and sequences $\{b_j^m\}$ ($1 \leqslant m \leqslant k$) of integers such that, passing to a subsequence if necessary,*

$$\left\| z_j - \sum_{m=1}^{k} T_{b_j^m} w_m \right\| \to 0 \quad \text{and} \quad \sum_{m=1}^{k} \Phi(w_m) = c.$$

PROOF (outline). We shall only very briefly sketch the argument which is exactly the same as in [58] (where a Schrödinger equation has been considered) or [30], see also [26]. By Proposition 3.3, $\{z_j\}$ is bounded, and it is nonvanishing by the argument of Proposition 3.5. Hence (3.10) is satisfied, and so is (3.11), where $\tilde{z}_j = T_{a_j^1} z_j$. It follows that $\tilde{z}_j \rightharpoonup w_1 \neq 0$ after passing to a subsequence and w_1 is a homoclinic. Let $v_j^1 := \tilde{z}_j - w_1$. Then one shows that $\{v_j^1\}$ is a (PS)-sequence such that $\Phi(v_j^1) \to c - \Phi(w_1)$. This argument is rather technical, and it is here (in the proof that $\Phi'(v_j^1) \to 0$ to be more precise) that the condition (3.19) plays a role. Moreover, there exists $\alpha > 0$ such that $\Phi(w) \geqslant \alpha$ for all critical points $w \neq 0$.

Indeed, otherwise we find a sequence of critical points $w_n \neq 0$ with $\Phi(w_n) \to 0$. But then, according to Proposition 3.3, $w_n \to 0$ which is impossible because (3.3) and the fact that $\psi'(z) = o(\|z\|)$ as $z \to 0$ imply $w = 0$ is the only critical point in some neighbourhood of 0.

Now we can repeat the same argument for v_j^1 and obtain after passing to a subsequence again that $\tilde{v}_j^1 := T_{a_j^2} v_j^1 \rightharpoonup w_2$ and $\Phi(v_j^2) \to c - \Phi(w_1) - \Phi(w_2)$, where $v_j^2 = \tilde{v}_j^1 - w_2$. Since $\alpha > 0$, after a finite number of steps

$$\Phi\left(v_j^k\right) \to \beta := c - \sum_{m=1}^{k} \Phi(w_m) \leqslant 0.$$

But then $\beta = 0$ and $v_j^k \to 0$. Since (up to a subsequence) $T_{b_j^k} v_j^k = z_j - \sum_{m=1}^{k} T_{b_j^m} w_m$, where $b_j^m = -(a_j^1 + \cdots + a_j^m)$, the conclusion follows. $\qquad\square$

PROOF OF THEOREM 3.9. Assuming that (3.1) has finitely many geometrically distinct homoclinics, we shall show that the hypotheses of Theorem 1.21 are satisfied thereby obtaining a contradiction.

Since $T_g \in O(2N) \cap Sp(2N)$ for $g \in \mathcal{G}$, the quadratic form (3.2) is $\mathcal{G}$-invariant, hence Φ is $\mathcal{G}$-invariant. It has been shown in the proof of Theorem 3.6 that ψ is weakly sequentially lower semicontinuous. This and Proposition 3.1 imply (i). We already know that (ii) holds. In order to verify (iii), we first note that there is an increasing sequence of $\mathcal{G}$-invariant subspaces $F_n \subset E^+$ such that $\dim F_n = n$ if $p = 2$ (in this case all subspaces are invariant) and $\dim F_n = 2n$ if $p > 2$. Since there exists a continuous projection $L^\mu(\mathbb{R}, \mathbb{R}^{2N}) \to F_n$, we obtain

$$\Phi(z) \leqslant \frac{1}{2}\|z^+\|^2 - \frac{1}{2}\|z^-\|^2 - \delta\mu^{-1}\|z\|_\mu^\mu \leqslant \frac{1}{2}\|z^+\|^2 - \delta_0\|z^+\|_\mu^\mu - \frac{1}{2}\|z^-\|^2$$

for some $\delta_0 > 0$ and all $z \in E_n$. The right-hand side above tends to $-\infty$ as $\|z\| \to \infty$ because $\dim F_n < \infty$.

It remains to verify (iv). Choose a unique point in each $\mathbb{Z}$-orbit of homoclinics and denote the set of all such points by $\mathcal{F}$. According to our assumption, $\mathcal{F}$ is finite. Changing the b_j^m:s if necessary we may assume $w_m \in \mathcal{F}$ in Proposition 3.10. For a given positive integer l, let

$$[\mathcal{F}, l] := \left\{ \sum_{m=1}^{k} T_{a_m} w_m : 1 \leqslant k \leqslant l, \ a_m \in \mathbb{Z}, \ w_m \in \mathcal{F} \right\}.$$

If $I \subset (0, \infty)$ is a compact interval and l is large enough, then $[\mathcal{F}, l]$ is a $(PS)_I$-attractor according to Proposition 3.10. Finally, that

$$\inf\left\{ \|z^+ - w^+\| : z, w \in [\mathcal{F}, l], \ z^+ \neq w^+ \right\} > 0 \tag{3.20}$$

is a consequence of the result below. $\qquad\square$

PROPOSITION 3.11. *Let $\mathcal{F}$ be a finite set of points in E. Then (3.20) holds.*

Here it is not assumed that $\mathcal{F}$ is a set whose points have any special property. The proof is straightforward though rather tedious and may be found in [26, Proposition 1.55]. In [26] this result is proved for z and w (and not z^+ and w^+), however, the argument is exactly the same in our case.

Theorem 3.9, for $p = 2$ (even Hamiltonian), is due to Ding and Girardi [30]. They have allowed A to be time-dependent and 0 to be the left endpoint of a gap of the spectrum of $-J(\mathrm{d}/\mathrm{d}t) - A(t)$. A result similar to Theorem 3.9 but allowing much more general (also infinite) groups has been obtained by Arioli and Szulkin [6]. Subsequently the superquadraticity condition in [6] has been weakened by Xu [103] by means of a truncation argument mentioned in the preceding section.

3.4. *Multibump solutions and relation to the Bernoulli shift*

It is known by Melnikov's theory that certain integrable Hamiltonian systems having 0 as a hyperbolic point can be perturbed in such a way that the stable and unstable manifold at 0 intersect transversally. This in turn implies that there exists a compact set, invariant with respect to the Poincaré mapping and conjugate to the Bernoulli shift (these notions will be defined later), see Palmer [77], or [47], [100] for a more comprehensive account of the subject. However, in general it is not an easy task to decide whether the intersection is transversal. In this section we shall see that sometimes under conditions which are weaker than transversality it is still possible to show the existence of an invariant set which is *semiconjugate* to the Bernoulli shift.

Consider the Hamiltonian system (3.1), with H satisfying $(\widetilde{H}_1)$–$(\widetilde{H}_4)$ and suppose that z_0 is a homoclinic solution. Let $\chi \in C^\infty(\mathbb{R}, [0, 1])$ be a function such that $\chi(t) = 1$ for $|t| \leqslant 1/8$ and supp $\chi \subset (-1/4, 1/4)$. If $\chi_j(t) := \chi(t/j)$, one easily verifies that $\{\chi_j z_0\}$ is a (PS)-sequence, $\Phi(\chi_j z_0) \to \Phi(z_0) = c$ and $\mathrm{supp}(\chi_j z_0) \subset (-j/4, j/4)$. Let $w_j(t) := \chi_j(t)z_0(t) + \chi_j(t - j)z_0(t - j)$. Since $\chi_j z_0$ and $\chi_j(\cdot - j)z_0(\cdot - j)$ have disjoint supports, it follows that $\{w_j\}$ is a (PS)-sequence such that $\Phi(z_j) \to 2c$. One can therefore expect that under suitable conditions there is a large j and a homoclinic solution

$$z(t) = \chi_j(t)z_0(t) + \chi_j(t - j)z_0(t - j) + \tilde{v}(t) = z_0(t) + z_0(t - j) + v(t)$$

such that $\|\tilde{v}\|_\infty$ (and hence also $\|v\|_\infty$) is small compared to $\|z_0\|_\infty$. We shall call this z a 2-bump solution. In a similar way one can look for k-bump solutions with $k > 2$.

Suppose now (3.1) has a homoclinic solution z_0 and for some ε reasonably small, say $\varepsilon \leqslant \frac{1}{2}\|z_0\|_\infty$, there exists $M > 0$ such that for any $k \in \mathbb{N}$ and any sequence of integers $a_1 < a_2 < \cdots < a_k$ satisfying $a_j - a_{j-1} \geqslant M$ for all j, there is a homoclinic solution

$$z(t) = \sum_{j=1}^{k} z_0(t - a_j) + v(t), \tag{3.21}$$

where $\|v\|_\infty \leqslant \varepsilon$ (so z is a k-bump solution). We emphasize that M is independent of k here. Existence of k-bump solutions for a class of superquadratic second order Hamiltonian systems has been shown by Coti Zelati and Rabinowitz [26]. However, in [26] M may depend on k. In [83] Séré has shown that under appropriate conditions on H, there is $M = M(\varepsilon)$ (independent of k) such that homoclinic solutions of the form (3.21) exist. More precisely, he has assumed that $H \in C^2(\mathbb{R}^{2N}, \mathbb{R})$ satisfies $(\widetilde{H}_1)$–$(\widetilde{H}_4)$, G is convex in z and

$$\delta_1 |z|^s \leqslant G(z, t) \leqslant \delta_2 |z|^s, \qquad sG(z, t) \leqslant z \cdot G_z(z, t)$$

for all z, t and some $\delta_1, \delta_2 > 0$, $s > 2$. Let $s' = s/(s - 1)$. Since G is convex, one can use Clarke's duality principle in order to construct a dual functional $\Psi \in C^1(L^{s'}(\mathbb{R}, \mathbb{R}^{2N}), \mathbb{R})$ such that there is a one-to-one correspondence between critical points $u \neq 0$ of Ψ and homoclinic solutions z of (3.1) ([25], see also [33]). The functional Ψ is better behaved than Φ; in particular, it is not strongly indefinite and under the conditions specified above it has the mountain pass geometry near 0. This fact has been employed in [25] in order to obtain a homoclinic.

Let c be the mountain pass level for Ψ, or more precisely, let

$$c := \inf_{h \in \Gamma} \max_{\tau \in [0,1]} \Psi\big(h(\tau)\big),$$

where

$$\Gamma := \big\{ h \in C\big([0, 1], L^{s'}(\mathbb{R}, \mathbb{R}^{2N})\big) \colon h(0) = 0, \ \Psi\big(h(1)\big) < 0 \big\}.$$

Since 0 is a strict local minimum of Ψ and Ψ is unbounded below, $c > 0$.

THEOREM 3.12 (Séré [83]). *Suppose that H, G satisfy the hypotheses above and there is $c' > c$ such that the set $\mathcal{K}_{c'} := \{u \in \Psi^{c'} \colon \Psi'(u) = 0\}$ is countable. Then for each $\varepsilon > 0$ there exists $M = M(\varepsilon)$ such that to every choice of integers $a_1 < a_2 < \cdots < a_k$ satisfying $a_j - a_{j-1} \geqslant M$ there corresponds a homoclinic solution z of* (3.1) *given by* (3.21).

The countability of $\mathcal{K}_{c'}$ is a sort of nondegeneracy condition. A similar (but stronger) condition has also been employed in [26]. Consider the Hamiltonian system with H given by (3.17). If z_0 is a homoclinic, u_0 corresponds to z_0 and $c = \Psi(u_0)$, then $\mathcal{K}_{c'}$ is uncountable (see (3.18)). On the other hand, there are no multibump solutions in this case. Therefore in general it is necessary to assume some kind of nondegeneracy. Note also that since autonomous systems are invariant with respect to time-translations by a for any $a \in \mathbb{R}$, the countability condition can never be satisfied in this case. However, as has been shown by Bolotin and Rabinowitz [21], autonomous systems may have multibumps.

The proof of Theorem 3.12, in particular the construction of M independent of k, is lengthy and very technical. Therefore we omit it and refer the reader to [83]. To our knowledge no multibump results are known for first order Hamiltonian systems with nonconvex G (except when a reduction to a second order system like in [26] can be made).

From now on we assume that (3.1) has k-bump solutions of the form (3.21) for any k and M is independent of k. Choose $a \geqslant M$ and let

$$z_j(t) := z_0(t - aj), \quad j \in \mathbb{Z}.$$

We claim that, given any sequence $\{s_j\}$ of 0's and 1's, there exists a solution

$$z(t) = \sum_{j \in \mathbb{Z}} s_j z_j(t) + v(t)$$

of (3.1) such that $\|v\|_\infty \leqslant \varepsilon$. Note that if $s_j = 1$ for infinitely many $j's$, then z has infinitely many bumps and is *not* a homoclinic. By our assumption, for any positive integer m we can find a solution

$$z^m(t) = \sum_{j=-m}^{m} s_j z_j(t) + v^m(t)$$

with $\|v^m\|_\infty \leqslant \varepsilon$. Since 0 is a hyperbolic point and $z_0(t) \to 0$ as $|t| \to \infty$, z_0 decays to 0 exponentially (this follows by exponential dichotomy [77], see also [25] or [31]). Therefore $\|z^m\|_\infty$ is bounded uniformly in m, and by (3.1), the same is true of $\|\dot{z}^m\|_\infty$. Hence z^m, and a posteriori also v^m, are uniformly bounded in $H^1_{\mathrm{loc}}(\mathbb{R}, \mathbb{R}^{2N})$. Since the embedding $H^1_{\mathrm{loc}}(\mathbb{R}, \mathbb{R}^{2N}) \hookrightarrow L^\infty_{\mathrm{loc}}(\mathbb{R}, \mathbb{R}^{2N})$ is compact and $\|v^m\|_\infty \leqslant \varepsilon$, $v^m \to v$ in $L^\infty_{\mathrm{loc}}(\mathbb{R}, \mathbb{R}^{2N})$ for some v with $\|v\|_\infty \leqslant \varepsilon$. It follows that the corresponding function z is a weak solution of (3.1) and $z \neq 0$. Moreover, $z \in H^1_{\mathrm{loc}}(\mathbb{R}, \mathbb{R}^{2N}) \cap C^2(\mathbb{R}, \mathbb{R}^{2N})$ (recall H is of class C^2).

Let now

$$\Sigma_2 := \{0, 1\}^{\mathbb{Z}} = \left\{ s = \{s_j\}_{j \in \mathbb{Z}} : \ s_j \in \{0, 1\} \right\}$$

be the set of doubly infinite sequences of 0's and 1's, endowed with the metric

$$d(s, \tilde{s}) := \sum_{j \in \mathbb{Z}} 2^{-|j|} |s_j - \tilde{s}_j|.$$

The space (Σ_2, d) is easily seen to be compact, totally disconnected and perfect (it is in fact homeomorphic to the Cantor set). The mapping $\sigma \in C(\Sigma_2, \Sigma_2)$ given by

$$\left(\sigma(s) \right)_j = s_{j+1}$$

is called the *Bernoulli shift on two symbols*. It is often considered as a prototype of a chaotic map. In particular, it has a countable infinity of periodic orbits, an uncountable infinity of nonperiodic orbits, a dense orbit, and it exhibits sensitive dependence on initial conditions. The details may be found, e.g., in Wiggins [100, Chapter 2].

Let

$$Z := \left\{ z \in L^\infty(\mathbb{R}, \mathbb{R}^{2N}) : \ z(t) = \sum_{j \in \mathbb{Z}} s_j z_j(t) + v(t), \ s_j \in \{0, 1\}, \ \|v\|_\infty \leqslant \varepsilon \right\}$$

and $I_j := [a(j - \frac{1}{2}), a(j + \frac{1}{2})]$. In Z we introduce a metric d by setting

$$d(z, 0) = \sum_{j \in \mathbb{Z}} 2^{-|j|} \big(s_j \|z_j\|_\infty + \|v\|_{L^\infty(I_j, \mathbb{R}^{2N})} \big). \tag{3.22}$$

Since z_0 decays exponentially, the topology induced by d coincides with the L^∞_{loc}-topology on Z as one readily verifies. Using this, the compactness of Σ_2 and of the embedding $H^1_{\mathrm{loc}}(\mathbb{R}, \mathbb{R}^{2N}) \hookrightarrow L^\infty_{\mathrm{loc}}(\mathbb{R}, \mathbb{R}^{2N})$, it follows that the set

$$X := \big\{ z \in Z : z \text{ is a solution of (3.1)} \big\}$$

is compact.

As before, let $T_a : X \to X$ be the mapping given by $(T_a z)(t) = z(t + a)$ and let $f_a : \mathbb{R}^{2N} \to \mathbb{R}^{2N}$ be the Poincaré (or time-a) mapping defined by $f_a(z_0) = z(a)$ where $z = z(t)$ is the unique solution of (3.1) satisfying the initial condition $z(0) = z_0$. Finally, let $Ev : X \to \mathbb{R}^{2N}$ be the evaluation mapping, $Ev(z) := z(0)$, and let $I := Ev(X)$. Since Ev is continuous and injective, it is a homeomorphism between X and I, and it is easy to verify that the diagram

$$\begin{array}{ccc} I & \xrightarrow{\;f_a\;} & I \\[1mm] {\scriptstyle Ev} \big\uparrow & & \big\uparrow {\scriptstyle Ev} \\[1mm] X & \xrightarrow[\;T_a\;]{} & X \end{array} \tag{3.23}$$

is commutative. Note in particular that the set I is invariant with respect to the Poincaré mapping f_a.

THEOREM 3.13. *There exists a continuous surjective mapping $g : I \to \Sigma_2$ such that the diagram*

$$\begin{array}{ccc} I & \xrightarrow{\;f_a\;} & I \\[1mm] {\scriptstyle g} \big\downarrow & & \big\downarrow {\scriptstyle g} \\[1mm] \Sigma_2 & \xrightarrow[\;\sigma\;]{} & \Sigma_2 \end{array}$$

is commutative.

If a continuous surjective mapping g as above exists, we shall say that $f_a : I \to I$ is *semiconjugate* to $\sigma : \Sigma_2 \to \Sigma_2$, and f_a will be called *conjugate* to σ if g is a homeomorphism.

PROOF. For $z = \sum_{j \in \mathbb{Z}} s_j z_j + v \in X$ we define $\varphi(z) = s = \{s_j\}_{j \in \mathbb{Z}}$. Then φ is a continuous mapping from X onto Σ_2 (cf. (3.22)) and the diagram

$$
\begin{array}{ccc}
X & \xrightarrow{\ T_a\ } & X \\
\varphi \downarrow & & \downarrow \varphi \\
\Sigma_2 & \xrightarrow{\ \sigma\ } & \Sigma_2
\end{array}
\qquad (3.24)
$$

commutes. Now the conclusion follows from (3.23) and (3.24) upon setting $g := \varphi \circ (Ev)^{-1}$. $\qquad\qquad\square$

REMARK 3.14. It follows from the definition of topological entropy $h(.)$ (see, e.g., [47, Definition 5.8.3] or [88, Definition 5.8.4]) and the uniform continuity of g that $h(f_a) \geqslant h(\sigma)$ (cf. [88, Exercise 5.8.1.B]). It is well known that $h(\sigma) > 0$. Hence $f_a|_I$, and therefore also the time-1-mapping f_1, has positive entropy. More precisely, $h(\sigma) = \log 2$ and $h(f_1) \geqslant (\log 2)/a$ according to [88, Example 5.8.1 and Theorem 5.8.4]. The same conclusion about the entropy may also be found in [83], where a different (in a sense, dual) argument has been used.

The approach presented in this section is taken from a work in progress by W. Zou and the second author. They study a certain second order system and hope to show that, in addition to the result of Theorem 3.13, to each m-periodic sequence $s \in \Sigma_2$ there corresponds a $z \in X$ which has period ma.

References

[1] A. Abbondandolo, *Morse Theory for Hamiltonian Systems*, Research Notes in Mathematics, Vol. 425, Chapman & Hall/CRC Press, Boca Raton, FL (2001).

[2] R.A. Adams, *Sobolev Spaces*, Academic Press, New York (1975).

[3] H. Amann and E. Zehnder, *Nontrivial solutions for a class of nonresonance problems and applications to nonlinear differential equations*, Ann. Scuola Norm. Sup. Pisa (4) **7** (1980), 539–603.

[4] H. Amann and E. Zehnder, *Periodic solutions of asymptotically linear Hamiltonian systems*, Manuscripta Math. **32** (1980), 149–189.

[5] A. Ambrosetti and V. Coti Zelati, *Periodic Solutions of Singular Lagrangian Systems*, Birkhäuser (1993).

[6] G. Arioli and A. Szulkin, *Homoclinic solutions of Hamiltonian systems with symmetry*, J. Differential Equations **158** (1999), 291–313.

[7] A. Bahri and H. Berestycki, *Existence of forced oscillations for some nonlinear differential equations*, Comm. Pure Appl. Math. **37** (1984), 403–442.

[8] A. Bahri and H. Berestycki, *Forced vibrations of superquadratic Hamiltonian systems*, Acta Math. **152** (1984), 143–197.

[9] T. Bartsch, *A simple proof of the degree formula for $\mathbb{Z}/p$-equivariant maps*, Math. Z. **212** (1993), 285–292.

[10] T. Bartsch, *Topological Methods for Variational Problems with Symmetries*, Springer-Verlag, Berlin (1993).

[11] T. Bartsch, *A generalization of the Weinstein–Moser theorems on periodic orbits of a Hamiltonian system near an equilibrium*, Ann. Inst. H. Poincaré Anal. Non Linéaire **14** (1997), 691–718.

[12] T. Bartsch and Y.H. Ding, *Homoclinic solutions of an infinite-dimensional Hamiltonian system*, Math. Z. **240** (2002), 289–310.

[13] T. Bartsch and Y.H. Ding, *Deformation theorems on non-metrizable vector spaces and applications to critical point theory*, Preprint.

[14] T. Bartsch and Z.-Q. Wang, *Periodic solutions of even Hamiltonian systems on the torus*, Math. Z. **224** (1997), 65–76.

[15] T. Bartsch and Z.-Q. Wang, *Periodic solutions of spatially periodic, even Hamiltonian systems*, J. Differential Equations (1997), 103–128.

[16] T. Bartsch and M. Willem, *Periodic solutions of non-autonomous Hamiltonian systems with symmetries*, J. Reine Angew. Math. **451** (1994), 149–159.

[17] V. Benci, *A geometrical index for the group S^1 and some applications to the study of periodic solutions of ordinary differential equations*, Comm. Pure Appl. Math. **34** (1981), 393–432.

[18] V. Benci, *On critical point theory of indefinite functionals in the presence of symmetries*, Trans. Amer. Math. Soc. **274** (1982), 533–572.

[19] V. Benci and P.H. Rabinowitz, *Critical point theorems for indefinite functionals*, Invent. Math. **52** (1979), 241–273.

[20] H. Berestycki, J.M. Lasry, G. Mancini and B. Ruf, *Existence of multiple periodic orbits on star-shaped Hamiltonian surfaces*, Comm. Pure Appl. Math. **38** (1985), 253–289.

[21] S.V. Bolotin and P.H. Rabinowitz, *A variational construction of chaotic trajectories for a Hamiltonian system on a torus*, Boll. Unione Mat. Ital. Ser. B (8) **1** (1998), 541–570.

[22] G. Cerami, *Un criterio di esistenza per i punti critici su varietà illimitate*, Istit. Lombardo Accad. Sci. Lett. Rend. A **112** (1978), 332–336.

[23] K.C. Chang, *Infinite Dimensional Morse Theory and Multiple Solution Problem*, Birkhäuser, Boston (1993).

[24] C. Conley and E. Zehnder, *The Birkhoff–Lewis fixed point theorem and a conjecture of V. Arnold*, Invent. Math. **73** (1983), 33–49.

[25] V. Coti Zelati, I. Ekeland and E. Séré, *A variational approach to homoclinic orbits in Hamiltonian systems*, Math. Ann. **288** (1990), 133–160.

[26] V. Coti Zelati and P.H. Rabinowitz, *Homoclinic orbits for second order Hamiltonian systems possessing superquadratic potentials*, J. Amer. Math. Soc. **4** (1991), 693–727.

[27] N. Dancer and S. Rybicki, *A note on periodic solutions of autonomous Hamiltonian systems emanating from degenerate stationary solutions*, Differential Integral Equations **12** (1999), 147–160.

[28] M. Degiovanni and L. Olian Fannio, *Multiple periodic solutions of asymptotically linear Hamiltonian systems*, Nonlinear Anal. **26** (1996), 1437–1446.

[29] Y.H. Ding, *Infinitely many homoclinic orbits for a class of Hamiltonian systems with symmetry*, Chinese Ann. Math. Ser. B **19** (1998), 167–178.

[30] Y.H. Ding and M. Girardi, *Infinitely many homoclinic orbits of a Hamiltonian system with symmetry*, Nonlinear Anal. **38** (1999), 391–415.

[31] Y.H. Ding and M. Willem, *Homoclinic orbits of a Hamiltonian system*, Z. Angew. Math. Phys. **50** (1999), 759–778.

[32] I. Ekeland, *Une théorie de Morse pour les systèmes hamiltoniens convexes*, Ann. Inst. H. Poincaré Anal. Non Linéaire **1** (1984), 19–78.

[33] I. Ekeland, *Convexity Methods in Hamiltonian Mechanics*, Springer-Verlag, Berlin (1990).

[34] I. Ekeland and L. Lassoued, *Multiplicité des trajectoires fermées de systèmes hamiltoniens convexes*, Ann. Inst. H. Poincaré Anal. Non Linéaire **4** (1987), 307–335.

[35] I. Ekeland and J.M. Lasry, *On the number of periodic trajectories for a Hamiltonian flow on a convex energy surface*, Ann. Math. **112** (1980), 283–319.

[36] E.R. Fadell and P.H. Rabinowitz, *Generalized cohomological index theories for Lie group actions with an application to bifurcation questions for Hamiltonian systems*, Invent. Math. **45** (1978), 139–174.

[37] P. Felmer, *Periodic solutions of spatially periodic Hamiltonian systems*, J. Differential Equations **98** (1992), 143–168.

[38] P. Felmer, *Periodic solutions of "superquadratic" Hamiltonian systems*, J. Differential Equations **102** (1993), 188–207.

[39] P. Felmer and Z.-Q. Wang, *Multiplicity for symmetric indefinite functionals and application to Hamiltonian and elliptic systems*, Topol. Methods Nonlinear Anal. **12** (1998), 207–226.

[40] A. Fonda and J. Mawhin, *Multiple periodic solutions of conservative systems with periodic nonlinearity*, Differential Equations and Applications, Vols. I and II, Ohio Univ. Press, Athens, OH (1989).

[41] K. Fukaya and K. Ono, *Arnold conjecture and Gromov–Witten invariants*, Topology **38** (1999), 933–1048.

[42] G. Fournier, D. Lupo, M. Ramos and M. Willem, *Limit relative category and critical point theory*, Dynamics Reported (New Series), Vol. 3, C.K.R.T. Jones, U. Kirchgraber and H.O. Walther, eds., Springer-Verlag, Berlin (1994), pp. 1–24.

[43] V.L. Ginzburg, *An embedding $S^{2n-1} \to \mathbb{R}^{2n}$, $2n - 1 \geqslant 7$, whose Hamiltonian flow has no periodic trajectories*, Internat. Math. Res. Notices 1995, 83–98.

[44] V.L. Ginzburg and B.Z. Gürel, *A C^2-smooth counterexample to the Hamiltonian Seifert conjecture in $\mathbb{R}^4$*, Ann. Math. **158** (2003), 953–976.

[45] V.L. Ginzburg and E. Kerman, *Periodic orbits of Hamiltonian flows near symplectic extrema*, Pacific J. Math. **206** (2002), 69–91.

[46] M. Girardi and M. Matzeu, *Solutions of minimal period for a class of nonconvex Hamiltonian systems and applications to the fixed energy problem*, Nonlinear Anal. **10** (1986), 371–382.

[47] J. Guckenheimer and P. Holmes, *Nonlinear Oscillations, Dynamical Systems, and Bifurcations of Vector Fields*, Springer-Verlag, New York (1983).

[48] Y.X. Guo, *Nontrivial periodic solutions for asymptotically linear Hamiltonian systems with resonance*, J. Differential Equations **175** (2001), 71–87.

[49] M. Herman, *Examples of compact hypersurfaces in $\mathbb{R}^{2p}$, $2p \geqslant 6$*, Preprint (1994).

[50] H. Hofer, K. Wysocki and E. Zehnder, *The dynamics on three-dimensional strictly convex energy surfaces*, Ann. Math. **148** (1998), 197–289.

[51] H. Hofer and E. Zehnder, *Periodic solutions on hypersurfaces and a result by C. Viterbo*, Invent. Math. **90** (1987), 1–9.

[52] H. Hofer and E. Zehnder, *Symplectic Invariants and Hamiltonian Dynamics*, Birkhäuser, Basel (1994).

[53] M. Izydorek, *A cohomological Conley index in Hilbert spaces and applications to strongly indefinite problems*, J. Differential Equations **170** (2001), 22–50.

[54] M. Izydorek, *Equivariant Conley index in Hilbert spaces and applications to strongly indefinite problems*, Nonlinear Anal. **51** (2002), 33–66.

[55] L. Jeanjean, *On the existence of bounded Palais–Smale sequences and application to a Landesman–Lazer type problem set on $\mathbf{R}^N$*, Proc. Roy. Soc. Edinburgh Sect. A **129** (1999), 787–809.

[56] M.A. Krasnosel'skii, *Topological Methods in the Theory of Nonlinear Integral Equations*, Pergamon Press, Oxford (1964).

[57] W. Kryszewski and A. Szulkin, *An infinite dimensional Morse theory with applications*, Trans. Amer. Math. Soc. **349** (1987), 3181–3234.

[58] W. Kryszewski and A. Szulkin, *Generalized linking theorem with an application to a semilinear Schrödinger equation*, Adv. Differential Equations **3** (1998), 441–472.

[59] G. Li and A. Szulkin, *An asymptotically periodic Schrödinger equation with indefinite linear part*, Commun. Contemp. Math. **4** (2002), 763–776.

[60] S.J. Li and J.Q. Liu, *Some existence theorems on multiple critical points and their applications*, Kexue Tongbao **17** (1984), 1025–1027 (in Chinese).

[61] S.J. Li and J.Q. Liu, *Morse theory and asymptotic linear Hamiltonian system*, J. Differential Equations **78** (1989), 53–73.

[62] S.J. Li and A. Szulkin, *Periodic solutions for a class of nonautonomous Hamiltonian systems*, J. Differential Equations **112** (1994), 226–238.

[63] Y.Q. Li, *A limit index theory and its applications*, Nonlinear Anal. **25** (1995), 1371–1389.

[64] E.H. Lieb and M. Loss, *Analysis*, Amer. Math. Soc., Providence, RI (1997).

[65] P.L. Lions, *The concentration compactness principle in the calculus of variations. The locally compact case. Part II*, Ann. Inst. H. Poincaré Anal. Non Linéaire **1** (1984), 223–283.

[66] J.Q. Liu, *A generalized saddle point theorem*, J. Differential Equations **82** (1989), 372–385.

[67] C.G. Liu, Y. Long and C. Zhu, *Multiplicity of closed characteristics on symmetric convex hypersurfaces in $\mathbb{R}^{2n}$*, Math. Ann. **323** (2002), 201–215.

[68] Y. Long, *Maslov-type index, degenerate critical points, and asymptotically linear Hamiltonian systems*, Sci. China Ser. A **33** (1990), 1409–1419.

[69] Y. Long, *Index Theory for Symplectic Paths with Applications*, Birkhäuser, Basel (2002).

[70] Y. Long and X. Xu, *Periodic solutions for a class of nonautonomous Hamiltonian systems*, Nonlinear Anal. 41 (2000), 455–463.

[71] Y. Long and C. Zhu, *Closed characteristics on compact convex hypersurfaces in* $\mathbf{R}^{2n}$, Ann. Math. 155 (2002), 317–368.

[72] A.M. Lyapunov, *Problème général de la stabilité du mouvement*, Ann. Fac. Sci. Toulouse Math. (6) **2** (1907), 203–474.

[73] J. Mawhin and M. Willem, *Critical Point Theory and Hamiltonian Systems*, Springer-Verlag, New York (1989).

[74] D. McDuff and D. Salamon, *Introduction to Symplectic Topology*, Oxford University Press, New York (1998).

[75] K.R. Meyer and G.R. Hall, *Introduction to Hamiltonian Dynamical Systems and the N-Body Problem*, Springer-Verlag (1992).

[76] J. Moser, *Periodic orbits near an equilibrium and a theorem by A. Weinstein*, Comm. Pure Appl. Math. **29** (1976), 727–747.

[77] K.J. Palmer, *Exponential dichotomies and transversal homoclinic points*, J. Differential Equations **55** (1984), 225–256.

[78] P.H. Rabinowitz, *Periodic solutions of Hamiltonian systems*, Comm. Pure Appl. Math. **31** (1978), 157–184.

[79] P.H. Rabinowitz, *Periodic solutions of large norm of Hamiltonian systems*, J. Differential Equations **50** (1983), 33–48.

[80] P.H. Rabinowitz, *Minimax Methods in Critical Point Theory with Applications to Differential Equations*, CBMS Reg. Conf. Ser. Math., Vol. 65, Amer. Math. Soc., Providence, RI (1986).

[81] P.H. Rabinowitz, *Variational methods for Hamiltonian systems*, Handbook of Dynamical Systems, Vol. 1A, B. Hasselblatt and A. Katok, eds., Elsevier (2002), 1091–1127.

[82] S. Secchi and C.A. Stuart, *Global bifurcation of homoclinic solutions of Hamiltonian systems*, Discrete Contin. Dynam. Systems 9 (2003), 1493–1518.

[83] E. Séré, *Looking for the Bernoulli shift*, Ann. Inst. H. Poincaré Anal. Non Linéaire 10 (1993), 561–590.

[84] E.H. Spanier, *Algebraic Topology*, Springer-Verlag, New York (1966).

[85] M. Struwe, *Variational Methods*, Springer-Verlag, Berlin (1990).

[86] M. Struwe, *Existence of periodic solutions of Hamiltonian systems on almost every energy surface*, Bol. Soc. Brasil Mat. (N.S.) **20** (1990), 49–58.

[87] C.A. Stuart, *Bifurcation into spectral gaps*, Bull. Belg. Math. Soc., Supplement, 1995.

[88] W. Szlenk, *An Introduction to the Theory of Smooth Dynamical Systems*, Polish Scientific Publishers, Warsaw; Wiley, New York (1984).

[89] A. Szulkin, *Morse theory and existence of periodic solutions of convex Hamiltonian systems*, Bull. Soc. Math. France **116** (1988), 171–197.

[90] A. Szulkin, *A relative category and applications to critical point theory for strongly indefinite functionals*, Nonlinear Anal. **15** (1990), 725–739.

[91] A. Szulkin, *Bifurcation for strongly indefinite functionals and a Liapunov type theorem for Hamiltonian systems*, Differential Integral Equations 7 (1994), 217–234.

[92] A. Szulkin, *Index theories for indefinite functionals and applications*, Topological and Variational Methods for Nonlinear Boundary Value Problems, P. Drábek, ed., Pitman Research Notes in Mathematics Series, Vol. 365, Harlow, Essex (1997), 89–121.

[93] A. Szulkin and W. Zou, *Infinite dimensional cohomology groups and periodic solutions of asymptotically linear Hamiltonian systems*, J. Differential Equations **174** (2001), 369–391.

[94] A. Szulkin and W. Zou, *Homoclinic orbits for asymptotically linear Hamiltonian systems*, J. Funct. Anal. **187** (2001), 25–41.

[95] K. Tanaka, *Homoclinic orbits in a first order superquadratic Hamiltonian system: convergence of subharmonic orbits*, J. Differential Equations **94** (1991), 315–339.

[96] C. Viterbo, *A proof of the Weinstein conjecture in* $\mathbb{R}^{2n}$, Ann. Inst. H. Poincaré Anal. Non Linéaire **4** (1987), 337–356.

[97] C. Viterbo, *Equivariant Morse theory for starshaped Hamiltonian systems*, Trans. Amer. Math. Soc. **311** (1989), 621–655.

[98] A. Weinstein, *Normal modes for nonlinear Hamiltonian systems*, Invent. Math. **20** (1973), 47–57.

[99] A. Weinstein, *Periodic orbits for convex Hamiltonian systems*, Ann. Math. **108** (1978), 507–518.

[100] S. Wiggins, *Global Bifurcations and Chaos*, Springer-Verlag, New York (1988).

[101] M. Willem, *Lectures on Critical Point Theory*, Trabalho de Mat., Vol. 250, Fundação Univ. Brasília, Brasília (1983).

[102] M. Willem, *Minimax Theorems*, Birkhäuser, Boston (1996).

[103] X. Xu, *Homoclinic orbits for first order Hamiltonian systems possessing super-quadratic potentials*, Nonlinear Anal. **51** (2002), 197–214.

CHAPTER 3

Differential Equations on Closed Sets[*],[†]

Ovidiu Cârjă

Faculty of Mathematics, "Al. I. Cuza" University of Iaşi, Romania

Ioan I. Vrabie

Faculty of Mathematics, "Al. I. Cuza" University of Iaşi, Romania and
"O. Mayer" Mathematics Institute of the Romanian Academy, Iaşi, Romania

Contents

1. Introduction 149
   Notations 155
2. Preliminaries 156
   2.1. Brezis–Browder ordering principle 156
   2.2. Projections 158
   2.3. Tangent cones 159
   2.4. The proximal normal cone 161
   2.5. Clarke's tangent cone 163
3. Problems of viability 166
   3.1. Nagumo's viability theorem 166
   3.2. Proof of the necessity 167
   3.3. Existence of approximate solutions 167
   3.4. Convergence of approximate solutions 170
   3.5. Existence of noncontinuable solutions 172
   3.6. Viability of the relative closure 174
   3.7. Comparison and viability 175
   3.8. Viable preordered subsets 176
4. Problems of invariance 178
   4.1. Preliminary facts 178
   4.2. Sufficient conditions for local invariance 179
   4.3. Viability and comparison imply exterior tangency 180

[*]Partially supported by CNCSIS Grant Code 1345/2003.

[†]The writing of this material was facilitated by numerous and very fruitful discussions with Professor Corneliu Ursescu and by a careful reading of the whole manuscript by Professor Mihai Necula. The authors take this opportunity to express their warmest thanks to both of them.

HANDBOOK OF DIFFERENTIAL EQUATIONS
Ordinary Differential Equations, volume 2
Edited by A. Cañada, P. Drábek and A. Fonda
© 2005 Elsevier B.V. All rights reserved

147

4.4. Sufficient conditions for invariance. Revisited 183
4.5. When tangency implies exterior tangency? 187
4.6. Local invariance and monotonicity 189
5. Carathéodory solutions 189
   5.1. A Lebesgue type derivation theorem 189
   5.2. Characterizations of Carathéodory viability 192
   5.3. Sufficient conditions for Carathéodory local invariance 197
   5.4. Sufficient conditions via generalized distance 198
6. Differential inclusions 200
   6.1. Multifunctions 200
   6.2. Viability with respect to a multifunction 201
   6.3. Existence of ε-approximate solutions 203
   6.4. Convergence of the ε-approximate solutions 206
   6.5. Noncontinuable solutions 207
   6.6. Viable preordered subsets 210
   6.7. Local invariance. Sufficient conditions 210
   6.8. Local invariance. Necessary conditions 212
7. Applications 214
   7.1. Invariance with respect to parametrized multifunctions 214
   7.2. Differentiability along trajectories 215
   7.3. Liapunov functions 216
   7.4. Hukuhara's theorem 219
   7.5. Kneser's theorem 221
   7.6. The characteristics method for a first order PDE 222
8. Notes and comments 227
   8.1. The upper semicontinuous case 227
   8.2. The case of Carathéodory mappings 228
   8.3. The lower semicontinuous case 228
   8.4. The semilinear single-valued case 229
   8.5. The semilinear multivalued case 231
   8.6. The nonlinear perturbed single-valued case 233
   8.7. The multivalued perturbed nonlinear case 233
   8.8. Applications 233
Appendix 234
References 235

1. Introduction

One of the most important problems in the Theory of Differential Equations is that of local existence. This consists in checking whether or not, for each initial datum, the Cauchy problem attached to a given differential equation has at least one solution. Unless otherwise specified, in that follows, we will assume that $\mathbb{R}^n$ is endowed with one of its equivalent norms $\| \cdot \|$. Sometimes, we will assume that $\| \cdot \|$ is the Euclidean norm, i.e., that it is defined by means of the usual inner product $\langle \cdot, \cdot \rangle$ as $\| \cdot \| = \sqrt{\langle \cdot, \cdot \rangle}$. In all these cases, either we will explicitly specify that, or we assume that it is implicitly notified by the simple use of the symbol $\langle \cdot, \cdot \rangle$. By a *domain* we understand a nonempty assumed and open subset $\mathbb{D} \subseteq \mathbb{R}^n$, and by $\mathbb{I}$ we denote an open interval in $\mathbb{R}$. Further, we denote by $\mathbb{K}$ a nonempty subset in $\mathbb{R}^n$. Let $f : \mathbb{I} \times \mathbb{K} \to \mathbb{R}^n$ be a given function. We recall that, excepting some other special mentions to be done in due course, by a *solution* to the differential equation

$$u'(t) = f\big(t, u(t)\big) \tag{1.1.1}$$

we mean a function $u : \mathbb{J} \to \mathbb{K}$, with $\mathbb{J}$ a non-degenerate interval included in $\mathbb{I}$, and satisfying the equality (1.1.1) for all $t \in \mathbb{J}$. Clearly, whenever f is continuous, all solutions to (1.1.1) are of class C^1. The first significant local existence result is due to Cauchy and refers to the case in which $\mathbb{K} = \mathbb{D}$ and f is a C^1-function. This was extended by Lipschitz to the class of all functions f satisfying the homonymous condition, and by Peano [80] to general continuous functions. More precisely, Peano [80] proved:

THEOREM 1.1.1 (Peano [80]). *If $\mathbb{D}$ is a domain and $f : \mathbb{I} \times \mathbb{D} \to \mathbb{R}^n$ is a continuous function, then, for every $(\tau, \xi) \in \mathbb{I} \times \mathbb{D}$, there exists $T \in \mathbb{I}$, $T > \tau$, such that the Cauchy problem*

$$\begin{cases} u'(t) = f\big(t, u(t)\big), \\ u(\tau) = \xi \end{cases} \tag{1.1.2}$$

has at least one solution $u : [\tau, T] \to \mathbb{D}$.

On the other hand, there are situations in which, instead of a domain $\mathbb{D}$, one has to consider a set $\mathbb{K}$ which contains non-interior points, say for instance when the state u of a certain system must evolve within a given closed subset $\mathbb{K}$ in $\mathbb{R}^n$. These considerations lead to the concept of *viability* of a set $\mathbb{K}$ with respect to a given function f. More precisely, we say that the subset $\mathbb{K} \subseteq \mathbb{D}$ is *right viable* with respect to the function $f : \mathbb{I} \times \mathbb{D} \to \mathbb{R}^n$ if for each $(\tau, \xi) \in \mathbb{I} \times \mathbb{K}$ there exists $T \in \mathbb{I}$, $T > \tau$, such that the differential equation (1.1.1) has at least one solution $u : [\tau, T] \to \mathbb{K}$ satisfying $u(\tau) = \xi$. In order to be consistent with the usual procedure, i.e. $g(t, u) = (1, f(t, u))$, which reduces the nonautonomous case to the autonomous one, in the latter situation, i.e. when $f : \mathbb{D} \to \mathbb{R}^n$, we say that $\mathbb{K}$ is *right viable* with respect to f if for each $\xi \in \mathbb{K}$ there exists $T > 0$ such that (1.1.1) has at least one solution $u : [0, T] \to \mathbb{K}$ satisfying $u(0) = \xi$. Sometimes, when we will be interested to get solutions defined at the left of τ, we will speak about *left viability*. Moreover, for the sake of simplicity, whenever no confusion may occur, we will use the term *viable* to design right viable. Let us notice that $\mathbb{K}$ is right viable with respect to f if and only if $\mathbb{K}$ is left viable with respect to $-f$. The next simple characterization of right viable subsets with respect to a given—possibly discontinuous—function which does not depend on the t-variable is almost obvious.

PROPOSITION 1.1.1. *Let* $\mathbb{D}$ *be a domain and* $f : \mathbb{D} \to \mathbb{R}^n$. *Then* $\mathbb{K} \subseteq \mathbb{D}$ *is right (left) viable with respect to* f *if and only if it is the union of a certain family of right (left) trajectories of* (1.1.1).

If $\mathbb{K}$ is open, then it is viable with respect to any continuous function f, and this is nothing else that the celebrated Peano's local existence theorem 1.1.1 just mentioned. By contrary, if $\mathbb{K}$ is not open, in general, $\mathbb{K}$ may fail to be viable with respect to any continuous f, as we can see from the simple example below.

EXAMPLE 1.1.1. Let us consider the plane $\mathbb{K} = \{(u_1, u_2, u_3); \ u_3 = 1\}$ and the function $f : \mathbb{K} \to \mathbb{R}^3$, defined by $f(u_1, u_2, u_3) = (u_2 + u_3, -u_1, -u_1)$ for every $(u_1, u_2, u_3) \in \mathbb{K}$. Then, if ξ is the projection of the origin on this plane, i.e. $\xi = (0, 0, 1)$, the problem (1.1.2) has no local solution. Indeed, assuming by contradiction that there exists such a solution $u : [0, T] \to \mathbb{K}$, we have $\langle u'(t), u(t) \rangle = \langle f(u(t)), u(t) \rangle = 0$ and therefore $\|u(t)\| = \|\xi\| = 1$ for every $t \in [0, T]$. Hence $u(t)$ lies on the sphere of center 0 and radius 1 which has only one point in common with $\mathbb{K}$, namely ξ. Then, necessarily $u(t) = \xi$ for every $t \in [0, T]$, which is impossible, because, in this case, one should have $u_1(t) = 0$ and $u_1'(t) = u_2(t) + u_3(t) = 1$ for every $t \in [0, T]$. This contradiction can be eliminated only if (1.1.2) has no local solution.

Thus, whenever $\mathbb{K}$ is not open, one has to compensate the lack of this crucial topological assumption by something else, as less restrictive as possible. In order to understand what extra-condition we have to impose, some comments are needed. More precisely, let us consider for the moment that $\mathbb{D}$ is a domain, $f : \mathbb{D} \to \mathbb{R}^n$ is continuous (and does not depend on t) and $U : \mathbb{D} \to \mathbb{R}$ is a function of class C^1 with $\nabla U(\xi) \neq 0$ on $\mathbb{D}$. Then, if U is a prime integral for the system $u'(t) = f(u(t))$, which amounts of saying that f is parallel to the tangent plane to each surface of constant level $\Sigma_\eta = \{\xi \in \mathbb{D}; \ U(\xi) = U(\eta)\}$ at every point of this surface, i.e.

$$\sum_{i=1}^{n} f(\xi) \frac{\partial U}{\partial x_i}(\xi) = 0,$$

for each $\xi \in \Sigma_\eta$, then the restriction $f_{|\Sigma_\eta}$, of the function f to any surface Σ_η, has the property that, for every $\xi \in \Sigma_\eta$, the equation $u'(t) = f_{|\Sigma_\eta}(u(t))$ has at least one local solution $u : [0, T] \to \Sigma_\eta$ satisfying $u(0) = \xi$. This condition constitutes a first step through a partial answer to the question: *what extra-conditions must satisfy the set* $\mathbb{K} \subseteq \mathbb{R}^n$ (*not necessarily a level set*) *and the continuous function* $f : \mathbb{K} \to \mathbb{R}^n$, *in order that, for every* $\xi \in \mathbb{K}$, *to exist at least one function of class* C^1, $u : [0, T] \to \mathbb{K}$, *such that* $u(0) = \xi$ *and* $u'(t) = f(u(t))$. However, the conditions offered by the result just mentioned in the case

$\mathbb{K} = \Sigma_\eta$ have three weak points. First, they ask f to be defined on the union of all surfaces Σ_η and not on a single one. Second, f must satisfy the mentioned "tangency condition" on each one of the surfaces of the family. Finally, the set $\mathbb{K}$ is in this case of a very specific type, namely it is a surface of constant level for a function $U : \mathbb{D} \to \mathbb{R}$, of class C^1 and satisfying $\nabla U(\xi) \neq 0$ for all $\xi \in \mathbb{D}$.

The first step in order to get an acceptable result in this direction would be expressed as in Theorem 1.1.2 below. See Vrabie [106, Theorem 6.1.2, p. 212].

THEOREM 1.1.2. *Let $\Sigma \subseteq \mathbb{D}$ be a regular surface and $f : \mathbb{D} \to \mathbb{R}^n$ a continuous function. The necessary and sufficient condition in order that, for every $\xi \in \Sigma$ to exist $T > 0$ and a function of class C^1, $u : [0, T] \to \Sigma$, such that $u(0) = \xi$ and $u'(t) = f(u(t))$ for every $t \in [0, T]$, is that for every $\eta \in \Sigma$, $f(\eta)$ be parallel to the tangent plane to Σ at η.*

As far as we know, the first general necessary and sufficient condition for viability has been discovered by Nagumo [70] in 1942 in the case when $\mathbb{K}$ is a closed, or merely locally closed subset in $\mathbb{R}^n$. We recall that $\mathbb{K}$ is *locally closed* if for each $\xi \in \mathbb{K}$ there exists $\rho > 0$ such that $B(\xi, \rho) \cap \mathbb{K}$ is closed. We notice that Nagumo used the term "*rechts zulässig*", i.e. *right admissible* to design viable, and proved the following fundamental characterization for viability:

THEOREM 1.1.3 (Nagumo [70]). *Let $\mathbb{K} \subseteq \mathbb{D}$ be locally closed and let $f : \mathbb{I} \times \mathbb{D} \to \mathbb{R}^n$ be continuous. A necessary and sufficient condition in order that $\mathbb{K}$ be viable with respect to f is that, for each $(t, \xi) \in \mathbb{I} \times \mathbb{K}$, $f(t, \xi)$ be tangent to $\mathbb{K}$ at ξ in the sense of Bouligand–Severi, i.e.*

$$\liminf_{h \downarrow 0} \frac{1}{h} \operatorname{dist}\big(\xi + hf(t, \xi); \mathbb{K}\big) = 0. \tag{1.1.3}$$

Here and thereafter, $\operatorname{dist}(\eta; \mathbb{K})$ denotes the distance from $\eta \in \mathbb{R}^n$ to the subset $\mathbb{K}$ in $\mathbb{R}^n$, i.e. $\operatorname{dist}(\eta; \mathbb{K}) = \inf\{\|\eta - \mu\|; \ \mu \in \mathbb{K}\}$.

Since (1.1.3) is obviously satisfied at each interior point of $\mathbb{K}$, Peano's theorem 1.1.1 is a direct consequence of Nagumo's theorem 1.1.3. Clearly, (1.1.3) is invariant with respect to equivalent norms on $\mathbb{R}^n$. Moreover, this essentially metric condition can be described, and therefore defined, only by means of linear topological concepts. See, for instance, Ursescu [97].

We notice that Nagumo's result (or variants of it) has been rediscovered independently, in the late sixties and early seventies, by Yorke [108,109], Crandall [38] and Hartman [55] among others. More precisely, Yorke [108] uses viability (*weak positive invariance* in his terminology) in order to prove necessary and sufficient conditions for stability by means of Liapunov functions, as well as to give very simple and elegant proofs for both Hukuhara and Kneser's theorems referring to the solution funnel. We notice that Yorke [108] analyzed the case of subsets $\mathbb{K}$ which are *closed relative to* $\mathbb{D}$, i.e., for which there exists a closed set $\mathbb{C} \subseteq \mathbb{R}^n$ such that $\mathbb{K} = \mathbb{C} \cap \mathbb{D}$. Nevertheless, as Remark 1.1.1 below shows, all his results hold true for locally closed sets as well.

REMARK 1.1.1. If $\mathbb{K} \subseteq \mathbb{D}$ is closed relative to $\mathbb{D}$, then it is locally closed. Conversely, if $\mathbb{K} \subseteq \mathbb{D}$ is locally closed, then there exists an open neighborhood $\mathbb{V} \subseteq \mathbb{D}$ of $\mathbb{K}$ such that $\mathbb{K}$ is closed relative to $\mathbb{V}$.

It should be emphasized that, from another point of view, Yorke [108] considered the strictly more general case in which, instead of a cylindrical set $\mathbb{I} \times \mathbb{D}$, f is defined on a non-cylindrical one $\mathcal{D} \subseteq \mathbb{R} \times \mathbb{R}^n$, situation requiring a more delicate analysis. Crandall [38] considers the case when $\mathbb{D}$ is arbitrary, $\mathbb{K} \subseteq \mathbb{D}$ is locally closed and $f : \mathbb{D} \to \mathbb{R}^n$ is continuous. He shows that a sufficient condition for $\mathbb{K}$ to be viable (*forward invariant* in his terminology) with respect to f is (1.1.3). Hartman [55] proves essentially the same result for $\mathbb{D}$ open, $\mathbb{K}$ closed relative to $\mathbb{D}$ and $f : \mathbb{D} \to \mathbb{R}^n$ continuous, and shows in addition that (1.1.3) is necessary for the viability of $\mathbb{K}$ with respect to f.

An extension of Nagumo's viability theorem to Carathéodory functions f was proved by Ursescu [98] by using Scorza Dragoni's theorem.

At this point, we can easily see that viability is independent of the values of f on $\mathbb{D} \setminus \mathbb{K}$, and therefore, in the study of such kind of problems there is no need for f to be defined "outside" $\mathbb{K}$. So, very often in that follows, we will speak about the viability of a subset $\mathbb{K} \subseteq \mathbb{R}^n$ with respect to a function f defined either on $\mathbb{I} \times \mathbb{D}$, or merely on $\mathbb{I} \times \mathbb{K}$. This is no longer true if we consider the problem of *local invariance* to be defined below. Namely, let $f : \mathbb{I} \times \mathbb{D} \to \mathbb{R}^n$, where $\mathbb{D}$ is a domain and let $\mathbb{K} \subseteq \mathbb{D}$. The subset $\mathbb{K}$ is *locally right invariant* with respect to f if for each $(\tau, \xi) \in \mathbb{I} \times \mathbb{K}$ and each solution $u : [\tau, c] \to \mathbb{D}$, $c \in \mathbb{I}$, $c > \tau$, of (1.1.1), satisfying the initial condition $u(\tau) = \xi$, there exists $T \in (\tau, c]$ such that we have $u(t) \in \mathbb{K}$ for each $t \in [\tau, T]$. It is *right invariant* if it satisfies the above condition of local invariance with $T = c$.

As in the case of right viability, whenever we will consider solutions defined at the left of τ we will speak about *local left invariance*. Again, for the sake of simplicity, if no confusion can occur, we will use the term *local invariant* to design local right invariant. We notice that if $\mathbb{K}$ is viable with respect to f, and (1.1.1) has the uniqueness property on $\mathbb{I} \times \mathbb{D}$, then $\mathbb{K}$ is locally invariant with respect to f. Furthermore, if $\mathbb{K}$ is locally invariant with respect to f, and the Cauchy problem (1.1.2) has the local existence property, which happens, for instance, when f is continuous, then $\mathbb{K}$ is viable with respect to f. However, in general, local invariance does not imply viability simply because the local invariance of a given set $\mathbb{K}$ could be a consequence of the lack of local existence for (1.1.2) for some $\xi \in \mathbb{K}$. The following example is instructive in this respect.

EXAMPLE 1.1.2. Let $f : \mathbb{R} \to \mathbb{R}$ be defined by

$$f(\xi) = \begin{cases} -1 & \text{if } \xi \geqslant 0, \\ 1 & \text{if } \xi < 0. \end{cases}$$

First we notice that the Cauchy problem $u'(t) = f(u(t))$, $u(0) = 0$ has no right solution. See Vrabie [106, Example 2.2.1, p. 57]. Then $\mathbb{K} = \{0\}$ is locally invariant, but not viable, with respect to f.

A completion of Proposition 1.1.1 is:

PROPOSITION 1.1.2. *Let $\mathbb{D}$ be a domain and let $f : \mathbb{D} \to \mathbb{R}^n$. If $\mathbb{K}$ is the union of all right (left) trajectories of (1.1.1) in $\mathbb{D}$, issuing from a given subset $\mathbb{C} \subseteq \mathbb{D}$, then $\mathbb{K}$ is locally right (left) invariant with respect to f. In particular, the subset $\mathbb{K} \subseteq \mathbb{D}$ is locally right (left) invariant with respect to the continuous function f if and only if it is the union of all right (left) trajectories of (1.1.1) issuing from $\mathbb{K}$.*

We also notice the following simple characterization of local right (left) invariance of arbitrary subsets with respect to a possibly discontinuous function.

PROPOSITION 1.1.3 (Yorke [108]). *Let $\mathbb{D}$ be a domain and $f : \mathbb{I} \times \mathbb{D} \to \mathbb{R}^n$. The subset $\mathbb{K} \subseteq \mathbb{D}$ is locally right invariant with respect to f if and only if $\mathbb{D} \setminus \mathbb{K}$ is locally left invariant with respect to f.*

It should be noticed that Proposition 1.1.3 cannot be extended to handle right viability, even if f is continuous, as the simple example below shows.

EXAMPLE 1.1.3. For $n = 1$, $\mathbb{K} = (-1, 1)$ is viable with respect to the function $f : \mathbb{R} \to \mathbb{R}$, $f \equiv 1$, but, nevertheless, $\mathbb{R} \setminus \mathbb{K}$ is neither right nor left viable with respect to f.

Sufficient conditions for local invariance were obtained by using some usual uniqueness assumptions ensuring that once a solution to (1.1.1) lies in $\mathbb{K}$, there is no other one issuing from the same initial point and which leaves $\mathbb{K}$ "immediately". Typical examples of this kind were obtained by Brezis [18], Bony [13], Redheffer [85] and Martin [66]. Brezis [18] analyzes the case when $\mathbb{D}$ is open, $\mathbb{K} \subseteq \mathbb{D}$ is relatively closed and f is locally Lipschitz, and proves that (1.1.3) with "lim" instead of "lim inf" is necessary and sufficient for $\mathbb{K}$ to be local invariant, *"flow invariant"* in his terminology, with respect to f. By means of a general concept of normal to a given set, and using a tangency condition expressed in the terms of this concept, Bony [13] get sufficient conditions for invariance. More precisely, let $\mathbb{K}$ be a given subset in $\mathbb{R}^n$. Let $\xi \in \mathbb{K}$ be such that there exists a sphere $B(\eta, \rho)$ containing ξ on its boundary but whose interior has empty intersection with $\mathbb{K}$. Then, the vector $v = \eta - \xi$ is *a metric normal to $\mathbb{K}$ at ξ*. A similar concept has been introduced and studied subsequently by Mordukhovich [69]. Coming back to local invariance we have:

THEOREM 1.1.4 (Bony [13]). *If $\mathbb{D}$ is a domain, $\mathbb{K}$ is relatively closed in $\mathbb{D}$, $f : \mathbb{D} \to \mathbb{R}^n$ is Lipschitz, and*

$$\langle v, f(\xi) \rangle \leqslant 0 \tag{1.1.4}$$

for each v which is metric normal to $\mathbb{K}$ at ξ, then $\mathbb{K}$ is locally invariant with respect to f.

We notice that unlike (1.1.3) which is invariant with respect to equivalent Banach norms, (1.1.4) is not invariant with respect to equivalent Euclidean norms. Redheffer [85] extends the main results in both Brezis [18] and Bony [13], by proving:

THEOREM 1.1.5 (Redheffer [85]). *If $\mathbb{D}$ is a domain, $\mathbb{K}$ is relatively closed in $\mathbb{D}$, $f : \mathbb{D} \to \mathbb{R}^n$ is continuous and there exists a uniqueness function[1] $\omega : \mathbb{R}_+ \to \mathbb{R}$ such that*

$$\langle \xi - \eta, f(\xi) - f(\eta) \rangle \leqslant \|\xi - \eta\| \omega(\|\xi - \eta\|), \tag{1.1.5}$$

for all $\xi, \eta \in \mathbb{D}$ and (1.1.4) is satisfied, then $\mathbb{K}$ is locally invariant with respect to f.

THEOREM 1.1.6 (Redheffer [85]). *If $\mathbb{D}$ is a domain, $\mathbb{K}$ is relatively closed in $\mathbb{D}$, $f : \mathbb{D} \to \mathbb{R}^n$ is continuous and there exists a uniqueness function $\omega : \mathbb{R}_+ \to \mathbb{R}$ such that (1.1.5) is satisfied for all $\xi, \eta \in \mathbb{D}$ and (1.1.3) is satisfied at all points $\xi \in \mathbb{K}$ at which there exists at least one metric normal to $\mathbb{K}$, then $\mathbb{K}$ is locally invariant with respect to f.*

It is interesting to notice that, in all Theorems 1.1.4, 1.1.5 and 1.1.6, it may happened that $\mathbb{K}$ fails to have metric normal vectors at many of its points, at which, in spite of the fact that it is by no means evident that some solution could not escape from $\mathbb{K}$ through such a point, no condition is imposed. The explanation of this apparently strange situation consists in that, every point at which $\mathbb{K}$ fails to have metric normal vectors is an interior-like point of $\mathbb{K}$, simply because the Bony's tangent cone to $\mathbb{K}$ at such a point is all of $\mathbb{R}^n$. Finally, Martin [66] analyzes the special case when f is continuous and dissipative. For other invariance results see [3,18,25,27,38,65,66,108]. Necessary conditions, which are expressed in the terms of a tangency condition of the type (1.1.3), require also some uniqueness hypotheses. In a slightly different spirit, more closely related to dynamical systems than to differential equations, the local invariance problem was studied by Ursescu [99,100]. The main idea in [100] was to consider from the very beginning that a given abstract evolution operator which stands for the set of "all solutions" satisfies a certain tangency condition coupled with a uniqueness hypothesis. It should be mentioned that, in this general context, there is no need of a "right-hand side" f of the associated differential equation—if any.

Unlike the above mentioned approaches, Cârjă et al. [24] consider the classical differential equation (1.1.1) and look for general sufficient and even necessary conditions for local invariance expressed only in terms of f, $\mathbb{K}$ and $\mathbb{D}$, but not in the terms of the panel of solutions to (1.1.1). The conditions there obtained, although by means of a comparison function—see definition below—allow (1.1.1) to have multiple solutions in $\mathbb{K}$.

We recall that a function $\omega : \mathbb{I} \times [0, a) \to \mathbb{R}$ is a *comparison function* if $\omega(t, 0) = 0$ for each $t \in \mathbb{I}$, and for each $[\tau, T) \subseteq \mathbb{I}$, the only continuous solution $x : [\tau, T) \to [0, a)$ to the differential inequality $[D_+ x](t) \leqslant \omega(t, x(t))$ for $t \in [\tau, T)$, satisfying $x(\tau) = 0$, is the null function.

Cârjă et al. [24] show that, if there exists an open neighborhood $\mathbb{V} \subseteq \mathbb{D}$ of $\mathbb{K}$ such that f satisfies the surprisingly simple "exterior tangency" condition

$$\liminf_{h \downarrow 0} \frac{1}{h} \left[\mathrm{dist}\big(\xi + hf(t, \xi); \mathbb{K}\big) - \mathrm{dist}(\xi; \mathbb{K}) \right] \leqslant \omega\big(t, \mathrm{dist}(\xi; \mathbb{K})\big) \tag{1.1.6}$$

---

[1]We recall that $\omega : \mathbb{R}_+ \to \mathbb{R}$ is a *uniqueness function* if the only continuous function $\phi : [0, \delta) \to \mathbb{R}_+$ satisfying $[D^- \phi](t) \leqslant \omega(\phi(t))$, $[D^+ \phi](t) \leqslant \omega(\phi(t))$ for each $t \in (0, \delta)$ and $\phi(0) = 0$ is the null function. Here D^- and D^+ are the upper left and right Dini derivatives.

for each $(t, \xi) \in \mathbb{I} \times \mathbb{V}$, where ω is a certain comparison function, then $\mathbb{K}$ is locally invariant with respect to f. This condition reduces to the classical Nagumo's tangency condition when applied to $\xi \in \mathbb{K}$, and this simply because, at each such point $\xi \in \mathbb{K}$, $\operatorname{dist}(\xi; \mathbb{K}) = 0$. So, we can easily see that, whenever $\mathbb{K}$ is open, (1.1.6) is automatically satisfied for the choice $\mathbb{V} = \mathbb{K}$. More than this, they prove that, in many situations, the condition above is even necessary for local invariance. The philosophy of this result rests on the simple observation that the local invariance is equivalent to the "$(\mathbb{D}, \mathbb{K})$-separating uniqueness" property defined below, while (1.1.6) implies $(\mathbb{D}, \mathbb{K})$-separating uniqueness and, even viability, if f is continuous. More precisely, we say that (1.1.1) has the $(\mathbb{D}, \mathbb{K})$-*separating uniqueness property* if, for each $(\tau, \xi) \in \mathbb{I} \times \mathbb{D}$ and every solutions $u, v : [\tau, T] \to \mathbb{D}$ of (1.1.1), satisfying $u(\tau) = v(\tau) = \xi$, there exists $c \in (\tau, T]$ such that both $u([\tau, c))$ and $v([\tau, c))$ are included either in $\mathbb{D} \setminus \mathbb{K}$, or in $\mathbb{K}$. In fact, if f is continuous, the Nagumo's tangency condition (1.1.3) combined with $(\mathbb{D}, \mathbb{K})$-separating uniqueness is nothing else than a simple rephrasing of the local invariance property. Indeed, *if $\mathbb{K} \subseteq \mathbb{D} \subseteq \mathbb{R}^n$, with $\mathbb{K}$ closed and $\mathbb{D}$ open, and $f : \mathbb{I} \times \mathbb{D} \to \mathbb{R}^n$ is continuous, then $\mathbb{K}$ is locally invariant with respect to f if and only if (1.1.3) is satisfied and (1.1.1) has the $(\mathbb{D}, \mathbb{K})$-separating uniqueness property.*

Notations

$B(\xi, \rho)$	— the closed ball in $\mathbb{R}^n$ centered at ξ and of radius $\rho > 0$
$\mathcal{B}_{\mathbb{K}}(\xi)$	— the Bony's tangent cone to $\mathbb{K}$ at ξ
$\mathcal{C}_{\mathbb{K}}(\xi)$	— $\{\eta \in \mathbb{R}^n;\ \lim_{h \downarrow 0, \mu \to \xi} \frac{1}{h} \operatorname{dist}(\mu + h\eta; \mathbb{K}) = 0\}$, i.e. the Clarke's tangent cone to $\mathbb{K}$ at ξ
$\mathbb{D}$	— a domain, i.e. a nonempty and open subset in $\mathbb{R}^n$
$\frac{\mathrm{d}^+}{\mathrm{d}s}$	— the right derivative
$\operatorname{dist}(\xi; \mathbb{K})$	— $\inf\{\|\xi - \eta\|;\ \eta \in \mathbb{K}\}$, i.e. the distance between $\xi \in \mathbb{R}^n$ and $\mathbb{K} \subseteq \mathbb{R}^n$
$\mathcal{F}_{\mathbb{K}}(\xi)$	— $\{\eta \in \mathbb{R}^n;\ \lim_{h \downarrow 0} \frac{1}{h} \operatorname{dist}(\xi + h\eta; \mathbb{K}) = 0\}$, i.e. the Federer's tangent cone to $\mathbb{K}$ at ξ
$F_{\tau, \xi}$	— $\{(s, u(s));\ s \geqslant \tau,\ u \in \mathcal{S}(\tau, \xi)\}$
$F_{\tau, \xi}(t)$	— $\{u(t);\ u \in \mathcal{S}(\tau, \xi)\}$
$\mathbb{I}$	— a nonempty and open interval in $\mathbb{R}$
$\mathbb{J}$	— a nonempty and non-degenerate, i.e. with nonempty interior interval in $\mathbb{R}$
$\mathbb{K}$	— a nonempty subset in $\mathbb{R}^n$
$L^p(\mathbb{S}; \mathbb{R}^n)$	— the space of all equivalence classes, with respect to the λ almost everywhere equality, of measurable functions from $\mathbb{S}$ to $\mathbb{R}^n$ whose norms are p-Lebesgue integrable over $\mathbb{S}$
$L^1_{\mathrm{loc}}(\mathbb{I})$	— the space of all equivalence classes, with respect to the λ almost everywhere equality, of measurable functions from $\mathbb{I}$ to $\mathbb{R}$ whose norms are Lebesgue integrable over each compact subset in $\mathbb{I}$
l.s.c.	— lower semicontinuous
$\mathbb{N}$	— the set of natural numbers, i.e. $0, 1, \ldots n, \ldots$
$\mathbb{N}^*$	— the set of strictly positive natural numbers, i.e. $1, 2, \ldots n, \ldots$
$\mathcal{N}_{\mathbb{K}}(\xi)$	— the cone of normals in the sense of Bony to $\mathbb{K}$ at ξ

$(\mathcal{N}_{\mathbb{K}}(\xi))^*$	— the conjugate of $\mathcal{N}_{\mathbb{K}}(\xi)$, i.e. the set of all $\eta \in \mathbb{R}^n$ satisfying $\langle \eta, \nu \rangle \leqslant 0$ for each $\nu \in \mathcal{N}_{\mathbb{K}}(\xi)$
ν	— a vector which is normal in the sense of Bony to $\mathbb{K}$ at ξ
$\mathbb{R}^n$	— the linear space over $\mathbb{R}$ of all n-tuples of real numbers
$\mathbb{R}$	— the set of real numbers
$\overline{\mathbb{R}}$	— $\mathbb{R} \cup \{-\infty, +\infty\}$, i.e. the extended set of real numbers
$\mathcal{S}(\tau, \xi)$	— the set of all noncontinuable solutions u of $u'(t) = f(t, u(t))$ satisfying $u(\tau) = \xi$
$\mathcal{T}_{\mathbb{K}}(\xi)$	— $\{\eta \in \mathbb{R}^n;\ \liminf_{h \downarrow 0} \frac{1}{h} \operatorname{dist}(\xi + h\eta; \mathbb{K}) = 0\}$, i.e. the Bouligand–Severi's tangent cone to $\mathbb{K}$ at ξ
u.s.c.	— upper semicontinuous
$[x, y]_+$	— the right directional derivative of the norm calculated at x in the direction y

2. Preliminaries

2.1. *Brezis–Browder ordering principle*

The goal of this subsection is to prove a general and very simple principle concerning preorder relations which unifies a number of various results in nonlinear functional analysis, principle due to Brezis and Browder [19]. We notice that this is an ordering principle similar to Zorn's lemma, but based on the *axiom of dependent choice* which, as shown by Feferman [42], turns out to be strictly weaker than *the axiom of choice*. For easy reference we recall:

THE AXIOM OF DEPENDENT CHOICE. *Let S be a nonempty set and let $\mathcal{R} \subseteq S \times S$ be a binary relation with the property that, for each $\xi \in S$, the set $\{\eta \in S;\ \xi \mathcal{R} \eta\}$ is nonempty. Then, for each $\xi \in S$, there exists a sequence $(\xi_m)_m$ in S such that $\xi_0 = \xi$ and $\xi_m \mathcal{R} \xi_{m+1}$ for each $m \in \mathbb{N}$.*

We notice that, in its turn, the axiom of dependent choice implies *the axiom of countable choice* stated below, which is sufficient to prove that a lot of remarkable properties in Real Analysis can be described by means of sequences. We emphasize that the axiom of dependent choice is "far enough" from the axiom of countable choice, as shown recently by Howard and Rubin [57].

THE AXIOM OF COUNTABLE CHOICE. *Let S be a nonempty set and let $\mathcal{F} = \{\mathbb{F}_m;\ m \in \mathbb{N}\}$ be a countable family of nonempty subsets in S. Then, there exists a sequence $(\xi_m)_m$ with the property that $\xi_m \in \mathbb{F}_m$ for each $m \in \mathbb{N}$.*

Finally, we notice that we preferred this framework simply because the results based on the axiom of dependent choice remain true no matter which initial assumption we make, i.e. no matter if we assume that either the axiom of choice, or its negation, holds true.

Let S be a nonempty set. A binary relation $\preceq \subseteq S \times S$ is a *preorder on S* if it is reflexive, i.e. $\xi \preceq \xi$ for each $\xi \in S$, and transitive, i.e. $\xi \preceq \eta$ and $\eta \preceq \zeta$ imply $\xi \preceq \zeta$.

DEFINITION 2.1.1. Let $\mathcal{S}$ be a nonempty set, $\preceq$ a preorder on $\mathcal{S}$ and let $\mathcal{M} : \mathcal{S} \to \mathbb{R} \cup \{+\infty\}$ be an increasing function. An $\mathcal{M}$-*maximal element* is an element $\bar{\xi} \in \mathcal{S}$ satisfying $\mathcal{M}(\xi) = \mathcal{M}(\bar{\xi})$, for every $\xi \in \mathcal{S}$ with $\bar{\xi} \preceq \xi$.

THEOREM 2.1.1 (Brezis–Browder ordering principle [19]). *Let $\mathcal{S}$ be a nonempty set, $\preceq$ a preorder on $\mathcal{S}$ and let $\mathcal{M} : \mathcal{S} \to \mathbb{R} \cup \{+\infty\}$ be a function. Suppose that:*

    (i) *for any increasing sequence $(\xi_k)_k$ in $\mathcal{S}$, there exists some $\eta \in \mathcal{S}$ such that $\xi_k \preceq \eta$ for all $k \in \mathbb{N}$;*

    (ii) *the function $\mathcal{M}$ is increasing.*

Then, for each $\xi_0 \in \mathcal{S}$ there exists an $\mathcal{M}$-maximal element $\bar{\xi} \in \mathcal{S}$ satisfying $\xi_0 \preceq \bar{\xi}$.

PROOF. Suppose first that the function $\mathcal{M}$ is bounded from above. Let us consider a fixed element $\xi_0 \in \mathcal{S}$ and let us construct (inductively) an increasing sequence $(\xi_k)_k$ as follows: if ξ_k is given, let us consider the set $\mathcal{S}_k = \{\xi \in \mathcal{S}; \ \xi_k \preceq \xi\}$ and let us denote by $\beta_k = \sup\{\mathcal{M}(\xi); \ \xi \in \mathcal{S}_k\}$. If ξ_k satisfies the conclusion, we have nothing to prove. If not, then $\beta_k > \mathcal{M}(\xi_k)$, and so we get ξ_{k+1} such that $\xi_k \preceq \xi_{k+1}$ and

$$\mathcal{M}(\xi_{k+1}) > \beta_k - \frac{\beta_k - \mathcal{M}(\xi_k)}{2}. \tag{2.1.1}$$

We have thus constructed an increasing sequence (here is the point where we have used the axiom of dependent choices) $(\xi_k)_k$ with the property that the sequence $(\mathcal{M}(\xi_k))_k$ is strictly increasing. By the assumption (i), $(\xi_k)_k$ is bounded above in $\mathcal{S}$, i.e., there exists $\bar{\xi} \in \mathcal{S}$ such that $\xi_k \preceq \bar{\xi}$ for every $k \in \mathbb{N}$. We show that this $\bar{\xi}$ satisfies the conclusion. Suppose by contradiction that there exists $\eta \in \mathcal{S}$ such that $\bar{\xi} \preceq \eta$ and $\mathcal{M}(\bar{\xi}) < \mathcal{M}(\eta)$. Since $\mathcal{M}$ is bounded from above, the sequence $(\mathcal{M}(\xi_k))_k$ is bounded from above, and thus convergent. Moreover $\lim_{k \to \infty} \mathcal{M}(\xi_k) \leq \mathcal{M}(\bar{\xi})$ and $\eta \in \mathcal{S}_k$ for each $k \in \mathbb{N}$. Therefore we have $\beta_k \geq \mathcal{M}(\eta)$. From (2.1.1), we deduce

$$2\mathcal{M}(\xi_{k+1}) - \mathcal{M}(\xi_k) \geq \beta_k \geq \mathcal{M}(\eta),$$

for all $k \in \mathbb{N}$. Passing to the limit as $k \to \infty$, we obtain $\mathcal{M}(\bar{\xi}) \geq \mathcal{M}(\eta)$, which is a contradiction. This achieves the proof under the extra-condition that $\mathcal{M}$ is bounded above.

Consider now the general case, and let us define the auxiliary function $\mathcal{M}_1 : \mathcal{S} \to (-\frac{\pi}{2}, \frac{\pi}{2}]$ by

$$\mathcal{M}_1(\xi) = \begin{cases} \arctan\big(\mathcal{M}(\xi)\big) & \text{if } \mathcal{M}(\xi) < +\infty, \\ \dfrac{\pi}{2} & \text{if } \mathcal{M}(\xi) = +\infty. \end{cases}$$

The function $\mathcal{M}_1$ is increasing and bounded from above. Therefore there exists an element $\bar{\xi} \in \mathcal{S}$ which verifies the conclusion with $\mathcal{M}_1$ instead of $\mathcal{M}$. But $\arctan \mathcal{M}(\bar{\xi}) = \arctan \mathcal{M}(\xi)$ implies $\mathcal{M}(\bar{\xi}) = \mathcal{M}(\xi)$, which completes the proof in the general case. $\quad\square$

In its original formulation of Brezis–Browder ordering principle, it is assumed that $\mathcal{M}$ is bounded from above. In order to handle a larger class of applications this condition has been dropped in Cârjă and Ursescu [25], by obtaining the very slight extension here presented. A simple inspection of the proof shows that the conclusion of Theorem 2.1.1 remains true if (i) is replaced by the weaker condition:

    (j) *For any increasing sequence $(\xi_k)_k$ in $\mathcal{S}$ with the property that the sequence $(\mathcal{M}(\xi_k))_k$ is strictly increasing, there exists some $\eta \in \mathcal{S}$ such that $\xi_k \preceq \eta$ for all $k \in \mathbb{N}$.*

2.2. *Projections*

We begin with:

DEFINITION 2.2.1. A subset $\mathbb{K} \subseteq \mathbb{R}^n$ is *locally closed* if for every $\xi \in \mathbb{K}$ there exists $\rho > 0$ such that $\mathbb{K} \cap B(\xi, \rho)$ is closed.

REMARK 2.2.1. Obviously every closed set is locally closed. Furthermore, if $\mathbb{D}$ is a given open subset in $\mathbb{R}^n$, then every relatively closed subset $\mathbb{K}$ in $\mathbb{D}$ is locally closed too. Indeed, if $\mathbb{K}$ is relatively closed, then $\mathbb{K} = \overline{\mathbb{K}} \cap \mathbb{D}$ and so, for every $\xi \in \mathbb{K}$ there exists $\rho > 0$ such that $B(\xi, \rho) \subseteq \mathbb{D}$. Consequently $\mathbb{K} \cap B(\xi, \rho) = \overline{\mathbb{K}} \cap \mathbb{D} \cap B(\xi, \rho) = \overline{\mathbb{K}} \cap B(\xi, \rho)$ which is closed, and this proves the assertion. Thus, each open subset $\mathbb{K}$ in $\mathbb{R}^n$ is locally closed. There exist however locally closed sets which are neither open, nor closed, nor even closed relatively to $\mathbb{D}$, as for example $\mathbb{K} \subseteq \mathbb{D} \subseteq \mathbb{R}^2$ defined by $\mathbb{D} = \{(x, y) \in \mathbb{R}^2;\ x > 0\}$ and $\mathbb{K} = \{(x, y) \in \mathbb{R}^2;\ 0 < x < 1, y = 0\}$. The set $\mathbb{K}$, which is in fact a line segment in the plane xOy, is locally closed but is neither open, nor closed, nor even closed relatively to $\mathbb{D}$.

    We say that $\xi \in \mathbb{R}^n$ *has projection on* $\mathbb{K}$ if there exists $\eta \in \mathbb{K}$ such that $\|\xi - \eta\| = \mathrm{dist}(\xi; \mathbb{K})$. Any $\eta \in \mathbb{K}$ enjoying the above property is called *a projection* of ξ on $\mathbb{K}$, and the set of all projections of ξ on $\mathbb{K}$ is denoted by $\Pi_{\mathbb{K}}(\xi)$.

DEFINITION 2.2.2. An open neighborhood $\mathbb{V}$ of $\mathbb{K}$, with $\Pi_{\mathbb{K}}(\xi) \neq \emptyset$ for each $\xi \in \mathbb{V}$, is called a *proximal neighborhood* of $\mathbb{K}$. If $\mathbb{V}$ is a proximal neighborhood of $\mathbb{K}$, then every single-valued selection, $\pi_{\mathbb{K}} : \mathbb{V} \to \mathbb{K}$, of $\Pi_{\mathbb{K}}$, i.e. $\pi_{\mathbb{K}}(\xi) \in \Pi_{\mathbb{K}}(\xi)$ for each $\xi \in \mathbb{V}$, is a *projection subordinated to* $\mathbb{V}$.

    The next lemma, proved in Cârjă and Ursescu [25, Lemma 18], shows that for each locally closed set $\mathbb{K}$, all the points which are sufficiently close to $\mathbb{K}$ do have projections, i.e. each locally closed set $\mathbb{K}$ has one proximal neighborhood.

LEMMA 2.2.1. *Let $\mathbb{K}$ be locally closed. Then the set of all $\xi \in \mathbb{R}^n$ such that $\Pi_{\mathbb{K}}(\xi)$ is nonempty is a neighborhood of $\mathbb{K}$.*

PROOF. Let $\xi \in \mathbb{K}$. Since $\mathbb{K}$ is locally closed, there exists $\rho > 0$ such that $\mathbb{K} \cap B(\xi, \rho)$ is closed. To complete the proof, it suffices to show that, for each $\eta \in \mathbb{K}$ satisfying $\|\xi - \eta\| < \rho/2$, $\Pi_{\mathbb{K}}(\eta)$ is nonempty. Indeed, for each η as above, there exists a sequence $(\zeta_k)_k$ in $\mathbb{K}$ such that $(\|\zeta_k - \eta\|)_k$ converges to $\mathrm{dist}(\eta; \mathbb{K})$. We can suppose, by taking a subsequence if necessary, that $(\zeta_k)_k$ converges to a point $\zeta \in \mathbb{R}^n$. So, we have $\mathrm{dist}(\eta; \mathbb{K}) = \|\zeta - \eta\|$. Clearly, $\|\zeta_k - \xi\| \leqslant \|\zeta_k - \eta\| + \|\eta - \xi\|$ for all $k \in \mathbb{N}$, and consequently $\|\zeta - \xi\| \leqslant \mathrm{dist}(\eta; \mathbb{K}) + \|\eta - \xi\| \leqslant 2\|\eta - \xi\| < \rho$. Finally, $\|\zeta_k - \eta\| < \rho$ for all $k \in \mathbb{N}$ sufficiently large. Therefore $\zeta_k \in \mathbb{K} \cap B(\xi, \rho)$, and since the latter is closed, it follows that $\zeta \in \mathbb{K}$. Thus $\Pi_{\mathbb{K}}(\eta)$ is nonempty, and this completes the proof. $\qquad\square$

2.3. *Tangent cones*

The next tangency concept was introduced independently by Bouligand [16] and Severi [89].

DEFINITION 2.3.1. Let $\mathbb{K} \subseteq \mathbb{R}^n$ and $\xi \in \mathbb{K}$. The vector $\eta \in \mathbb{R}^n$ is *tangent in the sense of Bouligand–Severi* to the set $\mathbb{K}$ at the point ξ if

$$\liminf_{h \downarrow 0} \frac{1}{h} \mathrm{dist}(\xi + h\eta; \mathbb{K}) = 0.$$

PROPOSITION 2.3.1. *The set $\mathcal{T}_{\mathbb{K}}(\xi)$ of all vectors which are tangent in the sense of Bouligand–Severi to the set $\mathbb{K}$ at the point ξ is a closed cone.*

PROOF. Let $\xi \in \mathbb{K}$. According to Definition 2.3.1, $\eta \in \mathcal{T}_{\mathbb{K}}(\xi)$ if

$$\liminf_{t \downarrow 0} \frac{1}{t} \mathrm{dist}(\xi + t\eta; \mathbb{K}) = 0.$$

Let $s > 0$ and let us observe that

$$\liminf_{t \downarrow 0} \frac{1}{t} \mathrm{dist}(\xi + ts\eta; \mathbb{K}) = s \liminf_{t \downarrow 0} \frac{1}{ts} \mathrm{dist}(\xi + ts\eta; \mathbb{K})$$

$$= s \liminf_{\tau \downarrow 0} \frac{1}{\tau} \mathrm{dist}(\xi + \tau\eta; \mathbb{K}) = 0.$$

Hence $s\eta \in \mathcal{T}_{\mathbb{K}}(\xi)$. In order to complete the proof, it remains to show that $\mathcal{T}_{\mathbb{K}}(\xi)$ is a closed set. To this aim let $(\eta_k)_{k \in \mathbb{N}^*}$ be a sequence of elements in $\mathcal{T}_{\mathbb{K}}(\xi)$, convergent to η. We have

$$\frac{1}{t} \mathrm{dist}(\xi + t\eta; \mathbb{K}) \leqslant \frac{1}{t} \|t(\eta - \eta_k)\| + \frac{1}{t} \mathrm{dist}(\xi + t\eta_k; \mathbb{K})$$

$$= \|\eta - \eta_k\| + \frac{1}{t} \mathrm{dist}(\xi + t\eta_k; \mathbb{K})$$

for every $k \in \mathbb{N}^*$. So $\liminf_{t \downarrow 0} \frac{1}{t} \mathrm{dist}(\xi + t\eta; \mathbb{K}) \leqslant \|\eta - \eta_k\|$ for every $k \in \mathbb{N}^*$. Since $\lim_{k \to \infty} \|\eta - \eta_k\| = 0$, it follows that $\liminf_{t \downarrow 0} \frac{1}{t} \mathrm{dist}(\xi + t\eta; \mathbb{K}) = 0$, which achieves the proof. $\qquad\square$

Following Bouligand [16], the cone $\mathcal{T}_{\mathbb{K}}(\xi)$ is called *the contingent cone* to the set $\mathbb{K}$ at the point ξ.

PROPOSITION 2.3.2. *A vector $\eta \in \mathbb{R}^n$ belongs to the cone $\mathcal{T}_{\mathbb{K}}(\xi)$ if and only if for every $\varepsilon > 0$ there exist $h \in (0, \varepsilon)$ and $p_h \in B(0, \varepsilon)$ with the property*

$$\xi + h(\eta + p_h) \in \mathbb{K}.$$

PROOF. Obviously $\eta \in \mathcal{T}_{\mathbb{K}}(\xi)$ if and only if, for every $\varepsilon > 0$ there exists $h \in (0, \varepsilon)$ and $z_h \in \mathbb{K}$ such that $\frac{1}{h}\|\xi + h\eta - z_h\| \leqslant \varepsilon$. Now, let us define $p_h = \frac{1}{h}(z_h - \xi - h\eta)$, and let us observe that we have both $\|p_h\| \leqslant \varepsilon$, and $\xi + h(\eta + p_h) = z_h \in \mathbb{K}$, thereby completing the proof. $\qquad\square$

A simple but useful consequence is:

COROLLARY 2.3.1. *A vector $\eta \in \mathbb{R}^n$ belongs to the cone $\mathcal{T}_{\mathbb{K}}(\xi)$ if and only if there exist two sequences $(h_m)_m$ in $\mathbb{R}_+$ and $(p_m)_m$ in $\mathbb{R}^n$ with $h_m \downarrow 0$ and $p_m \to 0$ as $m \to \infty$, and such that $\xi + h_m(\eta + p_m) \in \mathbb{K}$ for each $m \in \mathbb{N}$.*

REMARK 2.3.1. We notice that, if ξ is an interior point of the set $\mathbb{K}$, then $\mathcal{T}_{\mathbb{K}}(\xi) = \mathbb{R}^n$. Indeed, in this case there exists $\rho > 0$ such that $B(\xi, \rho) \subset \mathbb{K}$ and, therefore, for $t > 0$ sufficiently small, $\xi + t\eta \in B(\xi, \rho) \subseteq \mathbb{K}$. Obviously, for such numbers $t > 0$, we have $\text{dist}(\xi + t\eta; \mathbb{K}) = 0$, from where it follows the condition in Definition 2.3.1.

An interesting consequence of Corollary 2.3.1 is given below.

THEOREM 2.3.1. *Let $\mathbb{K}_1, \mathbb{K}_2 \subseteq \mathbb{R}^n$ be locally closed. If $\xi \in \mathbb{K}_1 \cap \mathbb{K}_2$ is an interior point of $\mathbb{K}_1 \cup \mathbb{K}_2$, then we have*

$$\mathcal{T}_{\mathbb{K}_1 \cap \mathbb{K}_2}(\xi) = \mathcal{T}_{\mathbb{K}_1}(\xi) \cap \mathcal{T}_{\mathbb{K}_2}(\xi).$$

PROOF. Obviously, for each $\xi \in \mathbb{K}_1 \cap \mathbb{K}_2$, $\mathcal{T}_{\mathbb{K}_1 \cap \mathbb{K}_2}(\xi) \subseteq \mathcal{T}_{\mathbb{K}_1}(\xi) \cap \mathcal{T}_{\mathbb{K}_2}(\xi)$. To prove that, whenever, in addition, ξ is in the interior of $\mathbb{K}_1 \cup \mathbb{K}_2$, the converse inclusion holds true, let $\xi \in \mathbb{K}_1 \cap \mathbb{K}_2$ and let $\eta \in \mathcal{T}_{\mathbb{K}_1}(\xi) \cap \mathcal{T}_{\mathbb{K}_2}(\xi)$. By Corollary 2.3.1, there exist four sequences $(h_m)_m$, $(\tilde{h}_m)_m$ in $\mathbb{R}_+$, $(p_m)_m$ and $(\tilde{p}_m)_m$ in X with $h_m \downarrow 0$, $\tilde{h}_m \downarrow 0$, $p_m \to 0$ and $\tilde{p}_m \to 0$ as $m \to \infty$, and such that $\xi + h_m(\eta + p_m) \in \mathbb{K}_1$ and $\xi + \tilde{h}_m(\eta + \tilde{p}_m) \in \mathbb{K}_2$ for each $m \in \mathbb{K}$. Now, if we assume further that ξ is in the interior of $\mathbb{K}_1 \cup \mathbb{K}_2$, there exists $\rho > 0$ such that $B(\xi, \rho) \subset \mathbb{K}_1 \cup \mathbb{K}_2$. Since $\mathbb{K}_1$ and $\mathbb{K}_2$ are locally closed, diminishing $\rho > 0$ if necessary, we may assume that both $\mathbb{K}_1 \cap B(\xi, \rho)$ and $\mathbb{K}_2 \cap B(\xi, \rho)$ are closed. Let $m_0 \in \mathbb{N}$ be such that, for each $m \geqslant m_0$, we have both $\xi_m = \xi + h_m(\eta + p_m) \in B(\xi, \rho)$ and $\tilde{\xi}_m = \xi + \tilde{h}_m(\eta + \tilde{p}_m) \in B(\xi, \rho)$. As a consequence, if $m \geqslant m_0$, the line segment $[\xi_m, \tilde{\xi}_m] \subseteq B(\xi, \rho) \subseteq \mathbb{K}_1 \cup \mathbb{K}_2$. Since $B(\xi, \rho)$ is connected, while $B(\xi, \rho) \cap \mathbb{K}_1$ and $B(\xi, \rho) \cap \mathbb{K}_2$ are closed, there exists $\eta_m \in [\xi_m, \tilde{\xi}_m] \cap \mathbb{K}_1 \cap \mathbb{K}_2$. Since $\eta_m \in [\xi_m, \tilde{\xi}_m]$, there exists $\theta_m \in [0, 1]$ such that $\eta_m = (1 - \theta_m)\xi_m + \theta_m \tilde{\xi}_m$. So, denoting

$$t_m = (1 - \theta_m)h_m + \theta_m \tilde{h}_m \quad \text{and} \quad q_m = \frac{(1 - \theta_m)h_m}{t_m} p_m + \frac{\theta_m \tilde{h}_m}{t_m} \tilde{p}_m,$$

we have

$$\eta_m = \xi + t_m(\eta + q_m) \in \mathbb{K}_1 \cap \mathbb{K}_2.$$

Finally, observing that $t_m \downarrow 0$ and $q_m \to 0$, and using Corollary 2.3.1, we get the conclusion. $\qquad\square$

In the case when $\mathbb{K}_1, \mathbb{K}_2$ are closed subsets in a normed vector space, a similar result was obtained by Quincampoix [84, Corollary 2.3].

The next tangency concept was introduced by Federer [41]. See also Girsanov [49].

DEFINITION 2.3.2. Let $\mathbb{K} \subseteq \mathbb{R}^n$ and $\xi \in \mathbb{K}$. The vector $\eta \in \mathbb{R}^n$ is *tangent in the sense of Federer* to the set $\mathbb{K}$ at the point ξ if

$$\lim_{h \downarrow 0} \frac{1}{h} \operatorname{dist}(\xi + h\eta; \mathbb{K}) = 0.$$

The set of all points $\eta \in \mathbb{R}^n$ which are tangent in the sense of Federer to $\mathbb{K}$ at ξ is denoted by $\mathcal{F}_{\mathbb{K}}(\xi)$. This set is a cone which clearly is included in $\mathcal{T}_{\mathbb{K}}(\xi)$. A not necessarily metrical tangency concept in general topological vector spaces which, in $\mathbb{R}^n$, reduces to that one of Federer [41], is due to Ursescu [95]. As we shall show next, for a continuous function $f : \mathbb{K} \to \mathbb{R}^n$, the following surprising equivalence holds true: $f(\xi) \in \mathcal{T}_{\mathbb{K}}(\xi)$ for each $\xi \in \mathbb{K}$ if and only if $f(\xi) \in \mathcal{F}_{\mathbb{K}}(\xi)$ for each $\xi \in \mathbb{K}$, and this in spite of the fact that $\mathcal{F}_{\mathbb{K}}(\xi) \neq \mathcal{T}_{\mathbb{K}}(\xi)$.

2.4. *The proximal normal cone*

We begin by recalling:

DEFINITION 2.4.1. Let $\xi \in \mathbb{K}$. We say that $v \in \mathbb{R}^n$ is *metric normal to* $\mathbb{K}$ *at* ξ, if there exist $\eta \in \mathbb{R}^n$ and $\rho > 0$ such that $B(\eta, \rho)$ contains ξ on its boundary, its interior has empty intersection with $\mathbb{K}$, and $v = \eta - \xi$.

In Fig. 1(a) is illustrated a point ξ at which there is no metric normal vector. We notice that we have $\mathcal{T}_{\mathbb{K}}(\xi) = \mathbb{R}^n$. For the case in which there is at least one normal vector to $\mathbb{K}$ at a point ξ, see Fig. 1(b).

DEFINITION 2.4.2. The *proximal normal cone to* $\mathbb{K}$ *at* $\xi \in \mathbb{K}$ is the set of all $\zeta \in \mathbb{R}^n$ of the form $\zeta = \lambda v$, where v is metric normal to $\mathbb{K}$ at ξ and $\lambda \geqslant 0$, whenever such a metric normal v exists, and $\{0\}$ if there is no metric normal to $\mathbb{K}$ at ξ. We denote this cone by $\mathcal{N}_{\mathbb{K}}(\xi)$.

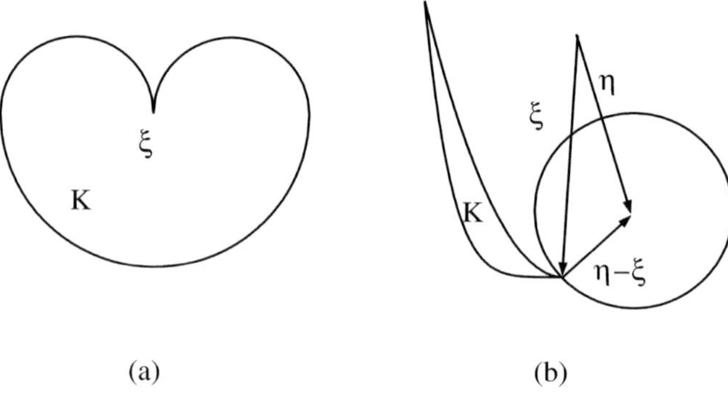

(a)   (b)

Fig. 1.

The use of the term "cone" in Definition 2.4.2, is justified by the simple observation that, for each $\xi \in \mathbb{K}$, $\mathcal{N}_{\mathbb{K}}(\xi)$ is a cone in the usual sense, i.e., for each $\zeta \in \mathcal{N}_{\mathbb{K}}(\xi)$ and $\lambda > 0$, we have $\lambda\zeta \in \mathcal{N}_{\mathbb{K}}(\xi)$.

The next tangency concept was introduced by Ursescu [101]. Let now $\mathbb{S}_{\mathbb{K}}(\xi)$ be the set of all $\eta \in \mathbb{R}^n$ such that $\eta - \xi$ is a metric normal to $\mathbb{K}$ at ξ whenever such a metric normal exists, and $\mathbb{S}_{\mathbb{K}}(\xi) = \{\xi\}$ otherwise. Let $\eta \in \mathbb{S}_{\mathbb{K}}(\xi)$, and let $\mathbb{E}(\xi, \eta) = \{\zeta \in \mathbb{R}^n;\ \|\eta - \zeta\| \geqslant \|\eta - \xi\|\}$. Since $\mathbb{K} \subseteq \mathbb{E}(\xi, \eta)$, for each $\eta \in \mathcal{S}_{\mathbb{K}}(\xi)$, we have

$$\mathcal{T}_{\mathbb{K}}(\xi) \subseteq \mathcal{B}_{\mathbb{K}}(\xi), \tag{2.4.1}$$

where

$$\mathcal{B}_{\mathbb{K}}(\xi) = \bigcap_{\eta \in \mathbb{S}_{\mathbb{K}}(\xi)} \mathcal{T}_{\mathbb{E}(\xi,\eta)}(\xi).$$

One may easily see that $\mathcal{B}_{\mathbb{K}}(\xi)$ is a cone in $\mathbb{R}^n$.

DEFINITION 2.4.3. The set $\mathcal{B}_{\mathbb{K}}(\xi)$, defined as above, is the *Bony tangent cone to* $\mathbb{K}$ *at* ξ, and its elements are the *tangents in the sense of Bony* to $\mathbb{K}$ at $\xi \in \mathbb{K}$.

Let $\|\cdot\|$ be a given norm on $\mathbb{R}^n$. If $x, y \in \mathbb{R}^n$, we denote by $[x, y]_+$ the *right directional derivative of the norm* calculated at x in the direction y, i.e.

$$[x, y]_+ = \lim_{h \downarrow 0} \frac{\|x + hy\| - \|x\|}{h}.$$

If $\|\cdot\|$ is the Euclidean norm, we have

$$[x, y]_+ = \begin{cases} \dfrac{\langle x, y \rangle}{\|x\|} & \text{if } x \neq 0, \\ \|y\| & \text{if } x = 0. \end{cases}$$

REMARK 2.4.1. Taking into account the definitions of both $\mathbb{E}(\xi, \eta)$ and $[\cdot, \cdot]_+$, we easily deduce that $\zeta \in \mathcal{T}_{\mathbb{E}(\xi,\eta)}(\xi)$ if and only if $[\xi - \eta, \zeta]_+ \geqslant 0$. Therefore, $\zeta \in \mathcal{B}_{\mathbb{K}}(\xi)$ if and only if $[-\nu, \zeta]_+ \geqslant 0$ for each ν which is a metric normal to $\mathbb{K}$ at ξ. In particular, when $\mathbb{R}^n$ is endowed with the Euclidean norm, we easily deduce that, for each $\xi \in \mathbb{K}$ and each $\eta \in \mathbb{S}_{\mathbb{K}}(\xi)$, $\mathcal{T}_{\mathbb{E}(\xi,\eta)}(\xi)$ is a closed half-space having the exterior normal $\eta - \xi$. Therefore, in this case, we have

$$\mathcal{B}_{\mathbb{K}}(\xi) = \left(\mathcal{N}_{\mathbb{K}}(\xi)\right)^*,$$

where $\left(\mathcal{N}_{\mathbb{K}}(\xi)\right)^*$ is the so-called *conjugate cone* of $\mathcal{N}_{\mathbb{K}}(\xi)$, i.e.

$$\left(\mathcal{N}_{\mathbb{K}}(\xi)\right)^* = \left\{\eta \in \mathbb{R}^n; \ \langle \nu, \eta \rangle \leqslant 0, \ \text{for each } \nu \in \mathcal{N}_{\mathbb{K}}(\xi)\right\}.$$

REMARK 2.4.2. If there is no metric normal vector to $\mathbb{K}$ at ξ, we may easily see that $\mathcal{B}_{\mathbb{K}}(\xi) = \mathbb{R}^n$. See also Fig. 1(a).

2.5. *Clarke's tangent cone*

We are now ready to study another useful tangency concept defined by Clarke [32].

DEFINITION 2.5.1. Let $\mathbb{K} \subseteq \mathbb{R}^n$ and $\xi \in \mathbb{K}$. The vector $\eta \in \mathbb{R}^n$ is *tangent in the sense of Clarke* to the set $\mathbb{K}$ at the point ξ if

$$\lim_{\substack{h \downarrow 0 \\ \mu \to \xi; \mu \in \mathbb{K}}} \frac{1}{h} \operatorname{dist}(\mu + h\eta; \mathbb{K}) = 0.$$

We denote by $\mathcal{C}_{\mathbb{K}}(\xi)$ the set of all vectors $\eta \in \mathbb{R}^n$ which are tangent to $\xi \in \mathbb{K}$ in the sense of Clarke. It is not difficult to check out that $\mathcal{C}_{\mathbb{K}}(\xi)$ is a closed convex cone.

REMARK 2.5.1. One may easily see that, for each $\mathbb{K}$ and each $\xi \in \mathbb{K}$, we have

$$\mathcal{C}_{\mathbb{K}}(\xi) \subseteq \mathcal{F}_{\mathbb{K}}(\xi) \subseteq \mathcal{T}_{\mathbb{K}}(\xi) \subseteq \mathcal{B}_{\mathbb{K}}(\xi).$$

We notice that, whenever $\mathbb{K}$ admits a classical tangent space at $\xi \in \mathbb{K}$, all tangent cones previously introduced coincide with that tangent space. This happens for instance if $\mathbb{K}$ is a C^1 curve, or surface or, even if $\mathbb{K}$ is an m-dimensional C^1 manifold. However, if $\mathbb{K}$ is not smooth enough locally around ξ, the inclusions $\mathcal{C}_{\mathbb{K}}(\xi) \subseteq \mathcal{C}_{\mathbb{K}}(\xi) \subseteq \mathcal{T}_{\mathbb{K}}(\xi) \subseteq \mathcal{B}_{\mathbb{K}}(\xi)$ may be strict, as the following example shows.

EXAMPLE 2.5.1 (Necula [73]). Let $\mathbb{K} \subseteq \mathbb{R}^2$ be defined as $\mathbb{K} = \mathbb{K}_1 \cup \mathbb{K}_2$, where

$$\mathbb{K}_1 = \left\{(x, y); \ (x, y) \in \mathbb{R}^2, \ y \leqslant |x|\right\}$$

and

$$\mathbb{K}_2 = \big\{(0, 1/2^m);\; m \in \mathbb{N}\big\}$$

and let $\xi = (0, 0) \in \mathbb{K}$. Then, we have

$$\begin{cases} \mathcal{C}_{\mathbb{K}}(\xi) = \{0\}, \\ \mathcal{F}_{\mathbb{K}}(\xi) = \mathbb{K}_1, \\ \mathcal{T}_{\mathbb{K}}(\xi) = \mathbb{K}_1 \cup \big\{(0, y);\; y \geqslant 0\big\}, \\ \mathcal{B}_{\mathbb{K}}(\xi) = \mathbb{R}^2. \end{cases}$$

For a multifunction[2] $F : \mathbb{K} \rightsquigarrow \mathbb{R}^n$, we define

$$\liminf_{\substack{\xi \to \xi_0 \\ \xi \in \mathbb{K}}} F(\xi) = \Big\{\eta \in \mathbb{R}^n;\; \lim_{\substack{\xi \to \xi_0 \\ \xi \in \mathbb{K}}} \mathrm{dist}\big(\eta, F(\xi)\big) = 0\Big\}.$$

LEMMA 2.5.1. *Let us assume that the norm* $\|\cdot\|$ *on* $\mathbb{R}^n$ *satisfies*

$$[x, y]_+ = -[-x, y]_+$$

for each $x, y \in \mathbb{R}^n$.[3] *Let* $\mathbb{K} \subseteq \mathbb{R}^n$ *be locally closed. Then, for each* $\xi_0 \in \mathbb{K}$, *we have*

$$\liminf_{\xi \to \xi_0;\, \xi \in \mathbb{K}} \mathcal{B}_{\mathbb{K}}(\xi) = \mathcal{C}_{\mathbb{K}}(\xi_0). \tag{2.5.1}$$

PROOF. We begin by showing that

$$\liminf_{\xi \to \xi_0;\, \xi \in \mathbb{K}} \mathcal{B}_{\mathbb{K}}(\xi) \subseteq \mathcal{C}_{\mathbb{K}}(\xi_0). \tag{2.5.2}$$

Let $\eta \neq 0$, $\eta \in \liminf_{\xi \to \xi_0;\, \xi \in \mathbb{K}} \mathcal{B}_{\mathbb{K}}(\xi)$. It follows that, for each $\varepsilon > 0$, there exists $\theta > 0$ such that, for each $\phi \in \mathbb{K} \cap B(\xi_0, \theta)$, we have

$$B(\eta, \varepsilon) \cap \mathcal{B}_{\mathbb{K}}(\phi) \neq \emptyset. \tag{2.5.3}$$

Take a sufficiently small θ so that, for all $\xi \in \mathbb{K} \cap B(\xi_0, \frac{\theta}{4})$ and $t \in [0, \frac{\theta}{4\|\eta\|}]$, we have $\Pi_{\mathbb{K}}(\xi + t\eta) \neq \emptyset$. By virtue of Lemma 2.2.1, this is always possible. With ξ and t as above, let us define $g(t) = \mathrm{dist}(\xi + t\eta; \mathbb{K})$. In order to prove that $\eta \in \mathcal{C}_{\mathbb{K}}(\xi_0)$, it suffices to show that $g(t) \leqslant \varepsilon t$ for each $t \in [0, \frac{\theta}{4\|\eta\|}]$. Further, since $g(0) = 0$, it suffices to show that $g'(t) \leqslant \varepsilon$, whenever $g'(t)$ exists. To this aim, take $\phi \in \Pi_{\mathbb{K}}(\xi + t\eta)$, and let us observe that we have $\phi \in \mathbb{K} \cap B(\xi_0, \theta)$. Indeed,

$$\|\phi - \xi_0\| \leqslant \|\xi + t\eta - \phi\| + \|\xi + t\eta - \xi_0\| \leqslant 2\|\xi + t\eta - \xi_0\| \leqslant \theta,$$

---

[2] A function $F : \mathbb{K} \to \mathcal{P}(\mathbb{R}^n)$, where $\mathcal{P}(\mathbb{R}^n)$ is the class of all subsets of $\mathbb{R}^n$, is called *multifunction* and is denoted by $F : \mathbb{K} \rightsquigarrow \mathbb{R}^n$.

[3] This happens if $\|\cdot\|$ is Gâteaux differentiable.

as claimed. Now, for a sufficiently small $h > 0$, we obtain

$$g(t+h) - g(t) \leqslant \|\xi + t\eta + h\eta - \phi\| - \|\xi + t\eta - \phi\|.$$

Dividing by h and letting $h \downarrow 0$, we get

$$g'(t) \leqslant [\xi + t\eta - \phi, \eta]_+. \tag{2.5.4}$$

Taking into account that $\phi \in \Pi_{\mathbb{K}}(\xi + t\eta)$, from Definition 2.4.1, we deduce that $\xi + t\eta - \phi$ is metric normal to $\mathbb{K}$ at ϕ. In view of (2.5.3), there exists $w \in \mathcal{B}_{\mathbb{K}}(\phi)$ with $\|w - \eta\| \leqslant \varepsilon$. From Remark 2.4.1, we conclude

$$[-\xi - t\eta + \phi, w]_+ \geqslant 0.$$

According to the hypothesis, we have

$$[\xi + t\eta - \phi, \eta]_+ \leqslant [\xi + t\eta - \phi, \eta - w]_+ + [\xi + t\eta - \phi, w]_+ \leqslant \varepsilon.$$

From this inequality and (2.5.4), we get $g'(t) \leqslant \varepsilon$, as claimed. Thus (2.5.2) holds. Since $\mathcal{T}_{\mathbb{K}}(\xi) \subseteq \mathcal{B}_{\mathbb{K}}(\xi)$, to complete the proof, it suffices to show that

$$\mathcal{C}_{\mathbb{K}}(\xi_0) \subseteq \liminf_{\xi \to \xi_0; \xi \in \mathbb{K}} \mathcal{T}_{\mathbb{K}}(\xi).$$

So, let $\eta \in \mathcal{C}_{\mathbb{K}}(\xi_0)$ and let $\xi_m \in \mathbb{K}$ with $\xi_m \to \xi_0$. The idea is to find $\eta_m \in \mathcal{T}_{\mathbb{K}}(\xi_m)$ with $\eta_m \to \eta$. This would imply that $\eta \in \liminf_{\xi \to \xi_0; \xi \in \mathbb{K}} \mathcal{T}_{\mathbb{K}}(\xi)$. To this end, let us observe that, since $\eta \in \mathcal{C}_{\mathbb{K}}(\xi_0)$, for every $\varepsilon > 0$, there exist $m_\varepsilon \in \mathbb{N}$ and $h_\varepsilon > 0$ such that, for all $0 < h < h_\varepsilon$ and $m \geqslant m_\varepsilon$, we have

$$\mathrm{dist}(\xi_m + h\eta; \mathbb{K}) < h\varepsilon.$$

Fix m as above and take $\mu_m^h \in \mathbb{K}$ with $\|\mu_m^h - \xi_m - h\eta\| < h\varepsilon$. Let us consider

$$\eta_m^h = \frac{1}{h}\left(\mu_m^h - \xi_m\right).$$

Since $\|\eta_m^h - \eta\| \leqslant \varepsilon$, $\{\eta_m^h; \ h < 0 < h_\varepsilon\}$ is bounded and therefore it has a limit point η_m as $h \downarrow 0$. In its turn, η_m satisfies $\|\eta_m - \eta\| \leqslant \varepsilon$. A simple computational argument shows that $\eta_m \in \mathcal{T}_{\mathbb{K}}(\xi_m)$ and this completes the proof. $\qquad\square$

REMARK 2.5.2. From Lemma 2.5.1, and Remarks 2.4.1 and 2.5.1, we easily deduce

$$\liminf_{\xi \to \xi_0; \xi \in \mathbb{K}} \mathcal{F}_{\mathbb{K}}(\xi) = \liminf_{\xi \to \xi_0; \xi \in \mathbb{K}} \mathcal{T}_{\mathbb{K}}(\xi) = \liminf_{\xi \to \xi_0; \xi \in \mathbb{K}} \mathcal{B}_{\mathbb{K}}(\xi) = \mathcal{C}_{\mathbb{K}}(\xi_0). \tag{2.5.5}$$

Concerning the equalities (2.5.5), as far as we know, the main and most difficult part, i.e. the equality of the second term to the last one has been proved independently by Ursescu [96] and by Cornet [36] in $\mathbb{R}^n$. See also Treiman [93] for the proof of the inclusion $\liminf_{\xi \to \xi_0; \xi \in \mathbb{K}} \mathcal{T}_{\mathbb{K}}(\xi) \subseteq \mathcal{C}_{\mathbb{K}}(\xi_0)$ in general Banach spaces. We emphasize that there are examples showing that the converse inclusion does not hold in infinite-dimensional Banach spaces. See, for instance, Treiman [93]. We notice that Ursescu [96] proves a characterization of $\mathcal{C}_{\mathbb{K}}(\xi)$ in general Banach spaces, pertaining also an immediate proof of Treiman's main result in [93].

PROPOSITION 2.5.1. *Let* $\mathbb{K} \subseteq \mathbb{R}^n$ *be locally closed and let* $f : \mathbb{K} \to \mathbb{R}^n$ *be continuous. Then, the following conditions are equivalent*:
    (i) *for each* $\xi \in \mathbb{K}$, $f(\xi) \in \mathcal{C}_{\mathbb{K}}(\xi)$,
    (ii) *for each* $\xi \in \mathbb{K}$, $f(\xi) \in \mathcal{T}_{\mathbb{K}}(\xi)$,
    (iii) *for each* $\xi \in \mathbb{K}$, $f(\xi) \in \mathcal{B}_{\mathbb{K}}(\xi)$.
In general, if $\mathcal{G} : \mathbb{K} \rightsquigarrow \mathbb{R}^n$ *is such that* $\mathcal{C}_{\mathbb{K}}(\xi) \subseteq \mathcal{G}(\xi) \subseteq \mathcal{B}_{\mathbb{K}}(\xi)$ *for each* $\xi \in \mathbb{K}$, *then each one of the conditions above is equivalent to*:
    (iv) *for each* $\xi \in \mathbb{K}$, $f(\xi) \in \mathcal{G}(\xi)$.

PROOF. In view of Remark 2.5.1, it suffices to show that (i) is equivalent to (iii). But this easily follows from Lemma 2.5.1 and this completes the proof.   □

We notice that the equivalences in Proposition 2.5.1 were called to our attention by Ursescu [101].

3. Problems of viability

3.1. *Nagumo's viability theorem*

We begin with some background material we will need subsequently.

DEFINITION 3.1.1. Let $\mathbb{K} \subseteq \mathbb{R}^n$ be nonempty and $f : \mathbb{I} \times \mathbb{K} \to \mathbb{R}^n$. The subset $\mathbb{K}$ is *viable* with respect to f if for every $(\tau, \xi) \in \mathbb{I} \times \mathbb{K}$ there exists $T \in \mathbb{I}$, $T > \tau$, such that (1.1.1) has at least one solution, $u : [\tau, T] \to \mathbb{K}$, satisfying $u(\tau) = \xi$.

We can now proceed to the main result in this section.

THEOREM 3.1.1 (Nagumo [70]). *Let* $\mathbb{K} \subset \mathbb{R}^n$ *be a nonempty and locally closed set and let* $f : \mathbb{I} \times \mathbb{K} \to \mathbb{R}^n$ *be a continuous function. The necessary and sufficient condition in order that* $\mathbb{K}$ *be viable with respect to* f *is that, for every* $(t, \xi) \in \mathbb{I} \times \mathbb{K}$, $f(t, \xi) \in \mathcal{T}_{\mathbb{K}}(\xi)$.

3.2. *Proof of the necessity*

PROOF. In order to prove the necessity let $(t, \xi) \in \mathbb{I} \times \mathbb{K}$. Then, there exist $T \in \mathbb{I}$, $T > t$, and a function $u : [t, T] \to \mathbb{K}$, satisfying $u(t) = \xi$ and $u'(s) = f(s, u(s))$ for every $s \in [t, T]$. Consequently, we have

$$\lim_{h \downarrow 0} \frac{1}{h} \left\| \xi + hf(t, \xi) - u(t + h) \right\| = \lim_{h \downarrow 0} \left\| f(t, u(t)) - \frac{u(t + h) - u(t)}{h} \right\| = 0.$$

But, this relation shows that, for every $(t, \xi) \in \mathbb{I} \times \mathbb{K}$, $f(t, \xi) \in \mathcal{T}_{\mathbb{K}}(\xi)$ and the proof of the necessity is complete.

We notice that, in fact, we have proved much more than claimed by the necessity part of Theorem 3.1.1. Namely, we deduced:

THEOREM 3.2.1. *If $\mathbb{K}$ is viable with respect to $f : \mathbb{I} \times \mathbb{K} \to \mathbb{R}^n$, then, for each $(t, \xi) \in \mathbb{I} \times \mathbb{K}$, we have*

$$\lim_{h \downarrow 0} \frac{1}{h} \, \mathrm{dist}\big(\xi + hf(t, \xi); \mathbb{K}\big) = 0.$$

We notice that in Theorem 3.2.1 neither $\mathbb{K}$ is assumed to be locally closed, nor f to be continuous.

3.3. *Existence of approximate solutions*

PROOF OF THEOREM 3.1.1 (Continued). For the sake of simplicity, we divide the proof of the sufficiency into three steps. In the first one we shall prove the existence of a family of approximate solutions for the Cauchy problem (1.1.2) defined on intervals of the form $[\tau, c]$, with $c \in \mathbb{I}$. In the second step we will show that the problem (1.1.2) admits such approximate solutions, all defined on an interval $[\tau, T]$ independent of the "approximation order". Finally, in the last step, we shall prove the uniform convergence on $[\tau, T]$ of a sequence of such approximate solutions to a solution of the problem (1.1.2).

Let $(\tau, \xi) \in \mathbb{I} \times \mathbb{K}$ be arbitrary and let us choose $\rho > 0$, $M > 0$ and $T \in \mathbb{I}$, $T > \tau$, such that $B(\xi, \rho) \cap \mathbb{K}$ be closed,

$$\left\| f(t, x) \right\| \leqslant M \tag{3.3.1}$$

for every $t \in [\tau, T]$ and $x \in B(\xi, \rho) \cap \mathbb{K}$, and

$$(T - \tau)(M + 1) \leqslant \rho. \tag{3.3.2}$$

The existence of these three numbers is ensured by the fact that $\mathbb{K}$ is locally closed (from where it follows the existence of $\rho > 0$), by the continuity of f which implies its boundedness on $[\tau, T] \times B(\xi, \rho)$, and so the existence of $M > 0$, and by the fact that $T \in \mathbb{I}$, $T > \tau$, can be chosen as close to τ as we wish. We will show that, once fixed an $\varepsilon \in (0, 1)$ and the three numbers above, there exist three functions: $\sigma : [\tau, T] \to [\tau, T]$, nondecreasing, $g : [\tau, T] \to \mathbb{R}^n$, Riemann integrable and $u : [\tau, T] \to \mathbb{R}^n$, continuous, such that:

    (i) $t - \varepsilon \leqslant \sigma(t) \leqslant t$ for every $t \in [\tau, T]$;

    (ii) $\|g(t)\| \leqslant \varepsilon$ for every $t \in [\tau, T]$;

    (iii) $u(\sigma(t)) \in B(\xi, \rho) \cap \mathbb{K}$ for every $t \in [\tau, T]$ and $u(T) \in B(\xi, \rho) \cap \mathbb{K}$;

    (iv) u satisfies

$$u(t) = \xi + \int_\tau^t f\big(\sigma(s), u\big(\sigma(s)\big)\big)\, \mathrm{d}s + \int_\tau^t g(s)\, \mathrm{d}s$$

for every $t \in [\tau, T]$.

For the sake of simplicity, in all that follows, we will say that such a triple (σ, g, u) is an *ε-approximate solution* to the Cauchy problem (1.1.2) on the interval $[\tau, T]$. $\square$

The first step.   Let $\tau \in \mathbb{I}$, $\xi \in \mathbb{K}$ and let $\rho > 0$, $M > 0$ and $T \in \mathbb{I}$, $T > \tau$, be fixed as above. Let $\varepsilon \in (0, 1)$. We begin by showing the existence of an ε-approximate solution on an interval $[\tau, c]$ with $c \in (\tau, T]$. Since for every $(t, \xi) \in \mathbb{I} \times \mathbb{K}$, $f(t, \xi) \in \mathcal{T}_\mathbb{K}(\xi)$, from Proposition 2.3.2, it follows that there exists $c \in (\tau, T]$, $c - \tau \leqslant \varepsilon$ and $p \in \mathbb{R}^n$ with $\|p\| \leqslant \varepsilon$ such that $\xi + (c - \tau)f(\tau, \xi) + (c - \tau)p \in \mathbb{K}$. At this point, we can define the functions $\sigma : [\tau, c] \to [\tau, c]$, $g : [\tau, c] \to \mathbb{R}^n$ and $u : [\tau, c] \to \mathbb{R}^n$ by

$$\begin{cases} \sigma(t) = \tau & \text{for } t \in [\tau, c], \\ g(t) = p & \text{for } t \in [\tau, c], \\ u(t) = \xi + (t - \tau)f(\tau, \xi) + (t - \tau)p & \text{for } t \in [\tau, c]. \end{cases}$$

One can readily see that the triple (σ, g, u) is an ε-approximate solution to the Cauchy problem (1.1.2) on the interval $[\tau, c]$. Indeed the conditions (i), (ii) and (iv) are obviously fulfilled, while (iii) follows from (3.3.1), (3.3.2) and (i). Indeed, let us observe that $u(\sigma(t)) = \xi$, and therefore we have $u(\sigma(t)) \in B(\xi, \rho) \cap \mathbb{K}$ for every $t \in [\tau, c]$. Obviously $u(c) \in \mathbb{K}$. Moreover, by (3.3.1) and (3.3.2), we have

$$\|u(c) - \xi\| \leqslant (c - \tau)\|f(\tau, \xi)\| + (c - \tau)\|p\| \leqslant (T - \tau)(M + 1) \leqslant \rho$$

for every $t \in [\tau, c]$. Thus (iii) is also satisfied.

The second step.   Now, we will prove the existence of an ε-approximate solution defined on the whole interval $[\tau, T]$. To this aim we shall make use of Brezis–Browder Theorem 2.1.1, as follows. Let $\mathcal{S}$ be the set of all ε-approximate solutions to the problem (1.1.2) having the domains of definition of the form $[\tau, c]$ with $c \in (\tau, T]$. On $\mathcal{S}$ we define the relation "$\preceq$" by $(\sigma_1, g_1, u_1) \preceq (\sigma_2, g_2, u_2)$ if the domain of definition $[\tau, c_1]$ of the first triple is included in the domain of definition $[\tau, c_2]$ of the second triple, and the two ε-approximate solutions coincide on the common part of the domains. Obviously "$\preceq$" is a preorder relation on $\mathcal{S}$. Let us show first that each increasing sequence $((\sigma_m, g_m, u_m))_{m \in \mathbb{N}}$ is bounded from above. Indeed, let $((\sigma_m, g_m, u_m))_m$ be an increasing sequence, and let $c^* = \lim_m c_m$, where $[\tau, c_m]$ denotes the domain of definition of (σ_m, g_m, u_m). Clearly, $c^* \in (\tau, T]$. We will show that there exists at least one element, $(\sigma^*, g^*, u^*) \in \mathcal{S}$, defined on $[\tau, c^*]$ and satisfying $(\sigma_m, g_m, u_m) \preceq (\sigma^*, g^*, u^*)$ for each $m \in \mathbb{N}$. In order to do this, we have to prove first that there exists $\lim_m u_m(c_m)$. For each $m, k \in \mathbb{N}$, $m \leqslant k$, we have $u_m(s) = u_k(s)$ for all $s \in [\tau, c_m]$. Taking into account (iii), (iv) and (3.3.1), we deduce

$$\left\| u_m(c_m) - u_k(c_k) \right\| \leqslant \int_{c_m}^{c_k} \left\| f\big(\sigma_k(s), u_k\big(\sigma_k(s)\big)\big) \right\| \mathrm{d}s + \int_{c_m}^{c_k} \left\| g_k(s) \right\| \mathrm{d}s$$

$$\leqslant (M + \varepsilon)|c_k - c_m|$$

for every $m, k \in \mathbb{N}$, which proves that there exists $\lim_{m \to \infty} u_m(c_m)$. Since for every $m \in \mathbb{N}$, $u_m(c_m) \in B(\xi, \rho) \cap \mathbb{K}$, and the latter is closed, it readily follows that $\lim_{m \to \infty} u_m(c_m) \in B(\xi, \rho) \cap \mathbb{K}$. Furthermore, because all the functions in the set $\{\sigma_m;\ m \in \mathbb{N}\}$ are nondecreasing, with values in $[\tau, c^*]$, and satisfy $\sigma_m(c_m) \leqslant \sigma_p(c_p)$ for every $m, p \in \mathbb{N}$ with $m \leqslant p$, there exists $\lim_{m \to \infty} \sigma_m(c_m)$ and this limit belongs to $[\tau, c^*]$. This shows that we can define the triple of functions $(\sigma^*, g^*, u^*) : [\tau, c^*] \to [\tau, c^*] \times \mathbb{R}^n \times \mathbb{R}^n$ by

$$\sigma^*(t) = \begin{cases} \sigma_m(t) & \text{for } t \in [\tau, c_m],\, m \in \mathbb{N}, \\ \lim_{m \to \infty} \sigma_m(c_m) & \text{for } t = c^*, \end{cases}$$

$$g^*(t) = \begin{cases} g_m(t) & \text{for } t \in [\tau, c_m],\, m \in \mathbb{N}, \\ 0 & \text{for } t = c^*, \end{cases}$$

$$u^*(t) = \begin{cases} u_m(t) & \text{for } t \in [\tau, c_m],\, m \in \mathbb{N}, \\ \lim_{m \to \infty} u_m(c_m) & \text{for } t = c^*. \end{cases}$$

One can easily see that (σ^*, g^*, u^*) is an ε-approximate solution which is a majorant for $((\sigma_m, g_m, u_m))_m$. Let us define the function $\mathcal{M} : \mathcal{S} \to \mathbb{R} \cup \{+\infty\}$ by $\mathcal{M}((\sigma, g, u)) = c$, where $[\tau, c]$ is the domain of definition of (σ, g, u). Clearly $\mathcal{M}$ satisfies the hypotheses of Brezis–Browder Theorem 2.1.1. Then, $\mathcal{S}$ contains at least one $\mathcal{M}$-maximal element $(\bar{\sigma}, \bar{g}, \bar{u})$, defined on $[\tau, \bar{c}]$. In other words, if $(\tilde{\sigma}, \tilde{g}, \tilde{u}) \in \mathcal{S}$, defined on $[\tau, \tilde{c}]$, satisfies $(\bar{\sigma}, \bar{g}, \bar{u}) \preceq (\tilde{\sigma}, \tilde{g}, \tilde{u})$, then we necessarily have $\bar{c} = \tilde{c}$. We will show next that $\bar{c} = T$. Indeed, let us assume by contradiction that $\bar{c} < T$. Then, taking into account the fact that $\bar{u}(\bar{c}) \in B(\xi, \rho) \cap \mathbb{K}$, we deduce that

$$\left\| \bar{u}(\bar{c}) - \xi \right\| \leqslant \int_\tau^{\bar{c}} \left\| f\big(\bar{\sigma}(s), \bar{u}\big(\bar{\sigma}(s)\big)\big) \right\| \mathrm{d}s + \int_\tau^{\bar{c}} \left\| \bar{g}(s) \right\| \mathrm{d}s \leqslant (\bar{c} - \tau)(M + \varepsilon)$$

$$\leqslant (\bar{c} - \tau)(M + 1) < (T - \tau)(M + 1) \leqslant \rho.$$

Then, as $\bar{u}(\bar{c}) \in \mathbb{K}$ and $f(\bar{c}, \bar{u}(\bar{c})) \in \mathcal{T}_{\mathbb{K}}(\bar{u}(\bar{c}))$, there exist $\delta \in (0, T - \bar{c})$, $\delta \leqslant \varepsilon$ and $p \in \mathbb{R}^n$ such that $\|p\| \leqslant \varepsilon$ and $\bar{u}(\bar{c}) + \delta f(\bar{c}, \bar{u}(\bar{c})) + \delta p \in \mathbb{K}$. From the inequality above, it follows that we can diminish δ if necessary, in order to have $\|\bar{u}(\bar{c}) + \delta[f(\bar{c}, \bar{u}(\bar{c})) + p] - \xi\| \leqslant \rho$. Let us define the functions $\sigma : [\tau, \bar{c} + \delta] \to [\tau, \bar{c} + \delta]$ and $g : [\tau, \bar{c} + \delta] \to \mathbb{R}^n$ by

$$\sigma(t) = \begin{cases} \bar{\sigma}(t) & \text{for } t \in [\tau, \bar{c}], \\ \bar{c} & \text{for } t \in (\bar{c}, \bar{c} + \delta], \end{cases}$$

$$g(t) = \begin{cases} \bar{g}(t) & \text{for } t \in [\tau, \bar{c}], \\ p & \text{for } t \in (\bar{c}, \bar{c} + \delta]. \end{cases}$$

Clearly, g is Riemann integrable on $[\tau, \bar{c} + \delta]$ and $\|g(t)\| \leqslant \varepsilon$ for every $t \in [\tau, \bar{c} + \delta]$. In addition, for every $t \in [\tau, \bar{c} + \delta]$, $\sigma(t) \in [\tau, \bar{c}]$, and therefore $\bar{u}(\sigma(t))$ is well-defined and belongs to the set $B(\xi, \rho) \cap \mathbb{K}$. Accordingly, we can define $u : [\tau, \bar{c} + \delta] \to \mathbb{R}^n$ by

$$u(t) = \xi + \int_\tau^t f\big(\sigma(s), \bar{u}(\sigma(s))\big) \, ds + \int_\tau^t g(s) \, ds$$

for every $t \in [\tau, \bar{c} + \delta]$. Clearly u coincides with $\bar{u}$ on $[\tau, \bar{c}]$ and then it readily follows that u, σ and g satisfy all the conditions in (i) and (ii). In order to prove (iii) and (iv), let us observe that

$$u(t) = \begin{cases} \bar{u}(t) & \text{for } t \in [\tau, \bar{c}], \\ u(\bar{c}) + (t - \bar{c})f\big(\bar{c}, \bar{u}(\bar{c})\big) + (t - \bar{c})p & \text{for } t \in (\bar{c}, \bar{c} + \delta]. \end{cases}$$

Then u satisfies the equation in (iv). Since

$$u\big(\sigma(t)\big) = \begin{cases} \bar{u}\big(\bar{\sigma}(t)\big) & \text{for } t \in [\tau, \bar{c}], \\ \bar{u}(\bar{c}) & \text{for } t \in [\bar{c}, \bar{c} + \delta], \end{cases}$$

it follows that $u(\sigma(t)) \in B(\xi, \rho) \cap \mathbb{K}$. Furthermore, from the choice of δ and p, we have both $u(\bar{c} + \delta) = \bar{u}(\bar{c}) + \delta f(\bar{c}, \bar{u}(\bar{c})) + \delta p \in \mathbb{K}$, and

$$\big\|u(\bar{c} + \delta) - \xi\big\| = \big\|\bar{u}(\bar{c}) + \delta f\big(\bar{c}, \bar{u}(\bar{c})\big) + \delta p - \xi\big\| \leqslant \rho$$

and consequently u satisfies (iii). Thus $(\sigma, g, u) \in \mathcal{S}$. Furthermore, since $(\bar{\sigma}, \bar{g}, \bar{u}) \preceq (\sigma, g, u)$ and $\bar{c} < \bar{c} + \delta$, it follows that $(\bar{\sigma}, \bar{g}, \bar{u})$ is not an $\mathcal{M}$-maximal element. But this is absurd. This contradiction can be eliminated only if each maximal element in the set $\mathcal{S}$ is defined on $[\tau, T]$.

3.4. *Convergence of approximate solutions*

The third step.   Let $(\varepsilon_k)_{k \in \mathbb{N}}$ be a sequence from $(0, 1)$ decreasing to 0 and let $((\sigma_k, g_k, u_k))_{k \in \mathbb{N}}$ be a sequence of ε_k-approximate solutions defined on $[\tau, T]$. From (i) and (ii), it follows that

$$\lim_{k \to \infty} \sigma_k(t) = t \quad \text{and} \quad \lim_{k \to \infty} g_k(t) = 0 \tag{3.4.1}$$

uniformly on $[\tau, T]$. On the other hand, from (iii), (iv) and (3.3.2), we have

$$\|u_k(t)\| \leqslant \|u_k(t) - \xi\| + \|\xi\|$$

$$\leqslant \int_\tau^T \|f(\sigma_k(s), u_k(\sigma_k(s)))\| \, \mathrm{d}s + \int_\tau^T \|g_k(s)\| \, \mathrm{d}s + \|\xi\|$$

$$\leqslant (T - \tau)(M + 1) + \|\xi\| \leqslant \rho + \|\xi\|$$

for every $k \in \mathbb{N}$ and every $t \in [\tau, T]$. Hence, the sequence $(u_k)_{k \in \mathbb{N}}$ is uniformly bounded on $[\tau, T]$. Again from (iv), we have

$$\|u_k(t) - u_k(s)\| \leqslant \left| \int_s^t \|f(\sigma_k(\tau), u_k(\sigma_k(\tau)))\| \, \mathrm{d}\tau \right| + \left| \int_s^t \|g_k(\tau)\| \, \mathrm{d}\tau \right|$$

$$\leqslant (M + 1)|t - s|$$

for every $t, s \in [\tau, T]$. Consequently the sequence $(u_k)_{k \in \mathbb{N}}$ is equicontinuous on $[\tau, T]$. From Arzelà–Ascoli theorem—see Vrabie [106, Theorem 8.2.1, p. 320]—it follows that, at least on a subsequence, $(u_k)_{k \in \mathbb{N}}$ is uniformly convergent on $[\tau, T]$ to a function $u : [\tau, T] \to \mathbb{R}^n$. Taking into account of (iii), (3.4.1) and of the fact that $B(\xi, \rho) \cap \mathbb{K}$ is closed, we deduce that $u(t) \in B(\xi, \rho) \cap \mathbb{K}$ for every $t \in [\tau, T]$. Passing to the limit in the equation

$$u_k(t) = \xi + \int_\tau^t f(\sigma_k(\tau), u_k(\sigma_k(s))) \, \mathrm{d}s + \int_\tau^t g_k(s) \, \mathrm{d}s$$

and taking into account (3.4.1), we deduce that

$$u(t) = \xi + \int_\tau^t f(s, u(s)) \, \mathrm{d}s$$

for every $t \in [\tau, T]$, which achieves the proof of the theorem. $\qquad\square$

From Remark 2.3.1 combined with Theorem 3.1.1, we deduce Peano's local existence theorem 1.1.1. We mention that Theorem 1.1.2 too is a direct consequence of Theorem 3.1.1 combined with the observation below.

REMARK 3.4.1. Let $\mathbb{D}$ be a nonempty and open subset in $\mathbb{R}^n$ and let $U : \mathbb{D} \to \mathbb{R}$ be a function of class C^1 with $\nabla U(\xi) \neq 0$ on $\mathbb{D}$. Let $c \in \mathbb{R}$ be such that $\mathbb{K} = \{\xi \in \Omega; \ U(\xi) = c\}$ is nonempty. Then $\eta \in \mathbb{R}^n$ is tangent to $\mathbb{K}$ at the point $\xi \in \mathbb{K}$ if and only if $\langle \eta, \nabla U(\xi) \rangle = 0$. In other words, in this case, $\mathcal{T}_{\mathbb{K}}(\xi)$ coincides with the set of vectors in the tangent plane to $\mathbb{K}$ at ξ. Indeed, let us observe that a vector $\eta \in \mathcal{T}_{\mathbb{K}}(\xi)$ if and only if there exists a function $u : [0, 1] \to \mathbb{K}$ with $u(0) = \xi$, differentiable at $t = 0$, with $u'(0) = \eta$, and such that

$$\lim_{t \downarrow 0} \frac{1}{t} \|\xi + t\eta - u(t)\| = 0.$$

But, in the particular case of the set $\mathbb{K}$ considered, this relation holds if and only if $\langle \eta, \nabla U(\xi) \rangle = 0$, which achieves the proof.

Summing up, and using Proposition 2.5.1, we deduce:

THEOREM 3.4.1. *Let $\mathbb{K}$ be a nonempty and locally closed subset in $\mathbb{R}^n$ and let $f : \mathbb{I} \times \mathbb{K} \to \mathbb{R}^n$ be continuous. Then the following conditions are equivalent:*
    (i) *for every $(t, \xi) \in \mathbb{I} \times \mathbb{K}$, $f(t, \xi) \in \mathcal{C}_{\mathbb{K}}(\xi)$,*
    (ii) *for every $(t, \xi) \in \mathbb{I} \times \mathbb{K}$, $f(t, \xi) \in \mathcal{T}_{\mathbb{K}}(\xi)$,*
    (iii) *for every $(t, \xi) \in \mathbb{I} \times \mathbb{K}$, $f(t, \xi) \in \mathcal{B}_{\mathbb{K}}(\xi)$,*
    (iv) *the set $\mathbb{K}$ is viable with respect to f.*
In general, if $\mathcal{G} : \mathbb{K} \rightsquigarrow \mathbb{R}^n$ is such that $\mathcal{C}_{\mathbb{K}}(\xi) \subseteq \mathcal{G}(\xi) \subseteq \mathcal{B}_{\mathbb{K}}(\xi)$ for each $\xi \in \mathbb{K}$, then each one of the conditions above is equivalent to
    (v) *for each $\xi \in \mathbb{K}$, $f(t, \xi) \in \mathcal{G}(\xi)$.*

3.5. *Existence of noncontinuable solutions*

In this section we will prove some results concerning the existence of noncontinuable, or even global solutions to (1.1.1). We recall that a solution $u : [\tau, T) \to \mathbb{K}$ to (1.1.1) is called *noncontinuable*, if there is no other solution $v : [\tau, \widetilde{T}) \to \mathbb{K}$ of the same equation, with $T < \widetilde{T}$ and satisfying $u(t) = v(t)$ for all $t \in [\tau, T)$. The solution u is called *global* if $T = \sup \mathbb{I}$. The next theorem follows from Brezis–Browder Theorem 2.1.1.

THEOREM 3.5.1. *Let $\mathbb{K} \subseteq \mathbb{R}^n$ be nonempty and let $f : \mathbb{I} \times \mathbb{K} \to \mathbb{R}^n$ be a possibly discontinuous function. Then, the following conditions are equivalent:*
    (i) *$\mathbb{K}$ is viable with respect to f,*
    (ii) *for each $(\tau, \xi) \in \mathbb{I} \times \mathbb{K}$ there exists at least one noncontinuable solution $u : [\tau, T) \to \mathbb{K}$ of (1.1.1), satisfying $u(\tau) = \xi$.*

PROOF. Clearly (ii) implies (i). To prove that (i) implies (ii) it suffices to show that every solution u can be continued up to a noncontinuable one. To this aim, we will make use of Brezis–Browder Theorem 2.1.1. Let $\mathcal{S}$ be the set of all solutions to (1.1.1). On $\mathcal{S}$ which, by virtue of (i), is nonempty, we define the binary relation "$\preceq$" by $u \preceq v$ if the domain $[\tau, T_v)$ of v is larger that the domain $[\tau, T_u)$ of u, i.e. $T_u \leqslant T_v$, and $u(t) = v(t)$ for all $t \in [\tau, T_u)$. Clearly "$\preceq$" is a preorder on $\mathcal{S}$. Next, let $(u_m)_m$ be an increasing sequence in $\mathcal{S}$, and let us denote by $[\tau, T_m)$ the domain of definition of u_m. Let $T^* = \lim_{m \to \infty} T_m$, which can be finite, or not, and let us define $u^* : [\tau, T^*) \to \mathbb{K}$ by $u^*(t) = u_m(t)$ for each $t \in [\tau, T_m)$. Since $(T_m)_m$ is increasing and $u_m(t) = u_k(t)$ for each $m \leqslant k$ and each $t \in [\tau, T_m)$, u^* is well-defined and belongs to $\mathcal{S}$. Moreover, u^* is a majorant of $(u_m)_m$. Thus each increasing sequence in $\mathcal{S}$ is bounded from above. Moreover, the function $\mathcal{M} : \mathcal{S} \to \mathbb{R} \cup \{+\infty\}$, defined by $\mathcal{M}(v) = T_v$, for each $v \in \mathcal{S}$, is monotone, and therefore we are in the hypotheses of Theorem 2.1.1. Accordingly, for $u \in \mathcal{S}$, there exists at least one element $\bar{u} \in \mathcal{S}$ with $u \preceq \bar{u}$ and, in addition, $\bar{u} \preceq \tilde{u}$ implies $T_{\bar{u}} = T_{\tilde{u}}$. But this means that $\bar{u}$ is noncontinuable, and, of course, that it extends u. The proof is complete. $\qquad\square$

THEOREM 3.5.2. *Let* $\mathbb{K}$ *be a nonempty and locally closed subset in* $\mathbb{R}^n$, *and let* $f : \mathbb{I} \times \mathbb{K} \to \mathbb{R}^n$ *be a continuous function. Then a necessary and sufficient condition in order that for each* $(\tau, \xi) \in \mathbb{I} \times \mathbb{K}$ *there exists at least one noncontinuable solution to* (1.1.1) *satisfying* $u(\tau) = \xi$ *is one of the five equivalent conditions in Theorem* 3.4.1.

PROOF. In view of Theorem 3.4.1, each one of the first five conditions is equivalent to the viability of $\mathbb{K}$ with respect to f. The conclusion follows from Theorem 3.5.1, and the proof is complete. $\qquad\square$

In order to obtain global existence, some extra growth conditions on f are needed. We discuss below a very natural one, introduced in a more general framework in Vrabie [102, Definition 3.2.1, p. 95].

DEFINITION 3.5.1. A function $f : \mathbb{I} \times \mathbb{K} \to \mathbb{R}^n$ is called *positively sublinear* if there exists a norm $\| \cdot \|$ on $\mathbb{R}^n$ such that, for each $T \in \mathbb{I}$, there exist $a > 0$, $b \in \mathbb{R}$ and $c > 0$ satisfying

$$\left\| f(t, \xi) \right\| \leqslant a \|\xi\| + b$$

for each $(t, \xi) \in \mathbb{K}_+^c(f)$, where

$$\mathbb{K}_+^c(f) = \left\{ (t, \xi) \in \mathbb{I} \times \mathbb{K};\ t \leqslant T,\ \|\xi\| > c \text{ and } \left[\xi, f(t, \xi) \right]_+ > 0 \right\}.$$

As concerns the existence of global solutions we have:

THEOREM 3.5.3. *Let* $\mathbb{K}$ *be a nonempty and closed subset in* $\mathbb{R}^n$, *and let* $f : \mathbb{I} \times \mathbb{K} \to \mathbb{R}^n$ *be a continuous and positively sublinear function. Then a necessary and sufficient condition in order that for each* $(t, \xi) \in \mathbb{I} \times \mathbb{K}$ *there exists at least one global solution to* (1.1.1) *satisfying* $u(\tau) = \xi$ *is each one of the five equivalent conditions in Theorem* 3.4.1.

PROOF. Clearly, each one of the five conditions is necessary. To complete the proof it suffices to show that, whenever $\mathbb{K}$ is viable with respect to f, then, for each $(\tau, \xi) \in \mathbb{I} \times \mathbb{K}$, there exists at least one global solution $u : [\tau, T) \to \mathbb{K}$ to (1.1.1) satisfying $u(\tau) = \xi$. To this aim, let $(\tau, \xi) \in \mathbb{I} \times \mathbb{K}$ and let $u : [\tau, T) \to \mathbb{K}$ be a noncontinuable solution to (1.1.1) satisfying $u(\tau) = \xi$. We will show that $T = \sup \mathbb{I}$. To this aim, let us assume the contrary, i.e., that $T < \sup \mathbb{I}$. In particular, this means that $T < +\infty$. Since $u'(s) = f(s, (u(s))$ for all $s \in [\tau, T)$, we deduce

$$\left[u(s), u'(s) \right]_+ = \left[u(s), f\left(s, u(s)\right) \right]_+.$$

Since $[u(s), u'(s)]_+ = \frac{\mathrm{d}^+}{\mathrm{d}s}(\|u(s)\|)$ for $s \in [\tau, T)$, where $\| \cdot \|$ is the norm whose existence is ensured by Definition 3.5.1, integrating from τ to t the last equality, we get successively

$$\left\| u(t) \right\| = \|\xi\| + \int_{\tau}^{t} \left[u(s), f\left(s, u(s)\right) \right]_+ \mathrm{d}s$$

$$= \|\xi\| + \int_{\{\tau \leqslant s \leqslant t; \|u(s)\| \leqslant c\}} \left[u(s), f\big(s, u(s)\big) \right]_+ \mathrm{d}s$$

$$+ \int_{\{\tau \leqslant s \leqslant t; \|u(s)\| > c\}} \left[u(s), f\big(s, u(s)\big) \right]_+ \mathrm{d}s$$

$$\leqslant \|\xi\| + c\lambda\big(\{ s \in [\tau, t]; \, \|u(s)\| \leqslant c \}\big)$$

$$+ \int_{\{\tau \leqslant s \leqslant t; \|u(s)\| > c\}} \left[a\|u(s)\| + b \right] \mathrm{d}s$$

$$\leqslant \|\xi\| + (b + c)(T - \tau) + a \int_{\tau}^{t} \|u(s)\| \, \mathrm{d}s.$$

Here, as usual, λ denotes the Lebesgue measure on $\mathbb{R}$. Thanks to Gronwall's inequality—see Vrabie [106, Lemma 1.5.2, p. 46]—u is bounded on $[\tau, T)$ and, since $T < +\infty$, it follows that $\{ f(t, u(t)); \, t \in [\tau, T) \}$ is bounded. Therefore u is globally Lipschitz on $[\tau, T)$ and accordingly there exists $\lim_{t \uparrow T} u(t) = u^*$. Since $\mathbb{K}$ is closed and $T < \sup \mathbb{I}$, it follows that $(T, u^*) \in \mathbb{I} \times \mathbb{K}$. Using this observation and recalling that $\mathbb{K}$ is viable with respect to f, we conclude that u can be continued to the right of T. But this is absurd, because u is noncontinuable. This contradiction can be eliminated only if $T = \sup \mathbb{I}$, and this achieves the proof. $\square$

As, whenever $\mathbb{K}$ is compact, each continuous function $f : \mathbb{I} \times \mathbb{K} \to \mathbb{R}^n$ is positively sublinear, from Theorem 3.5.3 it readily follows:

COROLLARY 3.5.1. *Let $\mathbb{K}$ be a nonempty and compact subset of $\mathbb{R}^n$, and let $f : \mathbb{I} \times \mathbb{K} \to \mathbb{R}^n$ be continuous. Then a necessary and sufficient condition in order that for each $(\tau, \xi) \in \mathbb{I} \times \mathbb{K}$ there exists at least one solution, $u : [\tau, T) \to \mathbb{K}$ to (1.1.1), satisfying $T = \sup \mathbb{I}$ and $u(\tau) = \xi$, is any one of the five equivalent conditions in Theorem 3.4.1.*

3.6. *Viability of the relative closure*

PROPOSITION 3.6.1 (Roxin [86]). *Let $\mathbb{D} \subseteq \mathbb{R}^n$ be open, let $\mathbb{K} \subseteq \mathbb{D}$ and let $f : \mathbb{I} \times \mathbb{D} \to \mathbb{R}^n$ be continuous. If $\mathbb{K}$ is viable with respect to f, then its closure relative to $\mathbb{D}$, $\overline{\mathbb{K}}^{\mathbb{D}}$, is also viable with respect to f.*

PROOF. Let $\tau \in \mathbb{I}$ and let $(\xi_k)_k$ be a sequence in $\mathbb{K}$ convergent to $\xi \in \overline{\mathbb{K}}^{\mathbb{D}}$. Since $\mathbb{K}$ is viable with respect to f, there exists a sequence $(u(\cdot, \xi_k))_k$ of $\mathbb{K}$-valued noncontinuable solutions to (1.1.1) satisfying $u(\tau, \xi_k) = \xi_k$, for $k = 1, 2, \dots$. It is well known that the intersection of the domains of this sequence contains a nontrivial interval $[\tau, T]$. See, for instance, Vrabie [106, Lemma 3.2.1, p. 107]. Moreover, diminishing T if necessary, we may assume that there exists $\rho > 0$ such that $u(t, \xi_k) \in B(\xi, \rho) \subseteq \mathbb{D}$, for all $k \in \mathbb{N}$, and $t \in [\tau, T]$. By a compactness argument involving Arzelà–Ascoli theorem, we conclude that, on a subsequence at least, we have $\lim_{k \to \infty} u(\cdot, \xi_k) = u(\cdot, \xi)$ uniformly on $[\tau, T]$. Thus $u(t, \xi) \in \overline{\mathbb{K}}^{\mathbb{D}}$ for all $t \in [\tau, T]$, and this completes the proof. $\square$

3.7. *Comparison and viability*

The next theorem, called to our attention by Ursescu [101], gives a characterization of the viability of an epigraph of a certain function in the terms of a differential inequality. Similar results can be found in Clarke et al. [33, p. 266]. Throughout, we denote by $[D_+x](t)$ the *right lower Dini derivative* of the function x at t, i.e.

$$[D_+x](t) = \liminf_{h\downarrow 0} \frac{x(t+h) - x(t)}{h}.$$

THEOREM 3.7.1. *Let $\omega : \mathbb{I} \times \mathbb{R}_+ \to \mathbb{R}$ and $v : [\tau, T) \to \mathbb{R}_+$ be continuous, with $[\tau, T) \subseteq \mathbb{I}$. Then*

$$\mathrm{epi}(v) = \big\{(t, \eta); \ v(t) \leqslant \eta, \ t \in [\tau, T)\big\}$$

is viable with respect to $(t, y) \mapsto (1, \omega(t, y))$ if and only if v satisfies

$$[D_+v](t) \leqslant \omega\big(t, v(t)\big) \tag{3.7.1}$$

for each $t \in [\tau, T)$.

PROOF. *Sufficiency.* It suffices to show that the set $\{(t, v(t)); \ t \in [\tau, T)\}$, included in the boundary $\partial\mathrm{epi}(v)$ of $\mathrm{epi}(v)$, satisfies the Nagumo's tangency condition (1.1.3). From (3.7.1) it follows that

$$\left[D_+\left(v(\cdot) - \int_\tau^{\cdot} \omega\big(s, v(s)\big)\,ds\right)\right](t) \leqslant 0$$

for each $t \in [\tau, T)$. Thus, in view of a classical result in Hobson [56, p. 365], we necessarily have that $t \mapsto v(t) - \int_\tau^t \omega(s, v(s))\,ds$ is non-increasing on $[\tau, T]$. So, for each $t \in [\tau, T)$ and $h > 0$ such that $t + h < T$, we have

$$\left(t + h, v(t) + \int_t^{t+h} \omega\big(s, v(s)\big)\,ds\right) \in \mathrm{epi}(v),$$

and therefore

$$\mathrm{dist}\big((t, v(t)) + h\big(1, \omega\big(t, v(t)\big)\big); \mathrm{epi}(v)\big)$$

$$\leqslant \left\|(t, v(t)) + h\big(1, \omega\big(t, v(t)\big)\big) - \left(t + h, v(t) + \int_t^{t+h} \omega\big(s, v(s)\big)\,ds\right)\right\|$$

$$= \left|h\omega\big(t, v(t)\big) - \int_t^{t+h} \omega\big(s, v(s)\big)\,ds\right|.$$

Dividing by $h > 0$ and passing to $\liminf$ for $h \downarrow 0$ we get (1.1.3) and this completes the proof of the sufficiency.

Necessity. Let us assume that epi(v) is viable with respect to the function $(t, y) \mapsto (1, \omega(t, y))$, let $t \in [\tau, T)$, and let (s, x) be a solution to $s' = 1$, $x'(s) = \omega(s, x(s))$, satisfying the initial conditions $s(0) = t$ and $x(0) = v(t)$, and which remains in epi(v). We have

$$\frac{v(t+h) - v(t)}{h} \leqslant \frac{x(h) - x(0)}{h}.$$

Accordingly

$$[D_+v](t) \leqslant \omega\big(s(0), x(0)\big) = \omega\big(t, v(t)\big),$$

and this achieves the proof of the necessity. $\qquad\square$

DEFINITION 3.7.1. A function $\omega : \mathbb{I} \times [0, a) \to \mathbb{R}$, $0 < a \leqslant +\infty$, is a *comparison* function if $\omega(t, 0) = 0$ for each $t \in [0, a)$, and for each $[\tau, T) \subseteq \mathbb{I}$, the only continuous function $x : [\tau, T) \to [0, a)$, satisfying

$$\begin{cases} [D_+x](t) \leqslant \omega\big(t, x(t)\big) & \text{for all } t \in [\tau, T), \\ x(\tau) = 0, \end{cases}$$

is the null function.

COROLLARY 3.7.1. *Let $\omega : \mathbb{I} \times \mathbb{R}_+ \to \mathbb{R}$ be continuous and such that, for each $\tau \in \mathbb{I}$, the Cauchy problem $y'(t) = \omega(t, y(t))$, $y(\tau) = 0$ has only the null solution. Then ω is a comparison function.*

PROOF. Let $v : [\tau, T] \to \mathbb{R}_+$ be any solution to (3.7.1). By Theorem 3.7.1, epi(v) is viable with respect to $(t, y) \mapsto (1, \omega(t, y))$. So, the unique solution $y : [\tau, T) \to \mathbb{R}_+$ of the Cauchy problem $y'(t) = \omega(t, y(t))$, $y(\tau) = 0$ satisfies $0 \leqslant v(t) \leqslant y(t) = 0$. $\qquad\square$

3.8. *Viable preordered subsets*

Let us assume now that $\mathbb{K}$ is a nonempty subset in $\mathbb{R}^n$, and let "$\preceq$" be a *preorder* on $\mathbb{K}$, i.e. a reflexive and transitive binary relation. For our later purposes, it is convenient to identify "$\preceq$" with the multifunction $\mathcal{P} : \mathbb{K} \rightsquigarrow \mathbb{K}$, defined by

$$\mathcal{P}(\xi) = \{\eta \in \mathbb{K};\ \xi \preceq \eta\}$$

for each $\xi \in \mathbb{K}$, and called also a preorder. We say that a preorder "$\preceq$", or $\mathcal{P}$ is *closed* if "$\preceq$" is a closed subset in $\mathbb{R}^n \times \mathbb{R}^n$. Let $f : \mathbb{I} \times \mathbb{K} \to \mathbb{R}^n$, and let us consider the differential equation (1.1.1). We say that "$\preceq$", or $\mathcal{P}$, is *viable with respect to f* if for each $(\tau, \xi) \in \mathbb{I} \times \mathbb{K}$, there exist $[\tau, T] \subseteq \mathbb{I}$ and a solution $u : [\tau, T] \to \mathbb{R}^n$ of (1.1.1) satisfying $u(\tau) = \xi$, $u(t) \in \mathbb{K}$ for each $t \in [\tau, T]$ and u is "$\preceq$"-monotone on $[\tau, T]$, i.e., for each $\tau \leqslant s \leqslant t \leqslant T$, we have $u(s) \preceq u(t)$. The next lemma in Cârjă and Ursescu [25] is the main tool in our forthcoming analysis.

LEMMA 3.8.1. *Let $f : \mathbb{I} \times \mathbb{K} \to \mathbb{R}^n$ be continuous and let $\mathcal{P}$ be a preorder on $\mathbb{K}$. If $\mathcal{P}$ is viable with respect to f then, for each $\xi \in \mathbb{K}$, $\mathcal{P}(\xi)$ is viable with respect to f. If $\mathcal{P}$ is closed in $\mathbb{R}^n \times \mathbb{R}^n$ and, for each $\xi \in \mathbb{K}$, $\mathcal{P}(\xi)$ is viable with respect to f, then $\mathcal{P}$ is viable with respect to f.*

PROOF. Clearly, if $\mathcal{P}$ is viable with respect to f, then, for all $\xi \in \mathbb{K}$, $\mathcal{P}(\xi)$ is viable with respect to f.

Now, if $\mathcal{P}$ is closed, then, for each $\xi \in \mathbb{K}$, $\mathcal{P}(\xi)$ is a fortiori closed. Let us assume that, for each $\xi \in \mathbb{K}$, $\mathcal{P}(\xi)$ is viable with respect to f. Let $(\tau, \xi) \in \mathbb{I} \times \mathbb{K}$. We shall show that there exist $[\tau, T] \subseteq \mathbb{I}$ and at least one solution $u : [\tau, T] \to \mathbb{K}$ of (1.1.1), with $u(\tau) = \xi$, and such that $u([s, T]) \subseteq \mathcal{P}(u(s))$ for each $s \in [\tau, T]$. To this aim, we proceed in several steps.

In the first step, we note that, by standard qualitative arguments—see, for instance, Vrabie [106, Lemma 3.2.1, p. 107]—one can show that there exists $T > \tau$, $T \in \mathbb{I}$, such that for every noncontinuable solution $u : [\tau, T_m) \to \mathbb{K}$ to (1.1.1) with $u(\tau) = \xi$ we have $T < T_m$. Since $\mathcal{P}(\xi)$ is viable with respect to f, there exists a solution $u : [\tau, T] \to \mathbb{K}$ of (1.1.1) with $u(\tau) = \xi$ and $u([\tau, T]) \subseteq \mathcal{P}(\xi)$.

In the second step, we remark that, for every solution $v : [\tau, T] \to \mathbb{K}$ to (1.1.1), with $v(\tau) = \xi$ and $v([\tau, T]) \subseteq \mathcal{P}(\xi)$, and for every $v \in [\tau, T)$, there exists a solution $w : [\tau, T] \to \mathbb{K}$ to (1.1.1) such that w equals v on $[\tau, v]$ and $w([v, T]) \subseteq \mathcal{P}(w(v))$.

In the third step, we observe that, thanks to the first two steps, for every nonempty and finite subset S of $[\tau, T)$, with $\tau \in S$, there exists a solution $u : [\tau, T] \to \mathbb{K}$ of (1.1.1) satisfying both $u(\tau) = \xi$ and $u([s, T]) \subseteq \mathcal{P}(u(s))$ for all $s \in S$.

In the fourth step, we consider a sequence $(S_k)_{k \in \mathbb{N}}$ of nonempty finite subsets of $[\tau, T)$ such that $\tau \in S_k$, $S_k \subseteq S_{k+1}$ for each $n \in \mathbb{N}$, and the set $S = \cup_{k \in \mathbb{N}} S_k$ is dense in $[\tau, T]$. For example, we can take

$$S_k = \left\{ \tau + \frac{i}{2^k}(T - \tau); \ i \in \left\{ 0, 1, \ldots, 2^k - 1 \right\} \right\}.$$

Further, we shall make use of the third step to get a sequence of solutions $(u_k : [\tau, T] \to \mathbb{K})_k$ to (1.1.1), satisfying $u_k(\tau) = \xi$ and such that $u_k([s, T]) \subseteq \mathcal{P}(u_k(s))$ for each $k \in \mathbb{N}$ and each $s \in S_k$. Now, by virtue of Arzelà–Ascoli theorem, we can assume, taking a subsequence if necessary, that the sequence $(u_k)_k$ converges uniformly on $[\tau, T]$ to a solution $u : [\tau, T] \to \mathbb{K}$ of (1.1.1). Clearly $u(\tau) = \xi$.

In the fifth step, we show that $u([s, T]) \subset \mathcal{P}(u(s))$ for all $s \in S$. Indeed, given s as above, there exists $k \in \mathbb{N}$ such that $s \in S_k$. Then $s \in S_m$ and $u_m([s, T]) \subseteq \mathcal{P}(u_m(s))$ for all $m \in \mathbb{N}$ with $k \leqslant m$. At this point, the closedness of the graph of $\mathcal{P}$ shows that $u([s, T]) \subset \mathcal{P}(u(s))$.

In the sixth and final step, taking into account that S is dense in $[\tau, T]$, u is continuous on $[\tau, T]$ and the graph of $\mathcal{P}$ is closed, we conclude that the preceding relation holds for every $s \in [\tau, T]$, and this completes the proof. $\quad\square$

THEOREM 3.8.1. *Let $\mathcal{P}$ be a closed preorder on $\mathbb{K}$ and let $f : \mathbb{I} \times \mathbb{K} \to \mathbb{R}^n$ be continuous. Then a necessary and sufficient condition in order that $\mathcal{P}$ be viable with respect to f is the tangency condition below:*

$$f(t, \xi) \in \mathcal{T}_{\mathcal{P}(\xi)}(\xi)$$

for each $(t, \xi) \in \mathbb{I} \times \mathbb{K}$.

PROOF. The proof follows immediately from Lemma 3.8.1.   □

4. Problems of invariance

This section follows very closely Cârjă et al. [24].

4.1. *Preliminary facts*

Let $\mathbb{D}$ be a domain in $\mathbb{R}^n$, $\mathbb{K} \subseteq \mathbb{D}$ a locally closed subset, and let us consider the ordinary differential equation (1.1.1), where $f : \mathbb{I} \times \mathbb{D} \to \mathbb{R}^n$ is a given function.

DEFINITION 4.1.1. The subset $\mathbb{K}$ is *locally invariant* with respect to f if for each $(\tau, \xi) \in \mathbb{I} \times \mathbb{K}$ and each solution $u : [\tau, c] \to \mathbb{D}$, $c \in \mathbb{I}$, $c > \tau$, of (1.1.1), satisfying the initial condition $u(\tau) = \xi$, there exists $T \in (\tau, c]$ such that $u(t) \in \mathbb{K}$ for each $t \in [\tau, T]$. It is *invariant* if it satisfies the above condition of local invariance with $T = c$.

The relationship between viability and local invariance is clarified in

REMARK 4.1.1. If f is continuous on $\mathbb{I} \times \mathbb{D}$ and $\mathbb{K}$ is locally invariant with respect to f, then $\mathbb{K}$ is viable with respect to f. The converse of this assertion is no longer true, as we can see from the following example.

EXAMPLE 4.1.1. Let $\mathbb{D} = \mathbb{R}$, $\mathbb{K} = \{0\}$ and let $f : \mathbb{R} \to \mathbb{R}$ be defined by $f(u) = 3\sqrt[3]{u^2}$ for every $u \in \mathbb{R}$. Then $\mathbb{K}$ is viable with respect to f but $\mathbb{K}$ is not locally invariant with respect to f, because the differential equation $u'(t) = f(u(t))$ has at least two solutions which satisfy $u(0) = 0$, i.e. $u \equiv 0$ and $v(t) = t^3$.

A simple necessary and sufficient condition of invariance is stated below.

THEOREM 4.1.1. *Let $\mathbb{D}$ be a domain, $\mathbb{K} \subseteq \mathbb{D}$ a nonempty and locally closed subset and $f : \mathbb{I} \times \mathbb{D} \to \mathbb{R}^n$ a continuous function with the property that the associated Cauchy problem has the uniqueness property. Then, a necessary and sufficient condition in order that $\mathbb{K}$ be invariant with respect to f is that, for every $(t, \xi) \in \mathbb{I} \times \mathbb{K}$, $f(t, \xi) \in \mathcal{T}_{\mathbb{K}}(\xi)$.*

PROOF. The conclusion follows from Theorem 3.1.1 and Remark 4.1.1.   □

Theorem 4.1.1 says that, in general, if $\mathbb{K}$ is viable with respect to $f_{|_{\mathbb{I} \times \mathbb{K}}}$ and (1.1.1) has the uniqueness property, then $\mathbb{K}$ is locally invariant with respect to f. The preceding example shows that this is no longer true if we assume that $\mathbb{K}$ is viable with respect to f and merely $u'(t) = f_{|_{\mathbb{I} \times \mathbb{K}}}(t, u(t))$ has the uniqueness property.

REMARK 4.1.2. Moreover, if $f : \mathbb{I} \times \mathbb{D} \to \mathbb{R}^n$ is continuous and there exists one and only one point $\xi \in \mathbb{D}$ such that the differential equation (1.1.1) has at least two solutions u and v satisfying $u(\tau) = v(\tau) = \xi$, then, $\mathbb{K} = \{u(t); \ t \in [\tau, T]\}$ is viable with respect to f but it is not locally invariant with respect to f.

The next example reveals another interesting fact about local invariance. It shows that the local invariance of $\mathbb{K}$ with respect to f can take place even if $u'(t) = f_{|\mathbb{I} \times \mathbb{K}}(t, u(t))$ has not the uniqueness property.

EXAMPLE 4.1.2. Let $\mathbb{K} = \{(x, y) \in \mathbb{R}^2; \ y \geqslant 0\}$ and let $f : \mathbb{R}^2 \to \mathbb{R}^2$ be defined by

$$f\big((x, y)\big) = \begin{cases} (1, 0) & \text{if } (x, y) \in \mathbb{R}^2 \setminus \mathbb{K}, \\ \big(1, 3\sqrt[3]{y^2}\big) & \text{if } (x, y) \in \mathbb{K}. \end{cases}$$

Obviously $\mathbb{K}$ is locally invariant with respect to $f_{|\mathbb{K}}$ but $u'(t) = f_{|\mathbb{K}}(u(t))$ has not the uniqueness property. The latter assertion follows from the remark that, from each point, $(x, 0)$ (on the boundary of $\mathbb{K}$), we have at least two solutions to $u'(t) = f(u(t))$, $u(t) = (t + x, 0)$ and $v(t) = (t + x, t^3)$ satisfying $u(0) = v(0) = (x, 0)$.

4.2. *Sufficient conditions for local invariance*

Our first sufficient condition for local invariance says that, whenever there exists an open neighborhood $\mathbb{V} \subseteq \mathbb{D}$ of $\mathbb{K}$ such that f satisfies the "exterior tangency" condition

$$\liminf_{h \downarrow 0} \frac{1}{h}\big[\mathrm{dist}\big(\xi + hf(t, \xi); \mathbb{K}\big) - \mathrm{dist}(\xi; \mathbb{K})\big] \leqslant \omega\big(t, \mathrm{dist}(\xi; \mathbb{K})\big) \tag{4.2.1}$$

for each $(t, \xi) \in \mathbb{I} \times \mathbb{V}$, where ω is a comparison function in the sense of Definition 3.7.1, then $\mathbb{K}$ is locally invariant with respect to f. More precisely, we have:

THEOREM 4.2.1. *Let $\mathbb{K} \subseteq \mathbb{D} \subseteq \mathbb{R}^n$, with $\mathbb{K}$ locally closed and $\mathbb{D}$ open, and let $f : \mathbb{I} \times \mathbb{D} \to \mathbb{R}^n$. If (4.2.1) is satisfied, then $\mathbb{K}$ is locally invariant with respect to f.*

PROOF. Let $\mathbb{V} \subseteq \mathbb{D}$ be the open neighborhood of $\mathbb{K}$ whose existence is ensured by (4.2.1), and let $\omega : \mathbb{I} \times [0, a) \to \mathbb{R}$ the corresponding comparison function. Let $\xi \in \mathbb{K}$ and let $u : [\tau, c] \to \mathbb{V}$ be any solution to (1.1.1) satisfying $u(\tau) = \xi$. Diminishing c if necessary, we may assume that there exists $\rho > 0$ such that $B(\xi, \rho) \cap \mathbb{K}$ is closed, $u(t) \in B(\xi, \rho/2)$ and, in addition, $\mathrm{dist}(u(t); \mathbb{K}) < a$ for each $t \in [\tau, c]$. Let $g : [\tau, c] \to \mathbb{R}_+$ be defined by $g(t) = \mathrm{dist}(u(t); \mathbb{K})$ for each $t \in [\tau, c]$. Let $t \in [\tau, c)$ and $h > 0$ with $t + h \in [\tau, c]$. We have

$$g(t + h) = \mathrm{dist}\big(u(t + h); \mathbb{K}\big)$$

$$\leqslant h \left\| \frac{u(t + h) - u(t)}{h} - u'(t) \right\| + \mathrm{dist}\big(u(t) + hu'(t); \mathbb{K}\big).$$

Therefore

$$\frac{g(t+h) - g(t)}{h} \leqslant \alpha(h) + \frac{\mathrm{dist}(u(t) + hu'(t); \mathbb{K}) - \mathrm{dist}(u(t); \mathbb{K})}{h},$$

where

$$\alpha(h) = \left\| \frac{u(t+h) - u(t)}{h} - u'(t) \right\|.$$

Since $u'(t) = f(t, u(t))$ and $\lim_{h \downarrow 0} \alpha(h) = 0$, passing to the inf-limit in the inequality above for $h \downarrow 0$, and taking into account that $\mathbb{V}$, $\mathbb{K}$, and f satisfy (4.2.1), we get

$$[D_+ g](t) \leqslant \omega\big(t, g(t)\big)$$

for each $t \in [\tau, c)$. So, $g(t) \equiv 0$ which means that $u(t) \in \overline{\mathbb{K}} \cap B(\xi, \rho/2)$. But $\overline{\mathbb{K}} \cap B(\xi, \rho/2) \subseteq \mathbb{K} \cap B(\xi, \rho)$, and this achieves the proof. $\qquad\square$

REMARK 4.2.1. Clearly, (4.2.1) is satisfied with $\omega = \omega_f$, where the function $\omega_f : \mathbb{I} \times [0, a) \to \overline{\mathbb{R}}$, $a = \sup_{\xi \in \mathbb{V}} \mathrm{dist}(\xi; \mathbb{K})$ is defined by

$$\omega_f(t, x) = \sup_{\substack{\xi \in \mathbb{V} \\ \mathrm{dist}(\xi; \mathbb{K}) = x}} \liminf_{h \downarrow 0} \frac{1}{h}\Big[\mathrm{dist}\big(\xi + hf(t, \xi); \mathbb{K}\big) - \mathrm{dist}(\xi; \mathbb{K})\Big] \qquad (4.2.2)$$

for each $(t, x) \in \mathbb{I} \times [0, a)$.

So, Theorem 4.2.1 can be reformulated as:

THEOREM 4.2.2. *Let* $\mathbb{K} \subseteq \mathbb{D} \subseteq \mathbb{R}^n$, *with* $\mathbb{K}$ *locally closed and* $\mathbb{D}$ *open, and let* $f : \mathbb{I} \times \mathbb{D} \to \mathbb{R}^n$. *If there exists an open neighborhood* $\mathbb{V}$ *of* $\mathbb{K}$ *with* $\mathbb{V} \subseteq \mathbb{D}$ *such that* ω_f *defined by* (4.2.2) *is a comparison function, then* $\mathbb{K}$ *is locally invariant with respect to* f.

4.3. *Viability and comparison imply exterior tangency*

DEFINITION 4.3.1. *Let* $\mathbb{K} \subseteq \mathbb{D} \subseteq \mathbb{R}^n$. *We say that a function* $f : \mathbb{I} \times \mathbb{D} \to \mathbb{R}^n$ *has the comparison property with respect to* $(\mathbb{D}, \mathbb{K})$ *if there exist a proximal neighborhood* $\mathbb{V} \subseteq \mathbb{D}$ *of* $\mathbb{K}$, *one projection* $\pi_{\mathbb{K}} : \mathbb{V} \to \mathbb{K}$ *subordinated to* $\mathbb{V}$, *and one comparison function* $\omega : \mathbb{I} \times [0, a) \to \mathbb{R}$, *with* $a = \sup_{\xi \in \mathbb{V}} \mathrm{dist}(\xi; \mathbb{K})$, *such that*

$$\big[\xi - \pi_{\mathbb{K}}(\xi), f(t, \xi) - f\big(t, \pi_{\mathbb{K}}(\xi)\big)\big]_+ \leqslant \omega\big(t, \big\|\xi - \pi_{\mathbb{K}}(\xi)\big\|\big) \qquad (4.3.1)$$

for each $(t, \xi) \in \mathbb{I} \times \mathbb{V}$.

Let us observe that (4.3.1) is automatically satisfied for each $(t, \xi) \in \mathbb{I} \times \mathbb{K}$, and therefore, in Definition 4.3.1, we have only to assume that (4.3.1) holds for each $(t, \xi) \in \mathbb{I} \times [\mathbb{V} \setminus \mathbb{K}]$.

DEFINITION 4.3.2. The function $f : \mathbb{I} \times \mathbb{D} \to \mathbb{R}^n$ is called:

(i) $(\mathbb{D}, \mathbb{K})$-*Lipschitz* if there exist a proximal neighborhood $\mathbb{V} \subseteq \mathbb{D}$ of $\mathbb{K}$, a subordinated projection $\pi_{\mathbb{K}} : \mathbb{V} \to \mathbb{K}$, and $L > 0$, such that

$$\left\| f(t, \xi) - f\big(t, \pi_{\mathbb{K}}(\xi)\big) \right\| \leqslant L \left\| \xi - \pi_{\mathbb{K}}(\xi) \right\|$$

for each $(t, \xi) \in \mathbb{I} \times [\mathbb{V} \setminus \mathbb{K}]$;

(ii) $(\mathbb{D}, \mathbb{K})$-*dissipative* if there exist a proximal neighborhood $\mathbb{V} \subseteq \mathbb{D}$ of $\mathbb{K}$, and a projection, $\pi_{\mathbb{K}} : \mathbb{V} \to \mathbb{K}$, subordinated to $\mathbb{V}$, such that

$$\big[\xi - \pi_{\mathbb{K}}(\xi), f(t, \xi) - f\big(t, \pi_{\mathbb{K}}(\xi)\big) \big]_{+} \leqslant 0$$

for each $(t, \xi) \in \mathbb{I} \times [\mathbb{V} \setminus \mathbb{K}]$.

REMARK 4.3.1. We notice that, if we assume that (4.3.1), or either of the conditions (i), or (ii) in Definition 4.3.2 is satisfied for ξ replaced by ξ_1 and $\pi_{\mathbb{K}}(\xi)$ replaced by ξ_2 with $\xi_1, \xi_2 \in \mathbb{V}$, as considered in Kenmochi and Takahashi [60], then, for each $[\tau, T] \subseteq \mathbb{I}$ and $\xi \in \mathbb{K}$, there exists at most one solution $u : [\tau, T] \to \mathbb{K}$ to (1.1.1) satisfying $u(\tau) = \xi$. On contrary, in this more general frame, it may happen that, for certain (or for all) $[\tau, T] \subseteq \mathbb{I}$ and $\xi \in \mathbb{K}$, (1.1.1) have at least two solutions $u, v : [\tau, T] \to \mathbb{K}$ satisfying $u(\tau) = v(\tau) = \xi$.

Let $\mathbb{V}$ be a proximal neighborhood of $\mathbb{K}$, and let $\pi_{\mathbb{K}} : \mathbb{V} \to \mathbb{K}$ be a projection subordinated to $\mathbb{V}$. If $f : \mathbb{I} \times \mathbb{V} \to \mathbb{K}$ is a function with the property that, for each $t \in \mathbb{I}$ and $\eta \in \mathbb{K}$, the restriction of $f(t, \cdot)$ to the "segment"

$$\mathbb{V}_{\eta} = \big\{ \xi \in \mathbb{V} \setminus \mathbb{K}; \ \pi_{\mathbb{K}}(\xi) = \eta \big\}$$

is dissipative, then f is $(\mathbb{D}, \mathbb{K})$-dissipative.

It is easy to see that if f is either $(\mathbb{D}, \mathbb{K})$-Lipschitz, or $(\mathbb{D}, \mathbb{K})$-dissipative, then it has the comparison property with respect to $(\mathbb{D}, \mathbb{K})$. We notice that there are examples showing that there exist functions f which, although neither $(\mathbb{D}, \mathbb{K})$-Lipschitz, nor $(\mathbb{D}, \mathbb{K})$-dissipative, do have the comparison property. Moreover, there exist functions which, although $(\mathbb{D}, \mathbb{K})$-Lipschitz, are not Lipschitz on $\mathbb{D}$, as well as, functions which although $(\mathbb{D}, \mathbb{K})$-dissipative, are not dissipative on $\mathbb{D}$. In fact, these two properties describe merely the local behavior of f at the interface between $\mathbb{K}$ and $\mathbb{D} \setminus \mathbb{K}$. We include below two examples: the first one of an $(\mathbb{D}, \mathbb{K})$-Lipschitz function which is not locally Lipschitz, and the second one of a function which, although non-dissipative, is $(\mathbb{D}, \mathbb{K})$-dissipative. We notice that both examples refer to the autonomous case.

EXAMPLE 4.3.1. The graph of an $(\mathbb{D}, \mathbb{K})$-Lipschitz function $f : \mathbb{R} \to \mathbb{R}$ which is not Lipschitz is illustrated in Fig. 2. Here $\mathbb{K} = [a, b]$ and $\mathbb{D}$ is any open subset in $\mathbb{R}$ including $\mathbb{K}$.

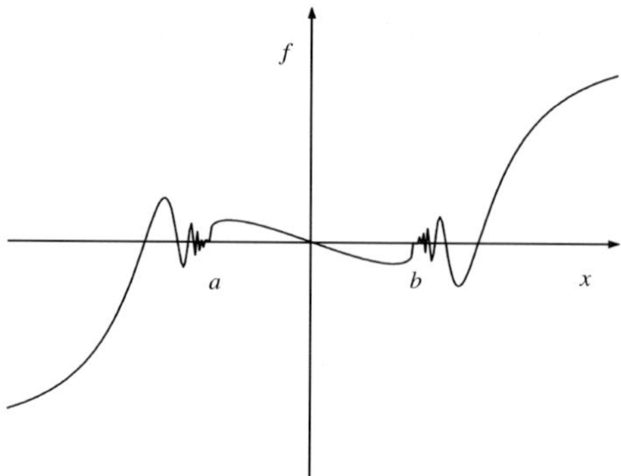

Fig. 2.

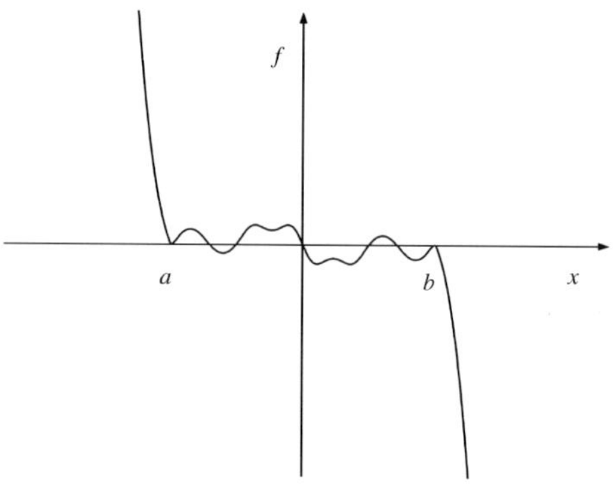

Fig. 3.

EXAMPLE 4.3.2. The graph of a function $f : \mathbb{R} \to \mathbb{R}$ which is $(\mathbb{D}, \mathbb{K})$-dissipative but not dissipative is illustrated in Fig. 3. This time, $\mathbb{K}$ is either $(-\infty, \beta]$, or $[\alpha, +\infty)$, or $[\alpha, \beta]$ with $\alpha \leqslant a \leqslant b \leqslant \beta$, and $\mathbb{D}$ is any open subset in $\mathbb{R}$ including $\mathbb{K}$.

We begin with:

THEOREM 4.3.1. *Let $\mathbb{K} \subseteq \mathbb{D} \subseteq \mathbb{R}^n$, with $\mathbb{K}$ locally closed and $\mathbb{D}$ open, and let $f : \mathbb{I} \times \mathbb{D} \to \mathbb{R}^n$. If f has the comparison property with respect to $(\mathbb{D}, \mathbb{K})$, and (1.1.3) is satisfied, then (4.2.1) holds true.*

PROOF. Let $\mathbb{V} \subseteq \mathbb{D}$ be the open neighborhood of $\mathbb{K}$ as in Definition 4.3.1, let $\xi \in \mathbb{V}$ and $[t, T) \subseteq \mathbb{I}$. Let $\rho > 0$ and let $\pi_{\mathbb{K}}$ be the selection of $\Pi_{\mathbb{K}}$ as in Definition 4.3.1. Let $h > 0$ with $t + h \in [t, T]$. Taking into account that $\|\xi - \pi_{\mathbb{K}}(\xi)\| = \mathrm{dist}(\xi; \mathbb{K})$, we have

$$\mathrm{dist}\big(\xi + hf(t, \xi); \mathbb{K}\big) - \mathrm{dist}(\xi; \mathbb{K})$$
$$\leqslant \big\| \xi - \pi_{\mathbb{K}}(\xi) + h\big[f(t, \xi) - f\big(t, \pi_{\mathbb{K}}(\xi)\big)\big]\big\|$$

$$- \left\| \xi - \pi_{\mathbb{K}}(\xi) \right\| + \mathrm{dist}\big(\pi_{\mathbb{K}}(\xi) + hf\big(t, \pi_{\mathbb{K}}(\xi)\big); \mathbb{K}\big).$$

Dividing by h, passing to the inf-limit for $h \downarrow 0$, and using (1.1.3), we get

$$\liminf_{h \downarrow 0} \frac{1}{h}\Big[\mathrm{dist}\big(\xi + hf(t, \xi); \mathbb{K}\big) - \mathrm{dist}(\xi; \mathbb{K})\Big]$$

$$\leq \big[\xi - \pi_{\mathbb{K}}(\xi),\, f(t, \xi) - f\big(t, \pi_{\mathbb{K}}(\xi)\big)\big]_{+}$$

$$\leq \omega\big(t, \left\| \xi - \pi_{\mathbb{K}}(\xi) \right\|\big).$$

But this inequality shows that (4.2.1) holds, and this completes the proof. $\square$

In the specific case in which f is continuous, we have:

THEOREM 4.3.2. *Let* $\mathbb{K} \subseteq \mathbb{D} \subseteq \mathbb{R}^n$, *with* $\mathbb{K}$ *locally closed and* $\mathbb{D}$ *open, and let* $f : \mathbb{I} \times \mathbb{D} \to \mathbb{R}^n$ *be continuous. Let us assume that* f *has the comparison property with respect to* $(\mathbb{D}, \mathbb{K})$, *and one of the four conditions below is satisfied:*
   (i) *for every* $(t, \xi) \in \mathbb{I} \times \mathbb{K}$, $f(t, \xi) \in \mathcal{C}_{\mathbb{K}}(\xi)$,
   (ii) *for every* $(t, \xi) \in \mathbb{I} \times \mathbb{K}$, $f(t, \xi) \in \mathcal{T}_{\mathbb{K}}(\xi)$,
   (iii) *for every* $(t, \xi) \in \mathbb{I} \times \mathbb{K}$, $f(t, \xi) \in \mathcal{B}_{\mathbb{K}}(\xi)$,
   (iv) *the set* $\mathbb{K}$ *is viable with respect to* f.
Then (4.2.1) *is also satisfied.*
   In general, if $\mathcal{G} : \mathbb{K} \rightsquigarrow \mathbb{R}^n$ *satisfies* $\mathcal{C}_{\mathbb{K}}(\xi) \subseteq \mathcal{G}(\xi) \subseteq \mathcal{B}_{\mathbb{K}}(\xi)$ *for each* $\xi \in \mathbb{K}$ *and:*
   (v) *for every* $(t, \xi) \in \mathbb{I} \times \mathbb{K}$, $f(t, \xi) \in \mathcal{G}(\xi)$,
then (4.2.1) *is satisfied too.*

PROOF. The conclusion follows from Theorem 3.4.1. $\square$

4.4. *Sufficient conditions for invariance. Revisited*

The next sufficient condition for invariance follows from Theorems 4.2.1 and 4.3.2.

THEOREM 4.4.1. *Let* $\mathbb{K} \subseteq \mathbb{D} \subseteq \mathbb{R}^n$, *with* $\mathbb{K}$ *locally closed and* $\mathbb{D}$ *open, and let* $f : \mathbb{I} \times \mathbb{D} \to \mathbb{R}^n$. *If* f *has the comparison property with respect to* $(\mathbb{D}, \mathbb{K})$, *and satisfies one of the conditions* (i), (ii), (iii), *or* (v) *in Theorem* 4.3.2, *then* $\mathbb{K}$ *is locally invariant with respect to* f.

Now, let $\mathbb{K} \subseteq \mathbb{R}^n$ and let $\mathbb{V}$ be an open neighborhood of $\mathbb{K}$.

DEFINITION 4.4.1. *A function* $g : \mathbb{V} \to \mathbb{K}$ *is a proximal generalized distance if:*
   (i) g *is Lipschitz continuous on bounded subsets in* $\mathbb{V}$,
   (ii) $g(\xi) = 0$ *if and only if* $\xi \in \mathbb{K}$.

If $\mathbb{K}$ is closed, a typical example of proximal generalized distance is offered by $g(\xi) = \alpha(\mathrm{dist}(\xi; \mathbb{K}))$, where $\alpha : [0, +\infty) \to [0, +\infty)$ is Lipschitz on bounded subsets, $\alpha(0) = 0$ and $\alpha(r) \neq 0$ if $r \neq 0$, while $\mathrm{dist}(\xi; \mathbb{K})$ is the usual distance from ξ to $\mathbb{K}$. We notice that if there exists a proximal generalized distance $g : \mathbb{V} \to [0, +\infty)$, then $\mathbb{K}$ is locally closed. Indeed, since $\mathbb{K} = \{\xi \in \mathbb{V};\ g(\xi) = 0\}$ and g is continuous, $\mathbb{K}$ is relatively closed in $\mathbb{V}$. But $\mathbb{V}$ is open and thus $\mathbb{K}$ is locally closed, as claimed.

THEOREM 4.4.2. *Let $\mathbb{K} \subseteq \mathbb{D} \subseteq \mathbb{R}^n$, and let $f : \mathbb{I} \times \mathbb{D} \to \mathbb{R}^n$. If there exist an open neighborhood $\mathbb{V}$ of $\mathbb{K}$, with $\mathbb{V} \subseteq \mathbb{D}$, a proximal generalized distance $g : \mathbb{V} \to \mathbb{R}_+$ and a comparison function $\omega : \mathbb{I} \times [0, a) \to \mathbb{R}$ such that*

$$\liminf_{h\downarrow 0} \frac{1}{h}\big[g\big(\xi + hf(t, \xi)\big) - g(\xi)\big] \leqslant \omega\big(t, g(\xi)\big) \tag{4.4.1}$$

for each $(t, \xi) \in \mathbb{I} \times \mathbb{V}$, then $\mathbb{K}$ is locally invariant with respect to f.

PROOF. The proof follows closely that one of Theorem 4.2.1, with the mention that here one has to use the obvious inequality $g(\lambda) \leqslant g(\eta) + L\|\lambda - \eta\|$ for each $\lambda, \eta \in B(\xi, \rho) \cap \mathbb{V}$, where $L > 0$ is the Lipschitz constant of g on $B(\xi, \rho) \cap \mathbb{V}$. $\square$

In order to obtain a simple, but useful, extension of Theorem 4.2.1, some observations are needed. Namely, if $g : \mathbb{V} \to [0, +\infty)$ is a generalized distance, we may consider the generalized tangency condition

$$\liminf_{h\downarrow 0} \frac{1}{h} g\big(\xi + hf(t, \xi)\big) = 0 \tag{4.4.2}$$

for each $(t, \xi) \in \mathbb{I} \times \mathbb{K}$, and one may ask whether this implies viability, whenever, of course f is continuous. The answer to this question is in the negative as the simple example below shows.

EXAMPLE 4.4.1. Let $\mathbb{K}$ be locally closed, let $\mathbb{V}$ be any open neighborhood of $\mathbb{K}$ and let $g : \mathbb{V} \to [0, +\infty)$ be defined as $g(\xi) = \mathrm{dist}^2(\xi; \mathbb{K})$ for each $\xi \in \mathbb{V}$. Further, let $f : \mathbb{I} \times \mathbb{K} \to \mathbb{R}^n$ be a continuous function such that $\mathbb{K}$ is not viable with respect to f. We can always find such a function whenever $\mathbb{K}$ is not open. Now, since

$$g\big(\xi + hf(t, \xi)\big) \leqslant \big\|\xi + hf(t, \xi) - \xi\big\|^2 \leqslant h^2 \big\|f(t, \xi)\big\|^2$$

for each $(t, \xi) \in \mathbb{I} \times \mathbb{K}$, (4.4.2) is trivially satisfied. So, the generalized tangency condition (4.4.2) does not imply the viability of $\mathbb{K}$ with respect to f.

This example shows that if g is a proximal generalized distance, and g^2 satisfies (4.4.2), it may happen that g does not satisfy (4.4.2). Therefore it justifies why, in the next result, we assume explicitly that (4.4.2) holds true, even though it is automatically satisfied by g^2.

THEOREM 4.4.3. *Let* $\mathbb{K} \subseteq \mathbb{D} \subseteq \mathbb{R}^n$, *and let* $f : \mathbb{I} \times \mathbb{D} \to \mathbb{R}^n$. *If there exist an open neighborhood* $\mathbb{V}$ *of* $\mathbb{K}$, *with* $\mathbb{V} \subseteq \mathbb{D}$, *and a proximal generalized distance* $g : \mathbb{V} \to \mathbb{R}_+$ *satisfying* (4.4.2) *and such that*

$$\liminf_{h \downarrow 0} \frac{1}{2h} \left[g^2(\xi + h f(t, \xi)) - g^2(\xi) \right] \leqslant g(\xi) \omega(t, g(\xi)) \tag{4.4.3}$$

for each $(t, \xi) \in \mathbb{I} \times \mathbb{V}$, *then* $\mathbb{K}$ *is locally invariant with respect to* f.

PROOF. We have only to observe that, in the presence of (4.4.2), (4.4.3) and (4.4.1) are equivalent, and to apply Theorem 4.4.2. $\qquad\square$

Using Theorem 4.4.3, we will prove next some other sufficient conditions for invariance expressed in the terms of a generalized Lipschitz projection. Namely, a subset $\mathbb{K}$ is a *Lipschitz retract* if there exist an open neighborhood $\mathbb{V}$ of $\mathbb{K}$ and a Lipschitz continuous map, $r : \mathbb{V} \to \mathbb{K}$, with $r(\xi) = \xi$ if and only if $\xi \in \mathbb{K}$. The function r as above is a *generalized Lipschitz projection*. For each Lipschitz retract $\mathbb{K}$, one can define a proximal generalized distance, $g : \mathbb{V} \to [0, +\infty)$, by $g(\xi) = \|r(\xi) - \xi\|$ for all $\xi \in \mathbb{V}$. Consequently, each Lipschitz retract is locally closed. Moreover, each open subset $\mathbb{K}$ is Lipschitz retract (take $\mathbb{V} = \mathbb{K}$ and r the identity). Another simple example of a Lipschitz retract is given by a closed subset $\mathbb{K}$ which has an open neighborhood $\mathbb{V}$ for which there exists a single-valued continuous projection $\pi_{\mathbb{K}} : \mathbb{V} \to \mathbb{K}$, i.e. $\mathrm{dist}(\xi; \mathbb{K}) = \|\pi_{\mathbb{K}}(\xi) - \xi\|$ for each $\xi \in \mathbb{V}$. In the latter case we say that $\mathbb{K}$ is *a proximate retract*. It should be noticed that the class of Lipschitz retract subsets is strictly larger than that of proximate retracts as the simple example below shows.

EXAMPLE 4.4.2. Take $\mathbb{R}^2$, endowed with the usual Hilbert structure and let us observe that the set

$$\mathbb{K} = \left\{ (x, y) \in \mathbb{R}^2;\ y \leqslant |x| \right\},$$

although Lipschitz retract, is not a proximate retract. Indeed, let $\mathbb{V} = \mathbb{R}^2$, and let $r((x, y))$ be defined, either as (x, y) if $(x, y) \in \mathbb{K}$, or as $(x, |x|)$ if $(x, y) \in \mathbb{V} \setminus \mathbb{K}$. It is easy to verify that r is a generalized Lipschitz projection with Lipschitz constant $\sqrt{2}$, and thus $\mathbb{K}$ is a Lipschitz retract. However, $\mathbb{K}$ is not a proximate retract since any selection $\pi_{\mathbb{K}}$ of the projection $\varPi_{\mathbb{K}}$ is discontinuous at each point $(0, y)$, with $y > 0$.

We emphasize that all the results which will follow can be reformulated to hold also for locally Lipschitz retracts, i.e. for those subsets $\mathbb{K}$ satisfying: *for each* $\xi \in \mathbb{K}$ *there exists* $\rho > 0$ *such that* $B(\xi; \rho) \cap \mathbb{K}$ *is Lipschitz retract*, but for the sake of simplicity we confined ourselves to the simpler case of Lipschitz retracts. First, let $\mathbb{K}$ be Lipschitz retract with the corresponding generalized Lipschitz projection $r : \mathbb{V} \to \mathbb{K}$. In the next two results, we will assume that the norm on $\mathbb{R}^n$ is defined by means of an inner product, i.e. $\|x\|^2 = \langle x, x \rangle$ for each $x \in \mathbb{R}^n$.

THEOREM 4.4.4. *Let $\mathbb{D} \subseteq \mathbb{R}^n$ be open and let $f : \mathbb{I} \times \mathbb{D} \to \mathbb{R}^n$. Let us assume that $\mathbb{K} \subseteq \mathbb{D}$ is Lipschitz retract with generalized Lipschitz projection $r : \mathbb{V} \to \mathbb{K}$ satisfying*

$$\liminf_{h \downarrow 0} \frac{1}{h} \left\| r\big(\xi + h f(t, \xi)\big) - \xi - h f(t, \xi) \right\| = 0 \tag{4.4.4}$$

for each $(t, \xi) \in \mathbb{I} \times \mathbb{K}$. Assume, in addition, that there exists a comparison function $\omega : \mathbb{I} \times [0, a) \to \mathbb{R}$, with $a = \sup_{\xi \in \mathbb{V}} \| r(\xi) - \xi \|$ such that

$$\liminf_{h \downarrow 0} \frac{1}{h} \big\langle r\big(\xi + h f(t, \xi)\big) - r(\xi) - h f(t, \xi), r(\xi) - \xi \big\rangle$$

$$\leqslant \| r(\xi) - \xi \| \omega\big(t, \| r(\xi) - \xi \| \big) \tag{4.4.5}$$

for each $(t, \xi) \in \mathbb{I} \times \mathbb{V}$. Then $\mathbb{K}$ is locally invariant with respect to f.

PROOF. Let us define $g(\xi) = \| r(\xi) - \xi \|$ for each $\xi \in \mathbb{V}$, and let $L > 0$ be the Lipschitz constant of r. Let us observe that

$$g^2(\xi + h\eta) - g^2(\xi)$$
$$= \big\langle r(\xi + h\eta) - (\xi + h\eta) - \big(r(\xi) - \xi\big), r(\xi + h\eta) - (\xi + h\eta) + \big(r(\xi) - \xi\big) \big\rangle$$
$$= \big\langle r(\xi + h\eta) - r(\xi), r(\xi + h\eta) + r(\xi) - 2\xi \big\rangle$$
$$\quad - h\big\langle \eta, 2r(\xi + h\eta) - 2\xi \big\rangle + h^2 \| \eta \|^2$$
$$= \big\| r(\xi + h\eta) - r(\xi) \big\|^2 + 2\big\langle r(\xi + h\eta) - r(\xi), r(\xi) - \xi \big\rangle$$
$$\quad - h\big\langle \eta, 2r(\xi + h\eta) - 2\xi \big\rangle + h^2 \| \eta \|^2$$
$$\leqslant \big(L^2 + 1\big)h^2 \| \eta \|^2 + 2\big\langle r(\xi + h\eta) - r(\xi), r(\xi) - \xi \big\rangle - 2h\big\langle \eta, r(\xi + h\eta) - \xi \big\rangle.$$

Therefore,

$$\liminf_{h \downarrow 0} \frac{1}{2h} \big[g^2(\xi + h\eta) - g^2(\xi) \big]$$

$$\leqslant \liminf_{h \downarrow 0} \frac{1}{h} \big\langle r(\xi + h\eta) - r(\xi), r(\xi) - \xi \big\rangle - \big\langle \eta, r(\xi) - \xi \big\rangle.$$

Since, by (4.4.4), g satisfies (4.4.2), taking $\eta = f(t, \xi)$ and using (4.4.5) and Theorem 4.4.3, we get the conclusion. $\qquad\square$

A consequence of Theorem 4.4.4 is stated below.

THEOREM 4.4.5. *Let $\mathbb{K} \subseteq \mathbb{D}$, with $\mathbb{D}$ open, and let $f : \mathbb{I} \times \mathbb{D} \to \mathbb{R}^n$. Let us assume that $\mathbb{K}$ is a Lipschitz retract with the generalized Lipschitz projection $r : \mathbb{V} \to \mathbb{K}$ satisfying (4.4.4).*

*Let us assume in addition that, for each $t \in \mathbb{I}$ and $\xi \in \mathbb{V}$, there exists the directional deriv-
ative, $r'(\xi)[f(t,\xi)]$, of r, at ξ in the direction $f(t,\xi)$, and*

$$\langle r'(\xi)[f(t,\xi)] - f(t,\xi), r(\xi) - \xi \rangle \leqslant \|r(\xi) - \xi\| \omega(t, \|r(\xi) - \xi\|), \qquad (4.4.6)$$

*where $\omega : \mathbb{I} \times [0, a) \to \mathbb{R}$ is a comparison function, and $a = \sup_{\xi \in \mathbb{V}} \|r(\xi) - \xi\|$. Then $\mathbb{K}$ is
locally invariant with respect to f.*

PROOF. It is easy to see that, in this specific case, (4.4.6) is equivalent to (4.4.5) and this
completes the proof. $\qquad\square$

REMARK 4.4.1. Let $\mathbb{K}$ be a Lipschitz retract subset and let $r : \mathbb{V} \to \mathbb{K}$ be the corresponding generalized Lipschitz projection. Let $f : \mathbb{I} \times \mathbb{K} \to \mathbb{R}^n$ be a continuous function, let
$a = \sup_{\xi \in \mathbb{V}} \|r(\xi) - \xi\|$, and let us define the function $\omega : \mathbb{I} \times [0, a) \to \overline{\mathbb{R}}_+$ by $\omega(t, 0) = 0$
and

$$\omega(t, x) = \sup_{\substack{\xi \in \mathbb{V} \\ \|r(\xi) - \xi\| = x}} \frac{\langle r'(\xi)[f(t,\xi)] - f(t,\xi), r(\xi) - \xi \rangle}{\|r(\xi) - \xi\|} \qquad (4.4.7)$$

for each $(t, x) \in \mathbb{I} \times (0, a)$. From Theorem 4.4.5, it follows that $\mathbb{K}$ is invariant with respect
to f if (4.4.4) is satisfied and ω, defined by (4.4.7), is a comparison function. Furthermore,
if $\mathbb{K} \subseteq \mathbb{R}^n$ is a closed linear subspace in $\mathbb{R}^n$, and r is the projection of $\mathbb{R}^n$ on $\mathbb{K}$, then r
is linear, and $r'(\xi)[\eta] = r(\eta)$ for each $\xi, \eta \in \mathbb{R}^n$. So, in this case, the condition (4.4.4) is
equivalent to $f(\mathbb{I} \times \mathbb{K}) \subseteq \mathbb{K}$. Take $\mathbb{V} = \{\xi \in \mathbb{R}^n; \ \text{dist}(\xi; \mathbb{K}) < \rho\}$, for some fixed $\rho > 0$,
and let us observe that the function ω, defined by (4.4.7), is given by $\omega(t, 0) = 0$ and

$$\omega(t, x) = \sup_{\substack{\xi \in \mathbb{V} \\ \|r(\xi) - \xi\| = x}} \frac{\langle r(f(t,\xi)) - f(t,\xi), r(\xi) - \xi \rangle}{\|r(\xi) - \xi\|} \qquad (4.4.8)$$

for each $(t, x) \in \mathbb{I} \times (0, \rho)$. Hence, if $f(\mathbb{I} \times \mathbb{K}) \subseteq \mathbb{K}$, and ω defined by (4.4.8) is a comparison function, then $\mathbb{K}$ is invariant with respect to (1.1.1).

4.5. *When tangency implies exterior tangency?*

Next, we will prove that, in special circumstances, the tangency condition (1.1.3) for a
function $f : \mathbb{I} \times \mathbb{K} \to \mathbb{R}$ comes from the exterior tangency condition (4.2.1) for a suitably
defined extension $\tilde{f} : \mathbb{I} \times \mathbb{D} \to \mathbb{R}$ of f. More precisely, we have:

THEOREM 4.5.1. *Let $f : \mathbb{I} \times \mathbb{K} \to \mathbb{R}^n$ be a given function satisfying (1.1.3). If $\mathbb{V} \subseteq \mathbb{R}^n$
is a proximal neighborhood of $\mathbb{K}$ and $r : \mathbb{V} \to \mathbb{K}$ is a projection subordinated to $\mathbb{V}$, then
$\tilde{f} : \mathbb{I} \times \mathbb{V} \to \mathbb{R}^n$, defined by $\tilde{f}(t, \cdot) = f(t, r(\cdot))$ satisfies (4.2.1).*

PROOF. Let $\xi \in \mathbb{V}$ and $h > 0$. We have

$$\mathrm{dist}\big(\xi + h\tilde{f}(t,\xi); \mathbb{K}\big) - \mathrm{dist}(\xi; \mathbb{K})$$

$$\leqslant \big\|\xi - r(\xi)\big\| + \mathrm{dist}\big(r(\xi) + hf\big(t, r(\xi)\big); \mathbb{K}\big) - \big\|\xi - r(\xi)\big\|$$

$$= \mathrm{dist}\big(r(\xi) + hf\big(t, r(\xi)\big); \mathbb{K}\big).$$

Dividing by $h > 0$ and passing to $\liminf$ for $h \downarrow 0$, we get

$$\liminf_{h \downarrow 0} \frac{1}{h}\big[\mathrm{dist}\big(\xi + h\tilde{f}(t,\xi); \mathbb{K}\big) - \mathrm{dist}(\xi; \mathbb{K})\big] \leqslant 0.$$

So, (4.2.1) holds true with $\omega \equiv 0$, and the proof is complete. $\qquad\square$

It should be noticed that, the conclusion of Theorem 4.5.1 is no longer true if we are looking for a continuous extension $\tilde{f}$ of a continuous function f satisfying (1.1.3), as the next example shows.

EXAMPLE 4.5.1. This example is adapted from Aubin and Cellina [3, p. 203]. Let $\mathbb{K}_1 = \{(x, 3\sqrt[3]{x^2}); \ x \in \mathbb{R}_+\}$, $\mathbb{K}_2 = \{(x, 3\sqrt[3]{x^2}); \ x \in \mathbb{R}_-\}$ and let $\mathbb{K} = \mathbb{K}_1 \cup \mathbb{K}_2$. If $\xi \in \mathbb{K}_1$, we define $f(\xi)$ as the unit clockwise oriented tangent vector to $\mathbb{K}_1$ at ξ, and if $\xi \in \mathbb{K}_2$, we define $f(\xi)$ as the unit counterclockwise oriented tangent vector to $\mathbb{K}_1$ at ξ. Of course, $f((0,0)) = (0,1)$. Thus $f : \mathbb{K} \to \mathbb{R}^2$ is continuous and $f(\xi) \in \mathcal{T}_{\mathbb{K}}(\xi)$ for each $\xi \in \mathbb{K}$. By virtue of Theorem 1.1.3, $\mathbb{K}$ is viable with respect to f. Let $\tilde{f}$ be any continuous extension of f to an open neighborhood $\mathbb{V}$ of the origin. We may assume that for each $v \in \mathbb{V}$, $\tilde{f}_2(v) \geqslant \frac{1}{2}$.

In fact, the equation $u'(t) = f(u(t))$ subjected to $u(0) = (0,0)$ has two local solutions $u, v : [0, \delta] \to \mathbb{K}$, with $u([0,\delta]) \subseteq \mathbb{K}_1$ and $v([0,\delta]) \subseteq \mathbb{K}_2$. Diminishing $\delta > 0$, we may assume that no solution to $u'(t) = \tilde{f}(u(t))$, $u(0) = (0,0)$, can escape from $\mathbb{V}$. Now, if we assume that $\mathbb{K}$ is invariant with respect to $\tilde{f}$, we have

$$F_{0,(0,0)}(\delta) = \big\{u(\delta); \ u'(t) = \tilde{f}\big(u(t)\big), \ \text{for all } t \in [0, \delta], \ u(0) = (0,0)\big\} \subseteq \mathbb{K},$$

and by virtue of a classical result due to Kneser—see Theorem 7.5.1—we know that $F_{0,(0,0)}(\delta)$ is connected, and therefore, we conclude that there exists at least one solution $w : [0, \delta] \to \mathbb{K}$ of $u'(t) = \tilde{f}(u(t))$, $u(0) = (0,0)$ with $w(\delta) = (0,0)$. But this is impossible, because $w_2(\delta) \geqslant \frac{1}{2}\delta$.

However, if f is continuous and $\mathbb{K}$ is smooth enough, by the very same proof we deduce:

THEOREM 4.5.2. *Let $f : \mathbb{I} \times \mathbb{K} \to \mathbb{R}^n$ be a continuous function satisfying (1.1.3). If there exists a proximal neighborhood $\mathbb{V} \subseteq \mathbb{R}^n$ of $\mathbb{K}$, and a continuous projection $r : \mathbb{V} \to \mathbb{K}$ subordinated to $\mathbb{V}$, then f can be extended to a continuous function $\tilde{f} : \mathbb{I} \times \mathbb{V} \to \mathbb{R}^n$ satisfying (4.2.1).*

4.6. *Local invariance and monotonicity*

We say that the preorder "$\preceq$" is *locally invariant with respect to* f if for each $(\tau, \xi) \in \mathbb{I} \times \mathbb{K}$, each solution $u : [\tau, c] \to \mathbb{D}$, $c \in (\tau, b]$, of (1.1.1) satisfying $u(\tau) = \xi$, there exists $T \in (\tau, c]$ such that $u(t) \in \mathbb{K}$ for each $t \in [\tau, T]$ and u is "$\preceq$"-monotone on $[\tau, T]$, i.e., for each $\tau \leqslant s \leqslant t \leqslant T$, we have $u(s) \preceq u(t)$. We recall that, for each $\xi \in \mathbb{K}$, $\mathcal{P}(\xi) = \{\eta \in \mathbb{K}; \ \xi \preceq \eta\}$.

REMARK 4.6.1. The preorder "$\preceq$" is locally invariant with respect to f if and only if for each $(\tau, \xi) \in \mathbb{I} \times \mathbb{K}$, each solution $u : [\tau, c] \to \mathbb{D}$, $c \in (\tau, b]$, of (1.1.1) satisfying $u(\tau) = \xi$, there exists $T \in (\tau, c]$ such that, for each $s \in [\tau, T]$ and $t \in [s, T]$, we have $u(t) \in \mathcal{P}(u(s))$.

In contrast with Lemma 3.8.1, the next lemma is almost obvious.

LEMMA 4.6.1. *The preorder "$\preceq$" is locally invariant with respect to f if and only if for each $\xi \in \mathbb{K}$, $\mathcal{P}(\xi)$ is locally invariant with respect to f.*

COROLLARY 4.6.1. *If for each $\xi \in \mathbb{K}$ there exists an open neighborhood $\mathbb{V} \subseteq \mathbb{D}$ of ξ and a comparison function $\omega : \mathbb{I} \times [0, a) \to \mathbb{R}$, with $a = \sup_{\eta \in \mathbb{V}} \mathrm{dist}(\eta; \mathcal{P}(\xi))$, and such that*

$$\liminf_{h \downarrow 0} \frac{1}{h} \big[\mathrm{dist}\big(\eta + hf(t, \eta); \mathcal{P}(\xi)\big) - \mathrm{dist}\big(\eta; \mathcal{P}(\xi)\big) \big] \leqslant \omega\big(t, \mathrm{dist}\big(\eta; \mathcal{P}(\xi)\big)\big)$$

for each $(t, \eta) \in \mathbb{I} \times \mathbb{V}$, then "$\preceq$" is locally invariant with respect to f.

COROLLARY 4.6.2. *If, for each $\xi \in \mathbb{K}$, f has the comparison property with respect to $(\mathbb{D}, \mathcal{P}(\xi))$ and, for each $(t, \eta) \in \mathbb{I} \times \mathcal{P}(\xi)$, we have*

$$\liminf_{h \downarrow 0} \frac{1}{h} \, \mathrm{dist}\big(\eta + hf(t, \eta); \mathcal{P}(\xi)\big) = 0,$$

then "$\preceq$" is locally invariant with respect to f.

5. Carathéodory solutions

5.1. *A Lebesgue type derivation theorem*

In this section we extend some of the previously established results to a more general case allowing the function f to be discontinuous with respect to the time variable. This case is very important in the study of some control problems to be analyzed in the sequel, when even starting with a continuous function $(t, v, u) \mapsto f(t, v, u)$, due to the discontinuities of the optimal control $t \mapsto v^*(t)$, we have to consider a very irregular right-hand side $(t, u) \mapsto f(t, v^*(t), u)$ which is only measurable with respect to t. First we recall:

DEFINITION 5.1.1. A function $f : \mathbb{I} \times \mathbb{K} \to \mathbb{R}^n$ is *of Carathéodory type*, if it satisfies the three conditions below.

(H$_1$) For every $\xi \in \mathbb{K}$ the function $f(\cdot, \xi)$ is measurable on $\mathbb{I}$.

(H$_2$) For almost every $t \in \mathbb{I}$ the function $f(t, \cdot)$ is continuous on $\mathbb{K}$.

(H$_3$) For every $m > 0$, there exists a function $\ell_m \in L^1_{\text{loc}}(\mathbb{I})$ such that $\| f(t, u)\| \leqslant \ell_m(t)$ for almost every $t \in \mathbb{I}$ and for all $u \in B(0, m) \cap \mathbb{K}$.

DEFINITION 5.1.2. A function $u : [\tau, T] \to \mathbb{K}$ is a *Carathéodory solution* to the differential equation (1.1.1) if u is absolutely continuous on $[\tau, T]$ and satisfies $u'(t) = f(t, u(t))$ a.e. for $t \in [\tau, T]$.

DEFINITION 5.1.3. We say that $\mathbb{K}$ is *Carathéodory viable* with respect to $f : \mathbb{I} \times \mathbb{K} \to \mathbb{R}^n$ if for each $(\tau, \xi) \in \mathbb{I} \times \mathbb{K}$ there exist $T \in \mathbb{I}$, $T > \tau$ and a Carathéodory solution $u : [\tau, T] \to \mathbb{K}$ to (1.1.1), satisfying $u(\tau) = \xi$.

Clearly, if $\mathbb{K}$ is viable with respect to f, then it is Carathéodory viable, but not conversely. In order to prove some necessary and sufficient conditions for Carathédory viability, we recall first a Lebesgue type derivation theorem due to Scorza Dragoni [88]. For more general results see Frankowska et al. [45] and Cârjă and Monteiro Marques [21].

THEOREM 5.1.1. *Assume* (H$_1$), (H$_2$) *and* (H$_3$). *Then there exists a negligible subset* $\mathbb{Z}$ *of* $\mathbb{I}$ *such that, for every* $t \in \mathbb{I} \setminus \mathbb{Z}$ *and every* $\xi \in \mathbb{K}$, *one has*

$$\lim_{h \downarrow 0} \frac{1}{h} \int_t^{t+h} f(s, u(s))\, ds = f(t, \xi) \tag{5.1.1}$$

for all continuous functions $u : \mathbb{I} \to \mathbb{K}$ *with* $u(t) = \xi$.

In order to prove Theorem 5.1.1, we recall for easy reference a specific form of a Lusin type continuity result due to Scorza Dragoni [87]. For more general results see Berliocchi and Lasry [10] and Kucia [63].

Here and thereafter, λ denotes the usual Lebesgue measure on $\mathbb{R}$.

THEOREM 5.1.2 (Scorza Dragoni [87]). *Let* $f : \mathbb{I} \times \mathbb{K} \to \mathbb{R}^n$ *be a function that satisfies* (H$_1$) *and* (H$_2$). *Then, for each* $\varepsilon > 0$, *there exists a closed set* $\mathbb{A} \subseteq \mathbb{I}$ *such that* $\lambda(\mathbb{I} \setminus \mathbb{A}) < \varepsilon$ *and the restriction of* f *to* $\mathbb{A} \times \mathbb{K}$ *is continuous.*

We note that the conclusion of Theorem 5.1.2 holds also true if $\mathbb{I}$ is replaced by any Lebesgue measurable subset in $\mathbb{R}$, while $\mathbb{K}$ and $\mathbb{R}^n$ are replaced by two separable metric spaces X and respectively Y. We can now proceed to the proof of Theorem 5.1.1.

PROOF OF THEOREM 5.1.1. Since $\mathbb{I}$ is a countable union of finite length intervals, it suffices to consider the case when $\mathbb{I}$ is of finite length. For each $\gamma > 0$, we shall obtain a set $\mathbb{L}_\gamma \subset \mathbb{I}$, with $\lambda(\mathbb{I} \setminus \mathbb{L}_\gamma) < \gamma$, and such that (5.1.1) holds for all $t \in \mathbb{L}_\gamma$. Finally, since $\lambda(\mathbb{I} \setminus \mathbb{L}_\gamma) < \gamma$, it will suffice to consider $\mathbb{Z} = \cap_m (\mathbb{I} \setminus \mathbb{L}_{1/m})$.

Let $\gamma > 0$ and let us observe that, by virtue of Theorem 5.1.2, it follows that there exists a compact set $\mathbb{A}_\gamma \subseteq \mathbb{I}$ such that $\lambda(\mathbb{I} \setminus \mathbb{A}_\gamma) < \gamma$, and the restriction of f to $\mathbb{A}_\gamma \times \mathbb{K}$ is continuous.

We define $\mathbb{L}_\gamma \subseteq \mathbb{A}_\gamma$ as the set of density points of $\mathbb{A}_\gamma$ which are also Lebesgue points of the functions $\tilde{\ell}_m : \mathbb{I} \to \mathbb{R}$, given by $\tilde{\ell}_m(t) = \ell_m(t)\chi_{\mathbb{I}\setminus\mathbb{A}_\gamma}(t)$, where ℓ_m is given by (H$_3$), $m = 1, 2, \ldots$. It is known that $\lambda(\mathbb{L}_\gamma) = \lambda(\mathbb{A}_\gamma)$ and, by definition, for $t \in \mathbb{L}_\gamma$ we have

$$\lim_{t\in\mathbb{J};\lambda(\mathbb{J})\to 0} \frac{\lambda(\mathbb{A}_\gamma \cap \mathbb{J})}{\lambda(\mathbb{J})} = 1,$$
$$\lim_{t\in\mathbb{J};\lambda(\mathbb{J})\to 0} \frac{1}{\lambda(\mathbb{J})} \int_{\mathbb{J}} \left| \tilde{\ell}_m(s) - \tilde{\ell}_m(t) \right| \, \mathrm{d}s = 0, \tag{5.1.2}$$

where $\mathbb{J}$ denotes intervals of positive length.

Let $t \in \mathbb{L}_\gamma$. Consider a continuous function $u : \mathbb{I} \to \mathbb{K}$, and denote by $\xi = u(t)$. Then, there is $m \geqslant 1$ such that $\|u(\theta)\| < m$ for all $\theta \in [t, t+\delta]$, where $\delta > 0$ is sufficiently small. Let $\varepsilon > 0$ be arbitrary. We can further assume that, for all $\theta \in \mathbb{A}_\gamma \cap [t, t+\delta]$,

$$\left\| f\big(\theta, u(\theta)\big) - f(t, \xi) \right\| \leqslant \frac{\varepsilon}{3}. \tag{5.1.3}$$

By taking a smaller δ if necessary, in view of (5.1.2), we can also ensure that

$$\frac{1}{s} \int_{[t,t+s]\setminus\mathbb{A}_\gamma} \ell_m(\theta) \, \mathrm{d}\theta \leqslant \frac{\varepsilon}{3}, \tag{5.1.4}$$

for every $s \in (0, \delta)$, and also

$$\frac{\lambda([t, t+s] \setminus \mathbb{A}_\gamma)}{s} \left\| f(t, \xi) \right\| \leqslant \frac{\varepsilon}{3} \tag{5.1.5}$$

for every $s \in (0, \delta)$. Then, by (5.1.3), for $s \in (0, \delta)$, we have

$$\frac{1}{s} \int_{[t,t+s]\cap\mathbb{A}_\gamma} \left\| f\big(\theta, u(\theta)\big) - f(t, \xi) \right\| \, \mathrm{d}\theta \leqslant \frac{\varepsilon}{3} \frac{\lambda([t, t+s] \cap \mathbb{A}_\gamma)}{s} \leqslant \frac{\varepsilon}{3},$$

while by (5.1.4) and (5.1.5) we have

$$\frac{1}{s} \int_{[t,t+s]\setminus\mathbb{A}_\gamma} \left\| f\big(\theta, u(\theta)\big) - f(t, \xi) \right\| \, \mathrm{d}\theta$$
$$\leqslant \frac{1}{s} \int_{[t,t+s]\setminus\mathbb{A}_\gamma} \big(\ell_m(\theta) + \left\| f(t, \xi) \right\|\big) \, \mathrm{d}\theta$$
$$\leqslant \frac{1}{s} \int_{[t,t+s]\setminus\mathbb{A}_\gamma} \ell_m(\theta) \, \mathrm{d}\theta + \frac{\lambda([t, t+s] \setminus \mathbb{A}_\gamma)}{s} \left\| f(t, \xi) \right\| \leqslant 2\frac{\varepsilon}{3}.$$

Finally, we have

$$\left\| \frac{1}{s} \int_t^{t+s} f\big(\theta, u(\theta)\big) \, d\theta - f(t, \xi) \right\| \leqslant \varepsilon,$$

for all $s \in (0, \delta)$ and this completes the proof. $\qquad\square$

Using similar arguments we can prove:

THEOREM 5.1.3. *Assume* (H_1), (H_2) *and* (H_3). *Then there exists a negligible subset* $\mathbb{Z}$ *of* $\mathbb{I}$ *such that, for every* $t \in \mathbb{I} \setminus \mathbb{Z}$ *and every* $\xi \in \mathbb{K}$, *one has*

$$\lim_{h \downarrow 0} \frac{1}{h} \int_t^{t+h} f\big(s, v(h)\big) \, ds = f(t, \xi)$$

for all functions $v : [0, +\infty) \to \mathbb{K}$ *satisfying* $\lim_{h \downarrow 0} v(h) = \xi$.

PROOF. The proof proceeds similarly with that one of Theorem 5.1.1 up to and including the sentence: "Let $t \in \mathbb{L}_\gamma$." We continue as follows. There is $m \geqslant 1$ such that $\|v(h)\| < m$ for $h \in (0, \delta)$ where $\delta > 0$ is sufficiently small. Let $\varepsilon > 0$ be arbitrary. We can assume that, for all $\theta \in \mathbb{A}_\gamma \cap [t, t + \delta]$ and $h \in (0, \delta)$,

$$\left\| f\big(\theta, v(h)\big) - f(t, \xi) \right\| \leqslant \frac{\varepsilon}{3}.$$

By taking a smaller δ if necessary, in view of (5.1.2), we can also ensure that (5.1.4) and (5.1.5) hold. From now on, the proof follows the very same way as that one of Theorem 5.1.1, with the observation that, here, $f(\theta, u(\theta))$ should be replaced by $f(\theta, v(h))$, wherever it appears. $\qquad\square$

5.2. *Characterizations of Carathéodory viability*

We are now ready to prove the main characterizations of Carathéodory viability.

THEOREM 5.2.1. *Suppose that* $\mathbb{K}$ *is locally closed and* $f : \mathbb{I} \times \mathbb{K} \to \mathbb{R}^n$ *is of Carathéodory type. Then the following conditions are equivalent:*
    (i) *there exists a negligible set* $\mathbb{Z} \subset \mathbb{I}$ *such that for every* $t \in \mathbb{I} \setminus \mathbb{Z}$ *and for every* $\xi \in \mathbb{K}$,
        $f(t, \xi) \in \mathcal{C}_{\mathbb{K}}(\xi)$,
    (ii) *there exists a negligible set* $\mathbb{Z} \subset \mathbb{I}$ *such that for every* $t \in \mathbb{I} \setminus \mathbb{Z}$ *and for every* $\xi \in \mathbb{K}$,
        $f(t, \xi) \in \mathcal{T}_{\mathbb{K}}(\xi)$,
    (iii) *there exists a negligible set* $\mathbb{Z} \subseteq \mathbb{I}$ *such that for every* $t \in \mathbb{I} \setminus \mathbb{Z}$ *and for every* $\xi \in \mathbb{K}$,
        $f(t, \xi) \in \mathcal{B}_{\mathbb{K}}(\xi)$,
    (iv) *for each* $\xi \in \mathbb{K}$ *there exists a negligible set* $\mathbb{Z}_\xi \subseteq \mathbb{I}$ *such that for every* $t \in \mathbb{I} \setminus \mathbb{Z}_\xi$,
        $f(t, \xi) \in \mathcal{C}_{\mathbb{K}}(\xi)$,

(v) *for each $\xi \in \mathbb{K}$ there exists a negligible set $\mathbb{Z}_\xi \subseteq \mathbb{I}$ such that for every $t \in \mathbb{I} \setminus \mathbb{Z}_\xi$,*
$$f(t, \xi) \in \mathcal{T}_\mathbb{K}(\xi),$$
(vi) *for each $\xi \in \mathbb{K}$ there exists a negligible set $\mathbb{Z}_\xi \subseteq \mathbb{I}$ such that for every $t \in \mathbb{I} \setminus \mathbb{Z}_\xi$,*
$$f(t, \xi) \in \mathcal{B}_\mathbb{K}(\xi),$$
(vii) *$\mathbb{K}$ is Carathéodory viable with respect to f.*

In general, if $\mathcal{G} : \mathbb{K} \rightsquigarrow \mathbb{R}^n$ is such that $\mathcal{C}_\mathbb{K}(\xi) \subseteq \mathcal{G}(\xi) \subseteq \mathcal{B}_\mathbb{K}(\xi)$ for each $\xi \in \mathbb{K}$, then each one of the conditions below:

(viii) *there exists a negligible set $\mathbb{Z} \subseteq \mathbb{I}$ such that for every $t \in \mathbb{I} \setminus \mathbb{Z}$ and for every $\xi \in \mathbb{K}$,*
$$f(t, \xi) \in \mathcal{G}(\xi),$$
(ix) *for each $\xi \in \mathbb{K}$ there exists a negligible set $\mathbb{Z}_\xi \subseteq \mathbb{I}$ such that for every $t \in \mathbb{I} \setminus \mathbb{Z}_\xi$,*
$$f(t, \xi) \in \mathcal{G}(\xi),$$

is equivalent to each one of the seven conditions above.

In order to prove Theorem 5.2.1, we need two auxiliary lemmas.

LEMMA 5.2.1. *Let $\mathbb{K} \subseteq \mathbb{R}^n$ be locally closed, let $f : \mathbb{I} \times \mathbb{K} \to \mathbb{R}^n$ be a function of Carathéodory type, let $[\tau, T] \subseteq \mathbb{I}$, $\xi \in \mathbb{K}$, and let $\rho > 0$ be such that $\mathbb{K} \cap B(\xi, \rho)$ is closed and there exists $\ell \in L^1(\tau, T)$ such that $\|f(s, \eta)\| \leqslant \ell(s)$ a.e. for $s \in [\tau, T]$ and for all $\eta \in \mathbb{K} \cap B(\xi, \rho)$. Then the family*

$$\mathcal{F} = \big\{ f(\cdot, \eta); \ \eta \in \mathbb{K} \cap B(\xi, \rho) \big\}$$

is compact in $L^1(\tau, T; \mathbb{R}^n)$.

PROOF. Let $(f(\cdot, \eta_m))_m$ be a sequence in $\mathcal{F}$. Since for all $m \in \mathbb{N}$, we have $\eta_m \in \mathbb{K} \cap B(\xi, \rho)$ and the latter is compact, there exist a subsequence of $(\eta_m)_m$, denoted for simplicity again by $(\eta_m)_m$, and $\eta \in \mathbb{K} \cap B(\xi, \rho)$, such that $\lim_m \eta_m = \eta$. It readily follows that $\lim_m f(s, \eta_m) = f(s, \eta)$ a.e. for $s \in [\tau, T]$. Since $\|f(s, \eta_m)\| \leqslant \ell(s)$ a.e. for $s \in [\tau, T]$ and for all $m \in \mathbb{N}$, by virtue of Lebesgue dominated convergence theorem, we deduce that

$$\lim_m \int_\tau^T \big\| f(s, \eta_m) - f(s, \eta) \big\| \, \mathrm{d}s = 0,$$

and this achieves the proof. $\qquad\square$

LEMMA 5.2.2. *Let $\mathbb{K} \subseteq \mathbb{R}^n$ be locally closed, let $f : \mathbb{I} \times \mathbb{K} \to \mathbb{R}^n$ be a function of Carathéodory type, let $[\tau, T] \subseteq \mathbb{I}$ with $T < \sup \mathbb{I}$, let $h > 0$ with $T + h < \sup \mathbb{I}$ and let $f_h : [\tau, T] \times \mathbb{K} \to \mathbb{R}^n$ be defined by*

$$f_h(t, \xi) = \frac{1}{h} \int_t^{t+h} f(s, \xi) \, \mathrm{d}s$$

for each $(t, \xi) \in [\tau, T] \times \mathbb{K}$. Then f_h is continuous on $[\tau, T] \times \mathbb{K}$.

PROOF. Let $(t, \xi) \in [\tau, T] \times \mathbb{K}$. Since $\mathbb{K}$ is locally closed and f satisfies (H_3), there exist $\rho > 0$ and $\ell \in L^1(\tau, T)$ such that $K \cap B(\xi, \rho)$ is closed and $\|f(s, \eta)\| \leqslant \ell(s)$ a.e. for

$s \in [\tau, T]$ and for all $\eta \in K \cap B(\xi, \rho)$. Then, for each θ with $[t + \theta, t + h + \theta] \subset \mathbb{I}$, and each $\eta \in K \cap B(\xi, \rho)$, we have

$$
\begin{aligned}
\big\| f_h(t + \theta, \eta) &- f_h(t, \xi) \big\| \\
&= \left\| \frac{1}{h} \int_{t+\theta}^{t+\theta+h} f(s, \eta)\,ds - \frac{1}{h} \int_t^{t+h} f(s, \xi)\,ds \right\| \\
&\leqslant \frac{1}{h} \int_t^{t+h} \big\| f(s + \theta, \eta) - f(s, \eta) \big\|\,ds + \frac{1}{h} \int_t^{t+h} \big\| f(s, \eta) - f(s, \xi) \big\|\,ds.
\end{aligned}
$$

In view of Lemma 5.2.1 and Theorem A.1.1, we have

$$
\lim_{\theta \to 0} \int_t^{t+h} \big\| f(s + \theta, \eta) - f(s, \eta) \big\|\,ds = 0
$$

uniformly for $\eta \in \mathbb{K} \cap B(\xi, \rho)$. Next, using condition (H_2) in Definition 5.1.1, the fact that $\| f(s, \eta) \| \leqslant \ell(s)$ a.e. for $s \in [t, t + h]$ and for all $\eta \in \mathbb{K} \cap B(\xi, \rho)$, by virtue of Lebesgue dominated convergence theorem, we deduce that

$$
\lim_{\eta \to \xi} \int_t^{t+h} \big\| f(s, \eta) - f(s, \xi) \big\|\,ds = 0,
$$

relation which, along with the preceding one, shows that

$$
\lim_{(\theta, \eta) \to (0, \xi)} \big\| f_h(t + \theta, \eta) - f_h(t, \xi) \big\| = 0,
$$

and this completes the proof. $\qquad\qquad\qquad\qquad\qquad\qquad\qquad\qquad\qquad\qquad\qquad$ $\square$

We can now proceed to the proof of Theorem 5.2.1.

PROOF OF THEOREM 5.2.1. In view of Theorem 3.4.1, (i) (ii) and (iii) are equivalent. Furthermore (i) implies (iv), (iv) implies (v) and (v) implies (vi). In order to show that the first seven conditions are equivalent, it remains to prove that (vi) implies (vii) and (vii) implies (ii). Let us prove that (vi) implies (vii). To this aim, let $\mathbb{V}$ be an open neighborhood of $\mathbb{K}$ such that $\Pi_{\mathbb{K}}(\xi) \neq \emptyset$ for each $\xi \in \mathbb{V}$. The existence of the set $\mathbb{V}$, enjoying the specified properties, is ensured by Lemma 2.2.1. Let $\ell \in L^1_{\mathrm{loc}}(\mathbb{I})$ and let $(\tau, \xi) \in \mathbb{I} \times \mathbb{K}$ be arbitrary. We claim that we can choose $\rho > 0$ and $T \in \mathbb{I}$ with $\tau < T < \sup \mathbb{I}$, such that $B(\xi, \rho) \cap \mathbb{K}$ is closed, $B(\xi, \rho) \subseteq \mathbb{V}$,

$$
\big\| f(t, u) \big\| \leqslant \ell(t) \tag{5.2.1}
$$

a.e. for $t \in [\tau, T]$ and for each $u \in B(\xi, \rho)$, and

$$
\int_\tau^T \ell(s)\,ds \leqslant \frac{\rho}{2}. \tag{5.2.2}
$$

Indeed, since $\mathbb{V}$ is open and $\mathbb{K}$ is locally closed, we can get a $\rho > 0$ with $B(\xi, \rho) \cap \mathbb{K}$ closed and $B(\xi, \rho) \subseteq \mathbb{V}$. Further, as f satisfies (H$_3$), it follows that there exists $\ell \in L^1_{\mathrm{loc}}(\mathbb{I})$ satisfying (5.2.1). Finally, diminishing T if necessary, we get (5.2.2) and $T < \sup \mathbb{I}$. Next, let $h > 0$ with $T + h < \sup \mathbb{I}$, and let us define $f_h : [\tau, T] \to \mathbb{R}^n$ by

$$f_h(t, \xi) = \frac{1}{h} \int_t^{t+h} f(s, \xi)\, \mathrm{d}s.$$

From Lemma 5.2.2, we know that f_h is continuous. On the other hand, since (vi) holds, we have $\langle f_h(t, \xi), \zeta \rangle \leqslant 0$, for all $t \in \mathbb{I}$ and $\zeta \in \mathcal{N}_{\mathbb{K}}(\xi)$, and thus the function f_h satisfies the condition (v) in Theorem 3.4.1 with $\mathcal{G} = \overline{\mathrm{co}}\, \mathcal{T}_{\mathbb{K}}(\xi)$. Therefore, for each $h > 0$, with $T + h < \sup \mathbb{I}$, $\mathbb{K}$ is viable with respect to f_h. This implies that, for any $h > 0$, $T + h < \sup \mathbb{I}$, there exists $T_h \in (\tau, T]$ and a solution $u_h : [\tau, T_h] \to \mathbb{K}$ of the differential equation $u'(t) = f_h(t, u(t))$ with $u_h(\tau) = \xi$. From Theorem 3.5.1, using (5.2.1) and (5.2.2), we conclude that we may choose $T_h = T$ for each $h > 0$, $T + h < \sup \mathbb{I}$. Therefore, for each sufficiently small $h > 0$, $u'(t) = f_h(t, u(t))$ has at least one solution $u_h : [\tau, T] \to B(\xi, \rho) \cap \mathbb{K}$ satisfying $u_h(\tau) = \xi$.

Take $h_m \downarrow 0$, and let $u_m = u_{h_m}$ be a solution to $u'(t) = f_m(t, u(t))$ with $f_m = f_{h_m}$ on $[\tau, T]$ and $u_m(\tau) = \xi$. We will prove that, for a certain function $u : [\tau, T] \to \mathbb{K}$, $\lim_m u_m(t) = u(t)$ uniformly for $t \in [\tau, T]$, and u is a Carathéodory solution to (1.1.1) satisfying $u(\tau) = \xi$. In order to do this, we will show first that $\{u_m;\ m \in \mathbb{N}\}$ satisfies the hypotheses of Arzelà–Ascoli theorem. Let $\mathbb{E} \subset [\tau, T]$ be a Lebesgue measurable set and let us observe that

$$\int_{\mathbb{E}} \left\| u'_m(t) \right\| \mathrm{d}t$$

$$= \int_{\mathbb{E}} \left\| f_{h_m}(t, u_m(t)) \right\| \mathrm{d}t = \int_{\mathbb{E}} \left\| \frac{1}{h_m} \int_t^{t+h_m} f(s, u_m(t))\, \mathrm{d}s \right\| \mathrm{d}t$$

$$= \int_{\mathbb{E}} \left\| \frac{1}{h_m} \int_0^{h_m} f(t + s, u_m(t))\, \mathrm{d}s \right\| \mathrm{d}t$$

$$\leqslant \frac{1}{h_m} \int_0^{h_m} \int_{\mathbb{E}} \left\| f(t + s, u_m(t)) \right\| \mathrm{d}t\, \mathrm{d}s \leqslant \frac{1}{h_m} \int_0^{h_m} \int_{\mathbb{E}} \ell(t + s)\, \mathrm{d}t\, \mathrm{d}s.$$

Since $\ell \in L^1_{\mathrm{loc}}(\mathbb{I})$, for each $\varepsilon > 0$ there exists $\delta(\varepsilon) > 0$ such that, for each $\mathbb{E} \subset [\tau, T + h_1]$ with $\lambda(\mathbb{E}) \leqslant \delta(\varepsilon)$, we have

$$\int_{\mathbb{E}} \ell(t)\, \mathrm{d}t \leqslant \varepsilon.$$

Since the Lebesgue measure is translation invariant, it follows that, for each $\mathbb{E} \subseteq [\tau, T]$ with $\lambda(\mathbb{E}) \leqslant \delta(\varepsilon)$, we have

$$\int_{\mathbb{E}} \ell(t + s)\, \mathrm{d}t \leqslant \varepsilon$$

for each $s \in [0, h_1]$. Summing up, we conclude that

$$\int_{\mathbb{E}} \left\| u_m'(t) \right\| dt \leq \frac{1}{h_m} \int_0^{h_m} \int_{\mathbb{E}} \ell(t+s) \, dt \, ds \leq \varepsilon$$

for each $\mathbb{E} \subseteq [\tau, T]$ with $\lambda(\mathbb{E}) \leq \delta(\varepsilon)$. So $\{u_m'; \ m \in \mathbb{N}\}$ is uniformly integrable. This implies that $\{u_m; \ m \in \mathbb{N}\}$ is equicontinuous. Moreover, since $u_m(t) \in \mathbb{K} \cap B(\xi, \rho)$ for each $t \in [\tau, T]$, if readily follows that $\{u_m; \ m \in \mathbb{N}\}$ is uniformly bounded. Therefore, there exists $u : [\tau, T] \to \mathbb{K} \cap B(\xi, \rho)$ such that, at least on a subsequence, denoted for simplicity again by $(u_m)_m$, we have $\lim_m u_m(t) = u(t)$ uniformly for $t \in [\tau, T]$. Clearly u is continuous. On the other hand, by Theorem 5.1.3, we have

$$\lim_{m \to \infty} \frac{1}{h_m} \int_s^{s+h_m} f\big(\theta, u_m(s)\big) \, d\theta = f\big(s, u(s)\big)$$

a.e. for $s \in [\tau, T]$. Since $\{u_m'; \ m \in \mathbb{N}\}$ is uniformly integrable, by virtue of Vitali's Theorem A.1.2, we conclude that

$$\lim_{m \to \infty} u_m(t) = \lim_{m \to \infty} \left(\xi + \int_\tau^t \frac{1}{h_m} \int_s^{s+h_m} f\big(\theta, u_m(s)\big) \, d\theta \, ds \right)$$
$$= \xi + \int_\tau^t f\big(s, u(s)\big) \, ds,$$

which shows that u is a Carathéodory solution to (1.1.1) satisfying $u(\tau) = \xi$. Hence $\mathbb{K}$ is Carathéodory viable with respect to f, and this shows that (vi) implies (vii).

Next, to prove that (vii) implies (ii), let $\mathbb{Z} \subset \mathbb{I}$ be a negligible set as in Theorem 5.1.1. Let $t \in \mathbb{I} \setminus \mathbb{Z}$, $\xi \in \mathbb{K}$ and let $u(\cdot)$ be a Carathéodory solution to $u'(s) = f(s, u(s))$ on $[t, t + T]$ with $u(t) = \xi$ and $u(s) \in \mathbb{K}$ for $s \in [t, t + h]$. We have

$$\frac{u(t+h) - u(t)}{h} = \frac{1}{h} \int_t^{t+h} f\big(s, u(s)\big) \, ds.$$

On the other hand, by (5.1.1), the right-hand side approaches $f(t, \xi)$ as $h \downarrow 0$. Accordingly

$$\lim_{h \downarrow 0} \frac{1}{h} \, \mathrm{dist}\big(\xi + h f(t, \xi); \mathbb{K}\big) \leq \lim_{h \downarrow 0} \left\| f(t, \xi) - \frac{u(t+h) - u(t)}{h} \right\| = 0$$

and so, $f(t, \xi) \in \mathcal{F}_{\mathbb{K}}(\xi) \subseteq \mathcal{T}_{\mathbb{K}}(\xi)$, which shows that (ii) holds.

Now, (viii) implies (iii) which implies (vii), and (vii) implies (i) which implies (viii). Finally, (ix) implies (vi) which implies (vii), and (vii) implies (iv) which implies (ix). The proof is complete. $\qquad\square$

5.3. *Sufficient conditions for Carathéodory local invariance*

The aim of this section is to extend the results concerning local invariance to the more general setting allowing the function f to be of Carathéodory type. Unlike the case of necessary and sufficient conditions for Carathéodory viability whose proofs differ essentially from those referring to viability, all the results which will follow are simple copies of their "C^1-solutions" counterparts we already presented when we dealt with local invariance. Therefore, we do not enter into details.

Let $\mathbb{D}$ be a domain in $\mathbb{R}^n$, $\mathbb{K} \subseteq \mathbb{D}$ a locally closed subset, and let us consider the ordinary differential equation (1.1.1), where $f : \mathbb{I} \times \mathbb{D} \to \mathbb{R}^n$ is a given function.

DEFINITION 5.3.1. The subset $\mathbb{K}$ is *Carathéodory locally invariant* with respect to f if for each $(\tau, \xi) \in \mathbb{I} \times \mathbb{K}$ and each Carathéodory solution $u : [\tau, c] \to \mathbb{D}$, $c \in \mathbb{I}$, $c > \tau$, to (1.1.1), satisfying the initial condition $u(\tau) = \xi$, there exists $T \in (\tau, c]$ such that we have $u(t) \in \mathbb{K}$ for each $t \in [\tau, T]$. It is *Carathéodory invariant* if it satisfies the above condition of Carathéodory local invariance with $T = c$.

REMARK 5.3.1. If $\mathbb{K}$ is Carathéodory locally invariant with respect to f, then $\mathbb{K}$ is locally invariant with respect to f. We notice that here we do have any extra-condition upon f as of being of Carathéodory type. It should be interesting to know if, under the latter assumption on f, the local invariance implies the Carathéodory local invariance.

The relationship between Carathéodory viability and Carathéodory local invariance is clarified in:

REMARK 5.3.2. If f is of Carathéodory type on $\mathbb{I} \times \mathbb{D}$ and $\mathbb{K}$ is Carathéodory locally invariant with respect to f, then $\mathbb{K}$ is Carathéodory viable with respect to f. The converse of this assertion is no longer true, as we already have seen in Example 4.1.1.

As in the case of local invariance, we have the following simple necessary and sufficient condition of Carathéodory local invariance.

THEOREM 5.3.1. *Let $\mathbb{D}$ be a domain, $\mathbb{K} \subseteq \mathbb{D}$ a nonempty and locally closed subset and $f : \mathbb{I} \times \mathbb{D} \to \mathbb{R}^n$ a Carathéodory type function with the property that (1.1.1) has the uniqueness property. Then, a necessary and sufficient condition in order that $\mathbb{K}$ be invariant with respect to f is each one of the nine conditions* (i)–(ix) *in Theorem* 5.2.1.

PROOF. The conclusion follows from the Carathéodory Local Existence Theorem combined with Remark 5.3.2 and the equivalence between (i) and (ix) in Theorem 5.2.1. $\square$

As in the case of local invariance, the Carathéodory local invariance of $\mathbb{K}$ with respect to f can take place even if $u'(t) = f_{|\mathbb{I} \times \mathbb{K}}(t, u(t))$ has not the uniqueness property. See Example 4.1.2.

We say that $\omega : \mathbb{I} \times [0, \rho) \to \mathbb{R}$ is a *Carathéodory comparison function* if, for each $[\tau, T) \subseteq \mathbb{I}$, the only absolutely continuous solution $x : [\tau, T) \to [0, \rho)$, of $[D_+ x](t) \leqslant \omega(t, x(t))$ a.e. for $t \in [\tau, T)$, satisfying $x(\tau) = 0$, is the null function.

Our first sufficient condition for Carathéodory local invariance is expressed by means of the "exterior tangency" condition: there exist a negligible subset $\mathbb{Z}$ in $\mathbb{I}$ and an open neighborhood $\mathbb{V} \subseteq \mathbb{D}$ of $\mathbb{K}$ such that

$$\liminf_{h \downarrow 0} \frac{1}{h}\left[\operatorname{dist}\big(\xi + hf(t,\xi); \mathbb{K}\big) - \operatorname{dist}(\xi; \mathbb{K})\right] \leqslant \omega\big(t, \operatorname{dist}(\xi; \mathbb{K})\big) \qquad (5.3.1)$$

for each $(t, \xi) \in (\mathbb{I} \setminus \mathbb{Z}) \times \mathbb{V}$, where ω is a certain Carathéodory comparison function.

The next result is a "Carathéodory" counterpart of Theorem 4.2.1.

THEOREM 5.3.2. *Let* $\mathbb{K} \subseteq \mathbb{D} \subseteq \mathbb{R}^n$, *with* $\mathbb{K}$ *locally closed and* $\mathbb{D}$ *open, and let* $f : \mathbb{I} \times \mathbb{D} \to \mathbb{R}^n$. *If* (5.3.1) *is satisfied, then* $\mathbb{K}$ *is Carathéodory locally invariant with respect to* f.

Theorem 5.3.2 can be reformulated as:

THEOREM 5.3.3. *Let* $\mathbb{K} \subseteq \mathbb{D} \subseteq \mathbb{R}^n$, *with* $\mathbb{K}$ *locally closed and* $\mathbb{D}$ *open, and let* $f : \mathbb{I} \times \mathbb{D} \to \mathbb{R}^n$. *If there exists an open neighborhood* V *of* $\mathbb{K}$ *with* $\mathbb{V} \subseteq \mathbb{D}$ *such that* ω_f *defined by* (4.2.2) *is a Carathéodory comparison function, then* $\mathbb{K}$ *is Carathéodory locally invariant with respect to* f.

DEFINITION 5.3.2. Let $\mathbb{K} \subseteq \mathbb{D} \subseteq \mathbb{R}^n$. We say that a function $f : \mathbb{I} \times \mathbb{D} \to \mathbb{R}^n$ has the *Carathéodory comparison property with respect to* $(\mathbb{D}, \mathbb{K})$ if there exist a negligible subset $\mathbb{Z}$ of $\mathbb{I}$, a proximal neighborhood $\mathbb{V} \subseteq \mathbb{D}$ of $\mathbb{K}$, one projection $\pi_\mathbb{K} : \mathbb{V} \to \mathbb{K}$ subordinated to $\mathbb{V}$, and a Carathéodory comparison function $\omega : \mathbb{I} \times [0, a) \to \mathbb{R}$, with $a = \sup_{\xi \in \mathbb{V}} \operatorname{dist}(\xi; \mathbb{K})$, such that

$$\left[\xi - \pi_\mathbb{K}(\xi), f(t,\xi) - f\big(t, \pi_\mathbb{K}(\xi)\big)\right]_+ \leqslant \omega\big(t, \|\xi - \pi_\mathbb{K}(\xi)\|\big)$$

for each $(t, \xi) \in (\mathbb{I} \setminus \mathbb{Z}) \times \mathbb{V}$.

THEOREM 5.3.4. *Let* $\mathbb{K} \subseteq \mathbb{D} \subseteq \mathbb{R}^n$, *with* $\mathbb{K}$ *locally closed and* $\mathbb{D}$ *open, and let* $f : \mathbb{I} \times \mathbb{D} \to \mathbb{R}^n$. *If* f *has the Carathéodory comparison property with respect to* $(\mathbb{D}, \mathbb{K})$, *and* $\mathbb{K}$ *is Carathéodory viable with respect* f, *then* (5.3.1) *holds true.*

Other sufficient conditions for invariance obtained from Theorem 5.2.1 are:

THEOREM 5.3.5. *Let* $\mathbb{K} \subseteq \mathbb{D} \subseteq \mathbb{R}^n$, *with* $\mathbb{K}$ *locally closed and* $\mathbb{D}$ *open, and let* $f : \mathbb{I} \times \mathbb{D} \to \mathbb{R}^n$ *be of Carathéodory type. Let us assume that* f *has the Carathéodory comparison property with respect to* $(\mathbb{D}, \mathbb{K})$, *and satisfies one of the nine equivalent tangency conditions* (i)–(ix) *in Theorem 5.2.1. Then* $\mathbb{K}$ *is Carathéodory locally invariant with respect to* f.

5.4. *Sufficient conditions via generalized distance*

Here we reconsider the sufficient conditions for local invariance obtained by means of a proximal generalized distance in the context of Carathéodory type functions.

THEOREM 5.4.1. *Let* $\mathbb{K} \subseteq \mathbb{D} \subseteq \mathbb{R}^n$, *and let* $f : \mathbb{I} \times \mathbb{D} \to \mathbb{R}^n$. *If there exist an open neighborhood* $\mathbb{V}$ *of* $\mathbb{K}$, *with* $\mathbb{V} \subseteq \mathbb{D}$, *a proximal generalized distance* $g : \mathbb{V} \to \mathbb{R}_+$, *a negligible subset* $\mathbb{Z}$ *in* $\mathbb{I}$ *and a Carathéodory comparison function* $\omega : \mathbb{I} \times [0, a) \to \mathbb{R}$ *such that*

$$\liminf_{h \downarrow 0} \frac{1}{h} \left[g\big(\xi + hf(t, \xi)\big) - g(\xi) \right] \leqslant \omega\big(t, g(\xi)\big)$$

for each $(t, \xi) \in (\mathbb{I} \setminus \mathbb{Z}) \times \mathbb{V}$, *then* $\mathbb{K}$ *is Carathéodory locally invariant with respect to* f.

Let $g : \mathbb{V} \to [0, +\infty)$ be a proximal generalized distance, and let us introduce the generalized Carathéodory tangency condition: *there exists a negligible subset* $\mathbb{Z}$ *of* $\mathbb{I}$ *such that*

$$\liminf_{h \downarrow 0} \frac{1}{h} g\big(\xi + hf(t, \xi)\big) = 0 \tag{5.4.1}$$

for each $(t, \xi) \in (\mathbb{I} \setminus \mathbb{Z}) \times \mathbb{K}$.

THEOREM 5.4.2. *Let* $\mathbb{K} \subseteq \mathbb{D} \subseteq \mathbb{R}^n$, *and let* $f : \mathbb{I} \times \mathbb{D} \to \mathbb{R}^n$. *If there exist an open neighborhood* $\mathbb{V}$ *of* $\mathbb{K}$, *with* $\mathbb{V} \subseteq \mathbb{D}$, *a proximal generalized distance* $g : \mathbb{V} \to \mathbb{R}_+$, *a negligible subset* $\mathbb{Z}$ *in* $\mathbb{I}$ *and a Carathéodory comparison function* $\omega : \mathbb{I} \times [0, a) \to \mathbb{R}$ *satisfying* (5.4.1) *and such that*

$$\liminf_{h \downarrow 0} \frac{1}{2h} \left[g^2\big(\xi + hf(t, \xi)\big) - g^2(\xi) \right] \leqslant g(\xi)\omega\big(t, g(\xi)\big)$$

for each $(t, \xi) \in (\mathbb{I} \setminus \mathbb{Z}) \times \mathbb{V}$, *then* $\mathbb{K}$ *is Carathéodory locally invariant with respect to* f.

THEOREM 5.4.3. *Let* $\mathbb{D} \subseteq \mathbb{R}^n$ *be open and let* $f : \mathbb{I} \times \mathbb{D} \to \mathbb{R}^n$. *Let us assume that* $\mathbb{K} \subseteq \mathbb{D}$ *is a Lipschitz retract with the generalized Lipschitz projection* $r : \mathbb{V} \to \mathbb{K}$, *and there exists a negligible subset* $\mathbb{Z}$ *of* $\mathbb{I}$ *satisfying*

$$\liminf_{h \downarrow 0} \frac{1}{h} \left\| r\big(\xi + hf(t, \xi)\big) - \xi - hf(t, \xi) \right\| = 0 \tag{5.4.2}$$

for each $(t, \xi) \in (\mathbb{I} \setminus \mathbb{Z}) \times \mathbb{K}$. *Assume further that there exists a Carathéodory comparison function* $\omega : \mathbb{I} \times [0, a) \to \mathbb{R}$, *with* $a = \sup_{\xi \in \mathbb{V}} \| r(\xi) - \xi \|$, *such that*

$$\liminf_{h \downarrow 0} \frac{1}{h} \big\langle r\big(\xi + hf(t, \xi)\big) - r(\xi) - hf(t, \xi), r(\xi) - \xi \big\rangle$$

$$\leqslant \| r(\xi) - \xi \| \omega\big(t, \| r(\xi) - \xi \|\big)$$

for each $(t, \xi) \in (\mathbb{I} \setminus \mathbb{Z}) \times \mathbb{V}$. *Then,* $\mathbb{K}$ *is Carathéodory locally invariant with respect to* f.

A consequence of Theorem 5.4.3 is stated below.

THEOREM 5.4.4. *Let $\mathbb{K} \subseteq \mathbb{D}$ with $\mathbb{D}$ open and let $f : \mathbb{I} \times \mathbb{D} \to \mathbb{R}^n$. Let us assume that $\mathbb{K}$ is a Lipschitz retract with the generalized Lipschitz projection $r : \mathbb{V} \to \mathbb{K}$ and there exists a negligible subset $\mathbb{Z}$ of $\mathbb{I}$ such that (5.4.2) is satisfied. Let us assume, in addition, that, for each $t \in \mathbb{I} \setminus \mathbb{Z}$ and $\xi \in \mathbb{V}$, there exists the directional derivative, $r'(\xi)[f(t, \xi)]$, of r, at ξ in the direction $f(t, \xi)$, and*

$$\langle r'(\xi)[f(t, \xi)] - f(t, \xi), r(\xi) - \xi \rangle \leqslant \|r(\xi) - \xi\| \omega\big(t, \|r(\xi) - \xi\|\big),$$

where $\omega : \mathbb{I} \times [0, a) \to \mathbb{R}$ is a Carathéodory comparison function, with

$$a = \sup_{\xi \in \mathbb{V}} \|r(\xi) - \xi\|.$$

Then, $\mathbb{K}$ is Carathéodory locally invariant with respect to f.

6. Differential inclusions

6.1. *Multifunctions*

In this section we extend some of the previously established results to a more general case allowing the function f to be multivalued. First we introduce two classes of multifunctions. Let $\mathbb{K}$ be a nonempty subset in $\mathbb{R}^n$ and let $F : \mathbb{K} \rightsquigarrow \mathbb{R}^n$ a given mapping with nonempty values.

DEFINITION 6.1.1. The multifunction $F : \mathbb{K} \rightsquigarrow \mathbb{R}^n$ is *upper semicontinuous (u.s.c.) at* $\xi \in \mathbb{K}$ if for every open neighborhood $\mathbb{V}$ of $F(\xi)$ there exists an open neighborhood $\mathbb{U}$ of ξ such that $F(\eta) \subseteq \mathbb{V}$ for each $\eta \in \mathbb{U} \cap \mathbb{K}$. We say that F is *upper semicontinuous (u.s.c.) on* $\mathbb{K}$ if it is u.s.c. at each $\xi \in \mathbb{K}$.

DEFINITION 6.1.2. The multifunction $F : \mathbb{K} \rightsquigarrow \mathbb{R}^n$ is *lower semicontinuous (l.s.c.) at* $\xi \in \mathbb{K}$ if for every open set $\mathbb{V}$ in $\mathbb{R}^n$ with $F(\xi) \cap \mathbb{V} \neq \emptyset$ there exists an open neighborhood $\mathbb{U}$ of ξ such that $F(\eta) \cap \mathbb{V} \neq \emptyset$ for each $\eta \in \mathbb{U} \cap \mathbb{K}$. We say that F is *lower semicontinuous (l.s.c.) on* $\mathbb{K}$ if it is l.s.c. at each $\xi \in \mathbb{K}$.

The next two lemmas will prove useful later.

LEMMA 6.1.1. *If $F : \mathbb{K} \rightsquigarrow \mathbb{R}^n$ is a nonempty and compact valued u.s.c. multifunction, then, for each compact subset $\mathbb{C}$ of $\mathbb{K}$, $\bigcup_{\xi \in \mathbb{C}} F(\xi)$ is compact. In particular, for each compact subset $\mathbb{C}$ of $\mathbb{K}$, there exists $M > 0$ such that $\|\eta\| \leqslant M$ for each $\xi \in \mathbb{C}$ and each $\eta \in F(\xi)$.*

PROOF. Let $\mathbb{C}$ be a compact subset in $\mathbb{K}$ and let $\{\mathbb{D}_\sigma ; \ \sigma \in \Gamma\}$ be an arbitrary open covering of $\bigcup_{\xi \in \mathbb{C}} F(\xi)$. Since F is compact valued, for each $\xi \in \mathbb{C}$ there exists $n(\xi) \in \mathbb{N}$ such that

$$F(\xi) \subseteq \bigcup_{1 \leqslant k \leqslant n(\xi)} \mathbb{D}_{\sigma_k}.$$

But F is u.s.c. and therefore there exists an open neighborhood $\mathbb{U}(\xi)$ of ξ such that

$$F\big(\mathbb{U}(\xi) \cap \mathbb{K}\big) \subseteq \bigcup_{1 \leqslant k \leqslant n(\xi)} \mathbb{D}_{\sigma_k}.$$

The family $\{\mathbb{U}(\xi);\ \xi \in \mathbb{C}\}$ is an open covering of $\mathbb{C}$. As $\mathbb{C}$ is compact, there exists a finite family $\{\xi_1, \xi_2, \ldots, \xi_p\}$ in $\mathbb{C}$ such that

$$F(\mathbb{C}) \subseteq \bigcup_{1 \leqslant j \leqslant p} F\big(\mathbb{U}(\xi_j) \cap \mathbb{K}\big) \subseteq \bigcup_{1 \leqslant j \leqslant p} \bigcup_{1 \leqslant k \leqslant n(\xi_j)} \mathbb{D}_{\sigma_k},$$

and this completes the proof. $\qquad\square$

LEMMA 6.1.2. *Let $F : \mathbb{K} \rightsquigarrow \mathbb{R}^n$ be a u.s.c., nonempty, convex, compact valued multifunction, and let $u_m : [0, T] \to \mathbb{K}$ and $f_m \in L^1(0, T; \mathbb{R}^n)$ be such that $f_m(t) \in F(u_m(t))$ for each $m \in \mathbb{N}$ and a.e. for $t \in [0, T]$. If $\lim_m u_m(t) = u(t)$ a.e. for $t \in [0, T]$ and $\lim_m f_m = f$ weakly in $L^1(0, T; \mathbb{R}^n)$, then $f(t) \in F(u(t))$ a.e. for $t \in [0, T]$.*

PROOF. By Mazur's theorem—Dunford–Schwartz [40, Theorem 6, p. 416]—there exists a sequence $(g_m)_m$ of convex combinations of $\{f_k;\ k \geqslant m\}$, i.e. $g_m \in \mathrm{co}\{f_m, f_{m+1}, \ldots\}$ for each $m \in \mathbb{N}$, which converges strongly in $L^1(0, T; \mathbb{R}^n)$ to f. By a classical result due to Lebesgue, we know that there exists a subsequence (g_{m_p}) of (g_m) which converges almost everywhere on $[0, T]$ to f. Denote by $\mathbb{T}$ the set of all $s \in [0, T]$ such that both $(g_{m_p}(s))_p$ and $(u_m(s))_m$ are convergent to $f(s)$ and to $u(s)$, respectively, and, in addition, $f_m(s) \in F(u_m(s))$ for each $m \in \mathbb{N}$. Clearly $[0, T] \setminus \mathbb{T}$ has null measure. Let $s \in \mathbb{T}$ and let $\mathbb{E}$ be an open half-space in $\mathbb{R}^n$ including $F(u(s))$. Since F is upper semicontinuous at $u(s)$ and $(u_m(s))_m$ converges to $u(s)$, there exists $m(\mathbb{E})$ belonging to $\mathbb{N}$, such that $F(u_m(s)) \subseteq \mathbb{E}$ for each $m \geqslant m(\mathbb{E})$. From the relation above, taking into account that $f_m(s) \in F(u_m(s))$ for each $m \in \mathbb{N}$ and a.e. for $s \in [0, T]$, we easily conclude

$$g_{m_p}(s) \in \overline{\mathrm{co}}\left(\bigcup_{m \geqslant m(\mathbb{E})} F\big(u_m(s)\big) \right)$$

for each $p \in \mathbb{N}$ with $m_p \geqslant m(\mathbb{E})$. Passing to the limit for $p \to +\infty$ in the relation above we deduce $f(s) \in \overline{\mathbb{E}}$. Since $F(u(s))$ is closed and convex, it is the intersection of all closed half-spaces which include it. So, inasmuch as $\mathbb{E}$ was arbitrary, we, finally, get $f(s) \in F(u(s))$ for each $s \in \mathbb{T}$ and this completes the proof. $\qquad\square$

6.2. *Viability with respect to a multifunction*

Let us consider the autonomous differential inclusion

$$u'(t) \in F\big(u(t)\big) \tag{6.2.1}$$

where $F : \mathbb{K} \rightsquigarrow \mathbb{R}^n$ is a given multifunction. By a *solution* of (6.2.1) we mean an absolutely continuous function $u : \mathbb{J} \to \mathbb{K}$, with $\mathbb{J}$ a proper interval, satisfying $u'(t) \in F(u(t))$ a.e. for $t \in \mathbb{J}$.

DEFINITION 6.2.1. We say that $\mathbb{K}$ is *right viable with respect to* F if for each $\xi \in \mathbb{K}$, there exists $T > 0$ such that (6.2.1) has at least one solution $u : [0, T] \to \mathbb{K}$ satisfying $u(0) = \xi$.

THEOREM 6.2.1. *Let $\mathbb{K}$ be a locally closed subset in $\mathbb{R}^n$ and $F : \mathbb{K} \rightsquigarrow \mathbb{R}^n$ a u.s.c. nonempty, closed, convex and bounded valued mapping. Then a necessary and sufficient condition in order that $\mathbb{K}$ be right viable with respect to F is the tangency condition*

$$F(\xi) \cap \mathcal{T}_{\mathbb{K}}(\xi) \neq \emptyset \tag{6.2.2}$$

for each $\xi \in \mathbb{K}$.

PROOF. *Necessity.* Since F is upper semicontinuous and has nonempty, closed, convex and bounded values, by virtue of Lemma 6.1.1, we conclude that F is locally bounded. Let $\xi \in \mathbb{K}$ and let $u : [0, T] \to \mathbb{K}$ be a solution to (6.2.1) satisfying $u(0) = \xi$. Let $h \in (0, T]$ and let us define

$$u_h = \frac{1}{h} \int_0^h u'(s) \, ds.$$

Since F is locally bounded, by diminishing $T > 0$ if necessary, we may assume that $\{u'(s); \ s \in [0, T]\}$ is bounded a.e. In other words, there exists $M > 0$ such that $\|u'(s)\| \leqslant M$ a.e. for $s \in [0, T]$. This means that the set $\{\|u_h\|; \ h \in (0, T]\}$ is bounded, also by M, and therefore there exists $h_m \downarrow 0$ such that $\eta = \lim_{m \to \infty} u_{h_m}$. Since F is upper semicontinuous at ξ and has convex compact values, we infer that $\eta \in F(\xi)$. Indeed, for each open neighborhood $\mathbb{V}$ of $F(\xi)$ there exists an open neighborhood $\mathbb{U}$ of ξ such that $F(\zeta) \subseteq \mathbb{V}$ for each $\zeta \in \mathbb{K} \cap \mathbb{U}$. Therefore, if $h \in (0, T]$ is sufficiently small, we have $u'(s) \in \mathbb{V}$ a.e. for $s \in (0, h]$. So, $u_h \in \overline{co}\, \mathbb{V}$. Accordingly $\eta \in \bigcap_{\mathbb{V} \in \mathcal{V}(F(\xi))} \overline{co}\, \mathbb{V}$, where $\mathcal{V}(F(\xi))$ stands for the set of open neighborhoods of $F(\xi)$. But $F(\xi)$ is convex, and hence $\bigcap_{\mathbb{V} \in \mathcal{V}(F(\xi))} \overline{co}\, \mathbb{V} = F(\xi)$. To complete the proof of the necessity, it suffices to show that $\eta \in \mathcal{T}_{\mathbb{K}}(\xi)$. To this aim, let us observe that $\xi + hu_h = u(h) \in \mathbb{K}$, and so $\text{dist}(\xi + hu_h; \mathbb{K}) = 0$. Thus

$$\text{dist}(\xi + h\eta; \mathbb{K}) \leqslant \text{dist}(\xi + hu_h; \mathbb{K}) + \|\xi + hu_h - (\xi + h\eta)\| \leqslant h\|u_h - \eta\|.$$

Since η is a limit point of u_h as $h \downarrow 0$, we deduce

$$\liminf_{h \downarrow 0} \frac{1}{h} \text{dist}(\xi + h\eta; \mathbb{K}) = 0$$

which shows that $\eta \in \mathcal{T}_{\mathbb{K}}(\xi)$. The proof of the necessity is complete. $\qquad\square$

REMARK 6.2.1. On may ask whether the equivalence between (i) and (ii) in Proposition 2.5.1 can be extended to this framework. Namely, one may ask on the equivalence of the two conditions below:

(j) for each $\xi \in \mathbb{K}$, $F(\xi) \cap \mathcal{C}_{\mathbb{K}}(\xi) \neq \emptyset$,

(jj) for each $\xi \in \mathbb{K}$, $F(\xi) \cap \mathcal{T}_{\mathbb{K}}(\xi) \neq \emptyset$.

The next simple example shows that (j) and (jj) are not equivalent. Namely, take $n = 2$, $\mathbb{K} = \{(x_1, x_2); \ |x_1| \geqslant x_2\}$ and $\mathbb{V} = \{(x_1, 1); \ |x_1| \leqslant 1\}$, and let us define $F : \mathbb{K} \rightsquigarrow \mathbb{R}^2$ by $F(\xi) = \mathbb{V}$, for each $\xi \in \mathbb{K}$. Clearly, F is both u.s.c. and l.s.c., with nonempty, convex, compact values and (jj) holds true. Nevertheless $F(0, 0) \cap \mathcal{C}_{\mathbb{K}}((0, 0)) = \emptyset$.

We recall that, whenever $F : \mathbb{K} \rightsquigarrow \mathbb{R}^n$ is l.s.c. at $\xi \in \mathbb{K}$, for each $\eta \in F(\xi)$, and each $(\xi_m)_m$ with $\lim_m \xi_m = \xi$, there exist $(\eta_m)_m$ with $\eta_m \in F(\xi_m)$ for each $m \in \mathbb{N}$, and such that $\lim_m \eta_m = \eta$. See, for instance, Cârjă [20, Theorem 2.4, p. 25]. So, from Remark 2.5.1 and Lemma 2.5.1, we deduce:

PROPOSITION 6.2.1. *If $F : \mathbb{K} \rightsquigarrow \mathbb{R}^n$ is l.s.c., then the following conditions are equivalent:*

(i) *for each $\xi \in \mathbb{K}$, $F(\xi) \subseteq \mathcal{C}_{\mathbb{K}}(\xi)$,*

(ii) *for each $\xi \in \mathbb{K}$, $F(\xi) \subseteq \mathcal{T}_{\mathbb{K}}(\xi)$,*

(iii) *for each $\xi \in \mathbb{K}$, $F(\xi) \subseteq \mathcal{B}_{\mathbb{K}}(\xi)$,*

In general, if $\mathcal{G} : \mathbb{K} \rightsquigarrow \mathbb{R}^n$ is such that $\mathcal{C}_{\mathbb{K}}(\xi) \subseteq \mathcal{G}(\xi) \subseteq \mathcal{B}_{\mathbb{K}}(\xi)$ for each $\xi \in \mathbb{K}$, then each one of the conditions above is equivalent to:

(iv) *for each $\xi \in \mathbb{K}$, $F(\xi) \subseteq \mathcal{G}(\xi)$.*

REMARK 6.2.2. The next simple example shows that the equivalence between (i) and (ii) in Proposition 6.2.1 is no longer true if F is u.s.c. Namely, take $n = 2$, $\mathbb{K} = \{(x_1, x_2); \ |x_1| \geqslant x_2\}$ and let us define $F : \mathbb{K} \rightsquigarrow \mathbb{R}^2$ by

$$F(\xi) = \begin{cases} \{(0, 0)\} & \text{if } \xi \in \mathbb{K} \setminus \{(0, 0)\}, \\ \mathbb{K} & \text{if } \xi = (0, 0). \end{cases}$$

Clearly, F is u.s.c. with nonempty, closed values and (ii) holds true. Nevertheless (i) does not hold.

6.3. *Existence of ε-approximate solutions*

The proof of the sufficiency consists in showing that (6.2.2) along with Brezis–Browder Theorem 2.1.1 imply that, for each $\xi \in \mathbb{K}$, there exists at least one sequence of "approximate" solutions to (6.2.1), defined on the same interval, $u_m : [0, T] \to \mathbb{R}^n$, satisfying $u_m(0) = \xi$ for each $m \in \mathbb{N}$, and such that $(u_m)_m$ converges in some sense to a solution u of (6.2.1) satisfying $u(0) = \xi$.

The next lemma represents an existence result concerning "approximate solutions" of (6.2.1) satisfying $u_m(0) = \xi$.

LEMMA 6.3.1. *Let $\mathbb{K}$ be a nonempty and locally closed subset in $\mathbb{R}^n$, and let $F : \mathbb{K} \rightsquigarrow \mathbb{R}^n$ be a nonempty valued mapping which is locally bounded. If $\mathbb{K}$ and F satisfy (6.2.2) then for* *each $\xi \in \mathbb{K}$ there exist $\rho > 0$, $T > 0$, $M > 0$ such that $\mathbb{K} \cap B(\xi, \rho)$ is closed, and for each $\varepsilon \in (0, 1]$ there exist four functions: $\sigma : [0, T] \to [0, T]$ nondecreasing, $f : [0, T] \to \mathbb{R}^n$ and $g : [0, T] \to \mathbb{R}^n$ measurable, and $u : [0, T] \to \mathbb{R}^n$ continuous, satisfying*

  (i) $s - \varepsilon \leqslant \sigma(s) \leqslant s, u(\sigma(s)) \in \mathbb{K} \cap B(\xi, \rho)$ *and* $f(s) \in F(u(\sigma(s)))$, *a.e. for* $s \in [0, T]$,
  (ii) $\|f(s)\| \leqslant M$ *a.e. for* $s \in [0, T]$,
  (iii) $u(T) \in \mathbb{K} \cap B(\xi, \rho)$,
  (iv) $\|g(s)\| \leqslant \varepsilon$ *a.e. for* $s \in [0, T]$

and

$$u(t) = \xi + \int_0^t f(s)\,\mathrm{d}s + \int_0^t g(s)\,\mathrm{d}s \tag{6.3.1}$$

for each $t \in [0, T]$.

PROOF. Let $\xi \in \mathbb{K}$ be arbitrary, and choose $\rho > 0$, $T > 0$ and $M > 0$ such that $\mathbb{K} \cap B(\xi, \rho)$ is closed,

$$\|\eta\| \leqslant M \tag{6.3.2}$$

for each $x \in \mathbb{K} \cap B(\xi, \rho)$ and $\eta \in F(x)$, and

$$T(M + 1) \leqslant \rho. \tag{6.3.3}$$

This is always possible since $\mathbb{K}$ is locally closed and F is locally bounded. Let $\varepsilon \in (0, 1]$. We start by showing how to define the functions f, g, σ, and u on a sufficiently small interval $[0, \delta]$, $\delta \leqslant T$, and then we will show how to extend them to the whole interval $[0, T]$. We recall that, in view of Proposition 2.3.2, there exist $\delta \in (0, \varepsilon]$, $\eta \in F(\xi)$ and $p \in \mathbb{R}^n$, all depending on ε, with $\|p\| \leqslant \varepsilon$, and satisfying $\xi + \delta(\eta + p) \in \mathbb{K}$. Diminishing δ if necessary, and using (6.3.2), we may also assume that $\|\delta(\eta + p)\| \leqslant \rho$. Let us define $u : [0, \delta] \to \mathbb{R}^n$ by

$$u(t) = \xi + t\eta + tp \tag{6.3.4}$$

for each $t \in [0, \delta]$. By virtue of (6.3.2), u, η and p satisfy

  (j) $\eta \in F(u(0))$,
  (jj) $\|\eta\| \leqslant M$,
  (jjj) $u(\delta) \in \mathbb{K} \cap B(\xi, \rho)$,
  (jv) $\|p\| \leqslant \varepsilon$.

Setting $f(s) = \eta$, $g(s) = p$, $\sigma(s) = 0$ for $s \in [0, \delta]$ and u defined by (6.3.4), from (j)–(jv), the fact that $\delta \in (0, \varepsilon]$ and (6.3.4), we can easily see that (σ, f, g, u) satisfies (i)–(iv) and (6.3.1) with T substituted by δ.

Next, we are going to show that, for each $\varepsilon \in (0, 1]$, there exists at least one 4-tuple (σ, f, g, u), whose domain is $[0, T]$, satisfying (i)–(iv) and (6.3.1). To this aim we shall use Brezis–Browder Theorem 2.1.1 as follows. Let $\mathcal{S}$ be the set of all 4-tuples (σ, f, g, u) defined on $[0, a]$ with $a \in (0, T]$ and satisfying (i)–(iv) and (6.3.1) on $[0, a]$. This set is clearly nonempty because (σ, f, g, u), defined as above, belongs to $\mathcal{S}$. On $\mathcal{S}$ we define a binary relation "$\preceq$" as follows. We say that (σ, f, g, u) defined on $[0, a]$ and $(\tilde{\sigma}, \tilde{f}, \tilde{g}, \tilde{u})$ defined on $[0, b]$ satisfy

$$(\sigma, f, g, u) \preceq (\tilde{\sigma}, \tilde{f}, \tilde{g}, \tilde{u})$$

if $a \leqslant b$, $\sigma(s) = \tilde{\sigma}(s)$, $f(s) = \tilde{f}(s)$ and $g(s) = \tilde{g}(s)$ a.e. for $s \in [0, a]$. Endowed with "$\preceq$", $\mathcal{S}$ is obviously a preordered set. Let $((\sigma_m, f_m, g_m, u_m))_m$ be a monotone sequence, and let us denote by $[0, a_m]$ the domain of $(\sigma_m, f_m, g_m, u_m)$. Let $a^* = \lim_m a_m$, where $[0, a_m]$ denotes the domain of definition of $(\sigma_m, f_m, g_m, u_m)$. Clearly $a^* \in (0, T]$. We will show that there exists $(\sigma^*, f^*, g^*, u^*) \in \mathcal{S}$, defined on $[0, a^*]$ and satisfying

$$(\sigma_m, f_m, g_m, u_m) \preceq (\sigma^*, f^*, g^*, u^*)$$

for each $m \in \mathbb{N}$. We will prove first that there exists $\lim_m u_m(a_m)$. For each $m, k \in \mathbb{N}$, $m \leqslant k$, we have $u_m(s) = u_k(s)$ for all $s \in [0, a_m]$, and therefore, taking into account (iii), (iv) and (6.3.1), we deduce

$$\left\| u_m(a_m) - u_k(a_k) \right\| \leqslant \int_{a_m}^{a_k} \left\| f_k(s) \right\| \mathrm{d}s + \int_{a_m}^{a_k} \left\| g_k(s) \right\| \mathrm{d}s$$

$$\leqslant (M + \varepsilon)|a_k - a_m|$$

for all $m, k \in \mathbb{N}$, $m \leqslant k$, which proves that there exists $\lim_{m \to \infty} u_m(a_m)$. Since for every $m \in \mathbb{N}$, $u_m(a_m) \in B(\xi, \rho) \cap \mathbb{K}$, and the latter is closed, it readily follows that $\lim_{m \to \infty} u_m(a_m) \in B(\xi, \rho) \cap \mathbb{K}$. In addition, because all the functions in the set $\{\sigma_m;\ m \in \mathbb{N}\}$ are nondecreasing, with values in $[0, a^*]$ and satisfy $\sigma_m(a_m) \leqslant \sigma_p(a_p)$ for every $m, p \in \mathbb{N}$ with $m \leqslant p$, there exists $\lim_{m \to \infty} \sigma_m(a_m)$ and this limit belongs to $[0, a^*]$. This allows us to define the 4-tuple $(\sigma^*, f^*, g^*, u^*) : [0, a^*] \to [0, a^*] \times \mathbb{R}^n \times \mathbb{R}^n \times \mathbb{R}^n$ by

$$\sigma^*(t) = \begin{cases} \sigma_m(t) & \text{for } t \in [0, a_m],\ m \in \mathbb{N}, \\ \lim_{m \to \infty} \sigma_m(a_m) & \text{for } t = a^*, \end{cases}$$

$$f^*(t) = f_m(t) \quad \text{a.e. for } t \in [0, a_m],\ m \in \mathbb{N},$$

$$g^*(t) = \begin{cases} g_m(t) & \text{for } t \in [0, a_m],\ m \in \mathbb{N}, \\ 0 & \text{for } t = a^*, \end{cases}$$

$$u^*(t) = \begin{cases} u_m(t) & \text{for } t \in [0, a_m],\ m \in \mathbb{N}, \\ \lim_{m \to \infty} u_m(a_m) & \text{for } t = a^*. \end{cases}$$

One can easily see that $(\sigma^*, f^*, g^*, u^*)$ is an ε-approximate solution which is a majorant for $((\sigma_m, f_m, g_m, u_m))_m$. Let us define $\mathcal{M} : \mathcal{S} \to \mathbb{R} \cup \{+\infty\}$ by $\mathcal{M}((\sigma, f, g, u)) = c$, where $[0, c]$ is the domain of definition of (σ, f, g, u). Clearly $\mathcal{M}$ satisfies the hypotheses of Brezis–Browder Theorem 2.1.1. Then, there exists an $\mathcal{M}$-maximal element

$(\bar{\sigma}, \bar{f}, \bar{g}, \bar{u}) \in \mathcal{S}$, defined on $[0, \bar{c}]$. This means that if $(\bar{\sigma}, \tilde{f}, \tilde{g}, \tilde{u}) \in \mathcal{S}$ is defined on $[0, \tilde{c}]$ and $(\bar{\sigma}, \bar{f}, \bar{g}, \bar{u}) \preceq (\tilde{\sigma}, \tilde{f}, \tilde{g}, \tilde{u})$, we have $\mathcal{M}((\bar{\sigma}, \bar{f}, \bar{g}, \bar{u})) = \mathcal{M}((\tilde{\sigma}, \tilde{f}, \tilde{g}, \tilde{u}))$. We will show next that $\bar{c} = T$. Indeed, let us assume by contradiction that $\bar{c} < T$. We have

$$\|\bar{u}(\bar{c}) - \xi\| \leqslant \int_0^{\bar{c}} \|\bar{f}(s)\| \, \mathrm{d}s + \int_0^{\bar{c}} \|\bar{g}(s)\| \, \mathrm{d}s \leqslant \bar{c}(M + \varepsilon)$$

$$\leqslant \bar{c}(M + 1) < \rho.$$

Then, as $\bar{u}(\bar{c}) \in \mathbb{K}$, by virtue of (6.2.2), there exists $f_{\bar{c}} \in F(\bar{u}(\bar{c})) \cap \mathcal{T}_{\mathbb{K}}(\bar{u}(\bar{c}))$. This means that, there exist $\delta \in (0, T - \bar{c})$, $\delta \leqslant \varepsilon$ and $p \in \mathbb{R}^n$ such that $\|p\| \leqslant \varepsilon$ and $\bar{u}(\bar{c}) + \delta f_{\bar{c}} + \delta p \in \mathbb{K}$. From the inequality above, it follows that we can diminish δ if necessary, in order to have $\|\bar{u}(\bar{c}) + \delta(f_{\bar{c}} + p) - \xi\| \leqslant \rho$. Let us define the functions $\sigma : [0, \bar{c} + \delta] \to [0, \bar{c} + \delta]$, $f : [0, \bar{c} + \delta] \to \mathbb{R}^n$ and $g : [0, \bar{c} + \delta] \to \mathbb{R}^n$ by

$$\sigma(t) = \begin{cases} \bar{\sigma}(t) & \text{for } t \in [0, \bar{c}], \\ \bar{c} & \text{for } t \in (\bar{c}, \bar{c} + \delta], \end{cases}$$

$$f(t) = \begin{cases} \bar{f}(t) & \text{a.e. for } t \in [0, \bar{c}], \\ f_{\bar{c}} & \text{a.e. for } t \in (\bar{c}, \bar{c} + \delta], \end{cases}$$

$$g(t) = \begin{cases} \bar{g}(t) & \text{for } t \in [0, \bar{c}], \\ p & \text{for } t \in (\bar{c}, \bar{c} + \delta]. \end{cases}$$

It is not difficult to see that σ is nondecreasing, f and g are measurable on $[0, \bar{c} + \delta]$ and $\|g(t)\| \leqslant \varepsilon$ for every $t \in [0, \bar{c} + \delta]$. We define $u : [0, \bar{c} + \delta] \to \mathbb{R}^n$ by

$$u(t) = \xi + \int_0^t f(s) \, \mathrm{d}s + \int_0^t g(s) \, \mathrm{d}s$$

for every $t \in [0, \bar{c} + \delta]$. Clearly u coincides with $\bar{u}$ on $[0, \bar{c}]$ and then it readily follows that σ, f, g and u satisfy all the conditions in (i) and (ii). In order to prove (iii) and (iv), let us observe that

$$u(t) = \begin{cases} \bar{u}(t) & \text{for } t \in [0, \bar{c}], \\ \bar{u}(\bar{c}) + (t - \bar{c}) f_{\bar{c}} + (t - \bar{c})p & \text{for } t \in [\bar{c}, \bar{c} + \delta]. \end{cases}$$

As $u(\bar{c} + \delta) = \bar{u}(\bar{c}) + \delta f_{\bar{c}} + \delta p \in \mathbb{K}$, from the choice of δ, we also have $u(\bar{c} + \delta) \in \mathbb{K} \cap B(\xi, \rho)$. So, although $(\bar{\sigma}, \bar{f}, \bar{g}, \bar{u})$ is $\mathcal{M}$-maximal, we have both $(\bar{\sigma}, \bar{f}, \bar{g}, \bar{u}) \preceq (\sigma, f, g, u)$, and $\mathcal{M}((\bar{\sigma}, \bar{f}, \bar{g}, \bar{u})) < \mathcal{M}((\sigma, f, g, u))$, which is absurd. This contradiction can be eliminated only if each maximal element in the set $\mathcal{S}$ is defined on $[0, T]$, and this completes the proof of Lemma 6.3.1. $\square$

6.4. *Convergence of the ε-approximate solutions*

DEFINITION 6.4.1. Let $\xi \in \mathbb{K}$ and $\varepsilon \in (0, 1]$. A 4-tuple (σ, f, g, u) satisfying (i)–(iv) and (6.3.1) is called an *ε-approximate solution* of (6.2.1).

We are now prepared to prove the sufficiency of Theorem 6.2.1.

PROOF. Let $\varepsilon_m \downarrow 0$ and, for each $m \in \mathbb{N}$, let us fix an ε_m-approximate solution $(\sigma_m, f_m, g_m, u_m)$ of (6.2.1) defined on $[0, T]$.

Since, by (iv), $\|g_m(t)\| \leqslant \varepsilon_m$ for each $m \in \mathbb{N}$ and $t \in [0, T]$, we have

$$\lim_{m \to \infty} \int_0^t g_m(s)\, ds = 0$$

uniformly for $t \in [0, T]$. Moreover, as for each $m \in \mathbb{N}$, f_m satisfies (ii), we may assume with no loss of generality (by extracting a subsequence if necessary) that there exists $f \in L^1(0, T; \mathbb{R}^n)$ such that

$$\lim_{m \to \infty} f_m = f$$

weakly in $L^1(0, T; \mathbb{R}^n)$. As a consequence, from (6.3.1), we infer that there exists $u : [0, T] \to \mathbb{R}^n$ such that

$$\lim_{m \to \infty} u_m(t) = u(t)$$

uniformly for $t \in [0, T]$. Also from (6.3.1) and the last three relations, we easily conclude that

$$u(t) = \xi + \int_0^t f(s)\, ds$$

for each $t \in [0, T]$. Recalling that, by (i), we have $s - \varepsilon_m \leqslant \sigma_m(s) \leqslant s$ for each $m \in \mathbb{N}$ and a.e. for $s \in [0, T]$, from the remarks above, we obtain

$$\lim_{m \to \infty} u_m\big(\sigma_m(s)\big) = u(s)$$

a.e. for $s \in [0, T]$. Furthermore, again by (i), $u_m(\sigma_m(s)) \in \mathbb{K} \cap B(\xi, \rho)$ a.e. for $s \in [0, T]$ and since $\mathbb{K} \cap B(\xi, \rho)$ is closed, we have $u(s) \in \mathbb{K}$ a.e. for $s \in [0, T]$. Since u is continuous, using once again the fact that $\mathbb{K} \cap B(\xi, \rho)$ is closed, we deduce that $u(s) \in \mathbb{K} \cap B(\xi, \rho)$ for each $s \in [0, T]$. Finally, since F is upper semicontinuous with nonempty, convex and compact values, by virtue of Lemma 6.1.2, we conclude that $f(s) \in F(u(s))$ a.e. for $s \in [0, T]$, and thus u is a solution to (6.2.1) and $u(0) = \xi$. The proof is complete. $\qquad\square$

6.5. *Noncontinuable solutions*

In this section we will prove some results concerning the existence of noncontinuable, or even global solutions to (6.2.1). We recall that a solution $u : [0, T) \to \mathbb{K}$ of (6.2.1) is called *noncontinuable* if there is no other solution $v : [0, \widetilde{T}) \to \mathbb{K}$ of (6.2.1), with $T < \widetilde{T}$, satisfying $u(t) = v(t)$ for all $t \in [0, T)$. The next theorem is a consequence of Brezis–Browder Theorem 2.1.1.

THEOREM 6.5.1. *Let $\mathbb{K}$ be a nonempty and locally closed subset in $\mathbb{R}^n$, and let $F : \mathbb{K} \rightsquigarrow \mathbb{R}^n$ be a u.s.c., nonempty, closed, convex and bounded valued mapping. Then, a necessary and sufficient condition in order that for each $\xi \in \mathbb{K}$ to exist at least one noncontinuable solution to (6.2.1) satisfying $u(0) = \xi$ is the tangency condition (6.2.2).*

PROOF. The necessity is an easy consequence of Theorem 6.2.1. As concerns the sufficiency, let $\xi \in \mathbb{K}$ and let $u : [0, T) \to \mathbb{K}$ a solution to (6.2.1) satisfying $u(0) = \xi$. The existence of this solution is guaranteed by Theorem 6.2.1. To complete the proof it suffices to show that u can be continued up to a noncontinuable one. To this aim, we will make use of Brezis–Browder Theorem 2.1.1. Let $\mathcal{S}$ be the set of all solutions to (6.2.1), defined at least and coinciding with u on $[0, T)$. On $\mathcal{S}$, we define the binary relation "$\preceq$" by $u \preceq v$ if the domain $[0, T_v)$ of v is larger that the domain $[0, T_u)$ of u, i.e. $T_u \leqslant T_v$, and $u(t) = v(t)$ for all $t \in [0, T_u)$. Clearly "$\preceq$" is a preorder on $\mathcal{S}$. Next, let $(u_m)_m$ be an increasing sequence in $\mathcal{S}$, and let us denote by $[0, T_m)$ the domain of definition of u_m. Let $T^* = \lim_{m \to \infty} T_m$, which can be finite, or not, and let us define $u^* : [0, T^*) \to \mathbb{K}$ by $u^*(t) = u_m(t)$ for each $t \in [0, T_m)$. Since $(T_m)_m$ is increasing and $u_m(t) = u_k(t)$ for each $m \leqslant k$ and each $t \in [0, T_m)$, u^* is well-defined and belongs to $\mathcal{S}$. Moreover, u^* is a majorant of u_m for each $m \in \mathbb{N}$. Thus each increasing sequence in $\mathcal{S}$ is bounded from above. Moreover, the function $\mathcal{M} : \mathcal{S} \to \mathbb{R} \cup \{+\infty\}$, defined by $\mathcal{M}(v) = T_v$, for each $v \in \mathcal{S}$, is monotone, and therefore we are in the hypotheses of Theorem 2.1.1. Accordingly, there exists at least one $\mathcal{M}$-maximal element $\bar{u} \in \mathcal{S}$. But this means that $\bar{u}$ is noncontinuable, and, of course, that it extends u. The proof is complete. $\qquad \square$

We need next a multivalued counterpart of Definition 3.5.1.

DEFINITION 6.5.1. A mapping $F : \mathbb{K} \rightsquigarrow \mathbb{R}^n$ is called *positively sublinear* if there exist a norm $\| \cdot \|$ on $\mathbb{R}^n$, $a > 0$, $b \in \mathbb{R}$, and $c > 0$ such that

$$\sup\{\|\eta\|; \ \eta \in F(\xi)\} \leqslant a\|\xi\| + b$$

for each $\xi \in \mathbb{K}_+^c(F)$, where

$$\mathbb{K}_+^c(F) = \left\{\xi \in \mathbb{K}; \ \|\xi\| > c \ \text{and} \ \sup_{\eta \in F(\xi)} [\xi, \eta]_+ > 0\right\}.$$

As concerns the existence of global solutions to (6.2.1), we have:

THEOREM 6.5.2. *Let $\mathbb{K} \subseteq \mathbb{R}^n$ be nonempty and closed, and let $F : \mathbb{K} \rightsquigarrow \mathbb{R}^n$ be a u.s.c., nonempty, closed, convex and bounded valued mapping which is positively sublinear. Then, a necessary and sufficient condition in order that for each $\xi \in \mathbb{K}$ there exists at least one global solution to (6.2.1) satisfying $u(0) = \xi$ is the tangency condition (6.2.2).*

PROOF. Let $\xi \in \mathbb{K}$ and let $u : [0, T) \to \mathbb{K}$ be a noncontinuable solution to (6.2.1) satisfying $u(0) = \xi$. We will show that $T = +\infty$. To this aim, let us assume the contrary, i.e., that $T < +\infty$. Since $u'(s) \in F(u(t))$ a.e. for $s \in [0, T)$, we deduce

$$\big[u(s), u'(s)\big]_+ \leqslant \sup_{\eta \in F(u(s))} \big[u(s), \eta\big]_+ .$$

Observing that $[u(s), u'(s)] = \frac{\mathrm{d}}{\mathrm{d}s}(\|u(s)\|)$ a.e. for $s \in [0, T)$, where $\|\cdot\|$ is the norm in Definition 6.5.1, integrating from 0 to t the last equality, we obtain

$$\big\|u(t)\big\| \leqslant \|\xi\| + \int_0^t \sup_{\eta \in F(u(s))} \big[u(s), \eta\big]_+ \, \mathrm{d}s .$$

Denoting by $M = \sup\{\|\eta\|; \ \eta \in F(v), \ \|v\| \leqslant c\}$ and by λ the Lebesgue measure on $\mathbb{R}$, by the positive sublinearity of F and Lemma 6.1.1 we get

$$\big\|u(t)\big\| \leqslant \|\xi\| + \int_{\{s \leqslant t; \|u(s)\| \leqslant c\}} \sup_{\eta \in F(u(s))} \big[u(s), \eta\big]_+ \, \mathrm{d}s$$

$$+ \int_{\{s \leqslant t; \|u(s)\| > c\}} \sup_{\eta \in F(u(s))} \big[u(s), \eta\big]_+ \, \mathrm{d}s$$

$$\leqslant \|\xi\| + M\lambda\big(\{s \leqslant t; \ \|u(s)\| \leqslant c\}\big) + \int_{\{s \leqslant t; \|u(s)\| > c\}} \big[a\|u(s)\| + b\big] \, \mathrm{d}s$$

$$\leqslant \|\xi\| + (b + M)T + a \int_0^t \big\|u(s)\big\| \, \mathrm{d}s .$$

So, by Gronwall's Inequality, we deduce that u is bounded on $[0, T)$. Again from Lemma 6.1.1, it follows that the set $\{\eta; \ \eta \in F(u(t)), \ t \in [0, T)\}$ is bounded. Therefore u is globally Lipschitz on $[0, T)$ and accordingly there exists $\lim_{t \uparrow T} u(t) = u^*$. Since $\mathbb{K}$ is closed, it follows that $u^* \in \mathbb{K}$. Using this observation and recalling that (6.2.2) is satisfied, we conclude that u can be continued to the right of T, which is absurd as long as u is noncontinuable. This contradiction can be eliminated only if $T = +\infty$, and this completes the proof. $\qquad\square$

Noticing that, in view of Lemma 6.1.1, whenever $\mathbb{K}$ is compact, each upper semicontinuous, compact valued mapping $F : \mathbb{K} \rightsquigarrow \mathbb{R}^n$ is bounded, and thus positively sublinear, from Theorem 6.5.2, we deduce:

COROLLARY 6.5.1. *Let $\mathbb{K} \subseteq \mathbb{R}^n$ be nonempty and compact, and let $F : \mathbb{K} \rightsquigarrow \mathbb{R}^n$ be a u.s.c., nonempty, closed, convex and bounded valued mapping. Then, a necessary and sufficient condition in order that for each $\xi \in \mathbb{K}$ to exist at least one global solution, $u : [0, +\infty) \to \mathbb{K}$ of (6.2.1), satisfying $u(0) = \xi$ is the tangency condition (6.2.2).*

6.6. *Viable preordered subsets*

As in Section 3.8, let us consider $\mathbb{K}$ a nonempty subset in $\mathbb{R}^n$, and let "$\preceq$" be a preorder on $\mathbb{K}$, identified with the multifunction $\mathcal{P} : \mathbb{K} \rightsquigarrow \mathbb{K}$, defined by

$$\mathcal{P}(\xi) = \{\eta \in \mathbb{K};\ \xi \preceq \eta\}$$

for each $\xi \in \mathbb{K}$, and called also a preorder. Let $F : \mathbb{K} \rightsquigarrow \mathbb{R}^n$ be a multifunction. We say that "$\preceq$", or $\mathcal{P}$, is *viable with respect to F* if for each $\xi \in \mathbb{K}$, there exist $T > 0$ and a solution $u : [0, T] \to \mathbb{R}^n$ of the differential inclusion (6.2.1) satisfying $u(\tau) = \xi$, $u(t) \in \mathbb{K}$ for each $t \in [0, T]$ and u is "$\preceq$"-monotone on $[\tau, T]$, i.e., for each $\tau \leqslant s \leqslant t \leqslant T$, we have $u(s) \preceq u(t)$. An autonomous multivalued counterpart of Lemma 3.8.1 is proved below. See Cârjă and Ursescu [25].

LEMMA 6.6.1. *Let $F : \mathbb{K} \rightsquigarrow \mathbb{R}^n$ and let $\mathcal{P}$ be a preorder on $\mathbb{K}$. If $\mathcal{P}$ is viable with respect to F then, for each $\xi \in \mathbb{K}$, $\mathcal{P}(\xi)$ is viable with respect to F. If $\mathcal{P}$ is closed in $\mathbb{R}^n \times \mathbb{R}^n$ and, for each $\xi \in \mathbb{K}$, $\mathcal{P}(\xi)$ is viable with respect to F, then $\mathcal{P}$ is viable with respect to F.*

The proof follows the very same lines as that one of Lemma 3.8.1, and therefore we do not enter into details.

THEOREM 6.6.1. *Let $\mathcal{P}$ be a closed preorder on $\mathbb{K}$ and let $F : \mathbb{K} \rightsquigarrow \mathbb{R}^n$ be u.s.c. with nonempty, compact and convex values. Then a necessary and sufficient condition in order that $\mathcal{P}$ be viable with respect to F is the tangency condition below:*

$$F(\xi) \cap \mathcal{T}_{\mathcal{P}(\xi)}(\xi) \neq \emptyset$$

for each $\xi \in \mathbb{K}$.

PROOF. The proof follows immediately from Lemma 6.6.1. $\qquad\qquad\qquad\qquad\square$

6.7. *Local invariance. Sufficient conditions*

Let $\mathbb{D}$ be a domain in $\mathbb{R}^n$, $\mathbb{K} \subseteq \mathbb{D}$ a locally closed subset, and let us consider the differential inclusion (6.2.1), where $F : \mathbb{D} \rightsquigarrow \mathbb{R}^n$ is a given multifunction.

DEFINITION 6.7.1. The subset $\mathbb{K}$ is *locally invariant* with respect to F if for each $\xi \in \mathbb{K}$ and each solution $u : [0, c] \to \mathbb{D}$, $c > 0$, of (6.2.1), satisfying the initial condition $u(0) = \xi$, there exists $T \in (0, c]$ such that we have $u(t) \in \mathbb{K}$ for each $t \in [0, T]$. It is *invariant* if it satisfies the local invariance condition above with $T = c$.

REMARK 6.7.1. If F is upper semicontinuous and nonempty, convex and compact valued on $\mathbb{D}$ and $\mathbb{K}$ is locally invariant with respect to F, then $\mathbb{K}$ is viable with respect to F. The converse of this assertion is no longer true, as we already have seen, even in the single-valued case. See Example 4.1.1.

Our first sufficient condition for local invariance is expressed in terms of the exterior tangency condition below, extending its single-valued counterpart (1.1.6), i.e., there exists an open neighborhood $\mathbb{V}$ of $\mathbb{K}$, with $\mathbb{V} \subseteq \mathbb{D}$, such that

$$\liminf_{h\downarrow 0} \frac{1}{h}\big[\operatorname{dist}(\xi + h\eta; \mathbb{K}) - \operatorname{dist}(\xi; \mathbb{K})\big] \leqslant \omega\big(\operatorname{dist}(\xi; \mathbb{K})\big) \qquad (6.7.1)$$

for each $\xi \in \mathbb{V}$ and each $\eta \in F(\xi)$, where ω is an "autonomous" comparison function, i.e. a comparison function which does not depend on t. The main result in this section is:

THEOREM 6.7.1. *Let $\mathbb{K} \subseteq \mathbb{D} \subseteq \mathbb{R}^n$, with $\mathbb{K}$ locally closed and $\mathbb{D}$ open, and let $F : \mathbb{D} \rightsquigarrow \mathbb{R}^n$. If (6.7.1) is satisfied, then $\mathbb{K}$ is locally invariant with respect to F.*

PROOF. Let $\mathbb{V} \subseteq \mathbb{D}$ be the open neighborhood of $\mathbb{K}$ whose existence is ensured by (6.7.1) and let $\omega : [0, a) \to \mathbb{R}$ the corresponding comparison function. Let $\xi \in \mathbb{K}$ and let $u : [0, c] \to \mathbb{V}$ be any local solution to (6.2.1) satisfying $u(0) = \xi$. Diminishing c if necessary, we may assume that there exists $\rho > 0$ such that $B(\xi, \rho) \cap \mathbb{K}$ is closed, $u(t) \in B(\xi, \rho/2)$ and, in addition, $\operatorname{dist}(u(t); \mathbb{K}) < a$ for each $t \in [0, c]$. Let $g : [0, c] \to \mathbb{R}_+$ be defined by $g(t) = \operatorname{dist}(u(t); \mathbb{K})$ for each $t \in [0, c]$. Let us observe that g is absolutely continuous on $[0, c]$. Let $t \in [0, c)$ be such that both $u'(t)$ and $g'(t)$ exist and $u'(t) \in F(u(t))$, and let $h > 0$ with $t + h \in [0, c]$. We have

$$g(t + h) = \operatorname{dist}\big(u(t + h); \mathbb{K}\big)$$

$$\leqslant h \left\| \frac{u(t + h) - u(t)}{h} - u'(t) \right\| + \operatorname{dist}\big(u(t) + hu'(t); \mathbb{K}\big).$$

Therefore

$$\frac{g(t + h) - g(t)}{h} \leqslant \alpha(h) + \frac{\operatorname{dist}(u(t) + hu'(t); \mathbb{K}) - \operatorname{dist}(u(t); \mathbb{K})}{h},$$

where

$$\alpha(h) = \left\| \frac{u(t + h) - u(t)}{h} - u'(t) \right\|.$$

Since $\lim_{h\downarrow 0} \alpha(h) = 0$, passing to the inf-limit for $h \downarrow 0$ and taking into account that $\mathbb{V}$, $\mathbb{K}$ and F satisfy (6.7.1), we get

$$[D_+ g](t) \leqslant \omega\big(g(t)\big)$$

a.e. for $t \in [0, c)$. So, $g(t) \equiv 0$ which means that $u(t) \in \overline{\mathbb{K}} \cap B(\xi, \rho/2)$ for all $t \in [0, c)$. But $\overline{\mathbb{K}} \cap B(\xi, \rho/2) \subset \mathbb{K} \cap B(\xi, \rho)$ for each $t \in [0, c)$, and this completes the proof. $\qquad\square$

REMARK 6.7.2. One may easily see that (6.7.1) is satisfied with $\omega = \omega_F$, where the function $\omega_F : [0, a) \to \overline{\mathbb{R}}$, $a = \sup_{\xi \in \mathbb{V}} \mathrm{dist}(\xi; \mathbb{K})$, is defined by

$$\omega_F(x) = \sup_{\substack{\xi \in \mathbb{V} \\ \mathrm{dist}(\xi; \mathbb{K}) = x}} \sup_{\eta \in F(\xi)} \liminf_{h \downarrow 0} \frac{1}{h} \big[\mathrm{dist}(\xi + h\eta; \mathbb{K}) - \mathrm{dist}(\xi; \mathbb{K}) \big] \tag{6.7.2}$$

for each $x \in [0, a)$.

So, Theorem 6.7.1 can be reformulated as:

THEOREM 6.7.2. *Let $\mathbb{K} \subseteq \mathbb{D} \subseteq \mathbb{R}^n$, with $\mathbb{K}$ locally closed and $\mathbb{D}$ open, and let $F : \mathbb{D} \leadsto \mathbb{R}^n$. If there exists an open neighborhood V of $\mathbb{K}$ with $\mathbb{V} \subseteq \mathbb{D}$ such that ω_F defined by (6.7.2) is a comparison function, then $\mathbb{K}$ is locally invariant with respect to F.*

6.8. *Local invariance. Necessary conditions*

We begin with the following simple necessary condition for local invariance.

THEOREM 6.8.1. *Let $\mathbb{K} \subseteq \mathbb{D} \subseteq \mathbb{R}^n$, with $\mathbb{K}$ nonempty and $\mathbb{D}$ open, and let $F : \mathbb{D} \leadsto \mathbb{R}^n$ be l.s.c. with nonempty, convex and compact values. If $\mathbb{K}$ is locally invariant with respect to F, then, for each $\xi \in \mathbb{K}$, $F(\xi) \subseteq \mathcal{C}_{\mathbb{K}}(\xi)$.*

PROOF. Let $\xi \in \mathbb{K}$ and $\eta \in F(\xi)$. Since F is l.s.c. with nonempty, closed and convex values, by the Michael' selection theorem [67,68], there exists a continuous function $f : \mathbb{D} \to \mathbb{R}^n$ such that $f(\xi) = \eta$ and $f(x) \in F(x)$ for each $x \in \mathbb{D}$. As $\mathbb{D}$ is open, by Peano's local existence theorem 1.1.1, there exists at least one solution $u : [0, T] \to \mathbb{D}$ of the equation $u'(t) = f(u(t))$ satisfying $u(0) = \xi$. Clearly $u'(0) = f(u(0)) = \eta$ and, in addition, u is a solution to the differential inclusion (6.2.1). Since $\xi \in \mathbb{K}$ and the latter is locally invariant with respect to F, there exists $0 < a \leqslant T$ such that $u(t) \in \mathbb{K}$ for all $t \in [0, a]$. Now, repeating the same arguments as in the proof of Theorem 3.2.1—Section 3.2—we conclude that $\eta = u'(0) \in \mathcal{F}_{\mathbb{K}}(\xi)$. The conclusion follows from Proposition 6.2.1, and the proof is complete. $\qquad\square$

Next, we rephrase some concepts we introduced in the single-valued case.

DEFINITION 6.8.1. *Let $\mathbb{K} \subseteq \mathbb{D} \subseteq \mathbb{R}^n$. A multifunction $F : \mathbb{D} \leadsto \mathbb{R}^n$ has the comparison property with respect to $(\mathbb{D}, \mathbb{K})$ if there exist a proximal neighborhood $\mathbb{V} \subseteq \mathbb{D}$ of $\mathbb{K}$, one projection $\pi_{\mathbb{K}} : \mathbb{V} \to \mathbb{K}$ subordinated to $\mathbb{V}$, and one comparison function*

$$\omega : [0, a) \to \mathbb{R}, \quad \text{with } a = \sup_{\xi \in \mathbb{V}} \mathrm{dist}(\xi; \mathbb{K}),$$

such that

$$\sup_{\eta \in F(\xi)} \inf_{\eta_\pi \in F(\pi_K(\xi))} \big[\xi - \pi_{\mathbb{K}}(\xi), \eta - \eta_\pi \big]_+ \leqslant \omega\big(\| \xi - \pi_{\mathbb{K}}(\xi) \| \big) \tag{6.8.1}$$

for each $\xi \in \mathbb{V}$.

A condition similar to that one in Definition 6.8.1, with ξ replaced by ξ_1 and $\pi_{\mathbb{K}}(\xi)$ replaced by ξ_2 with $\xi_1, \xi_2 \in \mathbb{V}$, has been used previously by Cârjă and Ursescu [25]. As in the single-valued case, here, (6.8.1) is also automatically satisfied for each $\xi \in \mathbb{K}$, and therefore, in Definition 6.8.1, we have only to assume that (6.8.1) holds for each $\xi \in \mathbb{V} \setminus \mathbb{K}$.

DEFINITION 6.8.2. The multifunction $F : \mathbb{D} \rightsquigarrow \mathbb{R}^n$ is called:

   (i) $(\mathbb{D}, \mathbb{K})$-*Lipschitz* if there exist a proximal neighborhood $\mathbb{V} \subseteq \mathbb{D}$ of $\mathbb{K}$, a subordinated projection $\pi_{\mathbb{K}} : \mathbb{V} \to \mathbb{K}$, and $L > 0$, such that

$$\sup_{\eta \in F(\xi)} \inf_{\eta_\pi \in F(\pi_K(\xi))} \| \eta - \eta_\pi \| \leqslant L \| \xi - \pi_{\mathbb{K}}(\xi) \|$$

   for each $\xi \in \mathbb{V} \setminus \mathbb{K}$;

  (ii) $(\mathbb{D}, \mathbb{K})$-*dissipative* if there exist a proximal neighborhood $\mathbb{V} \subset \mathbb{D}$ of $\mathbb{K}$, and a projection, $\pi_{\mathbb{K}} : \mathbb{V} \to \mathbb{K}$, subordinated to $\mathbb{V}$, such that

$$\sup_{\eta \in F(\xi)} \inf_{\eta_\pi \in F(\pi_K(\xi))} \left[\xi - \pi_{\mathbb{K}}(\xi), \eta - \eta_\pi \right]_+ \leqslant 0$$

   for each $\xi \in \mathbb{V} \setminus \mathbb{K}$.

A strictly more restrictive Lipschitz condition, with ξ replaced by ξ_1 and $\pi_K(\xi)$ by ξ_2, with ξ_1, ξ_2 belonging to $\mathbb{D}$, has been considered first by Filippov [43]. In a very same spirit as Filippov [43], Kobayashi [62] has used a dissipative type condition strictly restrictive than that one in Definition 6.8.2. It is easy to see that if F is either $(\mathbb{D}, \mathbb{K})$-Lipschitz, or $(\mathbb{D}, \mathbb{K})$-dissipative, then it has the comparison property with respect to $(\mathbb{D}, \mathbb{K})$. We notice that there are examples showing that there exist multifunctions F which, although neither $(\mathbb{D}, \mathbb{K})$-Lipschitz, nor $(\mathbb{D}, \mathbb{K})$-dissipative, have the comparison property. See, for instance, the "single-valued" Examples 4.3.1 and 4.3.2.

THEOREM 6.8.2. *Let* $\mathbb{K} \subseteq \mathbb{D} \subseteq \mathbb{R}^n$, *with* $\mathbb{K}$ *locally closed and* $\mathbb{D}$ *open, and let* $F : \mathbb{D} \rightsquigarrow \mathbb{R}^n$ *be a multifunction. If F has the comparison property with respect to* $(\mathbb{D}, \mathbb{K})$, *and*

$$\liminf_{h \downarrow 0} \frac{1}{h} \operatorname{dist}(\xi + h\eta; \mathbb{K}) = 0 \tag{6.8.2}$$

for each $\xi \in \mathbb{K}$ *and each* $\eta \in F(\xi)$, *then* (6.7.1) *holds true.*

PROOF. Let $\mathbb{V} \subseteq \mathbb{D}$ be the open neighborhood of $\mathbb{K}$ as in Definition 6.8.1, let $\xi \in \mathbb{V}$ and $\eta \in F(\xi)$. Let $\rho > 0$ and let $\pi_{\mathbb{K}}$ be the selection of $\Pi_{\mathbb{K}}$ as in Definition 6.8.1. Let $h \in (0, T]$. Since $\| \xi - \pi_{\mathbb{K}}(\xi) \| = \operatorname{dist}(\xi; \mathbb{K})$, we have

$$\operatorname{dist}(\xi + h\eta; \mathbb{K}) - \operatorname{dist}(\xi; \mathbb{K})$$

$$\leqslant \left\| \xi - \pi_{\mathbb{K}}(\xi) + h[\eta - \zeta] \right\| - \left\| \xi - \pi_{\mathbb{K}}(\xi) \right\| + d\left(\pi_{\mathbb{K}}(\xi) + h\zeta; \mathbb{K} \right),$$

for each $\zeta \in F(\pi_{\mathbb{K}}(\xi))$.

Dividing by h, passing to the $\liminf$ for $h \downarrow 0$, and using (6.8.2), we get

$$\liminf_{h \downarrow 0} \frac{1}{h}\left[\operatorname{dist}(\xi + h\eta; \mathbb{K}) - \operatorname{dist}(\xi; \mathbb{K})\right] \leqslant \left[\xi - \pi_{\mathbb{K}}(\xi), \eta - \zeta\right]_{+}.$$

Since $\zeta \in F(\pi_{\mathbb{K}}(\xi))$ is arbitrary, we have

$$\liminf_{h \downarrow 0} \frac{1}{h}\left[\operatorname{dist}(\xi + h\eta; \mathbb{K}) - \operatorname{dist}(\xi; \mathbb{K})\right] \leqslant \inf_{\zeta \in F(\pi_{\mathbb{K}}(\xi))}\left[\xi - \pi_{\mathbb{K}}(\xi), \eta - \zeta\right]_{+}.$$

Therefore

$$\liminf_{h \downarrow 0} \frac{1}{h}\left[\operatorname{dist}(\xi + h\eta; \mathbb{K}) - \operatorname{dist}(\xi; \mathbb{K})\right] \leqslant \omega\big(\|\xi - \pi_{\mathbb{K}}(\xi)\|\big).$$

But this inequality shows that (6.7.1) holds, and this completes the proof.   □

COROLLARY 6.8.1. *Let $\mathbb{K} \subseteq \mathbb{D} \subseteq \mathbb{R}^{n}$, with $\mathbb{K}$ locally closed and $\mathbb{D}$ open, and let $F : \mathbb{D} \rightsquigarrow \mathbb{R}^{n}$ be a multifunction. If F has the comparison property with respect to $(\mathbb{D}, \mathbb{K})$ and, for each $\xi \in \mathbb{K}$, $F(\xi) \subseteq \mathcal{T}_{\mathbb{K}}(\xi)$, then $\mathbb{K}$ is local invariant with respect to F.*

Corollary 6.8.1 and Proposition 6.2.1 yield:

COROLLARY 6.8.2. *Let $\mathbb{K} \subseteq \mathbb{D} \subseteq \mathbb{R}^{n}$, with $\mathbb{K}$ locally closed and $\mathbb{D}$ open, and let $F : \mathbb{D} \rightsquigarrow \mathbb{R}^{n}$ be l.s.c. with nonempty values. If, F has the comparison property with respect to $(\mathbb{D}, \mathbb{K})$ and, for each $\xi \in \mathbb{K}$, $F(\xi) \subseteq \mathcal{B}_{\mathbb{K}}(\xi)$, then $\mathbb{K}$ is local invariant with respect to F.*

7. Applications

7.1. *Invariance with respect to parametrized multifunctions*

Let us consider the parametrized multifunction $F : \mathbb{D} \rightsquigarrow \mathbb{R}^{n}$,

$$F(\xi) = \big\{ f(\xi, v); \ v \in \mathbb{V} \big\},$$

where $\mathbb{D}$ is a nonempty and open subset in $\mathbb{R}^{n}$, $\mathbb{V}$ is a nonempty and compact subset in $\mathbb{R}^{n}$ and $f : \mathbb{D} \times \mathbb{R}^{n} \to \mathbb{R}^{n}$ is a continuous function. Let $\mathbb{K}$ be a locally closed subset of $\mathbb{D}$ and let us assume that F has convex values. We assume further that, for each measurable function $v : [0, a] \to \mathbb{V}$ and for each $\xi \in \mathbb{D}$, there exists $T \in (0, a]$ such that the equation $u'(t) = f(u(t), v(t))$ has a unique Carathéodory solution $u : [0, T] \to \mathbb{R}^{n}$ satisfying $u(0) = \xi$.

THEOREM 7.1.1. *Under the above circumstances, the following conditions are equivalent*:
     (i) *for each $\xi \in \mathbb{K}$ and $v \in \mathbb{V}$, $f(\xi, v) \in \mathcal{C}_{\mathbb{K}}(\xi)$,*
     (ii) *for each $\xi \in \mathbb{K}$ and $v \in \mathbb{V}$, $f(\xi, v) \in \mathcal{T}_{\mathbb{K}}(\xi)$,*

(iii) *for each $\xi \in \mathbb{K}$ and $v \in \mathbb{V}$, $f(\xi, v) \in \mathcal{B}_{\mathbb{K}}(\xi)$,*

(iv) *$\mathbb{K}$ is locally invariant with respect to F.*

In general, if $\mathcal{G} : \mathbb{K} \rightsquigarrow \mathbb{R}^n$ is such that $\mathcal{C}_{\mathbb{K}}(\xi) \subseteq \mathcal{G}(\xi) \subseteq \mathcal{B}_{\mathbb{K}}(\xi)$ for each $\xi \in \mathbb{K}$, then each one of the conditions above is equivalent to

(v) *for each $\xi \in \mathbb{K}$, $f(\xi, v) \in \mathcal{G}(\xi)$.*

PROOF. In view of Proposition 2.5.1, it suffices to show that (ii) is equivalent to (iv). Let us assume that (ii) holds true and let $u : [0, T] \to \mathbb{D}$ be a solution to $u'(t) \in F(u(t))$ with $u(0) = \xi \in \mathbb{K}$. We have to show that there exists $\tau \in (0, T]$ such that $u(t) \in \mathbb{K}$ for all $t \in [0, \tau]$. By a classical result of Filippov, there exists a measurable function $v : [0, T] \to \mathbb{V}$ such that

$$u'(t) = f\big(u(t), v(t)\big) \tag{7.1.1}$$

a.e. for $t \in [0, T]$. By the uniqueness property, taking into account (ii) and using Theorem 5.2.1, we deduce that there exists $\tau \in (0, T]$ such that the unique solution to (7.1.1) satisfies $u(t) \in \mathbb{K}$ for all $t \in [0, \tau]$. Thus (ii) implies (iv).

Let us suppose now that (iv) holds true. Fix $\xi \in \mathbb{K}$ and $v \in \mathbb{V}$. The equation $u'(t) = f(u(t), v)$ with $u(0) = \xi$ has a unique solution $u : [0, T] \to \mathbb{D}$. Since $\mathbb{K}$ is locally invariant with respect to F, there exists $\tau \in (0, T]$ such that we have $u(t) \in \mathbb{K}$ for all $t \in [0, \tau]$. As $u'(0) = f(\xi, v)$, it readily follows that $f(\xi, v) \in \mathcal{F}_{\mathbb{K}}(\xi) \subseteq \mathcal{T}_{\mathbb{K}}(\xi)$, and thus (iv) implies (ii). This completes the proof. $\qquad\square$

7.2. *Differentiability along trajectories*

In this section we will show how viability can be used in order to obtain sufficient conditions for asymptotic stability via Liapunov functions. Let us consider the autonomous differential equation

$$u'(t) = g\big(u(t)\big), \tag{7.2.1}$$

where $g : \mathbb{R}^n \to \mathbb{R}^n$ is continuous. Let $V : \mathbb{R}^n \to \mathbb{R}$, and let us define

$$V^*(\xi) = \limsup_{h \downarrow 0} \frac{1}{h}\big[V\big(\xi + hg(\xi)\big) - V(\xi)\big]$$

and

$$V_*(\xi) = \liminf_{h \downarrow 0} \frac{1}{h}\big[V\big(\xi + hg(\xi)\big) - V(\xi)\big].$$

If $V^*(\xi) = V_*(\xi)$, we denote this common value by $\dot{V}(\xi)$ and we note that, if V is differentiable, we have $\dot{V}(\xi) = \langle \operatorname{grad} V(\xi), g(\xi)\rangle$. The next result will prove useful in the sequel.

THEOREM 7.2.1 (Yoshizawa [110]). *If $V : \mathbb{R}^n \to \mathbb{R}$ is locally Lipschitz, $u : [0, T) \to \mathbb{R}^n$ is any solution to (7.2.1) and $t \in [0, T)$, then*

$$V^*\big(u(t)\big) = \limsup_{h\downarrow 0} \frac{1}{h}\big[V\big(u(t+h)\big) - V\big(u(t)\big)\big].$$

PROOF. We have

$$V^*\big(u(t)\big) = \limsup_{h\downarrow 0} \frac{1}{h}\big[V\big(u(t)\big) + hg\big(u(t)\big) - V\big(u(t)\big)\big]$$

$$\leqslant \limsup_{h\downarrow 0} \frac{1}{h}\left[V\left(u(t) + \int_t^{t+h} g\big(u(s)\big)\,\mathrm{d}s\right) - V\big(u(t)\big)\right]$$

$$+ \limsup_{h\downarrow 0} \frac{1}{h}\left[-V\left(u(t) + \int_t^{t+h} g\big(u(s)\big)\,\mathrm{d}s\right)\right.$$

$$\left. + V\big(u(t)\big) + hg\big(u(t)\big)\right].$$

Since

$$\left|\limsup_{h\downarrow 0} \frac{1}{h}\left[-V\left(u(t) + \int_t^{t+h} g\big(u(s)\big)\,\mathrm{d}s\right) + V\big(u(t)\big) + hg\big(u(t)\big)\right]\right|$$

$$\leqslant L \limsup_{h\downarrow 0} \frac{1}{h}\left\|\int_t^{t+h} g\big(u(s)\big)\,\mathrm{d}s - hg\big(u(t)\big)\right\| = 0,$$

where $L > 0$ is the Lipschitz constant of V on a suitably chosen neighborhood of $u(t)$, we deduce

$$V^*\big(u(t)\big) \leqslant \limsup_{h\downarrow 0} \frac{1}{h}\big[V\big(u(t+h)\big) - V\big(u(t)\big)\big].$$

Similarly, we get

$$V^*\big(u(t)\big) \geqslant \limsup_{h\downarrow 0} \frac{1}{h}\big[V\big(u(t+h)\big) - V\big(u(t)\big)\big]$$

and this completes the proof.   □

7.3. *Liapunov functions*

DEFINITION 7.3.1. We say that 0 is *stable* for (7.2.1) if for each $\varepsilon > 0$ there exists $\delta(\varepsilon) \in (0, \varepsilon)$ such that for each $\xi \in \mathbb{R}^n$, $\|\xi\| \leqslant \delta(\varepsilon)$, each solution u of (7.2.1), satisfying $u(0) = \xi$, is defined on $[0, +\infty)$, and $\|u(t)\| \leqslant \varepsilon$ for all $t \geqslant 0$.

Clearly, if 0 is stable for (7.2.1), $\{0\}$ is both viable and locally invariant with respect to g, and thus $g(0) = 0$. In other words, if 0 is stable for (7.2.1), then $u \equiv 0$ is necessarily a solution to (7.2.1), and this is the only one issuing from 0.

DEFINITION 7.3.2. We say that V has positive gradient at $v \in \mathbb{R}$ if

$$\liminf_{\substack{\mathrm{dist}(\xi;\mathbb{K})\downarrow 0 \\ \xi \notin \mathbb{K}}} \frac{V(\xi) - v}{\mathrm{dist}(\xi;\mathbb{K})} > 0,$$

where $\mathbb{K} = V^{-1}((-\infty, v])$.

If V is of class C^1 and $\|\,\mathrm{grad}\,V\,\|$ is bounded from below on $V^{-1}(\{v\})$ by a constant $c > 0$, then V has positive gradient at v.

PROPOSITION 7.3.1. *Let* $g : \mathbb{R}^n \to \mathbb{R}^n$ *and* $V : \mathbb{R}^n \to \mathbb{R}$ *be continuous, and let* $\mathbb{K} = V^{-1}((-\infty, v])$. *If* V *has positive gradient at* v *and* $V^*(\xi) \leqslant 0$ *whenever* $V(\xi) = v$, *then* $\mathbb{K}$ *is viable with respect to* g.

PROOF. Let us assume by contradiction that $\mathbb{K}$ is not viable with respect to g. In view of Theorem 3.1.1, this means that there exists $\xi \in \partial\mathbb{K}$ such that $g(\xi)$ is not tangent in the sense of Federer to $\mathbb{K}$ at ξ. So, there exist $\gamma > 0$ and a sequence $h_m \downarrow 0$ such that

$$\gamma < \frac{1}{h_m}\,\mathrm{dist}\big(\xi + h_m g(\xi); \mathbb{K}\big)$$

for all $m \in \mathbb{N}$. As $\xi \in \partial\mathbb{K}$, it follows that $V(\xi) = v$. Furthermore, since V has positive gradient at v, there exists $\nu > 0$ such that

$$\nu < \frac{V(\xi + h_m g(\xi)) - v}{\mathrm{dist}(\xi + h_m g(\xi); \mathbb{K})} < \frac{1}{h_m \gamma}\big[V\big(\xi + h_m g(\xi)\big) - v\big]$$

for each $m \in \mathbb{N}$. Hence

$$0 < \nu \leqslant \frac{1}{\gamma}\limsup_{h\downarrow 0} \frac{V(\xi + hg(\xi)) - V(\xi)}{h} = \frac{1}{\gamma} V^*(\xi),$$

thereby contradicting the hypothesis that $V^*(\xi) \leqslant 0$. This contradiction can be eliminated only if $\mathbb{K}$ is viable with respect to g, and this completes the proof. $\square$

THEOREM 7.3.1 (Yorke [108]). *Assume that (7.2.1) has the uniqueness property. Then 0 is stable for (7.2.1) if and only if there exist a continuous function* $V : \mathbb{R}^n \to [0, +\infty)$ *and a sequence* $v_m \downarrow 0$ *such that* $V(x) = 0$ *if and only if* $x = 0$, *and, for every* $m \in \mathbb{N}$ *for which* $V(x) = v_m$, *we have* $V^*(x) \leqslant 0$, *and, for each* $m \in \mathbb{N}$, V *has positive gradient at* v_m.

PROOF. We denote by $u(\cdot, \xi) : [0, +\infty) \to \mathbb{R}^n$ the unique solution of (7.2.1) satisfying $u(0, \xi) = \xi$.

Sufficiency. Let us assume that such a sequence $(v_m)_m$ and function V exist. Let $\mathbb{K}_m = V^{-1}([0, v_m])$. Since (7.2.1) has the uniqueness property, by Proposition 7.3.1, it follows that $\mathbb{K}_m$ is both viable and locally invariant with respect to g. Let $\varepsilon > 0$ and let us define both $\mu_*(\varepsilon) = \inf_{\|\xi\|=\varepsilon} V(\xi)$ and $\mu^*(\varepsilon) = \sup_{\|\xi\|=\varepsilon} V(\xi)$. Then, $\mu_*(\varepsilon) > 0$. For any $\varepsilon > 0$, choose $m = m(\varepsilon)$ such that $v_m < \mu_*(\varepsilon)$, and choose $\delta(\varepsilon) \in (0, \varepsilon)$ such that $\mu^*(\delta) < v_m$ for each $\delta \in (0, \delta(\varepsilon))$. Let $\xi \in \mathbb{R}^n$ with $\|\xi\| \leqslant \delta(\varepsilon)$. Clearly, $\xi \in \mathbb{K}_m$. If there exists $t > 0$ such that $\|u(t, \xi)\| > \varepsilon$, then there exists $t_0 \in (0, t)$ such that $\|u(t_0, \xi)\| = \varepsilon$. On the other hand, by the choice of m, we have $u(t_0, \xi) \notin \mathbb{K}_m$, thereby contradicting the local invariance of $\mathbb{K}_m$ with respect to g. This contradiction can be eliminated only if $\|u(t, \xi)\| < \varepsilon$ for all $t \geqslant 0$, and this completes the proof of the sufficiency.

Necessity. Suppose that 0 is stable for (7.2.1). Let $\delta(\cdot)$ the function in Definition 7.3.1, choose $\varepsilon_1 > 0$, and let us define inductively $\varepsilon_{m+1} = \delta(\varepsilon_m)/2$. Clearly, $\varepsilon_m \downarrow 0$. Let us denote by

$$\mathbb{K}_m^0 = \left\{ u(t, \xi); \ \|\xi\| \leqslant \delta(\varepsilon_m), \ \text{and } t \geqslant 0 \right\}.$$

Since $\mathbb{K}_m^0$ contains only points reached at positive time by solutions starting in $B(0, \delta(\varepsilon_m))$, it follows that $\mathbb{K}_m^0$ is viable with respect to g. By Proposition 3.6.1, we deduce that $\mathbb{K}_m = \overline{\mathbb{K}_m^0}$ is viable with respect to g. Since $u'(t) = g(u(t))$ has the uniqueness property, it follows that $\mathbb{K}_m$ is in fact invariant with respect to g. For each $m = 1, 2, \ldots$, let

$$v_m = \sum_{i=m+1}^{\infty} \varepsilon_i.$$

Clearly, $v_m = \frac{1}{2} \sum_{i=m+1}^{\infty} \delta(\varepsilon_{i-1}) \leqslant \varepsilon_m$ for $m = 1, 2, \ldots$. Let $\mathbb{K}_0 = \mathbb{R}^n$ and $v_0 = +\infty$. We have $\cdots \subseteq \mathbb{K}_2 \subseteq \mathbb{K}_1 \subseteq \mathbb{K}_0$ and therefore, we may define $N(\xi) = \max\{m; \ \xi \in \mathbb{K}_m\}$. Notice that we want to have $V(\xi) = v_m$ for all $\xi \in \partial \mathbb{K}_m$. Define $V(0) = 0$ and

$$V(\xi) = \min\left\{ v_{N(\xi)}, v_{N(\xi)+1} + \text{dist}(\xi; \mathbb{K}_{N(\xi)+1}) \right\}$$

for $\xi \neq 0$. We begin by proving that V is continuous. First, let us observe that, for each $m \in \mathbb{N}$, the function N is constant on $\mathbb{K}_m \setminus \mathbb{K}_{m+1}$. Accordingly the restriction of V to $\mathbb{K}_m \setminus \mathbb{K}_{m+1}$ is continuous, and so the function V itself is continuous at each interior point of $\mathbb{K}_m \setminus \mathbb{K}_{m+1}$. Next, we show that V is continuous on $\partial \mathbb{K}_m$ for each $m \in \mathbb{N}$. Let ξ be arbitrary in $\partial \mathbb{K}_m$. We have both $\|\xi\| \in [\delta(\varepsilon_m), \varepsilon_m)$ and $\sup_{\eta \in \mathbb{K}_{m+1}} \|\eta\| \leqslant \varepsilon_{m+1}$. Therefore, $N(\xi) = m$ and $\text{dist}(\xi; \mathbb{K}_{m+1}) > \delta(\varepsilon_m) - \varepsilon_{m+1} = \delta(\varepsilon_m)/2$. Hence we have $v_{N(\xi)+1} + \text{dist}(\xi; \mathbb{K}_{N(\xi)+1}) > v_{N(\xi)}$ and $V(\xi) = v_m$. Choose $\xi_i \notin \mathbb{K}_m$ for $i = 1, 2, \ldots$, with $\xi_i \to \xi$. Then, $\lim_{i \to \infty} \text{dist}(\xi_i; \mathbb{K}_m) = 0$, and

$$\lim_{i \to \infty} V(\xi_i) = v_m = V(\xi),$$

which shows that V is continuous at ξ. It remains to prove that V is continuous at 0. Take $\xi_i \to 0$. Then $N(\xi_i) \to \infty$ and $V(\xi_i) \leqslant v_{N(\xi_i)}$. But $v_m \to 0$ as $m \to \infty$, so that

$$\lim_{\xi \to 0} V(\xi) = 0 = V(0).$$

Thus V is continuous on $\mathbb{R}^n$.

Finally, we show that $V(\xi) = v_m$ implies $V^*(\xi) \leqslant 0$. Indeed, if for some $m \in \mathbb{N}$ we have $V(\xi) = v_m$, then $\xi \in \mathbb{K}_m$, $\xi + hg(\xi) \in \mathbb{K}_{m-1}$ for $h > 0$ sufficiently small, and

$$V\big(\xi + hg(\xi)\big) \leqslant v_m + \operatorname{dist}\big(\xi + hg(\xi); \mathbb{K}_m\big).$$

Since $\mathbb{K}_m$ is viable with respect to g, by virtue of Theorem 3.2.1, we have $g(\xi) \in \mathcal{T}_{\mathbb{K}_m}(\xi)$, and therefore

$$V^*(\xi) \leqslant \limsup_{h \downarrow 0} \frac{1}{h} \operatorname{dist}\big(\xi + hg(\xi); \mathbb{K}_m\big) = 0.$$

If $0 < \operatorname{dist}(\xi; \mathbb{K}_m) \leqslant v_m - v_{m-1} = \varepsilon_m$, then $V(\xi) = v_m + \operatorname{dist}(\xi; \mathbb{K}_m)$, so that V has positive gradient at each v_m, thereby completing the proof. $\qquad\square$

7.4. *Hukuhara's theorem*

In this section, by using viability and invariance techniques, we will prove two celebrated results concerning the funnel of solutions, results due to Hukuhara [47] and Kneser [61]. Let $f : \mathbb{I} \times \mathbb{D} \to \mathbb{R}^n$ be a continuous function, and let us consider the non-autonomous differential equation (1.1.1). Let $(\tau, \xi) \in \mathbb{I} \times \mathbb{D}$, and let us denote by $\mathcal{S}(\tau, \xi)$ the set of all noncontinuable solutions u of (1.1.1) satisfying $u(\tau) = \xi$.

DEFINITION 7.4.1. The *right solution funnel* through $(\tau, \xi) \in \mathbb{I} \times \mathbb{D}$, $F_{\tau,\xi}$, is defined by

$$F_{\tau,\xi} = \big\{(s, u(s)); \; s \geqslant \tau, \; u \in \mathcal{S}(\tau, \xi)\big\}.$$

If $t \geqslant \tau$, we define the *t-cross section* of $F_{\tau,\xi}$ by

$$F_{\tau,\xi}(t) = \big\{u(t); \; u \in \mathcal{S}(\tau, \xi)\big\}.$$

The next compactness result will prove useful in that follows.

PROPOSITION 7.4.1. *Let* $(\tau, \xi) \in \mathbb{I} \times \mathbb{D}$ *and let* $t > \tau$ *be such that, for each* $u \in \mathcal{S}(\tau, \xi)$, *$u(t)$ is defined. Then*

$$F_{\tau,\xi}\big([\tau, t]\big) = \big\{(s, u(s)); \; s \in [\tau, t], \; u \in \mathcal{S}(\tau, \xi)\big\}$$

is compact.

PROOF. Let $(t_m, u_m(t_m))_m$ be an arbitrary sequence in $F_{\tau,\xi}([\tau,t])$, with $u_m : (a_m, b_m) \to$ $\mathbb{D}$ for each $m \in \mathbb{N}$. We may assume with no loss of generality that $\lim_m t_m = s$. Obviously, $s \in [\tau, t] \subseteq (a_m, b_m)$ for every $m \in \mathbb{N}$, and therefore, by Hartman [54, Theorem 3.2, p. 26], there exists at least one subsequence of $(u_m)_m$, denoted for simplicity again by $(u_m)_m$, and $u \in \mathcal{S}(\tau, \xi)$, with $\lim_m u_m = u$ uniformly on $[\tau, t]$. But this shows that $\lim_m (t_m, u_m(t_m)) = (s, u(s))$. Since $(s, u(s)) \in F_{\tau,\xi}([\tau, t])$, the proof is complete. $\qquad\Box$

From Proposition 7.4.1, we deduce:

COROLLARY 7.4.1. *The set $F_{\tau,\xi}$ is locally closed.*

THEOREM 7.4.1 (Hukuhara [47]). *Let $(\tau, \xi) \in \mathbb{I} \times \mathbb{D}$ and let $t > \tau$ be such that, for each $u \in \mathcal{S}(\tau, \xi)$, $u(t)$ is defined. Then, for each $\eta \in \partial F_{\tau,\xi}(t)$ there exists a solution v such that $v(s) \in \partial F_{\tau,\xi}(s)$ for all $s \in [\tau, t]$.*

In order to prove Theorem 7.4.1, we need:

THEOREM 7.4.2 (Yorke [108]). *If $\mathbb{D}$ is a domain, $\mathbb{K}_1, \mathbb{K}_2 \subseteq \mathbb{D}$ are locally closed and viable with respect to f, and if $\mathbb{K}_1 \cup \mathbb{K}_2 = \mathbb{D}$, then $\mathbb{K}_1 \cap \mathbb{K}_2$ is viable with respect to f.*

PROOF. The conclusion is a consequence of Theorem 2.3.1 and Nagumo's viability theorem 3.1.1. $\qquad\Box$

REMARK 7.4.1. A result similar to Theorem 7.4.2 holds true trivially in the case of local invariance. More precisely, if $\mathbb{D} \subseteq \mathbb{R}^n$ is open, $\mathbb{K}_1, \mathbb{K}_2 \subseteq \mathbb{D}$ are locally closed and locally invariant with respect to f, then $\mathbb{K}_1 \cap \mathbb{K}_2$ is locally invariant with respect to f.

Let $\tau \in \mathbb{I}$ be fixed, let us denote by $\mathcal{D} = \{s \in \mathbb{I};\ s > \tau\} \times \mathbb{D}$, and let us define $\mathcal{F} : \mathcal{D} \to \mathbb{R} \times \mathbb{R}^n$, by $\mathcal{F}(t, \xi) = (1, f(t, \xi))$ for each $(t, \xi) \in \mathcal{D}$. Throughout, we denote by $\partial_{\mathcal{D}} F_{\tau,\xi}$ the boundary of $F_{\tau,\xi}$ relative to $\mathcal{D}$, i.e. $\partial_{\mathcal{D}} F_{\tau,\xi} = \overline{(\mathcal{D} \setminus F_{\tau,\xi})}^{\mathcal{D}} \cap \overline{\mathcal{D} \cap F_{\tau,\xi}}^{\mathcal{D}}$. We will deduce Theorem 7.4.1 from a slightly more general result, i.e. Theorem 7.4.3 below.

THEOREM 7.4.3 (Yorke [108]). *For each $(\tau, \xi) \in \mathbb{I} \times \mathbb{D}$, the set $\partial_{\mathcal{D}} F_{\tau,\xi}$ is left viable with respect to $\mathcal{F}$.*

PROOF. Let us observe that (1.1.1) can be equivalently written as

$$w'(t) = \mathcal{F}(w(t)),$$

where $\mathcal{F}$ is defined as above, and $w = (s, u)$. By the definition of $F_{\tau,\xi}$, we easily deduce that $\mathcal{D} \cap F_{\tau,\xi}$ is right viable and right locally invariant with respect to $\mathcal{F}$, and hence, by Propositions 1.1.1–1.1.3, it follows that $\mathcal{D} \setminus F_{\tau,\xi}$ is both left viable and left locally invariant with respect to $\mathcal{F}$. So, thanks to Proposition 3.6.1, we conclude that $\mathcal{K}_1 = \overline{\mathcal{D} \setminus F_{\tau,\xi}}^{\mathcal{D}}$ is left viable with respect to $\mathcal{F}$. Further, also by definition, $\mathcal{D} \cap F_{\tau,\xi}$ is left viable with respect to $\mathcal{F}$, and again by Proposition 3.6.1, it follows that $\mathcal{K}_2 = \overline{\mathcal{D} \cap F_{\tau,\xi}}^{\mathcal{D}}$ is left viable with respect to $\mathcal{F}$. Since $\mathcal{K}_1 \cup \mathcal{K}_2 = \mathcal{D}$, by Remark 1.1.1 and Theorem 7.4.2, we conclude that $\mathcal{K}_1 \cap \mathcal{K}_2 = \partial_{\mathcal{D}} F_{\tau,\xi}$ is left viable with respect to $\mathcal{F}$, and this completes the proof. $\qquad\square$

We may now proceed to the proof of Theorem 7.4.1.

PROOF OF THEOREM 7.4.1. First, let us observe that thanks to Proposition 7.4.1, it follows that $\partial_{\mathcal{D}} F_{\tau,\xi} \subseteq F_{\tau,\xi}$. Hence, in view of Theorem 7.4.3, we know that, for each $(t, u(t)) \in \partial_{\mathcal{D}} F_{\tau,\xi}$, there exists at least one solution $v : [\theta, t] \to \mathbb{D}$, with $\tau \leqslant \theta < t$, $v(t) = u(t)$ and such that $(s, v(s)) \in \partial_{\mathcal{D}} F_{\tau,\xi}(s)$ for each $s \in [\theta, t]$. But, by virtue of Proposition 7.4.1, $F_{\tau,\xi}([\tau, t])$ is compact, and therefore a simple maximality argument shows that we can always extend such a solution to $[\tau, t]$, and this achieves the proof. $\qquad\square$

REMARK 7.4.2. Theorem 7.4.3 implies that, for each $(\tilde{t}, \tilde{y}) \in \partial_{\mathcal{D}} F_{\tau,\xi}$, there exists at least one noncontinuable solution, $v(\cdot) : (\sigma, \tilde{t}] \to \mathbb{R}^n$, of (1.1.1), such that $(s, v(s)) \in \partial_{\mathcal{D}} F_{\tau,\xi}$ for all $s \in (\sigma, \tilde{t}] \cap [\tau, \tilde{t}]$. If $t > \tau$ is chosen as in Theorem 7.4.1 and $\tilde{t} \in [\tau, t]$, then by virtue of both Theorem 7.4.3 and Proposition 7.4.1, it follows that $[\tau, \tilde{t}] \subset (\sigma, \tilde{t}]$.

REMARK 7.4.3. We notice that

$$\partial F_{\tau,\xi}(t) \subseteq (\partial_{\mathcal{D}} F_{\tau,\xi})(t) = \left\{ v \in \mathbb{R}^n; \ (t, v) \in \partial_{\mathcal{D}} F_{\tau,\xi} \right\}$$

and the inclusion can be strict. So, Theorem 7.4.3 is more general that Theorem 7.4.1 because it considers all of $\partial_{\mathcal{D}} F_{\tau,\xi}$.

7.5. *Kneser's theorem*

We conclude this section with the celebrated theorem of Kneser.

THEOREM 7.5.1 (Kneser [61]).   *Let $(\tau, \xi) \in \mathbb{I} \times \mathbb{D}$ and let $t > \tau$ be such that, for each $u \in \mathcal{S}(\tau, \xi)$, $u(t)$ is defined. Then $F_{\tau,\xi}(t)$ is connected.*

PROOF. Let us assume by contradiction that $F_{\tau,\xi}(t)$ is not connected. Then there exist two nonempty subsets $\mathbb{C}_1$, $\mathbb{C}_2$ with $F_{\tau,\xi}(t) = \mathbb{C}_1 \cup \mathbb{C}_2$ but $\mathbb{C}_1 \cap \overline{\mathbb{C}_2} = \overline{\mathbb{C}_1} \cap \mathbb{C}_2 = \emptyset$. Let $\mathbb{K}_1$ be the union of all right noncontinuable trajectories[4] of (1.1.1) whose corresponding solutions v, either are not defined at t, or, if defined, satisfy

$$\mathrm{dist}\big(v(t); \mathbb{C}_1\big) \leqslant \mathrm{dist}\big(v(t); \mathbb{C}_2\big). \tag{7.5.1}$$

Similarly, we define $\mathbb{K}_2$ by reversing the inequality (7.5.1). By virtue of Proposition 1.1.1, both $\mathbb{K}_1$ and $\mathbb{K}_2$ are right viable with respect to $\mathcal{F}$. In addition, $\mathbb{K}_1$ and $\mathbb{K}_2$ are locally closed in $\mathbb{I} \times \mathbb{D}$ and $\mathbb{K}_1 \cup \mathbb{K}_2 = \mathbb{I} \times \mathbb{D}$. So, by Proposition 3.6.1, it follows that both $\overline{\mathbb{K}_1}^{\mathcal{D}}$ and $\overline{\mathbb{K}_2}^{\mathcal{D}}$ are

---

[4] We recall that a trajectory of (1.1.1) is a set of the form $\{v(t); \ t \in \mathbb{I}_v\}$, where $v : \mathbb{I}_v \to \mathbb{D}$ is a solution of (1.1.1).

right viable with respect to $\mathcal{F}$ and, of course, locally closed. In view of Theorem 7.4.2, we conclude that $\mathbb{K} = \overline{\mathbb{K}_1}^{\mathcal{D}} \cap \overline{\mathbb{K}_2}^{\mathcal{D}}$ is right viable with respect to $\mathcal{F}$. As $(\tau, \xi) \in \mathbb{K}$, there exists a noncontinuable solution v such that $(s, v(s)) \in \mathbb{K}$ for each s in the domain of v. By the choice of t, $v(t)$ is defined and belongs to $\mathbb{C}_1 \cup \mathbb{C}_2$. Since $(t, v(t)) \in \mathbb{K}$, $\operatorname{dist}(v(t); \mathbb{C}_1) = \operatorname{dist}(v(t); \mathbb{C}_2)$ which must be 0. But this is absurd because $v(t)$ would be either in $\mathbb{C}_1 \cap \overline{\mathbb{C}_2}$, or in $\overline{\mathbb{C}_1} \cap \mathbb{C}_2$. This contradiction can be eliminated only if $F_{\tau, \xi}(t)$ is connected as claimed. $\qquad\qquad\qquad\qquad\qquad\qquad\qquad\qquad\qquad\qquad\qquad\qquad\qquad\qquad\square$

7.6. *The characteristics method for a first order PDE*

Let Ω be a nonempty open subset of $\mathbb{R}^n$, let $H : \Omega \times \mathbb{R} \rightsquigarrow \mathbb{R}^n \times \mathbb{R}$ be a multifunction with nonempty values, and consider the first order partial differential equation

$$\inf_{(u,v) \in H(x, w(x))} \left(Dw(x)(u) - v \right) = 0, \tag{7.6.1}$$

where D denotes the differentiability concept of Severi [90]. We recall that a function $w : \Omega \to \mathbb{R}$ is *Severi differentiable* at a point $x \in \Omega$ if, for every $u \in \mathbb{R}^n$, there exists the finite limit

$$Dw(x)(u) = \lim_{\substack{s \downarrow 0 \\ p \to 0}} \left(\frac{1}{s} \right) \left(w\left(x + s(u + p) \right) - w(x) \right),$$

called the *Severi differential of w at x in the direction u*. If w is Severi differentiable at x, the function $u \mapsto Dw(x)(u)$ from $\mathbb{R}^n$ to $\mathbb{R}$ is the *Severi differential of w at x*. For details see Ursescu [94]. We have

$$\operatorname{graph}\left(Dw(x) \right) = \mathcal{T}_{\operatorname{graph}(w)}\left(x, w(x) \right).$$

For $u \in \mathbb{R}^n$, let us consider the extended real numbers:

$$\underline{D}w(x)(u) = \liminf_{\substack{s \downarrow 0 \\ p \to 0}} \left(\frac{1}{s} \right) \left(w\left(x + s(u + p) \right) - w(x) \right),$$

$$\overline{D}w(x)(u) = \limsup_{\substack{s \downarrow 0 \\ p \to 0}} \left(\frac{1}{s} \right) \left(w\left(x + s(u + p) \right) - w(x) \right).$$

We have the equalities:

$$\operatorname{epi}\left(\underline{D}w(x) \right) = \mathcal{T}_{\operatorname{epi}(w)}\left(x, w(x) \right),$$

$$\operatorname{hyp}\left(\overline{D}w(x) \right) = \mathcal{T}_{\operatorname{hyp}(w)}\left(x, w(x) \right),$$

where epi stands for the epigraph and hyp for the hypograph. To conclude, w is differentiable at x if and only if $\underline{D}w(x)(u)$ and $\overline{D}w(x)(u)$ are finite and equal to each other for all $u \in \mathbb{R}^n$, while $\overline{D}w(x)(u) \leqslant \underline{D}w(x)(u)$ for all $u \in \mathbb{R}^n$.

The epigraph and hypograph equalities above show that $u \mapsto \underline{D}w(x)(u)$ and $u \mapsto \overline{D}w(x)(u)$ are l.s.c. and u.s.c., respectively.

By a solution to Eq. (7.6.1) we mean a differentiable function $w : \Omega \to \mathbb{R}$ which satisfies equality (7.6.1) for all $x \in \Omega$. Equation (7.6.1) is worth noting since it contains at least two important particular cases: the quasilinear, first order partial differential equations (Goursat [51], Perron [81], Kamke [59], Carathéodory [29], Courant and Hilbert [37]); the Bellman equations (Bellman [7–9], Pontryagin et al. [83], Boltyanskii [11,12], Gonzales [50], Hájek [53], Cesari [30], Clarke and Vinter [35], and many others). Among other particular cases of Eq. (7.6.1) we mention also the eikonal equation (see Ishii [58]). Let us consider further the ordinary differential inclusion

$$\bigl(X'(t), Y'(t)\bigr) \in H\bigl(X(t), Y(t)\bigr), \tag{7.6.2}$$

whose solutions we label as characteristics with regard to Eq. (7.6.1). The theory for the inclusion (7.6.2) is well developed (see Section 6.1 above) and we use it to start developing an existence theory for Eq. (7.6.1). We take here as a model the classical characteristics method and characterize the solutions w of (7.6.1) by means of the behavior of the functions w along solutions (X, Y) of (7.6.2). Since the inclusion (7.6.2) is an autonomous one, we consider only solutions $(X, Y) : [0, T) \to \mathbb{R}^n \times \mathbb{R}$ where $0 < T \leqslant \infty$. In order to characterize the solutions w of (7.6.1), we need the following conditions:

(C_1) *For every $x \in \Omega$, there exists a solution $(X, Y) : [0, T) \to \mathbb{R}^n \times \mathbb{R}$ of the inclusion (7.6.2), with $(X(0), Y(0)) = (x, w(x))$, such that, for every $s \in (0, T)$, $w(X(s)) \leqslant Y(s)$;*

(C_2) *For every $x \in \Omega$, for every solution $(X, Y) : [0, T) \to \mathbb{R}^n \times \mathbb{R}$ of the inclusion (7.6.2), with $(X(0), Y(0)) = (x, w(x))$, and for every $s \in (0, T)$, $Y(s) \leqslant w(X(s))$.*

Now we are ready to state a first result concerning Eq. (7.6.1).

THEOREM 7.6.1. *Let $H : \Omega \times \mathbb{R} \rightsquigarrow \mathbb{R}^n \times \mathbb{R}$ be both u.s.c. and l.s.c. with nonempty, compact and convex values, and let $w : \Omega \to \mathbb{R}$ be differentiable and such that H has the comparison property with respect to $(\Omega \times \mathbb{R}, \mathrm{hyp}(w))$. Then w is a solution to Eq. (7.6.1) if and only if it satisfies conditions (C_1) and (C_2).*

As far as we know, the differentiability concept of Severi is the least restrictive for which the characterization above holds true. Every other differentiability concepts used in the literature devoted to particular cases of Eq. (7.6.1) implies the classical Fréchet, referred better as to Stolz–Young–Fréchet–Hadamard, differentiability. And, w is Fréchet differentiable at x if and only if both w is Severi differentiable at x, and the Severi differential $Dw(x)$ is linear on $\mathbb{R}^n$. In addition, Eq. (7.6.1) admits a natural substitute which dispenses with any differentiability restriction, and which can be still characterized by using (C_1)

and (C_2). The substitute for Eq. (7.6.1) consists of the couple of first order partial differential inequalities below:

$$\inf_{(u,v)\in H(x,w(x))} \left(\underline{D}w(x)(u) - v\right) \leqslant 0, \tag{7.6.3}$$

$$0 \leqslant \inf_{(u,v)\in H(x,w(x))} \left(\overline{D}w(x)(u) - v\right), \tag{7.6.4}$$

called *generalized Bellman equation*. A *solution to inequality* (7.6.3) (or (7.6.4)) is a function $w : \Omega \to \mathbb{R}$ which satisfies inequality (7.6.3) (or (7.6.4)) for all $x \in \Omega$. Clearly a function w is a solution to (7.6.1) if and only if it is a differentiable solution to the couple (7.6.3) and (7.6.4). Since differentiability at a point implies continuity at that point, we conclude that Theorem 7.6.1 above is a natural corollary of Theorem 7.6.2 below.

THEOREM 7.6.2. *Let $H : \Omega \times \mathbb{R} \rightsquigarrow \mathbb{R}^n \times \mathbb{R}$ be both u.s.c. and l.s.c. with nonempty, compact and convex values, and let $w : \Omega \to \mathbb{R}$ be continuous such that H has the comparison property with respect to $(\Omega \times \mathbb{R}, \mathrm{hyp}(w))$. Then, w is a solution to (7.6.3) and (7.6.4) if and only if it satisfies conditions (C_1) and (C_2).*

We mention that all solutions to every variational problem satisfy both (C_1) and (C_2) (the Bellman "principle") with a suitable chosen H. Hence, under rather common hypotheses upon the components of a variational problem, its continuous solution satisfies inequalities (7.6.3) and (7.6.4) (the generalized Bellman "equation"). A typical example is the time optimal control problem associated to a control system and a target. More precisely, consider the multifunction $F : \mathbb{R}^n \rightsquigarrow \mathbb{R}^n$, the differential inclusion

$$X'(t) \in F\big(X(t)\big), \tag{7.6.5}$$

and fix a *target set* $\mathcal{T}$ (nonempty and closed). Let $\mathcal{R}$ be the *reachable set*, that is, the set of all initial points which can be transferred to $\mathcal{T}$ by trajectories of (7.6.5). For $x \in \mathcal{R}$ define $T(x)$ as the infimum of the transition times. The well-known Bellman equation for the optimal time problem is

$$-1 + \sup_{u \in F(x)} -DT(x)(u) = 0, \quad x \in \mathcal{R} \setminus \mathcal{T},$$

which is a particular case of (7.6.1), for the choice $H(x, y) = (F(x), \{-1\})$.

In its turn, Theorem 7.6.2 follows from the following anatomized variant of itself.

THEOREM 7.6.3. (a) *Let $H : \Omega \times \mathbb{R} \rightsquigarrow \mathbb{R}^n \times \mathbb{R}$ be u.s.c. with nonempty, compact and convex values. A continuous function $w : \Omega \to \mathbb{R}$ is a solution to inequality (7.6.3) if and only if satisfies condition (C_1).*

(b) *Let $H : \Omega \times \mathbb{R} \rightsquigarrow \mathbb{R}^n \times \mathbb{R}$ be l.s.c. with closed and convex values. Let $w : \Omega \to \mathbb{R}$ be a continuous function such that H has the comparison property with respect to $(\Omega \times \mathbb{R}, \mathrm{hyp}(w))$. Then w is a solution to inequality (7.6.4) if and only if it satisfies condition (C_2).*

PROOF. (a) Since $H(x, w(x))$ is compact and since $\underline{D}w(x)$ is l.s.c. on $\mathbb{R}^n$, it follows that "inf" can be replaced by "min" in (7.6.3), hence inequality (7.6.3) states that

$$\phi \neq H\big(x, w(x)\big) \cap \mathcal{T}_{\mathrm{epi}(w)}\big(x, w(x)\big). \tag{7.6.6}$$

Since w is continuous, (7.6.6) is equivalent to (7.6.7) below

$$\phi \neq H(x, t) \cap \mathcal{T}_{\mathrm{epi}(w)}(x, t) \tag{7.6.7}$$

for $t \geqslant w(x)$, and this simply because $\mathcal{T}_{\mathrm{epi}(w)}(x, t) = \mathbb{R}^n \times \mathbb{R}$ whenever $w(x) < t$. On the other hand, since w is continuous, condition (C_1) is equivalent to the viability of epi(w) with respect to H. But the set epi(w) is closed in $\Omega \times \mathbb{R}$, and therefore the conclusion is an immediate consequence of Theorem 6.2.1.

(b) The inequality in (7.6.4) states that $H(x, w(x)) \subseteq \mathcal{T}_{\mathrm{hyp}(w)}(x, w(x))$. Since w is continuous, we have $H(x, t) \subseteq \mathcal{T}_{\mathrm{hyp}(w)}(x, t)$ in case $w(x) > t$. On the other hand, condition (C_2) states that the set hyp(w) is invariant with respect to H. Since the set hyp(w) is closed in $\Omega \times \mathbb{R}$, the conclusion follows from Theorem 6.8.1 and Corollary 6.8.1. $\qquad\square$

A natural question is whether we can weaken the continuity property of the function w in Theorem 7.6.2. This question arises from the fact that epi(w) is closed even in the case when w is l.s.c. and hyp(w) is closed in case w is u.s.c. The answer is in the negative as the following examples in Cârjă and Ursescu [25] show. Consider first the inequality

$$\underline{D}w(x)\big(w(x)\big) - 1 \leqslant 0.$$

Here $H(x, y) = \{(y, 1)\}$, for all $(x, y) \in \mathbb{R} \times \mathbb{R}$ and the characteristic system $X'(s) = Y(s)$, $Y'(s) = 1$ has the solution $X(s) = X(0) + sY(0) + s^2/2$, $Y(s) = Y(0) + s$. The l.s.c. function $w : \mathbb{R} \to \mathbb{R}$ given by $w(x) = 0$ for $x = 0$ and by $w(x) = 1$ for $x \neq 0$ is a solution to the preceding inequality but does not satisfy the condition: *for every $x \in \mathbb{R}$ and for every* $s \in (0, +\infty)$, $w(x + sw(x) + s^2/2) \leqslant w(x) + s$ (take $x = 0$ and $s \in (0, 1)$ and observe $w(0 + sw(0) + s^2/2) = 1 > s = w(0) + s$). A second example is given by the differential inequality

$$0 \leqslant \overline{D}w(x)\big(w(x)\big) + 1.$$

Here $H(x, y) = \{(y, -1)\}$ for all $(x, y) \in \mathbb{R} \times \mathbb{R}$ and the characteristic system $X'(s) = Y(s)$, $Y'(s) = -1$ has the solutions $X(s) = X(0) + sy(0) - s^2/2$, $Y(s) = Y(0) - s$. The u.s.c. function $w : \mathbb{R} \to \mathbb{R}$ given by $w(x) = 0$ for $x = 0$ and by $w(x) = -1$ for $x \neq 0$ is a solution to the preceding inequality but does not satisfy the condition: *for every $x \in \mathbb{R}$ and for every $s \in (0, +\infty)$*, $w(x) - s \leqslant w(x + sw(x) - s^2/2)$ (take $x = 0$ and $s \in (0, 1)$ and observe $w(0) - s = -s > -1 = w(0) + sw(0) - s^2/2)$.

A condition which assures that (C_1) in part (a) of Theorem 7.6.3 holds for a l.s.c. function w, while (C_2) in part (b) holds for a u.s.c. function w is given below:

THEOREM 7.6.4. *Let $H : \Omega \times \mathbb{R} \rightsquigarrow \mathbb{R}^n \times \mathbb{R}$ be a nonempty and convex valued multifunction satisfying*

$$H(x, y_1) \subseteq H(x, y_2) \tag{7.6.8}$$

for each $x \in \Omega$ and $y_1, y_2 \in \mathbb{R}$, with $y_1 \leqslant y_2$. (a) Assume that H is u.s.c. with compact values. Then, a l.s.c. function $w : \Omega \to \mathbb{R}$ is a solution to the inequality (7.6.3) if and only if it satisfies (C_1).

(b) Assume that H is l.s.c. with closed values. Let $w : \Omega \to \mathbb{R}$ a u.s.c. function such that H has the comparison property with respect to $(\Omega \times \mathbb{R}, \mathrm{hyp}(w))$. Then, $w : \Omega \to \mathbb{R}$ is a solution to the inequality (7.6.4) if and only if it satisfies (C_2).

PROOF. The proof goes in the same spirit as that of Theorem 7.6.3. Indeed, in order to prove (a), since $\mathcal{T}_{\mathrm{epi}(w)}(x, w(x)) \subseteq \mathcal{T}_{\mathrm{epi}(w)}(x, t)$ if $w(x) \leqslant t$, (7.6.8) implies that (7.6.6) is equivalent to (7.6.7). In its turn, (7.6.6) is equivalent to (7.6.3). On the other hand, since (7.6.8) is satisfied, condition (C_1) is equivalent to the viability of $\mathrm{epi}(w)$ with respect to H. Indeed, if $w(x) < t$, and $(X, Y) : [0, T] \to \mathbb{R}^n \times \mathbb{R}$ is a solution to (7.6.2) with $(X(0), Y(0)) = (x, w(x))$, and $w(X(s)) \leqslant Y(s)$ for all $s \in [0, T]$, then the function $s \mapsto (X(s), Y(s) + t - w(x)) = (\overline{X}(s), \overline{Y}(s))$ is a solution to (7.6.2) with $(\overline{X}(0), \overline{Y}(0)) = (x, t)$ and satisfies $w(\overline{X}(s)) \leqslant \overline{Y}(s)$ for all $s \in [0, T]$. This completes the proof of (a). The proof of (b) goes in the very same spirit, and therefore we do not give details. $\qquad\square$

REMARK 7.6.1. The property (C_1) is related to the so-called *weakly decreasing systems* discussed in Clarke et al. [34, p. 211], which in turn are related to the Liapunov theory of stabilization. See also [34, p. 208]. The "monotone variant" of (C_1) labelled below as (C_3) is connected to the so-called *strongly decreasing systems* discussed also in [34, p. 217].

(C_3) *For each $x \in \Omega$ there exists a solution $(X, Y) : [0, T] \to \mathbb{R}^n \times \mathbb{R}$ of the differential inclusion (7.6.2) with $(X(0), Y(0)) = (x, w(x))$, such that the function $s \mapsto w(X(s)) - Y(s)$ is decreasing.*

We have:

THEOREM 7.6.5. *Let $H : \Omega \times \mathbb{R} \rightsquigarrow \mathbb{R}^n \times \mathbb{R}$ be a u.s.c., nonempty and convex valued multifunction. Let, us assume that (7.6.8) is satisfied. Then, a continuous function w satisfies (C_1) if and only if it satisfies (C_3).*

PROOF. On $\mathrm{epi}(w)$ we define a relation $\mathcal{P}$ by

$$\mathcal{P}(x, t) = \big\{(y, s); \ w(y) - s \leqslant w(x) - t\big\}.$$

Obviously, $\mathcal{P}$ is a preorder on $\mathrm{epi}(w)$ and condition (C_3) is equivalent to the viability of the preorder $\mathcal{P}$ with respect to H. Moreover, condition (C_1) is equivalent to the viability of $\mathcal{P}(x, t)$ with respect to H for each $(x, t) \in \mathrm{epi}(w)$. The conclusion follows from Lemma 6.6.1. $\qquad\square$

8. Notes and comments

8.1. *The upper semicontinuous case*

Since of its birth in 1942, viability theory emerged in several directions we will discuss sequentially below. The first viability result for the multivalued case is due to Bebernes and Schuur [6]. We mention that, in 1936, Zaremba [111] proved that if $F : \mathbb{D} \rightsquigarrow \mathbb{R}^n$ is u.s.c. with nonempty compact and convex values and $\mathbb{D}$ is open then, for each $\xi \in \mathbb{D}$, the differential inclusion (6.2.1) has at least one solution $u : [0, T] \to \mathbb{D}$ satisfying $u(0) = \xi$. It should be noticed that the concept of solution used by Zaremba in [111] is in the sense of the *contingent derivative*. More precisely, if $u : [0, T] \to \mathbb{R}^n$ is continuous and $t \in [0, T)$, the set

$$Du(t) = \left\{ \lim_{m \to \infty} \frac{u(t + t_m) - u(t)}{t_m}; \ t_m \downarrow 0 \right\}$$

is called the *contingent derivative* of u at t. We say that $u : [0, T] \to \mathbb{D}$ is a *contingent solution* of (6.2.1) if

$$\emptyset \neq Du(t) \subseteq F\big(u(t)\big) \tag{8.1.1}$$

for each $t \in [0, T)$. In 1961, Ważewski [107] proved that, if F is u.s.c. with nonempty, compact and convex values, u is a contingent solution to (6.2.1) if and only if u is a Carathéodory solution to (6.2.1). So, Zaremba's existence result in [111] is nothing than the multivalued counterpart of Peano's local existence theorem 1.1.1. In the same spirit, the viability result of Bebernes and Schuur is the multivalued version of Nagumo's viability theorem 1.1.3. Proposition 8.1.1 below, due to Ważewski [107], is in fact equivalent to the necessity part of Theorem 6.2.1.

PROPOSITION 8.1.1. *Let $\mathbb{K} \subseteq \mathbb{R}^n$ be nonempty and locally closed and let $F : \mathbb{K} \rightsquigarrow \mathbb{R}^n$ be u.s.c. with nonempty, convex, compact values. Then, for every $\xi \in \mathbb{K}$, and every solution $u : [0, T] \to \mathbb{K}$ to (6.2.1), with $u(0) = \xi$, there exist $\eta \in F(\xi)$ and a sequence $(t_m)_m$ in $(0, T)$ convergent to 0 such that the sequence $(\frac{1}{t_m}(u(t_m) - \xi))_m$ converges to η.*

It is interesting to notice that, by using the viability theory developed in Section 6.1 for the locally closed set $\mathbb{K} = \{(t, u(t)); \ t \in [0, T]\}$ and the multifunction $\{1\} \times F$, we can prove (see [22]) that condition (8.1.1) is also equivalent to each one of the following:

   (i) $Du(t) \cap F(u(t)) \neq \emptyset$ for each $t \in [0, T)$, or

   (ii) $\overline{\text{co}} \, Du(t) \cap F(u(t)) \neq \emptyset$ for each $t \in [0, T)$.

In 1981, Haddad [52] obtained the first result on viability of preorders. In fact, Haddad adapted the proof of viability of sets in order to obtain viability of preorders. Cârjă and Ursescu [25] showed that the viability, as well as the invariance of preorders can be completely described in terms of viability, or invariance of sets. See Section 3.8.

The structure of the set of viable solutions of a differential inclusion on a subset $\mathbb{K}$, for which $\Pi_{\mathbb{K}}$ has continuous selections, was studied by Plaskacz [82]. See also the references therein.

As far as the infinite dimensional case is concerned, i.e. the case in which instead of $\mathbb{R}^n$ we are considering an infinite-dimensional Banach space X, we mention the pioneering contribution of Gauthier [48]. First, he used Zorn's Lemma in order to get approximate solutions defined on an a priori given interval. Second, he used a sufficient weak tangency condition of the form: *for each $\xi \in \mathbb{K}$, there exist $\eta \in F(\xi)$, a sequence $(h_m)_m$ decreasing to 0, a sequence $(q_m)_m$ weakly convergent to 0 satisfying $\|q_m + \eta\| \leqslant 2\|\eta\|$ and $\xi + h_m(\eta + q_m) \in \mathbb{K}$ for each $m \in \mathbb{N}$.* This tangency is far from being necessary for the viability of $\mathbb{K}$. A necessary and sufficient condition for the viability of $\mathbb{K}$ in a more general setting has been obtained by Cârjă and Vrabie [27] by means of the so-called "bounded weak tangency condition".

8.2. *The case of Carathéodory mappings*

The first viability result in the case of a single-valued Carathéodory right-hand side is due to Ursescu [98]. Theorem 5.2.1 is an extension of Ursescu's result in [98] which contains only the equivalence between (ii) and (vi) in the above mentioned theorem.

Although not presented here, the Carathéodory case for differential inclusions is well developed and there exists a rather large literature on the subject. Among the first notable results in this direction we mention those of Tallos [92], Ledyaev [64], Frankowska et al. [45]. In all these papers, theorems of Scorza Dragoni type are the main tools. Results of the same kind as in Theorem 5.2.1, but for differential inclusions, can be found in Cârjă and Monteiro Marques [22] in the finite-dimensional case, and in Cârjă and Monteiro Marques [23] in the infinite-dimensional setting. There, a technique of approximation (as in Theorem 5.2.1) of the multifunction through the Aumann integral mean is used.

8.3. *The lower semicontinuous case*

The next example shows that the convexity condition on the values of F is essential in obtaining the viability of a locally closed set $\mathbb{K}$ with respect to a u.s.c. multifunction $F : \mathbb{K} \rightsquigarrow \mathbb{R}^n$ by means of the tangency condition $\mathcal{T}_{\mathbb{K}}(\xi) \cap F(\xi) \neq \emptyset$ for all $\xi \in \mathbb{K}$.

EXAMPLE 8.3.1 (Aubin and Cellina [3, p. 202]). Let $n = 2$, $\mathbb{K} = B(0, 1)$ and $F : \mathbb{K} \to \mathbb{R}^2$, defined by $F(\xi) = \{(-1, 0), (1, 0)\}$ for each $\xi \in \mathbb{K}$. Then, one may easily see that $\mathbb{K}$ is locally closed (in fact closed and convex), F is u.s.c., satisfies the tangency condition, but nevertheless $\mathbb{K}$ is not viable with respect to F.

The lack of convexity of the values of F can be however counterbalanced by an l.s.c. extra-assumption combined with the stronger tangency condition

$$F(\xi) \subseteq \mathcal{T}_{\mathbb{K}}(\xi) \tag{8.3.1}$$

for each $\xi \in \mathbb{K}$. More precisely, we have:

THEOREM 8.3.1 (Aubin and Cellina [3, p. 198]). *Let $\mathbb{K}$ be locally closed and let $F : \mathbb{K} \rightsquigarrow \mathbb{R}^n$ be both u.s.c. and l.s.c. with nonempty and closed values. If (8.3.1) is satisfied, then $\mathbb{K}$ is viable with respect to F. Moreover, x' is a regulated function.*[5]

We emphasize that even for the convex-valued case, l.s.c. and (8.3.1) do not ensure invariance (see the case of functions), but implies viability via the Michael's selection theorem [67,68].

8.4. *The semilinear single-valued case*

Another direction was to consider a larger framework in order to handle semilinear partial differential equations as well. To this aim, let us consider an infinite dimensional Banach space X with norm $\| \cdot \|$ and let $\mathcal{L}(X)$ be the space of all linear bounded operators from X to X, endowed with the usual *operator norm* $\| \cdot \|_{\mathcal{L}(X)}$. We recall that $\{S(t);\ t \geqslant 0\} \subseteq \mathcal{L}(X)$ is a C_0-*semigroup* if

    (i) $S(0) = I$;

    (ii) $S(t + s) = S(t)S(s)$ for all $t, s \geqslant 0$;

    (iii) $\lim_{h \downarrow 0} S(h)\xi = \xi$ for each $\xi \in X$.

The *infinitesimal generator* of $\{S(t);\ t \geqslant 0\}$ is the possibly unbounded linear operator $A : D(A) \subseteq X \rightarrow X$, defined by

$$\begin{cases} D(A) = \left\{ \xi \in X;\ \exists \lim_{h \downarrow 0} \dfrac{1}{h}\big[S(h)\xi - \xi\big] \right\}, \\[2mm] A\xi = \lim_{h \downarrow 0} \dfrac{1}{h}\big[S(h)\xi - \xi\big] \quad \text{for } \xi \in D(A). \end{cases}$$

Further, if $\{S(t);\ t \geqslant 0\}$ is a C_0-semigroup, $A : D(A) \subseteq X \rightarrow X$ is its infinitesimal generator and $\xi \in D(A)$, then the mapping $t \mapsto S(t - \tau)\xi$ is the unique classical, i.e., C^1, solution to the ordinary homogeneous differential equation

$$\begin{cases} u' = Au, \\ u(\tau) = \xi, \end{cases} \tag{8.4.1}$$

defined on $[\tau, +\infty)$. Moreover, since $D(A)$ is dense in X, it follows that, for each $\xi \in X$, the mapping $t \mapsto S(t - \tau)\xi$, which may fail to be C^1, can be approximated uniformly on compact subsets in $[\tau, +\infty)$ by classical solutions to problem (8.4.1). Thus, for each $\xi \in X$, the mapping above can be considered as a generalized solution of (8.4.1), called the *semigroup solution*. Furthermore, inspired from the variation of constants formula, we may define the so-called *mild, or C^0 solution*, $u : [\tau, a) \rightarrow X$, of the nonhomogeneous problem

$$\begin{cases} u' = Au + f(t), \\ u(\tau) = \xi, \end{cases}$$

---

[5]A function is *regulated* if it is uniform limit of step functions.

by

$$u(t) = S(t - \tau)\xi + \int_\tau^t S(t - s) f(s) \, ds$$

for each $t \in [\tau, a)$. Here $f : [\tau, a) \to X$ is a given continuous, or even locally integrable, function and $a > \tau$ is finite or not. So, if $\mathbb{D} \subseteq X$ is a nonempty and open subset in X, $\mathbb{I}$ is a given nonempty open interval, $f : \mathbb{I} \times \mathbb{D} \to X$, $\tau \in \mathbb{I}$ and $\xi \in \mathbb{D}$, we may consider the *semilinear differential equation*

$$\begin{cases} u' = Au + f(t, u), \\ u(\tau) = \xi. \end{cases} \tag{8.4.2}$$

As expected, a *mild, or C^0 solution* of (8.4.2) is a continuous function $u : [\tau, T) \to \mathbb{D}$, with $T \in \mathbb{I}$, and satisfying

$$u(t) = S(t - \tau)\xi + \int_\tau^t S(t - s) f\big(s, u(s)\big) \, ds \tag{8.4.3}$$

for each $t \in [\tau, T)$. For details on C_0-semigroups see Vrabie [105]. Further, if $\mathbb{K} \subseteq \mathbb{D}$ is locally closed, we redefine the concepts of right viability and right invariance of $\mathbb{K}$ with respect to $A + f$ by using mild or C^0 solutions. The main problem here is that, in general, $\mathbb{K} \cap D(A)$ is very narrow, even empty (we emphasize that $D(A)$ is only dense and could have empty interior), and therefore a necessary, or/and sufficient condition for viability of the type (1.1.3), i.e.

$$\liminf_{h \downarrow 0} \frac{1}{h} \operatorname{dist}\big(\xi + h[A\xi + f(t, \xi)]; \mathbb{K}\big) = 0, \tag{8.4.4}$$

for each $(t, \xi) \in \mathbb{I} \times [\mathbb{K} \cap D(A)]$, turns out to be unrealistic for nonsmooth semigroups. An example which justifies the remark above is that one when $\mathbb{K}$ is the trajectory of a nowhere differentiable right mild solution to (8.4.2), case in which $\mathbb{K} \cap D(A) = \emptyset$. The case of smooth semigroups (even nonlinear) was considered recently by Barbu and Pavel [5] by means of the tangency condition (8.4.4) with "lim" instead of "lim inf". It is the merit of Pavel [75] to observe that, in general, the really useful sufficient, and, very often, even necessary, condition for right viability which works in this case is

$$\lim_{h \downarrow 0} \frac{1}{h} \operatorname{dist}\big(S(h)\xi + h f(t, \xi); \mathbb{K}\big) = 0, \tag{8.4.5}$$

for each $(t, \xi) \in \mathbb{I} \times \mathbb{K}$. We notice that, whenever $\xi \in \mathbb{K} \cap D(A)$, (8.4.5) is equivalent to (8.4.4) which is a stronger form, i.e. with "lim inf" replaced by "lim", of the classical Nagumo's tangency condition (1.1.3) with $A + f$ instead of f. We re-emphasize that whenever $\mathbb{K}$ is not included in $D(A)$, or even $\mathbb{K} \cap D(A)$ is empty, the only tangency condition which could be of some use is (8.4.5). Namely, the main result in Pavel [75] is:

THEOREM 8.4.1 (Pavel [75]). *If $S(t)$ is compact for each $t > 0$ and $f : \mathbb{I} \times \mathbb{K} \to X$ is continuous, then a necessary and sufficient condition in order that $\mathbb{K}$ be right viable with respect to $(t, u) \mapsto Au + f(t, u)$ is (8.4.5).*

We notice that the compactness of $S(t)$ for each $t > 0$ is a parabolicity condition. The simplest nontrivial example of a C_0 semigroup satisfying this condition is that one occurring in the study of the heat flow. Namely, let $S(t) : L^2(0, \pi) \to L^2(0, \pi)$ be defined by

$$\left[S(t)\xi \right](x) = \sqrt{\frac{2}{\pi}} \sum_{k=1}^{\infty} a_k(\xi) e^{-k^2 t} \sin kx$$

for $x \in (0, \pi)$, where $a_k(\xi)$ are the Fourier coefficients of ξ with respect to the orthogonal system $\left\{ \sqrt{\frac{2}{\pi}} \sin x, \sqrt{\frac{2}{\pi}} \sin 2x, \ldots, \sqrt{\frac{2}{\pi}} \sin kx, \ldots \right\}$, i.e.

$$a_k(\xi) = \sqrt{\frac{2}{\pi}} \int_0^{\pi} \xi(y) \sin ky \, dy.$$

It is well known that this semigroup defines the $L^2(0, \pi)$-solutions to the one-dimensional heat equation

$$\begin{cases} u_t = u_{xx} & \text{for } (t, x) \in [0, +\infty) \times (0, \pi), \\ u(t, 0) = u(t, \pi) & \text{for } t \in [0, +\infty), \\ u(0, x) = \xi & \text{for } x \in (0, \pi). \end{cases}$$

Moreover, for each $t > 0$, $S(t)$ is a compact, in fact even a Hilbert–Schmidt operator, because it is the limit in the operator norm of a sequence of finite-dimensional range operators. Indeed, a simple example of such sequence $(S_m(t))_m$ is

$$\left[S_m(t)\xi \right](x) = \sqrt{\frac{2}{\pi}} \sum_{k=1}^{m} a_k(\xi) e^{-k^2 t} \sin kx$$

for $m = 1, 2, \ldots$ and each $\xi \in X$, where $a_k(\xi)$ are as above.

The extension of Pavel's Theorem 8.4.1 to Carathéodory perturbations has been considered by Cârjă and Monteiro Marques [21] for constant in time domains, and by Necula [71] for time dependent domains.

8.5. *The semilinear multivalued case*

As far as we know, the true semilinear and multivalued case, i.e. *A* linear unbounded and *F* multivalued, has been analyzed first by Pavel and Vrabie [78,79] by the end of seventies. A good source of references in this respect is Pavel [76]. Shi [91] considers the semilinear case (8.4.2) in which f is replaced by a multifunction, redefines the concept of viability by using strong solutions, i.e. continuous functions $u : [0, T) \to X$, $T \in (0, +\infty]$, which are absolutely continuous on $(0, T)$ and satisfy: $u'(t) = Au(t) + f(t)$ a.e. for $t \in [0, T)$, where $f \in L^\infty(0, T; X)$, $f(t) \in F(t, u(t))$ a.e. for $t \in (0, T)$, and proves a characterization of global viability, i.e. of right viability involving only strong solutions defined on $[0, +\infty)$.

THEOREM 8.5.1 (Shi [91]). *Let X be reflexive, $\mathbb{K}$ a compact subset of X, $F : X \rightsquigarrow X$ a nonempty, convex and compact valued upper semicontinuous mapping, and let $A : D(A) \subseteq X \to X$ be the infinitesimal generator of a differentiable C_0-semigroup $\{S(t); \ t \geqslant 0\}$ with $S(t)$ compact for all $t > 0$. Then a necessary and sufficient condition in order that $\mathbb{K}$ be right global viable with respect to F is the following tangency condition: for each $\xi \in \mathbb{K}$ there exist $\eta \in F(\xi)$, a sequence $(h_m)_m$ decreasing to 0 and a sequence $(p_m)_m$ strongly convergent to 0 such that*

$$S(h_m)\xi + h_m(\eta + p_m) \in \mathbb{K}$$

holds for each $m \in \mathbb{N}$.

As concerns sufficient conditions for global viability, we mention:

THEOREM 8.5.2 (Shi [91]). *Let X be reflexive, $\mathbb{K}$ a compact subset of X, $F : X \rightsquigarrow X$ a nonempty, bounded, closed and convex valued upper semicontinuous mapping, and let $A : D(A) \subseteq X \to X$ be the infinitesimal generator of a differentiable C_0-semigroup $\{S(t); \ t \geqslant 0\}$ with $S(t)$ compact for all $t > 0$. Then, a sufficient condition in order that $\mathbb{K}$ be right global viable with respect to F is the following tangency condition: for each $\xi \in \mathbb{K}$ and each $t > 0$ there exist $\eta \in F(\xi)$, a sequence $(h_m)_m$ decreasing to 0 and a sequence $(p_m)_m$ strongly convergent to 0 such that*

$$S(h_m)\xi + h_m\big(S(t)\eta + p_m\big) \in S(t)\mathbb{K}$$

holds for each $m \in \mathbb{N}$.

We note that the tangency condition in Theorem 8.5.2 is equivalent to:

$$AS(t)\xi + S(t)F(\xi) \subset \mathcal{T}_{S(t)\mathbb{K}}\big(S(t)\xi\big)$$

holds for each $\xi \in \mathbb{K}$ and each $t > 0$.

Clearly, in this case the general assumptions on $\mathbb{K}$ and F are significantly stronger than those in Pavel and Vrabie [78,79]. We note that for instance, in the infinite-dimensional setting, the compactness of $F(\xi)$ for each $\xi \in D$ is not satisfied if F is a superposition operator which is not single-valued. On the other hand, this "weakness" of the general setting of Shi [91] is well counterbalanced by the tangency condition which is quite close to its finite-dimensional counterpart.

The existence of monotone solutions has been considered in this context by Chiş-Şter [31].

8.6. *The nonlinear perturbed single-valued case*

Theorem 8.4.1 was partially extended by Vrabie [103] for the fully nonlinear case, i.e. when A m-dissipative. Namely, Vrabie [103] shows that, if A generates a nonlinear semi-group of nonexpansive compact, for $t > 0$, operators and $f : \mathbb{I} \times \mathbb{K} \to X$ is continuous on the locally closed subset $\mathbb{K}$, a sufficient condition for viability is

$$\lim_{h \downarrow 0} \frac{1}{h} \operatorname{dist}\big(u\big(t + h, t, \xi, f(t, \xi)\big); \mathbb{K}\big) = 0 \qquad (8.6.1)$$

uniformly for $(t, \xi) \in \mathbb{I} \times \mathbb{K}$, where $u(\cdot, t, \xi, \eta)$ denotes the unique mild solution to the problem $u'(s) \in Au(s) + \eta$ satisfying $u(t, t, \xi, \eta) = \xi$. Subsequent contributions in this context are due to Bothe [14] who allowed $\mathbb{K}$ to depend on t as well. In particular, in case $\mathbb{K}$ independent of t, Bothe [14] showed that the, possibly non uniform, tangency condition (8.6.1) satisfied for each $(t, \xi) \in \mathbb{I} \times \mathbb{K}$ is necessary and sufficient for viability. The case of Carathéodory perturbations defined on time dependent domains has been considered by Necula [72].

8.7. *The multivalued perturbed nonlinear case*

The more delicate case in which one allows A to be nonlinear, as well as f to be mul-tivalued, has been considered by Bressan and Staicu [17] who used the tangency con-dition proposed in Vrabie [103] after reducing the multivalued case to the single-valued one by means of a continuous selection argument. The case in which the multifunction F is strongly-weakly u.s.c. has been considered by Cârjă and Vrabie [28] by using a weak variant of the tangency condition in Vrabie [103], while the possibly nonconvex valued case has been analyzed by Necula and Vrabie [74] by using a selection theorem due to Fryszkowski [46].

8.8. *Applications*

The problem of finding Liapunov functions for differential inclusions is discussed in Aubin [1, Chapter 9], and Aubin and Cellina [3, Chapter 6]. As we have pointed out in Section 7.6, the characteristics method, when applied in control theory, is related to the dynamic pro-gramming method. Theorem 7.6.2 says that there is an equivalence between the Bellman equations (7.6.3) and (7.6.4) and the Bellman optimality conditions (C_1) and (C_2). If we want to study the uniqueness properties of the Bellman equation, we have to add appropri-ate boundary conditions and to use conditions (C_1) and (C_2) in order to get that a possible solution is necessarily the value function of the given control problem. The notion of so-lution for the couple of inequalities (7.6.3) and (7.6.4), defined in Section 7.6, is usually called *contingent solution*. It is interesting to notice that this kind of solution is equivalent with that of viscosity solution developed by Crandall and Lions [39]. However, the stan-dard technique to get uniqueness is different here. Among the first contributions in this area we mention Frankowska [44] and Cârjă and Ursescu [26]. See also the more recent article of Aubin [2] and the references therein. For other applications of viability and invariance techniques to Control Theory and Game Theory, we refer the reader to Aubin and Cellina [3, Chapter 5]. As concerns the existence theory for periodic problems via viability and/or invariance see Aubin and Cellina [3, Chapter 5] and Vrabie [104]. For applications to the existence of constrained solutions to nonlinear partial differential inclusions see Bothe [15].

Finally, it should be noticed that the theory of invariant sets pertains a very elegant approach to the study of orbital motions of a mass particle in a given force field. This was pointed out by means of a second order tangency concept by Pavel and Ursescu [77].

Appendix

THEOREM A.1.1. *Let $p \in [1, +\infty)$ and let $[\tau, T] \subseteq \mathbb{I}$. A subset $\mathcal{F}$ in $L^p_{\mathrm{loc}}(\mathbb{I}; \mathbb{R}^n)$ is relatively compact in $L^p(\tau, T; \mathbb{R}^n)$ if and only if it is bounded and*

$$\lim_{\theta \downarrow 0} \int_\tau^{T-\theta} \left\| f(s + \theta) - f(s) \right\| \, \mathrm{d}s = 0$$

uniformly for $f \in \mathcal{F}$.

See Vrabie [105, Theorem A.4.1, p. 305]. For the proof of the next result see Dunford and Schwartz [40, Theorem 15, p. 150].

THEOREM A.1.2 (Vitali). *Let $1 \leqslant p < +\infty$, let $\mathbb{S}$ be a Lebesgue measurable subset in $\mathbb{R}$, and let $(f_m)_m$ be a sequence in $L^p(\mathbb{S}; \mathbb{R}^n)$ converging almost everywhere to a function f. Then $f \in L^p(\mathbb{S}; \mathbb{R}^n)$ and*

$$\lim_m \| f_m - f \|_{L^p(\mathbb{S}; \mathbb{R}^n)} = 0$$

if and only if:
   (i) *$\{ \| f_m \|^p; \ m \in \mathbb{N} \}$ is uniformly integrable in $L^1(\mathbb{S}; \mathbb{R}^n)$;*
   (ii) *for each $\varepsilon > 0$ there exists a Lebesgue measurable set $\mathbb{E}_\varepsilon \subseteq \mathbb{S}$ with $\lambda(\mathbb{E}_\varepsilon) < +\infty$ and such that, for each $m \in \mathbb{N}$, we have*

$$\int_{\mathbb{S} \backslash \mathbb{E}_\varepsilon} \left\| f_m(s) \right\|^p \, \mathrm{d}s < \varepsilon.$$

We recall that a subset $\mathcal{F}$ in $L^1(\mathbb{S}; \mathbb{R}^n)$ is uniformly integrable if

$$\lim_{\lambda(\mathbb{E}) \downarrow 0} \int_\mathbb{E} \left\| f(s) \right\| \, \mathrm{d}s = 0$$

uniformly for $f \in \mathcal{F}$. We notice that whenever $\mathbb{S}$ is a finite length interval, the condition (ii) is automatically satisfied.

References

[1] J.P. Aubin, *Viability Theory*, Birkhäuser, Boston (1991).

[2] J.P. Aubin, *Boundary-value problems for systems of Hamilton–Jacobi–Bellman inclusions with constraints*, SIAM J. Control Optim. **41** (2002), 425–456.

[3] J.P. Aubin and A. Cellina, *Differential Inclusions*, Springer-Verlag, Berlin (1984).

[4] J.P. Aubin and H. Frankowska, *Set-Valued Analysis*, Birkhäuser (1990).

[5] V. Barbu and N.H. Pavel, *Flow-invariant closed sets with respect to nonlinear semigroup flows*, NoDEA Nonlinear Differential Equations Appl. **10** (2003), 57–72.

[6] W. Bebernes and I.D. Schuur, *The Ważewski topological method for contingent equations*, Ann. Mat. Pura Appl. **87** (1970), 271–278.

[7] R. Bellman, *Bottleneck problems and dynamic programming*, Proc. Natl. Acad. Sci. USA **39** (1953), 947–951.

[8] R. Bellman, *Bottleneck problems, functional equations and dynamic programming*, Econometrica **23** (1955), 73–87.

[9] R. Bellman, *Dynamic Programming*, Princeton Univ. Press, Princeton, NJ (1957).

[10] H. Berliocchi and J.M. Lasry, *Intégrandes normales et mesures paramétrées en calcul des variations*, Bull. Soc. Math. France **101** (1973), 129–184.

[11] V.G. Boltyanskii, *Sufficient optimality conditions* (in Russian), Dokl. Akad. Nauk SSSR **140** (1961), 994–997.

[12] V.G. Boltyanskii, *Sufficient conditions for optimality and the justification of the programming method* (in Russian), Izv. Akad. Nauk SSSR Ser. Mat. **28** (1964), 481–514.

[13] J.M. Bony, *Principe du maximum, inégalité de Harnack et unicité du problème de Cauchy pour les opérateurs elliptiques dégénérés*, Ann. Inst. Fourier (Grenoble) **19** (1969), 277–304.

[14] D. Bothe, *Flow invariance for perturbed nonlinear evolution equations*, Abstr. Appl. Anal. **1** (1996), 417–433.

[15] D. Bothe, *Reaction diffusion systems with discontinuities. A viability approach*, Nonlinear Anal. **30** (1997), 677–686.

[16] H. Bouligand, *Sur les surfaces dépourvues de points hyperlimités*, Ann. Soc. Polon. Math. **9** (1930), 32–41.

[17] A. Bressan and V. Staicu, *On nonconvex perturbations of maximal monotone differential inclusions*, Set-Valued Anal. **2** (1994), 415–437.

[18] H. Brezis, *On a characterization of flow-invariant sets*, Comm. Pure Appl. Math. **23** (1970), 261–263.

[19] H. Brezis and F. Browder, *A general principle on ordered sets in nonlinear functional analysis*, Adv. Math. **21** (1976), 355–364.

[20] O. Cârjă, *Unele metode de analiză funcțională neliniară*, MATRIXROM, Bucureşti (2003).

[21] O. Cârjă and M.D.P. Monteiro Marques, *Viability for nonautonomous semilinear differential equations*, J. Differential Equations **166** (2000), 328–346.

[22] O. Cârjă and M.D.P. Monteiro Marques, *Viability results for nonautonomous differential inclusions*, J. Convex Anal. **7** (2000), 437–443.

[23] O. Cârjă and M.D.P. Monteiro Marques, *Weak tangency, weak invariance and Carathédory mappings*, J. Dynam. Control Systems **8** (2002), 445–461.

[24] O. Cârjă, M. Necula and I.I. Vrabie, *Local invariance via comparison functions*, Electron. J. Differential Equations **2004**(50) (2004), 1–14.

[25] O. Cârjă and C. Ursescu, *The characteristics method for a first order partial differential equation*, An. Ştiin. Univ. Al. I. Cuza Iaşi, Secţ. I a Mat. **39** (1993), 367–396.

[26] O. Cârjă and C. Ursescu, *Viscosity solutions and partial differential inequalities*, Evolution Equations, Control Theory, and Biomathematics, Lecture Notes in Pure and Appl. Math., Vol. 155, Marcel Dekker, New York (1994), 39–44.

[27] O. Cârjă and I.I. Vrabie, *Some new viability results for semilinear differential inclusions*, NoDEA Nonlinear Differential Equations Appl. **4** (1997), 401–424.

[28] O. Cârjă and I.I. Vrabie, *Viability results for nonlinear perturbed differential inclusions*, Panamer. Math. J. **9** (1999), 63–74.

[29] C. Carathéodory, *Variationsrechnung und partielle Differentialgleichungen erster Ordnung*, Teubner, Leipzig–Berlin (1935).

[30] L. Cesari, *Optimization—Theory and Applications*, Springer, New York (1983).

[31] I. Chiş-Şter, *Existence of monotone solutions for semilinear differential inclusions*, NoDEA Nonlinear Differential Equations Appl. **6** (1999), 63–78.

[32] F.H. Clarke, *Generalized gradients and applications*, Trans. Amer. Math. Soc. **205** (1975), 247–262.

[33] F.H. Clarke, Yu.S. Ledyaev and R.J. Stern, *Invariance, monotonicity and applications*, Nonlinear Analysis, Differential Equations and Control, F.H. Clarke and R.J. Stern, eds., Kluwer Academic Publishers, Dordrecht (1999), 207–305.

[34] F.H. Clarke, Yu.S. Ledyaev, R.J. Stern and P.R. Wolenski, *Nonsmooth Analysis and Control Theory*, Springer-Verlag, Berlin (1998).

[35] F.H. Clarke and R.B. Vinter, *Local optimality conditions and lipschitzean solutions to the Hamilton–Jacobi equation*, SIAM J. Control Optim. **21** (1983), 856–870.

[36] B. Cornet, *Regularity properties of open tangent cones*, Nonlinear Analysis and Optimization, Louvain-la-Neuve, 1983, Math. Programming Stud. **30** (1987), 17–33.

[37] R. Courant and D. Hilbert, *Methoden der Mathematischen Physik*, Zweiter Band, Julius Springer, Berlin (1937).

[38] M.G. Crandall, *A generalization of Peano's existence theorem and flow-invariance*, Proc. Amer. Math. Soc. **36** (1972), 151–155.

[39] M.G. Crandall and P.L. Lions, *Viscosity solutions of Hamilton–Jacobi equations*, Trans. Amer. Math. Soc. **277** (1983), 1–42.

[40] N. Dunford and J.T. Schwartz, *Linear Operators*, Part I: *General Theory*, Interscience Publishers, New York (1958).

[41] H. Federer, *Curvature measures*, Trans. Amer. Math. Soc. **93** (1959), 418–491.

[42] S. Feferman, *Independence of the axiom of choice from the axiom of independent choices*, J. Symbolic Logic **29** (1967), 226–226.

[43] A.F. Filippov, *Classical solutions of differential equations with multi-valued right-hand side*, SIAM J. Control **5** (1967), 609–621.

[44] H. Frankowska, *Equations d'Hamilton–Jacobi contingentes*, C. R. Acad. Sci. Paris Sér. I Math. **304** (1987), 295–298.

[45] H. Frankowska, S. Plaskacz and T. Rzeżuchowski, *Measurable viability theorems and Hamilton–Jacobi–Bellman equations*, J. Differential Equations **116** (1995), 265–305.

[46] A. Fryszkowski, *Continuous selections for a class of non-convex multivalued maps*, Studia Math. **76** (1983), 163–174.

[47] M. Fukuhara[6], *Sur les systèmes des équations différentielles ordinaires II*, Japan. J. Math. **6** (1929), 269–299.

[48] S. Gautier, *Equations differentielles multivoques sur un fermé*, Publications de l'Université de Pau (1973).

[49] I.V. Girsanov, *Lekcii po Matematičeskoi Teorii Èkstremal'nyh Zadač*, Moscow University, Moscow (1970), (English translation: *Lecture on Mathematical Theory of Extremum Problems*, Lectures Notes in Econom. and Math. Systems, Vol. 67, Springer-Verlag, New York (1972)).

[50] R. Gonzales, *Sur l'existence d'une solution maximale de l'équation de Hamilton–Jacobi*, C. R. Acad. Sci. Paris Sér. I Math. **282** (1976), 1287–1290.

[51] E. Goursat, *Lecons sur l'Integration des Équation Partielles du Premier Ordre*, Hermann, Paris (1891).

[52] G. Haddad, *Monotone trajectories of differential inclusions and functional differential inclusions with memory*, Israel J. Math. **39** (1981), 83–100.

[53] O. Hájek, *On differentiability of the minimal time function*, Funkcial. Ekvac. **20** (1977), 97–114.

[54] P. Hartman, *Ordinary Differential Equations*, Wiley, New York (1964).

[55] P. Hartman, *On invariant sets and on a theorem of Ważewski*, Proc. Amer. Math. Soc. **32** (1972), 511–520.

[56] E.W. Hobson, *The Theory of Functions of a Real Variable and the Theory of Fourier's Series*, Vol. 1, third edition, Cambridge Univ. Press, Cambridge (1927).

[57] P. Howard and J.E. Rubin, *The Boolean prime ideal theorem plus countable choice do [does] not imply dependent choice*, Math. Logic Quart. **42** (1996), 410–420.

[58] H. Ishii, *A simple, direct proof of uniqueness for solutions of the Hamilton–Jacobi equations of eikonal type*, Proc. Amer. Math. Soc. **100** (1987), 247–251.

---

[6]The old English transliteration of the Japanese name Hukuhara.

[59] E. Kamke, *Differentialgleichungen reeler Funktionen*, Akademische Verlagsgesellschaft, Leipzig (1930).

[60] N. Kenmochi and T. Takahashi, *Nonautonomous differential equations in Banach spaces*, Nonlinear Anal. **4** (1980), 1109–1121.

[61] H. Kneser, Siz.-Ber. d. Preuss Akademie d. Wissensch., Phys.-Math. Klasse (1923), 171–174.

[62] J. Kobayashi, *Difference approximation of Cauchy problems for quasi-dissipative operators and generation of nonlinear semigroups*, J. Math. Soc. Japan **27** (1975), 640–665.

[63] A. Kucia, *Scorza Dragoni type theorems*, Fund. Math. **138** (1991), 197–203.

[64] Yu. Ledyaev, *Criteria for viability of trajectories of nonautonomous inclusions and their applications*, J. Math. Anal. Appl. **182** (1994), 165–188.

[65] V. Lupulescu and M. Necula, *Viability and local invariance for non-convex semilinear differential inclusions*, Nonlinear Funct. Anal. Appl. **9** (2004), 495–512.

[66] R.H. Martin Jr., *Differential equations on closed subsets of a Banach space*, Trans. Amer. Math. Soc. **179** (1973), 399–414.

[67] E. Michael, *Continuous selectors I*, Ann. of Math. **63** (1956), 361–382.

[68] E. Michael, *Continuous selectors II*, Ann. of Math. **64** (1956), 562–580.

[69] B.S. Mordukhovich, *Maximum principle in the problem of time optimal control with nonsmooth constraints*, J. Appl. Math. Mech. **40** (1976), 960–969.

[70] M. Nagumo, *Über die Lage der Integralkurven gewönlicher Differentialgleichungen*, Proc. Phys. Math. Soc. Japan **24** (1942), 551–559.

[71] M. Necula, *Viability of variable domains for nonautonomous semilinear differential equations*, Panamer. Math. J. **11** (2001), 65–80.

[72] M. Necula, *Viability of variable domains for differential equations governed by Carathéodory perturbations of nonlinear m-accretive operators*, An. Ştiinţ. Univ. Al. I. Cuza Iaşi, Secţ. I a Mat. **48** (2002), 41–60.

[73] M. Necula, Personal communication.

[74] M. Necula and I.I. Vrabie, Paper in preparation.

[75] N.H. Pavel, *Invariant sets for a class of semi-linear equations of evolution*, Nonlinear Anal. **1** (1977), 187–196.

[76] N.H. Pavel, *Differential Equations, Flow Invariance and Applications*, Res. Notes Math., Vol. 113, Pitman, Boston (1984).

[77] N.H. Pavel and C. Ursescu, *Flow-invariant sets for second order differential equations and applications in Mechanics*, Nonlinear Anal. **6** (1982), 35–74.

[78] N.H. Pavel and I.I. Vrabie, *Equations d'évolution multivoques dans des espaces de Banach*, C. R. Acad. Sci. Paris Sér. I Math. **287** (1978), 315–317.

[79] N.H. Pavel and I.I. Vrabie, *Semilinear evolution equations with multivalued right-hand side in Banach spaces*, An. Ştiinţ. Univ. Al. I. Cuza Iaşi Secţ. I a Mat. **25** (1979), 137–157.

[80] G. Peano, *Démonstration de l'intégrabilité des équations différentielles ordinaires*, Math. Ann. **37** (1890), 182–228.

[81] O. Perron, *Partieller Differentialgleichungssysteme im reellen Gebiet*, Math. Z. **27** (1928), 549–564.

[82] S. Plaskacz, *On the solution set for differential inclusions*, Boll. Unione Mat. Ital. Sez. A (7) **6** (1992), 387–394.

[83] L.S. Pontryagin, V.G. Boltyanskii, R.V. Gamkrelidze and E.F. Mishchenko, *The Mathematical Theory of Optimal Processes*, GIF-ML, Moscow (1961) (in Russian).

[84] M. Quincampoix, *Differential inclusions and target problems*, SIAM J. Control Optim. **30** (1992), 324–335.

[85] R. Redheffer, *The theorem of Bony and Brezis on flow-invariant sets*, Amer. Math. Monthly **79** (1972), 740–747.

[86] E. Roxin, *Stability in general control systems*, J. Differential Equations **1** (1965), 115–150.

[87] G. Scorza Dragoni, *Un teorema sulle funzioni continue rispetto ad una e misurabili rispetto ad un'altra variabile*, Rend. Sem. Mat. Univ. Padova **17** (1948), 102–106.

[88] G. Scorza Dragoni, *Una applicazione della quasicontinuità semiregolare delle funzioni misurabili rispetto ad una e continue rispetto ad un'altra variabile*, Atti Accad. Naz. Lincei Cl. Sci. Fis. Mat. Natur. Rend. Lincei (8) **12** (1952), 55–61.

[89] F. Severi, *Su alcune questioni di topologia infinitesimale*, Ann. Polon. Soc. Math. **9** (1930), 97–108.

[90] F. Severi, *Sulla differenziabilità totale delle funzioni di più variabili reali*, Ann. Mat. Pura Appl. **13** (1936), 1–35.

[91] S.Z. Shi, *Viability theorems for a class of differential operator inclusions*, J. Differential Equations **79** (1989), 232–257.

[92] P. Tallos, *Viability problems for nonautonomous differential inclusions*, SIAM J. Control Optim. **29** (1991), 253–263.

[93] J.S. Treiman, *Characterization of Clarke's tangent and normal cones in finite and infinite dimensions*, Nonlinear Anal. T.M.A. **7** (1983), 771–783.

[94] C. Ursescu, *Sur une généralisation de la notion de différentiabilité*, Atti Accad. Naz. Lincei Cl. Sci. Fis. Mat. Natur. Rend. Lincei (7) **54** (1973), 199–204.

[95] C. Ursescu, *Tangent sets and differentiable functions*, Math. Control Theory Proc. Conf. Zakopane, 1974, Banach Center Publications (1976), 151–155.

[96] C. Ursescu, *Probleme de tangenţă cu aplicaţii în teoria optimizării*, Internal Report, Grant 11996/25.09.1980, Iaşi (1981).

[97] C. Ursescu, *Tangent sets' calculus and necessary conditions for extremality*, SIAM J. Control Optim. **20** (1982), 563–574.

[98] C. Ursescu, *Carathéodory solutions of ordinary differential equations on locally closed sets in finite dimensional spaces*, Math. Japon. **31** (1986), 483–491.

[99] C. Ursescu, C_0-*semigroups, characteristics method, and generic differential equations*, An. Univ. Timişoara Ser. Mat.-Inform. **34** (1996), 283–290.

[100] C. Ursescu, *Evolutors and invariance*, Set-Valued Anal. **11** (2003), 69–89.

[101] C. Ursescu, Personal communication.

[102] I.I. Vrabie, *Compactness Methods for Nonlinear Evolutions*, second edition, Pitman Monographs Surveys Pure Appl. Math., Vol. 75, Longman, Harlow (1995).

[103] I.I. Vrabie, *Compactness methods and flow-invariance for perturbed nonlinear semigroups*, An. Ştiinţ. Univ. Al. I. Cuza Iaşi Secţ. I a Mat. **27** (1981), 117–125.

[104] I.I. Vrabie, *Periodic solutions for nonlinear evolution equations in a Banach space*, Proc. Amer. Math. Soc. **109** (1990), 653–661.

[105] I.I. Vrabie, C_0-*Semigroups and Applications*, North-Holland Math. Stud., Vol. 191 (2003).

[106] I.I. Vrabie, *Differential Equations. An Introduction to Basic Results, Concepts and Applications*, World Scientific, New Jersey (2004).

[107] T. Ważewski, *Sur un condition équivalent à l'équation au contingent*, Bull. Polon. Acad. Sci. Ser. Sci. Math. Astr. Phys. **9** (1961), 865–867.

[108] J.A. Yorke, *Invariance for ordinary differential equations*, Math. Systems Theory **1** (1967), 353–372.

[109] J.A. Yorke, *Invariance for ordinary differential equations: Correction*, Math. Systems Theory **2** (1968), 381.

[110] T. Yoshizawa, *Stability theory by Liapunov's second method*, The Mathematical Society of Japan (1966).

[111] S.C. Zaremba, *Sur les équations au paratingent*, Bull. Sci. Math. (2) **60** (1936), 139–160.

CHAPTER 4

Monotone Dynamical Systems

M.W. Hirsch[*]

Department of Mathematics, University of California, Berkeley, CA 94720-3840, USA

Hal Smith[†]

Department of Mathematics, Arizona State University, Tempe, AZ 85287-1804, USA

Contents

0. Introduction 241
1. Strongly order-preserving semiflows 243
   1.1. Definitions and basic results 243
   1.2. Nonordering of omega limit sets 249
   1.3. Local semiflows 251
   1.4. The limit set dichotomy 252
   1.5. Q is plentiful 255
   1.6. Stability in normally ordered spaces 258
   1.7. Stable equilibria in strongly ordered Banach spaces 261
   1.8. The search for stable equilibria 263
2. Generic convergence and stability 265
   2.1. The sequential limit set trichotomy 265
   2.2. Generic quasiconvergence and stability 271
   2.3. Improving the limit set dichotomy for smooth systems 273
   2.4. Generic convergence and stability 279
3. Ordinary differential equations 281
   3.1. The quasimonotone condition 282
   3.2. Strong monotonicity with linear systems 286
   3.3. Autonomous K-competitive and K-cooperative systems 289
   3.4. Dynamics of cooperative and competitive systems 291
   3.5. Smale's construction 294
   3.6. Invariant surfaces and the carrying simplex 295
   3.7. Systems in $\mathbb{R}^2$ 296
   3.8. Systems in $\mathbb{R}^3$ 297

[*]Supported by NSF Grant DMS 9700910.
[†]Supported by NSF Grant DMS 0107160.

HANDBOOK OF DIFFERENTIAL EQUATIONS
Ordinary Differential Equations, volume 2
Edited by A. Cañada, P. Drábek and A. Fonda

239

4. Delay differential equations 302
    4.1. The semiflow 302
    4.2. The quasimonotone condition 304
    4.3. Eventual strong monotonicity 308
    4.4. K is an orthant 310
    4.5. Generic convergence for delay differential equations 312
5. Monotone maps 313
    5.1. Background and motivating examples 313
    5.2. Definitions and basic results 316
    5.3. The order interval trichotomy 320
    5.4. Sublinearity and the cone limit set trichotomy 323
    5.5. Smooth strongly monotone maps 327
    5.6. Monotone planar maps 329
6. Semilinear parabolic equations 332
    6.1. Solution processes for abstract ODEs 332
    6.2. Semilinear parabolic equations 338
    6.3. Parabolic systems with monotone dynamics 347
References 348

0. Introduction

This chapter surveys a restricted but useful class of dynamical systems, namely, those enjoying a comparison principle with respect to a closed order relation on the state space. Such systems, variously called monotone, order-preserving or increasing, occur in many biological, chemical, physical and economic models.

The following notation will be used. $\mathbb{Z}$ denotes the set of integers; $\mathbb{N} = \{0, 1, \ldots\}$, the set of natural numbers; $\mathbb{N}_+$ is the set of positive integers, and $\mathbb{R}$ is the set of real numbers. For $u, v \in \mathbb{R}^n$ ($=$ Euclidean n-space), we write

$$
\begin{aligned}
u \leqslant v &\iff u_i \leqslant v_i, \\
u < v &\iff u_i \leqslant v_i, \quad u \neq v, \\
u \ll v &\iff u_i < v_i,
\end{aligned}
$$

where $i = 1, \ldots, n$. This relation $\leqslant$ is called the *vector order* in $\mathbb{R}^n$.

The prototypical example of monotone dynamics is a Kolmogorov model of cooperating species,

$$
\dot{x}_i = x_i G_i(x), \quad x_i \geqslant 0, \quad i = 1, \ldots, n \tag{0.1}
$$

in the positive orthant $\mathbb{R}^n_+ = [0, \infty)^n$, where $G : \mathbb{R}^n_+ \to \mathbb{R}^n$ is continuously differentiable. x_i denotes the population and G_i the *per capita* growth rate of species i. Cooperation means that an increase in any population causes an increase of the growth rates of all the other populations, modeled by the assumption that $\partial G_i / \partial x_j \geqslant 0$ for $i \neq j$. The right-hand side $F_i = x_i G_i$ of (0.1) then defines a *cooperative vector field* $F : \mathbb{R}^n \to \mathbb{R}^n$, meaning that $\partial F_i / \partial x_j \geqslant 0$ for $i \neq j$.

Assume for simplicity that solutions to Eq. (0.1) are defined for all $t \geqslant 0$. Let $\Phi = \{\Phi_t : \mathbb{R}^n_+ \to \mathbb{R}^n_+\}_{t \geqslant 0}$ denote the resulting semiflow in $\mathbb{R}^n_+$ that describes the evolution of states in positive time: the solution with initial value u is given by $x(t) = \Phi_t(u)$. The key to the long-term dynamics of cooperative vector fields is an important differential inequality due to Müller [148] and Kamke [91].

$$
u \leqslant v \text{ and } t \geqslant 0 \implies \Phi_t(u) \leqslant \Phi_t(v).
$$

In other words: *The maps Φ_t preserve the vector order.* A semiflow Φ with this property is called *monotone*. Monotone semiflows and their discrete-time counterparts, order-preserving maps, form the subject of Monotone Dynamics.

Returning to the biological setting, we may make the assumption that each species directly or indirectly affect all the others. This is modeled by the condition that the Jacobian matrices $G'(x)$ are irreducible. An extension of the Müller–Kamke theorem shows that in the *open orthant* $\operatorname{Int} \mathbb{R}^n$, the restriction of Φ is *strongly monotone*: If $u, v \in \operatorname{Int} \mathbb{R}^n$, then

$$
u < v \text{ and } t > 0 \implies \Phi_t(u) \ll \Phi_t(v).
$$

A semiflow with this property is *strongly monotone*.

Similar order-preserving properties are found in other dynamical settings, including delay differential equations and quasilinear parabolic partial differential equations. Typically the state space is a subset of a (real) Banach space Y with a distinguished closed cone $Y_+ \subset Y$. An order relation is introduced by $x \geqslant y \Leftrightarrow x - y \in Y_+$. When Y is a space of real valued functions on some domain, Y_+ is usually (but not always) the cone of functions with values in $\mathbb{R}_+ := [0, \infty)$. When $Y = \mathbb{R}^n$, the cooperative systems defined above use the cone $\mathbb{R}_+^n$.

Equations (0.1) model an ecology of competing species if $\partial G_i / \partial x_j \leqslant 0$ for $i \neq j$. The resulting vector field K with components $K_i = x_i G_i$ is not generally cooperative, but its negative $F = -K$ *is* cooperative. Many dynamical properties of the semiflow of K can be deduced from that of F, which is monotone.

We will see that the long-term behavior of monotone systems is severely limited. Typical conclusions, valid under mild restrictions, include the following:

- If all forward trajectories are bounded, the forward trajectory of almost every initial state converges to an equilibrium.
- There are no attracting periodic orbits other than equilibria, because every attractor contains a stable equilibrium.
- In $\mathbb{R}^3$, every compact limit set that contains no equilibrium is a periodic orbit that bounds an invariant disk containing an equilibrium.
- In $\mathbb{R}^2$, each component of any solution is eventually increasing or decreasing.

Other cones in $\mathbb{R}^n$ are also used, especially the orthants defined by restricting the sign of each coordinate. For example, a system of two competing species can be modeled by ODEs

$$\dot{y}_i = y_i H_i(y); \quad y_i \geqslant 0, \quad i = 1, 2$$

with $\partial H_i / \partial y_j < 0$ for $i \neq j$. The coordinate change $x_1 = y_1$, $x_2 = -y_2$ converts this into a cooperative system in the second orthant K defined by $x_1 \geqslant 0 \geqslant x_2$. This system is thus both competitive and cooperative, albeit for different cones. Not surprisingly, the dynamics are very simple.

In view of such powerful properties of cooperative vector fields, it would be useful to know when a given field F in an open set $D \subset \mathbb{R}^n$ can be made cooperative or competitive by changing coordinates. The following sufficient condition appears to be due to DeAngelis et al. [39]; see also Smith [193], Hirsch [74]. Assume the Jacobian matrices $[a_{ij}(x)] = F'(x)$ have the following two properties:

(1) (Sign stability) If $i \neq j$ then a_{ij} does not change sign in D;
(2) (Sign symmetry) $a_{ij} a_{ji} \geqslant 0$ in D.

Let Γ be the combinatorial labeled graph with nodes $1, \ldots, n$ and an edge e_{ij} joining i and j labeled $\sigma_{ij} \in \{+, -\}$ if and only if $i \neq j$ and there exists $p \in D$ such that $\operatorname{sgn} a_{ij}(p) = \sigma_{ij} \neq 0$. Then F is cooperative (respectively, competitive) relative to some orthant if and only if in every closed loop in Γ the number of negative labels is even (respectively, odd).

Order-preserving dynamics also occur in discrete time systems. Consider a nonautonomous Kolmogorov system $\dot{x}_i = x_i H_i(t, x)$, where the map $H := (H_1, \ldots, H_n) : \mathbb{R} \times \mathbb{R}^n \to \mathbb{R}^n$ has period $\tau > 0$ in t. Denote by $T : \mathbb{R}_+^n \to \mathbb{R}_+^n$ the Poincaré map, which to $x \in \mathbb{R}^n_+$ assigns $y(\tau)$ where $y(t)$ denotes the solution with initial value x. Then T is monotone provided the $\partial H_i / \partial x_j \geqslant 0$ for $i \neq j$, and strongly monotone in the open orthant when these matrices are also irreducible. Most of the results stated above have analogs for T.

Convergence and stability properties of several kinds of order-preserving semiflows are developed in Sections 1 and 2, in the setting of general ordered metric spaces. Section 3 treats ODEs whose flows preserve the order defined by a cone in $\mathbb{R}^n$. Delay differential equations are studied in Section 4. In Section 5 we present results on order-preserving maps. The final section applies the preceding results to second order quasilinear parabolic equations.

1. Strongly order-preserving semiflows

This section introduces the basic definitions and develops the main tools of monotone dynamics. Several results on density of quasiconvergent points are proved, and used to establish existence of stable equilibria.

1.1. *Definitions and basic results*

The setting is a semiflow $\Phi = \{\Phi_t\}_{0 \leqslant t < \infty}$ in a (partially) ordered metric space that preserves the weak order relation: $x \leqslant y$ implies $\Phi_t(x) \leqslant \Phi_t(y)$. Such semiflows, called *monotone*, have severely restricted dynamics; for example, in $\mathbb{R}^n$ with the vector ordering there cannot be stable periodic orbits other than equilibria. But for generic convergence theorems we need semiflows with the stronger property of being "strongly order preserving," together with mild compactness assumptions. In later sections we will see that these conditions are frequently encountered in applications. The centerpiece of this section is the Limit Set Dichotomy, a fundamental tool for the later theory.

1.1.1. *Ordered spaces*　Let Z be a metric space and $A, B \subset Z$ subsets. The closure of A is denoted by $\overline{A}$ and its interior by $\mathrm{Int}\, A$. The distance from A to B is defined as $\mathrm{dist}(A, B) :=$ $\inf_{a \in A, b \in B} d(a, b)$. When B is a singleton $\{b\}$ we may write this as $\mathrm{dist}(A, b) = \mathrm{dist}(b, A)$.

X always denotes an *ordered space*. This means X is endowed with a metric d and an order relation $\mathsf{R} \subset X \times X$. As usual we write $x \leqslant y$ to mean $(x, y) \in \mathsf{R}$, and the order relation is:

　(i)　reflexive: $x \leqslant x$ for all $x \in X$,

　(ii)　transitive: $x \leqslant y$ and $y \leqslant z$ implies $x \leqslant z$,

　(iii)　antisymmetric: $x \leqslant y$ and $y \leqslant x$ implies $x = y$.

In addition, the ordering is compatible with the topology in the following sense:

　(iv)　if $x_n \to x$ and $y_n \to y$ as $n \to \infty$ and $x_n \leqslant y_n$, then $x \leqslant y$.

This is just to say that R is a closed subset of $X \times X$.

We write $x < y$ if $x \leqslant y$ and $x \neq y$. Given two subsets A and B of X, we write $A \leqslant B$ $(A < B)$ when $x \leqslant y$ $(x < y)$ holds for each choice of $x \in A$ and $y \in B$. The relation $A \leqslant B$ *does not imply* "$A < B$ or $A = B$"!

The notation $x \ll y$ means that there are open neighborhoods U, V of x, y respectively such that $U \leqslant V$. Equivalently, (x, y) belongs to the interior of R. The relation $\ll$, sometimes referred to as the *strong ordering*, is transitive; in many cases it is empty. We write $x \geqslant y$ to mean $y \leqslant x$, and similarly for $>$ and $\gg$.

We call X an *ordered subspace* of an ordered space X' if $X \subset X'$, and the order and topology on X are inherited from X'. When this is this case, the relation $u < v$ for points $u, v \in X$ means the same thing whether u and v are considered as points of X, or points of X'. But there are simple examples for which $u \ll v$ is true in X', yet false in X.

Let X be an ordered space. The *lower boundary* of a set $U \subset X$ is the set of points x in the boundary of U such that every neighborhood of x contains a point $y \in U$ with $y > x$. The *upper boundary* of U is defined dually.

Two points $x, y \in X$ are *order related* if $x < y$ or $y < x$; otherwise they are *unrelated*. A subset of X is *unordered* if it does not contain order related points. The empty set and singletons are unordered.

The (closed) *order interval* determined by $u, v \in X$ is the closed set

$$[u, v] = [u, v]_X := \{x \in X : u \leqslant x \leqslant v\}$$

which may be empty. The *open order interval* is the open set

$$[[u, v]] = \{x \in X : u \ll x \ll v\}.$$

A subset of X is *order bounded* if it lies in an order interval, and *order convex* if it contains $[u, v]$ whenever it contains u and v.

A point $x \in X$ is *accessible from below* if there is a sequence $x_n \to x$ with $x_n < x$; such a sequence is said to approximate x from below. We define *accessible from above* dually, that is, by replacing $<$ with $>$. In most applications there is a dense open subset of points that are accessible from both above and below.

The *supremum* $\sup S$ of a subset $S \subset X$, if it exists, is the unique point a such that $a \geqslant S$ and $x \geqslant S \Rightarrow x \geqslant a$. The *infimum* $\inf S$ is defined dually, i.e., substituting $\leqslant$ for $\geqslant$. A *maximal element* of S is a point $a \in S$ such that $x \in S$ and $x \geqslant a$ implies $x = a$. A *minimal element* is defined dually.

The following basic facts are well known:

LEMMA 1.1. *Assume the ordered space X is compact.*
   (i) *Every sequence in X that is increasing or decreasing converges.*
   (ii) *If X is totally ordered, it contains a supremum and an infimum.*
   (iii) *X contains a maximal element and a minimal element.*

PROOF. (i) If p and q denote subsequential limits, then $p \leqslant q$ and $q \leqslant p$, hence $p = q$.

(ii) For each $x \in X$, the set $B_x := \{y \in X : y \geqslant x\}$ is compact, and every finite family of such sets has nonempty intersection because X is totally ordered. Therefore there exists $a \in \bigcap_x B_x$, and clearly $a = \sup X$. Similarly, $\inf X$ exists.

(iii) Apply (ii) to a maximal totally ordered subset (using Zorn's lemma). $\qquad\square$

An *ordered Banach space* is an ordered space whose underlying metric space is a Banach space Y, and such that the set $Y_+ = \{y \in Y: y \geqslant 0\}$ is a cone, necessarily closed and convex. Thus Y_+ is a closed subset of Y with the properties:

$$\mathbb{R}_+ \cdot Y_+ \subset Y_+, \qquad Y_+ + Y_+ \subset Y_+, \qquad Y_+ \cap (-Y_+) = \{0\}.$$

We always assume $Y_+ \neq \{0\}$.

When $\operatorname{Int} Y_+$ is nonempty we call Y a *strongly ordered* Banach space. In this case $x \ll y \Leftrightarrow y - x \in \operatorname{Int} Y_+$.

The most important examples of ordered Banach spaces are completions of normed vector spaces of real-valued functions on some set Ω, with the positive cone corresponding to nonnegative functions. This cone defines the *functional ordering*. The simplest case is obtained from $\Omega = \{1, 2, \ldots, n\}$: here $Y = \mathbb{R}^n$ and $Y_+ = \mathbb{R}^n_+$, the *standard cone* comprising vectors with all components nonnegative. For the corresponding *vector ordering*, $x \leqslant y$ means that $x_i \leqslant y_i$ for all i. Other function spaces are used in Sections 4 and 6.

When Y is an ordered Banach space, the notation $X \subset Y$ tacitly assumes that X is an ordered subspace of Y (but not necessarily a linear subspace).

A subset S of an ordered Banach space is *p-convex* if it contains the line segment spanned by u, v whenever $u, v \in S$ and $u < v$.

1.1.2. *Semiflows* All maps are assumed to be continuous unless the contrary is indicated.

A *semiflow* on X is a map $\Psi : \mathbb{R}_+ \times X \to X, (t, x) \mapsto \Psi_t(x)$ such that:

$$\Psi_0(x) = x, \qquad \Psi_t\big(\Psi_s(x)\big) = \Psi_{t+s}(x) \quad (t, s \geqslant 0, \ x \in X).$$

Thus Ψ can be viewed as a collection of maps $\{\Psi_t\}_{t \in \mathbb{R}_+}$ such that Ψ_0 is the identity map of X and $\Psi_t \circ \Psi_s = \Psi_{t+s}$, and such that $\Psi_t(x)$ is continuous in (t, x).

A *flow* in a space M is a continuous map $\Psi : \mathbb{R} \times M \to M$, written $\Psi(t, x) = \Psi_t(x)$, such that

$$\Psi_0(x) = x, \qquad \Psi_t\big(\Psi_s(x)\big) = \Psi_{t+s}(x) \quad (t, s \in \mathbb{R}, x \in X).$$

Restricting a flow to $\mathbb{R}_+ \times M$ gives a semiflow. A C^1 vector field F on a compact manifold M, tangent to the boundary, generates a *solution flow*, for which the trajectory of x is the solution $u(t)$ to the initial value problem $du/dt = F(u), u(0) = x$.

The *trajectory* of x is the map $[0, \infty) \to X, t \mapsto \Psi_t(x)$; the image of the trajectory is the *orbit* $O(x, \Psi)$, denoted by $O(x)$ when Ψ is understood. When $O(x) = \{x\}$ then x is an *equilibrium*. The set of equilibria is denoted by E.

x and its orbit are called T-*periodic* if $T > 0$ and $\Psi_T(x) = x$; such a T is a *period* of x. In this case $\Psi_{t+T}(x) = \Psi_t(x)$ for all $t \geqslant 0$, so $O(x) = \Psi([0, T] \times \{x\})$. A periodic point is *nontrivial* if it is not an equilibrium.

A set $A \subset X$ is *positively invariant* if $\Psi_t A \subset A$ for all $t \geqslant 0$. It is *invariant* if $\Psi_t A = A$ for all $t \geqslant 0$. Orbits are positively invariant and periodic orbits are invariant.

A set K is said to *attract* a set S if for every neighborhood U of K there exists $t_0 \geqslant 0$ such that $t > t_0 \Rightarrow \Psi_t(S) \subset U$; when $S = \{x\}$ we say K attracts x. An *attractor* is a nonempty invariant set L that attracts a neighborhood of itself. The union of all such neighborhoods is the *basin* of L. If the basin of an attractor L is all of X then L is a *global attractor*.

The *omega limit set* of $x \in X$ is

$$\omega(x) = \omega(x, \Psi) := \bigcap_{t \geqslant 0} \overline{\bigcup_{s \geqslant t} \Psi_s(x)}.$$

This set is closed and positively invariant. When $\overline{O(x)}$ is compact, $\omega(x)$ is nonempty, compact, invariant and connected and it attracts $\overline{O(x)}$ (see, e.g., Saperstone [175]).

A point $x \in X$ is *quasiconvergent* if $\omega(x) \subset E$; the set of quasiconvergent points is denoted by Q. We call x *convergent* when $\omega(x)$ is singleton $\{p\}$; in this case $\Phi_t(x) \to p \in E$. We sometimes signal this by the abuse of notation $\omega(x) \in E$. The set of convergent points is denoted by C.

When all orbit closures are compact and E is totally disconnected (e.g., countable), then $Q = C$; because in this case every omega limit set, being a connected subset of E, is a singleton. For systems of ordinary differential equations generated by smooth vector fields, the Kupka–Smale theorem gives generic conditions implying that E is discrete (see Peixoto [157]); but in concrete cases it is often difficult to verify these conditions.

1.1.3. *Monotone semiflows*   A map $f : X_1 \to X_2$ between ordered spaces is *monotone* if

$$x \leqslant y \quad \Longrightarrow \quad f(x) \leqslant f(y),$$

strictly monotone if

$$x < y \quad \Longrightarrow \quad f(x) < f(y),$$

and *strongly monotone* if

$$x < y \quad \Longrightarrow \quad f(x) \ll f(y).$$

Let Φ denote a semiflow in the ordered space X. We call Φ monotone or strictly monotone according as each map Φ_t has the corresponding property.

We call Φ *strongly order-preserving*, SOP for short, if it is monotone and whenever $x < y$ there exist open subsets U, V of x, y respectively, and $t_0 \geqslant 0$, such that

$$\Phi_{t_0}(U) \leqslant \Phi_{t_0}(V).$$

Monotonicity of Φ then implies that $\Phi_t(U) \leqslant \Phi_t(V)$ for all $t \geqslant t_0$.

We call Φ *strongly monotone* if

$$x < y, \ 0 < t \quad \Longrightarrow \quad \Phi_t(x) \ll \Phi_t(y)$$

and *eventually strongly monotone* if it is monotone and whenever $x < y$ there exists $t_0 > 0$ such that

$$t \geq t_0 \implies \Phi_t(x) \ll \Phi_t(y).$$

This property obviously holds when Φ is strongly monotone. We shall see in Section 6 that many parabolic equations generate SOP semiflows in function spaces that are not strongly ordered and therefore do not support strongly monotone semiflows.

Strong monotonicity was introduced in Hirsch [68,69], while SOP was proposed later by Matano [133,134] and modified slightly by Smith and Thieme [197,199]. We briefly explore the relation between these two concepts.

PROPOSITION 1.2. *If Φ is eventually strongly monotone, it is SOP. If X is an open subset of a Banach space Y ordered by a cone Y_+, Φ is SOP and the maps $\Phi_t : X \to X$ are open, then Φ is eventually strongly monotone. In particular, Φ is eventually strongly monotone provided Y is finite-dimensional, Φ is SOP and the maps Φ_t are injective.*

PROOF. If $x < y$ and Φ is eventually strongly monotone, then there exists $t_0 > 0$ such that $\Phi_{t_0}(x) \ll \Phi_{t_0}(y)$. Take neighborhoods $\widetilde{U}$ of $\Phi_{t_0}(x)$ and $\widetilde{V}$ of $\Phi_{t_0}(y)$ such that $\widetilde{U} < \widetilde{V}$. By continuity of Φ_{t_0}, there are neighborhoods U of x and V of y such that $\Phi_{t_0}(U) \subset \widetilde{U}$ and $\Phi_{t_0}(V) \subset \widetilde{V}$. Therefore, $\Phi_{t_0}(U) < \Phi_{t_0}(V)$ so Φ is SOP.

Suppose that $X \subset Y$ is open and ordered by Y_+ and Φ is SOP. If $x < y$ and U, V are open neighborhoods as in the definition of SOP, the inequality $\Phi_t(U) \leq \Phi_t(V)$ together with the fact that $\Phi_t(U)$ and $\Phi_t(V)$ are open in Y imply that $\Phi_t(x) \ll \Phi_t(y)$. $\qquad\square$

The following very useful result shows that the defining property of SOP semiflows, concerning points $x < y$, extends to a similar property for compact sets $K < L$:

LEMMA 1.3. *Assume Φ is SOP and K, L are compact subsets of X satisfying $K < L$. Then there exists real numbers $t_1 \geq 0$, $\epsilon > 0$ and neighborhoods U, V of K, L respectively such that*

$$t \geq t_1 \quad and \quad 0 \leq s \leq \epsilon \implies \Phi_{t+s}(U) \leq \Phi_t(V).$$

PROOF. Let $x \in K$. For each $y \in L$ there exist $t_y \geq 0$, a neighborhood U_y of x, and a neighborhood V_y of y such that $\Phi_t(U_y) \leq \Phi_t(V_y)$ for $t \geq t_y$ since Φ is strongly order preserving. $\{V_y\}_{y \in L}$ is an open cover of L, so we may choose a finite subcover: $L \subset \bigcup_{i=1}^{n} V_{y_i} := \widetilde{V}$ where $y_i \in L$, $1 \leq i \leq n$. Let $\widetilde{U}_x = \bigcap_{i=1}^{n} U_{y_i}$, which is a neighborhood of x, and let $\tilde{t}_x = \max_{1 \leq i \leq n} t_{y_i}$. Then $\Phi_t(\widetilde{U}_x) \subset \Phi_t(U_{y_i}) \leq \Phi_t(V_{y_i})$, so $\Phi_t(\widetilde{U}_x) \leq \Phi_t(V_{y_i})$ for $t \geq \tilde{t}_x$. It follows that

$$t \geq \tilde{t}_x \implies \Phi_t(\widetilde{U}_x) \leq \Phi_t(\widetilde{V}).$$

Extract a finite subcover $\{\widetilde{U}_{x_j}\}$ of K from the family $\{\widetilde{U}_x\}$. Setting $U := \bigcup_j \widetilde{U}_{x_j} \supset K$ and $t_1 := \max_{1 \leqslant j \leqslant m} \tilde{t}_{x_j}$, we have

$$t \geqslant t_1 \quad \Longrightarrow \quad \Phi_t(U) = \bigcup_j \Phi_t(\widetilde{U}_j) \leqslant \Phi_t(V).$$

In order to obtain the stronger conclusion of the lemma, note that for each $z \in K$ there exists $\epsilon_z > 0$ and a neighborhood U'_z of z such that $\Phi([0, \epsilon_z) \times W_z) \subset U$. Choose $z_1, \ldots, z_m$ in K so that $K \subset \bigcup_j U'_{z_j}$. Define $U' = \bigcup_j U'_{z_j}$ and $\epsilon = \min_j \{\epsilon_{z_j}\}$. If $x \in U'$ and $0 \leqslant s < \epsilon$ then $x \in U'_{z_j}$ for some j so $\Phi_s(x) \in U$. Thus $\Phi([0, \epsilon) \times U') \subset U$ so $\Phi_s(U') \subset U$. It follows that $\Phi_{t+s}(U') \subset \Phi_t(U) \leqslant \Phi_t(V)$ for $t \geqslant t_1, 0 \leqslant s < \epsilon$. $\qquad\square$

Several fundamental results in the theory of monotone dynamical systems are based on the following sufficient conditions for a solution to converge to equilibrium.

THEOREM 1.4 (Convergence Criterion). *Assume Φ is monotone, $x \in X$ has compact orbit closure, and $T > 0$ is such that $\Phi_T(x) \geqslant x$. Then $\omega(x)$ is an orbit of period T. Moreover, x is convergent if the set of such T is open and nonempty or Φ is SOP and $\Phi_T(x) > x$.*

PROOF. Monotonicity implies that $\Phi_{(n+1)T}(x) \geqslant \Phi_{nT}(x)$ for $n = 1, 2, \ldots$ and therefore $\Phi_{nT}(x) \to p$ as $n \to \infty$ for some p by the compactness of the orbit closure. By continuity,

$$\Phi_{t+T}(p) = \Phi_{t+T}\left(\lim_{n \to \infty} \Phi_{nT}(x)\right)$$
$$= \lim_{n \to \infty} \Phi_{(n+1)T+t}(x)$$
$$= \lim_{n \to \infty} \Phi_t\left(\Phi_{(n+1)T}(x)\right)$$
$$= \Phi_t(p)$$

for all $t \geqslant 0$. Hence p is T-periodic.

To prove $\omega(x) = O(p)$, suppose $t_j \to \infty$ and $\Phi_{t_j}(x) \to q \in \omega(x)$ as $j \to \infty$, and write $t_j = n_j T + r_j$ where n_j is a natural number and $0 \leqslant r_j < T$. By passing to a subsequence if necessary, we may assume that $r_j \to r \in [0, T]$. Taking limits as $j \to \infty$ and noting that $n_j \to \infty$, we have by continuity:

$$\lim \Phi_{t_j}(x) = \lim \Phi_{r_j}\left(\lim \Phi_{n_j T}(x)\right) = \lim \Phi_{r_j}(p) = \Phi_r(p) = q.$$

Therefore $\omega(x) \subset O(p)$, and the opposite inclusion holds because $p \in \omega(x)$. This proves the first assertion of the theorem.

Suppose $\Phi_t(x) \geqslant x$ for all t in a nonempty open interval $(T - \epsilon, T + \epsilon)$. The first assertion shows that $\omega(x)$ is an orbit $O(p)$ of period τ for every $\tau \in (T - \epsilon, T + \epsilon)$. All elements of $O(p)$ have the same set G of periods; G is closed under addition and contains $(T - \epsilon, T + \epsilon)$. If $0 \leqslant s < \epsilon$ and $t \geqslant 0$ then

$$\Phi_{t+s}(p) = \Phi_t\left(\Phi_s(p)\right) = \Phi_t\left(\Phi_{s+T}(p)\right) = \Phi_t(p).$$

Hence $[0, \epsilon) \subset G$ and therefore $G = \mathbb{R}_+$, which implies $p \in E$. This proves the second assertion.

If $\Phi_T(x) > x$ and Φ is SOP then there exist neighborhoods U of x and V of $\Phi_T(x)$ and $t_0 > 0$ such that $\Phi_{t_0}(U) \leqslant \Phi_{t_0}(V)$. It follows that $\Phi_{t_0}(x) \leqslant \Phi_{t_0+T+\epsilon}(x)$ for all ϵ sufficiently small. The previous assertion implies $\omega(x) = p \in E$. $\qquad\square$

1.2. *Nonordering of omega limit sets*

The next result is the first of several describing the order geometry of limit sets.

PROPOSITION 1.5 (Nonordering of Periodic Orbits). *A periodic orbit of a monotone semiflow is unordered.*

PROOF. Let x have minimal period $s > 0$ under a monotone semiflow Φ. Suppose $x \leqslant z \in O(x)$. By compactness of $O(x)$ there is a maximal $y \in O(x)$ such that $y \geqslant z \geqslant x$. By periodicity and monotonicity $y = \Phi_t(x) \leqslant \Phi_t(y)$, $t > 0$, hence $y = \Phi_t(y)$ by maximality. Therefore t is an integer multiple of s, so $x = \Phi_t(x) = y$, implying $x = z$. $\qquad\square$

The following result, which implies (1.5), is a broad generalization of the obvious fact that for ODEs in $\mathbb{R}$, nonconstant solutions are everywhere increasing or everywhere decreasing. Let $J \subset \mathbb{R}$ be an interval and $f : J \to X$ a map. A compact subinterval $[a, b] \subset J$ is *rising* for f provided $f(a) < f(b)$, and *falling* if $f(b) < f(a)$.

THEOREM 1.6. *A trajectory of a monotone semiflow cannot have both a rising interval and a falling interval.*

This originated in Hirsch [67], with improvements in Smith [194], Smith and Waltman [203]. An analog for maps is given in Theorem 5.4.

PROOF. Let Φ be a monotone semiflow in X and fix a trajectory $f : [0, \infty) \to X$, $f(t) := \Phi_t(x)$. Call an interval $[d, d']$ *weakly falling* if $f(d) \geqslant f(d')$. Monotonicity shows that when this holds, the *right translates* of $[d, d']$—the intervals $[d + u, d' + u]$ with $u \geqslant 0$—are also weakly falling.

Proceeding by contradiction, we assume f has a falling interval $[a, a + r]$ and a rising interval $[c, c + q]$. To fix ideas we assume $a \leqslant c$, the case $c \leqslant a$ being similar. Define

$$b := \sup\{t \in [c, c+q]: \ f(t) \leqslant f(c), \ s := c+q-b\}.$$

Then $[b, b + s]$ is a rising interval in $[c, c + q]$, and

$$b < t \leqslant b+s \quad \Longrightarrow \quad f(t) \not\leqslant f(b). \tag{1.1}$$

Claim 1: *No interval $[b - l, b]$ is weakly falling.* Assume the contrary. Then (i) $l > s$, and (ii) $[b - (l - s), b]$ is weakly falling. To see (i), observe that $f(b + l) \leqslant f(b)$ because

$[b, b+l]$ is a right translate of $[b-l, b]$; hence $l \leqslant s$ would entail $b < b+l \leqslant b+s$, contradicting (1.1) with $t = b+l$. To prove (ii), note that right translation of $[b-l, b]$ shows that $[b-l+s, b+s]$ is weakly falling, implying $f(b-(l-s)) \geqslant f(b+s) > f(b)$; hence $[b-(l-s), b]$ is falling. Repetition of this argument with l replaced by $l-s, l-2s, \dots$ leads by induction on n to the absurdity that $l - ns > s$ for all $n \in \mathbb{N}$.

Claim 2: $r > s$. For $f(b+r) \leqslant f(b)$ because $[b, b+r]$ is falling, as it is a right translate of $[a, a+r]$. Therefore $r > s$, for otherwise $b < b+r \leqslant b+s$ and (1.1) leads to a contradiction.

As $b+s \geqslant a+r$, we can translate $[a, a+r]$ to the right by $(b+s)-(a+r)$, obtaining the weakly falling interval $[b+s-r, b+s]$. Note that $b+s-r < b$ by Claim 2. From $f(b+s-r) \geqslant f(b+s) > f(b)]$ we conclude that $[b-(r-s), b]$ is falling. But this contradicts Claim 1 with $l = r-s$. $\qquad\square$

LEMMA 1.7. *An omega limit set for a monotone semiflow Φ cannot contain distinct points x, y having respective neighborhoods U, V such that $\Phi_r U \leqslant \Phi_r V$ for some $r \geqslant 0$.*

PROOF. We proceed by contradiction. Suppose there exist distinct points $x, y \in \omega(z)$ having respective neighborhoods U, V such that $\Phi_r U \leqslant \Phi_r V$ for some $r \geqslant 0$. Then $\omega(z)$ is not a periodic orbit, for otherwise from $\Phi_r(x) \leqslant \Phi_r(y)$ we infer $x \leqslant y$ and hence $x < y$, violating Nonordering of Periodic Orbits.

There exist real numbers $a < b < c$ be such that $\Phi_a(z) \in U$, $\Phi_b(z) \in V$, $\Phi_c(z) \in U$. Therefore the properties of r, U and V imply

$$\Phi_{a+r}(z) \leqslant \Phi_{b+r}(z), \qquad \Phi_{b+r}(z) \geqslant \Phi_{c+r}(z).$$

As $\omega(z)$ is not periodic, the semiflow is injective on the orbit of z; hence the order relations above are strict. But this contradicts Theorem 1.6. $\qquad\square$

It seems to be unknown whether omega limit sets of monotone semiflows must be unordered. This holds for SOP semiflows by the following theorem due to Smith and Thieme [197, Proposition 2.2]; the strongly monotone case goes back to Hirsch [66]. This result is fundamental to the theory of monotone semiflows:

THEOREM 1.8 (Nonordering of Omega Limit Sets). *Let $\omega(z)$ be an omega limit set for a monotone semiflow Φ.*
   (i) *No points of $\omega(z)$ are related by $\ll$.*
   (ii) *If $\omega(z)$ is a periodic orbit or Φ is SOP, no points of $\omega(z)$ are related by $<$.*

PROOF. Assume $x, y \in \omega(z)$. If $\omega(z)$ is a periodic orbit then x, y are unrelated (Proposition 1.5). If $x \ll y$ or $x < y$ and Φ is SOP, there are respective neighborhoods U, V of x, y such that $\Phi_r(U) \leqslant \Phi_r(V)$ for some $r \geqslant 0$; but this violates Lemma 1.7. $\qquad\square$

COROLLARY 1.9. *Assume Φ is SOP.*
   (i) *If an omega limit set has a supremum or infimum, it reduces to a single equilibrium.*

(ii) *If the equilibrium set is totally ordered, every quasiconvergent point with compact orbit closure is convergent.*

PROOF. Part (i) follows from Theorem 1.8(ii), since the supremum or infimum, if it exists, belongs to the limit set. Part (ii) is a consequence of (i). $\qquad\square$

1.3. *Local semiflows*

For simplicity we have assumed trajectories are defined for all $t \geqslant 0$, but there are occasions when we need the more general concept of a *local semiflow* in X. This means a map $\Psi : \Omega \to X$, with $\Omega \subset [0, \infty) \times X$ an open neighborhood of $\{0\} \times X$, such that the maps

$$\Psi_t : D_t \to X, \quad x \mapsto \Psi(t, x) \quad (0 \leqslant t < \infty)$$

satisfy the following conditions: D_t is an open, possibly empty set in X, Ψ_0 is the identity map of X, and $\Psi_{s+t} = \Psi_s \circ \Psi_t$ in the sense that $D_{s+t} = D_t \cap \Psi_t^{-1}(D_s)$ and $\Psi_{s+t}(x) = \Psi_s(\Psi_t(x))$ for $x \in D_{s+t}$.

The trajectory of x is defined as the map

$$I_x \to X, \quad t \mapsto \Psi_t(x), \quad \text{where } I_x = \{t \in \mathbb{R}_+ : x \in D_t\}.$$

The composition law implies I_x is a half open interval $[0, \tau_x)$; we call $\tau_x \in (0, \infty]$ the *escape time* of x. It is easy to see that every point with compact orbit closure has infinite escape time. Thus a local semiflow with compact orbit closures is a semiflow. In dealing with local semiflows we adopt the convention that the notations $\Psi_t(x)$ and $\Psi_t(U)$ carry the assumptions that $t \in I_x$ and $U \subset D_t$. The image of I_x under the trajectory of x is the orbit $O(x)$. The omega limit set $\omega(x)$ is defined as $\omega(x) = \bigcap_{t \in I_x} \overline{O(\Psi_t(x))}$.

A *local flow* is a map $\Theta : \Lambda \to X$ where $\Lambda \subset \mathbb{R} \times X$ is an open neighborhood of $\{0\} \times X$, and the (possibly empty) maps

$$\Theta_t : D_t \to X, \quad x \mapsto \Theta(t, x) \quad (-\infty \leqslant t < \infty)$$

satisfy the following conditions: Θ_0 is the identity map of $D_0 := X$, Θ_t is a homeomorphism of D_t onto D_{-t} with inverse Θ_{-t}, and

$$x \in (\Theta_s)^{-1} D_r \implies \Theta_r \circ \Theta_s(x) = \Theta_{r+s}(x).$$

Θ is a *flow* provided $D_t = X$ for all t.

The set $J_x := \{t \in \mathbb{R} : x \in D_t\}$ is an open interval around 0. The positive and negative *semiorbits* of x are the respective sets

$$\gamma^+(x) = \gamma^+(x, \Theta) := \{\Theta_t(x) : t \in J_x, t \geqslant 0\},$$

$$\gamma^-(x) = \gamma^-(x, \Theta) := \{\Theta_t(x) : t \in J_x, t \leqslant 0\}.$$

The *time-reversal* of Θ is the local flow $\widetilde{\Theta}$ defined by $\widetilde{\Theta}(t, x) = \Theta(-t, x)$.

The omega limit set $\omega(x)$ (for Θ) is defined to be $\omega(x) = \bigcap_{t \in I_x, t \geqslant 0} \overline{O(\Psi_t(x))}$. The *alpha limit set* $\alpha(x) = \alpha(x, \Theta)$ of x is defined as the omega limit set of x under the time-reversal of Θ^+.

Let F be a locally Lipschitz vector field F on a manifold M tangent along the boundary. Denote by $t \mapsto u(t; x)$ the maximally defined solution to $\dot{u} = F(u), u(0, x) = x$. There is a local flow Θ^F on M such that $\Theta_t(x) = u(t; x)$. The time-reversal of Θ^F is Θ^{-F}. When M is compact, Θ^F is a flow. If we assume that F, rather than being tangent to the boundary, is transverse inward, we obtain a local semiflow.

Our earlier results are readily adapted to monotone local semiflows. In particular, omega limit sets are unordered. Theorems 1.8 and 1.6 have the following extension:

THEOREM 1.10. *Let Φ be a monotone local semiflow.*
  (a) *No trajectory has both a rising and a falling interval.*
  (b) *No points of an omega limit set are related by $\ll$, or by $<$ if Φ is SOP.*
  (c) *The same holds for alpha limit sets provided Φ is a local flow.*

PROOF. The proofs of Theorems 1.6 and 1.8 also prove (a) and (b), and (c) follows by time reversal. $\qquad\square$

1.4. *The limit set dichotomy*

Throughout the remainder of Section 1 we adopt the following assumptions:

  (H) Φ is a strongly order preserving semiflow in an ordered space X, with every orbit closure compact.

Our goal now is to prove the important Limit Set Dichotomy:

If $x < y$ then either $\omega(x) < \omega(y)$, or $\omega(x) = \omega(y) \subset E$.

LEMMA 1.11 (Colimiting Principle). *Assume $x < y$, $t_k \to \infty$, $\Phi_{t_k}(x) \to p$ and $\Phi_{t_k}(y) \to p$ as $k \to \infty$. Then $p \in E$.*

PROOF. Choose neighborhoods U of x and V of y and $t_0 > 0$ such that $\Phi_{t_0}(U) \leqslant \Phi_{t_0}(V)$. Let $\delta > 0$ be so small that $\{\Phi_s(x): 0 \leqslant s \leqslant \delta\} \subset U$ and $\{\Phi_s(y): 0 \leqslant s \leqslant \delta\} \subset V$. Then $\Phi_s(x) \leqslant \Phi_r(y)$ whenever $t_0 \leqslant r, s \leqslant t_0 + \delta$. Therefore,

$$\Phi_{t_k - t_0}\big(\Phi_s(x)\big) \leqslant \Phi_{t_k - t_0}\big(\Phi_{t_0}(y)\big) = \Phi_{t_k}(y) \tag{1.2}$$

for all $s \in [t_0, t_0 + \delta]$ and all large k. As

$$\Phi_{t_k - t_0}\big(\Phi_s(x)\big) = \Phi_{s - t_0}\big(\Phi_{t_k}(x)\big) = \Phi_r\big(\Phi_{t_k}(x)\big),$$

where $r = s - t_0 \in [0, \delta]$ if $s \in [t_0, t_0 + \delta]$, we have

$$\Phi_r\big(\Phi_{t_k}(x)\big) \leqslant \Phi_{t_k}(y)$$

for large k and $r \in [0, \delta]$. Passing to the limit as $k \to \infty$ we find that $\Phi_r(p) \leqslant p$ for $0 \leqslant r \leqslant \delta$. If, in (1.2), we replace $\Phi_s(x)$ by $\Phi_{t_0}(x)$ and replace $\Phi_{t_0}(y)$ by $\Phi_s(y)$, and argue as above then we find that $p \leqslant \Phi_r(p)$ for $0 \leqslant r \leqslant \delta$. Evidently, $\Phi_r(p) = p$, $0 \leqslant r \leqslant \delta$ and therefore for all $r \geqslant 0$, so $p \in E$. $\qquad\square$

THEOREM 1.12 (Intersection Principle). *If $x < y$ then $\omega(x) \cap \omega(y) \subset E$. If $p \in \omega(x) \cap \omega(y)$ and $t_k \to \infty$, then $\Phi_{t_k}(x) \to p$ if and only if $\Phi_{t_k}(y) \to p$.*

PROOF. If $p \in \omega(x) \cap \omega(y)$ then there exists a sequence $t_k \to \infty$ such that $\Phi_{t_k}(x) \to p$ and $\Phi_{t_k}(y) \to q \in \omega(y)$, and $p \leqslant q$ by monotonicity. If $p < q$ then we contradict the Nonordering of Limit Sets since $p, q \in \omega(y)$. Hence $p = q$. The Colimiting Principle then implies $p \in E$. $\qquad\square$

The proof of the next result has been substantially simplified over previous versions.

LEMMA 1.13. *Assume $x < y$, $t_k \to \infty$, $\Phi_{t_k}(x) \to a$, and $\Phi_{t_k}(y) \to b$ as $k \to \infty$. If $a < b$ then $O(a) < b$ and $O(b) > a$.*

PROOF. The set $W := \{t \geqslant 0 \colon \Phi_t(a) \leqslant b\}$ contains 0 and is closed. We prove $W = [0, \infty)$ by showing that W is also open. Observe first that if $t \in W$, then $\Phi_t(a) < b$. For equality implies $b \in \omega(x) \cap \omega(y) \subset E$, and then the Intersection Principle entails $\Phi_{t_k}(x) \to b$, giving the contradiction $a = b$.

Suppose $\bar{t} \in W$ is positive. By SOP there are open sets U, V with $\Phi_{\bar{t}}(a) \in U$, $b \in V$ and $t_1 \geqslant 0$ such that $\Phi_t(U) \leqslant \Phi_t(V)$ for $t \geqslant t_1$. There exists $\delta \in (0, \bar{t}/2)$ such that $\Phi_s(a) \in U$ for $|s - \bar{t}| \leqslant \delta$, so we can find an integer $\kappa > 0$ such that $\Phi_s(\Phi_{t_k}(x)) \in U$ for $k \geqslant \kappa$. Choose $k_0 \geqslant \kappa$ such that $\Phi_{t_{k_0}}(y) \in V$. Then we have $\Phi_{t+s+t_{k_0}}(x) \leqslant \Phi_{t+t_{k_0}}(y)$ for $t \geqslant t_1$. Setting $t = t_k - t_{k_0}$ for large k in this last inequality yields $\Phi_{t_k+s}(x) \leqslant \Phi_{t_k}(y)$ for large k. Taking the limit as $k \to \infty$ we get $\Phi_s(a) \leqslant b$ for $|s - \bar{t}| \leqslant \delta$. A similar argument in the case $\bar{t} = 0$ considering only $s \in [0, \delta]$ gives the previous inequality for such s. Therefore, W is both open and closed so $W = [0, \infty)$. This proves $O(a) < b$, and $O(b) > a$ is proved dually. $\square$

LEMMA 1.14 (Absorption Principle). *Let $u, v \in X$. If there exists $x \in \omega(u)$ such that $x < \omega(v)$, then $\omega(u) < \omega(v)$. Similarly, if there exists $x \in \omega(u)$ such that $\omega(v) < x$, then $\omega(v) < \omega(u)$.*

PROOF. Apply Lemma 1.3 to obtain open neighborhoods U of x and V of $\omega(v)$ and $t_0 > 0$ such that

$$r \geqslant t_0 \implies \Phi_r(U) \leqslant \Phi_r(V),$$

hence $\Phi_r(U) \leqslant \omega(v)$ since $\omega(v)$ is invariant. As $x \in \omega(u)$, there exists $t_1 > 0$ such that $\Phi_{t_1}(u) \in U$. Hence for $\Phi_{t_0+t_1}(u) \leqslant \omega(v)$, and monotonicity implies that $\Phi_{s+t_0+t_1}(u) \leqslant$

$\omega(v)$ for all $s \geqslant 0$. This implies that $\omega(u) \leqslant \omega(v)$. If $z \in \omega(u) \cap \omega(v)$ then $z = \sup \omega(u) = \inf \omega(v)$, whence $\{z\} = \omega(u) = \omega(v)$ by Corollary 1.9(ii). But this is impossible since $x < \omega(v)$ and $x \in \omega(u)$, so we conclude that $\omega(u) < \omega(v)$. $\qquad\square$

LEMMA 1.15 (Limit Set Separation Principle). *Assume $x < y$, $a < b$ and there is a sequence $t_k \to \infty$ such that $\Phi_{t_k}(x) \to a$, $\Phi_{t_k}(y) \to b$. Then $\omega(x) < \omega(y)$.*

PROOF. By Lemma 1.13, $O(a) < b$, and therefore $\omega(a) \leqslant b$. If $b \in \omega(a)$ then Corollary 1.9 implies that $\omega(a) = b \in E$. Applying the Absorption Principle with $u = x$, $v = a$, $x = a$, we have $a \in \omega(x)$, $a < \omega(a) = b$ which implies that $\omega(x) < \omega(a)$. This is impossible as $\omega(a) \subset \omega(x)$. Consequently, $\omega(a) < b$. By the Absorption Principle again (with $u = a$, $v = y$), we have $\omega(a) < \omega(y)$. Since $\omega(a) \subset \omega(x)$, the Absorption Principle gives $\omega(x) < \omega(y)$. $\qquad\square$

We now prove the fundamental tool in the theory of monotone dynamics, stated for strongly monotone semiflows in Hirsch [66,68].

THEOREM 1.16 (Limit Set Dichotomy). *If $x < y$ then either*
   (a) $\omega(x) < \omega(y)$, *or*
   (b) $\omega(x) = \omega(y) \subset E$.
If case (b) holds and $t_k \to \infty$ then $\Phi_{t_k}(x) \to p$ if and only if $\Phi_{t_k}(y) \to p$.

PROOF. If $\omega(x) = \omega(y)$ then $\omega(x) \subset E$ by the Intersection Principle, Theorem 1.12, which also establishes the final assertion. If $\omega(x) \neq \omega(y)$ then we may assume that there exists $q \in \omega(y) \setminus \omega(x)$, the other case being similar. There exists $t_k \to \infty$ such that $\Phi_{t_k}(y) \to q$. By passing to a subsequence if necessary, we can assume that $\Phi_{t_k}(x) \to p \in \omega(x)$. Monotonicity implies $p \leqslant q$ and, in fact, $p < q$ since $q \notin \omega(x)$. By the Limit Set Separation Principle, $\omega(x) < \omega(y)$. $\qquad\square$

Among the many consequences of the Convergence Criterion is that a monotone semiflow in a strongly ordered Banach space cannot have a periodic orbit γ that is *attracting*, meaning that γ attracts all points in some neighborhood of itself (Hadeler [55], Hirsch [69]). The following consequence of the Limit Set Dichotomy implies the same conclusion for periodic orbits of SOP semiflows:

THEOREM 1.17. *Let γ be a nontrivial periodic orbit, some point of which is accessible from above or below. Then γ is not attracting.*

The accessibility hypothesis is used to ensure that there are points near p that are order-related to p but different from p. Some such hypothesis is required, as otherwise we could simply take $X = \gamma$, and then γ is attracting!

PROOF. Suppose $\gamma \subset W$ attracts an open set W. By hypothesis there exists $p \in \gamma$ and $x \in W$ such that $x > p$ or $x < p$ and $\omega(x) = \gamma$. To fix ideas we assume $x > p$. Then

$p \in \omega(x)$, so the Limit Set Dichotomy implies $p \in E$. Hence the contradiction that γ contains an equilibrium. $\qquad\square$

It turns out that the periodic orbits γ considered above are not only not attracting; they enjoy the strong form of instability expressed in the next theorem.

A set $K \subset X$ is *minimal* if it is nonempty, invariant, and every orbit it contains is dense in K.

THEOREM 1.18. *Let K be a compact minimal set that is not an equilibrium, some point of which is accessible from below or above. Then there exists $\delta > 0$ with the following property: Every neighborhood of K contains a point x comparable to some point of K, such that* $\mathrm{dist}(\Phi_t(x), K) > \delta$ *for all sufficiently large t.*

PROOF. We may assume there exists a sequence $\tilde{x}_n \to p \in K$ with $\tilde{x}_n > p$. Suppose there is no such δ. Then there exist a subsequence $\{x_n\}$ and points $y_n \in \omega(x_n)$ such that $y_n \to q \in K$. Minimality of K implies $\omega(p) = \omega(q) = K$. Since $x_n > p$, the Limit Set Dichotomy implies $\omega(x_n) \geqslant \omega(p)$; therefore $y_n \geqslant K$, so $q \geqslant K$. It follows that $q = \sup K$, and Corollary 1.9 implies the contradiction that K is a singleton. $\qquad\square$

A stronger form of instability for periodic orbits is given in Theorem 2.6.

1.5. *Q is plentiful*

One of our main goals is to find conditions that make quasiconvergent points generic in various senses. The first such results are due to Hirsch [66,73]; the result below is an adaptation of Smith and Thieme [199, Theorem 3.5].

We continue to assume Φ is an SOP semiflow with compact orbit closures.

A *totally ordered arc* is the homeomorphic image of a nontrivial interval $I \subset \mathbb{R}$ under a map $f : I \to X$ satisfying $f(s) < f(t)$ whenever $s, t \in I$ and $s < t$.

THEOREM 1.19. *If $J \subset X$ is a totally ordered arc, $J \setminus Q$ is at most countable.*

Stronger conclusions are obtained in Theorems 2.8 and 2.24.

The following global convergence theorem is adapted from Hirsch [73, Theorem 10.3].

COROLLARY 1.20. *Let Y be an ordered Banach space. Assume $X \subset Y$ is an open set, a closed order interval, or a subcone of Y_+. If $E = \{p\}$, every trajectory converges to p.*

PROOF. If X is open in Y, there exists a totally ordered line segment $J \subset X$ and quasiconvergent points $u, v \in J$ with $u < x < v$, by Theorem 1.19. Therefore $\Phi_t(u) \to p$ and $\Phi_t(v) \to p$, so monotonicity and closedness of the order relation imply $\Phi_t(x) \to p$.

If $X = [a, b]$, the trajectories of a and b converge to p by the Convergence Criterion 1.4, and the previous argument shows all trajectories converge to p. Similarly if X is a subcone of Y_+. $\qquad\square$

PROOF OF THEOREM 1.19. Let $W = \overline{\Phi([0, \infty) \times J)}$. Continuity of Φ implies that W is a separable metric space which is positively invariant under Φ. Therefore we may as well assume that X is a separable metric space.

We show that if $x \in J$ and

$$\inf\{\operatorname{dist}(\omega(x), \omega(y)): y \in J, y \neq x\} = 0,$$

then $x \in Q$. Choose a sequence $x_n \in J$, $x_n \neq x$ such that $\operatorname{dist}(\omega(x), \omega(x_n)) \to 0$. We may assume that $x_n < x$ for all n. Taking a subsequence, we conclude from the Limit Set Dichotomy: Either some $\omega(x_n) = \omega(x)$, or every $\omega(x_n) < \omega(x)$.

In the first case, $x \in Q$. In the second case, choose $y_n \in \omega(x_n)$, $z_n \in \omega(x)$ such that $d(y_n, z_n) \to 0$. After passing to subsequences, we assume $y_n, z_n \to z \in \omega(x)$. Because $y_n \leqslant \omega(x)$, we conclude that $z \leqslant \omega(x)$. As $z \in \omega(x)$, Corollary 1.9 implies $\omega(x) = \{z\}$. Hence $x \in Q$ in this case as well.

It follows that for every $x \in J \setminus Q$, there exists an open set U_x containing $\omega(x)$ such that $U_x \cap \omega(y) = \emptyset$ for every $y \in J \setminus \{x\}$. By the axiom of choice we get an injective mapping

$$J \setminus Q \to X, \quad x \mapsto p_x \in \omega(x) \subset U_x.$$

The separable metric space X has a countable base $\mathcal{B}$. A second application of the axiom of choice gives a map

$$J \setminus Q \to \mathcal{B}, \quad x \mapsto V_x \subset U_x, \quad p_x \in V_x.$$

This map is injective. For if x, y are distinct points of $J \setminus Q$, then $V_x \neq V_y$ because V_x, being contained in U_x, does not meet $\omega(y)$; but $p_y \in V_y \cap \omega(y)$. This proves $J \setminus Q$ is countable. $\qquad\square$

Let Y be an ordered Banach space and assume $X \subset Y$ is an ordered subspace (not necessarily linear). When Y is finite-dimensional, Theorem 1.19 implies $X \setminus Q$ has Lebesgue measure zero, hence almost every point is quasiconvergent. For infinite-dimensional Y there is an analogous result for Gaussian measures (Hirsch [73, Lemma 7.7]). The next result shows that in this case Q is also plentiful in the sense of category.

A subset of a topological space S is *residual* if it contains the intersection of countably many dense open subsets of S. When S is a complete metric space every residual set is dense by the Baire category theorem.

The assumption on X in the following result holds for many subsets of an ordered Banach space, including all convex sets and all sets with dense interior.

THEOREM 1.21. *Assume X is a subset of an ordered Banach space Y, and a dense open subset $X_0 \subset X$ is covered by totally ordered line segments. Then Q is residual in X.*

PROOF. It suffices to show that the set $Q_1 := Q \cup (Y \setminus X_0)$ is residual in Y. Note that $Y \setminus Q_1 = X_0 \setminus Q$. Let $L \subset Y$ be the 1-dimensional space spanned by some positive vector. Every translate $y + L$ meets $Y \setminus Q_1$ in a finite or countably infinite set by Theorem 1.19, hence $(y + L) \cap Q_1$ is residual in the line $y + L$. By the Hahn–Banach theorem there is a closed linear subspace $M \subset Y$ and a continuous linear isomorphism $F \colon Y \approx M \times L$ such that $F(x + L) = \{x\} \times L$ for each $x \in M$. Therefore $F(Q_1) \cap (\{x\} \times L)$ is residual in $\{x\} \times L$ for all $x \in X_0$, whence $F(Q_1)$ is residual in $M \times L$ by the Kuratowski–Ulam Theorem (Oxtoby [154]). This implies Q_1 is residual in Y. $\qquad\square$

Additional hypotheses seem to be necessary in order to prove density of Q in general ordered spaces. The next theorem obtains the stronger conclusion that Q has dense interior. A different approach will be explored in Section 2.

A point x is *doubly accessible from below* (respectively, above) if in every neighborhood of x there exist f, g with $f < g < x$ (respectively, $x < f < g$).

Consider the following condition on a semiflow satisfying (H):

(L) Either every omega limit set has an infimum in X and the set of points that are doubly accessible from below has dense interior, or every omega limit set has a supremum in X and the set of points that are doubly accessible from above has dense interior.

This holds when X is the Banach space of continuous functions on a compact set with the usual ordering, for then every compact set has a supremum and infimum, and every point is doubly accessible from above and below.

THEOREM 1.22. *If* (L) *holds, then* $X \setminus Q \subset \overline{\operatorname{Int} C}$, *and* $\operatorname{Int} Q$ *is dense.*

The proof is based on the following result. For $p \in E$ define $C(p) := \{z \in X \colon \omega(z) = \{p\}\}$. Note that $C = \bigcup_{p \in E} C(p)$.

LEMMA 1.23. *Suppose* $x \in X \setminus Q$ *and* $a = \inf \omega(x)$. *Then* $\omega(a) = \{p\}$ *with* $p < \omega(x)$, *and* $x \in \overline{\operatorname{Int} C(p)}$ *provided* x *is doubly accessible from below.*

PROOF. Fix an arbitrary neighborhood M of x. Note that $a < \omega(x)$ because $\omega(x)$ is unordered (Theorem 1.8). By invariance of $\omega(x)$ we have $\Phi_t a \leqslant \omega(x)$, hence $\Phi_t a \leqslant a$. Therefore the Convergence Criterion Theorem 1.4 implies $\omega(a)$ is an equilibrium $p \leqslant a$. Because $p < \omega(x)$, SOP yields a neighborhood N of $\omega(x)$ and $s \geqslant 0$ such that $p \leqslant \Phi_t N$ for all $t \geqslant s$. Choose $r \geqslant 0$ with $\Phi_t x \in N$ for $t \geqslant r$. Then $p \leqslant \Phi_t x$ if $t \geqslant r + s$. The set $V := (\Phi_{r+s})^{-1}(N) \cap M$ is a neighborhood of x in M with the property that $p \leqslant \Phi_t V$ for all $t \geqslant r + 2s$. Hence:

$$u \in V \quad \Longrightarrow \quad p \leqslant \omega(u). \tag{1.3}$$

Now assume x doubly accessible from below and fix $y_1, y \in V$ with $y_1 < y < x$. By the Limit Set Dichotomy $\omega(y) < \omega(x)$, because $\omega(x) \not\subset E$. By SOP we fix a neighborhood $U \subset V$ of y_1 and $t_0 > 0$ such that $\Phi_{t_0} u \leqslant \Phi_{t_0} y$ for all $u \in U$. The Limit Set Dichotomy implies $\omega(u) = \omega(y)$ or $\omega(u) < \omega(y)$; as $\omega(y) < \omega(x)$, we therefore have:

$$u \in U \quad \Longrightarrow \quad \omega(u) < \omega(x). \tag{1.4}$$

                                    M.W. Hirsch and H. Smith

For all $u \in U$, (1.4) implies $\omega(u) \leqslant \omega(a) = \{p\}$, while (1.3) entails $p \leqslant \omega(u)$. Hence $U \subset C(p) \cap M$, and the conclusion follows. $\qquad\square$

PROOF OF THEOREM 1.22. To fix ideas we assume the first alternative in (L), the other case being similar. Let X_0 denote a dense open set of points doubly accessible from below. Lemma 1.23 implies $X_0 \subset Q \cup \overline{\operatorname{Int} C} \subset Q \cup \overline{\operatorname{Int} Q}$, hence the open set $X_0 \setminus \overline{\operatorname{Int} Q}$ lies in Q. This prove $X_0 \setminus \overline{\operatorname{Int} Q} \subset \operatorname{Int} Q$, so $X_0 \setminus \overline{\operatorname{Int} Q} = \emptyset$. Therefore $\overline{\operatorname{Int} Q} \supset X_0$, hence $\overline{\operatorname{Int} Q} \supset \overline{X_0} = X$. $\qquad\square$

EXAMPLE 1.24. An example in Hirsch [73] shows that generic quasiconvergence and the Limit Set Dichotomy need not hold for a monotone semiflow that does not satisfy SOP. Let X denote the ordered Banach space $\mathbb{R}^3$ whose ordering is defined by the "ice-cream" cone $X_+ = \{x \in \mathbb{R}^3 \colon x_3 \geqslant \sqrt{x_1^2 + x_2^2}\}$. The linear system $x_1' = -x_2$, $x_2' = x_1$, $x_3' = 0$ generates a flow Φ with global period 2π which merely rotates points about the x_3-axis. Evidently X_+ is invariant, so linearity of Φ implies monotonicity. On the other hand, Φ is not strongly order preserving: If $a = (1, 0, 1)$ (or any other point on ∂Y_+ except the origin 0), SOP would require $\Phi_t(a) \gg 0$ for $t > 0$ because Φ_t is a homeomorphism, but this fails for all $t > 0$. The Limit Set Dichotomy fails to hold: For $a = (1, 0, 1)$ and $b = (2, 0, 2)$ it is easy to see that $a < b$ (for the ordering defined by X_+) and $\omega(a) \cap \omega(b) = \emptyset$, but $\omega(a) \not< \omega(b)$. As $E = C = Q = \{x \colon x_1 = x_2 = 0\}$ and most points belonging to periodic orbits of minimal period 2π, quasiconvergence is rare. In fact, the set of nonquasiconvergent points—the complement of the x_3-axis—is open and dense. It is not known whether there is a similar example with a polyhedral cone.

1.6. *Stability in normally ordered spaces*

We continue to assume the semiflow Φ is SOP with compact orbit closures.

The *diameter* of a set Z is $\operatorname{diam} Z := \sup_{x, y \in Z} d(x, y)$.

We now introduce some familiar stability notions. A point $x \in X$ is *stable* (relative to $R \subset X$) if for every $\epsilon > 0$ there exists a neighborhood U of x such that $\operatorname{diam} \Phi_t(U \cap R) < \epsilon$ for all $t \geqslant 0$. The set of stable points is denoted by S.

Suppose x_0 is stable. Then omega limit sets of nearby points are close to $\omega(x_0)$, and if all orbit closures are compact, the map $x \mapsto \omega(x)$ is continuous at x_0 for the Hausdorff metric on the space of compact sets.

x is *stable from above* (respectively, *from below*) if x is stable relative to the set of points $\geqslant x$ (resp., $\leqslant x$). The set of points stable from above (resp., below is denoted by S_+ (resp., S_-).

The *basin of x in R* is the union of all subsets of R of the form $V \cap R$ where $V \subset X$ is an open neighborhood of x such that

$$\lim_{t \to \infty} \operatorname{diam} \Phi_t(V \cap R) = 0.$$

Notice that $\omega(x) = \omega(y)$ for all y in the basin.

If the basin of x in R is nonempty, we say x is *asymptotically stable* relative to R. This implies x is stable relative to R. If x is asymptotically stable relative to X we say x is *asymptotically stable*. The set of asymptotically stable points is an open set denoted by A.

x is *asymptotically stable from above* (respectively, *below*) if it is asymptotically stable relative to the set of points $\geqslant x$ (resp., $\leqslant x$). The basin of x relative to this set is called the *upper* (resp., *lower*) basin of x. The set of such x is denoted by A_+ (resp., A_-).

Note that continuity of Φ shows that asymptotic stability relative to R implies stability relative to R. In particular, $A \subset S$, $A_+ \subset S_+$ and $A_- \subset S_-$.

These stability notions for x depend only on the topology of X, and not on the metric, provided the orbit of x has compact closure.

The metric space X is *normally ordered* if there exists a *normality constant* $\kappa > 0$ such that $d(x, y) \leqslant \kappa d(u, v)$ whenever $u, v \in X$ and $x, y \in [u, v]$. In a normally ordered space order intervals are bounded and the diameter of $[u, v]$ goes to zero with $d(u, v)$. Many common function spaces, including L^p spaces and the Banach space of continuous functions with the uniform norm, are normally ordered by the cone of nonnegative functions. But spaces whose norms involve derivatives are not normally ordered. Normality is required in order to wring the most out of the Sequential Limit Set Trichotomy. The propositions that follow record useful stability properties of SOP dynamics in normally ordered spaces.

PROPOSITION 1.25. *Assume X is normally ordered.*
  (a) $x \in S_+$ *(respectively, S_-) provided there exists a sequence $y_n \to x$ such that $y_n > x$ (resp., $y_n < x$) and $\lim_{n \to \infty} \sup_{t>0} d(\Phi_t(x), \Phi_t(y_n)) = 0$.*
  (b) $x \in S$ *provided $x \in S_+ \cap S_-$ and x is accessible from above and below.*
  (c) $x \in A$ *provided $x \in A_+ \cap A_-$ and x is accessible from above and below.*
  (d) *Suppose $a < b$ and $\omega(a) = \omega(b)$. Then $a \in A_+$ and $b \in A_-$. If $a < x < b$ then $x \in A$ and the basin of x includes $[a, b] \setminus \{a, b\}$.*

In particular, (d) shows that an equilibrium e is in A_+ if $x > e$ and $\Phi_t(x) \to e$ (provided X is normally ordered); and dually for A_-.

PROOF. We prove (a) for the case $y_n > x$. Given $\epsilon > 0$, choose m and t_0 so that

$$t > t_0 \quad \Longrightarrow \quad d\big(\Phi_t(x), \Phi_t(y_m)\big) < \epsilon.$$

By SOP there exists a neighborhood W of x and $t_1 > t_0$ such that

$$t > t_1, \ v \in W \quad \Longrightarrow \quad \Phi_t(v) < \Phi_t(y_m).$$

Fixing t_1, we shrink W to a neighborhood W_ϵ of x so that

$$0 < t \leqslant t_1, \ v \in W_\epsilon \quad \Longrightarrow \quad d\big(\Phi_t(x), \Phi_t(v)\big) < \kappa\epsilon,$$

where $\kappa > 0$ is the normality constant. If $x < v \in W_\epsilon$ and $t > t_1$ then $\Phi_t(x) \leqslant \Phi_t(v) \leqslant \Phi_t(y_m)$, and therefore

$$t > t_1, \ x < v \in W_\epsilon \quad \Longrightarrow \quad d\big(\Phi_t(x), \Phi_t(v)\big) \leqslant \kappa d\big(\Phi_t(x), \Phi_t(y_m)\big) \leqslant \kappa\epsilon.$$

Hence we have proved

$$0 < t < \infty, \ v \in W_\epsilon \quad \Longrightarrow \quad d\big(\Phi_t(x), \Phi_t(v)\big) < \kappa\epsilon.$$

As ϵ is arbitrary, this proves $x \in S_+$.

To prove (b), let $u_n, v_n \to x$ with $u_n < x < v_n$. Because $x \in S_+ \cap S_-$, for any $\epsilon > 0$ there exists $\delta > 0$ such that if $d(y, x) < \delta$ and $y < x$ or $y > x$, then $\sup_{t>0} d(\Phi_t(y), \Phi_t(x)) < \epsilon$. Choose k such that $d(u_k, x) < \delta$ and $d(v_k, x) < \delta$. By SOP there is a neighborhood W_ϵ of x such that $\Phi_t(u_k) \leqslant \Phi_t(W_\epsilon) \leqslant \Phi_t(v_k)$ for sufficiently large t. Normality implies that for such t,

$$\kappa^{-1} \operatorname{diam} \Phi_t(W_e) \leqslant d\big(\Phi_t(u_k), \Phi_t(v_k)\big)$$

$$\leqslant d\big(\Phi_t(u_k), \Phi_t(x)\big) + d\big(\Phi_t(x), \Phi_t(v_k)\big) < 2\epsilon.$$

As κ is constant and ϵ is arbitrary, this proves x is stable.

The proofs of (c) and (d) are similar. $\square$

PROPOSITION 1.26. *Assume X is normally ordered, $p \in E$, and $\{K_n\}$ is a sequence of nonempty compact invariant sets such that $K_n < p$ and $\operatorname{dist}(K_n, p) \to 0$. Then:*

(a) *p is stable from below.*

(b) *If z is such that $\omega(z) = p$, then z is stable from below.*

In particular, if p is the limit of a sequence of equilibria $< p$ then p is stable from below.

PROOF. (a) Given $\epsilon > 0$, fix m such that $\operatorname{dist}(K_m, p) < \epsilon$. By Lemma 1.3 there is a neighborhood W of p and $t_0 > 0$ such that $t > t_0 \Longrightarrow \Phi_t(W) \geqslant K_m$, and therefore

$$t > t_0, \ v \in W, \ p > v \quad \Longrightarrow \quad d\big(\Phi_t(p), \Phi_t(v)\big) \leqslant \kappa d\big(\Phi_t(p), \Phi_t(K_m)\big) \leqslant \kappa\epsilon.$$

Pick a neighborhood $W_\epsilon \subset W$ of p so small that

$$0 \leqslant t \leqslant t_0, \ v \in W_\epsilon \quad \Longrightarrow \quad d\big(\Phi_t(p), \Phi_t(v)\big) < \kappa\epsilon.$$

Then

$$0 \leqslant t < \infty, \ v \in W_\epsilon, \ v < p \quad \Longrightarrow \quad d\big(\Phi_t(p), \Phi_t(v)\big) < \kappa\epsilon.$$

This proves $p \in S_-$, because ϵ is arbitrary.

(b) Choose a neighborhood U of z and $t_1 \geqslant 0$ such that $\Phi_{t_1}(U) \subset W$. Assume $y \in U$, $y < z$. If $t \geqslant t_1 + t_0$, then $K_m \leqslant \Phi_t(y) \leqslant \Phi_t(z)$, and therefore by normality, $d(\Phi_t(y), \Phi_t(z)) \leqslant \kappa \operatorname{dist}(K_m, \Phi_t(z))$. As $\Phi_t(z) \to p$, there exists $t_2 \geqslant t_1 + t_0$ such that

$$t \geqslant t_2 \quad \Longrightarrow \quad d\big(\Phi_t(y), \Phi_t(z)\big) \leqslant \kappa \operatorname{dist}(K_m, p) < \kappa\epsilon.$$

Fix this t_2. By continuity of Φ there is a neighborhood $U_1 \subset U$ of z so small that

$$0 \leqslant t \leqslant t_2 \quad \Longrightarrow \quad d\big(\Phi_t(y), \Phi_t(z)\big) \leqslant \kappa\epsilon.$$

As ϵ is arbitrary, this implies $z \in S_-$. $\qquad\square$

1.7. *Stable equilibria in strongly ordered Banach spaces*

In spaces that are not normally ordered we cannot directly use the results of the previous subsection to characterize stable equilibria. For strongly monotone semiflows in strongly ordered Banach spaces we work around this by introducing a weaker norm that makes the order normal, and for which the semiflows are continuous and SOP. This permits use of the earlier results.

Let Y be a strongly ordered Banach space. The *order topology* on Y is the topology generated by open order intervals. An *order norm* on the topological vector space $\widehat{Y}$ is defined by fixing $u \gg 0$ and assigning to x the smallest ϵ such that $x \in [-\epsilon u, \epsilon u]$. It is easy to see that $\widehat{Y}$ is normally ordered by the order norm, with normality constant 1. Every order neighborhood of p in $\widehat{Y}$ contains $[p - \epsilon u, p + \epsilon u]$ for all sufficiently small numbers $\epsilon > 0$. For example, $Y = C^1([0, 1])$ with the usual C^1-norm and with Y_+ the cone of nonnegative functions is strongly ordered but not normally ordered; putting $u := 1$, the order norm becomes the usual supremum.

The induced topology on any subset $Z \subset Y$ is also referred to as the order topology, and the resulting topological space is denoted by $\widehat{Z}$. A neighborhood in $\widehat{Z}$ is an *order neighborhood*.

Every open subset of $\widehat{Z}$ is open in Z, i.e., the identity map of Z is continuous from Z to $\widehat{Z}$. Therefore $\widehat{Z} = Z$ as topological spaces when Z is compact. As shown below, if Ψ is a monotone local semiflow in Z, it is also a local semiflow in $\widehat{Z}$, denoted by $\widehat{\Psi}$. Evidently Ψ and $\widehat{\Psi}$ have the same orbits and the same invariant sets.

LEMMA 1.27. *Let Ψ be a monotone local semiflow in a subset X of a strongly ordered Banach space Y, that extends to a monotone local semiflow in an open subset of Y. Then:*
  (a) *$\widehat{\Psi}$ is a monotone local semiflow.*
  (b) *If Ψ is a strongly monotone, then $\widehat{\Psi}$ is SOP.*

PROOF. It suffices to prove (a) and (b) when X is open in Y, which condition is henceforth assumed.

$\widehat{\Psi}$ is monotone because Ψ is monotone. To prove continuity of $\widehat{\Psi}$, let $N = [[a, b]]_Y \cap X$ and $(t_0, x_0) \in \widehat{\Psi}^{-1}(N)$. As the latter is open in $\mathbb{R}_+ \times X$, there exists $\epsilon > 0$ and U, an open neighborhood of x_0 in X, such that

$$\big[(t_0 - \epsilon, t_0 + \epsilon) \cap \mathbb{R}_+\big] \times U \subset \widehat{\Psi}^{-1}(N).$$

We may choose $u, v \in U$ such that $x \in [[u, v]]_Y$. If $z \in [[u, v]]_Y \cap X$ and $|t - t_0| < \epsilon$ then by monotonicity and $u, v \in U$ we have $a \ll \widehat{\Psi}_t(u) \leqslant \widehat{\Psi}_t(z) \leqslant \widehat{\Psi}_t(v) \ll b$. Thus,

$$\big[(t_0 - \epsilon, t_0 + \epsilon) \cap \mathbb{R}_+\big] \times \big([[u, v]]_Y \cap X\big) \subset \widehat{\Psi}^{-1}(N),$$

proving the continuity of $\widehat{\Psi}$.

Assume $x, y \in X$, $x < y$ and let $t_0 > 0$ be given. By strong monotonicity of Ψ there are respective open neighborhoods $U, V \subset X$ of x, y such that $\Psi_{t_0}(U) \leqslant \Psi_{t_0}(V)$ (see Proposition 1.2). Choose $w, u, v, z \in X$ such that $u, w \in U$, $v, z \in V$ and

$$w \ll x \ll u, \qquad v \ll y \ll z$$

so that $[[w, u]]_Y \cap X$ and $[[v, z]]_Y \cap X$ are order neighborhoods in X of x, y respectively. Monotonicity of Ψ implies

$$\Psi_{t_0}\big([[w, u]]_Y \cap X\big) \leqslant \Psi_{t_0}\big([[v, z]]_Y \cap X\big). \qquad \square$$

An equilibrium p for $\Psi : \mathbb{R}^+ \times X \to X$ is *order stable* (respectively, *asymptotically order stable* if p is stable (respectively, asymptotically stable) for $\widehat{\Psi}$.

PROPOSITION 1.28. *Let Ψ be a monotone local semiflow in a subset X of a strongly ordered Banach space Y, that extends to a monotone local semiflow in some open subset of Y. Assume p is an equilibrium having a neighborhood W that is attracted to a compact set $K \subset X$. If p is order stable (respectively, asymptotically order stable), it is stable (respectively, asymptotically stable).*

PROOF. Suppose p is order stable and let U be a neighborhood of p. As $\widehat{K} = K$, there is a closed order neighborhood N_0 of p such that $N_0 \cap K \subset U \cap K$. By order stability there exists an order neighborhood N_1 of p such that $O(N_1) \subset N_0$. Compactness of $N_0 \cap K$ implies there is an open set $V \supset K$ there is an open set $V \supset K$ such that $N_0 \cap V \subset U$. Because K attracts W, there is a neighborhood $U_2 \subset W$ of p and $r \geqslant 0$ such that

$$t \geqslant r \quad \Longrightarrow \quad \Psi_t(U_2) \subset V.$$

By continuity of Ψ_r at $p = \Psi_r(p)$ there is a neighborhood $U_3 \subset U_2$ of p such that

$$0 \leqslant t \leqslant r \quad \Longrightarrow \quad \Psi_t(U_3) \subset V.$$

and thus $O(U_3) \subset V$. Therefore $N_1 \cap U_3$ is a neighborhood of p such that

$$O(N_1 \cap U_3) \subset O(N_1) \cap O(U_3) \subset N_0 \cap V \subset U.$$

This shows p is stable.

Assume p is asymptotically order stable and choose an order neighborhood $M \subset X$ of p that is attracted to p by $\widehat{\Psi}$. We show that $M \cap W$ is in the basin of p for Ψ. Consider arbitrary sequences $\{x_k\}$ in $M \cap W$ and $t_k \to \infty$ in $[0, \infty)$. Fix $u \gg 0$. By the choice of M there are positive numbers $\epsilon_k \to 0$ such that

$$p - \epsilon_k u \ll \Psi_{t_k}(x_k) \ll p + \epsilon_k u.$$

This implies $\Psi_{t_k}(x_k) \to p$ in X, because the order relation on X is closed and $\{\Psi_{t_k}(x_k)\}$ is precompact in X by the choice of W and compactness of K. $\qquad \square$

1.8. *The search for stable equilibria*

The following results illustrate the usefulness of a dense set of quasiconvergent points. Φ denotes a strongly order preserving semiflow in X; Hypothesis (H) of Section 1.4 is still in force.

PROPOSITION 1.29. *Assume Q is dense. Let $p, q \in E$ be such that $p < q$, p is accessible from above, and q is accessible from below. Then there exists $z \in X$ satisfying one of the following conditions:*
  (a) *$p < z < q$, and $\Phi_t(z) \to p$ or $\Phi_t(z) \to q$;*
  (b) *$p < z < q$ and $z \in E$;*
  (c) *$z > p$ and $p \in O(z)$, or $z < q$ and $q \in O(z)$.*

PROOF. By SOP there are open neighborhoods U, V of p, q respectively and $t_0 \geq 0$ such that $\Phi_t U \leq \Phi_t V$ for $t \geq t_0$. Choose sequences $x_n \to p$ in U and $y_n \to q$ in V with $p < x_n$, $y_n < q$. We assume $p \notin O(x_n)$ and $q \notin O(y_n))$, as otherwise (c) is satisfied. Then

$$t \geq t_0 \quad \Longrightarrow \quad p < \Phi_t(x_n) \leq \Phi_t(y_n) < q.$$

Choose open neighborhoods U_1, W, V_1 of $p, \Phi_{t_0}(y_1), q$ respectively such that for some $t_1 \geq t_0$:

$$t \geq t_1 \quad \Longrightarrow \quad \Phi_t(U_1) \leq \Phi_t(W) \leq \Phi_t(V_1).$$

Choose $w \in Q \cap W$ and a sequence $s_k \to \infty$, $s_k \geq t_1$ such that $\Phi_{s_k}(w) \to e \in E$. Fix m so large that $x_m \in U_1$, $y_m \in V_1$. Then for sufficiently large k,

$$p < \Phi_{s_k}(x_m) \leq \Phi_{s_k}(w) \leq \Phi_{s_k}(y_m) < q.$$

It follows that $p \leq e \leq q$. If $e = p$ or q then $\omega(\Phi_{s_k}(w)) = p$ or q by the Convergence Criterion 1.4, giving (a) with $z = \Phi_{s_k}(w)$. Therefore if (a) does not hold, (b) holds with $z = e$. $\qquad\square$

The assumption in Proposition 1.29 that Q is dense can be considerably weakened, for example, to p (or q) being interior to $\overline{Q}$: Assume $y_1 \in \overline{\mathrm{Int}\, Q}$ and set $w = \Phi_{s_k}(w_0)$, $w_0 \in (\mathrm{Int}\, Q) \cap \Phi_{s_k}^{-1}(W)$, etc. In fact, density of Q can be replaced with the assumption that p or q lies in the interior of the set $Q_\#$ of points x such that there is a sequence $x_i \to x$ with $\lim_{i \to \infty} \mathrm{dist}(\omega(x_i), E) = 0$. Clearly $Q_\#$ is closed and contains Q, so density of Q implies $Q_\# = X$.

THEOREM 1.30. *Suppose X is normally ordered and the following three conditions hold:*
  (a) *Q is dense;*
  (b) *if $e \in E$ and e is not accessible from above (below) then $e = \sup X$ ($e = \inf X$);*
  (c) *there is a maximal totally ordered subset $R \subset E$ that is nonempty and compact.*
Then R contains a stable equilibrium, an asymptotically stable equilibrium if R is finite.

PROOF. By Lemma 1.1, $\sup R$ ($\inf R$) exists and is a maximal (minimal) element of E. We first prove that every maximal equilibrium q is in A_+. This holds vacuously when $q = \sup X$. Suppose $q \neq \sup X$. If q is in the orbit of some point $> q$ then $q \in A_+$ by Proposition 1.25(d). Hence we can assume:

$$t \geqslant 0, \; y > q \implies \Phi_t(y) > q.$$

By hypothesis we can choose $y > q$. By SOP there is an open neighborhood U of q and $s > 0$ such that $\Phi_s(y) \geqslant \Phi_s(U)$. By hypothesis we can choose $z \in U$ such that $\Phi_s(y) \neq \Phi_s(z)$ and $z > q$. Set $x_2 = \Phi_s(y)$, $x_1 = \Phi_s z$. Then $x_2 > x_1 > q$, By SOP and the assumption above there is a neighborhood V_2 of x_2 and $t_0 \geqslant 0$ such that

$$t > t_0 \implies q < \Phi_t(x_1) \leqslant \Phi_t(V_2).$$

Choose $v \in V_2 \cap Q$. Then $q < \Phi_t(v)$ for $t \geqslant t_0$, hence $q \leqslant \omega(v) = \omega(\Phi_{t_0}(v)) \subset E$. Therefore $\Phi_t(v) \to q$ by maximality of q, so and Proposition 1.25(d) implies $q \in A_+$, as required. The dual argument shows that every minimal equilibria is in A_-.

Assumption (c) and previous arguments establish that $q = \sup R$ and $p = \inf R$ satisfy $p \leqslant q$ and $q \in A_+$, $p \in A_-$.

Suppose $p = q$; in this case we prove $q \in A$. As q is both maximal and minimal in E, we have $q \in A_+ \cap A_-$. If q is accessible from above and below then $q \in A$ by Proposition 1.25(b). If q is not accessible from above then by hypothesis $q = \sup X$, in which case the fact that $q \in A_-$ implies $q \in A$. Similarly, $q \in A$ if q is not accessible from below.

Henceforth we assume $p < q$. As R is compact and $R \cap S_- \neq \emptyset$ because $p \in R$, it follows that R contains the equilibrium $r := \sup(R \cap S_-)$. Note that $r \in S_-$, because this holds by definition of r if r is isolated in $\{r' \in R: r' \leqslant r\}$, and otherwise $r \in S_-$ by Proposition 1.26(a). If $r = q$ a modification of the preceding paragraph proves $q \in S$.

Henceforth we assume $r < q$; therefore r is accessible from above.

If r is not accessible from below then $r = p = \inf X$ so $r \in S$ and we are done; so we may as well assume r is accessible from below as well as from above. If r is the limit of a sequence of equilibria $> r$ then $r \in S_+$ by the dual of Proposition 1.26, hence $r \in S$ by Proposition 1.25(b). Therefore we can assume R contains a smallest equilibrium $r_1 > r$. Note that $r_1 \notin S_-$ by maximality of r. We apply Proposition 1.29 to r, r_1: among its conclusions, the only one possible here is that $z > r$ and $\Phi_t(z) \to r$ (and perhaps $r \in O(z)$). Therefore $r \in S_+$ by Proposition 1.25(a), whence $r \in S$ by 1.25(b). When R is finite, a modification of the preceding arguments proves $\max(R \cap A_-) \subset A$. $\qquad\square$

Assumption (b) in the Theorem 1.30 holds for many subsets X of an ordered Banach space Y, including open sets, subcones of Y_+, closed order intervals, and so forth. This result is similar to Theorem 10.2 of Hirsch [73], which establishes equilibria that are merely order stable, but does not require normality.

Assumption (c) holds when E is compact, and also in the following situation: $X \subset Y$ where Y is an L^p space, $1 \leqslant p < \infty$, and E is a nonempty, closed, and order bounded subset of X; then every order bounded increasing or decreasing sequence converges. If (c) holds and some Φ_t is real analytic with spatial derivatives that are compact and strongly positive operators, then R is finite. This follows from the statements and proofs of Lemma 3.3 and Theorem 2 in Jiang and Yu [90].

For related results on stable equilibria see Jiang [86], Mierczyński [138,139], and Hirsch [69].

THEOREM 1.31. *Let Φ be a semiflow in a subset X of a strongly ordered Banach space Y, that extends to a strongly monotone local semiflow in some open subset of Y. Assume hypotheses* (a), (b), (c) *of Theorem* 1.30 *hold, and every equilibrium has a neighborhood attracted to a compact set. Let $R \subset E$ be as in* 1.30(c). *Then R contains a stable equilibrium, and an asymptotically stable equilibrium when R is finite.*

PROOF. Our strategy is to apply Theorem 1.30 to the semiflow $\widehat{\Phi}$ in $\widehat{X}$ (see Section 1.7). Give $\widehat{X}$ the metric coming from an order norm on $\widehat{Y}$; this makes $\widehat{X}$ is normally ordered. Lemma 1.27 shows that $\widehat{\Phi}$ is SOP. Therefore R contains an equilibrium p that is stable for $\widehat{\Phi}$, by Theorem 1.30. This means p is order stable for Φ, whence Proposition 1.28 shows that p is stable for Φ. The final assertion follows similarly. $\qquad\square$

Stable equilibria are found under various assumptions in Theorems 2.9, 2.10, 2.11, 2.26, 3.14, 4.12.

2. Generic convergence and stability

2.1. *The sequential limit set trichotomy*

Throughout Section 2 we assume Hypothesis (H) of Section 1.4:

> Φ *is a strongly order preserving semiflow in an ordered space X, with all orbit closures compact.*

The main result is that the typical orbit of an SOP semiflow is stable and approaches the set E of equilibria. Existence of stable equilibria is established under additional compactness assumptions.

The index n runs through the positive integers.

A point x is *strongly accessible from below* (*respectively,* above) if there exists a sequence $\{y_n\}$ converging to x such that $y_n < y_{n+1} < x$ (resp., $y_n > y_{n+1} > x$). In this case we say $\{y_n\}$ *strongly approximates x from* below (resp., *from above*).

The sequence $\{x_n\}$ is *omega compact* if $\bigcup_n \omega(x_n)$ is compact.

Define sets $BC, AC \subset X$ as follows:

$$x \in BC \iff x \text{ is strongly accessible from below by an omega compact sequence,}$$

$$x \in AC \iff x \text{ is strongly accessible from above by an omega compact sequence.}$$

In this notation "B" stands for "below," "A" for above, and "C" for "compact."

We will also use the following condition on a set $W \subset X$:

(C) Every sequence $\{w_n\}$ in W that strongly approximates a point of W from below or above is omega compact.

This does not assert that any point is strongly accessible from below or above. But if every point of W is accessible from above and W satisfies (C), then $W \subset AC$; and similarly for BC.

The next two propositions imply properties stronger than (C). Recall that a map $f : X \to X$ is *completely continuous* provided $\overline{f(B)}$ is compact for every bounded set $B \subset X$; and f *conditionally completely continuous* provided $\overline{f(B)}$ is compact whenever B and $f(B)$ are bounded subsets of X.

The *orbit* of any set $B \subset X$ is $O(B) = \bigcup_t \Phi_t(B)$.

PROPOSITION 2.1. *Assume the following two conditions*:
   (a) *every compact set has a bounded orbit, and*
   (b) Φ_s *is conditionally completely continuous for some* $s > 0$.
If $L \subset X$ *is compact, then* $\overline{\bigcup_{x \in L} \omega(x)}$ *is compact and this implies* X *has property* (C).

PROOF. $O(L)$ is a bounded set by (a), and positively invariant, so (b) implies compactness of $\overline{\Phi_s(O(L))}$. As the latter set contains $\omega(x)$ for all $x \in L$, the first assertion is proved. The second assertion follows from precompactness of $\{x_n\}$. $\qquad \square$

PROPOSITION 2.2. *Assume* $W \subset X$ *has the following property: For every* $x \in W$ *there is a neighborhood* $U_x \subset X$ *and a compact set* M_x *that attracts every point in* U_x. *Then* $\overline{O(x)}$ *is compact for every* $x \in W$, *and* $\overline{\bigcup_{y \in U_x} \omega(y)}$ *is compact. If* $z_n \to x \in W$ *then* $\overline{\bigcup_n \omega(z_n)}$ *is compact, therefore* W *has property* (C).

PROOF. It is easy to see that $\overline{O(x)}$ is compact and $\overline{\bigcup_{y \in U_x} \omega(y)}$ is compact because it lies in M_x. Fix $k \geqslant 0$ such that $z_n \in U_x$ for all $n \geqslant k$. Then

$$\overline{\bigcup_n \omega(z_n)} = \bigcup_{1 \leqslant n \leqslant k} \omega(z_n) \cup M_x,$$

which is the union of finitely many compact sets, hence compact. Condition (C) follows trivially. $\qquad \square$

The key to stronger results on generic quasiconvergence and stability is the following result of Smith and Thieme [197]:

THEOREM 2.3 (Sequential Limit Set Trichotomy). *Let* $\{\tilde{x}_n\}$ *be an omega compact sequence strongly approximating* $z \in BC$ *from below. Then there is a subsequence* $\{x_n\}$ *such that exactly one of the following three conditions holds:*

(a) *There exists $u_0 \in E$ such that*

$$\omega(x_n) < \omega(x_{n+1}) < \omega(z) = \{u_0\}$$

and

$$\lim_{n \to \infty} \mathrm{dist}\big(\omega(x_n), u_0\big) = 0.$$

In this case $z \in C$.
(b) *There exists $u_1 = \sup\{u \in E\colon u < \omega(z)\}$ and*

$$\omega(x_n) = \{u_1\} < \omega(z).$$

In this case $z \in \overline{\mathrm{Int}\, C}$. Moreover z has a neighborhood W such that if $w \in W$, $w < z$ then $\Phi_t(w) \to u_1$ and $\Phi_t(w) > u_1$ for sufficiently large t.
(c) $\omega(x_n) = \omega(z) \subset E$.
In this case $z \in \overline{\mathrm{Int}\, Q}$. Moreover $\omega(w) = \omega(z) \subset E$ for every $w < z$ sufficiently near z.

Note that z is convergent in (a), and strongly accessible from below by convergent points in (b). In (c), z is quasiconvergent and strongly accessible from below by quasiconvergent points.

If $z \in AC$ there is an analogous dual result, obtained by reversing the order relation in X. Although we do not state it formally, we will use it below. If $z \in AC \cap BC$ then both results apply. See Proposition 3.6 in Smith and Thieme [197].

PROOF OF THEOREM 2.3. By the Limit Set Dichotomy 1.16, either there exists a positive integer j such that $\omega(\tilde{x}_n) = \omega(\tilde{x}_m)$ for all $m, n \geqslant j$, or else there exists a subsequence $\{\tilde{x}_{n_i}\}$ such that $\omega(\tilde{x}_{n_i}) < \omega(\tilde{x}_{n_{i+1}})$ for all i. Therefore there is a subsequence $\{x_n\}$ such that $\omega(x_n) < \omega(x_{n+1})$ for all n, or $\omega(x_n) = \omega(x_{n+1})$ for all n.

Case I: $\omega(x_n) < \omega(x_{n+1})$. We will see that (a) holds. The Limit Set Dichotomy 1.16 implies $\omega(x_n) \leqslant \omega(z)$. In fact, that $\omega(x_n) < \omega(z)$. Otherwise $\omega(x_k) \cap \omega(z) \neq \emptyset$ for some k, and the Limit Set Dichotomy implies the contradiction $\omega(x_k) = \omega(z) \geqslant \omega(x_{k+1}) > \omega(x_k)$.

Define $K = \bigcup \omega(x_n)$, a nonempty compact invariant set. Consider the set

$$\Lambda = \left\{ y\colon y = \lim_{n \to \infty} y_n,\ y_n \in \omega(x_n) \right\} \subset K.$$

Clearly Λ is invariant and closed, and compactness of K implies Λ is compact and nonempty. We show that Λ is a single equilibrium. Suppose $y, v \in \Lambda$, so that $y_n \to y$, $v_n \to v$ with $y_n, v_n \in \omega(x_n)$. Since $y_n < v_{n+1}$ and $v_n < y_{n+1}$, we have $y \leqslant v$ and $v \leqslant y$, so $v = y$. Thus we can set $\Lambda = \{u_0\}$, and invariance implies $u_0 \in E$.

The definition of Λ and compactness of K imply $\lim_{n \to \infty} \mathrm{dist}(\omega(x_n), u_0) = 0$. From $\omega(x_n) < \omega(x_{n+1}) < \omega(z)$ we infer

$$\omega(x_n) < u_0 \leqslant \omega(z).$$

If $u_0 \in \omega(z)$ then $\omega(z) = \{u_0\}$ by Corollary 1.9, yielding (a).

We show that $u_0 < \omega(z)$ gives a contradiction. Choose a neighborhood W of $\omega(z)$ and $t_0 \geq 0$ such that $u_0 \leq \Phi_t(W)$ for all $t \geq t_0$ (by Lemma 1.3). There exists $t_1 > 0$ such that $\Phi_{t_1}(z) \in W$, and by continuity of Φ_{t_1} there exists m such that $\Phi_{t_1}(x_m) \in W$. It follows that $u_0 \leq \Phi_t(x_m)$ for $t \geq t_0 + t_1$. As $u_0 \in E$, we have $u_0 \leq \omega(x_m)$. But this contradicts $\omega(x_m) < u_0$. Thus (a) holds in Case I.

Case II: $\omega(x_n) = \omega(x_{n+1}) \subset E$. Since $x_n < z$, the Limit Set Dichotomy implies that either $\omega(x_n) = \omega(z)$, which gives (c), or else $\omega(x_n) < \omega(z)$, which we now assume. Choose an equilibrium $u_1 \in \omega(x_1)$. By Lemma 1.3 there exists an open set W containing $\omega(z)$ and $t_0 \geq 0$ such that $u_1 \leq \Phi_t(W)$ for all $t \geq t_0$. Arguing as in Case I, we obtain $u_1 \leq \Phi_t(x_m)$ for some m and all large t. Since $u_1 \in \omega(x_m)$, it follows that $\omega(x_m) = u_1$ by Corollary 1.9, and therefore $\omega(x_n) = \{u_1\}$ as asserted in case (b). Finally, if $u \in E$ and $u < \omega(z)$, we argue as above that $\omega(x_m) \geq u$ for some m, which implies $u_1 \geq u$.

To prove $z \in \overline{\text{Int } Q}$, use SOP to obtain a neighborhood U_n of x_n such that $\Phi_t(x_{n-1}) \leq \Phi_t(U_n) \leq \Phi_t(x_{n+1})$ for all large t, implying $U_n \subset Q$. A similar argument proves the analogous assertion in (b). $\qquad\square$

The following addendum to the Sequential Limit Set Trichotomy provides important stability information. In essence, it associates various kinds of stable points to arbitrary elements $z \in BC$:

PROPOSITION 2.4. *Assume X is normally ordered. In cases* (a), (b) *and* (c) *of the Sequential Limit Set Trichotomy, the following statements are valid respectively:*
  (a) *z and u_0 are stable from below;*
  (b) *z is not stable from below, $\omega(z)$ is unstable from below, and u_1 is asymptotically stable from above;*
  (c) *z is asymptotically stable from below, and $z \in \overline{A}$.*

PROOF. (a) follows from Proposition 1.26(a) and (b).

(b) The first two assertions are trivial. To prove $u_1 \in A_+$, take $w = x_n$ for some large n in the last assertion of (b) and apply 1.25(d) with $a = u_1$.

(c) follows from 1.25(d), taking $b = z$. $\qquad\square$

We expect in real world systems that observable motions are stable trajectories. Our next result implies stable trajectories approach equilibria.

PROPOSITION 2.5. $S \cap (BC \cup AC) \subset Q$.

PROOF. When $z \in S \cap (BC \cup AC)$, only (a) and (c) of the Sequential Limit Set Trichotomy are possible, owing to continuity at z of the function $x \mapsto \omega(x)$. In both cases $z \in Q$. $\qquad\square$

The inclusion $S \subset Q$ suggests trajectories issuing from nonquasiconvergent points are unlikely to be observed; the next result implies that their limit sets are, not surprisingly, unstable. There are as many concepts of instability as there are of stability, but for our purposes the following very strong property suffices: A set $M \subset X$ is *unstable from above* provided there is an equilibrium $u > M$ such that $\omega(x) = \{u\}$ if $u > x > y$, $y \in M$. Such an equilibrium u is unique, and SOP implies it attracts all points $< u$ in some neighborhood of u. *Unstable from below* is defined dually.

THEOREM 2.6. *Assume $z \in BC \setminus Q$ (respectively, $z \in AC \setminus Q$). Then $\omega(z)$ is unstable from below (resp., above).*

PROOF. To fix ideas we assume $z \in BC \setminus Q$. Then there exists a sequence $x_n \to z$ and an equilibrium u_1 as in conclusion (b) of the Sequential Limit Set Trichotomy. Suppose $u_1 < x < y$, $y \in \omega(z)$. SOP implies there exist open sets W_x and W_y containing x and y, respectively, and $t_0 \geqslant 0$, such that $\Phi_t(W_x) \leqslant \Phi_t(W_y)$ for all $t \geqslant t_0$. As $\Phi_s(z) \in W_y$ for some large s, by continuity $\Phi_s(x_n) \in W_y$ for some large n. Thus $u_1 \leqslant \Phi_t(x) \leqslant \Phi_{t+s}(x_n)$ for all $t \geqslant t_0$. Letting $t \to \infty$ and using the fact that $\omega(x_n) = \{u_1\}$, we find that $\omega(x) = \{u_1\}$. $\qquad\square$

A set is *minimal* if it is nonempty, closed and invariant, and no proper subset has these three properties. Every positively invariant nonempty compact set contains a minimal set (by Zorn's Lemma). A minimal set containing more than one point is called *nontrivial*.

COROLLARY 2.7. *A compact, nontrivial minimal set M that meets BC (respectively, AC) is unstable from below (resp., above).*

PROOF. Suppose $z \in M \cap BC$. The assumptions on M imply $M = \omega(z)$ and $M \cap E = \emptyset$. Therefore $z \in BC \setminus Q$, and instability follows from Theorem 2.6. $\qquad\square$

When X is a convex subset of a vector space, an alternative formulation of Theorem 2.6 is that $\omega(z)$ belongs to the upper boundary of the basin of attraction of the equilibrium u_1. Corollary 2.7 implies that periodic orbits are unstable. Theorem 2.6 is motivated by Theorem 1.6 in Hirsch [79].

The following sharpening of Theorem 1.19, due to Smith and Thieme [199], is an immediate corollary of the Sequential Limit Set Trichotomy.

THEOREM 2.8. *If $J \subset X$ is a totally ordered arc having property (C), then $J \setminus Q$ is a discrete, relatively closed subset of J; hence it is countable, and finite when J is compact.*

PROOF. Every limit point z of $J \setminus Q$ is strongly accessible from above or below by a sequence $\{\tilde{x}_n\}$ in $J \setminus Q$. As Property (C) implies $J \subset BC \cup AC$, there is a sequence $\{x_n\}$ satisfying (a), (b) or (c) of Theorem 2.3 (or its dual result), all of which imply $x_n \in Q$. Thus $J \setminus Q$ contains none of its limit points, which implies the conclusion. $\qquad\square$

The following result sharpens Theorems 1.30 and 2.8:

PROPOSITION 2.9. *Assume X is normally ordered and every point is accessible from above and below. Let $J \subset X$ be a totally ordered compact arc having property (C), with* *endpoints $a < b$ such that $\omega(a)$ is an equilibrium stable from below and $\omega(b)$ is an equilibrium stable from above. Then J contains a point whose trajectory converges to a stable equilibrium.*

PROOF. Denote by C_s (respectively: C_+, C_-) the set of convergent points whose omega limits belong to S (resp.: to S_+, S_-). Then $C_+ \cap C_- = C_s$ by Proposition 1.25(b).

Set $\sup(J \cap C_-) = z \in J$.

Case 1: $z \notin C_-$. Then $z > a$. Choose a sequence $x_1 < x_2 < \cdots < z$ in $J \cap C_-$ such that $x_n \to z$. By the Sequential Limit Set Trichotomy 2.3 it suffices to consider the following three cases:

(a) There exists $u_0 \in E$ such that

$$\omega(x_n) < \omega(x_{n+1}) < \omega(z) = \{u_0\}.$$

This is not possible, because $u_0 \in S_-$ by Proposition 2.4(a), yielding the contradiction $z \in C_-$.

(b) There exists $u_1 = \sup\{u \in E : u < \omega(z)\}$, and for all n we have

$$\omega(x_n) = \{u_1\} < \omega(z).$$

Now 2.4(b) has $u_1 \in S_+$, hence $x_n \in C_+$. Therefore $x_n \in C_+ \cap C_- = C_s$, as required.

(c) $\omega(x_n) = \omega(z)$. This is not possible because $x_n \in C_-$ and $\omega(z) = \omega(x_n)$ implies the contradiction $z \in C_-$.

Thus (b) holds, validating the conclusion when $z \notin C_-$.

Case 2: $z \in C_-$. If $z = b$ then $z \in C_+ \cap C_- = C_s$ and there is nothing more to prove. Henceforth we assume $z < b$.

The closed subinterval $K \subset J$ with endpoints z, b satisfies the hypotheses of the theorem. Set $\inf(K \cap C_+) = w \in K$. The dual of the reasoning above shows that if $w \notin C_+$ then the conclusion of the theorem is true.

From now on we assume $w \in C_+$. If $w = z$ there is nothing more to prove, so we also assume $w > z$. Let $L \subset K$ be the closed subinterval with endpoints w and z. Let $\{\bar{x}_n\}$ be a sequence in L converging to w from below.

One of the conclusions (a), (b) or (c) of 2.3 holds. Referring to the corresponding parts of 2.4, we see in case (a) that $\omega(w)$ is an equilibrium $\bar{u}_0$ that is stable from below; but $w > z$, so this contradicts the definition of z. If (b) holds, $\omega(\bar{x}_n)$ is an equilibrium $\bar{u}_1$ stable from above. But $\bar{x}_n < w$, so this contradicts the definition of w. In case (c) we have for all n that $\omega(\bar{x}_n) = \omega(w)$, which is an equilibrium stable from above. But $\bar{x}_n < w$ for $n > 1$, again contradicting the definition of w. $\qquad\square$

In the following result the assumption on equilibria holds when Φ has a global compact attractor.

PROPOSITION 2.10. *Assume X is an open subset of a strongly ordered Banach space, Φ is strongly monotone, and every equilibrium has a neighborhood attracted to a compact set. Let $J \subset X$ be a totally ordered compact arc, with endpoints $a < b$ such that $\omega(a)$ is* *an equilibrium stable from below and $\omega(b)$ is an equilibrium stable from above. Then J contains a point whose trajectory converges to a stable equilibrium.*

PROOF. Apply Proposition 2.9 to the to the SOP semiflow $\widehat{\Phi}$ in the normally ordered space $\widehat{X}$ (see Section 1.7), to obtain an equilibrium p that is stable for $\widehat{\Phi}$. This means p is order stable for Φ, hence stable for Φ by Proposition 1.28. $\qquad\square$

COROLLARY 2.11. *Let X be a p-convex open set in an ordered Banach space Y. Assume Φ has a compact global attractor. Suppose that either Y is normally ordered, or Y is strongly ordered and Φ is strongly monotone. Then:*

(i) *There is a stable equilibrium.*

(ii) *Let $u, v \in X$ be such that $u < v$ and there exist real numbers $r, s > 0$ such that $u < \Phi_r(u), \Phi_s(v) < v$. Then there is a stable equilibrium in $[u, v]$.*

In case (ii) *with Y normally ordered, the hypothesis of a global attractor can be replaced the assumption that the line segment joining u to v from satisfies condition* (C).

PROOF. We first prove (ii). Monotonicity shows that $\omega(x) \subset [u, v]$ for all $x \in [u, v]$. The Convergence Criterion implies

$$\Phi_t(u) \to a \in E \cap [u, v], \qquad \Phi_t(v) \to b \in E \cap [u, v].$$

We claim that $a \in S_-$ and $b \in S_+$, and $a \in S$ is stable if $a = b$. When Y is normal this follows from Propositions 1.25(b) and (d), and it is easy to prove directly when Φ is strongly monotone. Suppose $a < b$. By p-convexity and Theorems 2.9 and 2.10, the line segment from a to b lies in $[u, v] \cap X$ and contains a point whose x such that $\omega(x)$ is a stable equilibrium z. As noted above, $z \in [u, v]$.

We prove (i) by finding u and v as in (ii). By Theorem 2.8 and compactness of the global attractor, there is a minimal equilibrium p and a maximal equilibrium $q > p$. As X is open, it contains a totally ordered line segment $J < p$. By Theorem 1.19 J contains a quasiconvergent point $u < p$. As $\omega(u) \leqslant p$, minimality of p implies $\Phi_t(u) \to p$. Similarly there exist $v > q$ with $\Phi_t(v) \to q$. It follows from SOP that $u < \Phi_r(u), \Phi_s(v) < v$ for some $r, s > 0$. $\qquad\square$

For strongly monotone semiflows, the existence of order stable equilibria in attractors was treated in Hirsch [68,69,73].

2.2. *Generic quasiconvergence and stability*

The following result adapted from Smith and Thieme [197] refines Theorems 1.22 and 1.21:

THEOREM 2.12. (i) $AC \cup BC \subset \overline{\operatorname{Int} Q} \cup C$. *Therefore if $AC \cup BC$ is dense, so is Q.*

(ii) $(\operatorname{Int} AC) \cup (\operatorname{Int} BC) \subset \overline{\operatorname{Int} Q}$. *Therefore if $(\operatorname{Int} AC) \cup (\operatorname{Int} BC)$ is dense, so is $\operatorname{Int} Q$.*

PROOF. Every $z \in BC$ is the limit of an omega compact sequence $x_1 < x_2 < \cdots$ such that (a), (b) or (c) of the Sequential Limit Set Trichotomy Theorem 2.3 holds, and $z \in \overline{\text{Int } Q} \cup C$ in each case; the proof for AC is similar.

To prove (ii), assume $z \in \text{Int } BC$. If (a) holds for every point of a neighborhood W of z, then $W \subset C$, whence $z \in \text{Int } Q$. If there is no such W, every neighborhood of z contains a point for which (b) or (c) holds, hence $z \in \overline{\text{Int } Q}$. Similarly for $z \in \text{Int } AC$. $\square$

The next result extends Theorems 8.10 and 9.6 of Hirsch [73] and Theorem 3.9 of Smith and Thieme [197]:

THEOREM 2.13. *Assume X is normally ordered and $\text{Int}(BC \cup AC)$ is dense. Then $A \cup \text{Int } C$ is dense.*

PROOF. We argue by contradiction. If $A \cup \text{Int } C$ is not dense, there exists an open set U such that

$$U \cap \overline{A} = \emptyset = U \cap \overline{\text{Int } C}.$$

Suppose $z \in U \cap BC$, and let $\{x_n\}$ be a sequence in U strongly approximating z from below. Conclusion (b) of the Sequential Limit Set Trichotomy 2.3 is not possible because $z \notin \overline{\text{Int } C}$, and conclusion (c) is ruled out because $z \notin \overline{A}$ (see Proposition 2.4(c)). Therefore conclusion (a) holds, which makes z convergent; likewise when $z \in U \cap AC$. Thus we have $C \supset U \cap (BC \cup AC)$, so $\text{Int } C \supset U \cap \text{Int}(BC \cup AC)$. But the latter set is nonempty by the density hypothesis, yielding the contradiction $U \cap \text{Int } C \neq \emptyset$. $\square$

The following theorem concludes that generic trajectories are not only quasiconvergent, but also stable. Its full force will come into play in the next subsection, under assumptions entailing a dense open set of convergent points.

THEOREM 2.14. *If X is normally ordered and $\text{Int}(BC \cap AC)$ is dense, then $\text{Int}(Q \cap S)$ is dense.*

PROOF. $\text{Int } Q$ is dense by Theorem 2.12. To prove density of $\text{Int } S$, it suffices to prove that if $z \in \text{Int}(BC \cap AC)$, then every open neighborhood U of z meets $\text{Int } S$. We can assume $z \notin \overline{A}$ because $A \subset \text{Int } S$. Let $\{x_n\}$ be an omega compact sequence strongly approximating z from below. Suppose (b) or (c) of the Sequential Limit Set Trichotomy 2.3 holds. Then $x_m \in U$ for $m \geq m_0$. Fix $m \geq m_0$. It follows from Proposition 1.25(d) (with $a = x_m$, $x = x_{m+1}$, $b = x_{m+2}$) that $x_{m+1} \in A$, hence $z \in \overline{A}$; this is proved similarly when $\{x_n\}$ strongly approximates z from above.

Henceforth we can assume z belongs to the open set $W = \text{Int}(BC \cap AC) \setminus \overline{A}$, and consequently that there are omega compact sequences $\{x_n\}$, $\{y_n\}$ strongly approximating z from below and above respectively, for which Theorem 2.3(a) and its dual hold respectively. Then Proposition 1.26 implies $z \in S_+ \cap S_-$, whence $z \in S$ by Proposition 1.25(b). Thus the open set W is contained in $\text{Int } S$, and we have proved $\text{Int } S$ is dense. It follows that $\text{Int } S \cap \text{Int } Q$ is dense. $\square$

2.3. *Improving the limit set dichotomy for smooth systems*

The aim now is to strengthen the Limit Set Dichotomy with additional hypotheses, especially smoothness, in order to obtain the following property:

(ILSD) A semiflow satisfies the *Improved Limit Set Dichotomy* if $x_1 < x_2$ implies that
either
  (a) $\omega(x_1) < \omega(x_2)$, or
  (b) $\omega(x_1) = \omega(x_2) = e \in E$.

We begin with some definitions.

Let X be a subset of the Banach space Y. A map $f : X \to Y$ is said to be *locally* C^1 *at* $p \in X$ if there exists a neighborhood U of p in X and a continuous *quasiderivative* map $f' : U \to L(Y)$, where $L(Y)$ is the Banach space of bounded operators on Y, such that

$$f(x) - f(x_0) = f'(x_0)(x - x_0) + \phi(x, x_0)|x - x_0|, \quad x, x_0 \in U$$

with $\phi(x, x_0) \to 0$ as $x \to x_0$. The following result gives a setting where the quasiderivative is uniquely determined by f. We denote the open ball in Y of center p and radius r by $B_Y(p, r) := \{y \in Y : |y - p| < r\}$.

LEMMA 2.15. *Let* $p \in X \subset Y$ *where* Y *is a strongly ordered Banach space. Assume* $f : X \to Y$ *is locally* C^1 *at* p, *and suppose that either* $B_Y(p, r) \cap Y_+ \subset X$ *or* $B_Y(p, r) \cap (-Y_+) \subset X$ *for some* $r > 0$. *Then* $f'(p)$ *is uniquely defined.*

PROOF. Suppose $B_Y(p, r) \cap Y_+ \subset X$, the other case being similar. Fix $w \gg 0$ and let $y \in Y$. As $w + y/n := k_n \geqslant 0$ for large n, $y = n(k_n - w)$ so $Y = Y_+ - Y_+$.
Assume

$$f(x) - f(p) = A(x - p) + \phi(x, p)|x - p| = B(x - p) + \psi(x, p)|x - p|,$$

where $A, B \in L(Y)$ and $\phi, \psi \to 0$ as $x \to p$ in X. It suffices to show that $Av = Bv$ for all $v \geqslant 0$. The segment $x = p + sv \in X$ for all small $s \geqslant 0$. Inserting it in the formula above, dividing by s, and letting $s \to 0$ yields the desired result. $\qquad\square$

Let Φ be a monotone semiflow on the subset X of the strongly ordered Banach space Y. Concerning X and the set of equilibria E, we assume the following condition on the pair (Y, X):

(OC) Either X is an order convex subset of Y or $E \subset \text{Int } X$. For each $e \in E$ there exists
$r > 0$ such that either $B_Y(e, r) \cap Y_+ \subset X$ or $B_Y(e, r) \cap (-Y_+) \subset X$.

This relatively minor restriction is automatically satisfied if X is an open set, an order interval, or the cone Y_+. The second assertion of (OC) trivially holds if $E \subset \text{Int } X$.

We will also assume the following two conditions hold for some $\tau > 0$. A compact, strongly positive linear operator is called a *Krein–Rutman* operator.

(M) $x_1 < x_2 \implies \Phi_\tau(x_1) \ll \Phi_\tau(x_2)$
(D*) Φ_τ is locally C^1 at each $e \in E$, with $\Phi'_\tau(e)$ a Krein–Rutman operator.

As motivation for (D*), consider the case that X is an open set in Y and Φ_τ is C^1. If $x \in X, y \in Y_+, h > 0$, and $x + hy \in X$, then $(\Phi_\tau(x+hy) - \Phi_\tau(x))/h \geqslant 0$ by monotonicity; on taking the limit as $h \to 0$, we get $\Phi'_\tau(x)y \geqslant 0$. Consequently, $\Phi'_\tau(x)Y_+ \subset Y_+$, and hence the assumption that $\Phi'_\tau(x)$ is strongly positive is not such a severe one. Typically, one usually must verify it anyway to prove that Φ_τ is strongly monotone.

Observe that (M) implies that Φ is strongly order preserving on X.

THEOREM 2.16 (Improved Limit Set Dichotomy). *Let Φ be a monotone semiflow on a subset X of the strongly ordered Banach space Y for which* (OC), (M), *and* (D*) *are satisfied. Then* (ILSD) *holds.*

In particular, (ILSD) holds if X is open, the semiflow Φ continuously differentiable and strongly monotone, and the derivative $\Phi'_t(e)$ is a Krein–Rutman operator at each $e \in E$.

Before giving the proof, we explore the spectral and dynamical implications of (D*).

2.3.1. *The Krein–Rutman theorem* The spectrum of a linear operator $A : Y \to Y$ is denoted by $\mathrm{Spec}(A)$. When A is compact (i.e., completely continuous), $\mathrm{Spec}(A)$ consists of a countable set of eigenvalues and perhaps 0, and the eigenvalues have no accumulation point except possibly 0.

Let $\rho(A)$ be the *spectral radius* of A, that is, $\rho(A) = \max\{|\lambda| : \lambda \in \mathrm{Spec}(A)\}$. Denote the null space of A by $N(A)$ and the range by $\mathrm{Im}(A)$.

The set $\mathrm{KR}(Y)$ of Krein–Rutman operators on Y is given the metric induced by the uniform norm.

THEOREM 2.17 (Krein–Rutman). *Let $A \in \mathrm{KR}(Y)$ and set $r = \rho(A)$. Then Y decomposes into a direct sum of two closed invariant subspaces Y_1 and Y_2 such that $Y_1 = N(A - rI)$ is spanned by $z \gg 0$ and $Y_2 \cap Y_+ = \{0\}$. Moreover, the spectrum of $A|Y_2$ is contained in the closed ball of radius $v < r$ in the complex plane.*

See Krein and Rutman [104], Takáč [214] or Zeidler [244] for proofs.

It follows that each $A \in \mathrm{KR}(Y)$ has a unique unit eigenvector $z(A) \in Y_+$, and $z(A) \in \mathrm{Int}\,Y_+$, $Az(A) = \rho(A)z(A)$.

LEMMA 2.18. *$\rho(A)$ and $z(A)$ are continuous functions of $A \in \mathrm{KR}(Y)$.*

PROOF. The upper semicontinuity of the spectral radius follows from the upper semi-continuity of the spectrum as a function of the operator (Kato [92]). The lower semicontinuity follows from the lower semicontinuity of isolated parts of the spectrum (Kato [92, Chapter IV, Theorem 3.1, Remark 3.3, Theorem 3.16]). Let P_A be the projection onto $N(A) - \rho(A)I)$ along $\text{Im}(A - \rho(A)I)$. Continuity of $A \mapsto P_A$ is proved in [92, Chapter IV, Theorem 3.16]. Let $A_n \to A$ in $\text{KR}(Y)$ and set $z_n = z(A_n)$, $z = z(A)$. Then $(I - P_A))z_n = (P_{A_n} - P_A)z_n \to 0$ as $n \to \infty$, and $\{P_A z_n\}$ is precompact, so $\{z_n\}$ is precompact. If $z_{n_i} \to u$ for some subsequence, then

$$P_A u = \lim A_{n_i} z_{n_i} = \lim \rho(A_{n_i}) z_{n_i} = \rho(A)u.$$

Uniqueness of the positive eigenvector for A (Theorem 2.17) implies $u = z$ and $z_n \to z$. $\square$

For technical reasons it is useful to employ a norm that is more compatible with the partial order. If $w \in \text{Int}\, Y_+$ is fixed, then the set $U = \{y \in Y : -w \ll y \ll w\}$ is an open neighborhood of the origin. Consequently, if $y \in Y$, then there exists $t_0 > 0$ such that $t_0^{-1} y \in U$, hence, $-t_0 w \ll y \ll t_0 w$. Define the w-norm by

$$\|y\|_w = \inf\{t > 0 : -tw \leqslant y \leqslant tw\}.$$

Since $w \in \text{Int}\, Y_+$, there exists $\delta > 0$ such that for all $y \in Y \setminus \{0\}$ we have $w \pm \delta \frac{y}{|y|} \in Y_+$. Thus

$$\|y\|_w \leqslant \delta^{-1}|y|$$

holds for all $y \in Y$, implying that the w-norm is weaker than the original norm. In fact, the two norms are equivalent if Y_+ is normal, but we will have no need for this result. See Amann [6] and Hirsch [73] for more results in this direction. It will be useful to renormalize the positive eigenvector $z(A)$ for $A \in \text{KR}(Y)$. The next result says this can be done continuously. Continuity always refers to the original norm topology on Y unless the contrary is explicitly stated.

LEMMA 2.19. *Let* $Z(A) = z(A)/\|z(A)\|_w$ *and* $\beta(A) = \sup\{\beta > 0 : Z(A) \geqslant \beta w\}$. *Then* $\beta(A) > 0$, $Z(A) \geqslant \beta(A)w$, *and the maps* $A \to Z(A)$ *and* $A \to \beta(A)$ *are continuous on* $\text{KR}(Y)$.

PROOF. Since the w-norm is weaker than the original norm, the map $A \mapsto \|z(A)\|_w$ is continuous. This implies that $Z(A)$ is continuous in A. It is easy to see that $\beta(A) > 0$. Let $\epsilon > 0$ satisfy $2\epsilon < \beta(A)$ and let $A_n \to A$ in $\text{KR}(Y)$. Then $-\epsilon w \leqslant Z(A) - Z(A_n) \leqslant \epsilon w$ for all large n by continuity of Z and because the w-norm is weaker than the original norm. Therefore, $Z(A_n) = Z(A_n) - Z(A) + Z(A) \geqslant (\beta(A) - \epsilon)w$, so $\beta(A_n) \geqslant \beta(A) - \epsilon$ for all large n. Similarly, $Z(A) = Z(A) - Z(A_n) + Z(A_n) \geqslant (\beta(A_n) - \epsilon)w$ for all large n, so $\beta(A) \geqslant \beta(A_n) - \epsilon$ for all large n. Thus, $\beta(A) - \epsilon \leqslant \beta(A_n) \leqslant \beta(A) + \epsilon$ holds for all large n, completing the proof. $\square$

The key to improving the Limit Set Dichotomy is to show that the omega limit set of a point x that is quasiconvergent but not convergent, is uniformly unstable in the linear approximation. The direction of greatest instability at $e \in \omega(x)$ is the positive direction $z(e) := z(\Phi_\tau'(e))$. The number $\rho(e) := \rho(\Phi_\tau'(e))$ gives a measure of the instability.

Nonordering of Limit Sets means that positive directions are, in some rough sense, "transverse" to the limit set. Thus our next result means that the limit set is uniformly unstable in a transverse direction.

LEMMA 2.20. *Assume* (D*). *Let x be quasiconvergent but not convergent. Then $\rho(e) > 1$ for all $e \in \omega(x)$.*

PROOF. Fix $e \in \omega(x)$. Since $\omega(x)$ is connected, e is the limit of a sequence $\{e_n\}$ in $\omega(x) \cap U \setminus \{e_0\}$, where U is the neighborhood of e in the definition of Φ_τ is locally C^1 at e. Then

$$e_0 - e_n = \Phi_\tau(e) - \Phi_\tau(e_n) = \Phi'_\tau(e)(e - e_n) + \mathrm{o}(|e - e_n|),$$

where $\mathrm{o}(|e - e_n|)/|e - e_n| \to 0$ as $n \to \infty$. Put $v_n = (e - e_n)/|e - e_n|$. Then

$$v_n = \Phi'_\tau(e)v_n + r_n, \quad r_n \to 0, \quad n \to \infty.$$

The compactness of $\Phi'_\tau(e)$ implies that v_n has a convergent subsequence v_{n_i}; passing to the limit along this subsequence leads to $v = \Phi'_\tau(e)v$ for some unit vector v. Thus $\rho(e) \geqslant 1$. If $\rho(e) = 1$, then the Krein–Rutman Theorem implies $v = rz(e)$ where $r = \pm 1$. Consequently,

$$(e - e_{n_i})/|e - e_{n_i}| \to rz(e)$$

as $i \to \infty$. It follows that $e \ll e_{n_i}$ or $e \gg e_{n_i}$ for all large i, contradicting the Nonordering of Limit Sets. $\qquad\qquad\square$

PROOF OF THEOREM 2.16. By the Limit Set Dichotomy (Theorem 1.16), it suffices to prove: If $x_1 < x_2$ and $\omega(x_1) = \omega(x_2) = K \subset E$, then K is a singleton. K is compact and connected, unordered by the Nonordering of Limit Sets, and consists of fixed points of Φ_τ. Arguing by contradiction, we assume K is not a singleton.

Set $v_n = \Phi_{n\tau}(x_1)$, $u_n = \Phi_{n\tau}(x_2)$. Then $\mathrm{dist}(K, u_n) \to 0$ and $\mathrm{dist}(K, v_n) \to 0$ as $n \to \infty$. Moreover (M) and the final assertion of the Limit Set Dichotomy imply

$$u_n - v_n \gg 0, \quad u_n - v_n \to 0.$$

Fix $w \gg 0$ and define real numbers

$$\alpha_n = \sup\{\alpha \in \mathbb{R}: \alpha \geqslant 0, \ \alpha w \leqslant u_n - v_n\}.$$

Then $\alpha_n > 0$ and $\alpha_n \to 0$.

To simplify notation, define $S: X \to X$ by $S(x) := \Phi_\tau(x)$. Choose $e_n \in K$ such that $v_n - e_n \to 0$ as $n \to \infty$. By Lemma 2.20, local smoothness of Φ_τ and compactness of K, there exists $r > 1$ such that $\rho(e) > r$ for all $e \in K$. Let $z_n = Z(e_n)$ be the normalized positive eigenvector for $S'(e_n) = \Phi'_\tau(e_n)$ so $\|z_n\|_w = 1$ and $z_n \leqslant w$. By Lemma 2.19, there exists $\epsilon > 0$ such that $\beta(e_n) \geqslant \epsilon$ for all n. In particular, $w \geqslant z_n \geqslant \epsilon w$ for all n.

Fix a positive integer l such that $r^l \epsilon > 1$.

For each $e \in K$, by (D*) we can choose an open neighborhood W_e of e in X and a continuous map $S' : W_e \to L(Y)$ such that for $x, x_0 \in W_e$ we have

$$Sx - Sx_0 = S'(x_0)(x - x_0) + \phi(x, x_0)|x - x_0|, \qquad \lim_{x \to x_0} \phi(x, x_0) = 0.$$

Putting $x_0 = e$ and estimating norms, one easily sees that there exists a convex open neighborhood $U_e \subset W_e$ of e such that $S^i(U_e) \subset W_e$ for $1 \leqslant i \leqslant l$. Furthermore, a simple induction argument implies that S^l is locally C^1 at e with quasiderivative

$$\left(S^l\right)' : U_e \to L(Y), \qquad \left(S^l\right)'(x) = S'\left(S^{l-1}x\right) \circ S'\left(S^{l-2}x\right) \circ \cdots \circ S'(x).$$

By compactness of K there is a finite subset $\{e_1, \ldots, e_v\} \subset K$ such that the sets U_{e_j} cover K. Set $U_j = U_{e_j}$, $W_j = W_{e_j}$. Then

$$K \subset \bigcup_{j=1}^{v} U_j, \qquad S^i(U_j) \subset W_j \quad (1 \leqslant i \leqslant l),$$

and for $z, z_0 \in U_j$

$$S^l(z) - S^l(z_0) = \left(S^l\right)'(z_0)(z - z_0) + \phi_{l,j}(z, z_0)|z - z_0|, \qquad \lim_{z \to z_0} \phi_{l,j}(z, z_0) = 0,$$

and the usual chain rule expresses $(S^l)'$ in terms of S'.

By (OC), either X is order convex in Y or $E \subset \operatorname{Int} X$. In the order convex case, from $v_n \ll v_n + \alpha_n w \leqslant u_n$ we infer that $v_n + s\alpha_n w \in X$ for all $s \in [0, 1]$. Since $v_n - e_n \to 0$, $v_n - u_n \to 0$, and $\alpha_n \to 0$, for sufficiently large n there exists $j(n) \in \{1, \ldots, v\}$ such that $U_{j(n)}$ contains the points v_n, u_n, e_n, and $v_n + s\alpha_n w$ for all $s \in [0, 1]$. When $E \subset \operatorname{Int} X$ the same conclusion holds, and we can take U_j, V_j to be open in Y.

Lemma 2.15 justifies the application of the fundamental theorem of calculus to the map $[0, 1] \to X$, $s \mapsto S^l(v_n + s\alpha_n w)$, leading to

$$S^l(v_n + \alpha_n w) - S^l(v_n) = \left(S^l\right)'(e_n)(\alpha_n w) + \alpha_n \delta_n$$

and

$$\delta_n = \int_0^1 \left[\left(S^l\right)'(v_n + \eta\alpha_n w) - \left(S^l\right)'(e_n)\right] w \, d\eta.$$

Using that $v_n + \alpha_n w - e_n \to 0$, K is compact, and $(S^l)'$ is continuous, it is easy to show that

$$\lim_{n \to \infty} \max_{0 \leqslant \eta \leqslant 1} \left|\left[\left(S^l\right)'(v_n + \eta\alpha_n w) - \left(S^l\right)'(e_n)\right] w\right| = 0.$$

It follows that $d_n := \|\delta_n\|_w \to 0$ as $n \to \infty$. Because $w \geqslant z_n \geqslant \epsilon w \gg 0$ and $\delta_n \geqslant -d_n w$, for sufficiently large n we have:

$$
\begin{aligned}
S^l(v_n + \alpha_n w) - S^l(v_n) &\geqslant \left[(S^l)'(e_n)\right]\alpha_n w - \alpha_n d_n w \\
&\geqslant \left[S'(e_n)\right]^l \alpha_n w - \alpha_n d_n w \\
&\geqslant r^l \alpha_n z_n - \alpha_n d_n w \\
&\geqslant (r^l \epsilon - d_n)\alpha_n w \\
&\geqslant \alpha_n w,
\end{aligned}
$$

and therefore

$$
u_{n+l} = S^l(u_n) \geqslant S^l(v_n + \alpha_n w) \geqslant S^l v_n + \alpha_n w = v_{n+l} + \alpha_n w.
$$

Thus $\alpha_n w \leqslant u_{n+l} - v_{n+l}$, so the definition of α_{n+l} implies $\alpha_{n+l} \geqslant \alpha_n > 0$ for all sufficiently large n. Therefore the sequence $\{\alpha_i\}_{i \in \mathbb{N}_+}$, which converges to 0, contains a nondecreasing positive subsequence $\{\alpha_{n+kl}\}_{k \in \mathbb{N}_+}$. This contradiction implies K is a singleton. $\square$

A drawback of the Improved Limit Set Dichotomy, Theorem 2.16, is that the topology on X comes from a strongly ordered Banach space $Y \supset X$, severely limiting its application to infinite-dimensional systems. The following extension permits use of (ILSD) in more general spaces:

PROPOSITION 2.21. *Let X^1, X^0 be ordered spaces such that $X^1 \subset X^0$ and the inclusion map $j : X^1 \hookrightarrow X^0$ is continuous and order preserving. For $k = 0, 1$ let Φ^k be a monotone semiflow on X^k with compact orbit closures. Assume for all $t > 0$ that Φ_t^0 maps X^0 continuously into X^1, and $\Phi_t^0 | X^1 = \Phi_t^1$. If (ILSD) holds for Φ^1, it also holds for Φ^0.*

PROOF. Denote the closure in X^k of any $S \subset X^k$ by $\mathsf{C}_k S$. For $k \in \{0, 1\}$ and $x \in X^k$, let $O_k(x)$ and $\omega_k(x)$ respectively denote the orbit and omega limit set of x.

The hypotheses imply that the compact set $\mathsf{C}_0 O_0(x)$, which is positively invariant for Φ^0, is mapped homeomorphically by Φ_1^0 onto $\mathsf{C}_1 O_1(y) \subset X^1$, which is positively invariant for Φ^1. As Φ^0 and Φ^1 coincide in X^1, we see that $\omega_0(x) = \omega_1(y)$ as compact sets. Hence Φ^0 and Φ^1 have the same collection of omega limit sets, which implies the conclusion. $\square$

THEOREM 2.22 (Improved Sequential Limit Set Trichotomy). *Assume (ILSD). Let $\{\tilde{x}_n\}$ be a sequence approximating $z \in BC$ from below, with $\bigcup_n \omega(\tilde{x}_n)$ compact. Then there is a subsequence $\{x_n\}$ such that exactly one of the following three conditions holds for all n:*
    (a) *There exists $u_0 \in E$ such that*

$$
\omega(x_n) < \omega(x_{n+1}) < \omega(z) = \{u_0\}
$$

    and

$$
\lim_{n \to \infty} \mathrm{dist}\big(\omega(x_n), u_0\big) = 0.
$$

(b) *There exists $u_1 = \sup\{u \in E: u < \omega(z)\}$, and*

$$\omega(x_n) = \{u_1\} < \omega(z).$$

In this case $z \in \overline{\mathrm{Int}\,C}$. Moreover z has a neighborhood W such that if $w \in W$, $w < z$ then $\Phi_t(w) \to u_1$ and $\Phi_t(w) > u_1$ for sufficiently large t.
(c$'$) *There exists $u_2 \in E$ such that $\omega(x_n) = \omega(x_0) = u_2$.*

Note that z is convergent in (a), strongly accessible from below by convergent points in (b), and convergent in (c$'$).

PROOF. Conclusions (a) and (b) are the same as in the Sequential Limit Set Trichotomy, Theorem 2.3. If 2.3(c) holds, then (c$'$) follows from (ILSD). $\quad\square$

PROPOSITION 2.23. *Assume (ILSD). If $x \in BC \setminus C$ then $\omega(x)$ is unstable from below. If $x \in AC \setminus C$ then $\omega(x)$ is unstable from above.*

PROOF. This is just Theorem 2.6 if $x \notin Q$. If $x \in BC \cap (Q \setminus C)$, we must have conclusion (b) of Theorem 2.22. This provides $u_1 \in E$ such that $\omega(x_n) = \{u_1\}$ for all n, and the remainder of the proof mimics that of Theorem 2.6. $\quad\square$

A consequence of Proposition 2.23 is that if $x \in BC \cap AC$ is nonconvergent, then $\omega(x)$ lies in both the upper boundary of the basin of attraction of an equilibrium u_0 and the lower boundary of the basin of attraction of an equilibrium v_0, where $u_0 < L < v_0$. Thus $\omega(x)$ forms part of a separatrix separating the basins of attraction of u_0 and v_0.

2.4. *Generic convergence and stability*

The following result concludes that the set C of convergent points is dense and open in totally ordered arcs:

THEOREM 2.24. *Assume (ILSD) and let $J \subset X$ be a totally ordered arc having property (C). Then $J \setminus C$ is a discrete, relatively closed subset of J; hence it is countable, and finite when J is compact.*

PROOF. The proof is like that of Theorem 2.8, using the Improved Limit Set Trichotomy 2.22 instead of the Sequential Limit Set Trichotomy 2.3. $\quad\square$

We can now prove the following generic convergence and stability results:

THEOREM 2.25. *Assume (ILSD).*
(a) *$AC \cup BC \subset \overline{\mathrm{Int}\,C} \cup C$. In particular, if $AC \cup BC$ is dense, so is $\mathrm{Int}\,C$ is dense.*
(b) *If $\mathrm{Int}(BC \cap AC)$ is dense and X is normally ordered, then $\mathrm{Int}(C \cap S)$ is dense.*

PROOF. The proof of (a) is similar to that of Theorem 2.12: take $p \in X \setminus \operatorname{Int} C$ and use the Improved Limit Set Trichotomy (Theorem 2.22), instead of the Limit Set Trichotomy, to show that $p \in \overline{\operatorname{Int} C} \cup C$. Conclusion (b) follows from (a) and Theorem 2.14. $\square$

THEOREM 2.26. *Assume X is a subset of a strongly ordered Banach space Y, and a dense open subset of X is covered by totally ordered line segments. Let* (M) *and* (D*) *hold. Then:*
   (a) *The set of convergent points has dense interior.*
   (b) *Suppose Y is normally ordered. Then the set of stable points has dense interior.*
   (c) *Assume Y is normally ordered; X is open or order convex or a subcone of Y_+; and every closed totally ordered subset of E is compact. Then there is a stable equilibrium, and an asymptotically stable equilibrium when E is finite.*

PROOF. The assumption in (a) implies $BC \cap AC$ has dense interior and condition (OC) holds. Therefore the Improved Limit Set Dichotomy (ILSD) holds by Theorem 2.16, so (a) and (b) follow from Theorem 2.25(a). Conclusion (c) is a consequence of (a) and Theorem 1.30. $\square$

As most orbits with compact closure converge to an equilibrium, it is natural to investigate the nature of the convergence. It might be expected that most trajectories converging to a stable equilibrium are eventually increasing or decreasing. We quote a theorem of Mierczyński that demonstrates this under quite general conditions for smooth strongly monotone dynamical systems, including cases when the equilibrium is not asymptotically stable in the linear approximation. Mierczyński assumes the following hypothesis:

(M$_1$) X is an open set in a strongly ordered Banach space Y. Φ is C^1 on $(0, \infty) \times X$ and strongly monotone, $\Phi'_t(x)$ is strongly positive for all $t > 0$, $x \in X$, and $\Phi'_1(x)$ is compact.

The following local trichotomy due to Mierczyński [138] builds on earlier work of Poláčik [161]:

THEOREM 2.27. *Assume* (M$_1$). *Then each equilibrium e satisfying $\rho(\Phi'_1(e)) \leqslant 1$ belongs to a locally invariant submanifold Σ_e of codimension one that is smooth and unordered and has the following property. If $\lim_{t\to\infty} \Phi_t(x) = e$, there exists $t_0 \geqslant 0$ such that one of the following holds as $t \to \infty$, $t \geqslant t_0$:*
   (i) $\Phi_t(x)$ *decreases monotonically to e;*
   (ii) $\Phi_t(x)$ *increases monotonically to e;*
   (iii) $\Phi_t(x) \in \Sigma_e$.

Mierczyński also provides further important information: The trajectories in cases (i) and (ii) lie in curves tangent at e to the one-dimensional principle eigenspace Y_1 of $\Phi'_1(e)$ described in the Krein–Rutman Theorem 2.17. The hypersurface Σ_e is locally unique in a neighborhood of e. Its tangent space is the closed complementary subspace Y_2, hence Σ_e is transverse to $z = z(\Phi'_1(e)) \gg 0$ at e. Strong monotonicity implies that when (i) or (ii) holds, e is asymptotically stable for the induced local flow in Σ_e, even when e is not stable.

2.4.1. *Background and related results*   Smith and Thieme [197,199] introduced the compactness hypothesis (C) and obtained the Sequential Limit Set Trichotomy. This tool streamlines many of the arguments and leads to stronger conclusions so the presentation here follows [197,199]. Takáč [210] extends the compactness hypothesis, which leads to additional stability concepts.

The results of Smith and Thieme [199] on generic convergence for SOP semiflows were motivated by earlier work of Poláčik [160], who obtained such results for abstract semilinear parabolic evolution systems assuming less compactness but more smoothness than Smith and Thieme.

The set A of asymptotically stable points can be shown to be dense under suitable hypotheses. See, e.g., Hirsch [73, Theorem 9.6]; Smith and Thieme [197, Theorems 3.13 and 4.1].

Hirsch [69] shows that if K is a nonempty compact, invariant set that attracts all points in some neighborhood of itself, then K contains an order-stable equilibrium.

It is not necessary to assume, as we have done here, that the semiflow is globally defined, that is, that trajectories are defined for all $t \geqslant 0$; many of the results adapt to local semiflows. See Hirsch [73], Smith and Thieme [199].

3. Ordinary differential equations

Throughout this section $\mathbb{R}^n$ is ordered by a cone K with nonempty interior. Our first objective is to explore conditions on a vector field that make the corresponding local semiflow monotone with respect to the order defined by K. It is convenient to work with time-dependent vector fields. We then investigate the long-term dynamics of autonomous vector fields f that are K-cooperative, meaning that K is invariant under the forward flow of the linearized system. These results are applied to competitive vector fields by the trick of time-reversal. In fairly general circumstances, limit sets of cooperative or competitive systems in $\mathbb{R}^n$ are invariant sets for systems in $\mathbb{R}^{n-1}$. This leads to particularly sharp theorems for $n = 2$ and 3.

A cone is *polyhedral* if it is the intersection of a finite family of closed half spaces. For example, the standard cone $\mathbb{R}^n_+$ is polyhedral, while the ice-cream cone is not.

The *dual cone* to K is the closed cone K^* in the dual space $(\mathbb{R}^n)^*$ of linear functions on $\mathbb{R}^n$, defined by

$$K^* = \left\{ \lambda \in \left(\mathbb{R}^n\right)^* \colon \lambda(K) \geqslant 0 \right\}.$$

To $\lambda \in K^*$ we associate the vector $a \in \mathbb{R}^n$ such that $\lambda(x) = \langle a, x \rangle$ where $\langle a, x \rangle$ denotes the standard inner product on $\mathbb{R}^n$. Under this association K^* is canonically identified with a cone in $\mathbb{R}^n$, namely, the set of vectors a such that a is normal to a supporting hyperplane H of K, and a and K lie in a common halfspace bounded by H.

We use the following simple consequence of general results on the separation of two closed convex sets:

$$x \in K \quad \Longleftrightarrow \quad \lambda(x) \geqslant 0 \quad (\lambda \in K^*).$$

See, e.g., Theorem 1.2.8 of Berman et al. [18].

PROPOSITION 3.1. *If $x \in K$, then $x \in \mathrm{Int}\, K$ if and only if $\lambda(x) > 0$ for all $\lambda \in K^* \setminus \{0\}$.*

PROOF. Suppose $x \in \mathrm{Int}\, K$, $\lambda \in K^* \setminus \{0\}$, and $v \in X$ satisfies $\lambda(v) \neq 0$. Then $x \pm \epsilon v \in K$ for sufficiently small $\epsilon > 0$, so

$$\lambda(x \pm \epsilon v) = \lambda(x) \pm \epsilon \lambda(v) \geqslant 0,$$

implying that $\lambda(x) > 0$.

To prove the converse, assume $\mu(x) > 0$ for all functionals μ in the compact set $\Gamma = \{\lambda \in K^*\colon \|\lambda\| = 1\}$. As $\inf\{\mu(x)\colon \mu \in \Gamma\} > 0$, continuity of the map $(x, \lambda) \mapsto \lambda(x)$ implies $\mu(y) > 0$ for all y in some neighborhood U of x and all $\mu \in \Gamma$. If $\lambda \in K^*$ then $\|\lambda\|^{-1}\lambda \in \Gamma$ and therefore $\lambda(y) > 0$ for all $y \in U$. This proves $U \subset K$. $\qquad\square$

An immediate consequence of Proposition 3.1 is that if $x \in \partial K$, then there exists a nontrivial $\lambda \in K^*$ such that $\lambda(x) = 0$.

3.1. *The quasimonotone condition*

Let $J \subset \mathbb{R}$ be a nontrivial open interval, $D \subset \mathbb{R}^n$ an open set and $f : J \times D \to \mathbb{R}^n$ a locally Lipschitz function. We consider the ordinary differential equation

$$x' = f(t, x). \tag{3.1}$$

For every $(t_0, x_0) \in J \times D$, the initial value problem $x(t_0) = x_0$ has a unique noncontinuable solution defined on an open interval $J(t_0, x_0) \subset \mathbb{R}$. We denote this solution by $t \mapsto x(t, t_0, x_0)$. The notation $x(t, t_0, x_0)$ will carry the tacit assumption that $(t_0, x_0) \in J \times D$ and $t \in J(t_0, x_0)$. For fixed s_0, t_0 the map $x_0 \mapsto x(s_0, t_0, x_0)$ is a homeomorphism between open subsets of $\mathbb{R}^n$, the inverse being $x_0 \mapsto x(t_0, s_0, x_0)$.

System (3.1) is called *monotone* if $x_0 \leqslant x_1 \Longrightarrow x(t, t_0, x_0) \leqslant x(t, t_0, x_1)$.

The time-dependent vector field $f : J \times D \to \mathbb{R}^n$ satisfies the *quasimonotone condition* in D if for all $(t, x), (t, y) \in J \times D$ and $\phi \in K^*$ we have:

(QM) $x \leqslant y$ and $\phi(x) = \phi(y)$ implies $\phi(f(t, x)) \leqslant \phi(f(t, y))$.

The quasimonotone condition was introduced by Schneider and Vidyasagar [177] for finite-dimensional, autonomous linear systems and used later by Volkmann [224] for nonlinear infinite-dimensional systems. The following result is inspired by a result of Volkmann [224] and work of W. Walter [227]. See also Uhl [221], Walcher [226].

THEOREM 3.2. *Assume f satisfies (QM) in D, $t_0 \in J$, and $x_0, x_1 \in D$. Let $\prec$ denote any one of the relations $\leqslant, <, \ll$. If $x_0 \prec x_1$ then $x(t, t_0, x_0) \prec x(t, t_0, x_1)$, hence (3.1) is monotone. Conversely, if (3.1) is monotone then f satisfies (QM).*

PROOF. Assume that $x(t, t_0, x_i)$, $i = 0, 1$ are defined for $t \in [t_0, t_1]$ and $x_0 \leqslant x_1$. Let $v \gg 0$ be fixed and define $x_\epsilon := x_1 + \epsilon v$ and $f_\epsilon(t, x) := f(t, x) + \epsilon v$ for $\epsilon > 0$. Denote by $x(t) := x(t, t_0, x_0)$ and let $y_\epsilon(t) := x(t, t_0, x_\epsilon, \epsilon)$ denote the solution of the initial value problem $x'(t) = f_\epsilon(t, x)$, $x(t_0) = x_\epsilon$. It is well known that $y_\epsilon(t)$ is defined on $[t_0, t_1]$ for all sufficiently small ϵ. We show that $x(t) \ll y_\epsilon(t)$ for $t_0 \leqslant t \leqslant t_1$ and all sufficiently small $\epsilon > 0$. If not, then as $x(t_0) \ll y_\epsilon(t_0)$, there would exist $\epsilon > 0$ and $s \in (t_0, t_1]$ such that $x(t) \ll y_\epsilon(t)$ for $t_0 \leqslant t < s$ and $y_\epsilon(s) - x(s) \in \partial K$. By Proposition 3.1, there exists a non-trivial $\phi \in K^*$ such that $\phi(y_\epsilon(s) - x(s)) = 0$ but $\phi(y_\epsilon(t) - x(t)) > 0$ for $t_0 \leqslant t < s$. It follows that

$$\frac{d}{dt}\left[\phi\big(y_\epsilon(t)\big) - \phi\big(x(t)\big)\right]\bigg|_{t=s} \leqslant 0,$$

hence

$$\phi\big(f\big(s, y_\epsilon(s)\big)\big) < \phi\big(f\big(s, y_\epsilon(s)\big)\big) + \epsilon\phi(v) = \phi\big(f_\epsilon\big(s, y_\epsilon(s)\big)\big) \leqslant \phi\big(f\big(s, x(s)\big)\big),$$

where the last inequality follows from the one above. On the other hand, by (QM) we have

$$\phi\big(f\big(s, y_\epsilon(s)\big)\big) \geqslant \phi\big(f\big(s, x(s)\big)\big).$$

This contradiction proves that $x(t) \ll y_\epsilon(t)$ for $t_0 \leqslant t \leqslant t_1$ and all small $\epsilon > 0$. Since $y_\epsilon(t) = x(t, t_0, x_\epsilon, \epsilon) \to x(t, t_0, x_1)$ as $\epsilon \to 0$, by taking the limit we conclude that $x(t, t_0, x_0) \leqslant x(t, t_0, x_1)$ for $t_0 \leqslant t \leqslant t_1$.

Fix t_0 and $t \in J(t_0, x_0)$. As the map $h : x_0 \mapsto x(t, t_0, x_0)$ is injective, from $x_0 < x_1$ we infer $x(t, t_0, x_0) < x(t, t_0, x_1)$. Note that $h(D \cap [x_0, x_1]) \subset [x(t, t_0, x_0), x(t, t_0, x_1)]$. Therefore the relation $x_0 \ll x_1$ implies $\text{Int}\, D \cap [x_0, x_1] \neq \emptyset$. Injectivity of h and invariance of domain implies $\text{Int}[x(t, t_0, x_0), x(t, t_0, x_1)] \neq \emptyset$, which holds if and only if $x(t, t_0, x_0) \ll x(t, t_0, x_1)$.

Conversely, suppose that (3.1) is monotone, $t_0 \in J$, $x_0, x_1 \in D$ with $x_0 \leqslant x_1$ and $\phi(x_0) = \phi(x_1)$ for some $\phi \in K^*$. Since $x(t, t_0, x_0) \leqslant x(t, t_0, x_1)$ for $t \geqslant t_0$ we conclude that $\frac{d}{dt}\phi[x(t, t_0, x_1) - x(t, t_0, x_0)]|_{t=t_0} \geqslant 0$, or $\phi(f(t_0, x_1)) \geqslant \phi(f(t_0, x_0))$. Thus (QM) holds. $\square$

Theorem 3.2 has been stated so as to minimize technical details concerning the domain $J \times D$ by assuming that J and D are open. In many applications, D is a closed set, for example, $D = K$ or $D = [a, b]$ where $a \ll b$. The proof can be modified to handle these (and other) cases. If $D = K$ and K is positively invariant for (3.1), the proof is unchanged because whenever $x \in D$ then $x + \epsilon v \in D$ for small positive ϵ, and because K is also positively invariant for the modified equation. If $D = [a, b]$, then the result follows by applying Theorem 3.2 to $f | J \times [[a, b]]$ and using continuity.

A set S is called *positively invariant* under (3.1) if $S \subset D$ and solutions starting in S stay in S, or more precisely:

$$(t_0, x_0) \in J \times S \quad \text{and} \quad t \in J(t_0, x_0), \, t \geqslant t_0 \quad \Longrightarrow \quad x(t, t_0, x_0) \in S.$$

It will be useful to have the following necessary and sufficient condition for invariance of K:

PROPOSITION 3.3. *The cone K is positively invariant under* (3.1) *if and only if $K \subset D$ and for each $t \in J$*
(P) $\lambda \in K^*, x \in \partial K, \lambda(x) = 0 \Longrightarrow \lambda(f(t,x)) \geqslant 0.$

PROOF. The proof that (P) implies positive invariance of K is similar to that of Theorem 3.2. Given $x_1 \in K$, we pass immediately to $x_\epsilon \gg x_1$ and the solution $y_\epsilon(t)$ of the perturbed equation defined in the proof of Theorem 3.2 and show that $y_\epsilon(t) \gg 0$ for $t_0 \leqslant t \leqslant t_1$ by an argument similar to the one used in the aforementioned proof. The result $x(t, t_0, x_1) \geqslant 0$ for $t \geqslant t_0$ is obtained by passage to the limit as $\epsilon \to 0$. The converse is also an easy modification of the converse argument given in the proof of Theorem 3.2. $\qquad\square$

Since we will have occasion to apply (P) to systems other than (3.1), it will be convenient to refer to (P) by saying that (P) holds for $f : J \times D \to \mathbb{R}^n$ where $K \subset D$. Hypothesis (P) says that the time-dependent vector field $f(t, x)$ points into K at points $x \in \partial K$.

Let $A(t)$ be a continuous $n \times n$ matrix-valued function defined on the interval J containing t_0 and consider the linear initial value problem for the matrix solution X:

$$X' = A(t)X, \quad X(t_0) = I. \tag{3.2}$$

Observe that (P) and (QM) are equivalent for linear systems; therefore we have:

COROLLARY 3.4. *The matrix solution $X(t)$ satisfies $X(t)K \subset K$ for $t \geqslant t_0$ if and only if for all $t \in J$,* (P) *holds for the function $x \to A(t)x$. In fact,* (P) *implies that $X(t)$ maps $K \setminus \{0\}$ and* Int K *into themselves for all $t > t_0$.*

A matrix A is *K-positive* if $A(K) \subset K$. Corollary 3.4 implies that $X(t)$ is K-positive for $t \geqslant t_0$ if (P) holds.

If for every $t \in J$, there exists $\alpha \in \mathbb{R}$ such that $A + \alpha I$ is K-positive, then (P) holds for A. Indeed, if $\lambda \in K^*$ satisfies $\lambda(x) = 0$ then application of λ to $(A + \alpha I)x \geqslant 0$ yields that $\lambda(A(t)x) \geqslant 0$. The converse is false for general cones but true for polyhedral cones by Theorem 8 of Schneider and Vidyasagar [177]. See also Theorem 4.3.40 of Berman and Neumann [18]. Lemmert and Volkmann [118] give the following example of a matrix

$$A = \begin{bmatrix} 0 & 0 & 1 \\ 0 & 0 & 0 \\ 1 & 0 & 0 \end{bmatrix}$$

which satisfies (P) for the ice-cream cone above but $A + \alpha I$ is not K-positive for any α.

Recall that the domain D is *p-convex* if for every $x, y \in D$ satisfying $x \leqslant y$ the line segment joining them also belongs to D. Let $\frac{\partial f}{\partial x}(t, x)$ be continuous on $J \times D$. We say that f (or system (3.1)) is K-*cooperative* if for all $t \in J$, $y \in D$, (P) holds for the function $x \to \frac{\partial f}{\partial x}(t, y)x$. By Corollary 3.4 applied to the variational equation

$$X'(t) = \frac{\partial f}{\partial x}\big(t, x(t, t_0, x_0)\big)X, \quad X(t_0) = I$$

we conclude that if f is K-cooperative then $X(t) = \frac{\partial x}{\partial x_0}(t, t_0, x_0)$ is K-positive.

THEOREM 3.5. *Let $\frac{\partial f}{\partial x}(t, x)$ be continuous on $J \times D$. Then* (QM) *implies that f is K-cooperative. Conversely, if D is p-convex and f is K-cooperative, then* (QM) *holds.*

PROOF. Suppose that (QM) holds, $x \in D$, $h \in \partial K$, and $\phi \in K^*$ satisfies $\phi(h) = 0$. Since $x \leqslant x + \epsilon h$ and $\phi(x) = \phi(x + \epsilon h)$ for small $\epsilon > 0$, (QM) implies that $\phi(f(t, x)) \leqslant \phi(f(t, x + \epsilon h))$. Hence,

$$0 \leqslant \phi\left(\frac{f(t, x + \epsilon h) - f(t, x)}{\epsilon}\right)$$

and the desired result holds on taking the limit $\epsilon \to 0$.

Conversely, suppose that f is K-cooperative and D is p-convex. If $x, y \in D$ satisfy $x \leqslant y$ and $\phi(x) = \phi(y)$ for some $\phi \in K^*$, then either $\phi = 0$ or $y - x \in \partial K$. Consequently

$$\phi\big(f(t, y) - f(t, x)\big) = \int_0^1 \phi\left(\frac{\partial f}{\partial x}(t, sy + (1 - s)x)(y - x)\right)ds \geqslant 0$$

because the integrand is nonnegative. $\quad\square$

If for each $(t, x) \in J \times D$ there exists α such that $(\frac{\partial f}{\partial x}(t, x) + \alpha I)$ is K-positive, then f is K-cooperative. This is implied by the remark following Corollary 3.4.

In the special case that $K = \mathbb{R}^n_+$, the cone of nonnegative vectors, it is easy to see by using the standard inner product that we may identify K^* with K. The quasimonotone hypothesis reduces to the Kamke–Müller condition [91,148]: $x \leqslant y$ and $x_i = y_i$ for some i implies $f_i(t, x) \leqslant f_i(t, y)$. This holds by taking $\phi(x) = \langle e_i, x \rangle$ (e_i is the unit vector in the x_i-direction) and noting that every $\phi \in K^*$ can be represented as a positive linear combination of these functionals. If f is differentiable, the Kamke–Müller condition implies

$$\frac{\partial f_i}{\partial x_j}(t, x) \geqslant 0, \quad i \neq j. \tag{3.3}$$

Conversely, if $\frac{\partial f}{\partial x}(t, x)$ is continuous on $J \times D$, (3.3) holds and D is p-convex, then the Kamke–Müller condition holds by an argument similar to the one used in the proof of the converse in Theorem 3.5.

Stern and Wolkowicz [206] give necessary and sufficient conditions for (P) to hold for matrix A relative to the ice-cream cone $K = \{x \in \mathbb{R}^n : x_1^2 + x_2^2 + \cdots + x_{n-1}^2 \leqslant x_n^2, x_n \geqslant 0\}$. Let Q denote the $n \times n$ diagonal matrix with first $n - 1$ entries 1 and last entry -1. Then

(QM) holds for A if and only if $QA + A^T Q + \alpha Q$ is negative semidefinite for some $\alpha \in \mathbb{R}$. Their characterization extends to other ellipsoidal cones.

3.2. *Strong monotonicity with linear systems*

In this section, all matrices are assumed to be square. Recall that the matrix A is strongly positive if $A(K \setminus \{0\}) \subset \operatorname{Int} K$. We introduce the following milder hypothesis on the matrix A, following Schneider and Vidyasagar [177]:

(ST) For all $x \in \partial K \setminus \{0\}$ there exists $v \in K^*$ such that $v(x) = 0$ and $v(Ax) > 0$.

The following result for the case of constant matrices was proved by Elsner [44], answering a question in [177]. Our proof follows that of Theorem 4.3.26 of Berman et al. [18].

THEOREM 3.6. *Let the linear system* (3.2) *satisfy* (P). *Then the fundamental matrix* $X(t_1)$ *is strongly positive for* $t_1 > t_0$ *if there exists s satisfying* $t_0 \leqslant s \leqslant t_1$ *such that* (ST) *holds for* $A(s)$.

PROOF. Observe that the set of all s such that (ST) holds for $A(s)$ is open. If the result is false, there exists $x > 0$ such that the solution of (3.2) given by $y(t) = X(t)x$ satisfies $y(t_1) \in \partial K \setminus \{0\}$. By Corollary 3.4, $y(t) > 0$ for $t \geqslant t_0$ and $y(t) \in \partial K$ for $t_0 \leqslant t \leqslant t_1$. Let $s \in (t_0, t_1]$ be such that (ST) holds for $A(s)$. Then there exists $v \in K^*$ such that $v(y(s)) = 0$ and $v(A(s)y(s)) > 0$. As $v \in K^*$ and $y(t) \in K$, $h(t) := v(y(t)) \geqslant 0$ for $t_0 \leqslant t \leqslant t_1$. But $h(s) = 0$ and $\frac{d}{dt}|_{t=s} h(t) = v(A(s)y(s)) > 0$ which, taken together, imply that $h(s - \delta) < 0$ for small positive δ, giving the desired contradiction. $\qquad \square$

If (3.2) satisfies (P) and if $x \in \partial K$ then for all $\phi \in K^*$ such that $\phi(x) = 0$ we have $\phi(A(t)x) \geqslant 0$. Hypothesis (ST) asserts that if $x \neq 0$ then $\phi(A(t)x) > 0$ for at least one such ϕ. Berman et al. [18] refer to (ST) (they include (P) in their definition) by saying that A is strongly K-subtangential; while we do not use this terminology, our notation is motivated by it.

An example in [18] shows that (P) and (ST) are not necessary for strong positivity. Let K be the ice-cream cone $K = \{x \in \mathbb{R}^3 \colon x_1^2 + x_2^2 \leqslant x_3^2, x_3 \geqslant 0\}$ and consider the constant coefficient system (3.2) with matrix A given by

$$A = \begin{bmatrix} 0 & 1 & 0 \\ -1 & -1 & 0 \\ 0 & 0 & 0 \end{bmatrix}.$$

An easy calculation shows that $(x_1^2 + x_2^2)' = -2x_2^2$ so it follows easily that K is positively invariant, hence (P) holds by Corollary 3.4. The solution satisfying $x(0) = (\cos(\theta), \sin(\theta), 1)^T \in \partial K$ satisfies $x(t) \in \operatorname{Int} K$ for $t > 0$ since the calculation above and the fact that $x_2(t)$ can have only simple zeros implies that $x_1^2 + x_2^2$ is strictly decreasing while x_3 remains unchanged. The linear functional v, defined $v(x) :=$

$(-\cos(\theta), -\sin(\theta), 1)x$ belongs to K^* by an easy calculation and satisfies $v(x(0)) = 0$. It is unique, up to positive scalar multiple, with these properties because K is smooth so its positive normal at a point is essentially unique. But $v(Ax(0)) = \sin^2(\theta)$ vanishes if $\theta = 0, \pi$. Therefore (ST) fails although $X(t)$ is strongly positive for $t > 0$.

Theorem 3.6 leads to the following result on strong monotonicity for the nonlinear system (3.1).

LEMMA 3.7. *Assume* D *is* p-*convex,* $\frac{\partial f}{\partial x}(t, x)$ *is continuous on* $J \times D$ *and* f *is* K-*cooperative. Let* $x_0, x_1 \in D$ *satisfy* $x_0 < x_1$ *and* $t > t_0$ *with* $t \in J(t_0, x_0) \cap J(t_0, x_1)$. *If there exists* y_0 *on the line segment joining* x_0 *to* x_1 *and* $r \in [t_0, t]$ *such that* (ST) *holds for* $\frac{\partial f}{\partial x}(r, x(r, t_0, y_0))$ *then*

$$x(t, t_0, x_0) \ll x(t, t_0, x_1).$$

PROOF. First, observe that for y_0 on the segment it follows that $t \in J(t_0, y_0)$. We apply the formula

$$x(t, t_0, x_1) - x(t, t_0, x_0) = \int_0^1 \frac{\partial x}{\partial x_0}(t, t_0, sx_1 + (1-s)x_0)(x_1 - x_0)\,ds,$$

where $X(t) = \frac{\partial x}{\partial x_0}(t, t_0, y_0)$ is the fundamental matrix for (3.2) corresponding to the matrix $A(t) = \frac{\partial f}{\partial x}(t, x(t, t_0, y_0))$. The left-hand side belongs to $K \setminus \{0\}$ if $x_0 < x_1$ by Theorems 3.5 and 3.2 but we must show it belongs to Int K. For this to be true, it suffices that for each $t > t_0$ there exists $s \in [0, 1]$ such that the matrix derivative in the integrand is strongly positive. In fact, this derivative is K-positive by Corollary 3.4 for all values of the arguments with $t \geqslant t_0$, so application of any nontrivial $\phi \in K^*$ to the integral gives a nonnegative numerical result. If there exists s as above, then the application of ϕ to the integrand gives a positive numerical result for all s' near s by continuity and Proposition 3.1 and hence the integral belongs to Int K by Proposition 3.1. By Theorem 3.6, $\frac{\partial x}{\partial x_0}(t, t_0, y_0)$ is strongly positive for $t > t_0$ if (ST) holds for $A(r) = \frac{\partial f}{\partial x}(r, x(r, t_0, y_0))$ for some $r \in [t_0, t]$. But this is guaranteed by our hypothesis. $\qquad\square$

THEOREM 3.8. D *is* p-*convex,* $\frac{\partial f}{\partial x}(t, x)$ *is continuous on* $J \times D$, *and* f *is* K-*cooperative. Suppose for every* $x_0, x_1 \in D$ *with* $x_0 < x_1$ *and* $t_0 \in J$, *there exists* y_0 *on the line segment joining the* x_i *such that* (ST) *holds for* $\frac{\partial f}{\partial x}(t_0, y_0)$. *If* $x_0, x_1 \in D$, $x_0 < x_1$, *and* $t > t_0$ *then*

$$t \in J(t_0, x_0) \cap J(t_0, x_1) \quad \Longrightarrow \quad x(t, t_0, x_0) \ll x(t, t_0, x_1).$$

PROOF. This is an immediate corollary of Lemma 3.7. $\qquad\square$

As the main hypothesis of Theorem 3.8 will be difficult to verify in applications, the somewhat stronger condition of irreducibility may be more useful because there is a large body of theory related to it [18,19]. We now introduce the necessary background. A closed subset F of K that is itself a cone is called *a face* of K if $x \in F$ and $0 \leqslant y \leqslant x$ (inequalities induced by K) implies that $y \in F$. For example, the faces of $K = \mathbb{R}^n_+$ are of the form $\{x \in \mathbb{R}^n_+ : x_i = 0, \ i \in I\}$ where $I \subset \{1, 2, \ldots, n\}$. For the ice-cream cone $K = \{x \in \mathbb{R}^n : x_1^2 + x_2^2 + \cdots + x_{n-1}^2 \leqslant x_n^2, x_n \geqslant 0\}$, the faces are the rays issuing from the origin and passing through its boundary vectors. A K-positive matrix A is K-*irreducible* if the only faces F of K for which $A(F) \subset F$ are $\{0\}$ and K. The following is a special case of Theorem 2.3.9 in Berman and Neumann [18]; see Berman and Plemmons [19] for proofs. These references contain additional related results.

THEOREM 3.9. *Let A be an $n \times n$ K-positive matrix. Then the following are equivalent*:
   (i) *A is K-irreducible*;
   (ii) *No eigenvector of A belongs to ∂K*;
   (iii) *A has exactly one unit eigenvector in K and it belongs to* Int K;
   (iv) *$(I + A)^{n-1}(K \setminus \{0\}) \subset$ Int K*.

The famous Perron–Frobenius Theory is developed for K-positive and K-irreducible matrices in the references above. In particular, the spectral radius of A is a simple eigenvalue of A with corresponding eigenvector described in (iii) above.

Below we require the simple observation that if A is K-positive, then the adjoint A^* is K^*-positive. Indeed, if $v \in K^*$ then $(A^*v)(x) = v(Ax) \geqslant 0$ for all $x \in K$ so $A^*v \in K^*$. The next result is adapted from Theorem 4.3.17 of Berman et al. [18].

PROPOSITION 3.10. *Let A be an $n \times n$ matrix and suppose that there exists $\alpha \in \mathbb{R}$ such that $B := A + \alpha I$ is K-positive. Then B is K-irreducible if and only if* (ST) *holds for A*.

PROOF. Suppose that $B = A + \alpha I$ is K-positive and (ST) holds. If $Ax = \lambda x$ for some $\lambda \in \mathbb{R}$ and nonzero vector $x \in \partial K$ then there exists $v \in K^*$ such that $v(x) = 0$ and $v(Ax) > 0$. But $v(Ax) = \lambda v(x) = 0$. Consequently, no eigenvector of B belongs to ∂K so by Theorem 3.9, B is K-irreducible.

Conversely, suppose that B is K-positive and K-irreducible. Let $x \in \partial K$, $x \neq 0$ and let $v \in K^*$ satisfy $v \neq 0$ and $v(x) = 0$. By Theorem 3.9, $C := B + I$ has the property that C^{n-1} is strongly positive so $v(C^{n-1}x) > 0$. As C is K-positive, $v(C^r x) \geqslant 0$ for $r = 1, 2, \ldots, n - 1$. Because $v(x) = 0$, we may choose $p \in \{1, 2, \ldots, n - 1\}$ such that $v(C^p x) > 0$ but $v(C^{p-1}x) = 0$. Let $\tilde{v} = (C^*)^{p-1}v$. Then $\tilde{v} \in K^*$, $\tilde{v}(x) = 0$ and $\tilde{v}(Cx) > 0$. But then A satisfies (ST) because $\tilde{v}(Ax) = \tilde{v}(Cx) > 0$. $\qquad\square$

Motivated by Proposition 3.10, we introduce the following hypothesis for matrix A.

(CI) There exists $\alpha \in \mathbb{R}$ such that $A + \alpha I$ is K-positive and K-irreducible.

In the special case that $K = \mathbb{R}^n_+$, $n \geqslant 2$, matrix A satisfies (CI) if and only if $a_{ij} \geqslant 0$ for $i \neq j$ and there is no permutation matrix P such that

$$P^T A P = \begin{bmatrix} B & 0 \\ C & D \end{bmatrix},$$

where B and D are square. This is equivalent to the assertion that the incidence graph of A is strongly connected. See Berman and Plemmons [19].

The following is an immediate consequence of Theorem 3.8.

COROLLARY 3.11. *D is p-convex, $\frac{\partial f}{\partial x}(t,x)$ is continuous on $J \times D$ and f is K-cooperative. Suppose that for every $x_0, x_1 \in D$ with $x_0 < x_1$ and $t_0 \in J$, there exists y_0 on the line segment joining the x_i such that* (CI) *holds for $\frac{\partial f}{\partial x}(t_0, y_0)$. If $x_0, x_1 \in D$, $x_0 < x_1$, and $t > t_0$ then*

$$t \in J(t_0, x_0) \cap J(t_0, x_1) \implies x(t, t_0, x_0) \ll x(t, t_0, x_1).$$

PROOF. If (CI) holds then, by Proposition 3.10, (ST) holds for $\frac{\partial f}{\partial x}(t,x)$, so the conclusion follows from Theorem 3.8. $\qquad\square$

Corollary 3.11 is an improvement of the restriction of Theorem 10 of Kunze and Siegel [111] to the case that K has nonempty interior; their results also treat the case that K has empty interior in $\mathbb{R}^n$ but nonempty interior in some subspace of $\mathbb{R}^n$. Walter [228] gives a sufficient condition for strong monotonicity relative to $K = \mathbb{R}_+^n$ which does not require f to be differentiable.

For polyhedral cones it can be shown that matrix A satisfies (P) and (ST) if and only if there exists $\alpha \in \mathbb{R}$ such that $A + \alpha I$ is K-positive and K-irreducible. See Theorem 4.3.40 of Berman et al. [18]. For the case of polyhedral cones, therefore, Corollary 3.11 and Theorem 3.8 are equivalent.

3.3. *Autonomous K-competitive and K-cooperative systems*

Our focus now is on the autonomous system of ordinary differential equations

$$x' = f(x), \tag{3.4}$$

where f is a vector field on an open subset $D \subset \mathbb{R}^n$; all vector fields are assumed to be continuously differentiable. We change our notation slightly to conform to more dynamical notation, denoting $x(t, 0, x_0)$ by $\Phi_t(x)$, where Φ denotes the dynamical system ($=$ local flow) in D generated by f discussed in Section 1. The notation $\Phi_t(x)$ carries the tacit assumption that $t \in I_x$, the open interval in $\mathbb{R}$ containing the origin on which the trajectory of x under Φ is defined. The *positive semiorbit* (respectively, (*negative semiorbit*) of x is $\gamma^+(x) := \{\Phi_t(x): t \in t \geqslant 0\}$ (respectively, $\gamma^-(x) := \{\Phi_t(x): t \leqslant 0\}$). The limit sets of x can be defined as

$$\omega(x) = \bigcap_{t \geqslant 0} \overline{\bigcup_{\tau \geqslant t} \Phi_\tau(x)}, \qquad \alpha(x) = \bigcap_{t \leqslant 0} \overline{\bigcup_{\tau \leqslant t} \Phi_\tau(x)}.$$

We call f and Eq. (3.4) *K-competitive* in D if the time-reversed system

$$x' = -f(x)$$

is K-cooperative. When K is the standard cone $\mathbb{R}^n_+$, f is competitive if and only if $\partial f_i/\partial x_j \leqslant 0$ for $i \neq j$. Therefore if f is K-competitive with local flow Φ, then $-f$ is K-cooperative with local flow $\widetilde{\Phi}$, where $\widetilde{\Phi}_t(x) = \Phi_{-t}(x)$; and conversely. Thus time-reversal changes K-competitive systems into K-cooperative ones, and vice-versa. This fact will be exploited repeatedly below.

In the remainder of Section 3 we assume $\mathbb{R}^n$ is ordered by a cone $K \subset \mathbb{R}^n$ with nonempty interior.

A map is *locally monotone* if every point in its domain has a neighborhood on which the map is monotone. A local flow or local semiflow Φ is locally monotone if Φ_t is a locally monotone map for all $t > 0$. *Locally strongly monotone* is defined similarly.

THEOREM 3.12. *Let f be a K-cooperative vector field in an open set $D \subset \mathbb{R}^n$, generating the local flow Φ. Then Φ is locally monotone, and monotone when D is p-convex.*

PROOF. If D is p-convex, monotonicity follows from Theorem 3.2 (with $f(t, x) := f(x)$). Suppose D is not p-convex. Denote the domain of Φ_t by D_t.

We first claim: For every $p \in D$ there exists $\tau > 0$ and a neighborhood $N \subset D_\tau$ such that $\Phi_t|N$ is monotone if $t \in [0, \tau]$. But this is obvious since by restricting f to a p-convex neighborhood of p, we can use Theorem 3.2.

Now fix $p \in D$ and let $J(0, p) \cap [0, \infty) = [0, r)$, $0 < r \leqslant \infty$. Let I_p be the set of all nonnegative $s \in [0, r)$ such that there is a neighborhood U_s of p, contained in D_s, such that $\Phi_t|U_s$ is monotone for each $t \in [0, s]$. The previous claim implies that $[0, \tau] \subset I_p$ and, by its definition, I_p is an interval. Furthermore, straightforward applications of the previous claim establish that I_p is both an open and a closed subset of $[0, r)$. It follows that $I_p = [0, r)$. $\square$

The next theorem gives a sufficient condition for strong monotonicity. Define $G(f)$ to be the set of $x \in D$ such that (ST) holds for $A = f'(x)$. Note that $x \in G(f)$ provided (CI) holds for $A = f'(x)$, by Proposition 3.10. If $K = \mathbb{R}^n_+$, a sufficient condition for $x \in G(f)$ is that $f'(x)$ is an irreducible matrix with nonnegative off-diagonal entries.

THEOREM 3.13. *Let f be a K-cooperative vector field in an open set $D \subset \mathbb{R}^n$, generating the local flow Φ. Assume $D \setminus G(f)$ does not contain any totally ordered line segment (which holds when $D \setminus G(f)$ is zero dimensional). Then Φ is locally strongly monotone, and strongly monotone when D is p-convex.*

PROOF. Suppose D is p-convex, in which case Φ is monotone by from Theorem 3.2. By Theorem 3.8, Φ is strongly monotone.

When D is not p-convex, Φ is locally monotone by Theorem 3.12, and the previous paragraph implies Φ is locally strongly monotone. $\square$

The proof of Theorem 3.13 can be adapted to cover certain nonopen domains D, such as an order interval, a closed halfspace, and the cone K; see the discussion following the proof of Theorem 3.2.

Theorem 3.8 implies that Φ is strongly monotone provided D is p-convex and f satisfies the autonomous version of condition (ST) of Section 3.2, namely:

(ST*) For all $u \in D, x \in \partial K \setminus \{0\}$ there exists $v \in K^*$ such that $v(x) = 0$ and $v(f'(u)x) > 0$.

Without p-convexity of D, condition (ST*) yields local strong monotonicity.

3.4. *Dynamics of cooperative and competitive systems*

We continue to assume $\mathbb{R}^n$ is ordered by a cone K having nonempty interior; all notions involving order refer to that defined by K. For this section, the terms "competitive" and "cooperative" are tacitly understood to mean "K-competitive" and "K-cooperative," and monotonicity refers to the ordering defined by K.

We first apply results from Section 2 to obtain a generic stable convergence theorem for cooperative vector fields.

Let Φ denote the local flow generated by a vector field f on $D \subset \mathbb{R}^n$. We assume D is p-convex throughout this section without further mention. When $\Phi_t(x)$ is defined for all $(t, x) \in [0, \infty) \times D$, as when all positive semiorbits have compact closure in D, the corresponding positive local semiflow Φ^+ is a semiflow. To Φ we associate C, S and E, denoting respectively the sets of convergent, stable and equilibrium points for Φ^+.

THEOREM 3.14. *Let f be a cooperative vector field on an open set $D \subset \mathbb{R}^n$, generating a local flow Φ such that:*
  (a) *Every positive semiorbit of Φ has compact closure in D;*
  (b) *Condition (ST*) above is satisfied, and $D = AC \cup BC$.*
Then Φ has the following properties:
  (i) *$C \cap S$ contains a dense open subset of D, consisting of points whose trajectories converge to equilibria;*
  (ii) *If E is compact there is a stable equilibrium, and an asymptotically stable equilibrium when E is finite.*

PROOF. Assumption (ST*) makes Φ strongly monotone. The hypothesis of Theorem 2.26, with $X = D$, is fulfilled: D is normally ordered and $D = BC \cup AC$. Therefore Theorem 2.26 implies the conclusion. $\qquad\square$

Theorem 3.14, like Theorem 3.13, holds for some more general domains D, including relatively open subsets of V where V denotes a closed halfspace, a closed order interval, or the cone K.

One of the main results of this subsection is that n-dimensional competitive and cooperative systems behave like general systems of one less dimension. Theorems 3.21 and 3.22 illustrate this principle for $n = 2$ in a very strong form. In higher dimensions the principle holds for compact limit sets. The key tool in proving this is the following result due to Hirsch [67]:

THEOREM 3.15. *A limit set of a competitive or cooperative system cannot contain two points related by $\ll$.*

PROOF. By time reversal, if necessary, we assume the system is cooperative, hence the local flow is monotone. Now apply Proposition 1.10. $\square$

A periodic orbit of a competitive or cooperative system is a limit set and consequently it cannot contain two points related by $\ll$. The following sharper result will be useful later:

PROPOSITION 3.16. *Nontrivial periodic orbit of a competitive or cooperative system cannot contain two points related by $<$.*

PROOF. By time-reversal we assume the system is cooperative, and in this case the conclusion follows from Proposition 1.10. $\square$

Let Φ, Ψ be flows in respective spaces A, B. We say Φ and Ψ are *topologically equivalent* if there is a homeomorphism $Q : A \to B$ that is a conjugacy between them, i.e., $Q \circ \Phi_t = \Psi_t \circ Q$ for all $t \in \mathbb{R}$. The relationship of topological equivalence is an equivalence relation on the class of flow; it formalizes the notion of "having the same qualitative dynamics."

A system of differential equations $y' = F(y)$, defined on $\mathbb{R}^k$, is called *Lipschitz* if F is Lipschitz. That is, there exists $K > 0$ such that $|F(y_1) - F(y_2)| \leq K|y_1 - y_2|$ for all $y_1, y_2 \in \mathbb{R}^k$. With these definitions, we can state a result of Hirsch [67] that follows directly from Theorem 3.15.

THEOREM 3.17. *The flow on a compact limit set of a competitive or cooperative system in $\mathbb{R}^n$ is topologically equivalent to a flow on a compact invariant set of a Lipschitz system of differential equations in $\mathbb{R}^{n-1}$.*

PROOF. Let L be the limit set, $v \gg 0$ be a unit vector and let H_v be the hyperplane orthogonal to v, i.e, $H_v := \{x : \langle x, v \rangle = 0\}$. The orthogonal projection Q onto H_v is given by $Qx = x - \langle x, v \rangle v$. By Theorem 3.15, Q is one-to-one on L (this could fail only if L contains two points that are related by $\ll$). Therefore, Q_L, the restriction of Q to L, is a Lipschitz homeomorphism of L onto a compact subset of H_v. We argue by contradiction to establish the existence of $m > 0$ such that $|Q_L x_1 - Q_L x_2| \geq m|x_1 - x_2|$ whenever $x_1 \neq x_2$ are points of L. If this were false, then there exists sequences $x_n, y_n \in L$, $x_n \neq y_n$ such that

$$\frac{|Q(x_n) - Q(y_n)|}{|x_n - y_n|} = \frac{|(x_n - y_n) - v\langle v, x_n - y_n \rangle|}{|x_n - y_n|} \to 0$$

as $n \to \infty$. Equivalently, $|w_n - v\langle v, w_n \rangle| \to 0$ as $n \to \infty$ where $w_n = x_n - y_n/|x_n - y_n|$. We can assume that $w_n \to w$ as $n \to \infty$ where $|w| = 1$. Then, $w = v\langle v, w \rangle$ and therefore, $\langle v, w \rangle^2 = 1$ so $w = \pm v$. But then $x_n - y_n/|x_n - y_n| \to \pm v$ as $n \to \infty$ and this implies that $x_n \ll y_n$ or $y_n \ll x_n$ for all large n, contradicting Theorem 3.15. Therefore, Q_L^{-1} is Lipschitz on $Q(L)$. Since L is a limit set, it is an invariant set for (3.4). It follows that the dynamical system restricted to L can be modeled on a dynamical system in H_v. In fact, if $y \in Q(L)$ then $y = Q_L(x)$ for a unique $x \in L$ and $\Psi_t(y) \equiv Q_L(\Phi_t(x))$ is a dynamical system on $Q(L)$ generated by the vector field

$$F(y) = Q_L\big(f\big(Q_L^{-1}(y)\big)\big)$$

on $Q(L)$. According to McShane [137], a Lipschitz vector field on an arbitrary subset of H_v can be extended to a Lipschitz vector field on all of H_v, preserving the Lipschitz constant. It follows that F can be extended to all of H_v as a Lipschitz vector field. It is easy to see that $Q(L)$ is an invariant set for the latter vector field. We have established the topological equivalence of the flow Φ on L with the flow Ψ on $Q(L)$. $Q(L)$ is a compact invariant set for the $(n-1)$-dimensional dynamical system on H_v generated by the extended vector field. $\qquad\square$

A consequence of Theorem 3.17 is that the flow on a compact limit set, L, of a competitive or cooperative system shares common dynamical properties with the flow of a system of differential equations in one less dimension, restricted to the compact, connected invariant set $Q(L)$. Notice, however, that L may be the limit set of a trajectory not in L, and therefore $Q(L)$ need not be a limit set.

On the other hand, the flow Ψ in a compact limit sets enjoys the topological property of *chain recurrence*, due to Conley [31,30], which will be important in the next subsection. The definition is as follows. Let A be a compact invariant set for the flow Φ. Given two points z and y in A and positive numbers ϵ and t, an (ϵ, t)-chain from z to y in A is an ordered set

$$\{z = x_1, x_2, \ldots, x_{m+1} = y; t_1, t_2, \ldots, t_m\}$$

of points $x_i \in A$ and times $t_i \geqslant t$ such that

$$\left|\Phi_{t_i}(x_i) - x_{i+1}\right| < \epsilon, \quad i = 1, 2, \ldots, m. \tag{3.5}$$

A is *chain recurrent* for Φ if for every $z \in A$ and for every $\epsilon > 0$ and $t > 0$, there is an (ϵ, t)-chain from z to z in A.

Conley proved that when A is compact and connected, a flow Φ in A is chain recurrent if and only if there are no attractors. This useful condition can be stated as follows: For every proper nonempty compact set $S \subset A$ and all $t > 0$, there exists $s > t$ such that $\Phi_s(S) \not\subset \operatorname{Int} S$.

Compactness of A implies that chain recurrence of the flow in A is independent of the metric, and thus holds for any topologically equivalent flow.

It is intuitively clear that, as Conley proved, flows in compact alpha and omega limit sets are chain recurrent. Indeed, orbit segments of arbitrarily long lengths through point x repeatedly pass near any point of $\omega(x) \cup \alpha(x)$. Of course these segments do not necessarily belong to $\omega(x)$; but by taking suitable limits of points in these segments, one can find enough (ϵ, t)-chains in $\omega(x)$ and $\alpha(x)$ to prove the flows in these sets chain recurrent. For a rigorous proof, see Smith [194].

3.5. *Smale's construction*

Smale [182] showed that it is possible to embed essentially arbitrary dynamics in a competitive or cooperative irreducible system. His aim was to warn population modelers that systems designed to model competition could have complicated dynamics. His result is also very useful for providing counterexamples to conjectures in the theory of monotone dynamics, since by time reversal his systems are cooperative. In this section, competitive and cooperative are with respect to the usual cone.

Smale constructed special systems of Kolmogorov type

$$x_i' = x_i M_i(x), \quad 1 \leqslant i \leqslant n, \tag{3.6}$$

in $\mathbb{R}_+^n$ where the M_i are smooth functions satisfying

$$\frac{\partial M_i}{\partial x_j} < 0 \tag{3.7}$$

for all i, j; all sums are understood to be from 1 to n. We refer to such systems as *totally competitive*. They are simple models of competition between n species, where M_i is interpreted as the *per capita* growth rate of species i.

Smale's object was to choose the M_i so that the standard $(n-1)$-simplex $\Sigma_n = \{x \in \mathbb{R}_+^n : \sum x_i = 1\}$ is an attractor in which arbitrary dynamics may be specified.

In order to generate a dynamical system on Σ_n, let H denote the tangent space to Σ_n, that is, $H = \{x \in \mathbb{R}^n : \sum x_i = 0\}$, and let $h : \Sigma_n \to H$ be a smooth vector field on Σ_n, meaning that all partial derivatives of h exist and are continuous on Σ_n. We also assume that $h = (h_1, h_2, \ldots, h_n)$ has the form $h_i = x_i g_i(x)$ where the g_i are smooth functions on Σ_n. Then the differential equation

$$x_i' = h_i(x), \quad 1 \leqslant i \leqslant n \tag{3.8}$$

generates a flow in $\mathbb{R}_+^n$ that leaves Σ_n invariant. The form of the h_i ensures that if $x_i(0) = 0$, then $x_i(t) \equiv 0$ so each lower dimensional simplex forming part of the boundary of Σ_n is invariant.

The goal is to construct a competitive system of the form (3.6) satisfying (3.7) such that its restriction to Σ_n is equivalent to (3.8). Let $p : [0, \infty) \to \mathbb{R}_+$ have continuous derivatives of all orders, be identically 1 in a neighborhood of $s = 1$, and vanish outside the interval $[1/2, 3/2]$. As g is a smooth vector field on Σ_n, it has a smooth extension to $\mathbb{R}_+^n$ which we denote by g in order to conserve notation. An example of such an extension is the map $x \mapsto P(\sum x_j)g(x/\sum x_j)/P(1)$, where $P(u) = \int_0^u p(s)\,ds$.

For $\eta > 0$, define

$$M_i(x) = 1 - S(x) + \eta p\left(\sum x_j\right) g_i(x), \quad 1 \leqslant i \leqslant n.$$

Then (3.7) holds for sufficiently small η since $p(\sum x_j)$ vanishes identically outside a compact subset of $\mathbb{R}^n_+$. Consider the system (3.6) with M as above. $\mathbb{R}^n_+$ is positively invariant; and the function $S(x) = \sum_i x_i$, evaluated along a solution $x(t)$ of (3.6), satisfies

$$\frac{\mathrm{d}}{\mathrm{d}t} S\big(x(t)\big) = S\big(x(t)\big)\big[1 - S\big(x(t)\big)\big]$$

since $\sum x_i g_i(x) = \sum h_i(x) = 0$. Consequently Σ_n, which is $S^{-1}(1) \cap \mathbb{R}^n_+$, is positively invariant. Moreover if $x(0) \in \mathbb{R}^n_+$ then $S(x(0)) \geqslant 0$. This implies $S(x(t)) \to 1$ as $t \to \infty$, unless $x(t) \equiv 0$, and Σ_n attracts all nontrivial solutions of (3.6) in $\mathbb{R}^n_+$. Restricted to Σ_n, (3.6) becomes

$$x_i' = \eta h_i(x), \quad 1 \leqslant i \leqslant n.$$

Therefore the dynamics of (3.6) restricted to Σ_n is equivalent, up to a change in time scale, to the dynamics generated by (3.8).

As noted above, Smale's construction has implications for cooperative and irreducible systems since the time-reversed system corresponding to (3.6) is cooperative and irreducible in $\mathrm{Int}\,\mathbb{R}^n_+$. Time-reversal makes the simplex a repellor for a cooperative system Φ in $\mathbb{R}^n_+$. Therefore every invariant set in the simplex is unstable for Φ. Each trajectory of Φ that is not in the simplex is attracted to the equilibrium at the origin or to the virtual equilibrium at ∞. The simplex is the common boundary between the basins of attraction of these two equilibria.

3.6. *Invariant surfaces and the carrying simplex*

It turns out that the essential features of Smale's seemingly very special construction are found in a large class of totally competitive Kolmogorov systems

$$x_i' = x_i M_i(x), \quad x \in \mathbb{R}^n_+. \tag{3.9}$$

Here and below i and j run from 1 to n. Let Φ denote the corresponding local flow. The *unit $(n-1)$ simplex* is $\Delta^{n-1} := \{x \in \mathbb{R}^n_+ : \sum x_i = 1\}$.

THEOREM 3.18. *Assume (3.9) satisfies the following conditions:*
   (a) $\frac{\partial M_i}{\partial x_j} < 0$;
   (b) $M_i(0) > 0$;
   (c) $M_i(x) < 0$ *for* $|x|$ *sufficiently large.*
Then there exists an invariant compact hypersurface $\Sigma \subset \mathbb{R}^n_+$ such that
   (i) Σ *attracts every point in* $\mathbb{R}^n_+ \setminus \{0\}$;
   (ii) $\Sigma \cap \mathrm{Int}\,\mathbb{R}^n_+$ *is a locally Lipschitz submanifold;*
   (iii) $\Sigma \cap \mathrm{Int}\,\mathbb{R}^n_+$ *is transverse to every line that is parallel to a nonnegative vector and meets $\Sigma \cap \mathrm{Int}\,\mathbb{R}^n_+$;*
   (iv) Σ *is unordered;*

(v) *Radial projection defines a homeomorphism $h : \Sigma \to \Delta^{n-1}$ whose inverse is locally Lipschitz on the open $(n-1)$-cell $\Delta^{n-1} \cap \operatorname{Int} \mathbb{R}^n_+$. There is a flow Ψ on Δ^{n-1} such that $\Phi_t | \Sigma = h \circ \Phi_t \circ h^{-1}$.*

COROLLARY 3.19. *If $n = 3$, every periodic orbit in $\mathbb{R}^3_+$ bounds an unordered invariant disk.*

Assumption (a) is the condition of total competition; (b) and (c) have plausible biological interpretations. The attracting hypersurface Σ, named the *carrying simplex* by M. Zeeman, is analogous to the carrying capacity K in the one-dimensional logistic equation $dx/dt = rx(K - x)$. One can define Σ either as the boundary of the set of points whose alpha limit set is the origin, or as the boundary of the compact global attractor. These sets coincide if and only if Σ is unique, in which case it uniformly attracts every compact set in $\mathbb{R}^n_+ \setminus \{0\}$. Uniqueness holds under mild additional assumptions on the maps M_i (Wang and Jiang [230]). The geometry, smoothness and dynamics of carrying simplices have been investigated by Benaïm [14], Brunovsky [21], Miercyński [140,143,141], Tineo [220], van den Driessche and M. Zeeman [223], Wang and Jiang [230], E. Zeeman [239], E. Zeeman and M. Zeeman [240–242], M. Zeeman [243].

Theorem 3.18 is proved in Hirsch [72] using a general existence theorem for invariant hypersurfaces, of which the following is a generalization:

THEOREM 3.20. *Let Φ be a strongly monotone local flow in a p-convex open set $D \subset \mathbb{R}^n$. If $L \subset D$ is a nonempty compact unordered invariant set, L lies in an unordered invariant hypersurface M that is a locally Lipschitz submanifold.*

IDEA OF PROOF. Define U to be the set of $x \in D$ such that $\Phi_t(x) \gg y$ for some $t > 0$, and some $y \in L$. Continuity implies U is open, and it is nonempty since it contains $z \in D$ where $z > y \in L$. It can be shown that the lower boundary of U in D (Section 1.1) is a hypersurface with the required properties, by arguments analogous to the proof of Theorem 3.17. $\qquad\square$

3.7. *Systems in $\mathbb{R}^2$*

Cooperative and competitive systems in $\mathbb{R}^2$ have particularly simple dynamics. Versions of the following result were proved in Hirsch [67], Theorem 2.7 and Smith [194], Theorem 3.2.2. It is noteworthy that in the next two theorems Φ does not need to be monotone, only locally monotone; hence p-convexity of D is not needed.

THEOREM 3.21. *Let $D \subset \mathbb{R}^2$ be an open set and $g : D \to \mathbb{R}^2$ a vector field that is cooperative or competitive for the standard cone. Let $y(t)$ a nonconstant trajectory defined on an open interval $I \subset \mathbb{R}$ containing 0. Then there exists $t_* \in I$ such that each coordinate $y_i(t)$ is nonincreasing or nondecreasing on each connected component of $I \setminus \{t_*\}$.*

PROOF. It suffices to prove that $y_i'(t)$ can change sign at most once. We assume g is cooperative, otherwise reversing time. Let Φ be local flow of g and set $X(t,x) = \frac{\partial \Phi}{\partial x}(t,x)$. The matrix-valued function $X(t,x)$ satisfies the variational equation

$$\frac{\partial}{\partial t} X(t,x) = \frac{\partial g}{\partial x}\big(\Phi(t,x)\big) \cdot X(t,x), \quad X(0,x) = I.$$

Cooperativity and Corollary 3.4 show that $X(t,x)$ has nonnegative entries for $t \geqslant 0$, i.e., matrix multiplication by $X(t,x)$ preserves the standard cone. The tangent vector $y'(t)$ to the curve $y(t)$, being a solution of the variational equation, satisfies $y'(t) = X(t,y(0))y'(0)$. Nonnegativity of $X(t,x)$ implies that if $y'(t_0)$ lies in the first or third quadrants, then $y'(t)$ stays in the same quadrant, and hence its coordinates have constant sign, for $t > t_0$. On the other hand if $y'(t)$ for $t \geqslant t_0$ is never in the first or third quadrants, its coordinates again have constant sign. (Note that $y'(t)$ cannot transit directly between quadrants 1 and 3, or 2 and 4, since it cannot pass through the origin.) We have shown that there is at most one $t_0 \in I$ at which $y'(t)$ changes quadrants. If such a t_0 exists, set $t_* = t_0$; otherwise let $t_* \in I$ be arbitrary. $\qquad\square$

Variants of the next result have been proved many times for Kolmogorov type population models (Albrecht et al. [1], Grossberg [53], Hirsch and Smale [80], Kolmogorov [97], Rescigno and Richardson [168], Selgrade [178]).

THEOREM 3.22. *Let g be a K-cooperative or K-competitive vector field in a domain $D \subset \mathbb{R}^2$. If $\gamma^+(x)$ (respectively, $\gamma^-(x)$) has compact closure in D, then $\omega(x)$ (respectively, $\alpha(x)$) is a single equilibrium.*

PROOF. For the standard cone, denoted here by P, this follows from Theorem 3.21. The general case follows by making a linear coordinate change $y = Tx$ mapping K onto the standard cone. Here T is any linear transformation that takes a basis for $\mathbb{R}^2$ contained in ∂P into the standard basis, which lies in ∂K. Then we have $u \leqslant_K v$ if and only if $Tu \leqslant_P Tv$; in other words, T is an order isomorphism. It follows that the system $x' = g(x)$ is K-cooperative (respectively, K-competitive) if and only if the system $y' = h(y) := Tg(T^{-1}y)$ is P-cooperative (respectively, P-competitive). Therefore T is a conjugacy between the local flows Φ, Ψ of the two dynamical systems, that is, $T \circ \Phi_t = \Psi_t \circ T$. Consequently the conclusion for P, proved above, implies the conclusion for K. $\qquad\square$

3.8. *Systems in $\mathbb{R}^3$*

The following Poincaré–Bendixson theorem for three-dimensional cooperative and competitive systems is the most notable consequence of Theorem 3.17. It was proved by Hirsch [76] who improved earlier partial results [67,187]. The following result from Smith [194] holds for arbitrary cones $K \subset \mathbb{R}^3$ with nonempty interior:

THEOREM 3.23. *Let g be a K-cooperative or K-competitive vector field in a p-convex domain $D \subset \mathbb{R}^3$. Then a compact limit set of g that contains no equilibrium points is a periodic orbit.*

PROOF. Let Φ denote the flow of the system, and L the limit set. By Theorem 3.17, the restriction of Φ to L is topologically equivalent to a flow Ψ, generated by a Lipschitz planar vector field, restricted to the compact, connected, chain recurrent invariant set $Q(L)$. Since L contains no equilibria neither does $Q(L)$. The Poincaré–Bendixson theorem implies that $Q(L)$ consists of periodic orbits and, possibly, entire orbits whose omega and alpha limit sets are periodic orbits contained in $Q(L)$. The chain recurrence of Ψ on $Q(L)$ will be exploited to show that $Q(L)$ consists entirely of periodic orbits.

Let $z \in Q(L)$ and suppose that z does not belong to a periodic orbit. Then $\omega(z)$ and $\alpha(z)$ are distinct periodic orbits in $Q(L)$. Let $\omega(z) = \gamma$ and suppose for definiteness that z belongs to the interior component, V, of $\mathbb{R}^2 \setminus \gamma$ so that $\Psi_t(z)$ spirals toward γ in V. The other case is treated similarly. Then γ is asymptotically stable relative to V. Standard arguments using transversals imply the existence of compact, positively invariant neighborhoods U_1 and U_2 of γ in V such that $U_2 \subset \mathrm{Int}_V U_1$, $z \notin U_1$ and there exists $t_0 > 0$ for which $\Psi_t(U_1) \subset U_2$ for $t \geq t_0$. Let $\epsilon > 0$ be such that the 2ϵ-neighborhood of U_2 in D is contained in U_1. Choose t_0 larger if necessary such that $\Psi_t(z) \in U_2$ for $t \geq t_0$. This can be done since $\omega(z) = \gamma$. Then any (ϵ, t_0)-chain in $Q(L)$ beginning at $x_1 = z$ satisfies $\Psi_{t_1}(x_1) \in U_2$ and, by (3.5) and the fact that the 2ϵ-neighborhood of U_2 is contained in U_1, it follows that $x_2 \in U_1$. As $t_2 > t_0$, it then follows that $\Psi_{t_2}(x_2) \in U_2$ and (3.5) again implies that $x_3 \in U_1$. Continuing this argument, it is evident that the (ϵ, t_0)-chain cannot return to z. There can be no (ϵ, t_0)-chain in $Q(L)$ from z to z and therefore we have contradicted that $Q(L)$ is chain recurrent. Consequently, every orbit of $Q(L)$ is periodic. Since $Q(L)$ is connected, it is either a single periodic orbit or an annulus consisting of periodic orbits. It follows that L is either a single periodic orbit or a cylinder of periodic orbits.

To complete the proof we must rule out the possibility that $Q(L)$ consists of an annulus of periodic orbits. We can assume that the system is cooperative. The argument will be separated into two cases: $L = \omega(x)$ or $L = \alpha(x)$.

If $L = \omega(x)$ consists of more than one periodic orbit then $Q(L)$ is an annulus of periodic orbits in the plane containing an open subset O. Then there exists $t_0 > 0$ such that $Q(\Phi_{t_0}(x)) \in O$. Let y be the unique point of L such that $Q(y) = Q(\Phi_{t_0}(x))$. $y = \Phi_{t_0}(x)$ cannot hold since this would imply that L is a single periodic orbit so it follows that either $y \ll \Phi_{t_0}(x)$ or $\Phi_{t_0}(x) \ll y$. Suppose that the latter holds, the argument is similar in the other case. Then there exists $t_1 > t_0$ such that $\Phi_{t_1}(x)$ is so near y that $\Phi_{t_0}(x) \ll \Phi_{t_1}(x)$. But then the Convergence Criterion from Chapter 1 implies that $\Phi_t(x)$ converges to equilibrium, a contradiction to our assumption that L contains no equilibria. This proves the theorem in this case.

If $L = \alpha(x)$ and $Q(L)$ consists of an annulus of periodic orbits, let $C \subset L$ be a periodic orbit such that $Q(L)$ contains C in its interior. $Q(C)$ separates $Q(L)$ into two components. Fix a and b in $L \setminus C$ such that $Q(a)$ and $Q(b)$ belong to different components of $Q(L) \setminus Q(C)$. Since $\Phi_t(x)$ repeatedly visits every neighborhood of a and b as $t \to -\infty$, $Q(\Phi_t(x))$ must cross $Q(C)$ at a sequence of times $t_k \to -\infty$. Therefore, there exist $z_k \in C$ such that $Q(z_k) = Q(\Phi_{t_k}(x))$ and consequently, as in the previous case, either $z_k \ll \Phi_{t_k}(x)$

or $\Phi_{t_k}(x) \ll z_k$ holds for each k. Passing to a subsequence, we can assume that either $z_k \ll \Phi_{t_k}(x)$ holds for all k or $\Phi_{t_k}(x) \ll z_k$ holds for all k. Assume the latter, the argument is essentially the same in the other case. We claim that for every $s < 0$ there is a point $w \in C$ such that $\Phi_s(x) > w$. For if $t_k < s$ then

$$\Phi_s(x) = \Phi_{s-t_k} \circ \Phi_{t_k}(x) < \Phi_{s-t_k}(z_k) \in C.$$

If $y \in L$ then $\Phi_{s_n}(x) \to y$ for some sequence $s_n \to -\infty$. By the claim, there exists $w_n \in C$ such that $\Phi_{s_n}(x) > w_n$. Passing to a subsequence if necessary, we can assume that $w_n \to w \in C$ and $y \geq w$. Therefore, every point of L is related by $\leq$ to some point of C.

The same reasoning applies to every periodic orbit $C' \subset L$ for which $Q(C')$ belongs to the interior of $Q(L)$: either every point of L is $\leq$ some point of C' or every point of L is $\geq$ some point of C'. Since there are three different periodic orbits in L whose projections are contained in the interior of $Q(L)$, there will be two of them for which the same inequality holds between points of L and points of the orbit. Consider the case that there are two periodic orbits C_1 and C_2 such that every point of L is $\leq$ some point of C_1 and $\leq$ some point of C_2. The case that the opposite relations hold is treated similarly. If $u \in C_1$ then it belongs to L so we can find $w \in C_2$ such that $u < w$ (equality can't hold since the points belong to different periodic orbits). But $w \in L$ so we can find $z \in C_1$ such that $w < z$. Consequently, $u, z \in C_1$ satisfy $u < z$, a contradiction to Proposition 3.16. This completes the proof. $\qquad\square$

A remarkable fact about three-dimensional competitive or cooperative systems on suitable domains is that the existence of a periodic orbit implies the existence of an equilibrium point inside a certain semi-invariant closed ball having the periodic orbit on its boundary. Its primary use is to locate equilibria, or conversely, to exclude periodic orbits. The construction below is adapted from Smith [187,194] where the case $K = \mathbb{R}^3_+$ was treated; here we treat the general case that K has nonempty interior. The terms "competitive" and "cooperative" will be used to mean K-competitive and K-cooperative for brevity. A related result appears in Hirsch [75]. Throughout the remainder of this section, the system is assumed to be defined on a p-convex subset D of $\mathbb{R}^3_+$.

We can assume the system is competitive. Let γ denote the periodic orbit and assume that there exist p, q with $p \ll q$ such that

$$\gamma \subset [p, q] \subset D. \tag{3.10}$$

Define

$$B = \left\{ x \in \mathbb{R}^3 : x \text{ is not related to any point } y \in \gamma \right\} = (\gamma + K)^c \cap (\gamma - K)^c.$$

Here we use the notation A^c for the complement of the subset A in $\mathbb{R}^3$. Observe that in defining B we ignored the domain D of (3.4), viewing γ as a subset of $\mathbb{R}^3$. Another way to define B is to express its complement as $B^c = (\gamma + K) \cup (\gamma - K)$.

A *3-cell* is a subset of $\mathbb{R}^3$ that is homeomorphic to the open unit ball.

THEOREM 3.24. *Let γ be a nontrivial periodic orbit of a competitive system in $D \subset \mathbb{R}^3$ and suppose that (3.10) holds. Then B is an open subset of $\mathbb{R}^3$ consisting of two connected components, one bounded and one unbounded. The bounded component, $B(\gamma)$, is a 3-cell contained in $[p, q]$. Furthermore, $B(\gamma)$ is positively invariant and its closure contains an equilibrium.*

Combining this result with Theorem 3.23 leads to the following dichotomy from Hirsch [75].

COROLLARY 3.25. *Assume the domain $D \subset \mathbb{R}^3$ of a cooperative or competitive system contains $[p, q]$ with $p \ll q$. Then one of the following holds:*
   (i) *$[p, q]$ contains an equilibrium;*
   (ii) *the forward and backward semi-orbits of every point of $[p, q]$ meet $D \setminus [p, q]$.*

PROOF. We take the system to be competitive, otherwise reversing time. Assume (ii) is false. Then $[a, b]$ contains a compact limit set L. If L is not a cycle, it contains an equilibrium by Theorem 3.23. If L is a cycle, (i) follows from Theorem 3.24. $\qquad\square$

PROOF SKETCH OF THEOREM 3.24. That B is open is a consequence of the fact that $\gamma + K$ and $\gamma - K$ are closed. We show that $B \cap D$ is positively invariant. If $x \in B \cap D$, $y \in \gamma$ and $t > 0$ then $\Phi_{-t}(y) \in \gamma$ so x is not related to it. Since the forward flow of a competitive system preserves the property of being unrelated, $\Phi_t(x)$ is unrelated to y. Therefore, $\Phi_t(x) \in B \cap D$.

As in the proof of Theorem 3.17, for $v > 0$, H_v denotes the hyperplane orthogonal to v and Q the orthogonal projection onto H_v along v. Q is one-to-one on γ so $Q(\gamma)$ is a Jordan curve in H_v. Let H_i and H_e denote the interior and exterior components of $H_v \setminus Q(\gamma)$. If $x \in Q^{-1}(Q(\gamma))$ then $Q(x) = Q(y)$ for some $y \in \gamma$ and therefore either $x = y$, $x \ll y$ or $y \ll x$. In any case, $x \notin B$. Hence,

$$B = \left(B \cap Q^{-1}(H_i)\right) \cup \left(B \cap Q^{-1}(H_e)\right).$$

Set $B(\gamma) = B \cap Q^{-1}(H_i)$.

Given $z \in H_i$, let $A_z^+ := \{s \in \mathbb{R} : z + sv \in \gamma + K\}$ and $A_z^- := \{s \in \mathbb{R} : z + sv \in \gamma - K\}$. A_z^+ clearly contains all large s by compactness of γ and it is closed because $\gamma + K$ is closed. If $s \in A_z^+$, there exists $y \in \gamma$ and $k \in K$ such that $z + sv = y + k$ so $z + (s + r)v = y + k + rv$, implying that $s + r \in A_z^+$ for all $r \geqslant 0$. It follows that $A_z^+ = [s_+(z), \infty)$, and similarly, $A_z^- = (-\infty, s_-(z)]$. If $s_-(z) \geqslant z_+(z)$ so $A_z^+ \cap A_z^-$ is nonempty, then there exists $s \in \mathbb{R}$, $k_i \in K$, and $y_i \in \gamma$ such that $z + sv = y_1 + k_1 = y_2 - k_2$. We must have $k_1 = k_2 = 0$ or else $y_2 > y_1$, a contradiction to Proposition 3.16, but then $z + sv = y_1$ so $z = Qy_1$ contradicting that $z \in H_i$. We conclude that $s_-(z) < z_+(z)$ and that $z + sv \in B(\gamma)$ if and only if $s_-(z) < s < s_+(z)$. It follows that

$$B(\gamma) = \left\{z + sv : z \in H_i, \ s \in \left(s_-(z), s_+(z)\right)\right\}.$$

It is easy to show that the maps $z \mapsto s_{\pm}(z)$ are continuous and satisfy $s_+(z) - s_-(z) \to 0$ as $z \to y \in \gamma$ and this implies that $B(\gamma)$ is a 3-cell. See the argument given in Smith [187, 194].

To prove $B(\gamma) \subset [p, q]$, we identify K^* as the set of x such that $\langle x, k \rangle \geqslant 0$ for all $k \in K$ (where $\langle x, k \rangle$ denotes inner product). Schneider and Vidyasagar [177] proved the elegant result that every vector x has a unique representation

$$x = k - w, \quad k \in K, \ w \in K^*, \ \langle w, k \rangle = 0.$$

Choose any $z \in B \cap (\mathbb{R}^3 \setminus [p, q])$ and write

$$z - p = k - w, \quad k \in K, w \in K^*, \ \langle w, k \rangle = 0,$$

$$q - z = k' - w', \quad k' \in K, w' \in K^*, \ \langle w', k' \rangle = 0.$$

Observe that $w > 0$, $w' > 0$ because $z \in B$.

Either $k > 0$ or $k' > 0$. For if $k = k' = 0$ then $q - p = -(w + w')$, so

$$0 \leqslant \langle w + w', q - p \rangle = -\|w + w'\|^2 \leqslant 0.$$

This entails $w + w' = 0$ and thus $p = q$, a contradiction.

We assume $k > 0$, as the case $k' > 0$ is similar, and even follows formally by replacing K with $-K$. Then $w > 0$. Consider the ray $R = \{z + tk: t \geqslant 0\}$. If $y \in \gamma$, then

$$\langle w, z + tk - y \rangle = \langle w, z - p \rangle + \langle w, p - y \rangle \leqslant \langle w, z - p \rangle = -\|w\|^2 < 0.$$

Because z and u are unrelated, there exists $u \in K^*$ such that $\langle u, z - y \rangle > 0$. So

$$\langle u, z + tk - y \rangle = \langle u, z - y \rangle + t(z, k) \geqslant \langle u, z - y \rangle > 0.$$

This shows that no point of R is related to any point of γ. Therefore R and hence z are in the unbounded component of B.

As $B(\gamma)$ is a connected component of the positively invariant set B, it is positively invariant. Consequently its closure is a positively invariant set homeomorphic to the closed unit ball in $\mathbb{R}^3$. It therefore contains an equilibrium by a standard argument using the Brouwer Fixed Point Theorem (see, e.g., Hale [57, Theorem I.8.2]). $\qquad\square$

If $B(\gamma)$ contains only nondegenerate equilibria $x_1, x_2, \ldots, x_m$, then standard topological degree arguments imply that m is odd and that $1 = \sum_{i=1}^{m} (-1)^{s_i}$ where $s_i \in \{0, 1, 2, 3\}$ is the number of positive eigenvalues of $Df(x_i)$. See Smith [187] for the proof and further information on equilibria in $B(\gamma)$.

There are many papers devoted to competitive Lotka–Volterra systems in $\mathbb{R}^3$, largely stimulated by the work of M. Zeeman. See for example [82,223,237,239,243,240,242] and references therein. The paper of Li and Muldowney [115] contains an especially nice application to epidemiology. Additional results for three-dimensional competitive and cooperative systems can be found in references [66,75–77,41,194,196,247,248].

The recent paper of Ortega and Sánchez [153] is noteworthy for employing a cone related to the ice-cream cone and observing that results for competitive systems are valid for general cones with nonempty interior. They show that if P is a symmetric matrix of dimension n having one positive eigenvalue λ_+ with corresponding unit eigenvector e_+, and $n-1$ negative eigenvalues, then (3.4) is monotone with respect to the order generated by the cone $K := \{x \in \mathbb{R}^n : \langle Px, x \rangle \geqslant 0, \ \langle x, e_+ \rangle \geqslant 0\}$ if and only if there exists a function $\mu : \mathbb{R}^n \to \mathbb{R}$ such that the matrix $P \cdot Df_x + (Df_x)^T \cdot P + \mu(x)P$ is positive semidefinite for all x. They use this result to show that one of the results of R.A. Smith [204] on the existence of an orbitally stable periodic orbit, in the special case $n = 3$, follows from the results for competitive systems. It is not hard to see that if (3.4) satisfies the conditions above then after a change of variables in (3.4), the resulting system is monotone with respect to the standard ice-cream cone.

For applications of competitive and cooperative systems, see for example Benaïm [15], Benaïm and Hirsch [16,17], Hirsch [69,74] Hofbauer and Sandholm [81], Hsu and Waltman [84], Smith [194,196], Smith and Waltman [202].

4. Delay differential equations

4.1. *The semiflow*

The aim of the present section is to apply the theory developed in Sections 1 and 2 to differential equations containing delayed arguments. Such equations are often referred to as delay differential equations or functional differential equations. Since delay differential equations contain ordinary differential equations as a special case, when all delays are zero, the treatment is quite similar to the previous section. The main difference is that a delay differential equation generally can't be solved backward in time and therefore there is not a well-developed theory of competitive systems with delays.

Delay differential equations generate infinite-dimensional dynamical systems and there are several choices of state space. We restrict attention here to equations with bounded delays and follow the most well-developed theory (see Hale and Verduyn Lunel [61]). If r denotes the maximum delay appearing in the equation, then the space $\mathcal{C} := C([-r, 0], \mathbb{R}^n)$ is a natural choice of state space. Given a cone K in $\mathbb{R}^n$, $\mathcal{C}_K$ contains the cone of functions which map $[-r, 0]$ into K. The section begins by identifying sufficient conditions on the right hand side of the delay differential equation for the semiflow to be monotone with respect to the ordering induced by this cone. This quasimonotone condition reduces to the quasimonotone condition for ordinary differential equations when no delays are present. Our main goal is to identify sufficient conditions for a delay differential equation to generate an eventually strongly monotone semiflow so that results from Sections 1 and 2 may be applied.

In order to motivate fundamental well-posedness issues for delay equations, it is useful to start with a consideration of a classical example that has motivated much research in the field (see, e.g., Krisztin et al. [105] and Hale and Verduyn Lunel [61]), namely the equation

$$x'(t) = -x(t) + h\big(x(t-r)\big), \quad t \geqslant 0, \tag{4.1}$$

where h is continuous and $r > 0$ is the delay. It is clear that $x(t)$ must be prescribed on the interval $[-r, 0]$ in order that it be determined for $t \geqslant 0$. A natural space of initial conditions is the space of continuous functions on $[-r, 0]$, which we denote by C, where $n = 1$ in this case. C is a Banach space with the usual uniform norm $|\phi| = \sup\{|\phi(\theta)|: -r \leqslant \theta \leqslant 0\}$. If $\phi \in C$ is given, then it is easy to see that the equation has a unique solution $x(t)$ for $t \geqslant 0$ satisfying

$$x(\theta) = \phi(\theta), \quad -r \leqslant \theta \leqslant 0.$$

If the state space is C, then we need to construct from the solution $x(t)$, an element of the space C to call the state of the system at time t. It should have the property that it uniquely determines $x(s)$ for $s \geqslant t$. The natural choice is $x_t \in C$, defined by

$$x_t(\theta) = x(t+\theta), \quad -r \leqslant \theta \leqslant 0.$$

Then, $x_0 = \phi$ and $x_t(0) = x(t)$.

The general autonomous functional differential equation is given by

$$x'(t) = f(x_t), \tag{4.2}$$

where $f : D \to \mathbb{R}^n$, D is an open subset of C and f is continuous. In the example above, f is given by $f(\phi) = -\phi(0) + h(\phi(-r))$ for $\phi \in C$. Observe that (4.2) includes the system of ordinary differential equations

$$x' = g(x),$$

where $g : \mathbb{R}^n \to \mathbb{R}^n$, as a special case. Simply let $f(\phi) = g(\phi(0))$ so that $f(x_t) = g(x_t(0)) = g(x(t))$.

It will always be assumed that (4.2), together with the initial condition $x_0 = \phi \in D$ has a unique, maximally defined solution, denoted by $x(t, \phi)$, on an interval $[0, \sigma)$. The state of the system is denoted by $x_t(\phi)$ to emphasize the dependence on the initial data. Uniqueness of solutions holds if, for example, f is Lipschitz on compact subsets of D (see Hale and Verduyn Lunel [61]). This holds, for example, if $f \in C^1(D)$ has locally bounded derivative. If uniqueness of solutions of initial value problems hold, then the map $(t, \phi) \to x_t(\phi)$ is continuous. Therefore, a (local) semiflow on D can be defined by

$$\Phi_t(\phi) = x_t(\phi). \tag{4.3}$$

In contrast to the case of ordinary differential equations, $x(t, \phi)$ cannot usually be defined for $t \leqslant 0$ as a solution of (4.2) and consequently, Φ_t need not be one-to-one.

It will be convenient to have notation for the natural embedding of $\mathbb{R}^n$ into C. If $x \in \mathbb{R}^n$, let $\hat{x} \in C$ be the constant function equal to x for all values of its argument. The set of equilibria for (4.2) is given by

$$E = \left\{ \hat{x} \in D\colon x \in \mathbb{R}^n \text{ and } f(\hat{x}) = 0 \right\}.$$

4.2. *The quasimonotone condition*

Given that C is a natural state space for (4.2), we now consider what sort of cones in C will yield useful order relations. The most natural such cones are those induced by cones in $\mathbb{R}^n$. Let K be a cone in $\mathbb{R}^n$ with nonempty interior and K^* denote the dual cone. All inequalities hereafter are assumed to be those induced on $\mathbb{R}^n$ by K. The cone K induces a cone C_K in the Banach space C defined by

$$C_K = \left\{ \phi \in C\colon \phi(\theta) \geqslant 0,\ -r \leqslant \theta \leqslant 0 \right\}.$$

It has nonempty interior in C given by $\operatorname{Int} C_K = \{ \phi \in C_K \colon \phi(\theta) \gg 0, \theta \in [-r, 0] \}$. The usual notation $\leqslant, <, \ll$ will be used for the various order relations on C generated by C_K. In particular, $\phi \leqslant \psi$ holds in C if and only if $\phi(s) \leqslant \psi(s)$ holds in $\mathbb{R}^n$ for every $s \in [-r, 0]$. The same notation will also be used for the various order relations on $\mathbb{R}^n$ but hopefully the context will alert the reader to the appropriate meaning. Cones in C that are not induced by a cone in $\mathbb{R}^n$ have also proved useful. See Smith and Thieme [198,200,194].

An immediate aim is to identify sufficient conditions on f for the semiflow Φ to be a monotone semiflow. The following condition should seem natural since it generalizes the condition (QM) for ordinary differential equations in the previous section. We refer to it here as the *quasimonotone condition*, (QMD) for short. "D" in the notation, standing for delay, is used so as not to confuse the reader with (QM) of the previous section. We follow this pattern in several definitions in this section.

(QMD) $\phi, \psi \in D$, $\phi \leqslant \psi$ and $\eta(\phi(0)) = \eta(\psi(0))$ for some $\eta \in K^*$, implies $\eta(f(\phi)) \leqslant \eta(f(\psi))$.

For the special case $K = \mathbb{R}^n_+$, (QMD) becomes:

$$\phi, \psi \in D,\ \phi \leqslant \psi \text{ and } \phi_i(0) = \psi_i(0) \quad \text{implies} \quad f_i(\phi) \leqslant f_i(\psi).$$

As in Section 3, it is convenient to consider the nonautonomous equation

$$x'(t) = f(t, x_t), \tag{4.4}$$

where $f \colon \Omega \to \mathbb{R}^n$ is continuous on Ω, an open subset of $\mathbb{R} \times C$. Given $(t_0, \phi) \in \Omega$, we write $x(t, t_0, \phi, f)$ and $x_t(t_0, \phi, f)$ for the maximally defined solution and state of the system at time t satisfying $x_{t_0} = \phi$. We assume this solution is unique, which will be the case if f is Lipschitz in its second argument on each compact subset of Ω. We drop the last argument f from $x(t, t_0, \phi, f)$ when no confusion over which f is being considered will result.

$f : \Omega \to \mathbb{R}^n$ is said to satisfy (QMD) if $f(t, \cdot)$ satisfies (QMD) on $\Omega_t \equiv \{\phi \in C : (t, \phi) \in \Omega\}$ for each t.

The next theorem not only establishes the desired monotonicity of the semiflow Φ but also allows comparisons of solutions between related functional differential equations. It generalizes Theorem 3.2 of Chapter 3 to functional differential equations and is a generalization of Proposition 1.1 of [190] and Theorem 5.1.1 of [194] where $K = \mathbb{R}^n_+$ is considered. The quasimonotone condition for delay differential equations seems first to have appeared in the work of Kunisch and Schappacher [109], Martin [128], and Ohta [152].

THEOREM 4.1. *Let $f, g : \Omega \to \mathbb{R}^n$ be continuous, Lipschitz on each compact subset of Ω, and assume that either f or g satisfies* (QMD). *Assume also that $f(t, \phi) \leqslant g(t, \phi)$ for all $(t, \phi) \in \Omega$. Then*

$$\phi, \psi \in \Omega_{t_0}, \ \phi \leqslant \psi, \ t \geqslant t_0, \quad \Longrightarrow \quad x(t, t_0, \phi, f) \leqslant x(t, t_0, \psi, g)$$

for all t for which both are defined.

PROOF. Assume that f satisfies (QMD), a similar argument holds if g satisfies (QMD). Let $e \in \mathbb{R}^n$ satisfy $e \gg 0$, $g_\epsilon(t, \phi) := g(t, \phi) + \epsilon e$ and $\psi_\epsilon := \psi + \epsilon \hat{e}$, for $\epsilon \geqslant 0$. If $x(t, t_0, \psi, g)$ is defined on $[t_0 - r, t_1]$ for some $t_1 > t_0$, then $x(t, t_0, \psi_\epsilon, g_\epsilon)$ is also defined on this same interval for all sufficiently small positive ϵ and

$$x(t, t_0, \psi_\epsilon, g_\epsilon) \to x(t, t_0, \psi, g), \quad \epsilon \to 0,$$

for $t \in [t_0, t_1]$ by Hale and Verduyn Lunel [61, Theorem 2.2.2]. We will show that $x(t, t_0, \phi, f) \ll x(t, t_0, \psi_\epsilon, g_\epsilon)$ on $[t_0 - r, t_1]$ for small positive ϵ. The result will then follow by letting $\epsilon \to 0$. If the assertion above were false for some ϵ, then applying the remark below Proposition 3.1, there exists $s \in (t_0, t_1]$ such that $x(t, t_0, \phi, f) \ll x(t, t_0, \psi_\epsilon, g_\epsilon)$ for $t_0 \leqslant t < s$ and $\eta(x(s, t_0, \phi, f)) = \eta(x(s, t_0, \psi_\epsilon, g_\epsilon))$ for some nontrivial $\eta \in K^*$. As $\eta(x(t, t_0, \phi, f)) < \eta(x(t, t_0, \psi_\epsilon, g_\epsilon))$ for $t_0 \leqslant t < s$, by Proposition 3.1, we conclude that $\frac{d}{dt}|_{t=s} \eta(x(s, t_0, \phi, f)) \geqslant \frac{d}{dt}|_{t=s} \eta(x(s, t_0, \psi_\epsilon, g_\epsilon))$. But

$$\frac{d}{dt}\bigg|_{t=s} \eta\big(x(s, t_0, \psi_\epsilon, g_\epsilon)\big) = \eta\big(g\big(s, x_s(t_0, \psi_\epsilon, g_\epsilon)\big)\big) + \epsilon \eta(e)$$

$$> \eta\big(f\big(s, x_s(t_0, \psi_\epsilon, g_\epsilon)\big)\big)$$

$$\geqslant \eta\big(f\big(s, x_s(t_0, \phi, f)\big)\big)$$

$$= \frac{d}{dt}\bigg|_{t=s} \eta\big(x(s, t_0, \phi, f)\big),$$

where the last inequality follows from (QMD). This contradiction implies that no such s can exist, proving the assertion. $\qquad\square$

In the case of the autonomous system (4.2), taking $f = g$ in Theorem 1.1 implies that $x_t(\phi) \leqslant x_t(\psi)$ for $t \geqslant 0$ such that both solutions are defined. In other words, the semi-flow Φ defined by (4.3) is monotone. In contrast to Theorem 3.2 of the previous section, if $\phi < \psi$ we cannot conclude that $x(t, \phi) < x(t, \psi)$ or $x_t(\phi) < x_t(\psi)$ since Φ_t is not generally one-to-one. A simple example is provided by the scalar equation (4.2) with $r = 1$ and $f(\phi) := \max \phi$, which satisfies (QMD). Let $\phi < \psi$ be strictly increasing on $[-1, -1/2]$, $\phi(-1) = \psi(-1) = 0$, $\phi(-1/2) = \psi(-1/2) = 1$, and $\phi(\theta) = \psi(\theta) = -2\theta$ for $-1/2 < \theta \leqslant 0$. It is easy to see that $x(t, \phi) = x(t, \psi)$ for $t \geqslant 0$.

It is useful to have sufficient conditions for the positive invariance of K. By this we mean that $t_0 \in J$ and $\phi \geqslant 0$ implies $x(t, t_0, \phi) \geqslant 0$ for all $t \geqslant t_0$ for which it is defined. The following result provides the expected necessary and sufficient condition. The proof is similar to that of Theorem 4.1; the result is the delay analog of Proposition 3.3.

THEOREM 4.2. *Assume that* $J \times K \subset \Omega$ *where* J *is an open interval. Then* K *is positively invariant for* (4.4) *if and only if for all* $t \in J$
   (PD) $\phi \geqslant 0$, $\lambda \in K^*$ *and* $\lambda(\phi(0)) = 0$ *implies* $\lambda(f(t, \phi)) \geqslant 0$
holds.

Let $L : J \to L(\mathcal{C}, \mathbb{R}^n)$ be continuous, where $L(\mathcal{C}, \mathbb{R}^n)$ denotes the space of bounded linear operators from $\mathcal{C}$ to $\mathbb{R}^n$, and consider the initial value problem for the linear nonautonomous functional differential equation

$$x' = L(t)x_t, \quad x_{t_0} = \phi. \tag{4.5}$$

Observing that (PD) and (QMD) are equivalent for linear systems, we have the following corollary.

COROLLARY 4.3. *Let* $x(t, t_0, \phi)$ *be the solution of* (4.5). *Then* $x(t, t_0, \phi) \geqslant 0$ *for all* $t \geqslant t_0$ *and all* $\phi \geqslant 0$ *if and only if for each* $t \in J$, (PD) *holds for* $L(t)$.

As in the case of ordinary differential equations, a stronger condition than (PD) for linear systems is that for every $t \in J$, there exists $\alpha \in \mathbb{R}$ such that $L(t)\phi + \alpha\phi(0) \geqslant 0$ whenever $\phi \geqslant 0$.

It is useful to invoke the Riesz Representation Theorem [171] in order to identify $L(t)$ with a matrix of signed Borel measures $\eta(t) = (\eta(t)_{ij})$:

$$L(t)\phi = \int_{-r}^{0} \mathrm{d}\eta(t)\phi. \tag{4.6}$$

The Radon–Nikodym decomposition of η_{ij} with respect to the Dirac measure δ with unit mass at 0 gives $\eta_{ij}(t) = a_{ij}(t)\delta + \tilde{\eta}_{ij}(t)$ where a_{ij} is a scalar and $\tilde{\eta}_{ij}(t)$ is mutually singular with respect to δ. In particular, the latter assigns zero mass to $\{0\}$. Therefore,

$$L(t)\phi = A(t)\phi(0) + \tilde{L}(t)\phi, \quad \tilde{L}(t)\phi := \int_{-r}^{0} \mathrm{d}\tilde{\eta}(t)\phi. \tag{4.7}$$

Continuity of the map $t \to A(t)$ follows from continuity of $t \to L(t)$. The decomposition (4.7) leads to sharp conditions for (PD) to hold for $L(t)$.

PROPOSITION 4.4. (PD) *holds for* $L(t)$ *if and only if*
  (a) $A(t)$ *satisfies* (P) *of Proposition 3.3, and*
  (b) $\tilde{L}(t)\phi \geqslant 0$ *whenever* $\phi \geqslant 0$.

PROOF. If (a) and (b) hold, $\phi \geqslant 0$, $\lambda \in K^*$ and $\lambda(\phi(0)) = 0$ then $\lambda(L(t)\phi) = \lambda(A(t)\phi(0))$ $+ \lambda(\tilde{L}(t)\phi) \geqslant 0$ because each summand on the right is nonnegative.

Conversely, if (PD) holds for $L(t)$, $v \in \partial K$, $\lambda \in K^*$, and $\lambda(v) = 0$, define $\phi_n(\theta) = e^{n\theta} v$ on $[-r, 0]$. Then $\phi_n \geqslant 0$ and ϕ_n converges point-wise to zero, almost everywhere with respect to $\tilde{\eta}(t)$. By (PD),

$$\lambda(L(t)\phi_n) = \lambda(A(t)v + \tilde{L}(t)\phi_n) \geqslant 0.$$

Letting $n \to \infty$, we get $\lambda(A(t)v) \geqslant 0$ implying that (P) holds for $A(t)$. Let $\phi \geqslant 0$ be given and define $\phi_n(\theta) = [1 - e^{n\theta}]\phi(\theta)$ on $[-r, 0]$, $n \geqslant 1$. ϕ_n converges point-wise to $\phi\chi$, where χ is the indicator function of the set $[-r, 0)$, and $\phi\chi = \phi$ almost everywhere with respect to $\tilde{\eta}(t)$. If $\lambda \in K^*$, then $\lambda(\phi_n(0)) = 0$ so applying (PD) we get $0 \leqslant \lambda(L(t)\phi_n) = \tilde{L}(t)\phi_n$. Letting $n \to \infty$ we get (b). $\qquad\square$

For the remainder of this section, we suppose that $\Omega = J \times D$ where J is a nonempty open interval and $D \subset C$ is open. Suppose that $\frac{\partial f}{\partial \phi}(t, \psi)$ exists and is continuous on $J \times D$ to $L(C, \mathbb{R}^n)$. In that case, $x(t, t_0, \phi)$ is continuously differentiable in its last argument and $y(t, t_0, \chi) = \frac{\partial x}{\partial \phi}(t, t_0, \phi)\chi$ satisfies the variational equation

$$y'(t) = \frac{\partial f}{\partial \phi}\big(t, x_t(t, \phi)\big)y_t, \quad y_{t_0} = \chi. \tag{4.8}$$

See Theorem 2.4.1 of Hale and Verduyn Lunel [61]. We say that f (or (4.4)) is K-*cooperative* if for all $(t, \chi) \in J \times D$ the function $\psi \to \frac{\partial f}{\partial \phi}(t, \chi)\psi$ satisfies (PD). By Corollary 4.3 applied to the variational equation we have the following analog of Theorem 3.5 for functional differential equations. The proof is essentially the same.

THEOREM 4.5. *Let* $\frac{\partial f}{\partial \phi}(t, \psi)$ *exist and be continuous on* $J \times D$. *If* (QMD) *holds for* (4.4), *then* f *is* K-*cooperative. Conversely, if* D *is* p-*convex and* f *is* K-*cooperative, then* (QMD) *holds for* f.

Consider the nonlinear system

$$x'(t) = g\big(x(t), x(t - r_1), x(t - r_2), \ldots, x(t - r_m)\big), \tag{4.9}$$

where $g(x, y^1, y^2, \ldots, y^m)$ is continuously differentiable on $\mathbb{R}^{(m+1)n}$ and $r_{j+1} > r_j > 0$. Then

$$\frac{\partial f}{\partial \phi}(\psi) = \frac{\partial g}{\partial x}(x, Y)\delta + \sum_k \frac{\partial g}{\partial y^k}(x, Y)\delta_{-r_k}, \tag{4.10}$$

where δ_{-r_k} is the Dirac measure with unit mass at $\{-r_k\}$ and $x = \psi(0)$, $y^k = \psi(-r_k)$ and $(x, Y) := (x, y^1, y^2, \ldots, y^m)$. By Theorem 4.5, Corollary 4.3, and Proposition 4.4, (QMD) holds if and only if for each (x, Y), $\frac{\partial g}{\partial x}(x, Y)$ satisfies condition (P) and $\frac{\partial g}{\partial y^k}(x, Y)$ is K-positive. If $K = \mathbb{R}^n_+$, the condition becomes $\frac{\partial g_i}{\partial x_j}(x, Y) \geqslant 0$, for $i \neq j$ and $\frac{\partial g_i}{\partial y^k_j}(x, Y) \geqslant 0$ for all i, j, k; if, in addition, $n = 1$ then $\frac{\partial g}{\partial y^k}(x, Y) \geqslant 0$ for all k suffices.

4.3. *Eventual strong monotonicity*

We begin by considering the linear system (4.5). The following hypothesis for the continuous map $L: J \to L(\mathcal{C}, \mathbb{R}^n)$ reduces to (ST) of the previous section when $r = 0$:

> (STD)   for all $t \in J$ and $\phi \geqslant 0$ with $\phi(0) \in \partial K$ satisfying one of the conditions
>   (a) $\phi(-r) > 0$ and $\phi(0) = 0$, or
>   (b) $\phi(s) > 0$ for $-r \leqslant s \leqslant 0$,
>   there exists $\nu \in K^*$ such that $\nu(\phi(0)) = 0$ and $\nu(L(t)\phi) > 0$.

The following result is the analog of Theorem 3.6 of the previous section for delay differential equations.

THEOREM 4.6. *Let linear system (4.5) satisfy* (PD) *and* (STD) *and let* $t_0 \in J$. *Then*

$$\phi > 0, \ t \geqslant t_0 + 2r \quad \Longrightarrow \quad x(t, t_0, \phi) \gg 0.$$

In particular, $x_t(t_0, \phi) \gg 0$ *for* $t \geqslant t_0 + 3r$.

PROOF. By Corollary 4.3, we have that $x(t) := x(t, t_0, \phi) \geqslant 0$ for all $t \geqslant t_0$ that belong to J. There exists $t_1 \in (t_0, t_0 + r)$ such that $x(t_1 - r) = \phi(t_1 - r) = x_{t_1}(-r) > 0$ since $\phi > 0$. If $x(t_1) = 0$, then (STD)(a) implies the existence of $\nu \in K^*$ such that $\nu(L(t_1)x_{t_1}) > 0$. As $\nu(x(t)) \geqslant 0$ for $t \geqslant t_0$ and $\nu(x(t_1)) = 0$ we conclude that $\frac{d}{dt}|_{t=t_1}\nu(x(t)) \leqslant 0$. But $\frac{d}{dt}|_{t=t_1}\nu(x(t)) = \nu(L(t_1)x_{t_1}) > 0$, a contradiction. Therefore, $x(t_1) > 0$.

Now, by (4.7)

$$x' = A(t)x + \tilde{L}(t)x_t$$

from which we conclude

$$x(t) = X(t, t_1)x(t_1) + \int_{t_1}^t X(t, r)\tilde{L}(r)x_r \, dr,$$

where $X(t, t_0)$ is the fundamental matrix for $y' = A(t)y$ satisfying $X(t_0, t_0) = I$. From (a) of Proposition 4.4 and Corollary 3.4, it follows that $X(t, t_0)$ is K positive for $t \geq t_0$. This, the fact that $x_r \geq 0$, and (b) of Proposition 4.4 imply that the integral belongs to K so we conclude that

$$x(t) \geq X(t, t_1)x(t_1) > 0, \quad t \geq t_1.$$

We claim that $x(t) \gg 0$ for $t \geq t_1 + r$. If not, there is a $t_2 \geq t_1 + r$ such that $x(t_2) = x_{t_2}(0) \in \partial K$ but $x_{t_2}(s) > 0$ for $-r \leq s \leq 0$. Then (STD) implies the existence of $v \in K^*$ such that $v(x(t_2)) = 0$ and $v(L(t_2)x_{t_2}) > 0$. Since $v(x(t)) \geq 0$ for $t \geq t_0$ we must have $\frac{d}{dt}|_{t=t_2} v(x(t)) \leq 0$. But $\frac{d}{dt}|_{t=t_2} v(x(t)) = v(L(t_2)x_{t_2}) > 0$, a contradiction. We conclude that $x(t) \gg 0$ for $t \geq t_1 + r$. $\square$

In a sense, (STD)(a) says that r has been correctly chosen; (STD)(b) is more fundamental. The next result gives sufficient conditions for it to hold.

PROPOSITION 4.7. *If $L(t)$ satisfies* (PD) *and either*
  (a) *$A(t)$ satisfies* (ST), *or*
  (b) $\phi > 0 \Longrightarrow \tilde{L}(t)\phi \gg 0$
then (STD)(b) *holds.*

PROOF. This is immediate from the definitions, the decomposition (4.7), Proposition 4.4, and the expression $v(L(t)\phi) = v(A(t)\phi(0)) + v(\tilde{L}(t)\phi)$. $\square$

Theorem 4.6 leads immediately to a result on eventual strong monotonicity for the nonlinear system (4.4) where we assume that $\Omega = J \times D$ as above.

THEOREM 4.8. *Let D be p-convex, $\frac{\partial f}{\partial \phi}(t, \psi)$ exist and be continuous on $J \times D$ to $L(\mathcal{C}, \mathbb{R}^n)$, and f be K-cooperative. Suppose that* (STD) *holds for $\frac{\partial f}{\partial \phi}(t, \psi)$, for each $(t, \psi) \in J \times D$. Then*

$$\phi_0, \phi_1 \in D, \ \phi_0 < \phi_1 \quad \Longrightarrow \quad x(t, t_0, \phi_0) \ll x(t, t_0, \phi_1)$$

for all $t \geq t_0 + 2r$ for which both solutions are defined.

PROOF. By Theorem 4.5, we have $x(t, t_0, \phi_0) \leq x(t, t_0, \phi_1)$ for $t \geq t_0$ for which both solutions are defined. We apply the formula

$$x(t, t_0, \phi_1) - x(t, t_0, \phi_0) = \int_0^1 \frac{\partial x}{\partial \phi}(t, t_0, s\phi_1 + (1-s)\phi_0)(\phi_1 - \phi_0) \, ds.$$

Here, for $\psi \in D$ and $\beta \in \mathcal{C}$, $y(t, t_0, \beta) := \frac{\partial x}{\partial \phi}(t, t_0, \psi)\beta$ satisfies the variational equation (4.5) where $\phi = \beta$ and $L(t) = \frac{\partial f}{\partial \phi}(t, x_t(t_0, \psi))$. See Theorem 2.4.1 of Hale and Verduyn Lunel [61]. The desired conclusion will follow if we show that $y(t, t_0, \beta) \gg 0$ for

$t \geqslant t_0 + 2r$ for $\psi = s\phi_1 + (1-s)\phi_0$ and $\beta = \phi_1 - \phi_0 > 0$. By Theorem 4.6, it suffices to show that $L(t)$ satisfies (PD) and (STD). But this follows from our hypotheses. $\square$

In the next result, Theorem 4.8 is applied to system (4.9). We make use of notation introduced below Theorem 4.5.

COROLLARY 4.9. *Let* $g : \mathbb{R}^{(m+1)n} \to \mathbb{R}^n$ *be continuously differentiable and satisfy*
  (a) $\frac{\partial g}{\partial x}(x, Y)$ *satisfies* (P) *for each* $(x, Y) \in Z$;
  (b) *for each* k, $\frac{\partial g}{\partial y^k}(x, Y)$ *is* K *positive*;
  (c) *either* $\frac{\partial g}{\partial x}(x, Y)$ *satisfies* (ST) *or some* $\frac{\partial g}{\partial y^k}(x, Y)$ *is strongly positive on* K.
Then the hypotheses of Theorem 4.8 hold for (4.9).

PROOF. Recalling (4.10), it is evident that (a) and (b) imply that (4.9) is K-cooperative. Hypothesis (c) and Proposition 4.7 imply that (STD) holds. $\square$

In the special case that (4.9) is a scalar equation, $m = 1$ and $K = \mathbb{R}_+$, then $\frac{\partial g}{\partial y}(x, y) > 0$ suffices to ensure an eventually strongly monotone semiflow.

4.4. *K is an orthant*

Our results can be improved in the case that K is a product cone such as $\mathbb{R}^n_+ = \prod_{i=1}^n \mathbb{R}_+$, i.e., an orthant. The following example illustrates the difficulty with our present set up.

$$x_1'(t) = -x_1(t) + x_2(t - 1/2),$$
$$x_2'(t) = x_1(t - 1) - x_2(t).$$

Observe that (PD) holds for the standard cone. For initial data, take $\phi = (\phi_1, \phi_2) \in C$ $(r = 1)$ where $\phi_1 = 0$ and $\phi_2(\theta) > 0$ for $\theta \in (-1, -2/3)$ and $\phi_2(\theta) = 0$ elsewhere in $[-1, 0]$. The initial value problem can be readily integrated by the method of steps of length $1/2$ and one sees that $x(t) = 0$ for all $t \geqslant -2/3$. In the language of semiflows, $\phi > 0$ yet $\Phi_t(\phi) = \Phi_t(0) = 0$ for all $t \geqslant 0$. The problem is that $C([-1, 0], \mathbb{R}^2)$ is not the optimal state space; a better one is the product space $X = C([-1, 0], \mathbb{R}) \times C([-1/2, 0], \mathbb{R})$. Obviously, an arbitrary cone in $\mathbb{R}^2$ will not induce a cone in the product space X.

For the remainder of this section we focus on the standard cone but the reader should observe that an analogous construction works for any orthant $K = \{x: (-1)^{m_i} x_i \geqslant 0\}$. Motivated by the example in the previous paragraph, let $r = (r_1, r_2, \dots, r_n) \in \mathbb{R}^n_+$ be a vector of delays, $R = \max r_i$ and define

$$C_r = \prod_{i=1}^n C([-r_i, 0], \mathbb{R}).$$

Note that we allow some delays to be zero. We write $\phi = (\phi_1, \phi_2, \ldots, \phi_n)$ for a generic point of C_r. C_r is a Banach space with the norm $|\phi| = \sum |\phi_i|$. Let

$$C_r^+ = \prod_{i=1}^{n} C\big([-r_i, 0], \mathbb{R}_+\big)$$

denote the cone of functions in C_r with nonnegative components. It has nonempty interior given by those functions with strictly positive components. As usual, we use the notation $\leqslant, <, \ll$ for the corresponding order relations on C_r induced by C_r^+. If $x_i(t)$ is defined on $[-r_i, \sigma)$, $1 \leqslant i \leqslant n$, $\sigma > 0$ then we may redefine $x_t \in C_r$ as $x_t = (x_t^1, x_t^2, \ldots, x_t^n)$ where $x_t^i(\theta) = x_i(t + \theta)$ for $\theta \in [-r_i, 0]$. Notice that now, the subscript signifying a particular component will be raised to a superscript when using the subscript "t" to denote a function.

If $D \subset C_r$ is open, J is an open interval and $f : J \times D \to \mathbb{R}^n$ is given, then the standard existence and uniqueness theory for the initial value problem associated with (4.4) is unchanged. Furthermore, Theorems 4.1 and 4.2, and Corollary 4.3 remain valid in our current setting where, of course, we need only make use of the coordinate maps $\eta(x) = x_i$, $1 \leqslant i \leqslant n$ in (QMD) and (PD). Our goal now is to modify (STD) so that we may obtain a result like Theorem 4.6 that applies to systems such as the example above. We begin by considering the linear system (4.5) where $L : J \to L(C_r, \mathbb{R}^n)$ is continuous and let $L_i(t)\phi := \langle e_i, L(t)\phi \rangle$, $1 \leqslant i \leqslant n$.

In our setting, $L(t)$ satisfies (PD) if and only if:

$$\phi \geqslant 0 \text{ and } \phi_i(0) = 0 \quad \text{implies} \quad L_i(t)\phi \geqslant 0.$$

THEOREM 4.10. *Let linear system* (4.5) *satisfy* (PD) *and*
    (i) $t \in J$, $r_j > 0$, $\phi \geqslant 0$, $\phi_j(-r_j) > 0 \Longrightarrow L_i(t)\phi > 0$ *for some* i;
    (ii) *for every proper subset* I *of* $N := \{1, 2, \ldots, n\}$, *there exists* $j \in N \setminus I$ *such that* $L_j(t)\phi > 0$ *whenever* $\phi \geqslant 0$, $\phi_i(s) > 0$, $-r_i \leqslant s \leqslant 0$, $i \in I$.
Then $x(t, \phi, t_0) \gg 0$ *if* $\phi > 0$ *for all* $t \geqslant t_0 + nR$.

PROOF. By (PD) and Corollary 4.3 we have $x(t) \geqslant 0$ for $t \geqslant t_0$. An application of the Riesz Representation Theorem and Radon–Nikodym Theorem implies that for $i = 1, 2, \ldots, n$, we have

$$L_i(t)\phi = a_i(t)\phi_i(0) + \sum_{j=1}^{n} \int_{-r_j}^{0} \phi_j(\theta) \, d_\theta \eta_{ij}(t, \theta) = a_i(t)\phi_i(0) + \bar{L}_i(t)\phi,$$

where $\eta_{ij}(t)$ is a positive Borel measure on $[-r_j, 0]$, $a_i(t) \in \mathbb{R}$ and $\bar{L}_i(t)\phi \geqslant 0$ whenever $\phi \geqslant 0$. Moreover, $t \to \eta_{ij}(t)$ and $t \to a_i(t)$ are continuous. See Smith [190,194] for details. The representation of L_i in terms of signed measures, $\bar{\eta}_{ij}$, is standard; (PD) implies that $\eta_{ij} := \bar{\eta}_{ij}$ must be positive for $i \neq j$ and that $\bar{\eta}_{ii}$ has the Lebesgue decomposition $\bar{\eta}_{ii} = a_i \delta + \eta_{ii}$ with respect to δ, the Dirac measure of unit mass at zero, and η_{ii} is a positive measure which is mutually singular with respect to δ.

If $x_i(t_1) > 0$ for some i and $t_1 > t_0$ then from $x_i'(t) = a_i(t)x_i(t) + \bar{L}_i(t)x_t \geqslant a_i(t)x_i(t)$ we conclude from standard differential inequality arguments that $x_i(t) > 0$ for $t \geqslant t_1$.

As $\phi > 0$, there exists j such that $\phi_j > 0$. If $r_j = 0$ then $x_j(t_0) > 0$; if $r_j > 0$ then $x_j(t_1 - r_j) > 0$ for some $t_1 \in (t_0, t_0 + r_j)$. In this case, it follows from (i) that $x_i'(t_1) = L_i(t_1)\phi > 0$ for some i and hence $x_i(t_1) > 0$. Hence, $x_i(t) > 0$ for $t \geqslant t_1$ by the previous paragraph. Applying (ii) with $I = \{i\}$ and $t = t_2 = t_1 + r_i$ we may find $k \neq i$ such that $x_k'(t_2) = L_k(t_2)x_{t_2} > 0$ because $x_{t_2}^i(s) > 0$, $-r_i \leqslant s \leqslant 0$. Therefore, we must have $x_k(t_2) > 0$ and hence $x_k(t) > 0$ for $t \geqslant t_2$. Obviously, we may continue in this manner until we have all components positive for $t \geqslant t_0 + nR$ as asserted. $\qquad\square$

Theorem 4.10 leads directly to a strong monotonicity result for the nonlinear nonautonomous delay differential equation (4.4) in the usual way. We extend the definition of K-cooperativity of f to our present setup with state space $\mathcal{C}_r$ exactly as before.

THEOREM 4.11. *Let $D \subset \mathcal{C}_r$ be p-convex, $\frac{\partial f}{\partial \phi}(t, \psi)$ exist and be continuous on $J \times D$ to $L(\mathcal{C}_r, \mathbb{R}^n)$, and f be K-cooperative. Suppose that for all $(t, \psi) \in J \times D$, $L(t) := \frac{\partial f}{\partial \phi}(t, \psi)$ satisfies the conditions of Theorem 4.10. Then*

$$\phi_0, \phi_1 \in D, \ \phi_0 < \phi_1, \ t \geqslant t_0 + nR \implies x(t, t_0, \phi_0) \ll x(t, t_0, \phi_1).$$

The biochemical control circuit with delays, modeled by the system

$$\begin{aligned} x_1'(t) &= g\big(x_n(t - r_n)\big) - \alpha_1 x_1(t), \\ x_j'(t) &= x_{j-1}(t - r_{j-1}) - \alpha_j x_j(t), \quad 2 \leqslant j \leqslant n \end{aligned} \tag{4.11}$$

with decay rates $\alpha_j > 0$ and delays $r_i \geqslant 0$ with $R > 0$ provides a good application of Theorem 4.11 which cannot be obtained by Theorem 4.8 if the delays are distinct. We assume the $g : \mathbb{R}_+ \to \mathbb{R}_+$ is continuously differentiable and $g' > 0$. Equation (4.11) is an autonomous system for which $\mathcal{C}_r^+$ is positively invariant by Theorem 4.2. See Smith [191, 194] for more on this application.

4.5. *Generic convergence for delay differential equations*

The aim of this section is to apply Theorem 4.8 and Theorem 4.11 to the autonomous delay differential equation (4.2) to conclude that the generic solution converges to equilibrium. To Φ, defined by (4.3), we associate C, S and E, denoting respectively the sets of convergent, stable and equilibrium points. The main result of this section is the following.

THEOREM 4.12. *Let $f \in C^1(D)$, (4.2) be cooperative on the p-convex open subset D of C or $\mathcal{C}_r$ and satisfy:*

(a) *The hypotheses of Theorem 4.8 or of Theorem 4.11 hold;*

(b) *Every positive semiorbit of Φ has compact closure in D and $D = AC \cup BC$.*

Then

(i) $C \cap S$ *contains a dense open subset of* D, *consisting of points whose trajectories converge to equilibria;*

(ii) *If* E *is compact there is a stable equilibrium, and an asymptotically stable equilibrium when* E *is finite.*

PROOF. For definiteness, suppose that (4.2) is cooperative on the p-convex open subset D of C and that the hypotheses of Theorem 4.8 hold. The other case is proved similarly. Assumption (a) ensures that Φ is eventually strongly monotone. Moreover, the derivative of $\Phi_t(\phi)$ with respect to ϕ exists and $\Phi'_\tau(\phi)\chi = y_\tau(t_0, \chi)$, where $y(t, t_0, \chi)$ is the solution of the variational equation (4.8). As our hypotheses ensure that $L(t) = \frac{\partial f}{\partial \phi}(x_t(\phi))$ satisfies (STD), we conclude from Theorem 4.6 that $\Phi'_\tau(\phi)$ is strongly positive for $\tau \geqslant 3r$. Compactness of $\Phi'_\tau(\phi): C \to C$ for $\tau \geqslant r$ follows from the fact that a bound for $y_\tau(t_0, \chi)$, uniform for χ belonging to a bounded set $B \subset C$, can be readily obtained so, using (4.8), we may also find a uniform bound for $y'(t, t_0, \chi)$, $\tau - r \leqslant t \leqslant \tau$. See, e.g., Hale [58, Theorem 4.1.1] for more detail.

The hypotheses of Theorem 2.26, with $X = D$, are fulfilled: D is normally ordered and $D = BC \cup AC$; while (M) and (D^*) hold as noted above. Therefore Theorem 2.26 implies the conclusion. $\qquad\square$

In the special case that (4.2) is scalar ($n = 1$) we note that the set E of equilibria is totally ordered in C_r or C so the set of quasiconvergent points coincides with the set of convergent points: $Q = C$. The classical scalar delay differential equation (4.1) has been thoroughly investigated in the case of monotone delayed feedback ($f(0) = 0$ and $f' > 0$) by Krisztin et al. [105]. They characterize the closure of the unstable manifold of the trivial solution in case it is three-dimensional and determine in remarkable detail the dynamics on this invariant set.

Smith and Thieme [198,200,194] introduce an exponential ordering, not induced by a cone in $\mathbb{R}^n$, that extends the scope of application of the theory described here. One of the salient results from this work is that a scalar delay equation for which the product of the delay r and the Lipschitz constant of f is smaller than e^{-1} generates an eventually strongly monotone semiflow with respect to the exponential ordering and therefore the generic orbit converges to equilibrium: the dynamics mimics that of the associated ordinary differential equation obtained by ignoring the delay. See also work of Pituk [159].

We have considered only bounded delays. Systems of delay differential equations with unbounded and even infinite delay are also of interest. See Wu [234] for extensions to such systems. Wu and Freedman [235] and Krisztin and Wu [106–108] extend the theory to delay differential equations of neutral type.

5. Monotone maps

5.1. *Background and motivating examples*

One of the chief motivations for the study of monotone maps is their importance in the study of periodic solutions to periodic quasimonotone systems of ordinary differential equations. See for example the monograph of Krasnosel'skii [99], the much cited paper of de Mottoni and Schiaffino [42], Hale and Somolinos [60], Smith [188,189], Liang and Jiang [121], and Wang and Jiang [229–231]. To fix ideas, let $f : \mathbb{R} \times \mathbb{R}^n \to \mathbb{R}^n$ be a locally Lipschitz function and consider the ordinary differential equation

$$x' = f(t, x). \tag{5.1}$$

As usual, denote by $x(t, t_0, x_0)$ the noncontinuable solution of the initial value problem $x(t_0) = x_0$, which for simplicity we assumed is defined for all t. If f is periodic in t of period one: $f(t + 1, x) = f(t, x)$ for all (t, x), then it is natural to consider the *period map* $T : \mathbb{R}^n \to \mathbb{R}^n$ defined by

$$T(x_0) = x(1, 0, x_0). \tag{5.2}$$

Its fixed points (periodic points) are in one-to-one correspondence with the periodic (subharmonic) solutions of (5.1). If K is a cone in $\mathbb{R}^n$ for which f satisfies the quasimonotone condition (QM), then it follows from Theorem 3.2 that T is a monotone map: $x \leqslant y$ implies $Tx \leqslant Ty$. Moreover, T has the important property, not shared with general monotone maps, that it is an orientation-preserving homeomorphism.

In a similar way, periodic solutions for second order parabolic partial differential equations with time-periodic data can be analyzed by considering period maps in appropriate function spaces. Here monotonicity comes from classical maximum principles. Hess [63] remains an up-to-date survey. See also Alikakos et al. [3] and Zhao [245]. Remarkable results are known for equations on a compact interval with standard boundary conditions. Chen and Matano [23] show that every forward (backward) bounded solution is asymptotic to a periodic solution; Brunovsky et al. [22] extend the result to more general equations. Chen et al. [24] give conditions for the period map to generate Morse–Smale dynamics and thus be structurally stable. Although monotonicity of the period map is an important consideration in these results, it is not the key tool. The fact that the number of zeros on the spatial interval of a solution of the linearized equation is non-increasing in time is far more important. See Hale [59] for a nice survey.

A different theme in order-preserving dynamics originates in the venerable subject of nonlinear elliptic and parabolic boundary value problems. The 1931 edition of Courant and Hilbert's famous book [34] refers to a paper of Bieberbach in *Göttingen Nachrichten*, 1912 dealing with the elliptic boundary value problem $\Delta u = e^u$ in Ω, $u|\partial\Omega = f$, in a planar region Ω. A solution is found by iterating a monotone map in a function space. Courant and Hilbert extended this method to a broad class of such problems. Out of this technique grew the method of "upper and lower solutions" (or "supersolutions and subsolutions") for solving, both theoretically and numerically, second order elliptic PDEs (see Amann [4], Keller and Cohen [95], Keller [93,94], Sattinger [176]). Krasnosel'skii and Zabreiko [101] trace the use of positivity in functional analysis—closely related to monotone dynamics— to a 1924 paper by Uryson [222] on concave operators. The systematic use of positivity in PDEs was pioneered Krasnosel'skii and Ladyzhenskaya [100] and Krasnosel'skii [98].

Amann [5] showed how a sequence $\{u_n\}$ of approximate solutions to an elliptic problem can be viewed as the trajectory $\{T^n u_0\}$ of u_0 under a certain monotone map T in a suitable function space incorporating the boundary conditions, with fixed points of T being solutions of the elliptic equation. The dynamics of T can therefore be used to investigate the equation. Thus when T is globally asymptotically stable, there is a unique solution; while if T has two asymptotically stable fixed points, in many cases degree theory yields a third fixed point. As Amann [6] emphasized, a few key properties of T—continuity, monotonicity and some form of compactness—allow the theory to be efficiently formulated in terms of monotone maps in ordered Banach spaces.

Many questions in differential equations are framed in terms of eigenvectors of linear and nonlinear operators on Banach spaces. The usefulness of operators that are positive in some sense stems from the theorem of Perron [158] and Frobenius [51], now almost a century old, asserting that for a linear operator on $\mathbb{R}^n$ represented by a matrix with positive entries, the spectral radius is a simple eigenvalue having a positive eigenvector, and all other eigenvalues have smaller absolute value and only nonpositive eigenvectors. In 1912 Jentzsch [85] proved the existence of a positive eigenfunction with a positive eigenvalue for a homogeneous Fredholm integral equation with a continuous positive kernel.

In 1935 the topologists Alexandroff and Hopf [2] reproved the Perron–Frobenius theorem by applying Brouwer's fixed-point theorem to the action of a positive $n \times n$ matrix on the space of lines through the origin in $\mathbb{R}^n_+$. This was perhaps the first explicit use of the dynamics of operators on a cone to solve an eigenvalue problem. In 1940 Rutman [173] continued in this vein by reproving Jentzsch's theorem by means of Schauder's fixed-point theorem, also obtaining an infinite-dimensional analog of Perron–Frobenius, known today as the Krein–Rutman theorem [104,214]. In 1957 G. Birkhoff [20] initiated the dynamical use of Hilbert's projective metric for such questions.

The dynamics of cone-preserving operators continues to play an important role in functional analysis; for a survey, see Nussbaum [149,150]. One outgrowth of this work has been a focus on purely dynamical questions about such operators; some of these results are presented below. Polyhedral cones in Euclidean spaces have lead to interesting quantitative results, including *a priori* bounds on the number of periodic orbits. For recent work see Lemmens et al. [117], Nussbaum [151], Krause and Nussbaum [102], and references therein.

Monotone maps frequently arise as mathematical models. For example, the discrete Lotka–Volterra competition model (see May and Oster [136]):

$$(u_{n+1}, v_{n+1}) = T(u_n, v_n)$$
$$:= \left(u_n \exp\left[r(1 - u_n - bv_n)\right], v_n \exp\left[s(1 - cu_n - v_n)\right]\right)$$

generates a monotone dynamical system relative to the fourth-quadrant cone only when the intrinsic rate of increase of each population is not too large ($r, s \leqslant 1$) and then only on the order interval $[0, r^{-1}] \times [0, s^{-1}]$ (Smith [192]). Fortunately in this case, every point in the first quadrant enters and remains in this order interval after one iteration. As is typical in ecological models, the Lotka–Volterra map is neither injective nor orientation-preserving or orientation-reversing. For monotone maps as models for the spread of a gene or an epidemic through a population, see Thieme [218], Selgrade and Ziehe [181], Weinberger [232], Liu [123] and the references therein.

5.2. *Definitions and basic results*

A continuous map $T : X \to X$ on the ordered metric space X is *monotone* if $x \leqslant y \Rightarrow Tx \leqslant Ty$, *strictly monotone* if $x < y \Rightarrow Tx < Ty$, *strongly monotone* if $x < y \Rightarrow Tx \ll Ty$, and *eventually strongly monotone* if whenever $x < y$, there exists $n_0 \geqslant 1$ such that $T^n x \ll T^n y$. We call T *strongly order-preserving* (*SOP*) if T is monotone, and whenever $x < y$ there exist respective neighborhoods U, V of x, y and $n_0 \geqslant 1$ such that $n \geqslant n_0 \Rightarrow T^n U \leqslant T^n V$.[1] As with semiflows, eventual strong monotonicity implies the strong order preserving property.

The orbit of x is $O(x) := \{T^n x\}_{n \geqslant 0}$, and the omega limit set of x is $\omega(x) := \bigcap_{k \geqslant 0} \overline{O(T^k x)}$. If $O(x)$ has compact closure, $\omega(x)$ is nonempty, compact, invariant (that is, $T\omega(x) = \omega(x)$) and *invariantly connected*. The latter means that $\omega(x)$ is not the disjoint union of two closed invariant sets [116].

If $T(x) = x$ then x is a *fixed point* or *equilibrium*. E denotes the set of fixed points. More generally, if $T^k x = x$ for some $k \geqslant 1$ we call x *periodic*, or k-*periodic*. The minimal such k is called the period of x (and $O(x)$).

Let Y denotes an ordered Banach space with order cone Y_+. A linear operator $A \in L(Y)$ is called *positive* if $A(Y_+) \subset Y_+$ (equivalently, A is a monotone map) and *strongly positive* if $A(Y_+ \setminus \{0\}) \subset \operatorname{Int} Y_+)$ (equivalently, A is a strongly monotone map).

The following result is useful for proving smooth maps monotone or strongly monotone:

LEMMA 5.1. *Let $X \subset Y$ be a p-convex set and $f : X \to Y$ a locally C^1 map with quasiderivative $h : U \to L(Y)$ defined on an open set $U \subset Y$. If the linear maps $h(x) \in L(Y)$ are positive (respectively, strongly positive) for all $x \in U$, then f is monotone (respectively, strongly monotone).*

PROOF. By p-convexity it suffices to prove that every $p \in X$ has a neighborhood N such that $f|N \cap X$ is monotone (respectively, strongly monotone). We take N to be an open ball in U centered at p. Suppose $p + z \in X \cap N$, $z > 0$. By p-convexity, $X \cap N$ contains the line segment from p to $p + z$. The definition (above Lemma 2.15) of locally C^1 implies that the map $g : [0, 1] \to Y, t \mapsto f(p + tz)$ is C^1 with $g'(t) = h(tz)z$. Therefore

$$f(p + z) - f(z) = g(0) - g(1) = \int_0^1 g'(t)\,dt = \int_0^1 h(tz)z\,dt.$$

Because $h(tz) \in L(Y)$ is positive and $z > 0$, we have $h(tz)z \in Y_+$, therefore $f(p + z) - f(p) \geqslant 0$. If the operators $h(tz)$ are strongly positive, $f(p + z) - f(p) \gg 0$. $\square$

PROPOSITION 5.2 (Nonordering of Periodic Orbits). *A periodic orbit of a monotone map is unordered.*

---

[1] Our use of "strongly order-preserving" conflicts with Dancer and Hess [38], who use these words to mean what we have defined as "strongly monotone". Our usage is consistent with that of several authors. Takáč [208,209] uses "strongly increasing" for our SOP.

PROOF. If not, there exists x in the orbit such that $T^k(x) > x$ for some $k > 0$. Induction on n shows that $T^{nk}(x) > x$ for all $n > 0$. But if x has period $m > 0$, induction on k proves that $T^{mk}(x) = x$. $\qquad\square$

LEMMA 5.3 (Monotone Convergence Criterion). *Assume T is monotone and $O(z)$ has compact closure. If $m \geqslant 1$ is such that $T^m z < z$ or $T^m z > z$ then $\omega(z)$ is an m-periodic orbit.*

PROOF. Consider first the case $m = 1$. Compactness of $\overline{O(z)}$ implies the decreasing sequence $\{T^k z\}$ converges to a point $p = \omega(x)$. Now suppose $m > 1$. Applying the case just proved to the map T^m, we conclude that $\{T^{km} z\}$ converges to a point $p = T^m(p)$. It follows that $\omega(z) = \{p, Tp, T^2 p, \dots, T^{m-1} p\}$. $\qquad\square$

Lemma 5.3 yields information on one-sided stability of compact limit sets when T is SOP; see Hirsch [70].

In order to state the following lemma succinctly, we call a set $J \subset \mathbb{N}$ an *interval* if it is nonempty and contains all integers between any two of its members. For $a, b \in \mathbb{N}$ we set $[a, b] = \{j \in \mathbb{N} : a \leqslant j \leqslant b\}$ (there will be no confusion with real intervals). Two intervals *overlap* if they have more than one point in common.

Let $J \subset \mathbb{N}$ be an interval and $f : J \to X$ be a map. A subinterval $[a, b] \subset J, a < b$ is *rising* if $f(a) < f(b)$, and *falling* if $f(b) < f(a)$.

THEOREM 5.4. *A trajectory of a monotone map cannot have both a rising interval and a falling interval.*

PROOF. Follows from Theorem 1.6. $\qquad\square$

LEMMA 5.5. *If T is monotone, $\omega(z)$ cannot contain distinct points having respective neighborhoods U, V such that $T^r(U) \leqslant T^r(V)$ for some $r \geqslant 0$.*

PROOF. Follows from Theorem 5.4 (see proof of Lemma 1.7). $\qquad\square$

The next result is fundamental to the theory of monotone maps:

THEOREM 5.6 (Nonordering Principle). *Let $\omega(z)$ be an omega limit set for a monotone map T.*
  (i) *No points of $\omega(z)$ are related by $\ll$.*
  (ii) *If $\omega(z)$ is a periodic orbit or T is SOP, no points of $\omega(z)$ are related by $<$.*

PROOF. Follows from Proposition 5.2 and Lemma 5.5 (see the proof of Theorem 1.8). $\quad\square$

Call x *convergent* if $\omega(x)$ is a fixed point, and *quasiconvergent* if $\omega(x) \subset E$. Just as for semiflows, Proposition 5.6 leads immediately to a convergence criterion:

COROLLARY 5.7. *Assume Φ is SOP.*

(i) *If an omega limit set has a supremum or infimum, it reduces to a single fixed point.*

(ii) *If the fixed point set is totally ordered, every quasiconvergent point with compact orbit closure is convergent.*

PROOF. Part (i) follows from Theorem 5.6(ii), since the supremum or infimum, if it exists, belongs to the limit set. Part (ii) is a consequence (i). $\square$

5.2.1. *Failure of the limit set dichotomy* We now point out a significant difference between strongly monotone maps and semiflows:

The Limit Set Dichotomy fails for strongly monotone maps.

Recall that for an SOP semiflow with compact orbit closures, the dichotomy (Theorem 1.16) states:

$$\text{If } a < b, \text{ either } \omega(a) < \omega(b) \text{ or } \omega(a) = \omega(b) \subset E.$$

Takáč [211, Theorem 3.10], gives conditions on strongly monotone maps under which $a < b$ implies that either $\omega(a) \cap \omega(b) = \emptyset$ or $\omega(a) = \omega(b)$. He also gives a counterexample showing that $\omega(a) \cap \omega(b) = \emptyset$ does not imply $\omega(a) < \omega(b)$, nor does $\omega(a) = \omega(b)$ imply that these limit sets consist of fixed points (they are period-two orbits in his example). However, the mapping in his example is defined on a disconnected space.

For any map T in a Banach space, having an asymptotically stable periodic point p of period > 1, the Limit Set Dichotomy as formulated above must fail: take a point $q > p$ so near to p that $O(p) = \omega(p) = \omega(q)$. Clearly $\omega(p)$, being a nontrivial periodic orbit, contains no fixed points. Thus the second assertion of the Limit Set Dichotomy fails in this case.

Dancer and Hess [38] gave a simple example in $\mathbb{R}^k$ for prime k of a strongly monotone map with an asymptotically stable periodic point of period k which we describe below. Therefore the second alternative of the Limit Set Dichotomy can be no stronger than that $\omega(a) = \omega(b)$ is a periodic orbit.

The Limit Set Dichotomy fails even for strictly monotone maps in $\mathbb{R}^2$. Let $f(x) = 2\arctan(x)$, let $a > 0$ be its unique positive fixed point, and note that $0 < f'(a) < 1$. Define $T_0 : \mathbb{R}^2 \to \mathbb{R}^2$ by $T_0(x, y) := (f(y), f(x))$. Then $E = \{(-a, -a), (0, 0), (a, a)\}$ since f has no points of period 2. The fixed points of T_0^2 are the nine points obtained by taking all pairings of $-a, 0, a$. An easy calculation shows that $\{(-a, a), (a, -a)\}$ is an asymptotically stable period-two orbit of T_0 because the Jacobian matrix of T_0^2 is $f'(a)^2$ times the identity matrix. T^0 is strictly monotone but not strongly monotone. Now consider the perturbations $T_\epsilon(x, y) := T_0(x, y) + (\epsilon x, \epsilon y)$. It is easy to see that T_ϵ is strongly monotone for $\epsilon > 0$; and by the implicit function theorem, for small $\epsilon > 0$, T_ϵ has an asymptotically stable period-two orbit $O(p_\epsilon)$ with p_ϵ near $(-a, a)$. As noted in [38], this example can be generalized to $\mathbb{R}^k$ for prime k.

Takáč [212] shows that linearly stable periodic points can arise for the period map associated with monotone systems of ordinary and partial differential equations. Other counterexamples for low-dimensional monotone maps can be found in Smith [192,195].

As we have shown, asymptotically stable periodic orbits that are not singletons can exist for monotone, even strongly monotone maps. Later we will show that the generic orbit of a smooth, dissipative, strongly monotone map converges to a periodic orbit. Here, we show that every attractor contains a stable periodic orbit.

Recall that a point p is *wandering* if there exists a neighborhood U of p and a positive integer n_0 such that $T^n(U) \cap U = \emptyset$ for $n > n_0$. The *nonwandering set* Ω, consisting of all points q that are not wandering, contains all limit sets. In the following, we assume that X is an open subset of the strongly ordered Banach space Y and $T : X \to X$ is monotone with compact orbit closures. The following result is adapted from Hirsch [71, Theorems 4.1, 6.3].

THEOREM 5.8. *If T is strongly monotone and K is a compact attractor, then K contains a stable periodic orbit.*

The proof relies on the following result that does not use strong monotonicity nor that K attracts uniformly:

THEOREM 5.9. *Let $p \in K$ be a maximal (resp., minimal) nonwandering point. Then p is periodic, and every neighborhood of p contains an open set $W \gg p$ (resp., $W \ll p$) such that $\omega(x) = O(p)$ for all $x \in W$.*

PROOF. Suppose K attracts the open neighborhood U of K and fix $y \gg p$, $y \in U$. Since p is nonwandering there exists a convergent sequence $x_i \to p$ and a sequence $n_i \to \infty$ such that $T^{n_i} x_i \to p$. For all large i, $x_i \leqslant y$. Passing to a subsequence, we assume that $T^{n_i} y \to q$. By monotonicity and $x_i \leqslant y$ for large i, we have $q \geqslant p$. But $q \in K \cap \Omega$ and the maximality of p requires $q = p$. Since $p \ll y$ and $T^{n_i} y \to p$ it follows that $T^m y \ll y$ for some m. Lemma 5.3 implies that $\omega(y)$ is an m-periodic orbit containing p. As this holds for every $y \gg p$, the result follows. $\square$

LEMMA 5.10. *Let $p, q \in K$ be fixed points such that $p \ll q$, p is order stable from below, and q is order stable from above. Then $K \cap [p, q]$ contains a stable equilibrium.*

PROOF. Let R be a maximal totally ordered set of fixed points in $K \cap [p, q]$. An argument similar to the one in the proof of Theorem 1.30 shows that the fixed point

$$e := \inf\{z \in E \cap R: z \text{ is order stable from above}\}$$

is order stable. That e is stable follows from the analog of Proposition 1.28. $\square$

PROOF OF THEOREM 5.8. Theorem 5.9 shows that some iterate T^n, $n \geqslant 0$ has fixed points p, q as in Lemma 5.10, which result therefore implies Theorem 5.8. $\square$

Jiang and Yu [90, Theorem 2] implies that if T is analytic, order compact with strongly positive derivative, then K must contain an asymptotically stable periodic orbit.

5.3. *The order interval trichotomy*

In this section we assume that X is a subset of an ordered Banach space Y with positive cone Y_+, with the induced order and topology. Much of the early work on monotone maps on ordered Banach spaces focused on the existence of fixed points for self maps of order intervals $[a, b]$ such that $a, b \in E$; see especially Amann [6]. The following result of Dancer and Hess [38], quoted without proof, is crucial for analyzing such maps.

Let u, v be fixed points of T. A doubly-infinite sequence $\{x_n\}_{n \in \mathbb{Z}}$ ($\mathbb{Z}$ is the set of all integers) in Y is called an *entire orbit from u to v* if

$$x_{n+1} = T(x_n), \quad \lim_{n \to -\infty} x_n = u, \quad \lim_{n \to \infty} x_n = v.$$

If $x_n \leqslant x_{n+1}$ (respectively, $x_n < x_{n+1}$), the entire orbit is *increasing* (respectively, *strictly increasing*). If $x_n \geqslant x_{n+1}$ (respectively, $x_n > x_{n+1}$), the entire orbit is *decreasing* (respectively, *strictly decreasing*). If the entire orbit $\{x_n\}$ is increasing but not strictly increasing, then $x_n = v$ for all sufficiently large n; and similarly for decreasing.

Consider the following hypothesis:

(G) $X = [a, b]$ where $a, b \in Y$, $a < b$ and $Ta = a$, $Tb = b$. The map $T : X \to X$ is monotone and $T(X)$ has compact closure in X.

THEOREM 5.11 (The Order Interval Trichotomy). *Under hypothesis* (G), *at least one of the following holds:*
   (a) *there is a fixed point c such that $a < c < b$;*
   (b) *there exists an entire orbit from a to b that is increasing, and strictly increasing if T is strictly monotone;*
   (c) *there exists an entire orbit from b to a that is decreasing, and strictly decreasing if T is strictly monotone.*

An extension of Theorem 5.11 to allow additional fixed points on the boundary of $[a, b]$ is carried out in Hsu et al. [83]. Wu et al. [236] weaken the compactness condition. See Hsu et al. [83], Smith [192], and Smith and Thieme [201] for applications to generalized two-species competition dynamics. For related results see Hess [63], Matano [133], Poláčik [162], Smith [184,194].

A fixed point q of T is *stable* if every neighborhood of q contains a positively invariant neighborhood of q. An immediate corollary of the Order Interval Trichotomy is:

COROLLARY 5.12. *Assume hypothesis* (G), *and let a and b be stable fixed points. Then there is a third fixed point in $[a, b]$.*

Corollary 5.14 establishes a third fixed point under different assumptions.

In general, more than one of the alternatives (a), (b), (c) may hold (see [83]). The following complement to the Order Interval Trichotomy gives conditions for exactly one to hold; (iii) is taken from Proposition 2.2 of [83].

Consider the following three conditions:

(a′) there is a fixed point c such that $a < c < b$;
(b′) there exists an entire orbit from a to b;
(c′) there exists an entire orbit from b to a.

PROPOSITION 5.13. *Assume hypothesis* (G).
   (i) *If T is strongly order-preserving, exactly one of* (a′), (b′), (c′) *can hold. More precisely: Assume $a < y < b$ and y has compact orbit closure. Then $\omega(y) = \{b\}$ if there is an entire orbit from a to b, while $\omega(y) = \{a\}$ if there is an entire orbit from b to a.*
   (ii) *If $a \ll b$, at most one of* (b′), (c′) *can hold.*
   (iii) *Suppose $a \ll b$, and $E \cap [a, b] \setminus \{a, b\} \neq \emptyset$ implies $E \cap [[a, b]] \setminus \{a, b\} \neq \emptyset$. Then at most one of* (a′), (b′), (c′) *can hold.*

PROOF. For (i), consider an entire orbit $\{x_n\}$ from a to b. There is a neighborhood U of a such that $T^k U \leqslant T^k y$ for sufficiently large k. Choose $x_j \in U$. Then $T^k x_j \leqslant T^k y \leqslant b$ for all large k. As $\lim_{k \to \infty} T^k x_j = b$ and the order relation is closed, b is the limit of every convergent subsequence of $\{T^k y\}$. The case of an entire orbit from b to a is similar.

In (ii), choose neighborhoods U, V of a, b respectively such that $U \ll V$. Fix j so that $x_j \in U$. If $y \in V$ then an argument similar to the proof of (i) shows that $\omega(y) = \{b\}$. Hence there cannot be an entire orbit from b to a, since it would contain a point of V.

Assume the hypothesis of (iii), and note that (ii) makes (b′) and (c′) incompatible. If (a′), there is a fixed point $c \in [[a, b]]$, and arguments similar to the proof of (ii) show that neither (b′) nor (c′) holds. $\square$

COROLLARY 5.14. *In addition to hypothesis* (G), *assume T is strongly order preserving with precompact image. If some trajectory does not converge, there is a third fixed point.*

PROOF. Follows from the Order Interval Trichotomy 5.11 and Proposition 5.13(i). $\square$

A number of authors have considered the question of whether *a priori* knowledge that every fixed point is stable implies the convergence of every trajectory. See Alikakos et al. [3], Dancer and Hess [38], Matano [133] and Takáč [209] for such results. The following theorem is adapted from [38].

A set $A \subset X$ is a *uniform global attractor* for the map $T : X \to X$ if $T(A) = A$ and $\mathrm{dist}(T^n x, A) \to 0$ uniformly in $x \in X$.

THEOREM 5.15. *Let $a, b \in Y$ with $a < b$. Assume $T : [a, b] \to [a, b]$ is strongly order preserving with precompact image, and every fixed point is stable. Then E is a totally ordered arc J that is a uniform global attractor, and every trajectory converges.*

PROOF. We first show that there exists a totally ordered arc of fixed points; this will not use the SOP property. $O(a)$ is an increasing sequence converging to the smallest fixed point in $[a, b]$. Similarly, $O(b)$ is a decreasing sequence converging to the largest fixed point in $[a, b]$. By renaming a and b as these fixed points, we may as well assume that $a, b \in E$. The stability hypothesis and Corollary 5.12 implies there is a fixed point c satisfying $a < c < b$.

The same reasoning applies to $[a, c]$ and $[c, b]$, and can be repeated indefinitely to show that every maximal totally ordered set of fixed points is compact and connected, hence an arc (Wilder [233, Theorem I.11.23]). Thus by Zorn's Lemma there is a totally ordered arc $J \subset E$ joining a to b.

Next we prove: *Every unordered compact invariant set K is a point of J.* This will not use precompactness of $T([a, b])$. Set $q = \inf\{e \in J\colon K \leqslant e\}$. It suffices to prove $q \in K$, for then K, being unordered, reduces to $\{q\}$. If $q \notin K$ then $q > k$. By SOP and invariance of K there is a are neighborhoods V of p and $n > 0$ such that $K = T^n(K) \leqslant T(V)$, hence $K \leqslant T^n(V \cap J) = V \cap J$. This gives the contradiction $K \leqslant \inf(V \cap J) < q$.

Every $\omega(x)$ is compact by the precompactness assumption, and unordered by the Nonordering Principle 5.6(ii). Total ordering of J therefore implies $\omega(x)$ is a point of J. This proves every trajectory converges.

To show that J is a global attractor, let N be the open ϵ-neighborhood of J for an arbitrary $\epsilon > 0$. The stability hypothesis implies N contains a positively invariant open neighborhood W of J. It suffices to prove $T^n(X) \subset W$ when n is sufficiently large. Convergence of all trajectories implies that for every $x \in X$ there exists an open neighborhood $U(x)$ of x and $n(x) > 0$ such that $T^n(x) \in W$ for all $n \geqslant n(x)$. Precompactness of $T(X)$ implies $T(X) \subset \bigcup U(x_i)$ for some finite set $\{x_i\}$. Hence $T^n(X) \subset W$ provided $n > \max\{n(x_i)\}$. $\quad\square$

If the map T in Theorem 5.15 is C^1 and strongly monotone, then E is a smooth totally ordered arc by a result of Takáč [211].

5.3.1. *Existence of fixed points*　　Dancer [37] obtained remarkable results concerning the dynamics of monotone maps with some compactness properties: Limit sets can always be bracketed between two fixed points, and with additional hypotheses these fixed points can be chosen to be stable. The next two theorems are adapted from [37].

A map $T : Y \to Y$ is *order compact* if it takes each order interval, and hence each order bounded set, into a precompact set.

THEOREM 5.16. *Let X be an order convex subset of Y. Assume that $T : X \to X$ is monotone and order compact, with every orbit having compact closure in X and every omega limit set order bounded. Then for all $z \in Y$ there are fixed points f, g such that $f \leqslant \omega(z) \leqslant g$.*

PROOF.　There exists $u \in X$ such that $u \geqslant \omega(z)$ because omega limit sets order bounded. Since $T(\omega(z)) = \omega(z)$, it follows that $\omega(z) \leqslant T^i u$ for all i, hence $\omega(z) \leqslant \omega(u)$. Similarly, there exists $s \in X$ such that $\omega(u) \leqslant \omega(s)$. The set $F := \{x \in Y\colon \omega(z) \leqslant x \leqslant \omega(s)\}$ is the intersection of closed order intervals, hence closed and convex, nonempty because it contains $\omega(u)$, and obviously order bounded. Moreover $F \subset X$ because X is order convex. Therefore $T(F)$ is defined and is precompact. Monotonicity of T and invariance of $\omega(z)$ and $\omega(s)$ imply $T(F) \subset F$. It follows from the Schauder fixed point theorem that there is a fixed point $g \in F$, and $g \geqslant \omega(z)$ as required. The existence of f is proved similarly.　$\square$

The cone Y_+ is *reproducing* if $Y = Y_+ - Y_+$. This holds for many function spaces whose norms do not involve derivatives. If Y_+ has nonempty interior, it is reproducing: any $x \in Y$

can be expressed as $x = \lambda e - \lambda(e - \lambda^{-1}x) \in Y_+ - Y_+$, where $e \gg 0$ is arbitrary and $\lambda > 0$ is a sufficiently large real number.

THEOREM 5.17. *Let* $X \subset Y$ *be order convex. Assume* $T : X \to X$ *is monotone, completely continuous, and order compact. Suppose orbits are bounded and omega limits sets are order bounded.*

(i) *For all* $z \in X$ *there are fixed points* f, g *such that* $f \leqslant \omega(z) \leqslant g$.

(ii) *Assume* Y_+ *is reproducing,* $X = Y$ *or* Y_+, *and* E *is bounded. Then there are fixed points* $e_M = \sup E$ *and* $e_m = \inf E$, *and all omega limit sets lie in* $[e_m, e_M]$. *Moreover, if* $x \leqslant e_m$ *then* $\omega(x) = \{e_m\}$, *while if* $x \geqslant e_M$ *then* $\omega(x) = \{e_M\}$.

(iii) *Assume* Y_+ *is reproducing,* $X = Y$ *or* Y_+, E *is bounded, and* T *is strongly order preserving. Suppose* $z_0 \in Y$ *is not convergent. Then there are three fixed points* $f < p < g$ *such that* $f < \omega(z_0) < g$. *If* T *is strongly monotone,* f *and* g *can be chosen to be stable.*

PROOF. We prove all assertions except for the stability in (iii). Complete continuity implies that every positively invariant bounded set is precompact. Therefore orbit closures are compact and omega limit sets are compact and nonempty, so (i) follows from Theorem 5.16.

To prove (ii), note that E is compact because it is bounded invariant and closed. Choose a maximal element $e_M \in E$ (Lemma 1.1). We must show that $e_M \geqslant e$ for every $e \in E$. Since the order cone is reproducing, $e_M - e = v - w$ with $v, w \geqslant 0$. Set $u := e + v + w$. Then $u \in X$, $u \geqslant e$, and $u \geqslant e_M$. Monotonicity implies $e_M = T^i e_M \leqslant T^i u$ for all $i \geqslant 0$, hence $e_M \leqslant \omega(u)$. By Theorem 5.16 there exists $g \in E$ such that $\omega(u) \leqslant g$. Hence $e_M \leqslant g$, whence $e_M = g$ by maximality. We now have $e_M \leqslant \omega(u) \leqslant g = e_M$, so $\omega(u) = \{e_M\}$. Monotonicity implies (as above) $e \leqslant \omega(u)$, therefore $e \leqslant e_M$ as required. This proves $e_M = \sup E$, and the dual argument proves $e_m = \inf E$. If $x \leqslant e_m$ then $\omega(x) \leqslant e_m$ by monotonicity; but $\omega(x) \geqslant e_m$ by (i), so $\omega(x) = \{e_m\}$. Similarly for the case $x \geqslant e_M$.

To prove the first assertion of (iii), note that $e_m < \omega(z) < e_M$ by (i) and the Nonordering Principle 5.6(ii). Monotonicity and order compactness of T imply $[e_m, e_M]$ is positively invariant with precompact image. As T is SOP, there is a third fixed point in $[e_m, e_M]$ by Corollary 5.14. $\qquad\square$

5.4. *Sublinearity and the cone limit set trichotomy*

Motivated by the problem of establishing the existence of periodic solutions of quasimonotone, periodic differential equations defined on the positive cone in R^n, Krasnosel'skii pioneered the dynamics of sublinear monotone self-mappings of the cone [99]. We will prove Theorem 5.20 below, adapted from the original finite-dimensional version of Krause and Ranft [103].

Let Y denote an ordered Banach space with positive cone Y_+. Denote the interior (possibly empty) of Y_+ by P. A map $T : Y_+ \to Y_+$ is *sublinear* (or "subhomogeneous") if

$$0 < \lambda < 1 \implies \lambda T(x) \leqslant T(\lambda x),$$

and *strongly sublinear* if

$$0 < \lambda < 1, \; x \gg 0 \quad \Longrightarrow \quad \lambda T(x) \ll T(\lambda x).$$

Strong sublinearity is the strong concavity assumption of Krasnosel'skii [99]. It can be verified by using the following result from that monograph:

LEMMA 5.18. *$T : P \to Y$ is strongly sublinear provided T is differentiable and $Tx \gg T'(x)x$ for all $x \gg 0$.*

PROOF. Let $F(s) = s^{-1}T(sx)$ for $s > 0$ and some fixed $x \gg 0$. Then $F'(s) = -s^{-2}T(sx) + s^{-1}T'(sx)x \ll 0$ by our hypothesis. So for $0 < \lambda < 1$, we have

$$\phi\big(Tx - \lambda^{-1}T(\lambda x)\big) = \phi\big(F(1)\big) - \phi\big(F(\lambda)\big) < 0$$

for every nontrivial $\phi \in Y^*_+$, the dual cone in Y^*, because $\frac{d}{ds}\phi(F(s)) < 0$. The desired conclusion follows from Proposition 3.1. $\qquad\square$

COROLLARY 5.19. *Assume Y is strongly ordered. A continuous map $T : Y_+ \to Y$ is sublinear provided T is differentiable in P and $Tx \geqslant T'(x)x$ for all $x \gg 0$.*

PROOF. By continuity it suffices to prove $T|P$ is sublinear. Fix $e \gg 0$. For each $\delta > 0$ the map $P \to Y$, $x \mapsto Tx + \delta e$ is strongly sublinear by Lemma 5.18. Sending δ to zero implies T is sublinear. $\qquad\square$

Krause and Ranft [103] have results establishing sublinearity of some iterate of T, which is an assumption used in Theorem 5.20 below.

The following theorem demonstrates global convergence properties for order compact maps that are monotone and sublinear in a suitably strong sense.

THEOREM 5.20 (Cone Limit Set Trichotomy). *Assume $T : Y_+ \to Y_+$ is continuous and monotone and has the following properties for some $r \geqslant 1$:*
  (a) *T^r is strongly sublinear;*
  (b) *$T^r x \gg 0$ for all $x > 0$;*
  (c) *T^r is order compact.*
Then precisely one of the following holds:
   (i) *each nonzero orbit is order unbounded;*
  (ii) *each orbit converges to 0, the unique fixed point of T;*
 (iii) *each nonzero orbit converges to $q \gg 0$, the unique nonzero fixed point of T.*

A key tool in the proofs of such results is Hilbert's projective metric and the related part metric due to Thompson [219]. We define the part metric $\mathrm{p}(x, y)$ here in a very limited way, as a metric on P (which is the "part"). For $x, y \gg 0$, define

$$\mathrm{p}(x, y) := \inf\{\rho > 0 : \mathrm{e}^{-\rho}x \ll y \ll \mathrm{e}^{\rho}y\}.$$

The family of open order intervals in P forms a base for the topology of the part metric. It is easy to see that the identity map of P is continuous from the original topology on P to that defined by the part metric.

When $Y = \mathbb{R}^n$ with vector ordering, with $P = \mathrm{Int}(\mathbb{R}^n_+)$, the part metric is isometric to the max metric on $\mathbb{R}^n$, defined by $d_{\max}(x, y) = \max_i |x_i - y_i|$, *via* the homeomorphism $\mathrm{Int}(\mathbb{R}^n_+) \approx \mathbb{R}^n$, $x \mapsto (\log x_1, \ldots, \log x_n)$. Restricted to compact sets in $\mathrm{Int}(\mathbb{R}^n_+)$, the part metric and the max metric are equivalent in the sense that there exist $\alpha, \beta > 0$ such that $\alpha\mathsf{p}(x, y) \leqslant |x - y|_{\max} \leqslant \beta\mathsf{p}(x, y)$.

The usefulness of the part metric in dynamics stems from the following result. Recall map T between metric spaces is a *contraction* if it has a Lipschitz constant < 1, and it is *nonexpansive* if it has Lipschitz constant 1. We say T is *strictly nonexpansive* if $\mathsf{p}(Tx, Ty) < \mathsf{p}(x, y)$ whenever $x \neq y$.

PROPOSITION 5.21. *Let $T : P \to P$ be a continuous, monotone, sublinear map.*
   (i) *T is nonexpansive for the part metric.*
   (ii) *If T is strongly sublinear, T is strictly nonexpansive for the part metric.*
   (iii) *If T is strongly monotone, $A \subset P$, and no two points of A are linearly dependent, then $T|_A$ is strictly nonexpansive for the part metric.*
   (iv) *Under the assumptions of (ii) or (iii), if $L \subset A$ is compact (in the norm topology) and $T(L) \subset L$, then the set $L_\infty = \bigcap_{n>0} T^n(L)$ is a singleton.*

PROOF. Fix distinct points $x, y \in A$ and set $e^{\mathsf{p}(x,y)} = \lambda > 1$, so that $\lambda^{-1}x \leqslant y \leqslant \lambda x$ and λ is the smallest number with this property. By sublinearity and monotonicity,

$$\lambda^{-1}Tx \leqslant T\left(\lambda^{-1}x\right) \leqslant Ty \leqslant T(\lambda x) \leqslant \lambda Tx \tag{5.3}$$

which implies $\mathsf{p}(Tx, Ty) \leqslant \mathsf{p}(x, y)$.

If T is strongly sublinear, the first and last inequalities in (5.3) can be replaced by $\ll$, which implies $\mathsf{p}(Tx, Ty) < \mathsf{p}(x, y)$.

When x and y are linearly independent, $\lambda^{-1}x < y < \lambda x$. If also T is strongly monotone, (5.3) is strengthened to

$$\lambda^{-1}Tx \leqslant T\left(\lambda^{-1}x\right) \ll Ty \ll T(\lambda x) \leqslant \lambda Tx$$

which also implies $\mathsf{p}(Tx, Ty) < \mathsf{p}(x, y)$.

To prove (iv), observe first that if L is compact in the norm metric, it is also compact in the part metric. In both (ii) and (iii) T reduces the diameter in the part metric of every compact subset of L. Since T maps L_∞ onto itself but reduces its part metric diameter, (iv) follows. $\qquad\square$

PROOF OF THE CONE LIMIT SET TRICHOTOMY 5.20. We first work under the assumption that $r = 1$. In this case Proposition 5.21 shows that every compact invariant set in P reduces to a fixed point, and there is at most one fixed point in P. It suffices to consider the orbits of points $x \in P$, by (b).

Suppose there is a fixed point $q \gg 0$. There exist numbers $0 < \lambda < 1 < \mu$ such that $x \in [\lambda q, \mu q] \subset P$. For all n we have

$$0 \ll \lambda q = \lambda T^n q \leqslant T^n(\lambda q) \leqslant T^n x \leqslant T^n(\mu q) \leqslant \mu T^n q = \mu q.$$

Hence $O(x) \subset [\lambda q, \mu q]$, so $O(Tx)$ lies in $T([\lambda p, \mu q])$, which is precompact by (c). Therefore $\omega(x)$ is a compact unordered invariant set in P. Proposition 5.21(iii) implies that $\omega(x) = \{q\}$. This verifies (iii).

Case I: If some orbit $O(y)$ is order unbounded, we prove (i). We may assume $y \gg 0$. There exists $0 < \gamma < 1$ such that $\gamma y \ll x$. Then $\gamma T^n y \leqslant T^n(\gamma y) \leqslant T^n x$, implying $O(x)$ is unbounded.

Case II: If $0 \in \omega(y)$ for some y, we prove (ii). We may assume $y \gg 0$. Fix $\mu > 1$ with $x \ll \mu y$. Then $0 \leqslant T^n x \leqslant T^n(\mu y) \leqslant \mu T^n y \to 0$. Therefore $\overline{O(x)}$ is compact and $T^n x \to 0$.

Case III: If the orbit closure $\overline{O(x)} \subset [a, b] \subset P$, then (iii) holds. For $\overline{O(x)}$ is compact by (c), so $\omega(x)$ is a nonempty compact invariant set. Because $\omega(x) \subset \overline{O(x)} \subset P$, Case I implies (iii).

Cases I, II and III cover all possibilities, so the proof for $r = 1$ is complete. Now assume $r > 1$. One of the statements (i), (ii), (iii) is valid for T^r in place of T. If (i) holds for T^r, it obviously holds for T. Assume (ii) holds for T^r. If $x > 0$ then $\omega(x) = \{0, T(0), \ldots, T^{r-1}(0)\}$. As this set is compact and T^r invariant, it reduces to $\{0\}$, verifying (ii) for T. A similar argument shows that if (iii) holds for T^r, it also holds for T. $\qquad\square$

The conclusion of the Cone Limit Trichotomy can fail for strongly monotone sublinear maps—simple linear examples in the plane have a line of fixed points. But the following holds:

THEOREM 5.22. *Assume:*
   (a) $T : Y_+ \to Y_+$ *is continuous, sublinear, strongly monotone, and order compact;*
   (b) *for each $x > 0$ there exists $r \in \mathbb{N}$ such that $T^r x \gg 0$.*
Then:
   (i) *either $O(x)$ is not order bounded for all $x > 0$, or $O(x)$ converges to a fixed point for all $x \geqslant 0$;*
   (ii) *the set of fixed points > 0 has the form $\{\lambda e: a \leqslant \lambda \leqslant b\}$ where $e \gg 0$ and $0 \leqslant a \leqslant b \leqslant \infty$.*

PROOF. Let $y > 0$ be arbitrary. If $O(y)$ is not order bounded, or $0 \in \omega(y)$, the proof of (i) follows Cases I and II in the proof of the Cone Limit Set Trichotomy 5.20. If $\overline{O(x)} \subset [a, b] \subset P$, then $\omega(y)$ is a compact invariant set in P, as in Case III of 5.20. As $\omega(y)$ is unordered, every pair of its elements are linearly independent. Therefore Proposition 5.21(iv) implies $\omega(y)$ reduces to a fixed point, proving (i). The same reference shows that all fixed points lie on a ray $R \subset Y_+$ through the origin, which must pass through some

$e \gg 0$ by (b). Suppose p, q are distinct fixed points and $0 \ll p \ll x <\ll q$. There exist unique numbers $0 < \mu < 1 < \nu$ such that $x = \mu p = \mu q$. Then

$$Tx \geqslant \mu T p = \mu p = x, \qquad Tx \leqslant \nu T q = \nu q = x$$

proving $Tx = x$. This implies (ii). $\qquad\qquad\square$

Papers related to sublinear dynamics and the part metric include Dafermos and Slemrod [35], Krause and Ranft [103], Krause and Nussbaum [102], Nussbaum [149,150], Smith [183], and Takáč [208,215]. For interesting applications of sublinear dynamics to higher order elliptic equations, see Fleckinger-Pellé and Takáč [45,46].

5.5. *Smooth strongly monotone maps*

Smoothness together with compactness allows one to settle questions of stability of fixed points and periodic points by examining the spectrum of the linearization of the mapping. Let $T : X \to X$ where X is an open subset of the ordered Banach space Y with cone Y_+ having nonempty interior in Y. Assume that T is a completely continuous, C^1 mapping with a strongly positive derivative at each point. Then T is strongly monotone by Lemma 5.1 and $T'(x)$ is a Krein–Rutman operator so the Krein–Rutman Theorem 2.17 holds for $T'(p)$, $p \in E$. Let ρ be the spectral radius of $T'(p)$, which the reader will recall is a simple eigenvalue which dominates all others in modulus and for which the generalized eigenspace is spanned by an eigenvector $v \gg 0$. Let V_1 be the span of v in Y. There is a complementing closed subspace V_2 such that $Y = V_1 \oplus V_2$ satisfying $T'(p)V_2 \subset V_2$ and $V_2 \cap Y_+ = \{0\}$. Let P denote the projection of Y onto V_2 along v. Finally, let τ denote the spectral radius of $T'(p)|V_2 : V_2 \to V_2$, which obviously satisfies $\tau < \rho$. Mierczyński [139] exploits this structure of the linearized mapping to obtain very detailed behavior of the orbits of points near p. In order to describe his results, define $K := \{x \in X : T^n x \to p\}$ to be the basin of attraction of p. Let $M_- := \{x \in X : T^{n+1} \ll T^n x, n \geqslant n_0, \text{ some } n_0\}$ be the set of eventually decreasing orbits, $M_+ := \{x \in X : T^n x \ll T^{n+1} x, n \geqslant n_0, \text{ some } n_0\}$ be the set of eventually increasing orbits, and $M := M_- \cup M_+$ be the set of eventually monotone (in the strong sense) orbits.

The following result is standard but nonetheless important.

THEOREM 5.23 (Principle of Linearized Stability). *If $\rho < 1$, there is a neighborhood U of p such that $T(U) \subset U$ and constants $c > 0$, $\kappa \in (\rho, 1)$ such that for each $x \in U$ and all n*

$$\left\| T^n x - p \right\| \leqslant c \kappa^n \| x - p \|.$$

In the more delicate case that $\rho \leqslant 1$, Mierczyński [139] obtains a smooth hypersurface C, which is an analog for T of the codimension-one linear subspace V_2 invariant under the linearized mapping $T'(p)$:

THEOREM 5.24. *If $\rho \leqslant 1$ there exists a codimension-one embedded invariant manifold $C \subset X$ of class C^1 having the following properties:*

    (i) *$C = \{x + Pw + R(w)v : w \in O\}$ where $R : O \to \mathbb{R}$ is a C^1 map defined on the relatively open subset O of V_2 containing 0, satisfying $R(0) = R'(0) = 0$. In particular, C is tangent to V_2 at p.*

    (ii) *C is unordered.*

    (iii) *$C = \{p \in X : \|T^n x - p\|/\kappa^n \to 0\} = \{x \in X : \|T^n x - p\|/\kappa^n$ is bounded$\}$, for any κ, $\tau < \kappa < \rho$. In particular, $C \subset K$.*

    (iv) *$K \setminus C = \{x \in K : \|T^n x - p\|/\kappa^n \to \infty\} = \{x \in K : \|T^n x - p\|/\kappa^n$ is unbounded$\}$, for any κ, $\tau < \kappa < \rho$.*

    (v) *$K \setminus C = K \cap M$.*

Conclusion (v) implies most orbits converging to p do so monotonically, but more can be said. Indeed, $K \cap M_+ = \{x \in K : (T^n x - p)/\|T^n x - p\| \to -v\}$ and a similar result for $K \cap M_-$ with v replacing $-v$ holds. The manifold C is a local version of the unordered invariant hypersurfaces obtained by Takáč in [209].

Corresponding to the space V_1 spanned by $v \gg 0$ for $T'(p)$, a locally forward invariant, one dimensional complement to the codimension one manifold C is given in the following result.

THEOREM 5.25. *There is $\epsilon > 0$ and a one-dimensional locally forward invariant C^1 manifold $W \subset B(p; \epsilon)$, tangent to V at p. If $\rho > 1$, then W is locally unique, and for each $x \in W$ there is a sequence $\{x_{-n}\} \subset W$ with $T x_{-n} = x_{-n+1}$, $x_0 = x$, and $\kappa^n \|x_{-n} - p\| \to 0$ for any κ, $1 < \kappa < \rho$.*

Here $B(p; \epsilon)$ is the open ϵ-ball centered at p. Local forward invariance of W means that $x \in W$ and $Tx \in B(x; \epsilon)$ implies $Tx \in W$. Related results are obtained by Smith [184]. In summary, the above results assert that the dynamical behavior of the nonlinear map T behaves near p like that of its linearization $T'(p)$. Obviously, the above results can be applied at a periodic point p of period k by considering the map T^k which has all the required properties.

Mierczyński [139] uses the results above to classify the convergent orbits of T. Similar results are obtained by Takáč in [210].

It is instructive to consider the sort of stable bifurcations that can occur from a linearly stable fixed point, or a linearly stable periodic point, for a one parameter family of mappings satisfying the hypotheses of the previous results, as the parameter passes through a critical value at which $\rho = 1$. The fact that there is a simple positive dominant eigenvalue of $(T^k)'(p)$ ensures that period-doubling bifurcations from a stable fixed point or from a stable periodic point, as a consequence of a real eigenvalue passing through -1, cannot occur. In a similar way, a Neimark–Sacker [113] bifurcation to an invariant closed curve cannot occur from a stable fixed or periodic point. These sorts of bifurcations can occur from unstable fixed or periodic points but then they will "be born unstable."

The generic orbit of a smooth strongly order preserving semiflow converges to fixed point but such a result fails to hold for discrete semigroups, i.e., for strongly order preserving mappings. Indeed, such mappings can have attracting periodic orbits of period exceeding one as we have seen. However, Tereščák [217], improving earlier joint work with Poláčik [164,165], and [65], has obtained the strongest result possible for strongly monotone, smooth, dissipative mappings.

THEOREM 5.26 (Tereščák, 1996). *Let* $T : Y \to Y$ *be a completely continuous, C^1, point dissipative map whose derivative is strongly positive at every point of the ordered Banach space Y having cone Y_+ with nonempty interior. Then there is a positive integer m and an open dense set $U \subset Y$ such that the omega limit set of every point of U is a periodic orbit with period at most m.*

The map P is *point dissipative* (see Hale [58]) provided there is a bounded set B with the property that for every $x \in X$, there is a positive integer $n_0 = n_0(x)$ such that $P^n x \in B$ for all $n \geqslant n_0$. We note that the hypothesis that $T'(x)$ is strongly positive implies that T is strongly monotone by Lemma 5.1.

5.6. *Monotone planar maps*

A remarkable convergence result for planar monotone maps was first obtained by de Mottoni and Schiaffino [42]. They focused on the period-map for the two-species, Lotka–Volterra competition system of ordinary differential equations with periodic coefficients. The full generality of their arguments was recognized and improved upon by Hale and Somolinos [60] and Smith [188,189,192]. We follow the treatment Smith in [192].

In addition to the usual order relations on $\mathbb{R}^2$, $\leqslant$, $<$, $\ll$, generated by $\mathbb{R}^2_+$, we have the "southeast ordering" ($\leqslant_K$), generated by the fourth quadrant $K = \{(u, v): u \geqslant 0, v \leqslant 0\}$. The map T is *cooperative* if it is monotone relative to $\leqslant$ and *competitive* if it is monotone relative to K.

Throughout this subsection, we assume that $T : A \to A$ is a continuous competitive map on the subset A of the plane. Further hypotheses concerning A will be made below. As noted above, all of the results have obvious analogs in the case of cooperative planar maps (just interchange cones). Competitive planar maps preserve the order relation $\leqslant_K$ by definition, but they also put constraints on the usual ordering, as we show below.

LEMMA 5.27. *Let $T : A \to A$ be a competitive map on $A \subset \mathbb{R}^2$. If $x, y \in A$ satisfy $Tx \ll Ty$, then either $x \ll y$ or $y \ll x$.*

PROOF. If neither $x \ll y$ nor $y \ll x$ hold, then $x \leqslant_K y$ or $y \leqslant_K x$ holds. But $x \leqslant_K y$ implies $Tx \leqslant_K Ty$ which is incompatible with $Tx \ll Ty$. A similar contradiction is obtained from $y \leqslant_K x$. $\qquad\square$

Lemma 5.27 suggests placing one of the following additional assumptions on T.

(O$_+$) If $x, y \in A$ and $Tx \ll Ty$, then $x \leqslant y$.
(O$_-$) If $x, y \in A$ and $Tx \ll Ty$, then $y \leqslant x$.

As we shall soon see, if T is orientation preserving, then (O_+) holds and if it is orientation reversing, then (O_-) holds. A sequence $\{x_n = (u_n, v_n)\} \subset \mathbb{R}^2$ is *eventually componentwise monotone* if there exists a positive integer N such that either $u_n \leqslant u_{n+1}$ for all $n \geqslant N$ or $u_{n+1} \leqslant u_n$ for all $n \geqslant N$ and similarly for v_n.

In the case of orientation-preserving maps, the following result was first proved by de Mottoni and Schiaffino [42] for the period map of a periodic competitive Lotka–Volterra system of differential equations.

THEOREM 5.28. *If T is a competitive map for which (O_+) holds then for all $x \in A$, $\{T^n x\}_{n \geqslant 0}$ is eventually component-wise monotone. If the orbit of x has compact closure in A, then it converges to a fixed point of T. If, instead, (O_-) holds then for all $x \in A$, $\{T^{2n} x\}_{n \geqslant 0}$ is eventually component-wise monotone. If the orbit of x has compact closure in A, then its omega limit set is either a period-two orbit or a fixed point.*

PROOF. We first note that if T is competitive and (O_-) holds then T^2 is competitive and (O_+) holds (use Lemma 5.27) so the second conclusion of the theorem follows from the first.

Suppose that (O_+) holds. If $T^n x \leqslant_K T^{n+1} x$ or $T^{n+1} x \leqslant_K T^n x$ holds for some $n \geqslant 1$, then it holds for all larger n so the conclusion is obvious. Therefore, we assume that this is not the case. It follows that for each $n \geqslant 1$ either (a) $T^n x \ll T^{n+1} x$ or (b) $T^{n+1} x \ll T^n x$. We claim that either (a) holds for all n or (b) holds for all n. Assume $x \ll Tx$ (the argument is similar in the other case). If the claim is false, then there is an $n \geqslant 1$ such that $x \ll Tx \ll \cdots \ll T^{n-1} x \ll T^n x$ but $T^{n+1} x \ll T^n x$. But (O_+) implies $T^n x \leqslant T^{n-1} x$ contradicting the displayed inequality. $\qquad\square$

Orbits may not converge to a fixed point if (O_-) holds. Consider the map $T : I \to I$ where $I = [-1, 1]^2$ and $T(u, v) = (-v, -u)$ reflects points through the line $v = -u$. It is easy to see using Lemma 5.1 that T is competitive and that (O_-) holds (see below). Fixed points of T lie on the above-mentioned line but all other points in I are period-two points.

The hypotheses (O_+) and (O_-) on T are global in nature and therefore can be difficult to check in specific examples. We now give sufficient conditions for them to hold that may be easier to verify in applications. *A contains order intervals* if $x, y \in A$ and $x \ll y$ implies that $[x, y] \subset A$. Clearly, $A = [a, b]$ contains order intervals. If $A \subset \mathbb{R}^2$ and $T : A \to \mathbb{R}^2$, we say that T is C^1 if for each $a \in A$ there is an open set U in $\mathbb{R}^2$ and a continuously differentiable function $F : U \to \mathbb{R}^2$ that coincides with T on $U \cap A$. We will have occasion to make certain hypotheses concerning $T'(x)$ even though it is not necessarily uniquely defined. What we mean by this is that there exists an F as above such that $T'(x) = DF(x)$ has the desired properties. This abuse of language will lead to no logical difficulties in the arguments below. In the applications, A will typically be $\mathbb{R}^2_+$ or some order interval $[a, b]$ where $a \ll b$ in which case T' is uniquely defined.

Consider the following hypothesis:

(H_+) (a) A contains order intervals and is p-convex with respect to $\leqslant_K$.
      (b) $\det T'(x) > 0$ for $x \in A$.

(c) $T'(x)(K) \subset K$ for $x \in A$.
(d) T is injective.

Hypothesis (H_-) is identical except the inequality is reversed in (b).

LEMMA 5.29. *If $T : A \to A$ satisfies (H_+), then T is competitive and (O_+) holds. If (H_-) holds, then T is competitive and (O_-) holds.*

PROOF. T is competitive by hypothesis (c) since A is p-convex with respect to $\leqslant_K$. Assuming that (H_+) holds, $x, y, Tx, Ty \in A$ and $Tx \ll Ty$, we will show that $x \ll y$. According to Lemma 5.27, the only alternative to $x \ll y$ is $y \ll x$ so we assume the latter for contradiction. Let a and b be the northwest and southeast corners of the rectangle $[y, x] \subset A$ so that $a \ll_K b$ and $[y, x] = [a, b]_K$. Since T is competitive on A, $T([y, x]) \subset [Ta, Tb]_K$ and $Tx \ll Ty$ implies that $Ta \ll_K Tb$. Consider the oriented Jordan curve forming the boundary of $[y, x]$ starting at a and going horizontally to x, then going vertically down to b, horizontally back to y and vertically up to a. As T is injective on A, the image of this curve is an oriented Jordan curve. Monotonicity of T implies that the image curve is contained in $[Ta, Tb]_K$, begins at Ta and moves monotonically with respect to $\leqslant_K$ (southwest) through Tx and then monotonically to Tb before moving monotonically (decreasing or northwest) from Tb through Ty and on to Ta. $(H_+)(b)$ implies that T is locally orientation preserving, so upon traversing the first half of the image curve from Ta to Ty to Tb, the curve must make a "right turn" at Tb before continuing on to Tx and to Ta. As the image curve cannot intersect itself, we see that $Tx \ll Ty$ cannot hold, a contradiction. $\square$

In specific examples it is often difficult to check that T is injective. It automatically holds if A is compact and connected and there exists $z \in T(A)$ such that the set $T^{-1}(z)$ is a single point. This is because the cardinality of $T^{-1}(w)$ is finite and constant for $w \in T(A)$ by Chow and Hale [26, Lemma 2.3.4].

The following is an immediate corollary of Theorem 5.28 and Lemma 5.29.

COROLLARY 5.30. *If $T : A \to A$ satisfies (H_+), then $\{T^n x\}$ is eventually component-wise monotone for every $x \in A$. In this case, if an orbit has compact closure in A, then it converges to a fixed point of T. If T satisfies (H_-), then $\{T^{2n} x\}$ is eventually component-wise monotone for every $x \in A$. In this case, if an orbit has compact closure in A, then its omega limit set is either a fixed point or a period-two orbit.*

As an application of Corollary 5.30, we recall the celebrated results of de Mottoni and Schiaffino [42] for the periodic Lotka–Volterra system

$$
\begin{aligned}
x' &= x\big[r(t) - a(t)x - b(t)y\big], \\
y' &= y\big[s(t) - c(t)x - b(t)y\big],
\end{aligned}
\tag{5.4}
$$

where r, s, a, b, c, d are periodic of period one and $a, b, c, d \geqslant 0$. The period map $T : \mathbb{R}_+^2 \to \mathbb{R}_+^2$, defined by (5.2) for (5.4), is strictly monotone relative to the fourth quadrant cone K by virtue of Theorem 3.5. Indeed, (5.4) is a competitive system relative to the cone $\mathbb{R}^2_+$ (the off-diagonal entries of the Jacobian $J = J(t, x, y)$ of the right-hand side are nonpositive), and every such system is monotone relative to K. Observe that $J + \alpha I$, for large enough $\alpha > 0$, has nonnegative diagonal entries so $(J + \alpha I)(u, v)^T \in K$ if $(u, v)^T \in K$ (i.e., $u \geqslant 0$, $v \leqslant 0$). T is strongly monotone relative to K in Int $\mathbb{R}^2_+$ if $b, c > 0$ by Corollary 3.11. Because T is injective and orientation preserving by Liouville's theorem, (H_+) holds. Orbits are seen to be bounded by simple differential inequality arguments, e.g., applied to $x' \leqslant x[r(t) - a(t)x]$. Consequently, by Corollary 5.30, all orbits $O(T)$ converge to a fixed point; equivalently, every solution of (5.4) is asymptotic to a period-one solution.

System (5.4) is most interesting when each species can survive in the absence of its competitor, i.e. the time average of r and s are positive. In that case, aside from the trivial fixed point $E_0 := (0, 0)$, there are unique fixed points of type $E_1 := (e, 0)$ and $E_2 := (0, f)$. Of course, $e, f > 0$ give initial data corresponding to the unique nontrivial one-periodic solutions of the scalar equations: $x' = x[r(t) - a(t)x]$ and $y' = y[s(t) - d(t)y]$. The dynamics of the period map for these equations is described by alternative (iii) of Theorem 5.20.

It is shown by de Mottoni and Schiaffino that there is a monotone, relative to K, T-invariant curve joining E_1 to E_2 which is the global attractor for the dynamics of T in $\mathbb{R}^2_+ \setminus \{E_0\}$. This work has inspired a very large amount of work on competitive dynamics. See Hale and Somolinos [60], Smith [188,189], Hess and Lazer [64], Hsu et al. [83], Smith and Thieme [201], Wang and Jiang [230,231,229], Liang and Jiang [121], Zanolin [238].

6. Semilinear parabolic equations

The purpose of this section is to analyze the monotone dynamics in a broad class of second order, semilinear parabolic equations.

For basic theory and further information on many topics we refer the reader to books of Amann [11], Henry [62] Cholewa and Dlotko [25], Hess [63], Lunardi [124] and Martin [125], the papers of Amann [7–10], and the survey article of Poláčik [163].

Solution processes for semilinear parabolic problems have been obtained by many authors; see for example [3,8,38,62,63,124,125,134,147,163,194,174,208,246]. We briefly outline the general procedure, due to Henry, with important improvements by Mora and Lunardi.

To balance the sometimes conflicting goals of order, topology and dynamics, the domain of a solution process must be chosen carefully. We rely on results of Mora [147], refined by Lunardi [124], for solution processes in Banach subspaces $C^k_B(\overline{\Omega}) \subset C^k(\overline{\Omega})$, $k = 0, 1$ determined by the boundary operator B.

6.1. *Solution processes for abstract ODEs*

If Y and X are spaces such that Y is a subset of X and the inclusion map $Y \to X$ is continuous, we write $Y \hookrightarrow X$. When Y and X are ordered Banach space structures, this notation tacitly states that Y is a linear subspace of X and $Y_+ = Y \cap X_+$.

The domain and range of any map h are denoted by $\mathsf{D}(h)$ and $\mathsf{R}(h)$.

6.1.1. *Processes*   Let Z be a topological space and set $\widehat{Z} := \{(t, t_0, z) \in \mathbb{R}_+ \times \mathbb{R}_+ \times Z \colon t \geqslant t_0\}$. A *process* in Z is a family $\Theta = \{\Theta_{t,t_0}\}_{0 \leqslant t_0 < t}$ of continuous maps

$$\Theta_{t,t_0} \colon D_{t,t_0} \to Z, \quad D_{t,t_0} \text{ open in } Z,$$

where the set $\{(t, t_0, z) \in \widehat{Z} \colon z \in D(t, t_0)\}$ is open in $\widehat{Z}$ containing $\{(t, t, z) \colon t \geqslant 0, z \in Z\}$, with the properties:

- the map $(t, t_0, z) \mapsto \Theta_{t,t_0}(z)$ is continuous from $\widehat{Z}$ to Z.
- the *cocycle identities* hold:

$$t \geqslant t_1 \geqslant t_0 \quad \Longrightarrow \quad \Theta_{t,t_1} \circ \Theta_{t_1,t_0} = \Theta_{t,t_0}, \quad \Theta_{t,t} = \text{identity map of } Z.$$

Equivalently: there is a local semiflow Λ on $\mathbb{R}_+ \times Z$ such that $\Lambda_t(t_0, u_0) = (t + t_0, \Theta_{t,t_0}(u_0))$. It follows that for each (t_0, z) there is a maximal $\tau := \tau(t_0, z) \in (t_0, \infty]$ such that $z \in D_{t,t_0}$ for all $t \in [t_0, \tau)$. The *trajectory* of (t_0, z) is the parametrized curve $[t_0, \tau) \to Z, t \mapsto \Theta_{t,t_0}(z)$, whose image is the *orbit* of (t_0, z). A subset $S \subset Z$ is *positively invariant* if it contains the orbit of every point in $\mathbb{R}_+ \times S$.

A trajectory is *global* if it is defined on $[t_0, \infty)$. The process is called global when all trajectories are global.

Let S be a space such that $S \hookrightarrow Z$. It may be that S is positively invariant under the process Θ, and the maps $\Theta_{t,t_0} \colon S \cap D(t, t_0) \to S$ are continuous respecting the topology on S and furthermore, the map $(t, t_0, s) \to \Theta_{t,t_0} s$ is continuous from $\widehat{S}$ to S. In this case these maps form the *induced process* Θ^S in S.

A process Θ in an ordered space is called (locally) monotone, SOP, Lipschitz, compact, and so forth, provided every map $\Theta_{t,t_0}, t > t_0$ has the corresponding property.

6.1.2. *Solution processes*   Let X be a Banach space. $\mathcal{A}$ denotes a linear operator (usually unbounded) in X with domain $\mathsf{D}(\mathcal{A}) \subset X$, that is *sectorial* in the following strong sense:

- $\mathcal{A}$ is a densely defined, closed operator generating an analytic semigroup $\{e^{t\mathcal{A}}\}_{t \geqslant 0}$ in $L(X)$, and the resolvent operators $(\lambda I - \mathcal{A})^{-1} \in L(X)$ are compact for sufficiently large $\lambda \geqslant 0$.

The latter property ensures that $e^{t\mathcal{A}}$ is compact for $t > 0$ [156, Theorem 2.3.3].

We make $\mathsf{D}(\mathcal{A})$ into a Banach space with the graph norm $\|x\|_{\mathsf{D}(\mathcal{A})} = \|x\| + \|\mathcal{A}x\|$, or any equivalent norm. Then $\mathcal{A} \colon \mathsf{D}(\mathcal{A}) \to X$ is bounded, and $\mathsf{D}(\mathcal{A}) \hookrightarrow X$.

For $0 \leqslant \alpha \leqslant 1$ we define the fractional power domain of $\mathcal{A}^\alpha$ to be $X^\alpha = X^\alpha(\mathcal{A}) := \mathsf{D}(\mathcal{A}^\alpha)$. Thus we have [62]

$$\mathsf{D}(\mathcal{A}) \hookrightarrow X^\alpha \hookrightarrow X, \quad \overline{\mathsf{D}(\mathcal{A})} = X.$$

Let $F \colon [0, \infty) \times X^\alpha \to X$ be a continuous map that is *locally Lipschitz in the second variable*, i.e.:

- $F|[0, \tau] \times B(r)$ has Lipschitz constant $L(\tau, r)$ in the second variable whenever $[0, \tau] \subset [0, \infty)$ and $B(r)$ is the closed ball of radius r in X^α.

Locally Hölder in the first variable is defined analogously. We say F is C^1 *in the second variable* if $\partial_w F(t, w)$ is continuous.

The data $(X, \mathcal{A}, F)$ determine the abstract initial value problem

$$\begin{cases} u'(t) = \mathcal{A}u(t) + F(t, u(t)) & (t > t_0), \\ u(t_0) = u_0 \in X. \end{cases} \tag{6.1}$$

A continuous curve $u : [t_0, \tau) \to X$, $t_0 < \tau \leqslant \infty$ is a (classical) *solution through* (t_0, u_0) if $u(t) \in \mathsf{D}(\mathcal{A})$ for $t_0 < t < \tau$ and (6.1) holds. It is well known (e.g., Lunardi [124, 4.1.2]) that every solution is also a *mild solution*, i.e., it satisfies the integral equation

$$u(t) = \mathrm{e}^{(t-t_0)\mathcal{A}} u_0 + \int_{t_0}^{t} \mathrm{e}^{(t-s)\mathcal{A}} F\big(s, u(s)\big) \, \mathrm{d}s \quad (t_0 \leqslant t < \tau). \tag{6.2}$$

Moreover, every mild solution is a solution provided F is locally Hölder in t (Lunardi [124, Proposition 7.1.3]).

A classical or mild solution is *maximal* if it does not extend to a classical or mild solution on a larger interval in $[t_0, \infty)$; it is then referred to as a *trajectory* at (t_0, u_0), and its image is an *orbit*. When such a trajectory is unique it is denoted by $t \mapsto u(t, t_0, u_0)$. In this case the *escape time* of (t_0, u_0) is $\tau(t_0, u_0) := \tau$. If $\tau = \infty$ the trajectory is called *global*.

The following basic result means that Eq. (6.1) is well-posed in a strong sense, and that solutions enjoy considerable uniformity and compactness.

THEOREM 6.1. *Let* $(t_0, u_0) \in \mathbb{R}_+ \times X^\alpha$. *There is a unique mild trajectory at* (t_0, u_0), *and it is a classical trajectory provided* $F(t, u)$ *is locally Hölder in* t. *If* $t_0 < t_1 < \tau(t_0, u_0)$, *there is a neighborhood* U *of* x_0 *in* X^α *and* $M > 0$ *such that*

$$\|u(t, t_0, u_1) - u(t, t_0, u_2)\|_{X^\alpha} \leqslant M \|u_1 - u_2\|_{X^\alpha}, \quad u_1, u_2 \in U.$$

There exist $C > 0$, $t_0 < t_1 < \tau(t_0, u_0)$, *a bounded neighborhood* N *of* u_0 *in* X *and a continuous map*

$$\Psi : [t_0, t_1] \times N \to X, \qquad (t, v) \mapsto u(t, t_0, v),$$

where $u(t, t_0, v)$ *is a mild solution, such that the following hold. If* $s, t \in (t_0, t_1]$, $0 \leqslant \alpha < 1$ *and* $v, w \in N$:

    (i) $\|\Psi(s, v) - \Psi(s, w)\| \leqslant C \|v - w\|$;

    (ii) $\|\Psi(s, v) - \Psi(s, w)\|_{X^\alpha} \leqslant (s - t_0)^{-\alpha} C \|v - w\|$;

    (iii) $\Psi([s, t_1] \times N)$ *is precompact in* X^α;

    (iv) $u(\cdot, t_0, v) : (t_0, t] \to X^\alpha$ *and* $u(\cdot, t_0, v) : [t_0, t] \to X$ *are continuous*;

    (v) *trajectories bounded in* X^α *are global*.

PROOF. Lunardi [124, Theorems 7.1.2, 7.1.3 and 7.1.10] proves the first assertion. Items (i), (ii) and (iv) follow from [124, Theorem 7.1.5], and (v) follows from Theorem 7.1.8 (see also Henry [62, 3.3.4]). Fix β with $\alpha < \beta < 1$. As N is bounded in X, $\Psi(s \times N)$ is bounded in X^β by (ii) (with α in (ii) replaced by β). Therefore $\Psi(s \times N)$ is precompact in X^α, and (iii) follows because Ψ defines a local semiflow on $\mathbb{R}_+ \times X^\alpha$. $\square$

Equation (6.1) induces a *solution process* Θ in X, defined by $\Theta_{t,t_0}(u_0) := u(t, t_0, u_0)$. Its restriction to X^α defines an induced solution process on that space. When Eq. (6.1) is autonomous, i.e., $F(t, u) = F(u)$, this solution process boils down to a local semiflow Φ in X^α, defined by $\Phi_t(t_0, u_0) = u(t + t_0, t_0, u_0)$.

When $F(t, u)$ has period $\lambda > 0$ in t, the solution process is λ-*periodic*: $\Theta_{t,t_0} \equiv \Theta_{t+\lambda, t_0+\lambda}$. In this case Θ reduces to a local semiflow on $\mathbf{S}^1 \times X^\alpha$, the dynamics of which are largely determined by the Poincaré map $T := \Theta_{\lambda,0}$ which maps an open subset of X^α continuously into X^α.

Let S be a set and Z a Banach space. We use expressions such as "S is bounded in Z" or "$S \subset Z$ is bounded" to mean $S \subset Z$ and $\sup_{u \in S} \|u\|_Z < \infty$. Note that S may also be unbounded in other Banach spaces.

A map defined on a metric space is *compact* if every bounded set in its domain has precompact image. It is *locally compact* if every point of the domain has a neighborhood with precompact image.

A Banach space Y is *adapted* to the data $(X, \mathcal{A}, F)$ if the following two conditions hold:

$$X^\alpha \hookrightarrow Y \hookrightarrow X \tag{6.3}$$

and the map $(t, u_0) \mapsto \Theta_{t,t_0} u_0$ from $[t_0, \tau) \times D(t, t_0) \cap Y$ to Y is continuous. The solution process Θ determines the *induced solution process* Θ^Y in Y. The domain of Θ^Y_{t,t_0} is the open subset $D^Y(t, t_0) := D_{t,t_0} \cap Y$ of Y.

Rather than work with fractional power spaces, one can assume that $F : [0, \infty) \times K \to X$ where K is a suitable subset of X. The subset $K \subset X$ is *locally closed* in the Banach space X if for each $x \in K$ there exists $r > 0$ such that $\{y \in K: \|x - y\| \leqslant r\}$ is closed in X. Closed and open subsets K of X are locally closed. Note that the following result gives existence and uniqueness of mild solutions while at the same time giving positive invariance. It is a special case of Theorems VIII.2.1 and VIII.3.1 in Martin [125]. Assumptions on the semigroup $e^{t\mathcal{A}}$ remain as above.

THEOREM 6.2. *Let K be a nonempty locally closed subset of a Banach space X and let $F : [0, \infty) \times K \to X$ be continuous and satisfy: For each $R > 0$ there are $L_R > 0$ and $\gamma \in (0, 1]$ such that for $x, y \in K$, $\|x\|, \|y\| \leqslant R, 0 \leqslant s, t \leqslant R$*

$$\left\| F(t, x) - F(s, y) \right\| \leqslant L_R \left(|t - s|^\gamma + \|x - y\| \right). \tag{6.4}$$

Suppose also that:
   (a) $e^{t\mathcal{A}}(K) \subset K$ *for all $t \geqslant 0$, and*
   (b) $\liminf_{h \searrow 0} \frac{1}{h} \operatorname{dist}(x + hF(t, x), K) = 0$ *for $(t, x) \in [0, \infty) \times K$.*
Then for each $(t_0, u_0) \in [0, \infty) \times K$, there is a unique classical trajectory $u(t, t_0, u_0)$ of (6.2) defined on a maximal interval $[t_0, \tau)$, and $u(t) \in K$ for $t_0 \leqslant t < \tau$.

This result is useful for parabolic systems when $X = C^k(\overline{\Omega})$, $k = 0, 1$ but not when $X = L^p(\Omega)$. The substitution operators are well-behaved in the former cases but require very stringent growth conditions for the latter; see Martin [125]. By virtue of the uniqueness assertions of Theorem 6.1 and Theorem 6.2, the solution processes given by the two results agree on K if (6.4) holds.

Hypothesis (a) is obviously required for the positive invariance of K in case $F = 0$. Hypothesis (b), called the *subtangential condition*, is easily seen to be a necessary condition for the positive invariance of K if $\mathcal{A} = 0$. See Martin [125, Theorem VI.2.1]. Both hypotheses are trivially satisfied if $K = X$.

The following result is a special case of [125, Proposition VIII.4.1]:

PROPOSITION 6.3. *Let $F : [0, \infty) \times X \to X$ be continuous and satisfy (6.4) with $K = X$ and let $u(t) = u(t, t_0, x_0)$ be the unique classical trajectory defined on a maximal interval $[t_0, \tau)$ guaranteed by Theorem 6.2. If $\tau < \infty$ then $\lim_{t \to \tau} \|u(t)\| = \infty$.*

6.1.3. *Monotone processes*   Given our interest in establishing monotonicity properties of solution processes induced by parabolic systems in various functions spaces, there are two approaches one may take. One is to establish the properties on spaces of smooth functions such as fractional power spaces X^α for $\alpha < 1$ near unity and then try to extend the monotonicity to larger spaces, e.g., $C^0(\overline{\Omega})$, by approximation. An alternative is to establish the monotonicity properties on the larger spaces first and then get corresponding properties on the smaller spaces by restriction. We give both approaches here, beginning with the former.

A process Θ is *very strongly order preserving* (= VSOP) if it is monotone and has the following property: Given $t_0 \geqslant 0$, $u > v$, and $\epsilon > 0$, there exist $s \in (t_0, t_0 + \epsilon]$ and neighborhoods U, V of u, v respectively such that

$$t \geqslant s \quad \Longrightarrow \quad \Theta_{t,t_0}(U \cap D_{t,t_0}) > \Theta_{t,t_0}(V \cap D_{t,t_0}).$$

This implies Θ is SOP and strictly monotone.

THEOREM 6.4. *Assume X is an ordered Banach space and $Y \hookrightarrow X$ an ordered Banach space such that Y is dense in X and the order cone $Y_+ := Y \cap X_+$ is dense in X_+. Let Θ be a process in X that induces a monotone process Θ^Y in Y. Then:*
 (a) *Θ is monotone.*
 (b) *Assume $R(\Theta_{t,t_0}) \subset Y$ for all $t > t_0 \geqslant 0$. Then Θ is strictly monotone if Θ^Y is strictly monotone, and Θ is VSOP provided Θ^Y is strongly monotone and $\Theta_{t,t_0} : D(t, t_0) \to Y$ is continuous for $t > t_0$.*

PROOF. (a) Fix u and $v > u$ in X. The closed line segment $\overline{uv}$ spanned by u and v is compact, hence there exists $\rho > t_0$ with $\overline{uv} \subset D_{t_0,\rho}$. By the density assumptions there exist convergent sequences $u_n \to u$, $v_n \to v$ in $D_{t_0,\rho}$ such that $u_n, v_n \in Y$ and $u_n < v_n$. As Θ^Y is induced from Θ, it follows that $u_n, v_n \in D^Y_{t_0,\rho}$. For all $t \in [t_0, \rho)$,

$$\Theta_{t,t_0}(u_n) = \Theta^Y_{t,t_0}(u_n) \leqslant \Theta^Y_{t,t_0}(v_n) = \Theta_{t,t_0}(v_n).$$

Taking limits as $n \to \infty$ proves $\Theta_{t,t_0}(u) \leqslant \Theta_{t,t_0}(v)$. Thus Θ is monotone.

(b) Assume now that Θ^Y is strictly monotone. We show that Θ is strictly monotone. Let $u(t), v(t)$ be local trajectories with $u(t_0) < v(t_0)$. If $r \in (t_0, t_1]$ is sufficiently near t_0, then $u(r), v(r)$ are distinct points of Y, and $u(r) < v(r)$ by (a). Hence $u(t_1) < v(t_1)$ by strict monotonicity of Θ^Y.

To prove Θ is VSOP, let $u(t), v(t)$ be as above with $u(t_0), v(t_0) \in D_{t,t_0}$. If $t_0 < s < r < t$, strict monotonicity implies $u(s) > v(s)$. These points are in Y, Θ^Y is strongly monotone, and Θ agrees with Θ^Y in Y. Therefore there are disjoint neighborhoods $U_1, V_1 \subset Y$ of $u(s), v(s)$ respectively, such that

$$\Theta_{r,s}(U_1 \cap D_{r,s}) \gg \Theta_{r,s}(V_1 \cap D_{r,s})$$

and strict monotonicity implies that

$$t > r \quad \Longrightarrow \quad \Theta_{t,r}(U_1 \cap D_{t,r}) > \Theta_{t,r}(V_1 \cap D_{t,r}). \tag{6.5}$$

As $\Theta_{r,t_0} : D(r, t_0) \to Y$ is continuous, we may define neighborhoods $U, V \subset X$ of $u(t_0), v(t_0)$ respectively by

$$U = \Theta_{r,t_0}^{-1}(U_1), \qquad V = \Theta_{r,t_0}^{-1}(V_1).$$

By (6.5) and the cocycle identities,

$$t > r \quad \Longrightarrow \quad \Theta_{t,t_0}(U \cap D_{t,t_0}) > \Theta_{t,t_0}(V \cap D_{t,t_0}). \qquad \square$$

Let X be an ordered Banach space with positive cone X_+ and K a locally closed subset. The mapping $F : K \to X$ is said to be *quasimonotone* (relative to X_+) if:

(QM) For all $(t, x), (t, y) \in [0, \infty) \times K$ satisfying $x \leqslant y$ we have:

$$\lim_{h \searrow 0} \frac{1}{h} \operatorname{dist}\big(y - x + h[F(t, y) - F(t, x)], X_+\big) = 0.$$

The next result is due to [125, Proposition VIII.6.1 and Lemma 6.3] (see also [129] in case of abstract delay differential equations).

THEOREM 6.5. *Assume the hypotheses of Theorem 6.2 hold, F is quasimonotone, and e^{tA} is a positive operator for $t \geqslant 0$. In addition, suppose one of the following:*
   (i) *K is open.*
   (ii) *$K + X_+ \subset K$.*
   (iii) *X is a Banach lattice and $K = [u, v]$ for some $u, v \in X \cup \{-\infty, \infty\}$, $u \leqslant v$.*
Then

$$x, y \in K, x \leqslant y \quad \Longrightarrow \quad u(t, x) \leqslant u(t, y) \quad \big(0 \leqslant t \leqslant \min\{\tau_x, \tau_y\}\big).$$

By $[-\infty, v]$, $v \in X$, is meant the set $\{x \in X: x \leqslant v\}$; similarly for other intervals involving $\pm\infty$. Of course, $-\infty \leqslant v \leqslant \infty$ for every $v \in X$. Observe that $K = [u, \infty]$ is covered by both (ii) and (iii).

REMARK 6.6. If F has the property that for each $x, y \in K$ with $x \leqslant y$, there exists $\lambda > 0$ such that $F(t, x) + \lambda x \leqslant F(t, y) + \lambda y$ then F is quasimonotone because

$$y - x + h\big[F(t, y) - F(t, x)\big] = (1 - \lambda h)(y - x)$$
$$+ h\big[F(t, y) + \lambda y - F(t, x) - \lambda x\big] \in X_+$$

when $h < \lambda^{-1}$.

REMARK 6.7. It is well-known that e^{tA} is a positive operator if and only if $(\lambda I - A)^{-1}$ is a positive operator for all large positive λ. See, e.g., [11, Theorem II 6.4.1] or [125, Proposition 7.5.3]. Indeed, if K is a closed convex subset of X, then $e^{tA}K \subset K$ if and only if $(\lambda I - A)^{-1}K \subset K$ for all large positive λ.

A Banach space X is a *Banach lattice* if for each $x, y \in X$, $x \vee y := \sup\{x, y\}$ exists and the norm is monotone in the sense:

$$|x| \leqslant |y| \quad \Longrightarrow \quad \|x\| \leqslant \|y\|,$$

where $|x|$ denotes the absolute value of x: $|x| := (-x) \vee x$ (see Vulikh [225]). Banach lattices are easy to work with due to simple formulas such as

$$\text{dist}(x, X_+) = \|x - x_+\| = \|x_-\|,$$

where $x_+ := x \vee 0$ and $x_- = -(-x)_+$. The requirement that X be a Banach lattice is a rather strong hypothesis which essentially restricts applicability to $X = L^p(\Omega), C^0(\overline{\Omega})$ or $C_0^0(\overline{\Omega})$. However, the latter two will be important for reaction–diffusion systems.

6.2. *Semilinear parabolic equations*

Let $\Omega \subset \mathbb{R}^n$ be the interior of a compact n-dimensional manifold with C^2 boundary $\partial\Omega$. We consider the semilinear system of m coupled equations ($1 \leqslant i \leqslant m$):

$$\begin{aligned}
\frac{\partial u_i}{\partial t} &= (A_i u_i)(t, x) + f_i(t, x, u, \nabla u) \quad &(x \in \Omega, \ t > t_0), \\
(B_i u_i)(t, x) &= 0 \quad &(x \in \partial\Omega, \ t > t_0), \\
u_i(t_0, x) &= v_{0,i}(x) \quad &(x \in \overline{\Omega}).
\end{aligned} \tag{6.6}$$

Here the unknown function is $u = (u_1, \ldots, u_m) : \overline{\Omega} \to \mathbb{R}^m$, and $\nabla u := (\nabla u_1, \ldots, \nabla u_m) \in (\mathbb{R}^n)^m$ lists the spatial gradients ∇u_i of the u_i, i.e., $\nabla u_i := (\frac{\partial u_i}{\partial x_1}, \ldots, \frac{\partial u_i}{\partial x_n})$. Each $A_i(x)$ is a second order, elliptic differential operator of the form

$$A_i(x) = \sum_{l,j=1}^{n} C_{lj}^i(x) \frac{\partial}{\partial x_l} \frac{\partial}{\partial x_j} + \sum_{j=1}^{n} b_j^i(x) \frac{\partial}{\partial x_j} \tag{6.7}$$

with uniformly continuous and bounded coefficients. Each $n \times n$ matrix $C^i(x) := [C^i_{lj}(x)]$ is assumed positive definite:

$$0 < \inf(\langle C^i(x)y, y\rangle) \quad (x \in \Omega, y \in \mathbb{R}^n, |y| = 1),$$

where $\langle \cdot, \cdot \rangle$ denotes the Euclidean inner product on $\mathbb{R}^n$.

The function

$$f := (f_1, \ldots, f_m) : \mathbb{R}_+ \times \overline{\Omega} \times \mathbb{R}^m \times \left(\mathbb{R}^n\right)^m \to \mathbb{R}^m$$

is continuous, and $f(t, x, u, \xi)$ is locally Lipschitz in $(u, \xi) \in \mathbb{R}^m \times (\mathbb{R}^n)^m$.

Each boundary operator B_i acts on sufficiently smooth functions $v : [t_0, \tau) \times \overline{\Omega} \to \mathbb{R}$ in one of the following ways, where $x \in \partial\Omega$:

Dirichlet: $(B_i v)(t, x) = v(t, x);$

Robin: $(B_i v)(t, x) = \gamma_i v(t, x) + \dfrac{\partial v}{\partial \xi_i}(t, x);$

Neumann: $(B_i v)(t, x) = \dfrac{\partial v}{\partial \xi_i}(t, x),$

where $\gamma_i : \overline{\Omega} \to [0, \infty)$ is continuously differentiable, and $\xi_i : \overline{\Omega} \to \mathbb{R}^n$ is a continuously differentiable vector field transverse to $\partial\Omega$ and pointing outward from Ω. Note that Neumann is a special case of Robin.

We rewrite (6.6) as an initial-boundary value problem for an unknown vector-valued function $u := (u_1, \ldots, u_m) : [t_0, \tau) \times \overline{\Omega} \to \mathbb{R}^m$,

$$\begin{aligned}
\frac{\partial u}{\partial t} &= (Au)(t, x) + f(t, x, u, \nabla u) & (x \in \Omega, \ t > t_0), \\
(Bu)(t, x) &= 0 & (x \in \partial\Omega, \ t > t_0), \\
u(t_0, x) &= u_0(x) & (x \in \overline{\Omega}),
\end{aligned} \tag{6.8}$$

where the operators $A := A_1 \times \cdots \times A_m$ and $B := B_1 \times \cdots \times B_m$ act componentwise on $u = (u_1, \ldots, u_m)$. By a *solution process* for Eq. (6.8) we mean a process in some function space on $\overline{\Omega}$, whose trajectories are solutions to (6.8).

Of special interest are *autonomous* systems, for which $f = f(x, u, \nabla u)$; and the *reaction–diffusion* systems, characterized by $f = f(t, x, u)$.

Assume $n < p < \infty$. To Eq. (6.8) we associate an abstract differential Eq. (6.1) in $L^p(\Omega, \mathbb{R}^m)$. The pair of operators (A_i, B_i) has a sectorial realization $\mathcal{A}_i$ in $L^p(\Omega)$ with domain $\mathcal{D}(\mathcal{A}_i) \hookrightarrow L^p(\Omega)$ (Lunardi [124, 3.1.3]). The operator $\mathcal{A} := \mathcal{A}_1 \times \cdots \times \mathcal{A}_m$ is sectorial on $X := L^p(\Omega, \mathbb{R}^m) = [L^p(\Omega)]^m$.

For $\alpha \in [0, 1)$ set $X^\alpha := X^\alpha(\mathcal{A})$. We choose α so that f defines a continuous *substitution operator*

$$F : \mathbb{R}_+ \times X^\alpha \to X, \qquad F(t, u)(x) := f\big(t, x, u(x), \nabla u(x)\big).$$

It suffices to take α

$$1 > \alpha > \frac{1}{2}\left(1 + \frac{n}{p}\right), \tag{6.9}$$

for then $X^\alpha \hookrightarrow C^1(\overline{\Omega}, \mathbb{R}^m)$ by the Sobolev embedding theorems.

The data $(\mathcal{A}, F)$ thus determine an abstract differential equation $u' = \mathcal{A}u + F(t, u)$ in X, whose trajectories $u(t)$ correspond to solutions $u(t, x) := u(t)(x)$ of (6.6). The assumptions on f make $F(t, u)$ locally Lipschitz in $u \in X^\alpha$.

By Theorem 6.1 and the Sobolev embedding theorem we have:

PROPOSITION 6.8. *Equation* (6.8) *defines a solution process* Θ *on* $X := L^p(\Omega, \mathbb{R}^m)$ *which induces a solution process in* X^β *for every* $\beta \in [0, 1)$ *with* $\beta \geqslant \alpha$.

We quote a useful condition for globality of a solution:

PROPOSITION 6.9. *Assume there are constants* $C > 0$ *and* $0 < \epsilon \leqslant 1$ *such that*

$$\left\| f(t, x, v, \xi) \right\| \leqslant C\left(1 + \|v\| + \|\xi\|^{2-\epsilon}\right) \quad \textit{for all } (t, x, v, \xi) \in \mathbb{R}_+ \times \overline{\Omega} \times S \times \mathbb{R}^n. \tag{6.10}$$

If $u : [t_0, \tau) \to L^p(\Omega, \mathbb{R}^m)$ *is a trajectory such that*

$$\limsup_{t \to \tau-} \left\| u(t) \right\|_{L^p(\Omega, \mathbb{R}^m)} < \infty \tag{6.11}$$

then $\tau = \infty$.

PROOF. Follows from Amann [9, Theorem 5.3(i)], taking the constants of that result to be $m = k = p_0 = \gamma_0 = 1, \kappa = s_0 = 0, \gamma_1 = 2 - \epsilon$. $\qquad\square$

Solutions $u : [t_0, \tau) \times \overline{\Omega} \to \mathbb{R}^m$ to (6.8) enjoy considerable smoothness. For example, if the data $\partial\Omega_i, f_i, A_i, B_i$ are smooth of class $C^{2+2\epsilon}, 0 < 2\epsilon < 1$, then $u \in C^{1+\epsilon, 2+2\epsilon}([t_1, t_2] \times \overline{\Omega}, \mathbb{R}^m)$ for all $t_0 < t_1 < t_2 < \tau$ (Lunardi [124, 7.3.3(iii)]).

While useful for many purposes, solution processes in the spaces X^α suffer from the drawback that X^α and its norm are defined implicitly, leaving unclear the domains of solutions and the meaning of convergence, stability, density and similar topological terms. In addition, the topology of X^α might be unsuitable for a given application. To overcome these difficulties we could appeal to results of Colombo and Vespri [29], Lunardi [124] and Mora [147], establishing induced processes in Banach spaces of continuous, smooth or L^p functions; or we can apply Theorem 6.2. We now define these spaces.

For $r \in \mathbb{N}$ let $C^r(\overline{\Omega})$ denotes the usual Banach space of C^r functions on $\overline{\Omega}$. Set

$$C_0^r(\overline{\Omega}) := \left\{ v \in C^r(\overline{\Omega}) : v|\partial\Omega = 0 \right\}.$$

With γ, ξ as in a Robin boundary operator and $r \geqslant 1$, define

$$C^r_{\gamma,\xi}\left(\overline{\Omega}\right) := \left\{ v \in C^r\left(\overline{\Omega}\right): \ \gamma(x)v(x) + \frac{\partial v}{\partial \xi}(x) = 0 \ (x \in \partial\Omega) \right\}.$$

It is not hard to show that:
- $C^0(\overline{\Omega})$, $C^1(\overline{\Omega})$ and $C^1_{\gamma,\xi}(\overline{\Omega})$ are strongly ordered, with $u \gg 0$ if and only if $u(x) > 0$ for all $x \in \overline{\Omega}$;
- $C^1_0(\overline{\Omega})$ is strongly ordered, with $u \gg 0$ if and only if $u(x) > 0$ for all $x \in \Omega$ and $\partial u/\partial \nu > 0$ where $\nu : \partial\Omega \to \mathbb{R}^n$ is the unit vector field inwardly normal to $\partial\Omega$;
- $C^0_0(\overline{\Omega})$ is not strongly ordered. Both $C^0_0(\overline{\Omega})$ and $C^0(\overline{\Omega})$ are Banach lattices.

In terms of the boundary operators B_i, for $k = 0, 1$ we define Banach spaces

$$C^k_{B_i}\left(\overline{\Omega}\right) := \begin{cases} C^k_0\left(\overline{\Omega}\right) & \text{if } B_i \text{ is Dirichlet,} \\ C^k_{\gamma,\xi}\left(\overline{\Omega}\right) & \text{if } B_i \text{ is Robin and } k = 1, \\ C^0\left(\overline{\Omega}\right) & \text{if } B_i \text{ is Robin and } k = 0. \end{cases}$$

Note that $C^1_{B_i}(\overline{\Omega})$ is strongly ordered, while $C^0_{B_i}(\overline{\Omega})$ is strongly ordered if and only if B_i is Robin; $C^0_{B_i}(\overline{\Omega})$ is a Banach lattice. The ordered Banach space

$$C^k_B\left(\overline{\Omega}, \mathbb{R}^m\right) := \Pi_i C^k_{B_i}\left(\overline{\Omega}\right),$$

with the product order cone, is strongly ordered if $k = 1$, or $k = 0$ and no B_i is Dirichlet. The order cone $L^p(\Omega, \mathbb{R}^m)_+$ is the subset of $L^p(\Omega, \mathbb{R}^m)$ comprising equivalence classes represented by functions $\Omega \to \mathbb{R}^m_+$. Note that $L^p(\Omega, \mathbb{R}^m)$ is normally ordered but not strongly ordered.

It is known that the pair of operators (A_i, B_i) has a sectorial realization $\mathcal{A}_i$ on $C^k(B_i)$ and therefore the product operator $\mathcal{A}$ is sectorial on $C^k_B(\overline{\Omega}, \mathbb{R}^m)$. See Corollary 3.1.24, Theorems 3.1.25, 3.1.26 in [124].

LEMMA 6.10. *For* $X = L^p(\Omega, \mathbb{R}^m)$ *or* $C^k_B(\overline{\Omega}, \mathbb{R}^m)$, *the analytic semigroup* $\mathrm{e}^{t\mathcal{A}}$ *is a positive operator for* $t \geqslant 0$ *with respect to the cone of componentwise nonnegative functions in* X.

PROOF. As noted in Remark 6.7, it suffices to show that $(\lambda I - \mathcal{A})^{-1}$ is positive for large $\lambda > 0$, or equivalently, that for each i and $f_i \geqslant 0$, the solution $g_i \in D(\mathcal{A}_i)$ of $f_i = \lambda g_i - \mathcal{A}_i g_i$ satisfies $g_i \geqslant 0$. The existence of g_i is not the issue but rather its positivity. Thus it boils down to $\lambda g_i - \mathcal{A}_i g_i \geqslant 0 \Longrightarrow g_i \geqslant 0$. But these follow from standard maximum principle arguments. See Lemma 3.1.4 in [155]. $\qquad\square$

With $X = L^p(\Omega, \mathbb{R}^m)$ and $\mathcal{A}$ and α as above, we have a chain of continuous inclusions of ordered Banach spaces

$$\mathsf{D}(\mathcal{A}) \hookrightarrow X^\alpha \hookrightarrow C^1_B\left(\overline{\Omega}, \mathbb{R}^m\right) \hookrightarrow C^0_B\left(\overline{\Omega}, \mathbb{R}^m\right) \hookrightarrow L^p\left(\Omega, \mathbb{R}^m\right),$$

with a solution process in $L^p(\Omega, \mathbb{R}^m)$ and an induced solution process in X^α.

PROPOSITION 6.11. *Let Θ be the solution process in $L^p(\Omega, \mathbb{R}^m)$ for Eq. (6.1) with $n < p < \infty$.*

    (a) *For all $t > t_0$, Θ_{t,t_0} maps D_{t,t_0} continuously into $C_B^1(\overline{\Omega}, \mathbb{R}^m)$.*

    (b) *Θ induces a solution process Θ^1 in $C_B^1(\overline{\Omega}, \mathbb{R}^m)$.*

    (c) *Θ induces a solution process Θ^0 in $C_B^0(\overline{\Omega}, \mathbb{R}^m)$ provided $f = f(t, x, u)$.*

PROOF. By uniqueness of solutions it suffices to establish induced solution processes in $C_B^1(\overline{\Omega}, \mathbb{R}^m) \hookrightarrow L^p(\Omega, \mathbb{R}^m)$, and in $C_B^0(\overline{\Omega}, \mathbb{R}^m) \hookrightarrow L^p(\Omega, \mathbb{R}^m)$ when $f = f(t, x, u)$. This is done in Lunardi [124, Proposition 7.3.3] for $m = 1$, and the general case is similar. Part (c) follows from Theorem 6.2. $\qquad\qquad\square$

Henceforth $\Theta^k, k \in \{0, 1\}$, denotes the process Θ^0 or Θ^1 as in Proposition 6.11.

6.2.1. *Dynamics in spaces X_Γ*   For any set $\Gamma \subset \mathbb{R}^m$ and $k = 0, 1$ define

$$
\begin{aligned}
X_\Gamma^k &:= \left\{ u \in C_B^k(\overline{\Omega}, \mathbb{R}^m) \colon u(\overline{\Omega}) \subset \Gamma \right\}, \\
X_\Gamma &:= \left\{ u \in L^p(\Omega, \mathbb{R}^m) \colon u(\Omega) \subset \Gamma \right\}.
\end{aligned}
\tag{6.12}
$$

A *rectangle* in $\mathbb{R}^m$ is a set of the form $J = J_1 \times \cdots \times J_m$ where each $J_i \subset \mathbb{R}$ is a non-degenerate closed interval. $\mathbb{R}^m$, $\mathbb{R}_+^m$ and closed order intervals $[a, b]$, $a \leqslant b$ are rectangles.

PROPOSITION 6.12. *Let $J := \Pi_{i=1}^m J_i$ be a rectangle in $\mathbb{R}^m$ such that either $0 \in J_i$ or B_i is Neumann, and the following hold for all $x \in \overline{\Omega}$, $u \in \partial J$:*

$$
f_i(t, x, u, 0) \geqslant 0 \quad \textit{if } u_i = \inf J_i, \qquad f_i(t, x, u, 0) \leqslant 0 \quad \textit{if } u_i = \sup J_i. \tag{6.13}
$$

Then:

    (i) *In the reaction–diffusion case, X_J is positively invariant for Θ and X_J^k is positively invariant for Θ^k $(k = 0, 1)$.*

    (ii) *Suppose $k = m = 1$ and $J \subset \mathbb{R}$ is an interval. Then X_J is positively invariant for Θ and X_J^1 is positively invariant for Θ^1.*

PROOF. For the reaction–diffusion case we sketch a proof that X_J^0 is Θ^0-positively invariant using Theorem 6.2. The proof that X_J is Θ-positively invariant follows from this since $\Theta_{t,t_0}(u)$ is the L^p limit $\lim_k \Theta_{t,t_0}^0(u_k)$ where $u_k \in X_J^0$ approximates $u \in X_J$ in L^p and the facts: $\Theta^0 = \Theta$ on X_J^0, a dense subset of the closed subset X_J. In order to verify the subtangential condition for X_J^0, it suffices to verify the subtangential condition for J:

$$
\liminf_{h \searrow 0} \frac{1}{h} \operatorname{dist}\left(u + hf(t, x, u), J\right) = 0 \tag{6.14}
$$

for each $(t, x, u) \in [0, \infty) \times \overline{\Omega} \times J$ by Martin [125, Proposition IX.1.1]. But (6.14) is a necessary condition for J to be positively invariant for the ODE

$$v' = f(t, x, v),$$

where x is a parameter. See, e.g., [125, Theorem VI.2.1]. It is well-known and easy to prove that condition (6.13) implies the positive invariance of J for the ODE (see, e.g., Proposition 3.3, Smith and Waltman [203, Proposition B.7], or Walter [227, Chapter II, Section 12, Theorem II]). It follows that (6.14) holds. Therefore the subtangential condition for X_J^0 holds. Finally, we must verify that $e^{tA}X_J^0 \subset X_J^0$ or, equivalently, that $e^{tA_i}C_{B_i}^0(\overline{\Omega}, J_i) \subset C_{B_i}^0(\overline{\Omega}, J_i)$. This follows from Remark 6.7 and standard maximum principle arguments. It also follows from standard comparison principles for parabolic equations. See, e.g., Pao [155, Lemma 2.1] or Smith [194, Corollary 2.4].

The case $k = m = 1$ is a special case of [227, Chapter IV, Section 25, Theorem II, Section 31, Corollaries IV and V]. $\square$

Consider the case that (6.6) is autonomous:

$$\frac{\partial u_i}{\partial t} = A_i u_i + f_i(x, u, \nabla u) \quad (x \in \overline{\Omega}, \ t > t_0),$$
$$B_i u_i = 0 \qquad\qquad (x \in \partial\Omega, \ t > t_0), \tag{6.15}$$

$i = 1, \ldots, m$. The solution processes $\Theta, \Theta^1, \Theta^0$ reduce to local semiflows.

We introduce a mild growth condition, trivially satisfied in the reaction–diffusion case:

$$\text{For each } s > 0 \text{ there exists } C(s) > 0 \text{ such that}$$
$$|v| \leqslant s \implies |f(x, v, \xi)| \leqslant C(s)\big(1 + |\xi|^{2-\epsilon}\big). \tag{6.16}$$

The following result gives sufficient conditions for solution processes in X_Γ to be global, and to admit compact global attractors:

PROPOSITION 6.13. *Assume system (6.15) satisfies (6.16). Let $\Gamma \subset \mathbb{R}^m$ be a nonempty compact set such that X_Γ is positively invariant for (6.15). Then:*
    (a) *There are solution semiflows Φ, Φ^1 in X_Γ, X_Γ^1 respectively. Φ^1 is compact.*
    (b) *Assume (6.15) is reaction–diffusion. Then there is also a solution semiflow Φ^0 in X_Γ^0. The semiflows Φ, Φ^0, Φ^1 are compact and order compact. There is a compact set $K \subset X_\Gamma^1$ which is the global attractor for all three semiflows.*

PROOF. (a) Let Γ lie in the open ball of radius $R > 0$ about the origin in $\mathbb{R}^m$ and let $h: \mathbb{R}^m \to \mathbb{R}^m$ be any smooth bounded function that agrees with the identity on the open ball of radius R. Define g by $g(x, v, \xi) = f(x, h(v), \xi)$. Every trajectory in X_Γ of (6.15) is also a trajectory of the analogous system in which f is replaced by g (compare Poláčik [163, pp. 842–843]). Nonlinearity g satisfies (6.16) with $C(s)$ constant so (6.10) holds. As

$$\limsup_{t \to \tau-} \|u(t)\|_{C^0(\overline{\Omega}, \mathbb{R}^m)} \leqslant R,$$

which implies (6.11), all trajectories are global by Proposition 6.9. Thus, the restrictions of Ψ, Ψ^1 in X and $C_B^1(\overline{\Omega}, \mathbb{R}^m)$ respectively to X_Γ and X_Γ^1 define semiflows Φ and Φ^1. As Ψ^1 is compact by Hale [58], Theorem 4.2.2, Φ^1 is compact because X_Γ^1 is closed in $C_B^1(\overline{\Omega}, \mathbb{R}^m)$.

(b) In the reaction–diffusion case a similar argument establishes a compact solution semiflow Φ^0 in X_Γ^0; and Φ^0 is order compact because order intervals in X_Γ^0 are bounded. To prove Φ^0 order compact, let N be an order interval in X_Γ. For every $t = 2s > 0$, Φ_s maps N continuously into an order interval N' of X_Γ^0. Precompactness in X_Γ of $\Phi_t N$ follows from the precompactness in X_Γ^0 of $\Phi_s^0 N'$, already established, and the continuous inclusion $\Phi_t N = \Phi_s^0 \circ \Phi_s N \subset \Phi_s^0 N'$.

To prove order compactness of Φ^1, let $N_1 \subset X_\Gamma^1$ be an order interval. N_1 is contained in an order interval N_0 of X_Γ^0. Let C^k denote closure in X_Γ^k. For all $t > 0$ we have $\mathsf{C}^1(\Phi_t^1 N_1) = \mathsf{C}^1(\Phi_t^0 N_1) \subset \mathsf{C}^1 \mathsf{C}^0(\Phi_t^0 N_0)$, and the latter set is compact because Φ^0 is order compact. This proves $\Phi_t^1 N_1$ is precompact in X_Γ^1.

X_Γ^0 is closed and bounded in X^0, hence $\Phi_t^0 X_\Gamma^0$ is precompact in X_Γ^0 for all $t > 0$ by (a). Therefore $K := \bigcap_{t>0} \overline{\Phi_t^0 X_\Gamma^0}$ is a compact global attractor for Φ^0. Similarly, K (with the same topology) is a compact global attractor for Φ.

We rely on the identity $\Phi_t^1 = \Phi_t^0 | X_\Gamma^1$ and continuity of $\Phi_t^0 : X_\Gamma^0 \to X_\Gamma^1$ for all $t > 0$. As K is invariant under Φ^0, it follows that K is a compact subset of X_Γ^1. To prove K a global attractor for Φ^1, it suffices to prove: For arbitrary sequences $\{x(i)\}$ in X_Γ^1, and $t(i) \to \infty$ in $\mathbb{R}_+$ with $t(i) > \epsilon > 0$, there is a sequence $i_k \to \infty$ in $\mathbb{N}$ such that $\{\Phi_{t(i_k)}^1 x(i_k)\}$ converges in X_Γ^1 to a point of K. Choose $\{i_k\}$ so that $\Phi_{t(i_k)-\epsilon}^0 (i_k)$ converges X_Γ^0 as $k \to \infty$ to $p \in K$; this is possible because K is a compact global attractor for Φ^0. Then $\Phi_{t(i_k)}^1 x(i_k) = \Phi_\epsilon^0 \circ \Phi_{t(i_k)-\epsilon}^0 x(i_k)$, which converges in X_Γ^1 as $k \to \infty$ to $\Phi_\epsilon^0 p \in K$. $\qquad\square$

EXAMPLE. Let the u_i denote the concentrations or densities of entities such as chemicals or species. Such quantities are inherently positive, so taking the state space to be $L^p(\Omega, \mathbb{R}_+^m)$ or $C_B^k(\overline{\Omega}, \mathbb{R}_+^m)$ is appropriate. We make the plausible assumption that sufficiently high density levels must decrease. Modeling this situation by (a) and (b) below, we get the following result.

PROPOSITION 6.14. *In Eq.* (6.15) *assume* $f = f(x, u)$ *and let the following hold for* $i = 1, \ldots, m$:

    (a) $f_i(x, u) \geqslant 0$ *if* $u_i = 0$;

    (b) *there exists* $\kappa > 0$ *such that* $f_i(x, u) < 0$ *if* $u_i \geqslant \kappa$.

Then for $k = 0, 1$ *solution processes in the order cones* $L^p(\Omega, \mathbb{R}_+^m)$, $C_B^k(\overline{\Omega}, \mathbb{R}_+^m)$ *are defined by semiflows* Φ, Φ^k *respectively; and there is a compact set* $K \subset X_{[0,\kappa]^m}^k$ *that is the global attractor for* Φ, Φ^0 *and* Φ^1.

PROOF. Proposition 6.12 and (a) proves $L^p(\Omega, \mathbb{R}_+^m)$ and $C_B^k(\overline{\Omega}, \mathbb{R}_+^m)$ are positively invariant under the solution process.

Consider the compact rectangles $J(c) := [0, c\kappa]^m \subset \mathbb{R}^m$, $c \geqslant 1$. Assumption (b) and Proposition 6.12 entail positive invariance of $X_{J(c)}$. Proposition 6.13 shows that there are solution semiflows in $X_{J(c)}$ and $X^k_{J(c)}$ having a compact global attractor $K_c \subset X^1_{J(c)}$ in common. As the $J(c)$ are nested and exhaust $\mathbb{R}^m_+$, these semiflows come from solution semiflows Φ, Φ^k as required. Moreover, all the attractors K_c coincide with the compact set $K := K_1 \subset X^1_{J(1)}$. It is easy to see that K is the required global attractor. $\square$

Results on global solutions and positively invariant sets can be found in many places. See for example Amann [9,10], Cholewa and Dlotko [25], Cosner [33], Lunardi [124], Poláčik [163], Smith [194], Smoller [205].

6.2.2. *Monotone solution processes for parabolic equations* We restrict attention here to monotonicity properties with respect to the standard point-wise and component-wise ordering of functions $\Omega \to \mathbb{R}^m$: $f \leqslant g$ if and only if $f_i(x) \leqslant g_i(x)$ for all x and all i. The natural ordering on $L^p(\Omega, \mathbb{R}^m)$ is defined on equivalence classes by the condition on representatives that $f_i(x) \leqslant g_i(x)$ almost everywhere.

Orderings induced by orthants in $\mathbb{R}^n$ other than the positive orthant can be handled easily by change of variables. See Mincheva [144] and [145] for results in the case of polyhedral cones in $\mathbb{R}^n$.

Consider the case $m = 1$ in Eq. (6.8).

THEOREM 6.15. *In Eq. (6.8), assume $m = 1$ and f is C^1. Then:*
(i) *Θ is VSOP on $L^p(\Omega, \mathbb{R}^m)$.*
(ii) *Θ^1 is strongly monotone in $C^1_B(\overline{\Omega})$.*
(iii) *If $f = f(t, x, u)$ the induced process Θ^0 on $C^0_B(\overline{\Omega})$ is VSOP, and strongly monotone if all boundary operators are Robin.*

PROOF. Let $u, v : [t_0, t_1] \times \overline{\Omega} \to \mathbb{R}$ be solutions with $v(t_0, x) - u(t_0, x) \geqslant 0$ for all x and > 0 for some x. Then $w := v - u$ is the solution to the problem

$$\frac{\partial w}{\partial t} = Aw + \sum_{j=1}^{n} b_j \frac{\partial w}{\partial x_j} + cw \quad (x \in \overline{\Omega},\ t > t_0),$$

$$Bw(t, x) = 0 \qquad\qquad (x \in \partial\Omega,\ t > t_0), \tag{6.17}$$

$$w(t_0, x) \geqslant 0, \quad w(t_0, x) \not\equiv 0 \quad (x \in \overline{\Omega})$$

where $b_j = b_j(t, x)$ and $c_j = c_j(t, x)$ are obtained as follows. Evaluate u, v and their spatial gradients at (t, x), and for $s \in [0, 1]$ set

$$Z(s) = (1 - s)(t, x, u, \nabla u) + s(t, x, v, \nabla v),$$

$$b(t, x) = \big(b_1(t, x), \ldots, b_n(t, x)\big) = \int_0^1 D_4 f\big(Z(s)\big)\, ds,$$

$$c(t, x) = \int_0^1 D_3 f\big(Z(s)\big)\, ds$$

where $D_4 f$ and $D_3 f$ denote respectively the derivatives of $f(t, x, y, \xi)$ with respect to $\xi \in \mathbb{R}^n$ and $y \in \mathbb{R}$. By Taylor's theorem

$$f(t, x, v, \nabla v) - f(t, x, u, \nabla u) = b(t, x)(\nabla u - \nabla v) + c(t, x)(u - v),$$

whence (6.17) follows.

The parabolic maximum principle and boundary point lemma ([194, Theorems 7.2.1, 7.2.2]) imply that the function $w(t_1, \cdot)$, considered as an element of $C_B^1(\overline{\Omega})$, is $\gg 0$. This proves (ii), and the first assertion of (iii) follows from Theorem 6.4(b). The proof of strong monotonicity for Robin boundary conditions is similar to the arguments given above. Part (i) follows from strong monotonicity of Θ^1, Theorem 6.4 and continuity of $\Theta_{t,t_0} : L^p(\Omega, \mathbb{R}^m) \to X^\alpha \hookrightarrow C_B^1(\overline{\Omega}, \mathbb{R}^m)$. $\qquad\square$

For $m \geq 2$ we impose further conditions on system (6.6) in order to have a monotone solution process: it must be of reaction–diffusion type, and the vector fields $f(t, x, \cdot)$ on $\mathbb{R}^m$ must be *cooperative*. In other words, $f(t, x, u)$ is C^1 in u and $\partial f_i / \partial u_j \geq 0$ for all $i \neq j$. (The latter condition holds vacuously if $m = 1$). When this holds then the system is called cooperative. If in addition, there exists $\bar{x} \in \Omega$ such that the $m \times m$ Jacobian matrix $[\partial f_i / \partial u_j(t, \bar{x}, u)]$ is irreducible for all (t, u), we call the system *cooperative and irreducible*

THEOREM 6.16. *If system (6.15) is cooperative, then $\Theta, \Theta^k, k = 0, 1$ are monotone. If the system is also irreducible, then:*
  (i) *Θ is VSOP on $L^p(\Omega, \mathbb{R}^m)$.*
  (ii) *Θ^1 is strongly monotone in $C_B^1(\overline{\Omega}, \mathbb{R}^m)$.*
  (iii) *Θ^0 is VSOP in $C_B^0(\overline{\Omega}, \mathbb{R}^m)$ and is strongly monotone when all boundary operators are Robin.*

PROOF. Monotonicity in $C_B^0(\overline{\Omega}, \mathbb{R}^m)$ follows directly from Theorem 6.5 and Remark 6.6. Indeed, let $u \leq v$ in $C_B^0(\overline{\Omega}, \mathbb{R}^m)$ and t be fixed. Then

$$\left[F(t, v) - F(t, u) + \lambda(v - u) \right](x)$$

$$= \int_0^1 \left(\frac{\partial f}{\partial u}(t, x, su(x) + (1 - s)v(x)) + \lambda I \right) ds (v - u)(x) \geq 0$$

for some $\lambda > 0$ and all $x \in \overline{\Omega}$ by cooperativity of f and compactness of $\overline{\Omega}$. This implies that (QM) holds. The positivity of e^{tA} follows from Lemma 6.10. Monotonicity of Θ in $L^p(\Omega, \mathbb{R}^m)$ follows from monotonicity of Θ^0 and Theorem 6.4.

The proof of VSOP and strong monotonicity for Robin boundary conditions in $C_B^0(\overline{\Omega}, \mathbb{R}^m)$ is like that of Theorem 6.15(i), exploiting the maximum principle for weakly coupled parabolic systems (Protter and Weinberger [166, Chapter 3, Theorems 13, 14, 15 and pp. 192, Remark (i)]). See Smith [194, Section 7.4] for a similar proof.

Monotonicity of Θ^1 follows from monotonicity of Θ^0. Strong monotonicity of Θ^1, in the case of Dirichlet boundary conditions, requires exploiting the maximum principle as in the previous case (the same references apply). VSOP of Θ follows from strong monotonicity of Θ^1, Theorem 6.4 and continuity of the composition $\Theta_{t,t_0}: L^p(\Omega, \mathbb{R}^m) \to X^\alpha \to C_B^1(\overline{\Omega}, \mathbb{R}^m)$. $\qquad\qquad\square$

6.3. *Parabolic systems with monotone dynamics*

We now treat autonomous systems (6.15) having monotone dynamics. Our goal is Theorem 6.17, a sample of the convergence and stability results derivable from the general theory.

In addition to the assumptions for (6.6), we require the following conditions to hold for the solution process Θ in $X := L^p(\Omega, \mathbb{R}^m)$, with p satisfying (6.9) and X_Γ^k defined in (6.12):

(SP) If $m \geqslant 2$ in system (6.15) then $f = f(x, u)$ and the system is cooperative and irreducible. $\Gamma \subset \mathbb{R}^m$ is a nonempty set, either an open set or the closure of an open set. The solution process induces semiflows Φ, Φ^1 in X_Γ, X_Γ^1 respectively, and Φ^0 in X_Γ^0 for the reaction–diffusion case. These semiflows are assumed to have compact orbit closures.

Simple conditions implying (SP) can be derived from Propositions 6.13.

The following statements follow from (SP), assertions about Φ^0 having the implied hypothesis $f = f(x, u)$:

- X_Γ^1 is dense in X_Γ^0 and in X_Γ.
- Φ and Φ^0 agree on X_Γ^0, and Φ, Φ^0 and Φ^1 agree on X_Γ^1.
- Φ_t (respectively, Φ_t^0) maps X_Γ (respectively, X_Γ^0) continuously into X_Γ^1 for $t > 0$ (Proposition 6.11).
- Φ, Φ^1 and Φ^0 have the same omega limit sets, compact attractors and equilibria.
- If Γ is open or order convex and $f(x, u, \xi)$ is C^1 in (u, ξ), the Improved Limit Set Dichotomy (ILSD) holds for Φ^1 by Theorem 2.16, and for Φ and Φ^0 by Proposition 2.21.
- If Γ is compact then Φ^1 is compact. In the reaction–diffusion case with Γ compact, Φ, Φ^1 and Φ^0 are compact and order compact, and a common compact global attractor (Propositions 6.13).
- Φ^1 is strongly monotone; Φ^0 is VSOP, and strongly monotone if all boundary operators are Robin; Φ is VSOP (Theorem 6.16).

The sets of quasiconvergent, convergent and stable points for any semiflow Ψ are denoted respectively by $Q(\Psi), C(\Psi), S(\Psi)$. References to intrinsic or extrinsic topology of these sets (e.g., closure, density) for Φ, Φ^1 or Φ^0 are to be interpreted in terms of the topology of the corresponding domain X_Γ, X_Γ^1 or X_Γ^0.

THEOREM 6.17. *If system (6.15) satisfies hypothesis (SP), then:*

(i) *The sets $Q(\Phi)$, $Q(\Phi^0)$ and $Q(\Phi^1)$ are residual.*

(ii) *Assume Γ is open or order convex and $f(x, u, \xi)$ is C^1 in (u, ξ). Then the sets $C(\Phi) \cap S(\Phi), C(\Phi^0) \cap S(\Phi^0)$ and $C(\Phi^1)$ have dense interiors.*

(iii) *Assume $f = f(x, u)$ and Γ is compact. Then the semiflows Φ, Φ^0, Φ^1 are compact and order compact, and they have a compact global attractor in common.*

(iv) *Assume Γ is open or order convex and E_Γ is compact. Then some $p \in E_\Gamma$ is stable for Φ. Every such p is also stable for Φ^1, and for Φ^0 in the reaction–diffusion case. When E_Γ is finite, the same holds for asymptotically stable equilibria.*

PROOF. (i) follows from Theorem 1.21.

(ii) for Φ and Φ^0 follows from Theorem 2.25(b). For Φ^1, (ii) follows from Theorem 2.26(a).

(iii) is a special case of Proposition 6.13(b).

In (iv), to find a $p \in E_\Gamma$ having the asserted stability properties for Φ, it suffices to verify the hypotheses of Theorem 1.30: (a) follows from (i), while (b) and (c) holds by the assumptions on Γ and compactness of E. Similarly for Φ^0 in the reaction–diffusion case.

To prove the stability properties for p under Φ^1, it suffices by Theorem 1.31 to show that p has a neighborhood in X_Γ^1 that is attracted to a compact set. By (i) and the assumptions on Γ, there are sequences $\{u_k\}, \{v_k\}$ in $Q(\Phi^1)$ converging to p in X_Γ, such that

$$u_k \leqslant u_{k+1} \leqslant p \leqslant v_{k+1} \leqslant v_k$$

and

$$p \neq \inf X_\Gamma \implies u_k < u_{k+1} < p, \qquad p \neq \sup X_\Gamma \implies p < v_{k+1} < v_k.$$

Replacing u_k, v_k by their images under Φ_{ϵ_k} for sufficiently small $\epsilon_k > 0$, we see from strong monotonicity of Φ^1 that we can assume:

$$p \neq \inf X_\Gamma \implies u_k \ll u_{k+1} \ll p, \qquad p \neq \sup X_\Gamma \implies p \ll v_{k+1} \ll v_k.$$

The sets $N_k := [[u_k, v_k]]_X \cap X_\Gamma$ are positively invariant and form a neighborhood basis at p in X_Γ.

Fix k_0 such that N_k is bounded in X_Γ^1 for all $k \geqslant k_0$. By Theorem 6.1(iii), for every $s > 0$ there exists $j \geqslant k_0$ such that $\Phi_s(N_j)$ is precompact in X^α, hence in X_Γ^1. Fix such numbers s and j and let P denote the closure of $\Phi_s(N_j)$ in X_Γ^1. Being compact and positively invariant, P contains the compact global attractor $K := \bigcap_{t>0} \overline{\Phi_t P}$ for the semiflow in $\Phi^1|P$. Then N_j is a neighborhood p in X_Γ^1 that is attracted under Φ^1 to K. $\qquad \square$

References

[1] F. Albrecht, H. Gatzke, A. Haddad and N. Wax, *The dynamics of two interacting populations*, J. Math. Anal. Appl. **46** (1974), 658–670.

[2] A. Alexandroff and H. Hopf, *Topologie*, Springer, Berlin (1935).

[3] N. Alikakos, P. Hess and H. Matano, *Discrete order-preserving semigroups and stability for periodic parabolic differential equations*, J. Differential Equations **82** (1989), 322–341.

[4] H. Amann, *On the existence of positive solutions of nonlinear elliptic boundary value problems*, Indiana Univ. Math. J. **21** (1971), 125–146.

[5] H. Amann, *On the number of solutions of nonlinear equations in ordered Banach spaces*, J. Funct. Anal. **11** (1972), 346–384.

[6] H. Amann, *Fixed point equations and nonlinear eigenvalue problems in ordered Banach spaces*, SIAM Rev. **18** (1976), 620–709.

[7] H. Amann, *Invariant sets and existence theorems for semilinear parabolic and elliptic systems*, Math. Anal. Appl. **65** (1978), 432–467.

[8] H. Amann, *Existence and regularity for semilinear parabolic evolution equations*, Ann. Scuola Norm. Sup. Pisa Cl. Sci. (4) **11** (1984), 593–676.

[9] H. Amann, *Global existence for semilinear parabolic systems*, J. Reine Angew. Math. **360** (1985), 47–83.

[10] H. Amann, *Quasilinear evolution equations and parabolic systems*, Trans. Amer. Math. Soc. **293** (1986), 191–227.

[11] H. Amann, *Linear and Quasilinear Parabolic Problems, Vol. 1: Abstract Linear Theory*, Monogr. Math., Vol. 89, Birkhäuser, Boston, MA (1995), xxxvi + 335 pp.

[12] S. Angenent, *Nonlinear analytic semiflows*, Proc. Roy. Soc. Edinburgh Sect. A **115** (1990), 91–107.

[13] C. Batty and C. Robinson, *Positive one-parameter semigroups on ordered spaces*, Acta Appl. Math. **2** (1984), 221–296.

[14] M. Benaïm, *On invariant hypersurfaces of strongly monotone maps*, J. Differential Equations **137** (1997), 302–319.

[15] M. Benaïm, *Convergence with probability one of stochastic approximation algorithms whose average is cooperative*, Nonlinearity **13** (2000), 601–616.

[16] M. Benaïm and M. Hirsch, *Adaptive processes, mixed equilibria and dynamical systems arising from fictitious plays in repeated games*, Games Econom. Behav. **29** (1999), 36–72.

[17] M. Benaïm and M. Hirsch, *Stochastic approximation with constant step size whose average is cooperative*, Ann. Appl. Probab. **9** (1999), 216–241.

[18] A. Berman, M. Neumann and R. Stern, *Nonnegative Matrices in Dynamic Systems*, Wiley, New York (1989).

[19] A. Berman and R. Plemmons, *Nonnegative Matrices in the Mathematical Sciences*, Academic Press, New York (1979).

[20] G. Birkhoff, *Extensions of Jentzsch's theorem*, Trans. Amer. Math. Soc. **85** (1957), 219–227.

[21] P. Brunovsky, *Controlling nonuniqueness of local invariant manifolds*, J. Reine Angew. Math. **446** (1994), 115–135.

[22] P. Brunovsky, P. Poláčik and B. Sandstede, *Convergence in general periodic parabolic equations in one space variable*, Nonlinear Anal. **18** (1992), 209–215.

[23] X.-Y. Chen and H. Matano, *Convergence, asymptotic periodicity, and finite-point blow-up in one-dimensional semilinear heat equations*, J. Differential Equations **78** (1989), 160–190.

[24] M. Chen, X. Chen and J.K. Hale, *Structural stability for time-periodic one-dimensional parabolic equations*, J. Differential Equations **96** (1992), 355–418.

[25] J. Cholewa and T. Dlotko, *Global Attractors in Abstract Parabolic Problems*, London Math. Soc. Lecture Note Ser., Vol. 278, Cambridge Univ. Press, Cambridge (2000).

[26] S.N. Chow and J.K. Hale, *Methods of Bifurcation Theory*, Springer, New York (1982).

[27] Ph. Clements, H. Heijmans, S. Angenent, C. van Duijn and B. de Pagter, *One-Parameter Semigroups*, CWI Monographs, Vol. 5, North-Holland, Amsterdam (1987).

[28] E. Coddington and N. Levinson, *Theory of Ordinary Differential Equations*, McGraw–Hill, New York (1955).

[29] F. Colombo and V. Vespri, *Generation of analytic semigroups in $W^{k,p}$*, Differential Integral Equations **9** (1996), 421–436.

[30] C. Conley, *Isolated Invariant Sets and the Morse Index*, CBMS Reg. Conf. Ser. Math., Vol. 38, Amer. Math. Soc., Providence, RI (1978).

[31] C. Conley, *The gradient structure of a flow: I*, IBM Research, RC 3939 (1972) (17806), Yorktown Heights, NY. Also, Ergodic Theory Dynam. Systems **8** (1988), 11–26.

[32] W. Coppel, *Stability and Asymptotic Behavior of Differential Equations*, Heath, Boston (1965).

[33] C. Cosner, *Pointwise a priori bounds for strongly coupled semilinear systems of parabolic partial differential equations*, Indiana Univ. Math. J. **30** (1981), 607–620.

[34] R. Courant and D. Hilbert, *Methoden der mathematischen Physik, Bd. 2*, Springer, Berlin (1937).

[35] C.M. Dafermos and M. Slemrod, *Asymptotic behaviour of nonlinear contractions semigroups*, J. Funct. Anal. **13** (1973), 97–106.

[36] E.N. Dancer, *Upper and lower stability and index theory for positive mappings and applications*, Nonlinear Anal. **17** (1991), 205–217.

[37] E. Dancer, *Some remarks on a boundedness assumption for monotone dynamical systems*, Proc. Amer. Math. Soc. **126** (1998), 801–807.

[38] E. Dancer and P. Hess, *Stability of fixed points for order-preserving discrete-time dynamical systems*, J. Reine Angew. Math. **419** (1991), 125–139.

[39] D.L. DeAngelis, W.M. Post and C.C. Travis, *Positive Feedback in Natural Systems*, Springer, Berlin (1986).

[40] K. Deimling, *Nonlinear Functional Analysis*, Springer, New York (1985).

[41] P. de Leenheer and H.L. Smith, *Virus dynamics: A global analysis*, SIAM J. Appl. Math. (2003).

[42] P. de Mottoni and A. Schiaffino, *Competition systems with periodic coefficients: A geometric approach*, J. Math. Biol. **11** (1982), 319–335.

[43] A. Eden, C. Foias, B. Nicolaenko and R. Temam, *Exponential Attractors for Dissipative Evolution Equations*, Wiley, New York (2000).

[44] L. Elsner, *Quasimonotonie und Ungleichungen in halbgeordneten Raumen*, Linear Algebra Appl. **8** (1974), 249–261.

[45] J. Fleckinger-Pellé and P. Takáč, *Uniqueness of positive solutions for nonlinear cooperative systems with the p-Laplacian*, Indiana Univ. Math. J. **43** (1994), 1227–1253.

[46] J. Fleckinger-Pellé and P. Takáč, *Unicité de la solution d'un système non linéaire strictement coopératif*, C. R. Acad. Sci. Paris Sér. I Math. **319** (1994), 447–450.

[47] A. Friedman, *Partial Differential Equations of Parabolic Type*, Prentice Hall, Englewood Cliffs, NJ (1964).

[48] A. Friedman, *Remarks on nonlinear parabolic equations*, Proc. Sympos. Applied Math., Vol. XVII, *Application of Nonlinear Partial Differential Equations in Mathematical Physics*, Amer. Math. Soc. (1965).

[49] P. Fife, *Mathematical Aspects of Reacting and Diffusing Systems*, Lecture Notes in Math., Vol. 28, Springer, New York (1979).

[50] P. Fife and M. Tang, *Comparison principles for reaction–diffusion systems: Irregular comparison functions and applications to questions of stability and speed of propagation of disturbances*, J. Differential Equations **40** (1981), 168–185.

[51] G. Frobenius, *Über aus nicht negativen Elementen*, Sitzungsber. Königl. Preuss. Akad. Wiss. (1912), 456–477.

[52] G. Greiner, *Zur Perron–Frobenius Theorie stark stetiger Halbgruppen*, Math. Z. **177** (1981), 401–423.

[53] S. Grossberg, *Competition, decision, and consensus*, J. Math. Anal. Appl. **66** (1978), 470–493.

[54] S. Grossberg, *Biological competition: Decision rules, pattern formation, and oscillations*, Proc. Nat. Acad. Sci. **77** (1980), 2338–2342.

[55] K. Hadeler and D. Glas, *Quasimonotone systems and convergence to equilibrium in a population genetics model*, J. Math. Anal. Appl. **95** (1983), 297–303.

[56] J.K. Hale, *Theory of Functional Differential Equations*, Springer, New York (1977).

[57] J.K. Hale, *Ordinary Differential Equations*, Krieger, Malabar, FL (1980).

[58] J.K. Hale, *Asymptotic Behavior of Dissipative Systems*, Amer. Math. Soc., Providence (1988).

[59] J.K. Hale, *Dynamics of a scalar parabolic equation*, Canad. Appl. Math. J. **5** (1997), 209–306.

[60] J.K. Hale and A.S. Somolinos, *Competition for fluctuating nutrient*, J. Math. Biol. **18** (1983), 255–280.

[61] J.K. Hale and S.M. Verduyn Lunel, *Introduction to Functional Differential Equations*, Springer, New York (1993).

[62] D. Henry, *Geometric Theory of Semilinear Parabolic Equations*, Lecture Notes in Math., Vol. 840, Springer, New York (1981).

[63] P. Hess, *Periodic-parabolic Boundary Value Problems and Positivity*, Longman Scientific and Technical, New York (1991).

[64] P. Hess and A.C. Lazer, *On an abstract competition model and applications*, Nonlinear Anal. **16** (1991), 917–940.

[65] P. Hess and P. Poláčik, *Boundedness of prime periods of stable cycles and convergence to fixed points in discrete monotone dynamical systems*, SIAM J. Math. Anal. **24** (1993), 1312–1330.

[66] M.W. Hirsch, *Convergence in ordinary and partial differential equations*, Notes for Colloquium Lectures at University of Toronto, August 23–26, 1982, Amer. Math. Soc., Providence, RI (1982).

[67] M.W. Hirsch, *Systems of differential equations which are competitive or cooperative. I: Limit sets*, SIAM J. Appl. Math. **13** (1982), 167–179.

[68] M.W. Hirsch, *Differential equations and convergence almost everywhere in strongly monotone semiflows*, Contemp. Math. **17** (1983), 267–285.

[69] M.W. Hirsch, *The dynamical systems approach to differential equations*, Bull. Amer. Math. Soc. **11** (1984), 1–64.

[70] M.W. Hirsch, *Systems of differential equations which are competitive or cooperative. II: Convergence almost everywhere*, SIAM J. Math. Anal. **16** (1985), 423–439.

[71] M.W. Hirsch, *Attractors for discrete-time monotone dynamical systems in strongly ordered spaces*, Geometry and Topology, J. Alexander and J. Harer, eds., Lecture Notes in Math., Vol. 1167, Springer, New York (1985), 141–153.

[72] M.W. Hirsch, *Systems of differential equations which are competitive or cooperative. III: Competing species*, Nonlinearity **1** (1988), 51–71.

[73] M.W. Hirsch, *Stability and convergence in strongly monotone dynamical systems*, J. Reine Angew. Math. **383** (1988), 1–53.

[74] M.W. Hirsch, *Convergent activation dynamics in continuous time neural networks*, Neural Networks **2** (1989).

[75] M.W. Hirsch, *Systems of differential equations that are competitive or cooperative. V: Convergence in 3-dimensional systems*, J. Differential Equations **80** (1989), 94–106.

[76] M.W. Hirsch, *Systems of differential equations that are competitive or cooperative. IV: Structural stability in three dimensional systems*, SIAM J. Math. Anal. **21** (1990), 1225–1234.

[77] M.W. Hirsch, *Systems of differential equations that are competitive or cooperative. VI: A local C^r closing lemma for 3-dimensional systems*, Ergodic Theory Dynam. Systems **11** (1991), 443–454.

[78] M.W. Hirsch, *Fixed points of monotone maps*, J. Differential Equations **123** (1995), 171–179.

[79] M.W. Hirsch, *Chain transitive sets for smooth strongly monotone dynamical systems*, Differential Equations Dynamical Systems **5** (1999), 529–543.

[80] M.W. Hirsch and S. Smale, *Differential Equations, Dynamical Systems, and Linear Algebra*, Acad. Press, New York (1974).

[81] J. Hofbauer and W. Sandholm, *On the global convergence of stochastic fictitious play*, Econometrica **70** (2002), 2265–2294.

[82] J. Hofbauer and J.W.-H. So, *Multiple limit cycles for three-dimensional competitive Lotka–Volterra equations*, Appl. Math. Lett. **7** (1994), 65–70.

[83] S.-B. Hsu, H. Smith and P. Waltman, *Competitive exclusion and coexistence for competitive systems on ordered Banach spaces*, Trans. Amer. Math. Soc. **348** (1996), 4083–4094.

[84] S.-B. Hsu and P. Waltman, *Analysis of a model of two competitors in a chemostat with an external inhibitor*, SIAM J. Applied Math. **52** (1992), 528–540.

[85] R. Jentzsch, *Über Integralgleichungen mit positiven Kern*, J. Reine Angew. Math. **141** (1912), 235–244.

[86] J. Jiang, *Attractors for strongly monotone flows*, J. Math. Anal. Appl. **162** (1991), 210–222.

[87] J. Jiang, *On the existence and uniqueness of connecting orbits for cooperative systems*, Acta Math. Sinica (N.S.) **8** (1992), 184–188.

[88] J. Jiang, *Three- and four-dimensional cooperative systems with every equilibrium stable*, J. Math. Anal. Appl. **18** (1994), 92–100.

[89] J. Jiang, *Five-dimensional cooperative systems with every equilibrium stable*, Differential Equations and Control Theory, Wuhan, 1994, Lecture Notes in Pure and Appl. Math., Vol. 176, Dekker, New York (1996), 121–127.

[90] J. Jiang and S. Yu, *Stable cycles for attractors of strongly monotone discrete-time dynamical systems*, J. Math. Anal. Appl. **202** (1996), 349–62.

[91] E. Kamke, *Zur Theorie der Systeme gewöhnlicher Differentialgleichungen II*, Acta Math. **58** (1932), 57–85.

[92] T. Kato, *Perturbation Theory for Linear Operators*, Springer, Berlin (1976).

[93] H.B. Keller, *Elliptic boundary value problems suggested by nonlinear diffusion processes*, Arch. Ration Mech. Anal **35** (1969), 363–381.

[94] H.B. Keller, *Positive solutions of some nonlinear eigenvalue problems*, J. Math. Mech. **19** (1969), 279–296.

[95] H.B. Keller and D.S. Cohen, *Some positone problems suggested by nonlinear heat generation*, J. Math. Mech. **16** (1967), 327–342.

[96] W. Kerscher and R. Nagel, *Asymptotic behavior of one-parameter semigroups of positive operators*, Acta Appl. Math. **2** (1984), 297–309.

[97] A. Kolmogorov, *Sulla teoria di Volterra della lotta per l'esistenza*, Giorno. Ist. Ital. Attuari **7** (1936), 74–80.

[98] M. Krasnosel'skii, *Positive Solutions of Operator Equations*, Groningen, Noordhoff (1964).

[99] M. Krasnosel'skii, *The Operator of Translation along Trajectories of Differential Equations*, Transl. Math. Monogr., Vol. 19, Amer. Math. Soc., Providence (1968).

[100] M.A. Krasnosel'skii and L. Ladyzhenskaya, *The structure of the spectrum of positive nonhomogeneous operators*, Trudy Moskov. Mat. Obshch. **3** (1954), 321–346 (in Russian).

[101] M.A. Krasnosel'skii and P.P. Zabreiko, *Geometric Methods of Nonlinear Analysis*, Springer, New York (1984).

[102] U. Krause and R.D. Nussbaum, *A limit set trichotomy for self-mappings of normal cones in Banach spaces*, Nonlinear Anal. **20** (1993), 855–870.

[103] U. Krause and P. Ranft, *A limit set trichotomy for monotone nonlinear dynamical systems*, Nonlinear Anal. **19** (1992), 375–392.

[104] M.G. Krein and M.A. Rutman, *Linear operators leaving invariant a cone in a Banach space*, Uspekhi Mat. Nauk **3** (1948), 3–95. AMS Translation No. 26, Amer. Math. Soc., Providence, RI (1950).

[105] T. Krisztin, H.-O. Walther and J. Wu, *Shape, Smoothness and Invariant Stratification of an Attracting Set for Delayed Monotone Positive Feedback*, Fields Inst. Monogr., Vol. 11, Amer. Math. Soc., Providence, RI (1999).

[106] T. Krisztin and J. Wu, *Monotone semiflows generated by neutral equations with different delays in neutral and retarded parts*, Acta Math. Univ. Comenian. (N.S.) **63** (1994), 207–220.

[107] T. Krisztin and J. Wu, *Asymptotic behaviors of solutions of scalar neutral functional differential equations*, Differential Equations Dynam. Systems **4** (1996), 351–366.

[108] T. Krisztin and J. Wu, *Asymptotic periodicity, monotonicity, and oscillations of solutions of scalar neutral functional differential equations*, J. Math. Anal. Appl. **199** (1996), 502–525.

[109] K. Kunisch and W. Schappacher, *Order preserving evolution operators of functional differential equations*, Boll. Un. Mat. Ital. B (6) (1979), 480–500.

[110] H. Kunze and D. Siegel, *Monotonicity with respect to closed convex cones I*, Dynam. Contin. Discrete Impuls. Systems **5** (1999), 433–449.

[111] H. Kunze and D. Siegel, *Monotonicity with respect to closed convex cones II*, Appl. Anal. **77** (2001), 233–248.

[112] H. Kunze and D. Siegel, *A graph theoretic approach to strong monotonicity with respect to polyhedral cones*, Positivity **6** (2002), 95–113.

[113] Y.A. Kuznetsov, *Elements of Applied Bifurcation Theory*, second edition, Springer (1998).

[114] L. Ladyzhenskaya, *Attractors for Semigroups and Evolution Equations*, Cambridge Univ. Press, Cambridge (1991).

[115] M. Li and J. Muldowney, *Global stability for the SEIR model in epidemiology*, Math. Biosci. **125** (1995), 155–164.

[116] J.P. LaSalle, *The Stability of Dynamical Systems*, SIAM, Philadelphia, PA (1976).

[117] B. Lemmens, R.D. Nussbaum and S.M. Verduyn Lunel, *Lower and upper bounds for ω-limit sets of nonexpansive maps*, Indag. Math. (N.S.) **12** (2001), 191–211.

[118] R. Lemmert and P. Volkmann, *On the positivity of semigroups of operators*, Comment. Math. Univ. Carolin. **39** (1998), 483–489.

[119] A.W. Leung, *Systems of Nonlinear Partial Differential Equations: Applications to Biology and Engineering*, Kluwer Academic, Dordrecht (1989).

[120] X. Liang and J. Jiang, *The classification of the dynamical behavior of 3-dimensional type K monotone Lotka–Volterra systems*, Nonlinear Anal. **51** (2002), 749–763.

[121] X. Liang and J. Jiang, *On the finite-dimensional dynamical systems with limited competition*, Trans. Amer. Math. Soc. **354** (2002), 3535–3554.

[122] R. Loewy and H. Schneider, *Positive operators on the n-dimensional ice-cream cone*, J. Math. Anal. Appl. **49** (1975), 375–392.

[123] R. Lui, *A nonlinear integral operator arising from a model in population genetics. IV: Clines*, SIAM J. Math. Anal. **17** (1986), 152–168.

[124] A. Lunardi, *Analytic Semigroups and Optimal Regularity in Parabolic Problems*, Progr. Nonlinear Differential Equations Appl., Vol. 16, Birkhäuser, Boston (1995).

[125] R.H. Martin, *Nonlinear Operators and Differential Equations in Banach Spaces*, Wiley, New York (1976).

[126] R.H. Martin, *Asymptotic stability and critical points for nonlinear quasimonotone parabolic systems*, J. Differential Equations **30** (1978), 391–423.

[127] R.H. Martin, *A maximum principle for semilinear parabolic systems*, Proc. Amer. Math. Soc. **14** (1979), 66–70.

[128] R.H. Martin, *Asymptotic behavior of solutions to a class of quasimonotone functional differential equations*, Proc. of Workshop on Functional Differential Equations and Nonlinear Semigroups, Pitman Res. Notes Math. Ser., Vol. 48, Pitman, Boston, MA (1981).

[129] R.H. Martin and H.L. Smith, *Abstract functional differential equations and reaction–diffusion systems*, Trans. Amer. Math. Soc. **321** (1990), 1–44.

[130] R.H. Martin and H.L. Smith, *Reaction–diffusion systems with time-delays: Monotonicity, invariance, comparison and convergence*, J. Reine Angew. Math. **413** (1991), 1–35.

[131] H. Matano, *Convergence of solutions of one-dimensional semilinear parabolic equations*, J. Math. Kyoto Univ. **18** (1978), 221–227.

[132] H. Matano, *Asymptotic behavior and stability of solutions of semilinear diffusion equations*, Publ. Res. Inst. Math. Sci. **19** (1979), 645–673.

[133] H. Matano, *Existence of nontrivial unstable sets for equilibriums of strongly order preserving systems*, J. Fac. Sci. Univ. Tokyo **30** (1984), 645–673.

[134] H. Matano, *Strongly order-preserving local semi-dynamical systems—theory and applications*, Semigroups, Theory and Applications, Vol. 1, H. Brezis, M.G. Crandall and F. Kappel, eds., Res. Notes Math., Vol. 141, Longman Scientific and Technical, London (1986), 178–185.

[135] H. Matano, *Strong comparison principle in nonlinear parabolic equations*, Nonlinear Parabolic Equations: Qualitative Properties of Solutions, L. Boccardo and A. Tesei, eds., Pitman Res. Notes Math. Ser., Longman Scientific and Technical, London (1987), 148–155.

[136] R. May and G. Oster, *Bifurcations and dynamic complexity in simple ecological models*, Amer. Naturalist **110** (1976), 573–599.

[137] E.J. McShane, *Extension of range of functions*, Bull. Amer. Math. Soc. (N.S.) **40** (1934), 837–842.

[138] J. Mierczyński, *On monotone trajectories*, Proc. Amer. Math. Soc. **113** (1991), 537–544.

[139] J. Mierczyński, *P-arcs in strongly monotone discrete-time dynamical systems*, Differential Integral Equations **7** (1994), 1473–1494.

[140] J. Mierczyński, *The C^1 property of carrying simplices for a class of competitive systems of ODEs*, J. Differential Equations **111** (1994), 385–409.

[141] J. Mierczyński, *On peaks in carrying simplices*, Colloq. Math. **81** (1999), 285–292.

[142] J. Mierczyński, *Smoothness of carrying simplices for three-dimensional competitive systems: A counterexample*, Dynam. Contin. Discrete Impuls. Systems **6** (1999), 147–154.

[143] J. Mierczyński, *A remark on M.W. Hirsch's paper "Chain transitive sets for smooth strongly monotone dynamical systems" [Dynam. Contin. Discrete Impuls. Systems 5 (1999), 529–543]*, Dynam. Contin. Discrete Impuls. Systems **7** (2000), 455–461.

[144] M. Mincheva, *Qualitative properties of reaction–diffusion systems in chemical reactions*, Ph.D. Thesis, University of Waterloo (2003).

[145] M. Mincheva and D. Siegel, *Comparison principles for reaction–diffusion systems with respect to proper polyhedral cones*, Dynam. Cont., Discrete and Impulsive Systems **10**, (2003) p. 477.

[146] K. Mischaikow, H.L. Smith and H.R. Thieme, *Asymptotically autonomous semiflows: Chain recurrence and Lyapunov functions*, Trans. Amer. Math. Soc. **347** (1995), 1669–1685.

[147] X. Mora, *Semilinear parabolic problems define semiflows on C^k spaces*, Trans. Amer. Math. Soc. **278** (1983), 21–55.

[148] M. Müller, *Über das Fundamenthaltheorem in der Theorie der gewöhnlichen Differentialgleichungen*, Math. Z. **26** (1926), 619–645.

[149] R.D. Nussbaum, *Hilbert's projective metric and iterated nonlinear maps*, Mem. Amer. Math. Soc. **391** (1988).

[150] R.D. Nussbaum, *Iterated nonlinear maps and Hilbert's projective metric, II*, Memoirs of the Amer. Math. Soc. **401** (1989).

[151] R.D. Nussbaum, *Estimates of the periods of periodic points for nonexpansive operators*, Israel J. Math. **76** (1991), 345–380.

[152] Y. Ohta, *Qualitative analysis of nonlinear quasimonotone dynamical systems described by functional differential equations*, IEEE Trans. Circuits Systems **28** (1981), 138–144.

[153] R. Ortega and L. Sánchez, *Abstract competitive systems and orbital stability in R^3*, Proc. Amer. Math. Soc. **128** (2000), 2911–2919.

[154] J. Oxtoby, *Measure and Category*, Grad. Texts in Math., Vol. 2. Springer, New York (1980).

[155] C.V. Pao, *Nonlinear Parabolic and Elliptic Equations*, Plenum, New York (1992).

[156] A. Pazy, *Semigroups of Linear Operators and Applications to Partial Differential Equations*, Springer, New York (1983).

[157] M. Peixoto, *On an approximation theorem of Kupka and Smale*, J. Differential Equations **3** (1967), 214–227.

[158] O. Perron, *Zur Theorie der Matrizen*, Math. Ann. **64** (1907), 248–263.

[159] M. Pituk, *Convergence to equilibrium in scalar nonquasimonotone functional differential equations*, J. Differential Equations **193** (2003), 95–130.

[160] P. Poláčik, *Convergence in smooth strongly monotone flows defined by semilinear parabolic equations*, J. Differential Equations **79** (1989), 89–110.

[161] P. Poláčik, *Domains of attraction of equilibria and monotonicity properties of convergent trajectories in parabolic systems admitting strong comparison principle*, J. Reine Angew. Math. **400** (1989), 32–56.

[162] P. Poláčik, *Existence of unstable sets for invariant sets in compact semiflows. Applications in order-preserving semiflows*, Comment. Math. Univ. Carolin. **31** (1990), 263–276.

[163] P. Poláčik, *Parabolic Equations: Asymptotic Behavior and Dynamics on Invariant Manifolds*, Handbook of Dynamical Systems, Vol. 2, B. Fiedler, ed., Elsevier, New York (2002).

[164] P. Poláčik and I. Tereščák, *Convergence to cycles as a typical asymptotic behavior in smooth strongly monotone discrete-time dynamical systems*, Arch. Ration. Mech. Anal. **116** (1992), 339–360.

[165] P. Poláčik and I. Tereščák, *Exponential separation and invariant bundles for maps in ordered Banach spaces with applications to parabolic equations*, J. Dynam. Differential Equations **5** (1993), 279–303.

[166] M.H. Protter and H.F. Weinberger, *Maximum Principles in Differential Equations*, Prentice–Hall, Englewood Cliffs, NJ (1967).

[167] G. Raugel, *Global Attractors in Partial Differential Equations*, Handbook of Dynamical Systems, Vol. 2, B. Fiedler, ed., Elsevier, New York (2002).

[168] A. Rescigno and I. Richardson, *The struggle for life, I: Two species*, Bull. Math. Biophysics **2** (1967), 377–388.

[169] C. Robinson, *Stability theorems and hyperbolicity in dynamical systems*, Rocky Mountain J. Math. **7** (1977), 425–437.

[170] J. Robinson, *Infinite-Dimensional Dynamical Systems*, Cambridge Univ. Press, Cambridge (2001).

[171] W. Rudin, *Real and Complex Analysis*.

[172] D. Ruelle, *Analyticity properties of characteristic exponents of random matrix products*, Adv. Math. **32** (1979), 68–80.

[173] M. Rutman, *Sur les opérateurs totalement continus linéaires laissant invariant un certain cône*, Rec. Math. (Mat. Sbornik) N.S. **8** (1940), 77–96.

[174] R. Sacker and G. Sell, *Lifting properties in skew-product flows with applications to differential equations*, Mem. Amer. Math. Soc. **190** (1977).

[175] S.H. Saperstone, *Semidynamical Systems in Infinite-Dimensional Spaces*, Appl. Math. Sci., Vol. 37, Springer, Berlin (1981).

[176] D. Sattinger, *Monotone methods in nonlinear elliptic and parabolic boundary value problems*, Indiana Univ. Math. J. **21** (1972), 979–1000.

[177] H. Schneider and M. Vidyasagar, *Cross-positive matrices*, SIAM J. Numer. Anal. **7** (1970), 508–519.

[178] J. Selgrade, *Asymptotic behavior of solutions to single loop positive feedback systems*, J. Differential Equations **38** (1980), 80–103.

[179] J. Selgrade, *A Hopf bifurcation in single loop positive feedback systems*, Quart. Appl. Math. (1982), 347–351.

[180] J. Selgrade, *On the existence and uniqueness of connecting orbits*, Nonlinear Anal. **7** (1983), 1123–1125.

[181] J. Selgrade and M. Ziehe, *Convergence to equilibrium in a genetic model with differential viability between sexes*, J. Math. Biol. **25** (1987), 477–490.

[182] S. Smale, *On the differential equations of species in competition*, J. Math. Biol. **3** (1976), 5–7.

[183] H.L. Smith, *Cooperative systems of differential equations with concave nonlinearities*, Nonlinear Anal. **10** (1986), 1037–1052.

[184] H.L. Smith, *Invariant curves for mappings*, SIAM J. Math. Anal. **17** (1986), 1053–1067.

[185] H.L. Smith, *On the asymptotic behavior of a class of deterministic models of cooperating species*, SIAM J. Appl. Math. **46** (1986), 368–375.

[186] H.L. Smith, *Competing subcommunities of mutualists and a generalized Kamke theorem*, SIAM J. Appl. Math. **46** (1986), 856–874.

[187] H.L. Smith, *Periodic orbits of competitive and cooperative systems*, J. Differential Equations **65** (1986), 361–373.

[188] H.L. Smith, *Periodic competitive differential equations and the discrete dynamics of competitive maps*, J. Differential Equations **64** (1986), 165–194.

[189] H.L. Smith, *Periodic solutions of periodic competitive and cooperative systems*, SIAM J. Math. Anal. **17** (1986), 1289–1318.

[190] H.L. Smith, *Monotone semiflows generated by functional differential equations*, J. Differential Equations **66** (1987), 420–442.

[191] H.L. Smith, *Oscillation and multiple steady states in a cyclic gene model with repression*, J. Math. Biol. **25** (1987), 169–190.

[192] H.L. Smith, *Planar competitive and cooperative difference equations*, J. Differ. Equations Appl. **3** (1988), 335–357.

[193] H.L. Smith, *Systems of ordinary differential equations which generate an order preserving flow. A survey of results*, SIAM Rev. **30** (1988), 87–113.

[194] H.L. Smith, *Monotone Dynamical Systems, an Introduction to the Theory of Competitive and Cooperative Systems*, Math. Surveys Monogr., Vol. 41, American Mathematical Society, Providence, RI (1995).

[195] H.L. Smith, *Complicated dynamics for low-dimensional strongly monotone maps*, Nonlinear Anal. **30** (1997), 1911–1914.

[196] H.L. Smith, *Dynamics of Competition*, Mathematics Inspired by Biology, Lecture Notes in Math., Vol. 1714, Springer (1999), 191–240.

[197] H.L. Smith and H.R. Thieme, *Quasi convergence for strongly ordered preserving semiflows*, SIAM J. Math. Anal. **21** (1990), 673–692.

[198] H.L. Smith and H.R. Thieme, *Monotone semiflows in scalar nonquasi-monotone functional differential equations*, J. Math. Anal. Appl. **150** (1990), 289–306.

[199] H.L. Smith and H.R. Thieme, *Convergence for strongly ordered preserving semiflows*, SIAM J. Math. Anal. **22** (1991), 1081–1101.

[200] H.L. Smith and H.R. Thieme, *Strongly order preserving semiflows generated by functional differential equations*, J. Differential Equations **93** (1991), 332–363.

[201] H.L. Smith and H.R. Thieme, *Stable coexistence and bi-stability for competitive systems on ordered Banach spaces*, J. Differential Equations **176** (2001), 195–222.

[202] H.L. Smith and P. Waltman, *A classification theorem for three-dimensional competitive systems*, J. Differential Equations **70** (1987), 325–332.

[203] H.L. Smith and P. Waltman, *The Theory of the Chemostat*, Cambridge Stud. Math. Biol., Vol. 13, Cambridge Univ. Press (1995).

[204] R.A. Smith, *Orbital stability for ordinary differential equations*, J. Differential Equations **69** (1987), 265–287.

[205] J. Smoller, *Shock Waves and Reaction Diffusion Equations*, Springer, New York (1983).

[206] R. Stern and H. Wolkowicz, *Exponential nonnegativity on the ice-cream cone*, SIAM J. Matrix Anal. Appl. **12** (1991), 160–165.

[207] H. Stewart, *Generation of analytic semigroups by strongly elliptic operators under general boundary conditions*, Trans. Amer. Math. Soc. **259** (1980), 299–310.

[208] P. Takáč, *Asymptotic behavior of discrete-time semigroups of sublinear, strongly increasing mappings with applications in biology*, Nonlinear Anal. **14** (1990), 35–42.

[209] P. Takáč, *Convergence to equilibrium on invariant d-hypersurfaces for strongly increasing discrete-time semigroups*, J. Math. Anal. Appl. **148** (1990), 223–244.

[210] P. Takáč, *Domains of attraction of generic ω-limit sets for strongly monotone semiflows*, Z. Anal. Anwendungen **10** (1991), 275–317.

[211] P. Takáč, *Domains of attraction of generic ω-limit sets for strongly monotone discrete-time semigroups*, J. Reine Angew. Math. **432** (1992), 101–173.

[212] P. Takáč, *Linearly stable subharmonic orbits in strongly monotone time-periodic dynamical systems*, Proc. Amer. Math. Soc. **115** (1992), 691–698.

[213] P. Takáč, *A construction of stable subharmonic orbits in monotone time-periodic dynamical systems*, Monatsh. Math. **115** (1993), 215–244.

[214] P. Takáč, *A short elementary proof of the Kreĭn–Rutman theorem*, Houston J. Math. **20** (1994), 93–98.

[215] P. Takáč, *Convergence in the part metric for discrete dynamical systems in ordered topological cones*, Nonlinear Anal. **26** (1996), 1753–1777.

[216] P. Takáč, *Discrete monotone dynamics and time-periodic competition between two species*, Differential Integral Equations **10** (1997), 547–576.

[217] I. Tereščák, *Dynamics of C^1 smooth strongly monotone discrete-time dynamical system*, Preprint, Comenius University, Bratislava (1994).

[218] H.R. Thieme, *Density dependent regulation of spatially distributed populations and their asymptotic speeds of spread*, J. Math. Biol. **8** (1979), 173–187.

[219] A.C. Thompson, *On certain contraction mappings in a partially ordered vector space*, Proc. Amer. Math. Soc. **14** (1963), 438–443.

[220] A. Tineo, *On the convexity of the carrying simplex of planar Lotka–Volterra competitive systems*, Appl. Math. Comput. **123** (2001), 93–108.

[221] R. Uhl, *Ordinary differential inequalities and quasimonotonicity in ordered topological vector spaces*, Proc. Amer. Math. Soc. **126** (1998), 1999–2003.

[222] P.S. Uryson, *Mittlehre Breite une Volumen der konvexen Körper im n-dimensionalen Räume*, Mat. Sb. **31** (1924), 477–486.

[223] P. van den Driessche and M. Zeeman, *Three-dimensional competitive Lotka–Volterra systems with no periodic orbits*, SIAM J. Appl. Math. **58** (1998), 227–234.

[224] P. Volkmann, *Gewöhnliche Differentialungleichungen mit quasimonoton wachsenden Funktionen in topologischen Vektorräumen*, Math. Z. **127** (1972), 157–164.

[225] B. Vulikh, *Introduction to the Theory of Partially Ordered Spaces*, Wolters-Noordhoff Scientific Publications, Groningen (1967).

[226] S. Walcher, *On cooperative systems with respect to arbitrary orderings*, J. Math. Anal. Appl. **263** (2001), 543–554.

[227] W. Walter, *Differential and Integral Inequalities*, Springer, Berlin (1970).

[228] W. Walter, *On strongly monotone flows*, Ann. Polon. Math. **LXVI** (1997), 269–274.

[229] Y. Wang and J. Jiang, *The general properties of discrete-time competitive dynamical systems*, J. Differential Equations **176** (2001), 470–493.

[230] Y. Wang and J. Jiang, *Uniqueness and attractivity of the carrying simplex for discrete-time competitive dynamical systems*, J. Differential Equations **186** (2002), 611–632.

[231] Y. Wang and J. Jiang, *The long-run behavior of periodic competitive Kolmogorov systems*, Nonlinear Anal. Real World Appl. **3** (2002), 471–485.

[232] H.F. Weinberger, *Long-time behavior of a class of biological models*, SIAM J. Math. Anal. **13** (1982), 353–396.

[233] R. Wilder, *Topology of Manifolds*, Amer. Math. Soc., Providence, RI (1949).

[234] J. Wu, *Global dynamics of strongly monotone retarded equations with infinite delay*, J. Integral Equations Appl. **4** (1992), 273–307.

[235] J. Wu and H. Freedman, *Monotone semiflows generated by neutral functional differential equations with application to compartmental systems*, Canad. J. Math. **43** (1991), 1098–1120.

[236] J. Wu, H.I. Freedman and R.K. Miller, *Heteroclinic orbits and convergence of order-preserving set-condensing semiflows with applications to integrodifferential equations*, J. Integral Equations Appl. **7** (1995), 115–133.

[237] D. Xiao and W. Li, *Limit cycles for the competitive three-dimensional Lotka–Volterra system*, J. Differential Equations **164** (2000), 1–15.

[238] F. Zanolin, *Permanence and positive periodic solutions for Kolmogorov competing species systems*, Results Math. **21** (1992), 224–250.

[239] E. Zeeman, *Classification of quadratic carrying simplices in two-dimensional competitive Lotka–Volterra systems*, Nonlinearity **15** (2002), 1993–2018.

[240] E.C. Zeeman and M.L. Zeeman, *On the convexity of carrying simplices in competitive Lotka–Volterra systems*, Differential Equations, Dynamical Systems, and Control Science, Lecture Notes in Pure and Appl. Math., Vol. 152, Dekker, New York (1994), 353–364.

[241] E.C. Zeeman and M.L. Zeeman, *An n-dimensional competitive Lotka–Volterra system is generically determined by the edges of its carrying simplex*, Nonlinearity **15** (2002), 2019–2032.

[242] E.C. Zeeman and M.L. Zeeman, *From local to global behavior in competitive Lotka–Volterra systems*, Trans. Amer. Math. Soc. **355** (2003), 713–734 (electronic).

[243] M. Zeeman, *Hopf bifurcations in competitive three-dimensional Lotka–Volterra systems*, Dynam. Stability Systems **8** (1993), 189–217.

[244] E. Zeidler, *Nonlinear Functional Analysis and Its Applications, I: Fixed Point Theorems*, Springer, New York (1986).

[245] X.-Q. Zhao, *Global attractivity and stability in some monotone discrete dynamical systems*, Bull. Austral. Math. Soc. **53** (1996), 305–324.

[246] X.-Q. Zhao and B.D. Sleeman, *Permanence in Kolmogorov competition models with diffusion*, IMA J. Appl. Math. **51** (1993), 1–11.

[247] H.-R. Zhu, *The existence of stable periodic orbits for systems of three-dimensional differential equations that are competitive*, Ph.D. Thesis, Arizona State University (1991).

[248] H.-R. Zhu and H.L. Smith, *Stable periodic orbits for a class of three-dimensional competitive systems*, J. Differential Equations **110** (1994), 143–156.

CHAPTER 5

Planar Periodic Systems of Population Dynamics

Julián López-Gómez[†]

Departamento de Matemática Aplicada, Universidad Complutense de Madrid, 28040-Madrid, Spain
E-mail: Lopez_Gomez@mat.ucm.es

Contents

1. Introduction 361
2. Basic preliminaries. The single logistic equation 364
   2.1. The differential operator $\frac{d}{dt} + V : \mathcal{C}_T^1 \to \mathcal{C}_T$ 364
   2.2. The single logistic equation 367
   2.3. Point-wise behaviour of the map $\rho \mapsto \theta_{[\rho\alpha,\beta]}$ 369
   2.4. Attractiveness of $\theta_{[\rho\alpha,\beta]}$ 375
3. Linearized stability of the semi-trivial solutions 376
4. The curves of change of stability of the semi-trivial positive solutions 378
   4.1. The species u and v compete ($b > 0$ and $c > 0$) 379
   4.2. The species v preys on u ($b > 0$ and $c < 0$) 386
   4.3. The species u and v cooperate ($b < 0$ and $c < 0$) 387
   4.4. The general case when b and c change of sign 388
5. The existence of coexistence states 392
   5.1. Abstract unilateral bifurcation from semi-trivial states 392
   5.2. The existence of coexistence states in the general case 401
   5.3. Local structure of $\mathfrak{C}_\mu^\lambda$ and $\mathfrak{C}_\lambda^\mu$ at $(\mu_0, \theta_{[\lambda\ell,a]}, 0)$ and $(\lambda_0, 0, \theta_{[\mu m,d]})$ 406
6. The symbiotic model ($b < 0$ and $c < 0$) 411
   6.1. Periodic systems of cooperative type 411
   6.2. Structure of the set of coexistence states 421
   6.3. Stability of the coexistence states along $\mathfrak{C}_\mu^\lambda$ and $\mathfrak{C}_\lambda^\mu$ 426
   6.4. Low and strong symbiosis effects 434
7. The competing species model ($b > 0$ and $c > 0$) 438
   7.1. Periodic systems of quasi-cooperative type 439
   7.2. Structure of the components $\mathfrak{C}_\mu^\lambda$ and $\mathfrak{C}_\lambda^\mu$ 443
   7.3. Stability of the coexistence states along $\mathfrak{C}_\mu^\lambda$ and $\mathfrak{C}_\lambda^\mu$ 448
   7.4. Some bifurcation diagrams 452
   7.5. Dynamics of (1.1) 454
8. The species v preys on u ($b > 0$ and $c < 0$) 455
   8.1. Characterizing the existence of coexistence states 456

[†]Supported by the Ministry of Science and Technology of Spain under grant REN2003-00707.

HANDBOOK OF DIFFERENTIAL EQUATIONS
Ordinary Differential Equations, volume 2
Edited by A. Cañada, P. Drábek and A. Fonda

359

8.2. Dynamics of (1.1) 458
References 459

Abstract

This chapter analyzes the dynamics of the positive solutions of a general class of planar periodic systems including those of Lotka–Volterra type and a more general class of models simulating symbiotic interactions within global competitive environments. It seems this is the first occasion where this problematic has been addressed within the context of periodic planar systems. Most of our mathematical analysis is focused towards the study of the existence of coexistence states and the problem of ascertaining the structure, multiplicity and stability of these coexistence states in purely symbiotic, and competitive, environments.

Keywords: Lotka–Volterra systems, symbiosis in competitive environments, coexistence states.

MSC: 34A34, 34C12, 92B05.

1. Introduction

Throughout this chapter, for any $T > 0$ we denote by $\mathcal{C}_T$ the Banach space of real T-periodic continuous functions endowed with the norm

$$\|u\|_{\mathcal{C}_T} := \max_{t \in \mathbb{R}} |u(t)| = \max_{t \in [0,T]} |u(t)|.$$

Similarly, $\mathcal{C}_T^1$ stands for the Banach space of real T-periodic functions of class $\mathcal{C}^1$ endowed with the norm

$$\|u\|_{\mathcal{C}_T^1} := \|u\|_{\mathcal{C}_T} + \|u'\|_{\mathcal{C}_T},$$

where $' = \frac{\mathrm{d}}{\mathrm{d}t}$. The Banach space $\mathcal{C}_T$ is ordered by its cone of non-negative functions

$$P := \left\{ u \in \mathcal{C}_T : u(t) \geqslant 0, \ t \in \mathbb{R} \right\}.$$

Note that

$$\operatorname{Int} P = \left\{ u \in \mathcal{C}_T : u(t) > 0, \ t \in \mathbb{R} \right\}.$$

Given $u, v \in \mathcal{C}_T$, we shall write $u \geqslant v$ if $u - v \in P$, $u > v$ if $u - v \in P \setminus \{0\}$, and $u \gg v$ if $u - v \in \operatorname{Int} P$.

The main goal of this chapter is analyzing the dynamics of

$$\begin{cases} u'(t) = \lambda\, \ell(t)u(t) - a(t)u^2(t) - b(t)u(t)v(t), \\ v'(t) = \mu\, m(t)v(t) - d(t)v^2(t) - c(t)u(t)v(t), \\ u(0) = x > 0, \quad v(0) = y > 0, \end{cases} \tag{1.1}$$

and, in particular, the existence, attractiveness and multiplicity of the component-wise non-negative T-periodic solutions of

$$\begin{cases} u'(t) = \lambda\ell(t)u(t) - a(t)u^2(t) - b(t)u(t)v(t), \\ v'(t) = \mu m(t)v(t) - d(t)v^2(t) - c(t)u(t)v(t), \end{cases} \tag{1.2}$$

where $\{\ell, m, a, b, c, d\} \subset \mathcal{C}_T$ satisfy

$$\ell > 0, \quad m > 0, \quad a > 0, \quad d > 0, \tag{1.3}$$

and $(\lambda, \mu) \in \mathbb{R}^2$ are regarded as two real parameters. System (1.2) exhibits three different types of non-negative T-periodic solution pairs. Namely, the *trivial state* $(0,0)$, the *semi-trivial positive states* $(u, 0)$ and $(0, v)$ with $u \gg 0$ and $v \gg 0$, and the *coexistence states*, which are the solution pairs $(u, v) \in \mathcal{C}_T^1 \times \mathcal{C}_T^1$ with $u \gg 0$ and $v \gg 0$. The interest of analyzing these states coming from the fact that in many circumstances they govern the dynamics of (1.1).

It should be noted that we are not imposing any sign restriction on the coupling coefficient functions $b(t)$ and $c(t)$. Thus, (1.2) is far from falling within the classical family of Lotka–Volterra periodic systems, since b and c might not have a constant sign. Actually, if given $h \in \mathcal{C}_T$ we denote

$$h^+ := \max\{h, 0\}, \quad h^- := h^+ - h, \quad I_+^h := \operatorname{Int} \operatorname{supp} h^+, \quad I_-^h := \operatorname{Int} \operatorname{supp} h^-,$$

and

$$I_0^h := \operatorname{Int} h^{-1}(0) = \mathbb{R} \setminus \left(\bar{I}_+^h \cup \bar{I}_-^h \right),$$

then, the species u and v interact according to the following patterns:

- u and v compete if $t \in I_+^b \cap I_+^c$—periods of time where $b(t) > 0$ and $c(t) > 0$.
- u and v cooperate if $t \in I_-^b \cap I_-^c$—periods of time where $b(t) < 0$ and $c(t) < 0$.
- u preys on v if $t \in I_-^b \cap I_+^c$, and v preys on u if $t \in I_+^b \cap I_-^c$—periods of time where $b(t)c(t) < 0$.
- u is free from the action of v if $t \in I_0^b$, and v is free from the action of u if $t \in I_0^c$—periods of time where $b(t)c(t) = 0$.

Consequently, within our general setting (1.1) allows all different types of interactions between the species u and v as time passes by.

Due to (1.3), we have that $\int_0^T a > 0$ and $\int_0^T d > 0$, and, hence, the change of variable

$$u := \frac{T}{\int_0^T a} U, \quad v := \frac{T}{\int_0^T d} V, \quad \Lambda := \frac{\lambda}{T} \int_0^T \ell, \quad M := \frac{\mu}{T} \int_0^T m,$$

transforms (1.2) into

$$\begin{cases} U'(t) = \Lambda \dfrac{\ell(t)}{\frac{1}{T} \int_0^T \ell} U(t) - \dfrac{a(t)}{\frac{1}{T} \int_0^T a} U^2(t) - \dfrac{b(t)}{\frac{1}{T} \int_0^T d} U(t)V(t), \\[3ex] V'(t) = M \dfrac{m(t)}{\frac{1}{T} \int_0^T m} V(t) - \dfrac{d(t)}{\frac{1}{T} \int_0^T d} V^2(t) - \dfrac{c(t)}{\frac{1}{T} \int_0^T a} U(t)V(t). \end{cases}$$

Consequently, without lost of generality, we will throughout assume that the following conditions are satisfied:

$$\frac{1}{T} \int_0^T \ell = \frac{1}{T} \int_0^T m = \frac{1}{T} \int_0^T a = \frac{1}{T} \int_0^T d = 1. \tag{1.4}$$

Therefore,

$$h = 1 \quad \text{if } h \in \{\ell, m, a, d\} \text{ is assumed to be constant.} \tag{1.5}$$

Throughout this chapter, given any function $h \in \mathcal{C}_T$ we shall denote

$$\hat{h} := \frac{1}{T} \int_0^T h(t)\, dt, \quad h_M := \max_{\mathbb{R}} h, \quad h_L := \min_{\mathbb{R}} h. \tag{1.6}$$

Using these notations, conditions (1.3) and (1.4) can be summarized as follows:

$$\{\ell, m, a, d\} \subset P \setminus \{0\}, \qquad \hat{\ell} = \hat{m} = \hat{a} = \hat{d} = 1. \tag{1.7}$$

This work is distributed as follows. In Section 2 we provide some basic preliminaries needed for the subsequent mathematical analysis. Basically, it collects some well known results for the single logistic equation. Section 3 ascertains the linearized stability character of the semi-trivial positive solutions of (1.2).

Section 4 analyzes the minimal complexity of the components of the (λ, μ)-plane determined by the curves of neutral stability of the semi-trivial states. Several different situation cases will be differentiated, according to the nature of the interactions between the species u and v. The mathematical analysis carried out in Section 4 is reminiscent from J. Eilbeck and J. López-Gómez [12], where attention was exclusively focused into the competing species model.

In Section 5 we give an abstract unilateral global bifurcation result for systems. The main theorem of Section 5 is new in its full generality, and it admits a number of applications to the search of coexistence states in wide classes of nonlinear elliptic boundary value problems and planar periodic systems. Here, we will apply it to get all available existence results concerning the existence of coexistence states of (1.2). We point out that the abstract result of Section 5 does not follow as a direct application of the unilateral theory developed by P.H. Rabinowitz [31], but from the updated unilateral theorem of [19].

Section 6 considers the symbiotic prototype model ($b < 0$ and $c < 0$) and uses the theory of monotone periodic systems to show that the set of coexistence states linking the surfaces of the semi-trivial states along their respective curves of neutral stability is a real analytic surface, and that the stability character of the coexistence states changes as one crosses any *turning point* along that surface, while it remains unchanged when an hysteresis point is passed by. The theory developed here goes back to [17], within the context of quasi-cooperative systems, though here we are considerably tidying up the mathematical analysis of [17] and adding all necessary technical details that were omitted there in. Some of our abstract results for linear periodic systems provide with substantial improvements of some of the results of M.A. Krasnosel'skii [16]. Most precisely, the characterization of the attractive character of the system in terms of the existence of a strict positive supersolution seems to be a completely new result. This theorem is reminiscent of the characterization of the strong maximum principle for second order linear elliptic operators found by J. López-Gómez and M. Molina-Meyer [21,20]. Section 7 adapts the mathematical analysis of Section 6 to the competing species model ($b > 0$ and $c > 0$), which possesses a quasi-cooperative structure.

Finally, in Section 8 we briefly collect some of the main results available for predator prey models ($b > 0$ and $c < 0$, or $b < 0$ and $c > 0$). In a forthcoming chapter, we will adapt the theory developed by J. López-Gómez and M. Molina-Meyer [22,23], in order to discuss the effects of strategic symbiosis in competitive environments in order to show that strategic symbiosis allows the species to avoid extinction, while, simultaneously—and quite strikingly—the productivity of the ecosystem grows.

For historical remarks and a precise account of some of the most important works available about these models, the reader is sent to the expository chapter of J. Mawhin [27,

28], which bring an excellent complement to the mathematical analysis carried out in this chapter.

Throughout this chapter, given two real Banach spaces, X and Y, we denote by $\mathcal{L}(X; Y)$ the space of linear continuous operators between X and Y, and given $T \in \mathcal{L}(X; Y)$, $N[T]$, $R[T]$, and $\operatorname{spr} T$ will stand for the null space (kernel), the image (rank), and the spectral radius of T, respectively. The identity in X will be denoted by I_X, and, given $R > 0$ and $x \in X$, $B_R(x)$ will stand for the ball of radius R centered at x.

2. Basic preliminaries. The single logistic equation

In this section we focus our attention into the single logistic equation

$$w'(t) = \rho\alpha(t)w(t) - \beta(t)w^2(t), \tag{2.1}$$

where $\alpha, \beta \in \mathcal{C}_T$ satisfy

$$\hat{\alpha} > 0, \quad \beta > 0, \tag{2.2}$$

and $\rho \in \mathbb{R}$ is regarded as a real parameter. This equation provides us with the semi-trivial states—positive—of (1.2) when $(\alpha, \beta) \in \{(\ell, a), (m, d)\}$.

2.1. *The differential operator* $\frac{d}{dt} + V : \mathcal{C}_T^1 \to \mathcal{C}_T$

We begin studying some useful properties of the differential operator

$$\frac{d}{dt} + V : \mathcal{C}_T^1 \longrightarrow \mathcal{C}_T, \quad w \mapsto w' + Vw, \tag{2.3}$$

where $V \in \mathcal{C}_T$. The following result collects its main positivity properties.

PROPOSITION 2.1. *Suppose* $V, f \in \mathcal{C}_T$ *and consider the periodic problem*

$$w' + Vw = f, \quad w \in \mathcal{C}_T^1. \tag{2.4}$$

Then, the following assertions are true:
  (a) *(2.4) has a unique solution for each* $f \in \mathcal{C}_T$ *if, and only if,* $\int_0^T V \neq 0$. *In such case, we denote by* $\psi_{[V,f]}$ *its unique solution.*
  (b) *In case* $\int_0^T V > 0$, $\psi_{[V,f]} \gg 0$ *if* $f > 0$, *and* $\psi_{[V,f]} \ll 0$ *if* $f < 0$.
  (c) *In case* $\int_0^T V < 0$, $\psi_{[V,f]} \ll 0$ *if* $f > 0$, *and* $\psi_{[V,f]} \gg 0$ *if* $f < 0$.
  (d) *In case* $\int_0^T V = 0$, *(2.4) possesses a solution if, and only if,*

$$\int_0^T e^{\int_0^s V} f(s)\, ds = 0. \tag{2.5}$$

In particular, (2.4) cannot admit a solution if either $f > 0$ or $f < 0$. Actually, under condition (2.5) any solution of

$$w' + Vw = f$$

must be T-periodic.

PROOF. For any $w_0 \in \mathbb{R}$, the unique solution of the Cauchy problem

$$\begin{cases} w' + Vw = f, \\ w(0) = w_0, \end{cases}$$

is given by

$$w(t) = e^{-\int_0^t V} \left[w_0 + \int_0^t e^{\int_0^s V} f(s)\, ds \right], \quad t \in \mathbb{R}, \tag{2.6}$$

and it is T-periodic if, and only if, $w(0) = w(T)$. Equivalently,

$$\left(e^{\int_0^T V} - 1\right) w_0 = \int_0^T e^{\int_0^s V} f(s)\, ds. \tag{2.7}$$

Thus, in case $\int_0^T V \neq 0$,

$$w_0 := \frac{\int_0^T e^{\int_0^s V} f(s)\, ds}{e^{\int_0^T V} - 1} \tag{2.8}$$

is the unique initial value to a T-periodic solution, while, if $\int_0^T V = 0$, then identity (2.7) implies (2.5).

As a result from this elementary analysis, substituting (2.8) into (2.6) and rearranging terms, we find that under assumption $\int_0^T V \neq 0$ the function

$$\begin{aligned}
\psi_{[V,f]}(t) &:= \frac{e^{-\int_0^t V}}{e^{\int_0^T V} - 1} \left[\int_0^T e^{\int_0^s V} f(s)\, ds + \left(e^{\int_0^T V} - 1\right) \int_0^t e^{\int_0^s V} f(s)\, ds \right] \\
&= \frac{e^{-\int_0^t V}}{e^{\int_0^T V} - 1} \left(\int_t^T e^{\int_0^s V} f(s)\, ds + e^{\int_0^T V} \int_0^t e^{\int_0^s V} f(s)\, ds \right), \quad t \in \mathbb{R},
\end{aligned}$$

provides us with the unique solution of (2.4). Note that, for any $V \in \mathcal{C}_T$, the auxiliary function

$$\Xi(t) := \int_t^T e^{\int_0^s V} f(s)\, ds + e^{\int_0^T V} \int_0^t e^{\int_0^s V} f(s)\, ds, \quad t \in \mathbb{R}, \tag{2.9}$$

satisfies

$$\Xi \gg 0 \quad \text{if } f > 0, \quad \text{and} \quad \Xi \ll 0 \quad \text{if } f < 0.$$

Consequently, in case $\int_0^T V > 0$ we have that, for each $t \in \mathbb{R}$,

$$\psi_{[V,f]}(t) = \frac{e^{-\int_0^t V}}{e^{\int_0^T V} - 1} \, \Xi(t) > 0$$

if $f > 0$, whereas $\psi_{[V,f]}(t) < 0$ if $f < 0$. Similarly, in case $\int_0^T V < 0$, we have that $\psi_{[V,f]} \gg 0$ if $f < 0$, and $\psi_{[V,f]} \ll 0$ if $f > 0$, which concludes the proof of parts (a), (b) and (c).

Finally, suppose $\int_0^T V = 0$. In this case, we already know that (2.5) is necessary for the existence of a solution to (2.4). Suppose (2.5). Then, for each $w_0 \in \mathbb{R}$, we have that (2.6) satisfies

$$w(T) = e^{-\int_0^T V}\left[w_0 + \int_0^T e^{\int_0^s V} f(s)\,\mathrm{d}s \right] = w_0 = w(0)$$

and, therefore, it provides us with a solution of (2.4). This concludes the proof. $\qquad\square$

As an immediate consequence from Proposition 2.1, in case $\int_0^T V \neq 0$ the linear differential operator (2.3) is a bijection. Clearly, it is continuous. So, it follows from the open mapping theorem that it defines a linear isomorphism between $\mathcal{C}_T^1$ and $\mathcal{C}_T$. Subsequently, we shall denote by J the canonical injection

$$J : \mathcal{C}_T^1 \to \mathcal{C}_T.$$

By Ascoli–Arzela's theorem, J is compact, i.e., it sends bounded sets of $\mathcal{C}_T^1$ into relatively compact subsets of $\mathcal{C}_T$. Therefore, the *resolvent operator*

$$\mathfrak{R}_V := J\left(\frac{\mathrm{d}}{\mathrm{d}t} + V\right)^{-1} : \mathcal{C}_T \to \mathcal{C}_T \tag{2.10}$$

is as well compact. Actually, it is *strongly positive* if $\int_0^T V > 0$, while it is *strongly negative* if $\int_0^T V < 0$, i.e.,

$$\mathfrak{R}_V\left(P \setminus \{0\}\right) \subset \operatorname{Int} P \quad \text{if } \int_0^T V > 0, \tag{2.11}$$

while,

$$\mathfrak{R}_V\left(P \setminus \{0\}\right) \subset -\operatorname{Int} P \quad \text{if } \int_0^T V < 0. \tag{2.12}$$

As a consequence from these positivity properties, one can easily get the main comparison properties associated to Problem (2.4). To state them we need the following elementary concepts.

DEFINITION 2.2. *Given* $V, f \in C_T$,
  (1) *a function* $w \in C_T^1$ *is said to be a subsolution (resp. strict subsolution) of* (2.4) *if* $w' + Vw \leqslant f$ (*resp.* $w' + Vw < f$);
  (2) *a function* $w \in C_T^1$ *is said to be a supersolution (resp. strict supersolution) of* (2.4) *if* $w' + Vw \geqslant f$ (*resp.* $w' + Vw > f$).

PROPOSITION 2.3. *Suppose* $V, f \in C_T$, *with* $\int_0^T V \neq 0$. *Then*:
  (a) *In case* $\int_0^T V > 0$, *any strict subsolution* w *of* (2.4) *satisfies* $w \ll \psi_{[V,f]}$, *whereas* $w \gg \psi_{[V,f]}$ *for any strict supersolution* w.
  (b) *In case* $\int_0^T V < 0$, *any strict subsolution* w *of* (2.4) *satisfies* $w \gg \psi_{[V,f]}$, *whereas* $w \ll \psi_{[V,f]}$ *for any strict supersolution* w.

PROOF. Suppose $\int_0^T V > 0$ and w is a strict subsolution of (2.4). Then, the auxiliary function p defined by

$$p := \left(\frac{\mathrm{d}}{\mathrm{d}t} + V \right) (\psi_{[V,f]} - w)$$

satisfies $p > 0$ and, due to (2.11),

$$\psi_{[V,f]} - w = \mathfrak{R}_V(p) \gg 0.$$

Similarly, when w is a strict supersolution of (2.4),

$$q := \left(\frac{\mathrm{d}}{\mathrm{d}t} + V \right) (w - \psi_{[V,f]}) > 0$$

and (2.11) implies

$$w - \psi_{[V,f]} = \mathfrak{R}_V(q) \gg 0,$$

which concludes the proof of part (a). The proof of part (b) follows the same scheme, though one should use (2.12), instead of (2.11). $\square$

2.2. *The single logistic equation*

The following result characterizes the existence of positive solutions of (2.1).

PROPOSITION 2.4. *Suppose* α, $\beta \in \mathcal{C}_T$ *satisfy* (2.2). *Then,* (2.1) *possesses a positive T-periodic solution if, and only if,* $\rho > 0$. *Moreover, it is unique if it exists and if we denote it by* $\theta_{[\rho\alpha,\beta]}$, *then,*

$$\theta_{[\rho\alpha,\beta]} = \frac{1}{\psi_{[\rho\alpha,\beta]}}, \tag{2.13}$$

where $\psi_{[\rho\alpha,\beta]}$ *stands for the unique positive solution of*

$$z' + \rho\alpha z = \beta, \quad z \in \mathcal{C}_T^1. \tag{2.14}$$

PROOF. Note that any positive solution $w \in \mathcal{C}_T^1$ of (2.1) must satisfy $w \gg 0$, since $w' = (\rho\alpha - \beta w)w$. Moreover, the change of variable $w = z^{-1}$ transforms (2.1) into the linear equation (2.14). Thus, the positive T-periodic solutions of (2.1) are in one-to-one correspondence with the positive solutions of problem (2.14).

Suppose $\rho \neq 0$. Then, (2.2) implies $\rho\hat{\alpha} \neq 0$ and, hence, thanks to Proposition 2.1(a), (2.14) possesses a unique solution, $\psi_{[\rho\alpha,\beta]}$. Moreover, thanks to Proposition 2.1(b)–(c), $\psi_{[\rho\alpha,\beta]} \gg 0$ if $\rho > 0$, while $\psi_{[\rho\alpha,\beta]} \ll 0$ if $\rho < 0$.

Finally, note that if $\rho = 0$, then, due to Proposition 2.1(d), (2.14) cannot admit a positive solution, since

$$\int_0^T e^{\rho \int_0^s \alpha} \beta(s)\,\mathrm{d}s = \int_0^T \beta(s)\,\mathrm{d}s > 0.$$

Combining these features concludes the proof. $\qquad\square$

We now analyze some important comparison properties between subsolutions, supersolutions and solutions of (2.1). First, we will introduce some basic concepts.

DEFINITION 2.5. Let $\rho \in \mathbb{R}$ and α, $\beta \in \mathcal{C}_T$. Then,
(1) a function $w \in \mathcal{C}_T^1$ is said to be a subsolution (resp. strict subsolution) of (2.1) if $w' \leqslant \rho\alpha w - \beta w^2$ (resp. $w' < \rho\alpha w - \beta w^2$);
(2) a function $w \in \mathcal{C}_T^1$ is said to be a supersolution (resp. strict supersolution) of (2.1) if $w' \geqslant \rho\alpha w - \beta w^2$ (resp. $w' > \rho\alpha w - \beta w^2$).

PROPOSITION 2.6. *Suppose* $\rho > 0$ *and* α, $\beta \in \mathcal{C}_T$ *satisfy* $\hat{\alpha} > 0$ *and* $\beta > 0$. *Then, any strict subsolution* $\underline{w} \gg 0$ *of* (2.1) *satisfies* $\underline{w} \ll \theta_{[\rho\alpha,\beta]}$, *while any strict supersolution* $\overline{w} \gg 0$ *satisfies* $\overline{w} \gg \theta_{[\rho\alpha,\beta]}$.

PROOF. Suppose $\underline{w} \gg 0$ is a strict subsolution of (2.1). Then, the function $\underline{z} := 1/\underline{w}$ is a strict supersolution of (2.14) and, hence, it follows from Proposition 2.3(a) that $\underline{z} \gg \psi_{[\rho\alpha,\beta]}$. Therefore,

$$\underline{w} \ll \frac{1}{\psi_{[\rho\alpha,\beta]}} = \theta_{[\rho\alpha,\beta]}.$$

Similarly, if $\overline{w} \gg 0$ is a strict supersolution of (2.1), then $\overline{z} := 1/\overline{w}$ is a strict subsolution of (2.14) and, hence, $\overline{z} \ll \psi_{[\rho\alpha,\beta]}$. Consequently, $\overline{w} \gg \theta_{[\rho\alpha,\beta]}$. $\qquad\square$

COROLLARY 2.7. *Let, for each* $j \in \{1, 2\}$, $\rho_j > 0$ *and* $\alpha_j, \beta_j \in \mathcal{C}_T$ *satisfying* $\alpha_j > 0$ *and* $\beta_j > 0$. *Suppose*

$$\rho_1 \leqslant \rho_2, \quad \alpha_1 \leqslant \alpha_2, \quad \beta_1 \geqslant \beta_2, \quad (\rho_1, \alpha_1, \beta_1) \neq (\rho_2, \alpha_2, \beta_2). \tag{2.15}$$

Then, $\theta_{[\rho_1,\alpha_1,\beta_1]} \ll \theta_{[\rho_2,\alpha_2,\beta_2]}$.

PROOF. Under conditions (2.15), $\theta_{[\rho_1,\alpha_1,\beta_1]} \gg 0$ is a strict subsolution of

$$w' = \rho_2\alpha_2 w - \beta_2 w^2$$

and the conclusion follows from Proposition 2.6. $\qquad\square$

2.3. *Point-wise behaviour of the map* $\rho \mapsto \theta_{[\rho\alpha,\beta]}$

In this section we regard to the unique positive solution of (2.1), $\theta_{[\rho\alpha,\beta]}$, as a map depending on the parameter $\rho > 0$ in order to analyze some sharp properties of the global bifurcation diagram of non-negative solutions of (2.1). Our main result reads as follows.

THEOREM 2.8. *Suppose* $\hat{\alpha} > 0$ *and* $\beta > 0$, *and set*

$$\Theta(\rho) := \theta_{[\rho\alpha,\beta]} \in \mathcal{C}_T^1, \quad \rho > 0. \tag{2.16}$$

Then, the map $(0, \infty) \to \mathcal{C}_T^1$, $\rho \mapsto \Theta(\rho)$, *is real analytic and it possesses the following asymptotic expansion at* $\rho = 0$

$$\Theta(\rho) = \frac{\hat{\alpha}}{\hat{\beta}}\left[\rho + \rho^2 I_2 + O(\rho^3)\right] \quad \text{as } \rho \downarrow 0, \tag{2.17}$$

where I_2 *stands for the function defined by*

$$I_2(t) := \int_0^t \alpha - \frac{\hat{\alpha}}{\hat{\beta}} \int_0^t \beta - \frac{1}{T}\left(\int_0^T\int_0^t \alpha - \frac{\hat{\alpha}}{\hat{\beta}} \int_0^T\int_0^t \beta\right), \quad t \in \mathbb{R}.$$

In particular,

$$\lim_{\rho\downarrow 0} \left\|\Theta(\rho)\right\|_{\mathcal{C}_T} = 0. \tag{2.18}$$

If, in addition, $\alpha > 0$, then the map $(0, \infty) \to \mathcal{C}_T$, $\rho \mapsto \Theta(\rho)$, is strongly increasing in the sense that $\Theta(\rho) \gg \Theta(\tilde{\rho})$ if $\rho > \tilde{\rho}$. Further, if $\alpha \gg 0$ and $\beta \gg 0$, then, for each $\rho > 0$,

$$\frac{\alpha_L}{\beta_M} \rho \leqslant \Theta(\rho) \leqslant \frac{\alpha_M}{\beta_L} \rho \tag{2.19}$$

and

$$\lim_{\rho \uparrow \infty} \frac{\Theta(\rho)}{\rho} = \frac{\alpha}{\beta} \quad in \ \mathcal{C}_T. \tag{2.20}$$

PROOF. Note that the solutions of (2.1) are the zeroes of the non-linear operator $\mathfrak{F} : \mathbb{R} \times \mathcal{C}_T^1 \to \mathcal{C}_T$ defined by

$$\mathfrak{F}(\rho, w) := w' - \rho \alpha w + \beta w^2, \tag{2.21}$$

which is real analytic. Fix $\rho_0 > 0$. Then,

$$\mathfrak{F}\big(\rho_0, \Theta(\rho_0)\big) = 0$$

and $D_u \mathfrak{F}(\rho_0, \Theta(\rho_0)) \in \mathcal{L}(\mathcal{C}_T^1; \mathcal{C}_T)$ is given through

$$D_u \mathfrak{F}\big(\rho_0, \Theta(\rho_0)\big) := \frac{\mathrm{d}}{\mathrm{d}t} + 2\beta \Theta(\rho_0) - \rho_0 \alpha. \tag{2.22}$$

Since $\Theta(\rho_0) \gg 0$ satisfies

$$\frac{\mathrm{d}\Theta(\rho_0)}{\mathrm{d}t} = \rho_0 \alpha \Theta(\rho_0) - \beta\big[\Theta(\rho_0)\big]^2,$$

division by $\Theta(\rho_0)$ and integration in $[0, T]$ gives

$$0 = \mathrm{Log}\, \frac{\Theta(\rho_0)(T)}{\Theta(\rho_0)(0)} = \int_0^T \big[\rho_0 \alpha - \beta \Theta(\rho_0)\big] = 0$$

and, hence,

$$\int_0^T \big[2\beta \Theta(\rho_0) - \rho_0 \alpha\big] = \int_0^T \beta(t) \Theta(\rho_0)(t)\, \mathrm{d}t > 0,$$

since $\beta > 0$. Thus, thanks to Proposition 2.1, $D_u \mathfrak{F}(\rho_0, \Theta(\rho_0))$ is a linear isomorphism between $\mathcal{C}_T^1$ and $\mathcal{C}_T$. Therefore, thanks to the implicit function theorem, there exist $\varepsilon \in (0, \rho_0)$, $\delta > 0$, and a unique real analytic map

$$W : (\rho_0 - \varepsilon, \rho_0 + \varepsilon) \to \big\{ w \in \mathcal{C}_T^1 : \big\| w - \Theta(\rho_0) \big\|_{\mathcal{C}_T^1} < \delta \big\}$$

such that $W(\rho_0) = \Theta(\rho_0)$, $\mathfrak{F}(\rho, W(\rho)) = 0$ if $|\rho - \rho_0| < \varepsilon$, and $w = W(\rho)$ if $\mathfrak{F}(\rho, w) = 0$ with $|\rho - \rho_0| < \varepsilon$ and $\|w - \Theta(\rho_0)\|_{\mathcal{C}_T^1} < \delta$. As $W(\rho_0) = \Theta(\rho_0) \gg 0$, necessarily $W(\rho) \gg 0$ for $\rho \sim \rho_0$, and, therefore, $W(\rho) = \Theta(\rho)$, by the uniqueness of the positive solution of (2.1). This shows that $\rho \mapsto \Theta(\rho)$ is real analytic.

Although the fact that the map $(0, \infty) \to \mathcal{C}_T$, $\rho \mapsto \Theta(\rho)$, is strictly increasing if $\alpha > 0$ is a consequence from Corollary 2.7, it should be noted that implicit differentiation in

$$\mathfrak{F}\big(\rho, \Theta(\rho)\big) = 0$$

gives rise to

$$\left(\frac{\mathrm{d}}{\mathrm{d}t} + 2\beta\Theta(\rho) - \rho\alpha\right)\frac{\mathrm{d}\Theta}{\mathrm{d}\rho}(\rho) = \alpha\Theta(\rho) > 0$$

and, hence,

$$\frac{\mathrm{d}\Theta}{\mathrm{d}\rho}(\rho) = \left(\frac{\mathrm{d}}{\mathrm{d}t} + 2\beta\Theta(\rho) - \rho\alpha\right)^{-1}\big(\alpha\Theta(\rho)\big) \gg 0,$$

because of (2.11).

To prove (2.17) we will use a celebrated device introduced by M.G. Crandall and P.H. Rabinowitz [6]. Note that

$$D_u\mathfrak{F}(0, 0) = \frac{\mathrm{d}}{\mathrm{d}t} : \mathcal{C}_T^1 \to \mathcal{C}_T$$

and, hence,

$$N\big[D_u\mathfrak{F}(0, 0)\big] = \mathrm{span}\,[1].$$

Moreover,

$$\mathcal{C}_T^1 = N\big[D_u\mathfrak{F}(0, 0)\big] \oplus Y, \quad Y := \left\{y \in \mathcal{C}_T^1 : \int_0^T y = 0\right\}.$$

Indeed, any $w \in \mathcal{C}_T^1$ admits a unique decomposition as

$$w = x + y, \quad x \in \mathbb{R}, \ y \in Y.$$

Actually,

$$x = \hat{w}, \quad y = w - \hat{w}.$$

Now, consider the auxiliary operator $\mathfrak{G} : \mathbb{R}^2 \times Y \to \mathcal{C}_T$ defined by

$$\mathfrak{G}(s, \rho, y) := y' - \rho\alpha(1 + y) + \beta s(1 + y)^2, \quad (s, \rho, y) \in \mathbb{R}^2 \times Y. \tag{2.23}$$

It should be noted that $\mathfrak{G}(s, \rho, y) = 0$ implies $\mathfrak{F}(\rho, s(1 + y)) = 0$. It is rather clear that $\mathfrak{G}$ is a real analytic operator such that $\mathfrak{G}(0, 0, 0) = 0$. Moreover, the operator $D_{(\rho, y)}\mathfrak{G}(0, 0, 0) : \mathbb{R} \times Y \to \mathcal{C}_T$ is given by

$$D_{(\rho, y)}\mathfrak{G}(0, 0, 0)(\rho, y) = y' - \rho\alpha, \quad (\rho, y) \in \mathbb{R} \times Y,$$

and, so, it is a linear isomorphism. Indeed, if for some $(\rho, y) \in \mathbb{R} \times Y$ one has that $y' = \rho\alpha$, then integrating in $[0, T]$ gives $\rho\hat{\alpha} = 0$ and, hence, $\rho = 0$, since $\hat{\alpha} > 0$. Thus, $y' = 0$ and, therefore, y is constant. Consequently, as it has zero average, we obtain that $y = 0$, which concludes the proof of the injectivity. Now, fix $f \in \mathcal{C}_T$. Then, it is straightforward to check that

$$y(t) := -\frac{\hat{f}}{\hat{\alpha}}\int_0^t \alpha + \int_0^t f - \frac{1}{T}\left(-\frac{\hat{f}}{\hat{\alpha}}\int_0^T\int_0^t \alpha + \int_0^T\int_0^t f\right), \quad t \in \mathbb{R},$$

provides us with the unique element of Y solving

$$D_{(\rho, y)}\mathfrak{G}(0, 0, 0)(\rho, y) = f, \quad \rho := -\frac{\hat{f}}{\hat{\alpha}}.$$

Therefore, thanks to the implicit function theorem, there exist $s_0 > 0$, $\delta > 0$, and a unique real analytic map

$$(\rho, y) : (-s_0, s_0) \to (-\delta, \delta) \times \left\{y \in Y : \|y\|_{\mathcal{C}_T^1} < \delta\right\}$$

such that $(\rho(0), y(0)) = (0, 0)$, $\mathfrak{G}(s, \rho(s), y(s)) = 0$ for each $s \in (-s_0, s_0)$, and $(\rho, y) = (\rho(s), y(s))$ if $\mathfrak{G}(s, \rho, y) = 0$ with $|s| < s_0$, $|\rho| < \delta$, and $\|y\|_{\mathcal{C}_T^1} < \delta$. In particular,

$$\mathfrak{F}\big(\rho(s), s(1 + y(s))\big) = 0, \quad |s| < s_0,$$

and, so, for sufficiently small $s > 0$,

$$(\rho, w) := \big(\rho(s), s(1 + y(s))\big)$$

provides us with a positive T-periodic solution of (2.1). Necessarily, for sufficiently small $s > 0$, we have that

$$s(1 + y(s)) = \Theta\big(\rho(s)\big) \tag{2.24}$$

because of the uniqueness of the positive solution of (2.1). Now, in order to find the lowest order terms of the local developments

$$\rho(s) = \sum_{n=1}^{\infty} \rho_n s^n, \quad y(s) = \sum_{n=1}^{\infty} s^n y_n, \quad y_n \in Y, \ n \geq 1, \tag{2.25}$$

we substitute (2.25) into the equation of the zeroes of $\mathfrak{G}$. Then, we get

$$\sum_{n=1}^{\infty} s^n y_n' = \left[\alpha \sum_{n=1}^{\infty} \rho_n s^n - s\beta \left(1 + \sum_{n=1}^{\infty} s^n y_n \right) \right] \left(1 + \sum_{n=1}^{\infty} s^n y_n \right)$$

and identifying terms of order one in both members gives

$$y_1' = \rho_1 \alpha - \beta.$$

Thus,

$$\rho_1 = \frac{\hat{\beta}}{\hat{\alpha}}, \quad y_1(t) = \rho_1 \int_0^t \alpha - \int_0^t \beta - \frac{1}{T}\left(\rho_1 \int_0^T \int_0^t \alpha - \int_0^T \int_0^t \beta \right)$$

and, therefore, (2.24) shows that

$$\rho(s) = \rho_1 s + O(s^2) \quad \text{and} \quad \Theta(\rho(s)) = s(1 + s\,y_1(t) + O(s^2)) \tag{2.26}$$

as $s \downarrow 0$. Thanks again to the implicit function theorem, one can eliminate s as a function of ρ at $\rho = 0$ from the identity

$$\rho_1 s + O(s^2) = \rho.$$

Actually,

$$s = \frac{1}{\rho_1}\rho + O(\rho^2) = \frac{\hat{\alpha}}{\hat{\beta}}\rho + O(\rho^2),$$

and, so, substituting this expansion into the second identity of (2.26) shows that

$$\Theta(\rho) = \frac{\hat{\alpha}}{\hat{\beta}}\left[\rho + \frac{\hat{\alpha}}{\hat{\beta}}y_1\rho^2 + O(\rho^3) \right]$$

as $\rho \downarrow 0$, which provides us with the asymptotic expansion (2.17).

Subsequently, we suppose that $\alpha \gg 0$ and $\beta \gg 0$. The global estimates (2.19) and (2.20) are consequences from Proposition 2.6. Indeed, setting $\Theta := \Theta(\rho)$, we have that

$$\rho\alpha_L\Theta - \beta_M\Theta^2 \leqslant \Theta' = \rho\alpha\Theta - \beta\Theta^2 \leqslant \rho\alpha_M\Theta - \beta_L\Theta^2,$$

and, hence, for each $\rho > 0$, $\Theta(\rho)$ is a positive subsolution of $w' = \rho\alpha_M w - \beta_L w^2$ and a positive supersolution of $w' = \rho\alpha_L w - \beta_M w^2$. Therefore, (2.19) follows from Proposition 2.6 by taking into account that $\frac{\alpha_M}{\beta_L}\rho$ and $\frac{\alpha_L}{\beta_M}\rho$ provide us with the unique T-periodic positive solutions of each of these autonomous equations.

Now, fix $\delta > 0$ sufficiently small so that $\frac{\alpha}{\beta} \gg \delta$, consider two functions $\underline{\theta}, \bar{\theta} \in C_T^1$ such that

$$\frac{\alpha}{\beta} - \delta < \underline{\theta} < \frac{\alpha}{\beta} - \frac{\delta}{2}, \qquad \frac{\alpha}{\beta} + \frac{\delta}{2} < \bar{\theta} < \frac{\alpha}{\beta} + \delta, \tag{2.27}$$

and set

$$\underline{w} := \rho\,\underline{\theta}, \qquad \overline{w} := \rho\,\bar{\theta}.$$

Then, for each sufficiently large $\rho > 0$, say $\rho \geqslant \rho_0$, we have that

$$\rho\alpha\overline{w} - \beta\overline{w}^2 = \rho^2\beta\bar{\theta}\left(\frac{\alpha}{\beta} - \bar{\theta}\right) < -\frac{\delta}{2}\rho^2\beta\bar{\theta} \leqslant \rho\bar{\theta}' = \overline{w}',$$

and

$$\rho\alpha\underline{w} - \beta\underline{w}^2 = \rho^2\beta\underline{\theta}\left(\frac{\alpha}{\beta} - \underline{\theta}\right) > \frac{\delta}{2}\rho^2\beta\underline{\theta} \geqslant \rho\underline{\theta}' = \underline{w}'.$$

Consequently, for each $\rho \geqslant \rho_0$, $\underline{w}$ and $\overline{w}$ provide with a strict subsolution and a strict supersolution of (2.1), respectively, and, hence, thanks to Proposition 2.6,

$$\underline{w} = \rho\,\underline{\theta} \ll \Theta(\rho) \ll \overline{w} = \rho\,\bar{\theta}.$$

Thus, it follows from (2.27) that, for each $\rho \geqslant \rho_0$,

$$\frac{\alpha}{\beta} - \delta \ll \frac{\Theta(\rho)}{\rho} \ll \frac{\alpha}{\beta} + \delta.$$

Therefore, passing to the limit as $\rho \uparrow \infty$, shows that

$$\frac{\alpha}{\beta} - \delta \leqslant \liminf_{\rho\uparrow\infty} \frac{\Theta(\rho)}{\rho} \leqslant \limsup_{\rho\uparrow\infty} \frac{\Theta(\rho)}{\rho} \leqslant \frac{\alpha}{\beta} + \delta.$$

As these inequalities are valid for any $\delta > 0$, the proof of (2.20) is concluded, as well as the proof of the theorem. $\qquad\square$

It should be noted that the strong monotonicity of $\Theta(\rho)$ is lost when α changes of sign. Indeed, suppose that, for some sufficiently small $\delta > 0$, $\beta(t) = 0$ and $\alpha(t) < 0$ for each $t \in [0, \delta]$. Then, for each $t \in [0, \delta]$ we have that

$$\theta_{[\rho\alpha,\beta]}(t) = e^{\rho\int_0^t \alpha}\theta_{[\rho\alpha,\beta]}(0).$$

Since $\int_0^t \alpha < 0$, the map $\rho \mapsto \theta_{[\rho\alpha,\beta]}(t)$ is decreasing for each $t \in (0, \delta]$.

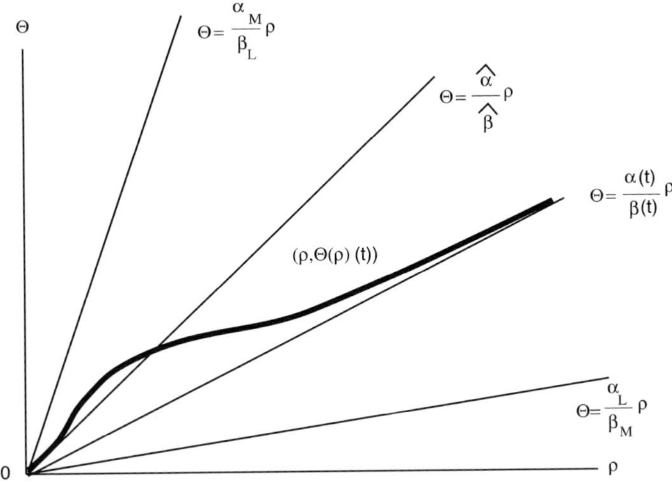

Fig. 1. Graph of $\rho \mapsto \Theta(\rho)(t)$ in case $\alpha \gg 0$ and $\beta \gg 0$.

In Fig. 1 we have represented the bifurcation diagram $(\rho, \Theta(\rho)(t))$ for some $t \in \mathbb{R}$ in the (ρ, Θ)-plane in the special, though important, case when $\alpha \gg 0$ and $\beta \gg 0$. Thanks to (2.19) the curve $\rho \mapsto \Theta(\rho)(t)$ must be confined in between the straight lines $\Theta = \frac{\alpha_L}{\beta_M}\rho$ and $\Theta = \frac{\alpha_M}{\beta_L}\rho$, and, due to (2.17) and (2.20), its tangents at $\rho = 0$ and $\rho = \infty$ are

$$\Theta = \frac{\hat{\alpha}}{\hat{\beta}}\,\rho \quad \text{and} \quad \Theta = \frac{\alpha(t)}{\beta(t)}\rho,$$

respectively.

2.4. *Attractiveness of* $\theta_{[\rho\alpha,\beta]}$

Note that the linearized equations of (2.1) at $w = 0$ and $w = \Theta(\rho)$ are given by

$$w' = \rho\alpha w \quad \text{and} \quad w' = \big[\rho\alpha - 2\beta\Theta(\rho)\big]w,$$

respectively, whose Floquet multipliers are

$$e^{\rho \int_0^T \alpha} \quad \text{and} \quad e^{\int_0^T [\rho\alpha - 2\beta\Theta(\rho)]}.$$

As a result, $w = 0$ is linearly stable if $\rho < 0$, linearly neutrally stable if $\rho = 0$ and linearly unstable if $\rho > 0$. Moreover, in the proof of Theorem 2.8 we have seen that, for each $\rho > 0$,

$$\int_0^T \big[\rho\alpha - \beta\Theta(\rho)\big] = 0, \tag{2.28}$$

and, hence,

$$e^{\int_0^T [\rho\alpha - 2\beta\Theta(\rho)]} = e^{-\int_0^T [\beta\Theta(\rho)]} < 1.$$

Therefore, $\Theta(\rho)$ is linearly stable. Actually, it is easy to see that 0 is a global attractor of (2.1) if $\rho \leqslant 0$, whereas $\Theta(\rho)$ attracts to all positive solutions of (2.1) if $\rho > 0$.

3. Linearized stability of the semi-trivial solutions

Thanks to Proposition 2.4, (1.2) possesses a semi-trivial state—throughout the remaining of the chapter, positive—of the form $(u, 0)$ if and only if $\lambda > 0$. Moreover, in such case, $(\theta_{[\lambda\ell,a]}, 0)$ is the unique semi-trivial state of this form. Similarly, $(0, \theta_{[\mu m,d]})$ provides us with the unique semi-trivial state of (2.1) of the form $(0, v)$, and it exists if and only if $\mu > 0$. The following result characterizes the stability of these states and the stability of the trivial state $(0, 0)$.

PROPOSITION 3.1. *The following assertions are true*:
   (a) *$(0, 0)$ is linearly stable if $\lambda < 0$ and $\mu < 0$, and it is linearly unstable if either $\lambda > 0$ or $\mu > 0$.*
   (b) *Suppose $\lambda > 0$. Then, $(\theta_{[\lambda\ell,a]}, 0)$ is linearly unstable if and only if*

$$\mu > \frac{1}{T} \int_0^T c(t)\theta_{[\lambda\ell,a]}(t)\,dt, \tag{3.1}$$

and linearly stable if and only if

$$\mu < \frac{1}{T} \int_0^T c(t)\theta_{[\lambda\ell,a]}(t)\,dt. \tag{3.2}$$

*Consequently, the **curve of change of stability** of $(\theta_{[\lambda\ell,a]}, 0)$ in the (λ, μ)-plane is given through its curve of neutral stability*

$$\mu = f(\lambda) := \frac{1}{T} \int_0^T c(t)\theta_{[\lambda\ell,a]}(t)\,dt. \tag{3.3}$$

   (c) *Suppose $\mu > 0$. Then, $(0, \theta_{[\mu m,d]})$ is linearly unstable if and only if*

$$\lambda > \frac{1}{T} \int_0^T b(t)\theta_{[\mu m,d]}(t)\,dt, \tag{3.4}$$

and linearly stable if and only if

$$\lambda < \frac{1}{T} \int_0^T b(t)\theta_{[\mu m,d]}(t)\,dt. \tag{3.5}$$

Consequently, the **curve of change of stability** *of* $(0, \theta_{[\mu m, d]})$ *in the* (λ, μ)*-plane is given through its curve of neutral stability*

$$\lambda = g(\mu) := \frac{1}{T} \int_0^T b(t)\theta_{[\mu m, d]}(t)\,dt. \tag{3.6}$$

PROOF. The linearization of (1.2) at any solution $(u_0, v_0) \in \mathcal{C}_T^1 \times \mathcal{C}_T^1$ is given by

$$\begin{cases} u' = (\lambda \ell - 2au_0 - bv_0)u - bu_0 v, \\ v' = -cv_0 u + (\mu m - 2dv_0 - cu_0)v. \end{cases} \tag{3.7}$$

When $(u_0, v_0) = (0, 0)$, (3.7) becomes into

$$\begin{cases} u' = \lambda \ell u, \\ v' = \mu m v, \end{cases}$$

whose Poincaré map is given by

$$\mathfrak{P}(x, y) = \begin{pmatrix} e^{\lambda T} & 0 \\ 0 & e^{\mu T} \end{pmatrix} \begin{pmatrix} x \\ y \end{pmatrix}, \quad (x, y) \in \mathbb{R}^2,$$

since $\hat{\ell} = \hat{m} = 1$. Thus, its Floquet multipliers, $e^{\lambda T}$ and $e^{\mu T}$, are less than one if $\lambda < 0$ and $\mu < 0$, while some of them is greater than one if either $\lambda > 0$ or $\mu > 0$, which concludes the proof of part (a).

When $(u_0, v_0) = (\theta_{[\lambda \ell, a]}, 0)$, (3.7) becomes into

$$\begin{cases} u' = (\lambda \ell - 2a\theta_{[\lambda \ell, a]})u - b\theta_{[\lambda \ell, a]}v, \\ v' = (\mu m - c\theta_{[\lambda \ell, a]})v, \end{cases} \tag{3.8}$$

and its associated Poincaré map is given by

$$\mathfrak{P}(x, y) = \begin{pmatrix} e^{\int_0^T (\lambda \ell - 2a\theta_{[\lambda \ell, a]})} & E \\ 0 & e^{\int_0^T (\mu m - c\theta_{[\lambda \ell, a]})} \end{pmatrix} \begin{pmatrix} x \\ y \end{pmatrix}, \quad (x, y) \in \mathbb{R}^2,$$

for some constant $E \in \mathbb{R}$ whose explicit knowledge is not important here. Thanks to (2.28),

$$\int_0^T (\lambda \ell - a\theta_{[\lambda \ell, a]}) = 0$$

and, hence,

$$0 < e^{\int_0^T (\lambda \ell - 2a\theta_{[\lambda \ell, a]})} = e^{-\int_0^T a(t)\theta_{[\lambda \ell, a]}(t)\,dt} < 1.$$

Consequently, the attractive character of $(\theta_{[\lambda\ell,a]}, 0)$ is given by the remaining Floquet multiplier

$$\nu := e^{\int_0^T (\mu m - c\theta_{[\lambda\ell,a]})} = e^{\mu T - \int_0^T c(t)\theta_{[\lambda\ell,a]}(t)\,dt}.$$

As $\nu > 1$ when (3.1) is satisfied, $\nu = 1$ if (3.3) is satisfied, and $0 < \nu < 1$ under condition (3.2), the proof of part (b) is concluded. Part (c) follows by symmetry. This ends the proof. $\qquad\square$

4. The curves of change of stability of the semi-trivial positive solutions

In this section we shall study some global properties of the curves of neutral stability (3.3) and (3.6) of the semi-trivial solutions of (1.2). Such analysis is imperative in order to ascertain the shape of each of the following regions within the parameter space (λ, μ):

$$\begin{aligned}
\mathbf{R}_{\mathrm{uu}} &:= \left\{(\lambda, \mu) \in (0, \infty)^2 : (\theta_{[\lambda\ell,a]}, 0) \text{ and } (0, \theta_{[\mu m,d]}) \text{ are l.u.}\right\}, \\
\mathbf{R}_{\mathrm{ss}} &:= \left\{(\lambda, \mu) \in (0, \infty)^2 : (\theta_{[\lambda\ell,a]}, 0) \text{ and } (0, \theta_{[\mu m,d]}) \text{ are l.s.}\right\},
\end{aligned} \tag{4.1}$$

$$\begin{aligned}
\mathbf{R}_{\mathrm{su}} &:= \left\{(\lambda, \mu) \in (0, \infty)^2 : (\theta_{[\lambda\ell,a]}, 0) \text{ is l.s., } (0, \theta_{[\mu m,d]}) \text{ is l.u.}\right\}, \\
\mathbf{R}_{\mathrm{us}} &:= \left\{(\lambda, \mu) \in (0, \infty)^2 : (\theta_{[\lambda\ell,a]}, 0) \text{ is l.u., } (0, \theta_{[\mu m,d]}) \text{ is l.s.}\right\},
\end{aligned} \tag{4.2}$$

$$\begin{aligned}
\mathbf{R}_{\mathrm{u}\star} &:= \left\{(\lambda, \mu) \in (0, \infty) \times (-\infty, 0] : (\theta_{[\lambda\ell,a]}, 0) \text{ is l.u.}\right\}, \\
\mathbf{R}_{\mathrm{s}\star} &:= \left\{(\lambda, \mu) \in (0, \infty) \times (-\infty, 0] : (\theta_{[\lambda\ell,a]}, 0) \text{ is l.s.}\right\},
\end{aligned} \tag{4.3}$$

$$\begin{aligned}
\mathbf{R}_{\star\mathrm{u}} &:= \left\{(\lambda, \mu) \in (-\infty, 0] \times (0, \infty) : (0, \theta_{[\mu m,d]}) \text{ is l.u.}\right\}, \\
\mathbf{R}_{\star\mathrm{s}} &:= \left\{(\lambda, \mu) \in (-\infty, 0] \times (0, \infty) : (0, \theta_{[\mu m,d]}) \text{ is l.s.}\right\},
\end{aligned} \tag{4.4}$$

where *l.u.* stands for *linearly unstable* and *l.s.* for *linearly stable*. As the global behaviour of the curves of neutral stability of the semi-trivial states is strongly based upon the nature of the model under study, i.e., under the sign properties of the interaction coefficient functions $b(t)$ and $c(t)$, in the subsequent discussion we will have to distinguish several different cases. We begin the section by giving a general result which is a straightforward consequence from Theorem 2.8.

THEOREM 4.1. *Suppose* $\alpha, \beta, \gamma \in \mathcal{C}_T$ *satisfy*

$$\alpha \gg 0, \quad \beta \gg 0, \quad \hat{\alpha} = \hat{\beta} = 1, \tag{4.5}$$

and consider the function

$$N(\rho) := \frac{1}{T} \int_0^T \gamma(t)\theta_{[\rho\alpha,\beta]}(t)\,dt, \quad \rho > 0. \tag{4.6}$$

Then,

$$N(\rho) = \hat{\gamma}\,\rho + \frac{1}{T}\int_0^T \gamma(t) I_2(t)\,dt\,\rho^2 + \mathrm{O}(\rho^3) \quad \text{as } \rho \downarrow 0, \tag{4.7}$$

where

$$I_2(t) := \int_0^t (\alpha - \beta) - \frac{1}{T}\int_0^T\int_0^t (\alpha - \beta), \quad t \in \mathbb{R}, \tag{4.8}$$

and

$$\lim_{\rho\uparrow\infty} \frac{N(\rho)}{\rho} = \frac{1}{T}\int_0^T \gamma(t)\frac{\alpha(t)}{\beta(t)}\,dt. \tag{4.9}$$

Moreover, $\rho \mapsto N(\rho)$ is increasing if $\gamma > 0$, decreasing if $\gamma < 0$, and, in general, is far from monotone if γ changes of sign. In such case, (4.9) can be either negative, or positive, or zero, according to the value of the average on its right-hand side.

PROOF. By (4.5), it follows from Theorem 2.8 that

$$\theta_{[\rho\alpha,\beta]} = \rho + I_2\,\rho^2 + \mathrm{O}(\rho^3) \quad \text{as } \rho \downarrow 0.$$

Thus,

$$N(\rho) := \frac{1}{T}\int_0^T \gamma(t)\theta_{[\rho\alpha,\beta]}(t)\,dt = \hat{\gamma}\rho + \frac{1}{T}\int_0^T \gamma(t) I_2(t)\,dt\rho^2 + \mathrm{O}(\rho^3),$$

which is (4.7). The validity of (4.9) is an immediate consequence from (2.20). The remaining assertions of the theorem are obvious, since $\theta_{[\rho\alpha,\beta]} \gg 0$. $\qquad\square$

It should be noted that, in general, nothing can be said about the concavity of $N(\rho)$ at $\rho = 0$. Indeed, since I_2 changes of sign if $I_2 \neq 0$, because $\hat{I}_2 = 0$, the average of the function γI_2 can have any sign we wish by choosing and adequate γ that may be taken either positive, or negative, or changing of sign, in strong contrast with the autonomous case, where $N(\rho) = \gamma\rho$ for every $\rho > 0$.

4.1. *The species u and v compete ($b > 0$ and $c > 0$)*

Besides (1.7), throughout this section we assume that

$$\ell \gg 0, \quad m \gg 0, \quad a \gg 0, \quad d \gg 0, \quad b > 0, \quad c > 0, \tag{4.10}$$

though most of our discussion can be easily adapted to cover much weaker requirements. Under assumptions (4.10), the curves of neutral stability of the semi-trivial states

$(\theta_{[\lambda\ell,a]}, 0)$ and $(0, \theta_{[\mu m,d]})$, (3.3) and (3.6), are increasing in λ and μ, respectively, and satisfy

$$f(\lambda) > 0, \quad g(\mu) > 0, \quad (\lambda, \mu) \in (0, \infty)^2. \tag{4.11}$$

It should be noted that these are the curves delimiting the regions defined by (4.1) and (4.2). Thanks to Theorem 4.1, the tangents of these curves at $(\lambda, \mu) = (0, 0)$ are given by

$$\mu = \hat{c}\lambda \quad \text{and} \quad \lambda = \hat{b}\mu, \tag{4.12}$$

respectively, as for the autonomous counterpart of (1.2)—where these are the curves of neutral stability of $(\lambda, 0)$ and $(0, \mu)$—and their tangents at infinity are given by

$$\mu = \alpha\lambda, \quad \alpha := \frac{1}{T}\int_0^T c(t)\frac{\ell(t)}{a(t)}\,dt,$$

$$\lambda = \beta\mu, \quad \beta := \frac{1}{T}\int_0^T b(t)\frac{m(t)}{d(t)}\,dt, \tag{4.13}$$

respectively. In strong contrast with the autonomous case—where (4.12) and (4.13) are the same straight lines—as a consequence from the non-autonomous character of (1.2), the *relative positions* of the straight lines (4.12) and (4.13) can be inter-exchanged at $(\lambda, \mu) = (0, 0)$ and at infinity, so entailing the existence of a crossing point between the curves of change of stability of the semi-trivial states; a feature of great significance from the point of view of the applications, as it might allow us designing environments where none of the competitors is driven to extinction by the other. Most precisely, there are admissible choices of the function coefficients a, b, c, d, ℓ and m, for which either

$$\hat{b}\hat{c} < 1 \quad \text{and} \quad \alpha\beta = \frac{1}{T}\int_0^T c(t)\frac{\ell(t)}{a(t)}\,dt \cdot \frac{1}{T}\int_0^T b(t)\frac{m(t)}{d(t)}\,dt > 1, \tag{4.14}$$

or else

$$\hat{b}\hat{c} > 1 \quad \text{and} \quad \alpha\beta = \frac{1}{T}\int_0^T c(t)\frac{\ell(t)}{a(t)}\,dt \cdot \frac{1}{T}\int_0^T b(t)\frac{m(t)}{d(t)}\,dt < 1. \tag{4.15}$$

In such circumstances, as the curves $\mu = f(\lambda)$, $\lambda > 0$, and $\lambda = g(\mu)$, $\mu > 0$, are real analytic, they must cross at most at a finite number of points—at least one—whose total sum of contact orders must be an odd integer number. To construct examples satisfying any of these requirements one can make the following choices:

$$a = d = 1, \quad \ell := 1 + \ell_1 \sin\left(\frac{2\pi}{T}\cdot\right), \quad m := 1 + m_1 \sin\left(\frac{2\pi}{T}\cdot\right), \tag{4.16}$$

for some constants $\ell_1, m_1 \in (-1, 1)$, and

$$b := b_0 + b_1 \sin\left(\frac{2\pi}{T}\cdot\right), \quad c := c_0 + c_1 \sin\left(\frac{2\pi}{T}\cdot\right), \tag{4.17}$$

for some constants

$$b_0 > 0, \quad c_0 > 0, \quad b_1 \in (-b_0, b_0), \quad c_1 \in (-c_0, c_0).$$

For these choices, we have that

$$\hat{b} = b_0, \quad \hat{c} = c_0,$$

and

$$\alpha = \frac{1}{T} \int_0^T c(t) \frac{\ell(t)}{a(t)} \, dt = c_0 + \frac{1}{2} c_1 \ell_1, \quad \beta = \frac{1}{T} \int_0^T b(t) \frac{m(t)}{d(t)} \, dt = b_0 + \frac{1}{2} b_1 m_1.$$

Thus, (4.14) and (4.15) are equivalent to

$$b_0 c_0 < 1 \quad \text{and} \quad \left(c_0 + \frac{1}{2} c_1 \ell_1 \right) \left(b_0 + \frac{1}{2} b_1 m_1 \right) > 1, \tag{4.18}$$

and

$$b_0 c_0 > 1 \quad \text{and} \quad \left(c_0 + \frac{1}{2} c_1 \ell_1 \right) \left(b_0 + \frac{1}{2} b_1 m_1 \right) < 1, \tag{4.19}$$

respectively. We now show that any of these conditions can be reached by an adequate choice of the several constants involved in their settings. Take, for sufficiently small $\varepsilon > 0$,

$$b_0 = c_0 = 1 - \varepsilon, \quad b_1 = c_1 = 1 - 2\varepsilon.$$

Then, the function $p(\varepsilon)$ defined by

$$p(\varepsilon) := \left[1 - \varepsilon + \frac{1}{2}(1 - 2\varepsilon)\ell_1 \right] \left[1 - \varepsilon + \frac{1}{2}(1 - 2\varepsilon)m_1 \right], \quad \varepsilon > 0,$$

satisfies

$$p(0) = 1 + \frac{1}{2}(\ell_1 + m_1) + \frac{1}{4}\ell_1 m_1$$

and, hence, $p(0) > 1$ if $\ell_1 > 0$ and $m_1 > 0$. Thus, if we choose $\ell_1, m_1 \in (0, 1)$, then, for sufficiently small $\varepsilon > 0$, we have that $p(\varepsilon) > 1$, by continuity, and, consequently, (4.18) is satisfied. On the contrary, choosing

$$b_0 = c_0 = 1 + \varepsilon, \quad b_1 = c_1 = 1,$$

the auxiliary function $n(\varepsilon)$ defined by

$$n(\varepsilon) := \left[1 + \varepsilon + \frac{1}{2}\ell_1 \right] \left[1 + \varepsilon + \frac{1}{2}m_1 \right], \quad \varepsilon > 0,$$

     J. López-Gómez

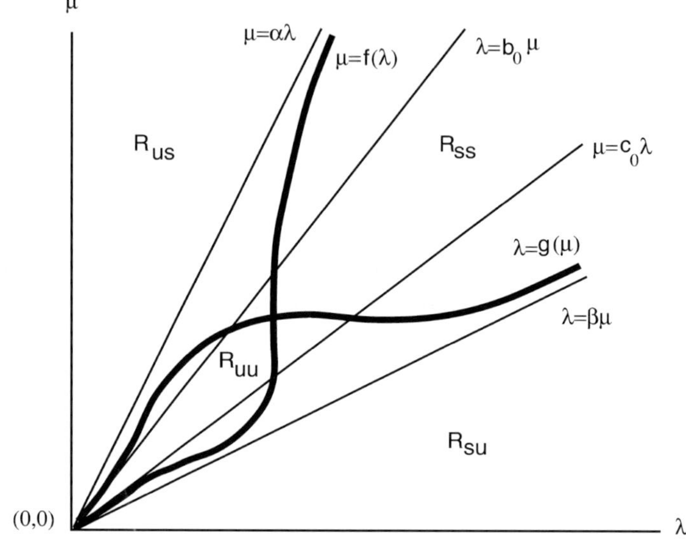

Fig. 2. The curves $\mu = f(\lambda)$ and $\lambda = g(\mu)$ in case (4.18).

satisfies

$$n(0) = 1 + \frac{1}{2}(\ell_1 + m_1) + \frac{1}{4}\ell_1 m_1$$

and, hence, $n(0) < 1$ if $\ell_1 < 0$ and $m_1 < 0$. Thus, if we choose $\ell_1, m_1 \in (-1, 0)$, then, for sufficiently small $\varepsilon > 0$, we have that $n(\varepsilon) < 1$, by continuity, and, consequently, (4.19) holds.

In Fig. 2 we have represented the curves of neutral stability of the semi-trivial states in an admissible situation satisfying (4.18). It should be noted that, for the previous choice, we have that, for each sufficiently small $\varepsilon > 0$,

$$\frac{1}{\beta} = \left(b_0 + \frac{1}{2}b_1 m_1\right)^{-1} = \left(1 - \varepsilon + \frac{1}{2}(1 - 2\varepsilon)m_1\right)^{-1} < c_0 = 1 - \varepsilon < 1$$

and

$$\alpha = c_0 + \frac{1}{2}c_1\ell_1 = 1 - \varepsilon + \frac{1}{2}(1 - 2\varepsilon)\ell_1 > \frac{1}{b_0} = \frac{1}{1 - \varepsilon} > 1$$

since $\ell_1 > 0$ and $m_1 > 0$. As illustrated by Fig. 2, the curves of change of stability of the semi-trivial states must cross at least once, and at most at a finite number of points whose total sum of contact orders must be odd. In the situations described by Fig. 2, one among all admissible ones, these curves divide the first quadrant into four different regions: the two regions enclosed by these curves, and the two exterior regions. In general,

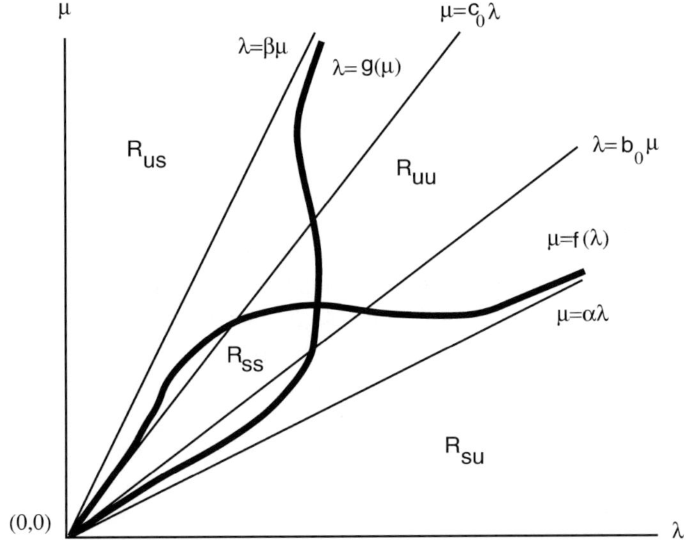

Fig. 3. The curves $\mu = f(\lambda)$ and $\lambda = g(\mu)$ in case (4.19).

the region enclosed between them consists of a finite number of components—one un-bounded, and the remaining bounded. Within the bounded components both semi-trivial states are—alternatively—either linearly unstable, or linearly stable. In Fig. 2, both are linearly unstable, because $\mu > f(\lambda)$ and $\lambda > g(\mu)$. Hence, this is the region R_{uu} defined in (4.1). Independently of the number of bounded components enclosed by the curves of neutral stability of the semi-trivial states, the unbounded component is always a subset of the region R_{ss} defined in (4.1), where both semi-trivial states are linearly stable. The re-maining regions correspond with the regions R_{su} and R_{us} defined in (4.2), where one of the semi-trivial states is linearly stable and the other linearly unstable. As in all competing species models, in this example we have that

$$R_{s*} = (0, \infty) \times (-\infty, 0], \quad R_{u*} = \emptyset,$$

$$R_{*s} = (-\infty, 0] \times (0, \infty), \quad R_{*u} = \emptyset.$$

In Fig. 3 we have represented an admissible situation adjusted to (4.19) for the choices previously analyzed. Now, for each sufficiently small $\varepsilon > 0$, we have that

$$\alpha := c_0 + \frac{1}{2}c_1\ell_1 = 1 + \varepsilon + \frac{1}{2}\ell_1 < \frac{1}{b_0} = \frac{1}{1+\varepsilon} < 1$$

and

$$\frac{1}{\beta} := \left(b_0 + \frac{1}{2}b_1m_1\right)^{-1} = \left(1 + \varepsilon + \frac{1}{2}m_1\right)^{-1} > c_0 = 1 + \varepsilon > 1,$$

since $\ell_1 < 0$ and $m_1 < 0$. As the relative positions between the curves of neutral stability of the semi-trivial states suffered a drastic change between the cases sketched in Figs. 2 and 3, in the latest case, within the bounded component enclosed by $\mu = f(\lambda)$ and $\lambda = g(\mu)$ both semi-trivial states must be linearly stable, while in the unbounded component both are linearly unstable. As it is easily realized, this inter-exchange between the relative positions of the curves of neutral stability of the semi-trivial states entails dramatic changes from the point of view of applications.

Although the previous examples exhibit a reminiscent behaviour from the autonomous model for (λ, μ) in a neighborhood of $(0, 0)$, in the sense that, for such range of (λ, μ) *low intensity competition* occurs if $\hat{b}\hat{c} < 1$, while *high intensity competition* occurs if $\hat{b}\hat{c} > 1$, these simple examples illustrate very well how temporal heterogeneities can affect the dynamics of (1.1), as for (λ, μ) sufficiently large the intensity of the competition is measured by α, instead of $\hat{c}$, and β, instead of $\hat{b}$. Actually, for (λ, μ) large, *low intensity competition* occurs if $\alpha\beta < 1$ (cf. Fig. 3), while *high intensity competition* occurs if $\alpha\beta > 1$ (cf. Fig. 2). The observation of this phenomenology goes back to [12].

In the previous examples, a further crossing point, perturbing from $(0, 0)$, might occur if $b_0 c_0 = 1$, as a result of the nature of the second order terms of the asymptotic expansion of (3.3) and (3.6) at $(0, 0)$, but we refrain of performing this sharper analysis here (cf. [13] for the corresponding analysis in the autonomous competing species model with diffusion).

It should be noted that most of the examples treated in the literature did not cover the previous two cases, but, instead, covered some very special situations where the function coefficients were chosen so that the curves of neutral stability of the semi-trivial states cannot meet in the interior of the first quadrant, as in those cases the method of sub- and supersolutions gives the existence of coexistence states in a rather simple way. Most precisely, let $t_L, t_M \in [0, T]$ such that

$$(\theta_{[\lambda\ell, a]})_j = \theta_{[\lambda\ell, a]}(t_j), \quad j \in \{L; M\}.$$

Then, since $\frac{d\theta_{[\lambda\ell, a]}}{dt}(t_j) = 0$, we have that

$$\theta_{[\lambda\ell, a]}(t_j) = \frac{\ell(t_j)}{a(t_j)} \lambda, \quad j \in \{L, M\},$$

and, hence,

$$\left(\frac{\ell}{a}\right)_L \lambda \leq \theta_{[\lambda\ell, a]} \leq \left(\frac{\ell}{a}\right)_M \lambda, \quad \lambda > 0. \tag{4.20}$$

By symmetry,

$$\left(\frac{m}{d}\right)_L \mu \leq \theta_{[\mu m, d]} \leq \left(\frac{m}{d}\right)_M \mu, \quad \mu > 0. \tag{4.21}$$

Thus, for each $\lambda > 0$,

$$\hat{c}\left(\frac{\ell}{a}\right)_L \lambda \leqslant f(\lambda) = \frac{1}{T}\int_0^T c(t)\theta_{[\lambda\ell,a]}(t)\,dt \leqslant \hat{c}\left(\frac{\ell}{a}\right)_M \lambda, \qquad (4.22)$$

and, for each $\mu > 0$,

$$\hat{b}\left(\frac{m}{d}\right)_L \mu \leqslant g(\mu) = \frac{1}{T}\int_0^T b(t)\theta_{[\mu m,d]}(t)\,dt \leqslant \hat{b}\left(\frac{m}{d}\right)_M \mu. \qquad (4.23)$$

Consequently, the curves of neutral stability (3.3) and (3.6) cannot meet at a point $(\lambda, \mu) \neq (0, 0)$ if either

$$\hat{b}\hat{c}\left(\frac{\ell}{a}\right)_M\left(\frac{m}{d}\right)_M < 1, \qquad (4.24)$$

or

$$\hat{b}\hat{c}\left(\frac{\ell}{a}\right)_L\left(\frac{m}{d}\right)_L > 1. \qquad (4.25)$$

Actually, in any of these cases the curves $\mu = f(\lambda)$ and $\lambda = g(\mu)$ divide the first quadrant into three regions and, necessarily, either $R_{ss} = \emptyset$ or $R_{uu} = \emptyset$ (cf. Fig. 4). As in the autonomous model, low intensity competition occurs under (4.24), while high intensity competition occurs if (4.25) is satisfied. It should be noted that (4.24) and (4.25) become into $\hat{b}\hat{c} < 1$ and $\hat{b}\hat{c} > 1$, respectively, if $\ell = a$ and $m = d$. These are the unique two possible situations exhibited by the autonomous model.

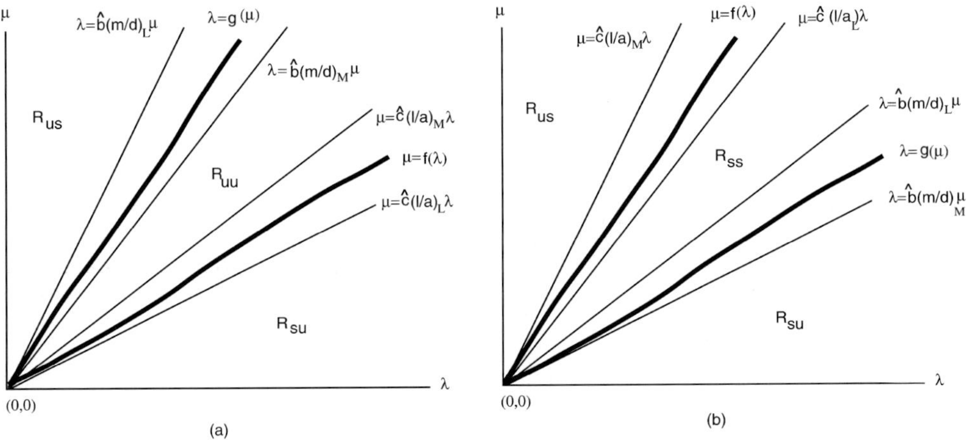

Fig. 4. Curves of neutral stability in cases (4.24) (a) and (4.25) (b).

4.2. *The species v preys on u ($b > 0$ and $c < 0$)*

Then, the curves of neutral stability of the semi-trivial states cannot meet at a point $(\lambda, \mu) \neq (0, 0)$, since

$$f(\lambda) = \frac{1}{T} \int_0^T c(t)\theta_{[\lambda\ell, a]}(t)\, dt < 0 \quad \text{for each } \lambda > 0,$$

and

$$g(\mu) = \frac{1}{T} \int_0^T b(t)\theta_{[\mu m, d]}(t)\, dt > 0 \quad \text{for each } \mu > 0.$$

Note that, as a consequence from Theorem 2.8, $\lambda \mapsto f(\lambda)$ is decreasing and $\mu \mapsto g(\mu)$ is increasing, since $\ell > 0$ and $m > 0$ (cf. Fig. 5). If we further assume that $\ell \gg 0$, $m \gg 0$, $a \gg 0$ and $d \gg 0$, then, besides the fact that $f(\lambda)$ and $g(\mu)$ are monotone, the tangents to $\mu = f(\lambda)$ and $\lambda = g(\mu)$ at $(0, 0)$ and at infinity are given by the straight lines (4.12) and (4.13), respectively.

Now, the half-plane $\lambda > 0$ is divided into four supplementary regions according to the existence and character of each of the semi-trivial states. The interior of the first quadrant is divided into the two regions R_{us} and R_{uu} (cf. (4.1) and (4.2)), and the quadrant $\lambda > 0$, $\mu \leqslant 0$, where (1.2) exclusively admits the semi-trivial state $(\theta_{[\lambda\ell, a]}, 0)$, is divided into the region $\mu > f(\lambda)$ R_{u*}, where $(\theta_{[\lambda\ell, a]}, 0)$ is linearly unstable, and the region $\mu < f(\lambda)$ R_{s*}, where $(\theta_{[\lambda\ell, a]}, 0)$ is linearly stable. Moreover,

$$R_{*s} = (-\infty, 0] \times (0, \infty), \quad R_{*u} = \emptyset, \quad R_{ss} = \emptyset.$$

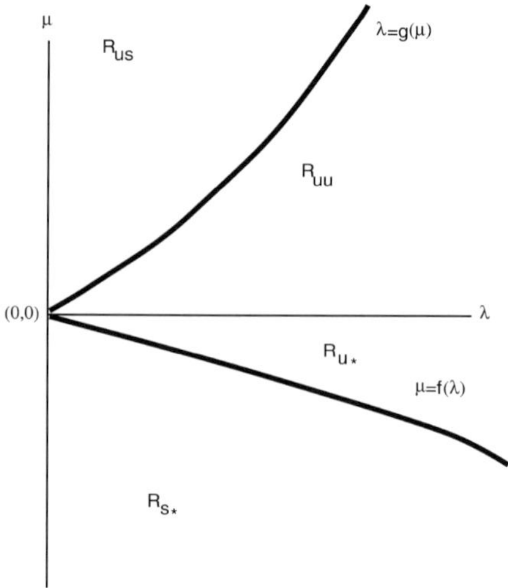

Fig. 5. Curves of neutral stability in the predator-prey model.

The configuration shown by Fig. 5 is reminiscent of the classical one for the autonomous model, where the curves of change of stability of the semi-trivial states $(\lambda, 0)$ and $(0, \mu)$ are $\mu = c\lambda$ and $\lambda = b\mu$, respectively. Actually, if $\ell = a$ and $m = d$, then $\mu = \hat{c}\lambda$ and $\lambda = \hat{b}\mu$ provide us with the tangents at $(0, 0)$ and at infinity to the curves $\mu = f(\lambda)$ and $\lambda = g(\mu)$.

4.3. *The species u and v cooperate ($b < 0$ and $c < 0$)*

In this case, $\lambda \mapsto f(\lambda)$ and $\mu \mapsto g(\mu)$ are decreasing and, in general, they look like shows Fig. 6. Now, the regions defined in (4.1)–(4.4) are given by the following identities:

$$R_{uu} = (0, \infty) \times (0, \infty), \quad R_{ss} = \emptyset, \quad R_{us} = \emptyset, \quad R_{su} = \emptyset,$$

$$R_{u*} = \big\{(\lambda, \mu): \lambda > 0, f(\lambda) < \mu \leqslant 0\big\}, \quad R_{s*} = \big\{(\lambda, \mu): \lambda > 0, \mu < f(\lambda)\big\},$$

$$R_{*u} = \big\{(\lambda, \mu): \mu > 0, g(\mu) < \lambda \leqslant 0\big\}, \quad R_{*s} = \{(\lambda, \mu): \mu > 0, \lambda < g(\mu)\}.$$

As in the predator-prey model, if we further assume $\ell \gg 0$, $m \gg 0$, $a \gg 0$ and $d \gg 0$, then, besides the fact that $f(\lambda)$ and $g(\mu)$ are monotone, the tangents to the curves of neutral stability $\mu = f(\lambda)$ and $\lambda = g(\mu)$ at $(0, 0)$ and at infinity are given by the straight lines (4.12) and (4.13), respectively. Similarly, the configuration shown by Fig. 6 is reminiscent of the classical one for the autonomous model, where the curves of change of stability of the semi-trivial states $(\lambda, 0)$ and $(0, \mu)$ are $\mu = c\lambda$ and $\lambda = b\mu$, respectively. Actually, if $\ell = a$ and $m = d$, then $\mu = \hat{c}\lambda$ and $\lambda = \hat{b}\lambda$ provide us with the tangents at $(0, 0)$ and at infinity to the curves $\mu = f(\lambda)$ and $\lambda = g(\mu)$.

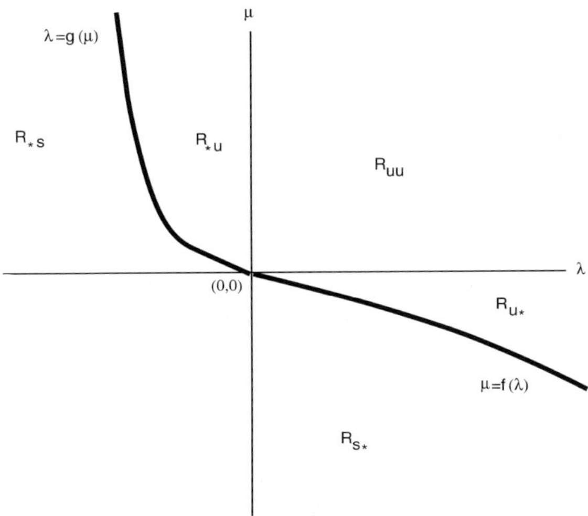

Fig. 6. Curves of neutral stability for the symbiotic model.

4.4. *The general case when b and c change of sign*

As for the competing species model, in order to simplify our analysis, besides (1.7) in this section we assume that

$$a \gg 0, \quad d \gg 0, \quad \ell \gg 0, \quad m \gg 0. \tag{4.26}$$

Then, as in Section 4.1, (4.12) and (4.13) provide us with the tangents at $(0,0)$ and at infinity of the curves of neutral stability of the semi-trivial states. As we are assuming that b and c change of sign in $[0, T]$, we have

$$I^b_+ := \operatorname{Int supp} b^+ \neq \emptyset, \quad I^b_- := \operatorname{Int supp} b^- \neq \emptyset,$$
$$I^c_+ := \operatorname{Int supp} c^+ \neq \emptyset, \quad I^c_- := \operatorname{Int supp} c^- \neq \emptyset,$$

and, hence, $f(\lambda)$ and $g(\mu)$ fail to be monotone. Nevertheless, since

$$\lim_{(c^-)_M \downarrow 0} f(\lambda) = \frac{1}{T} \int_0^T c^+(t)\theta_{[\lambda\ell,a]}(t)\,dt,$$

$$\tag{4.27}$$

$$\lim_{(b^-)_M \downarrow 0} g(\mu) = \frac{1}{T} \int_0^T b^+(t)\theta_{[\mu m,d]}(t)\,dt,$$

uniformly on compact subsets of (λ, μ), the curves of change of stability of the semi-trivial states approximate the corresponding curves of the competition model with interaction function coefficients b^+ and c^+, as $b^- \downarrow 0$ and $c^- \downarrow 0$. Therefore, on compact subsets of (λ, μ), the regions defined by (4.1) and (4.2) approximate the corresponding regions of the associated competition model. Similarly, since

$$\lim_{(c^+)_M \downarrow 0} f(\lambda) = -\frac{1}{T} \int_0^T c^-(t)\theta_{[\lambda\ell,a]}(t)\,dt,$$

$$\tag{4.28}$$

$$\lim_{(b^+)_M \downarrow 0} g(\mu) = -\frac{1}{T} \int_0^T b^-(t)\theta_{[\mu m,d]}(t)\,dt,$$

uniformly on compact subsets of (λ, μ), the curves of change of stability of the semi-trivial states approximate the corresponding curves of the symbiotic model with interaction function coefficients $-b^-$ and $-c^-$ as $b^+ \downarrow 0$ and $c^+ \downarrow 0$. Consequently, for (λ, μ) varying in any compact subset of $(0, \infty)^2$, the regions defined by (4.1) and (4.2) approximate the corresponding regions of the symbiotic model with interactions $-b^-$ and $-c^-$. Therefore, the general case when b and c change of sign can be regarded as a sort of intermediate situation between a competitive and a symbiotic model. Actually, the map $(B, C) : [0, 1] \to \mathcal{C}_T \times \mathcal{C}_T$ defined, for each $\varepsilon \in [0, 1]$, by

$$B(\varepsilon) := \varepsilon b^+ - (1 - \varepsilon)b^- + 2\varepsilon(1 - \varepsilon)\big(b^+ - b^-\big),$$

$$C(\varepsilon) := \varepsilon c^+ - (1 - \varepsilon)c^- + 2\varepsilon(1 - \varepsilon)\big(c^+ - c^-\big),$$

establishes a homotopy between (b^+, c^+), $(-b^-, -c^-)$ and (b, c). Indeed,

$$\big(B(0), C(0)\big) = \big(-b^-, -c^-\big), \qquad \big(B(1), C(1)\big) = \big(b^+, c^+\big),$$

and

$$\big(B(1/2), C(1/2)\big) = (b, c).$$

Consequently, (1.2) should exhibit an intermediate behavior between those exhibited by a competition and a symbiotic model.

In Fig. 7, we have represented two admissible curves $\mu = f(\lambda)$ and $\lambda = g(\mu)$ in the special case when conditions

$$\hat{b} > 0, \quad \hat{c} > 0, \quad \hat{b}\hat{c} < 1, \tag{4.29}$$

and

$$\alpha := \frac{1}{T} \int_0^T c(t) \frac{\ell(t)}{a(t)} \, \mathrm{d}t < 0, \quad \beta := \frac{1}{T} \int_0^T b(t) \frac{m(t)}{d(t)} \, \mathrm{d}t < 0 \tag{4.30}$$

are satisfied. Conditions (4.29) can be reached by taking b^+ and c^+ with $\hat{b}^+\hat{c}^+ < 1$ and choosing sufficiently small b^- and c^-. In order to get (4.30) it suffices to choose ℓ sufficiently small in I_+^c and m sufficiently small in I_+^b. This entails the birth rate of each of the species must be relatively slow during the time periods where the intensity of its aggressions on the other species reaches its highest level. Consequently, although the species compete, the model exhibits some sort of cooperative behaviour. Nevertheless, one should be very careful in choosing the adequate ranges of values of the parameters where this phenomenology occurs.

For the situation illustrated in Fig. 7, the first quadrant is divided into there regions: the region R_{uu}, where both semi-trivial states are linearly unstable, which is the region where $\mu > f(\lambda)$ and $\lambda > g(\mu)$, the region R_{su}, where $(\theta_{[\lambda\ell, a]}, 0)$ is linearly stable and $(0, \theta_{[\mu m, d]})$ is linearly unstable, which is the region where $\lambda > 0$ and $0 < \mu < f(\lambda)$, and the region R_{us}, where $(\theta_{[\lambda\ell, a]}, 0)$ is linearly unstable and $(0, \theta_{[\mu m, d]})$ is linearly stable, which is given by $\mu > 0$ and $0 < \lambda < g(\mu)$. In the second quadrant ($\lambda \leqslant 0$ and $\mu > 0$), $(0, \theta_{[\mu m, d]})$ is the unique semi-trivial state; it is linearly stable if $\lambda < g(\mu)$ and linearly unstable if $\lambda > g(\mu)$. In the fourth quadrant ($\lambda > 0$ and $\mu \leqslant 0$), $(\theta_{[\lambda\ell, a]})$ is the unique semi-trivial state; it is linearly stable if $\mu < f(\lambda)$ and linearly unstable if $\mu > f(\lambda)$. It should be noted that under conditions (4.29) and (4.30) the curves $\mu = f(\lambda)$ and $\lambda = g(\mu)$ might cross. In such case, their total number of crossing points, counting contact orders, must be even, because $\mu = f(\lambda)$ and $\lambda = g(\mu)$ are real analytic. Also, note that Fig. 7 shows a low-competitive-intensity behaviour around $(\lambda, \mu) = (0, 0)$ and a symbiotic type of behavior for sufficiently large λ and μ.

Now, by choosing b^+ and c^+ such that $\hat{b}^+\hat{c}^+ > 1$, $b^- \sim 0$, $c^- \sim 0$, $\ell \sim 0$ in I_+^c and $m \sim 0$ in I_+^b, one gets

$$\hat{b} > 0, \quad \hat{c} > 0, \quad \hat{b}\hat{c} > 1, \quad \alpha < 0, \quad \beta < 0. \tag{4.31}$$

                          J. López-Gómez

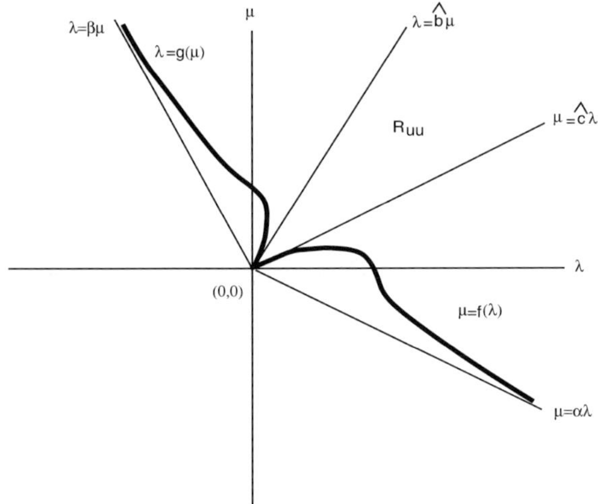

Fig. 7. Curves of neutral stability under conditions (4.29) and (4.30).

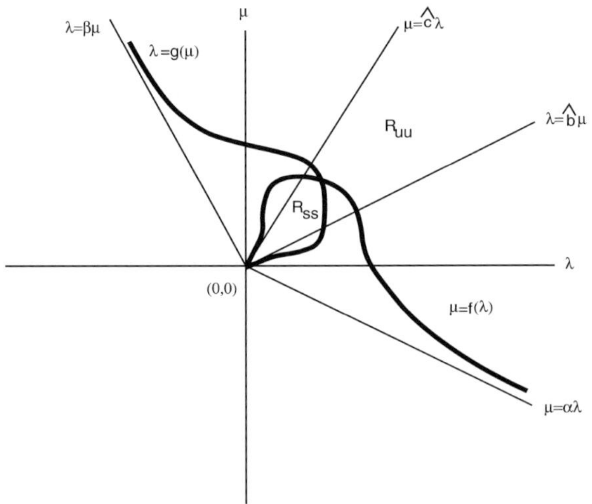

Fig. 8. Curves of neutral stability under condition (4.31).

Fig. 8 shows two admissible curves of neutral stability of semi-trivial states adjusted to (4.31). Now, these curves must cross at an odd number of points, counting contact orders, in such a way that both semi-trivial states are linearly unstable in the unbounded component of the enclosed region between them. Rather naturally, the model exhibits a high-intensity-competition behavior around $(0, 0)$, while it exhibits a genuine cooperative behavior for sufficiently large λ and μ; as far as to the structure of the regions defined in (4.1) and (4.2) concerns. It seems the first occasion that such a kind of behaviour has been observed had been in [22] within the context of reaction diffusion models.

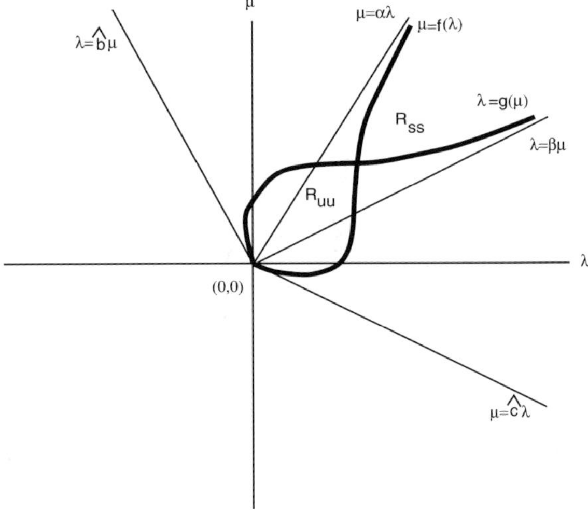

Fig. 9. Curves of neutral stability under condition (4.32).

Now, by choosing sufficiently small b^+ and c^+, $\ell \sim 0$ in I_-^c and sufficiently large in I_+^c, and $m \sim 0$ in I_-^b and sufficiently large in I_+^b, the following conditions can be reached

$$\hat{b} < 0, \quad \hat{c} < 0, \quad \alpha > 0, \quad \beta > 0, \quad \alpha\beta > 1. \tag{4.32}$$

Fig. 9 shows two admissible curves of neutral stability according to (4.32). As in case (4.31), the total number of crossing points between $\mu = f(\lambda)$ and $\lambda = g(\mu)$ must be odd, though, in case (4.32), R_{ss} includes the unbounded component enclosed by the two curves, while the bounded component containing $(0, 0)$ is a subset of R_{uu}, in strong contrast with the situation described by Fig. 8. Actually, in case (4.32), the model exhibits a genuine symbiotic behavior around $(0, 0)$, while it possesses a high-intensity-competitive behavior for sufficiently large values of λ and μ.

Finally, by choosing sufficiently small b^- and c^+, one can easily reach

$$\hat{b} > 0, \quad \hat{c} < 0, \quad \alpha < 0, \quad \beta > 0, \tag{4.33}$$

for instance, by taking $\ell = m = a = d = 1$. In such cases the curves $\mu = g(\lambda)$ and $\lambda = g(\mu)$ exhibit a behaviour quite reminiscent of the case when v preys on u (cf. Fig. 10); rather naturally, since these models can be regarded as perturbations from the case when $b^- = 0$ and $c^+ = 0$, where $b > 0$ and $c < 0$—v preys on u—though the symbiotic effects of the interaction might play a significant role for large values of λ and μ. In Fig. 10 we have represented the curves of change of stability of the semi-trivial states of (1.2) in one of the several admissible situation cases.

J. López-Gómez

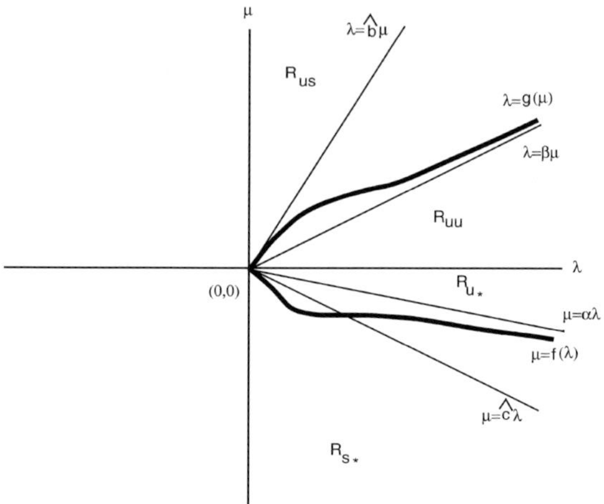

Fig. 10. Curves of neutral stability under condition (4.33).

5. The existence of coexistence states

In this section we are going to give a very general result about the existence of coexistence states of (1.2) for the general case where b and c change of sign. As a matter of fact, our result covers all admissible situations, as we are not imposing any sign restriction on b and c. The general methodology adopted in this section goes back to the pioneering chapter by J.M. Cushing [7], where one of the parameters, λ or μ, was fixed while the other was considered as a bifurcation parameter in order to get the coexistence states as emanating from one of the semi-trivial states in the context of competing species models. This idea has been extremely fruitful in a number of different contexts [19, Chapter 7].

Most precisely, in Section 5.1 we use the abstract unilateral theorem of [19, Chapter 6] to get an abstract global unilateral theorem for systems, which is the basic abstract result to get—as an easy consequence—our theorem about the existence of coexistence states, which will be given in Section 5.2. Our abstract result admits a number of applications in rather different contexts, but we refrain of giving more details here in. It should be noted that the well known, and very celebrated, unilateral theorems of P.H. Rabinowitz [31] are wrong as they are stated. Actually, there is an available—very recent—counterexample by E.N. Dancer [9]. In the subsequent subsections we shall consider each of the possible interactions separately, as in each special circumstance one can obtain sharper consequences from our abstract theorem.

5.1. *Abstract unilateral bifurcation from semi-trivial states*

Let X be an ordered Banach space whose positive cone P is normal and it has nonempty interior, and consider a nonlinear abstract equation of the form

$$\mathfrak{F}(\rho, x, y) := \mathfrak{L}(\rho)(x - x_0, y) + \mathfrak{R}(\rho, x, y) = 0, \quad (\rho, x, y) \in \mathbb{R} \times X^2, \qquad (5.1)$$

for some $x_0 \in P \setminus \{0\}$, where

(HL) $\mathfrak{K}(\rho) := I_{X^2} - \mathfrak{L}(\rho) \in \mathcal{L}(X^2)$, $\rho \in \mathbb{R}$, is a compact and continuous operator pencil with a discrete set of *singular values*, denoted by Σ. Since $\mathfrak{L}(\rho)$ is Fredlhom of index zero, $\sigma \in \Sigma$ if, and only if, σ is an *eigenvalue* of $\mathfrak{L}(\rho)$, i.e., if $\dim N[\mathfrak{L}(\sigma)] \geqslant 1$.

(HR) $\mathfrak{R} \in C(\mathbb{R} \times X^2; X^2)$ is compact on bounded sets and

$$\lim_{(x,y) \to (x_0, 0)} \frac{\mathfrak{R}(\rho, x, y)}{\|(x - x_0, y)\|} = 0$$

uniformly on compact intervals of $\mathbb{R}$.

(HP) The solutions of (5.1) satisfy the *strong maximum principle* in the sense that

$$(\rho, x, y) \in \mathbb{R} \times (P \setminus \{0\}) \times X \quad \text{and} \quad \mathfrak{F}(\rho, x, y) = 0 \quad \text{imply} \quad x \in \operatorname{Int} P,$$

and

$$(\rho, x, y) \in \mathbb{R} \times X \times (P \setminus \{0\}) \quad \text{and} \quad \mathfrak{F}(\rho, x, y) = 0 \quad \text{imply} \quad y \in \operatorname{Int} P,$$

where $\operatorname{Int} P$ stands for the interior of the cone P.

Subsequently, the space X^2 is viewed as an ordered Banach space with positive cone P^2, and, given any ordered Banach space E with positive cone W, and $e_1, e_2 \in E$, we write $e_1 > e_2$ if $e_1 - e_2 \in W \setminus \{0\}$, and $e_1 \gg e_2$ if $e_1 - e_2 \in \operatorname{Int} W$. As

$$\operatorname{Int} P^2 = \operatorname{Int} P \times \operatorname{Int} P,$$

it is said that (ρ, x, y) is a *positive solution* of (5.1), if (ρ, x, y) is a solution of (5.1) with $(x, y) > 0$ ($x \geqslant 0$, $y \geqslant 0$ and $(x, y) \neq (0, 0)$). Thanks to Hypothesis (HP), any positive solution (ρ, x, y) of (5.1) must have one of the following forms: either $x \gg 0$ and $y \gg 0$—*strongly positive*, or $x \gg 0$ and $y = 0$, or $x = 0$ and $y \gg 0$—*semi-trivial positive solutions*.

Under Hypotheses (HL) and (HR),

$$\mathfrak{F}(\rho, x_0, 0) = 0 \quad \text{for each } \rho \in \mathbb{R}, \tag{5.2}$$

and, due to Hypothesis (HP), $x_0 \gg 0$, since $x_0 > 0$. Moreover, since $\mathfrak{L}(\rho)$ is a linear isomorphism for any $\rho \in \mathbb{R} \setminus \Sigma$, the *fixed point index*

$$\operatorname{Ind}\big(0, \mathfrak{K}(\rho)\big)$$

is well defined—the topological degree of $\mathfrak{L}(\rho)$ in any bounded open set containing 0. Thus, the following *crossing number* $C : \Sigma \to \{-1, 0, 1\}$ can be introduced

$$C(\sigma) := \frac{1}{2} \lim_{\varepsilon \downarrow 0} \left[\operatorname{Ind}\big(0, \mathfrak{K}(\sigma + \varepsilon)\big) - \operatorname{Ind}\big(0, \mathfrak{K}(\sigma - \varepsilon)\big) \right], \quad \sigma \in \Sigma. \tag{5.3}$$

Note that $C(\sigma) \neq 0$ if, and only if, $\mathrm{Ind}\,(0, \mathfrak{K}(\rho))$ changes as ρ crosses σ, and that this occurs if, and only if, $\mathrm{Ind}\,((x_0, 0), \mathfrak{F}(\rho, \cdot, \cdot))$ changes as ρ crosses σ.

Our main result concerns the bounded components of strongly positive solutions of (5.1) emanating from the curve $(\rho, x, y) = (\rho, x_0, 0)$ at a singular value $\rho_0 \in \Sigma$ with— classical—geometric multiplicity one and $C(\rho_0) \in \{-1, 1\}$. By a *component* of strongly positive solutions of (5.1) it is meant a maximal (for the inclusion) relatively closed and connected subset of the set of strongly positive solutions of (5.1)—in $\mathbb{R} \times \mathrm{Int}\,P^2$. Thanks to [19, Theorem 6.2.1], Eq. (5.1) possesses a component emanating from $(\rho, x, y) = (\rho, x_0, 0)$ at ρ_0 if $C(\rho_0) \in \{-1, 1\}$. Such a component will be subsequently denoted by $\mathfrak{C}_0$. Throughout the remaining of this section, we shall denote by P_x and P_y the projection operators

$$P_x(x, y) = x, \quad P_y(x, y) = y, \quad (x, y) \in X^2. \tag{5.4}$$

The main abstract result of this section can now be stated as follows.

THEOREM 5.1. *Suppose* $\rho_0 \in \Sigma$ *satisfies* $C(\rho_0) \neq 0$,

$$N\big[\mathfrak{L}(\rho_0)\big] = \mathrm{span}\big[(\varphi_{0x}, \varphi_{0y})\big], \quad \varphi_{0y} \in P \setminus \{0\}, \tag{5.5}$$

and, for each $x \in X$, $P_y \mathfrak{K}(\rho_0)(x, \cdot)$ *is strongly positive, in the sense that*

$$P_y \mathfrak{K}(\rho_0)(x, \cdot)\big(P \setminus \{0\}\big) \subset \mathrm{Int}\,P. \tag{5.6}$$

Then,

$$\mathrm{spr}\,P_y \mathfrak{K}(\rho_0)(\varphi_{0x}, \cdot) = 1 \tag{5.7}$$

and there exists a component $\mathfrak{C}_0^P$ *of* $\mathfrak{C}_0 \cap [\mathbb{R} \times (\mathrm{Int}\,P)^2]$ *such that*

$$(\rho_0, x_0, 0) \in \bar{\mathfrak{C}}_0^P. $$

Suppose, in addition, that $\mathfrak{C}_0^P$ *is bounded in* $\mathbb{R} \times X^2$ *and that*

$$\mathrm{spr}\,P_y \mathfrak{K}(\rho_0)(x, \cdot) = 1 \quad \text{for each } x \in X. \tag{5.8}$$

Then, some of the following alternatives occurs:
    A1. *There exists* $(\rho_1, x_1) \in (\mathbb{R} \times P) \setminus \{(\rho_0, x_0)\}$ *such that* $(\rho_1, x_1, 0) \in \bar{\mathfrak{C}}_0^P$. *In such case, thanks to* (HP), $x_1 \gg 0$ *if* $x_1 > 0$.
    A2. *There exists* $(\rho_1, y_1) \in \mathbb{R} \times P$ *such that* $(\rho_1, 0, y_1) \in \bar{\mathfrak{C}}_0^P$. *In such case, thanks to* (HP), $y_1 \gg 0$ *if* $y_1 > 0$.

Furthermore, alternative A2 *occurs if* $(\rho, x, 0) \in \mathfrak{F}^{-1}(0)$ *implies* $x = x_0$, *and* ρ_0 *is the unique value of* ρ *for which one is an eigenvalue of* $P_y \mathfrak{K}(\rho)(x_0, \cdot)$ *to a positive eigenvector.*

PROOF. The set of *non-trivial solutions* of (5.1) is defined through

$$\mathfrak{S} := \mathfrak{F}^{-1}\big((0,0)\big) \setminus \big[(\mathbb{R} \setminus \Sigma) \times \{(x_0, 0)\}\big]. \tag{5.9}$$

Note that $(\rho, x, y) \in \mathfrak{S}$ if $\mathfrak{F}(\rho, x, y) = 0$ with $(x, y) \neq (x_0, 0)$, or $(\rho, x, y) = (\rho, x_0, 0)$ with $\rho \in \Sigma$. Since $C(\rho_0) \neq 0$, thanks to [19, Theorem 6.2.1] there exists a component of $\mathfrak{S}$, denoted by $\mathfrak{C}_0$, such that $(\rho_0, x_0, 0) \in \mathfrak{C}_0$. Subsequently, we suppose that

$$\varphi_0 := (\varphi_{0x}, \varphi_{0y}) \in X \times \operatorname{Int} P$$

has been normalized so that

$$\|\varphi_0\|_{X^2} = 1.$$

It should be noted that, thanks to (HL),

$$\mathfrak{K}(\rho_0)\varphi_0 = \varphi_0$$

and, hence,

$$P_y \mathfrak{K}(\rho_0)(\varphi_{0x}, \varphi_{0y}) = \varphi_{0y}.$$

Thus, it follows from (5.5) and (5.6) that $\varphi_{0y} \gg 0$. Moreover, thanks to (5.6), (5.7) is a consequence from Krein–Rutman theorem, since the spectral radius is the unique real eigenvalue associated with it there is a positive eigenvector.

Now, let Z be a closed subspace of X^2 such that

$$X^2 = N\big[\mathfrak{L}(\rho_0)\big] \oplus Z.$$

Thanks to Hahn–Banach's theorem, there exists

$$\varphi_0^* := \big(\varphi_{0x}^*, \varphi_{0y}^*\big) \in X' \times X'$$

for which

$$Z = \big\{(x, y) \in X^2 : \langle \varphi_0^*, (x, y) \rangle = 0\big\}, \quad \langle \varphi_0^*, \varphi_0 \rangle = 1,$$

where $\langle \cdot, \cdot \rangle$ stands for the duality between X^2 and $X' \times X'$. Now, adapting the theory developed by P.H. Rabinowitz [31], for each $\eta \in (0, 1)$ and sufficiently small $\varepsilon > 0$ we set

$$\begin{aligned}
\mathcal{Q}_{\varepsilon, \eta} := \big\{(\rho, x, y) \in \mathbb{R} \times X^2 : &|\rho - \rho_0| < \varepsilon, \\
&\big|\langle \varphi_0^*, (x - x_0, y) \rangle\big| > \eta \|(x - x_0, y)\|\big\}.
\end{aligned}$$

Since the mapping

$$(x, y) \mapsto \big|\langle \varphi_0^*, (x - x_0, y) \rangle\big| - \eta \|(x - x_0, y)\|$$

is continuous, $\mathcal{Q}_{\varepsilon,\eta}$ is an open subset of $\mathbb{R} \times X^2$ consisting of the two disjoint components $\mathcal{Q}_{\varepsilon,\eta}^+$ and $\mathcal{Q}_{\varepsilon,\eta}^-$ defined through

$$
\begin{aligned}
\mathcal{Q}_{\varepsilon,\eta}^+ &:= \big\{ (\rho, x, y) \in \mathbb{R} \times X^2 : |\rho - \rho_0| < \varepsilon, \\
&\qquad \langle \varphi_0^*, (x - x_0, y) \rangle > \eta \| (x - x_0, y) \| \big\}, \\
\mathcal{Q}_{\varepsilon,\eta}^- &:= \big\{ (\rho, x, y) \in \mathbb{R} \times X^2 : |\rho - \rho_0| < \varepsilon, \\
&\qquad \langle \varphi_0^*, (x - x_0, y) \rangle < -\eta \| (x - x_0, y) \| \big\}.
\end{aligned}
$$

The following result collects the main consequences from [19, Theorem 6.2.1, Lemma 6.4.1, Proposition 6.4.2]; it is entirely attributable to P.H. Rabinowitz [31].

THEOREM 5.2. *For each sufficiently small* $\delta > 0$,

$$
\mathfrak{C}_0 \cap B_\delta(\rho_0, x_0, 0) \subset \mathcal{Q}_{\varepsilon,\eta} \cup \big\{ (\rho_0, x_0, 0) \big\}
$$

and each of the sets

$$
\mathfrak{S} \setminus \big[\mathcal{Q}_{\varepsilon,\eta}^- \cap B_\delta(\rho_0, x_0, 0) \big] \quad and \quad \mathfrak{S} \setminus \big[\mathcal{Q}_{\varepsilon,\eta}^+ \cap B_\delta(\rho_0, x_0, 0) \big]
$$

contains a component, denoted by $\mathfrak{C}_0^+$ *and* $\mathfrak{C}_0^-$, *respectively, such that*

$$
(\rho_0, x_0, 0) \in \mathfrak{C}_0^+ \cap \mathfrak{C}_0^-
$$

and

$$
\mathfrak{C}_0 \cap B_\delta(\rho_0, x_0, 0) = \big(\mathfrak{C}_0^+ \cup \mathfrak{C}_0^- \big) \cap B_\delta(\rho_0, x_0, 0). \tag{5.10}
$$

Moreover, for each

$$
(\rho, x, y) \in \big[\mathfrak{C}_0 \setminus \big\{ (\rho_0, x_0, 0) \big\} \big] \cap B_\delta(\rho_0, x_0, 0),
$$

there exists a unique $(s, \tilde{x}, \tilde{y}) \in \mathbb{R} \times Z$ *such that* $\rho = \rho(s)$ *and*

$$
(x, y) = (x_0, 0) + s\varphi_0 + (\tilde{x}, \tilde{y}) = \big(x_0 + s\varphi_{0x} + \tilde{x}, \, s\varphi_{0y} + \tilde{y} \big) \tag{5.11}
$$

with $|s| > \eta \| (x - x_0, y) \|$. *Actually,*

$$
\rho = \rho_0 + \mathrm{o}(1) \quad and \quad (\tilde{x}, \tilde{y}) = \mathrm{o}(s), \quad as \; s \to 0. \tag{5.12}
$$

Note that if

$$
(\rho, x, y) \in \mathfrak{C}_0^+ \cap B_\delta(\rho_0, x_0, 0), \quad (x, y) \neq (x_0, 0),
$$

then (5.11) holds with $s > \eta \, \|(x - x_0, y)\| > 0$ and, hence,

$$s^{-1} y = \varphi_{0y} + s^{-1} \tilde{y}.$$

Thus, due to (5.12), for sufficiently small $s > 0$ we have that $s^{-1} y \in \operatorname{Int} P$, since $\varphi_{0y} \gg 0$, and, hence, $y \gg 0$. The fact that $x \gg 0$ is an easy consequence from (5.12) and the fact that $x_0 \gg 0$. Therefore, for any sufficiently small $\delta > 0$ we have that

$$\left[\mathfrak{C}_0^+ \setminus \{(\rho_0, x_0, 0)\} \right] \cap B_\delta(\rho_0, x_0, 0) \subset \mathbb{R} \times \operatorname{Int} P^2. \tag{5.13}$$

This shows the existence of the component $\mathfrak{C}_0^P$. Actually, $\mathfrak{C}_0^P$ is the maximal sub-continuum of $\mathfrak{C}_0^+$ in $\mathbb{R} \times \operatorname{Int} P^2$ such that

$$(\rho_0, x_0, 0) \in \bar{\mathfrak{C}}_0^P.$$

If $\mathfrak{C}_0^P$ is unbounded in $\mathbb{R} \times X^2$ the proof is concluded. So, for the remaining of the proof we will assume that $\mathfrak{C}_0^P$ is bounded.

The following result, which is [19, Theorem 6.4.3], provides us with an updated version of the very celebrated unilateral theorem of P.H. Rabinowitz [31, Theorem 1.27] which is necessary to conclude the proof of Theorem 5.1. It should be noted that [31, Theorem 1.27] is wrong as originally stated (cf. the detailed discussion carried out in [19, p. 180] and the counterexample of E.N. Dancer [9]).

THEOREM 5.3. *The component* $\mathfrak{C}_0^+$ *satisfies some of the following alternatives:*
1. $\mathfrak{C}_0^+$ *is unbounded in* $\mathbb{R} \times X^2$.
2. *There exists* $\rho_1 \in \Sigma \setminus \{\rho_0\}$ *such that* $(\rho_1, x_0, 0) \in \mathfrak{C}_0^+$.
3. *There exists* $(\rho, x, y) \in \mathfrak{C}_0^+$ *such that*

$$(\rho, x - x_0, y) \in \mathbb{R} \times \left(Z \setminus \{0\} \right).$$

Now, we shall prove that

$$X^2 = N\left[\mathfrak{L}(\rho_0)\right] \oplus R\left[\mathfrak{L}(\rho_0)\right]. \tag{5.14}$$

Since $\mathfrak{L}(\rho_0)$ is Fredholm of index zero, to show (5.14) it suffices to prove that $\varphi_0 \notin R[\mathfrak{L}(\rho_0)]$. This will be accomplished by contradiction. Suppose, there exists $(x, y) \in X^2$ such that

$$\mathfrak{L}(\rho_0)(x, y) = (x, y) - \mathfrak{K}(\rho_0)(x, y) = \varphi_0 = (\varphi_{0x}, \varphi_{0y}).$$

Then,

$$y - P_y \mathfrak{K}(\rho_0)(x, y) = \varphi_{0y} \gg 0. \tag{5.15}$$

Note that, thanks to (HL) and (5.6), the map

$$y \mapsto P_y \mathfrak{K}(\rho_0)(x, y)$$

defines a linear compact strongly order preserving operator. Moreover, by construction,

$$P_y \mathfrak{K}(\rho_0)(\varphi_{0x}, \varphi_{0y}) = \varphi_{0y}$$

and, due to (5.8),

$$\operatorname{spr} P_y \mathfrak{K}(\rho_0)(x, \cdot) = 1.$$

Let $\psi_x \gg 0$ denote the principal eigenfunction associated to $P_y \mathfrak{K}(\rho_0)(x, \cdot)$. Then, thanks to (5.15), for each $\beta > 0$ we have that

$$y + \beta \psi_x - P_y \mathfrak{K}(\rho_0)(x, y + \beta \psi_x) = \varphi_{0y}.$$

On the other hand, since

$$\lim_{\beta \uparrow \infty} \left(\beta^{-1} y + \psi_x \right) = \psi_x \gg 0,$$

it is apparent that, for each sufficiently large $\beta > 0$, one has that

$$y + \beta \psi_x \gg 0$$

provides us with a positive solution of (5.15). Thanks to H. Amann [4, Theorem 3.2], this is impossible. This contradiction concludes the proof of (5.14). Consequently, in order to apply Theorem 5.3, we can make the special choice

$$Z = R\big[\mathfrak{L}(\rho_0)\big], \tag{5.16}$$

which will be maintained throughout the rest of the proof.

Since $\mathfrak{C}_0^P \subset \mathfrak{C}_0^+$, some of the following alternatives occurs: either

$$\mathfrak{C}_0^P = \mathfrak{C}_0^+ \setminus \big\{(\rho_0, x_0, 0)\big\}, \tag{5.17}$$

or

$$\mathfrak{C}_0^P \text{ is a proper subset of } \mathfrak{C}_0^+ \setminus \big\{(\rho_0, x_0, 0)\big\}. \tag{5.18}$$

Suppose (5.17). Then, $\bar{\mathfrak{C}}_0^P = \mathfrak{C}_0^+$ satisfies some of the alternatives of Theorem 5.3. Alternative 1 cannot be satisfied, since $\bar{\mathfrak{C}}_0^P$ is compact. Suppose alternative 3 occurs. Then, thanks to the choice (5.16), there exists

$$(\rho, x, y) \in \mathfrak{C}_0^+ = \bar{\mathfrak{C}}_0^P \subset \mathbb{R} \times P^2$$

such that

$$(\rho, x - x_0, y) \in \mathbb{R} \times R\big[\mathfrak{L}(\rho_0)\big], \quad y \neq 0.$$

Since $(\rho, x, y) \in \mathbb{R} \times P^2$ with $y \neq 0$, necessarily $y > 0$ and, consequently, $y \gg 0$, by (HP). Thus, there exists $(\tilde{x}, \tilde{y}) \in X^2$ such that

$$\mathfrak{L}(\rho_0)(\tilde{x}, \tilde{y}) = (\tilde{x}, \tilde{y}) - \mathfrak{K}(\rho_0)(\tilde{x}, \tilde{y}) = (x - x_0, y)$$

and, taking projections,

$$\tilde{y} - P_y \mathfrak{K}(\rho_0)\big(\tilde{x}, \tilde{y}\big) = y \gg 0.$$

Arguing as above, this is impossible, because, due to (5.8),

$$\operatorname{spr} P_y \mathfrak{K}(\rho_0)(\tilde{x}, \cdot) = 1.$$

Thus, alternative 2 of Theorem 5.3 must be satisfied. Consequently, there exists $\rho_1 \in \Sigma \setminus \{\rho_0\}$ such that

$$(\rho_1, x_0, 0) \in \mathfrak{C}_0^+ = \bar{\mathfrak{C}}_0^P$$

and, therefore, alternative A1 of the theorem is satisfied.

Now, suppose (5.18), instead of (5.17). Then, since

$$\mathfrak{C}_0^+ \cap B_\delta(\rho_0, x_0, 0) = \big[\mathfrak{C}_0^P \cap B_\delta(\rho_0, x_0, 0)\big] \cup \big\{(\rho_0, x_0, 0)\big\}$$

for each sufficiently small $\delta > 0$, fixing one of these δ's, there exists

$$(\rho_1, x_1, y_1) \notin B_\delta(\rho_0, x_0, 0) \tag{5.19}$$

such that

$$(\rho_1, x_1, y_1) \in \mathfrak{C}_0^+ \cap \big(\mathbb{R} \times \partial P^2\big) \cap \partial \mathfrak{C}_0^P,$$

where, given any set U, ∂U stands for the boundary of U. Then,

$$\mathfrak{F}(\rho_1, x_1, y_1) = 0 \quad \text{and} \quad (x_1, y_1) \in P^2.$$

If $x_1 > 0$ and $y_1 > 0$, then, due to (HP), $x_1 \gg 0$ and $y_1 \gg 0$, which is impossible, since $(x_1, y_1) \in \partial P^2$. Thus, either $x_1 = 0$, or $y_1 = 0$. Suppose $y_1 = 0$. Then, thanks to (5.19), $(\rho_1, x_1) \neq (\rho_0, x_0)$, and, therefore, alternative A1 is satisfied again. Obviously, if $x_1 = 0$, then the alternative A2 is satisfied.

Finally, suppose that $(\rho, x, 0) \in \mathfrak{F}^{-1}(0)$ implies $x = x_0$, and ρ_0 is the unique value of ρ for which one is an eigenvalue of $P_y \mathfrak{K}(\rho)(x_0, \cdot)$ to a positive eigenvector. To conclude the proof of the theorem we proceed by contradiction. Assume alternative A1 is satisfied. Then, there exists $(\rho_1, x_1) \in (\mathbb{R} \times P) \setminus \{(\rho_0, x_0)\}$ such that $(\rho_1, x_1, 0) \in \bar{\mathfrak{C}}_0^P$. Thus, $x_1 = x_0$, since $(\rho_1, x_1, 0) \in \mathfrak{F}^{-1}(0)$, and, hence,

$$\rho_1 \neq \rho_0. \tag{5.20}$$

Let $\{(\gamma_n, u_n, v_n)\}_{n \geq 1}$ be any subsequence of $\mathfrak{C}_0^P$ such that

$$\lim_{n \to \infty} (\gamma_n, u_n, v_n) = (\rho_1, x_0, 0). \tag{5.21}$$

Then, for each $n \geq 1$, $v_n \gg 0$, and, setting,

$$\psi_n := \frac{v_n}{\|v_n\|},$$

we find from $\mathfrak{F}(\gamma_n, u_n, v_n) = 0$ that

$$\psi_n = P_y \mathfrak{K}(\gamma_n)(u_n, \psi_n) - \frac{P_y \mathfrak{R}(\gamma_n, u_n, v_n)}{\|v_n\|}$$

$$= P_y \mathfrak{K}(\rho_1)(u_n, \psi_n) + P_y \big[\mathfrak{K}(\gamma_n) - \mathfrak{K}(\rho_1)\big](u_n, \psi_n) - \frac{P_y \mathfrak{R}(\gamma_n, u_n, v_n)}{\|v_n\|}.$$

As $\{(u_n, \psi_n)\}_{n \geq 1}$ is bounded, since we are assuming that $\mathfrak{C}_0^P$ is bounded, we have

$$\lim_{n \to \infty} P_y \big[\mathfrak{K}(\gamma_n) - \mathfrak{K}(\rho_1)\big](u_n, \psi_n) = 0,$$

by the continuity of $\mathfrak{K}(\rho)$ in ρ. Moreover, thanks to (5.21), it follows from (HR) that

$$\lim_{n \to \infty} \frac{P_y \mathfrak{R}(\gamma_n, u_n, v_n)}{\|v_n\|} = 0.$$

Thus, for each $n \geq 1$, there exists $w_n \in X$ such that

$$\psi_n = P_y \mathfrak{K}(\rho_1)(x_0, \psi_n) + w_n$$

with $\lim_{n \to \infty} w_n = 0$. Consequently, by compactness, there exists a subsequence of $\{\psi_n\}_{n \geq 1}$, again labeled by n, for which the limit

$$\Psi := \lim_{n \to \infty} \psi_n \in X$$

exists and it satisfies

$$\Psi = P_y \mathfrak{K}(\rho_1)(x_0, \Psi). \tag{5.22}$$

Necessarily $\Psi \in P$ and $\|\Psi\| = 1$. Thus, $\Psi > 0$ and, hence, Ψ provides us with a positive eigenvector of $P_y \mathfrak{K}(\rho_1)(x_0, \cdot)$ associated with the eigenvalue 1. By our assumptions, this is impossible, unless $\rho_1 = \rho_0$, which contradicts (5.20). This concludes the proof. $\qquad\square$

5.2. *The existence of coexistence states in the general case*

Subsequently, we fix $\lambda > 0$ and consider μ as the main bifurcation parameter. Also, $X := \mathcal{C}_T$ will be regarded as an ordered Banach space by its cone of non-negative functions, P. Note that P is normal, since every order interval is bounded. Also, note that, given any constant $M > 0$—to be chosen later—$(u, v) \in \mathcal{C}_T^1 \times \mathcal{C}_T^1$ is a solution of (1.2) if and only if,

$$(u, v) \in \mathcal{C}_T \times \mathcal{C}_T \quad \text{and} \quad \mathfrak{F}(\mu, u, v) = 0, \tag{5.23}$$

where

$$\mathfrak{F} : \mathbb{R} \times \mathcal{C}_T \times \mathcal{C}_T \to \mathcal{C}_T \times \mathcal{C}_T$$

is the operator defined by

$$\mathfrak{F}(\mu, u, v) := \begin{pmatrix} u - \mathfrak{R}_M\big[(\lambda\ell + M)u - au^2 - buv\big] \\ v - \mathfrak{R}_M\big[(\mu m + M)v - dv^2 - cuv\big] \end{pmatrix} \tag{5.24}$$

and $\mathfrak{R}_M$ stands for the operator introduced in (2.10),

$$\mathfrak{R}_M := J\left(\frac{\mathrm{d}}{\mathrm{d}t} + M\right)^{-1} : \mathcal{C}_T \to \mathcal{C}_T.$$

Thanks to Proposition 2.1, $\mathfrak{R}_M$ is compact and strongly order preserving. Moreover, for each $\mu \in \mathbb{R}$ we have that

$$\mathfrak{F}(\mu, \theta_{[\lambda\ell, a]}, 0) = 0$$

and, hence, since $\mathfrak{F}$ is real analytic, the general assumptions (HL) and (HR) of Section 5.1 are satisfied by choosing

$$(\rho, x_0, 0) := (\mu, \theta_{[\lambda\ell, a]}, 0), \quad \mu \in \mathbb{R}. \tag{5.25}$$

Moreover, if $\mathfrak{F}(\mu, u, v) = 0$ with $u(t_0) > 0$ for some $t_0 \in \mathbb{R}$, then

$$u(t) = e^{\int_{t_0}^t (\lambda\ell - au - bv)} u(t_0) > 0$$

for each $t \in \mathbb{R}$, since

$$u' = (\lambda\ell - au - bv)u,$$

and, hence, $u \gg 0$. Similarly, $v \gg 0$ if it is somewhere positive. Therefore, $\mathfrak{F}$ satisfies Hypothesis (HP) as well.

Maintaining the notations introduced in Section 5.1, for the choice (5.25) we have that the linearization

$$\mathfrak{L}(\mu) := D_{(u,v)}\mathfrak{F}(\mu, \theta_{[\lambda\ell,a]}, 0)$$

is given, for each $(u, v) \in \mathcal{C}_T \times \mathcal{C}_T$, through

$$\mathfrak{L}(\mu)\begin{pmatrix} u \\ v \end{pmatrix} := \begin{pmatrix} u - \mathfrak{R}_M\big[(\lambda\ell + M - 2a\theta_{[\lambda\ell,a]})u - b\theta_{[\lambda\ell,a]}v\big] \\ v - \mathfrak{R}_M\big[(\mu m + M - c\theta_{[\lambda\ell,a]})v\big] \end{pmatrix} \tag{5.26}$$

and, therefore,

$$\mathfrak{K}(\mu)\begin{pmatrix} u \\ v \end{pmatrix} := \begin{pmatrix} \mathfrak{R}_M\big[(\lambda\ell + M - 2a\theta_{[\lambda\ell,a]})u - b\theta_{[\lambda\ell,a]}v\big] \\ \mathfrak{R}_M\big[(\mu m + M - c\theta_{[\lambda\ell,a]})v\big] \end{pmatrix}.$$

Suppose $(u, v) \in N[\mathfrak{L}(\mu)]$ for some $\mu \in \mathbb{R}$ and $(u, v) \neq (0, 0)$. Then, $u, v \in \mathcal{C}_T^1$, and

$$\begin{cases} u' = (\lambda\ell - 2a\theta_{[\lambda\ell,a]})u - b\theta_{[\lambda\ell,a]}v, \\ v' = (\mu m - c\theta_{[\lambda\ell,a]})v. \end{cases} \tag{5.27}$$

Suppose $v = 0$. Then, $u \neq 0$ and the second equation of (5.27) holds true. Moreover, the first one reduces to

$$u' = (\lambda\ell - 2a\theta_{[\lambda\ell,a]})u. \tag{5.28}$$

Thanks to (2.28), we have that

$$\int_0^T (\lambda\ell - a\theta_{[\lambda\ell,a]}) = 0$$

and, hence,

$$\int_0^T (\lambda\ell - 2a\theta_{[\lambda\ell,a]}) = -\int_0^T a(t)\theta_{[\lambda\ell,a]}(t)\,\mathrm{d}t < 0. \tag{5.29}$$

Therefore, (5.28) cannot admit a non-trivial T-periodic solution. This contradiction shows that, necessarily, $v \neq 0$, which implies

$$\mu = \mu_0 := \frac{1}{T}\int_0^T c(t)\theta_{[\lambda\ell,a]}(t)\,\mathrm{d}t, \tag{5.30}$$

since $\hat{m} = 1$. Actually, in this case, there exists a constant $A \neq 0$ such that

$$v = A\varphi_v, \qquad \varphi_v(t) := \mathrm{e}^{\int_0^t (\mu_0 m - c\theta_{[\lambda\ell,a]})}, \ t \in \mathbb{R}. \tag{5.31}$$

Thanks to (5.29), it follows from Proposition 2.1 that the problem

$$u' = (\lambda\ell - 2a\theta_{[\lambda\ell,a]})u - b\theta_{[\lambda\ell,a]}\varphi_v, \quad u \in C_T^1, \tag{5.32}$$

possesses a unique solution. Namely,

$$\varphi_u := -\Re_{2a\theta_{[\lambda\ell,a]}-\lambda\ell}(b\theta_{[\lambda\ell,a]}\varphi_v). \tag{5.33}$$

Consequently, $\mathfrak{L}(\mu)$ is an isomorphism if $\mu \neq \mu_0$, whereas

$$N\big[\mathfrak{L}(\mu_0)\big] = \mathrm{span}\,\big[(\varphi_u, \varphi_v)\big]. \tag{5.34}$$

In particular, using the notations introduced in Section 5.1, this entails $\Sigma = \{\mu_0\}$. Note that, since $\varphi_v \gg 0$, condition (5.5) is satisfied. Also, for each $u, v \in X = C_T$, we have that

$$P_v\Re(\mu_0)(u, v) = \Re_M\big[(\mu_0 m + M - c\theta_{[\lambda\ell,a]})v\big], \tag{5.35}$$

which is independent of u. Now, choosing M sufficiently large so that

$$\mu_0 m + M - c\theta_{[\lambda\ell,a]} \gg 0,$$

it follows from Proposition 2.1 that $P_v\Re(\mu_0)(u, v) \gg 0$ if $v > 0$. Therefore, (5.6) and (5.7) are satisfied. Actually, since $P_v\Re(\mu_0)(u, \cdot)$ is constant in u, condition (5.8) is as well satisfied. Furthermore, since $(\theta_{[\lambda\ell,a]}, 0)$ is the unique semi-trivial solution of (1.2) for each $\mu \in \mathbb{R}$ and 1 cannot be an eigenvalue of $P_v\Re(\mu)(0, \cdot)$, unless $\mu = \mu_0$—we have already seen it in the proof of (5.34)—to show that all assumptions of Theorem 5.1 are satisfied it remains to prove that $C(\mu_0) \neq 0$. To prove it, one can argue as follows. Note that $\mathfrak{L}(\mu)$ is analytic in μ and that, for each $u, v \in C_T$, we have

$$\frac{\mathrm{d}\mathfrak{L}}{\mathrm{d}\mu}(\mu)(u, v) = \big(0, -\Re_M(mv)\big).$$

We now show that the following holds:

$$\frac{\mathrm{d}\mathfrak{L}}{\mathrm{d}\mu}(\mu_0)(\varphi_u, \varphi_v) = \big(0, -\Re_M(m\varphi_v)\big) \notin R\big[\mathfrak{L}(\mu_0)\big]. \tag{5.36}$$

Indeed, if (5.36) is not true, then there exists $v \in C_T$ such that

$$-\Re_M(m\varphi_v) = v - \Re_M\big[(\mu_0 m + M - c\theta_{[\lambda\ell,a]})v\big].$$

Equivalently,

$$v' = (\mu_0 m - \theta_{[\lambda\ell,a]})v - m\varphi_v, \quad v \in C_T. \tag{5.37}$$

Thanks to (5.30), it follows from Proposition 2.1(d) that (5.37) does not admit a solution, since $m\varphi_v > 0$. This contradiction shows (5.36). It should be noted that (5.36) is the classical transversality condition of M.G. Crandall and P.H. Rabinowitz [6]. Since $\mathfrak{L}(\mu_0)$ is Fredholm of index zero, because it is a compact perturbation of the identity, it is apparent from (5.36) that

$$\frac{d\mathfrak{L}}{d\mu}(\mu_0)\left(N\left[\mathfrak{L}(\mu_0)\right]\right) \oplus R\left[\mathfrak{L}(\mu_0)\right] = \mathcal{C}_T \times \mathcal{C}_T. \tag{5.38}$$

Thus, μ_0 is a 1-transversal eigenvalue of the analytic family $\mathfrak{L}(\mu)$, $\mu \in \mathbb{R}$, in the sense of [19, Definition 4.2.1]. Hence, the algebraic multiplicity defined there in equals one, i.e.,

$$\chi\left[\mathfrak{L}(\mu); \mu_0\right] = 1,$$

and, therefore, thanks to [19, Theorem 5.3.1], μ_0 is a pole of order one of the resolvent operator $\mathfrak{L}^{-1}(\mu)$, $\mu \neq \mu_0$. Consequently, thanks to [19, Theorem 5.6.2], the crossing number $C(\mu_0)$ defined in (5.3) satisfies

$$C(\mu_0) \in \{-1, 1\}.$$

Therefore, Theorem 5.1 provides us with the following existence result.

THEOREM 5.4. *Fix $\lambda > 0$ and consider μ as the main bifurcation parameter. Then, Problem* (1.2) *possesses a component of coexistence states*

$$\mathfrak{C}_\mu^\lambda \subset \mathbb{R} \times \operatorname{Int} P \times \operatorname{Int} P$$

such that

$$(\mu_0, \theta_{[\lambda\ell,a]}, 0) \in \bar{\mathfrak{C}}_\mu^\lambda, \tag{5.39}$$

where μ_0 is defined through (5.30). *Moreover, if $\mathfrak{C}_\mu^\lambda$ is bounded, then there exists $\mu_* > 0$ such that*

$$\lambda = \frac{1}{T}\int_0^T b(t)\theta_{[\mu_* m,d]}(t)\,dt \tag{5.40}$$

and

$$(\mu_*, 0, \theta_{[\mu_* m,d]}) \in \bar{\mathfrak{C}}_\mu^\lambda. \tag{5.41}$$

Therefore, $\mathfrak{C}_\mu^\lambda$ links the semi-trivial states $(\mu_0, \theta_{[\lambda\ell,a]}, 0)$ and $(\mu_*, 0, \theta_{[\mu_* m,d]})$ if it is bounded.

PROOF. The existence of $\mathfrak{C}_\mu^\lambda$ bifurcating from the curve $(\mu, u, v) = (\mu, \theta_{[\lambda\ell,a]}, 0)$ at $\mu = \mu_0$ is an immediate consequence from Theorem 5.1.

Suppose $\mathfrak{C}_\mu^\lambda$ is bounded. Then, it satisfies alternative A2 of Theorem 5.1 and, hence, there exists $(\mu_*, v_*) \in \mathbb{R} \times P$ such that

$$(\mu_*, 0, v_*) \in \bar{\mathfrak{C}}_\mu^\lambda.$$

Thus, there exists a sequence of coexistence states of (1.2), say $\{(\mu_n, u_n, v_n)\}_{n \geqslant 1}$, such that

$$\lim_{n \to \infty} (\mu_n, u_n, v_n) = (\mu_*, 0, v_*).$$

For each $n \geqslant 1$, we have that

$$u_n = \mathfrak{R}_M\left[(\lambda\ell + M)u_n - au_n^2 - bu_n v_n\right].$$

Thus, setting

$$U_n := \frac{u_n}{\|u_n\|_{C_T}}, \quad n \geqslant 1,$$

gives

$$U_n = \mathfrak{R}_M\left[(\lambda\ell + M)U_n - au_n U_n - bv_n U_n\right], \quad n \geqslant 1. \tag{5.42}$$

Since $\{(\lambda\ell + M)U_n - au_n U_n - bv_n U_n\}_{n \geqslant 1}$ is bounded in C_T and $\mathfrak{R}_M$ is compact, there exists a subsequence of U_n, labeled again by n, such that

$$\lim_{n \to \infty} U_n = \Phi.$$

Necessarily, $\|\Phi\|_{C_T} = 1$, and, hence, $\Phi > 0$, since $u_n \gg 0, n \geqslant 1$. Moreover, passing to the limit as $n \to \infty$ in (5.42), we find that

$$\Phi = \mathfrak{R}_M\left[(\lambda\ell + M - bv_*)\Phi\right].$$

Thus, $\Phi \in C_T^1$, $\Phi > 0$, and

$$\Phi' = (\lambda\ell - bv_*)\Phi. \tag{5.43}$$

Suppose $v_* = 0$, then, we find from (5.43) that $\lambda\hat{\ell} = \lambda = 0$, which is impossible, since we are assuming $\lambda > 0$. Thus, $v_* > 0$. Actually, since $(\mu_*, 0, v_*)$ solves (1.2), necessarily

$$\mu_* > 0 \quad \text{and} \quad v_* = \theta_{[\mu_* m, d]} \gg 0.$$

Consequently, (5.43) becomes into

$$\Phi' = (\lambda\ell - b\theta_{[\mu_*m,d]})\Phi, \tag{5.44}$$

and, therefore,

$$\lambda = \frac{1}{T}\int_0^T b(t)\theta_{[\mu_*m,d]}(t)\,dt,$$

which concludes the proof of the theorem. $\qquad\square$

By symmetry, if $\mu > 0$ is fixed and $\lambda \in \mathbb{R}$ is regarded as the main bifurcation parameter, then the following result is satisfied.

THEOREM 5.5. *Fix $\mu > 0$ and consider λ as the main bifurcation parameter. Then, Problem (1.2) possesses a component of coexistence states*

$$\mathfrak{C}_\lambda^\mu \subset \mathbb{R} \times \operatorname{Int} P \times \operatorname{Int} P$$

such that

$$(\lambda_0, 0, \theta_{[\mu m,d]}) \in \bar{\mathfrak{C}}_\lambda^\mu, \quad \lambda_0 := \frac{1}{T}\int_0^T b(t)\theta_{[\mu m,d]}(t)\,dt. \tag{5.45}$$

Moreover, if $\mathfrak{C}_\lambda^\mu$ is bounded, then there exists $\lambda_* > 0$ such that

$$\mu = \frac{1}{T}\int_0^T c(t)\theta_{[\lambda_*\ell,a]}(t)\,dt \tag{5.46}$$

and

$$(\lambda_*, \theta_{[\lambda_*\ell,a]}, 0) \in \bar{\mathfrak{C}}_\lambda^\mu. \tag{5.47}$$

Therefore, $\mathfrak{C}_\lambda^\mu$ links the semi-trivial states $(\lambda_0, 0, \theta_{[\mu m,d]})$ and $(\lambda_*, \theta_{[\lambda_*\ell,a]}, 0)$ if it is bounded.

5.3. *Local structure of $\mathfrak{C}_\mu^\lambda$ and $\mathfrak{C}_\lambda^\mu$ at $(\mu_0, \theta_{[\lambda\ell,a]}, 0)$ and $(\lambda_0, 0, \theta_{[\mu m,d]})$*

The following result shows that, for each $\lambda > 0$, $\mathfrak{C}_\mu^\lambda$ consists of a real analytic curve of coexistence states of (1.2) in a neighborhood of

$$(\mu, u, v) = (\mu_0, \theta_{[\lambda\ell,a]}, 0).$$

Although this is a consequence from the celebrated theorem of M.G. Crandall and P.H. Rabinowitz [6], we shall give a self-contained proof—whose uniqueness part is based on Theorem 5.2—in order to calculate the bifurcation direction of the curve of coexistence states. Throughout this section, all notation introduced in Section 5.2 will be maintained.

THEOREM 5.6. *Fix $\lambda > 0$ and regard to $\mu \in \mathbb{R}$ as the main bifurcation parameter. Then, there exist $\varepsilon > 0$ and two real analytic maps*

$$\mu : (-\varepsilon, \varepsilon) \to \mathbb{R}, \qquad (u_1, v_1) : (-\varepsilon, \varepsilon) \to R\big[\mathfrak{L}(\mu_0)\big] \times R\big[\mathfrak{L}(\mu_0)\big], \qquad (5.48)$$

such that

$$\mu(0) = \mu_0 = \frac{1}{T} \int_0^T c(t)\,\theta_{[\lambda\ell,a]}(t)\,\mathrm{d}t, \quad \big(u_1(0), v_1(0)\big) = (0, 0), \qquad (5.49)$$

and, for each $s \in (-\varepsilon, \varepsilon)$,

$$\mathfrak{F}\big(\mu(s), \theta_{[\lambda\ell,a]} + s\big[\varphi_u + u_1(s)\big], s\big[\varphi_v + v_1(s)\big]\big) = 0, \qquad (5.50)$$

where φ_v and φ_u are the functions defined by (5.31) and (5.33), respectively. Moreover, there exists $\delta > 0$ such that, if B_δ^0 stands for the ball of radius δ centered at $(\mu_0, \theta_{[\lambda\ell,a]}, 0) \in \mathbb{R} \times C_T \times C_T$ and, for each $s \in (-\varepsilon, \varepsilon)$, we set

$$\big(u(s), v(s)\big) := \big(\theta_{[\lambda\ell,a]} + s\big[\varphi_u + u_1(s)\big], s\big[\varphi_v + v_1(s)\big]\big), \qquad (5.51)$$

then

$$\mathfrak{C}_\mu^\lambda \cap B_\delta^0 = \big\{\big(\mu(s), u(s), v(s)\big) : 0 < s < \varepsilon\big\} \cap B_\delta^0. \qquad (5.52)$$

Furthermore,

$$\frac{\mathrm{d}\mu}{\mathrm{d}s}(0) = \frac{1}{T} \int_0^T \big[c(t)\varphi_u(t) + d(t)\varphi_v(t)\big]\,\mathrm{d}t. \qquad (5.53)$$

PROOF. Consider the operator

$$\mathfrak{G} : \mathbb{R}^2 \times R\big[\mathfrak{L}(\mu_0)\big] \to C_T \times C_T$$

defined, for each $(s, \mu, u_1, v_1) \in \mathbb{R}^2 \times R[\mathfrak{L}(\mu_0)]$, by

$$\mathfrak{G}(s, \mu, u_1, v_1) := \begin{cases} s^{-1}\mathfrak{F}\big(\mu, \theta_{[\lambda\ell,a]} + s(\varphi_u + u_1), s(\varphi_v + v_1)\big), & \text{if } s \neq 0, \\ D_{(u,v)}\mathfrak{F}(\mu, \theta_{[\lambda\ell,a]}, 0)(\varphi_u + u_1, \varphi_v + v_1), & \text{if } s = 0. \end{cases}$$

The operator $\mathfrak{G}$ is real analytic and it satisfies

$$\mathfrak{G}(0, \mu_0, 0, 0) = D_{(u,v)}\mathfrak{F}(\mu_0, \theta_{[\lambda\ell,a]}, 0)(\varphi_u, \varphi_v) = \mathfrak{L}(\mu_0)(\varphi_u, \varphi_v) = (0, 0).$$

Moreover, for each $(\mu, u_1, v_1) \in \mathbb{R} \times R[\mathfrak{L}(\mu_0)]$,

$$D_{(\mu,u_1,v_1)}\mathfrak{G}(0, \mu_0, 0, 0)(\mu, u_1, v_1) = \lim_{h \to 0} \left[h^{-1}\mathfrak{G}(0, \mu_0 + h\mu, hu_1, hv_1) \right]$$

$$= \lim_{h \to 0} \left[h^{-1} D_{(u,v)}\mathfrak{F}(\mu_0 + h\mu, \theta_{[\lambda\ell,a]}, 0)(\varphi_u + hu_1, \varphi_v + hv_1) \right]$$

$$= \lim_{h \to 0} \left[h^{-1}\mathfrak{L}(\mu_0 + h\mu)(\varphi_u + hu_1, \varphi_v + hv_1) \right]$$

$$= \mathfrak{L}(\mu_0)(u_1, v_1) + \mu \frac{d\mathfrak{L}}{d\mu}(\mu_0)(\varphi_u, \varphi_v).$$

Thanks to the proof of Theorem 5.1, we already know that

$$N\big[\mathfrak{L}(\mu_0)\big] \oplus R\big[\mathfrak{L}(\mu_0)\big] = C_T \times C_T,$$

though this easily follows from the fact that $(\varphi_u, \varphi_v) \notin R[\mathfrak{L}(\mu_0)]$. Therefore,

$$\mathfrak{L}(\mu_0)|_{R[\mathfrak{L}(\mu_0)]} : R\big[\mathfrak{L}(\mu_0)\big] \to R\big[\mathfrak{L}(\mu_0)\big]$$

is a linear isomorphism and, hence, it follows from (5.36) that

$$D_{(\mu,u_1,v_1)}\mathfrak{G}(0, \mu_0, 0, 0) : \mathbb{R} \times R\big[\mathfrak{L}(\mu_0)\big] \to C_T \times C_T$$

is a linear isomorphism as well. Consequently, thanks to the implicit function theorem, there exist $\varepsilon > 0$ and the real analytic maps (5.48), satisfying (5.49), such that, for each $s \in (-\varepsilon, \varepsilon)$,

$$\mathfrak{G}\big(s, \mu(s), u_1(s), v_1(s)\big) = 0.$$

Moreover, if

$$\mathfrak{G}(s, \mu, u_1, v_1) = 0$$

for some (s, μ, u_1, v_1) sufficiently close to $(0, \mu_0, 0, 0)$, then

$$(s, \mu, u_1, v_1) = \big(s, \mu(s), u_1(s), v_1(s)\big).$$

It should be noted that, by the definition of $\mathfrak{G}$ itself, for each $s \in (-\varepsilon, \varepsilon)$, (5.50) is satisfied. Now, we will prove the uniqueness result.

Thanks to Theorem 5.2, there exists $\delta > 0$ such that, for each

$$(\mu, u, v) \in \big[\mathfrak{C}_\mu^\lambda \setminus \{(\mu_0, \theta_{[\lambda\ell,a]}, 0)\}\big] \cap B_\delta,$$

there exists a unique $(s, \tilde{u}, \tilde{v}) \in \mathbb{R} \times R[\mathfrak{L}(\mu_0)]$ such that

$$\mu = \mu(s), \quad (u, v) = (\theta_{[\lambda\ell,a]}, 0) + s(\varphi_u, \varphi_v) + \big(\tilde{u}, \tilde{v}\big), \tag{5.54}$$

with $s > \eta \, \|(u - \theta_{[\lambda\ell,a]}, v)\| > 0$. Moreover,

$$\mu = \mu_0 + o(1) \quad \text{and} \quad (\tilde{u}, \tilde{v}) = o(s) \quad \text{as } s \to 0. \tag{5.55}$$

So, setting

$$U(s) := s^{-1}\tilde{u}(s) = o(1), \quad V(s) := s^{-1}\tilde{v}(s) = o(1),$$

we have that $(U(s), V(s)) \in R[\mathfrak{L}(\mu_0)]$ and

$$\mathfrak{F}\big(\mu(s), \theta_{[\lambda\ell,a]} + s(\varphi_u + U(s)), s(\varphi_s + V(s))\big) = 0.$$

Thus,

$$\mathfrak{G}\big(s, \mu(s), U(s), V(s)\big) = 0$$

and, therefore, by the uniqueness obtained as an application of the implicit function theorem to $\mathfrak{G}$, if δ is taken sufficiently small, we have that

$$\big(U(s), V(s)\big) = \big(u_1(s), v_1(s)\big)$$

and, consequently,

$$(\mu, u, v) = \big(\mu(s), u(s), v(s)\big),$$

which concludes the proof of (5.52).

To prove (5.53), consider the second equation of (5.50). Then, after division by s, gives

$$\varphi_v + s\frac{dv_1}{ds}(0) + O(s^2)$$

$$= \mathfrak{R}_M\bigg\{\bigg[\bigg(\mu_0 + s\frac{d\mu}{ds}(0) + O(s^2)\bigg)m + M\bigg]\bigg(\varphi_v + s\frac{dv_1}{ds}(0) + O(s^2)\bigg)$$

$$- ds\bigg(\varphi_v + s\frac{dv_1}{ds}(0) + O(s^2)\bigg)^2 - c\big(\theta_{[\lambda\ell,a]} + s\varphi_u + O(s^2)\big)$$

$$\times \bigg(\varphi_v + s\frac{dv_1}{ds}(0) + O(s^2)\bigg)\bigg\}.$$

By definition,

$$\varphi_v = \mathfrak{R}_M\big\{(\mu_0 m + M - c\theta_{[\lambda\ell,a]})\varphi_v\big\}.$$

Thus, dividing by s the previous identity, passing to the limit as $s \to 0$, and rearranging terms yields to

$$\frac{dv_1}{ds}(0) = \mathfrak{R}_M\bigg[(\mu_0 m + M - c\theta_{[\lambda\ell,a]})\frac{dv_1}{ds}(0) + \frac{d\mu}{ds}(0)m\varphi_v - d\varphi_v^2 - c\varphi_u\varphi_v\bigg],$$

or, equivalently,

$$\frac{d}{dt}\frac{dv_1}{ds}(0) = (\mu_0 m - c\theta_{[\lambda\ell,a]})\frac{dv_1}{ds}(0) + \frac{d\mu}{ds}(0)m\varphi_v - d\varphi_v^2 - c\varphi_u\varphi_v.$$

Thus, thanks to Proposition 2.1(d),

$$\int_0^T e^{-\int_0^t (\mu_0 m - c\theta_{[\lambda\ell,a]})}\left(\frac{d\mu}{ds}(0)m(t)\varphi_v(t) - d(t)\varphi_v^2(t) - c(t)\varphi_u(t)\varphi_v(t)\right) dt = 0,$$

and, hence, (5.31) entails

$$\int_0^T \left(\frac{d\mu}{ds}(0)m(t) - d(t)\varphi_v(t) - c(t)\varphi_u(t)\right) dt = 0.$$

Therefore,

$$\frac{d\mu}{ds}(0) = \frac{1}{T}\int_0^T \left[c(t)\varphi_u(t) + d(t)\varphi_v(t)\right] dt,$$

since $\hat{m} = 1$. This concludes the proof. $\qquad\square$

By symmetry, for each $\mu > 0$, $\mathfrak{C}_\lambda^\mu$ consists of a real analytic curve of coexistence states of (1.2) in a neighborhood of $(\lambda, u, v) = (\lambda_0, 0, \theta_{[\mu m, d]})$. Actually, setting

$$\lambda_0 := \frac{1}{T}\int_0^T b(t)\theta_{[\mu m, d]}(t)\, dt, \qquad \tilde{\varphi}_u(t) := e^{\int_0^t (\lambda_0\ell - b\theta_{[\mu m, d]})}, \quad t \in \mathbb{R}, \qquad (5.56)$$

and

$$\tilde{\varphi}_v := -\mathfrak{R}_{2d\theta_{[\mu m, d]} - \mu m}(c\theta_{[\mu m, d]}\tilde{\varphi}_u), \qquad (5.57)$$

there exists $\delta > 0$ such that in the ball of radius $\delta > 0$ centered at $(\lambda_0, 0, \theta_{[\mu m, d]})$ the component $\mathfrak{C}_\lambda^\mu$ consists of a real analytic curve of the form

$$(\lambda, u, v) = \left(\lambda(s), s(\tilde{\varphi}_u + O(s)), \theta_{[\mu m, d]} + s(\tilde{\varphi}_v + O(s))\right), \quad s \sim 0, \ s > 0,$$

where $\lambda(0) = \lambda_0$ and

$$\frac{d\lambda}{ds}(0) = \frac{1}{T}\int_0^T \left[b(t)\tilde{\varphi}_v(t) + a(t)\tilde{\varphi}_u(t)\right] dt. \qquad (5.58)$$

6. The symbiotic model ($b < 0$ and $c < 0$)

In this section we analyze the structure of the set of coexistence states of (1.2) in the special case when $b < 0$ and $c < 0$. Our main result shows that, for each $\lambda > 0$ and $\mu > 0$, the components of coexistence states of (1.2) constructed in Section 5, $\mathfrak{C}_\lambda^\mu$ and $\mathfrak{C}_\mu^\lambda$, are real analytic curves. These curves fiber a real analytic surface entirely formed by coexistence states and linking the corresponding surfaces of semi-trivial solutions along their curves of neutral stability. Also, the stability of the coexistence states along these curves is completely ascertained. It turns out that the stability of the coexistence states along these curves changes at any turning point along the curve, while it remains unchanged when an hysteresis point is passed by. Some of the results of this section are substantial improvements of some results of M.A. Krasnosel'skii [16]. The theory developed in this section can be viewed as a development of many ideas and results going back to [17].

6.1. *Periodic systems of cooperative type*

In this section we study the linear periodic system

$$\begin{cases} u' = \alpha u + \beta v, \\ v' = \gamma u + \rho v, \end{cases} \tag{6.1}$$

where $\alpha, \beta, \gamma, \rho \in C_T$ satisfy

$$\int_0^T \alpha < 0, \quad \int_0^T \rho < 0, \quad \beta > 0, \quad \gamma > 0, \tag{6.2}$$

as well as its associated Cauchy problem

$$\begin{cases} u' = \alpha u + \beta v, \\ v' = \gamma u + \rho v, \\ u(0) = u_0, \quad v(0) = v_0. \end{cases} \tag{6.3}$$

The next result is the key theorem from which most of the results of this section will follow.

THEOREM 6.1. *Suppose* (6.1) *possesses a solution* $(u, v) \in C_T^1 \times C_T^1 \setminus \{(0, 0)\}$. *Then, either* $u \gg 0$ *and* $v \gg 0$, *or else* $u \ll 0$ *and* $v \ll 0$. *Therefore, all the coexistence states of* (1.2) *must be ordered, in the sense that, for any pair of coexistence states,* $(u_1, v_1) \neq (u_2, v_2)$, *either* $u_1 \gg u_2$ *and* $v_1 \gg v_2$, *or* $u_1 \ll u_2$ *and* $v_1 \ll v_2$.

PROOF. If $u = 0$, then $v' = \rho v$, and it follows from Proposition 2.1 that $v = 0$. Thus, $u \neq 0$, since $(u, v) \neq (0, 0)$. As the system is linear, without lost of generality we can assume that $u(t) > 0$ for some $t \in \mathbb{R}$. If $u \gg 0$, then $\gamma u > 0$ and, therefore,

$$v = \mathfrak{R}_{-\rho}(\gamma u) \gg 0,$$

which concludes the proof of the first thesis of the theorem. To show that $u \gg 0$ we will argue by contradiction. Suppose there are $t_0 < t_1$ such that

$$u(t_0) = u(t_1) = 0 \quad \text{and} \quad u(t) > 0 \quad \text{for each } t \in (t_0, t_1). \tag{6.4}$$

Then, for each $t \in \mathbb{R}$, we have that

$$u(t) = \int_{t_0}^{t} e^{\int_{\tau}^{t} \alpha} \beta(\tau) v(\tau) \, d\tau \tag{6.5}$$

and

$$v(t) = e^{\int_{t_0}^{t} \rho} v(t_0) + \int_{t_0}^{t} e^{\int_{\tau}^{t} \rho} \gamma(\tau) u(\tau) \, d\tau. \tag{6.6}$$

Suppose $v(t_0) \geq 0$. Then, thanks to (6.4) and (6.6), we find that, for each $t \in [t_0, t_1]$,

$$v(t) \geq e^{\int_{t_0}^{t} \rho} v(t_0) \geq 0,$$

since $\gamma u \geq 0$. Thus, (6.4) and (6.5) imply

$$0 = u(t_1) = \int_{t_0}^{t_1} e^{\int_{\tau}^{t_1} \alpha} \beta(\tau) v(\tau) \, d\tau$$

$$\geq \int_{t_0}^{\frac{t_0+t_1}{2}} e^{\int_{\tau}^{t_1} \alpha} \beta(\tau) v(\tau) \, d\tau$$

$$= \exp\left(\int_{\frac{t_0+t_1}{2}}^{t_1} \alpha\right) \int_{t_0}^{\frac{t_0+t_1}{2}} \exp\left(\int_{\tau}^{\frac{t_0+t_1}{2}} \alpha\right) \beta(\tau) v(\tau) \, d\tau$$

$$= \exp\left(\int_{\frac{t_0+t_1}{2}}^{t_1} \alpha\right) u\left(\frac{t_0 + t_1}{2}\right) > 0,$$

which is impossible. Hence, $v(t_0) < 0$. So, there is $\delta > 0$ such that $t_0 < t_0 + \delta < t_1$, for which $v(t) < 0$ if $t \in [t_0, t_0 + \delta]$. Thus, we find from (6.5) that $u(t) \leq 0$ for each $t \in [t_0, t_0 + \delta]$, which contradicts (6.4). Therefore, necessarily $u \gg 0$, which concludes the proof of the first statement.

Now, suppose $(u_1, v_1) \neq (u_2, v_2)$ are two coexistence states of (1.2). Then,

$$(u_2 - u_1)' = \lambda \ell (u_2 - u_1) - a(u_2 + u_1)(u_2 - u_1) - bu_2 v_2 + bu_1 v_1$$

$$= \left[\lambda \ell - a(u_2 + u_1) - bv_2\right](u_2 - u_1) - bu_1(v_2 - v_1).$$

By symmetry,

$$(v_2 - v_1)' = \left[\mu m - d(v_2 + v_1) - cu_2\right](v_2 - v_1) - cv_1(u_2 - u_1).$$

Thus, the pair

$$(\tilde{U}, V) := (u_2 - u_1, v_2 - v_1) \in \mathcal{C}_T^1 \times \mathcal{C}_T^1$$

satisfies

$$\begin{cases} U' = \left[\lambda\ell - a(u_2 + u_1) - bv_2\right]U - bu_1 V, \\ V' = \left[\mu m - d(v_2 + v_1) - cu_2\right]V - cv_1 U. \end{cases}$$

Since $u_1 \gg 0$, $v_1 \gg 0$, $b < 0$ and $c < 0$, necessarily $-bu_1 > 0$ and $-cv_1 > 0$. Moreover, since $u_2 \gg 0$ and $v_2 \gg 0$ satisfy

$$\frac{u_2'}{u_2} = \lambda\ell - au_2 - bv_2, \qquad \frac{v_2'}{v_2} = \mu m - dv_2 - cu_2,$$

integrating in $[0, T]$ gives

$$\int_0^T (\lambda\ell - au_2 - bv_2) = 0, \qquad \int_0^T (\mu m - dv_2 - cu_2) = 0. \tag{6.7}$$

Thus,

$$\int_0^T \left[\lambda\ell - a(u_2 + u_1) - bv_2\right] = -\int_0^T a(t)u_1(t)\,dt < 0$$

and

$$\int_0^T \left[\mu m - d(v_2 + v_1) - cu_2\right] = -\int_0^T d(t)v_1(t)\,dt < 0.$$

Therefore, by the first part of the theorem, either $U \gg 0$ and $V \gg 0$, or $U \ll 0$ and $V \ll 0$, which concludes the proof. $\qquad\square$

PROPOSITION 6.2. *Suppose* $(\bar{u}, \bar{v}) \in \mathcal{C}^1(\mathbb{R}_+) \times \mathcal{C}^1(\mathbb{R}_+)$, $\mathbb{R}_+ := [0, \infty)$, *satisfy*

$$\begin{cases} \bar{u}' \geqslant \alpha\bar{u} + \beta\bar{v}, \\ \bar{v}' \geqslant \gamma\bar{u} + \rho\bar{v}, \\ \bar{u}(0) > 0, \quad \bar{v}(0) > 0. \end{cases} \tag{6.8}$$

Then, $\bar{u}(t) > 0$ *and* $\bar{v}(t) > 0$ *for each* $t \geqslant 0$.

If $u_0 + v_0 > 0$ *and* $u_0 v_0 = 0$, *then there exists* $t_0 > 0$ *such that the solution* (u, v) *of* (6.3) *satisfies* $u(t) > 0$ *and* $v(t) > 0$ *for each* $t \geqslant t_0$.

PROOF. Setting

$$d := \bar{u}' - \alpha\bar{u} - \beta\bar{v} \geqslant 0, \qquad \delta := \bar{v}' - \gamma\bar{u} - \rho\bar{v} \geqslant 0,$$

and integrating the resulting system show that, for each $t \in \mathbb{R}_+$,

$$\bar{u}(t) = e^{\int_0^t \alpha} \bar{u}(0) + \int_0^t e^{\int_\tau^t \alpha} \left[\beta(\tau)\bar{v}(\tau) + d(\tau) \right] d\tau \tag{6.9}$$

and

$$\bar{v}(t) = e^{\int_0^t \rho} \bar{v}(0) + \int_0^t e^{\int_\tau^t \rho} \left[\gamma(\tau)\bar{u}(\tau) + \delta(\tau) \right] d\tau. \tag{6.10}$$

Now, suppose $\bar{u}(0) > 0$ and $\bar{v}(0) > 0$. Then, either $\bar{u}(t) > 0$ for each $t \in \mathbb{R}_+$, or there exists t^* such that

$$\bar{u}(t^*) = 0 \quad \text{and} \quad \bar{u}(t) > 0 \quad \text{for each } t \in [0, t^*). \tag{6.11}$$

Suppose (6.11) occurs. Then, since $\bar{u}(0) > 0$, $\bar{v}(0) > 0$, $\beta > 0$ and $d \geqslant 0$, it follows from (6.9) that there exists $\tilde{t} \in (0, t^*)$ such that

$$\bar{v}(\tilde{t}) = 0 \quad \text{and} \quad \bar{v}(t) > 0 \quad \text{for each } t \in [0, \tilde{t}),$$

which contradicts (6.10), since $\bar{v}(0) > 0$ and $\gamma \geqslant 0$, $\delta \geqslant 0$ and $\bar{u} \geqslant 0$ in $[0, \tilde{t}]$. Therefore, $\bar{u}(t) > 0$ for each $t \in \mathbb{R}_+$ and, thanks to (6.10), $\bar{v}(t) > 0$ for each $t \in \mathbb{R}_+$. This concludes the proof of the first part of the proposition.

Now, suppose $u_0 > 0$ and $v_0 = 0$, and let (u, v) denote the unique solution of (6.3). Thanks to the first part of the proposition, for each $\varepsilon > 0$ the solution $(u_\varepsilon, v_\varepsilon)$ of (6.1) with initial data $(u_\varepsilon(0), v_\varepsilon(0)) = (u_0, v_0 + \varepsilon)$ satisfies $u_\varepsilon(t) > 0$ and $v_\varepsilon(t) > 0$ for each $t \geqslant 0$. Thus, by continuous dependence with respect to the initial data, passing to the limit as $\varepsilon \downarrow 0$ we find that $u > 0$ and $v \geqslant 0$. Hence, for each $t \geqslant 0$,

$$u(t) = e^{\int_0^t \alpha} u_0 + \int_0^t e^{\int_\tau^t \alpha} \beta(\tau)v(\tau)\, d\tau \geqslant e^{\int_0^t \alpha} u_0 > 0.$$

If $v = 0$ in $\mathbb{R}_+$, then we find from the v-equation that $\gamma u = 0$ in $\mathbb{R}_+$, which is impossible since $\gamma > 0$ and $u \gg 0$. Thus, there exists $t_0 > 0$ such that $v(t_0) > 0$. Therefore, since $u(t_0) > 0$, by applying the first part of this theorem, we find that $u(t) > 0$ and $v(t) > 0$ for each $t \geqslant t_0$, which concludes the proof. $\qquad \square$

Subsequently, we denote by $U(t)$ the fundamental matrix of solutions of (6.1) such that $U(0) = I := I_{\mathbb{R}^2}$. Then, $U(T)$ provides us with the Poincaré map associated to (6.1). Thanks to Liouville's formula,

$$0 < \det U(T) = e^{\int_0^T (\alpha + \rho)} < 1,$$

since $\int_0^T (\alpha + \rho) < 0$. Thus, if we denote by ν_1 and ν_2 the two eigenvalues of $U(T)$, counting algebraic multiplicities, then

$$0 < \nu_1 \nu_2 < 1. \tag{6.12}$$

Actually, the following result is satisfied.

THEOREM 6.3. *The characteristic multipliers of* (6.1), v_1 *and* v_2, *are real and distinct, and satisfy* (6.12). *Moreover, if we order them so that* $0 < v_1 < v_2$, *then,* $0 < v_1 < 1$,

$$N\big[U(T) - v_1 I\big] = \mathrm{span}\big[(x_0, y_0)\big] \quad \text{with } x_0 < 0 \text{ and } y_0 > 0,$$

and

$$N\big[U(T) - v_2 I\big] = \mathrm{span}\big[(x_0, y_0)\big] \quad \text{with } x_0 > 0 \text{ and } y_0 > 0.$$

In particular, (6.1) *is asymptotically stable if* $v_2 < 1$, *neutrally stable if* $v_2 = 1$, *and unstable if* $v_2 > 1$. *Furthermore,*

$$\lambda_0 := -\frac{1}{T} \mathrm{Log}\, v_2 \tag{6.13}$$

provides us with the unique value of $\lambda \in \mathbb{R}$ *for which the eigenvalue problem*

$$\begin{cases} u' = \alpha u + \beta v + \lambda u, \\ v' = \gamma u + \rho v + \lambda v, \end{cases} \tag{6.14}$$

admits a solution $(u, v) \in \mathcal{C}_T^1 \times \mathcal{C}_T^1$ *with* $u \gg 0$ *and* $v \gg 0$.

PROOF. Suppose $u_0 > 0$, $v_0 > 0$, and let $(u(t), v(t))$ denote the unique solution of (6.3). Then, thanks to Proposition 6.2, $u(T) > 0$ and $v(T) > 0$, and, hence, $U(T)$ sends the interior of the first quadrant of the plane into itself. Setting

$$U(T) = \begin{pmatrix} u_{11} & u_{12} \\ u_{21} & u_{22} \end{pmatrix},$$

the invariance of the first quadrant entails $u_{ij} \geqslant 0$ for each $i, j \in \{1, 2\}$. On the other hand, the eigenvalues of $U(T)$ (v_1 and v_2) are the roots of

$$z^2 - (u_{11} + u_{22})z + u_{11}u_{22} - u_{12}u_{21} = 0,$$

which are given through

$$\frac{1}{2}\big(u_{11} + u_{22} \pm \sqrt{\Delta}\big),$$

where

$$\Delta := (u_{11} + u_{22})^2 - 4(u_{11}u_{22} - u_{12}u_{21}) = (u_{11} - u_{22})^2 + 4u_{12}u_{21} \geqslant 0.$$

Thus, both are real. Moreover, since

$$\det U(T) = u_{11}u_{22} - u_{12}u_{21} = v_1 v_2 > 0,$$

necessarily

$$u_{11} > 0, \quad u_{22} > 0, \quad v_1 + v_2 = u_{11} + u_{22}.$$

We now show that $u_{12} > 0$ and $u_{21} > 0$. Indeed, if $u_{12} = 0$, then

$$\{v_1, v_2\} = \{u_{11}, u_{22}\}$$

and, hence,

$$N[U(T) - u_{22}I] = \mathrm{span}[(0, 1)].$$

Consequently, the solution of (6.3) with initial data $(0, 1)$, say (u, v), satisfies $u(nT) = 0$ for each $n \geqslant 1$, which contradicts the second part of Proposition 6.2. This contradiction shows that $u_{12} > 0$. Similarly, $u_{21} > 0$ follows. Thus, $\Delta > 0$ and, therefore, $v_1 \neq v_2$, say $0 < v_1 < v_2$. Now, note that

$$\min\{u_{11}, u_{22}\} < v_2 = \frac{1}{2}\left(u_{11} + u_{22} + \sqrt{\Delta}\right).$$

In case $u_{11} = \min\{u_{11}, u_{22}\}$, $(u_{12}/(v_2 - u_{11}), 1)$ provides us with an eigenvector of $U(T)$ associated to v_2, while, in case $u_{22} = \min\{u_{11}, u_{22}\}$, $(1, u_{21}/(v_2 - u_{22}))$ provides us with an eigenvector of $U(T)$ associated to v_2. In both situations the two components of the eigenvector are positive. Similarly, in case $u_{11} = \min\{u_{11}, u_{22}\}$, $(1, u_{21}/(v_1 - u_{22}))$ is an eigenvector of $U(T)$ associated to v_1 whose second component is negative, since

$$v_1 = u_{11} + u_{22} - v_2 < u_{22}.$$

By symmetry, in case $u_{22} = \min\{u_{11}, u_{22}\}$, $(u_{12}/(v_1 - u_{11}), 1)$ is an eigenvector of $U(T)$ associated to v_1 whose second component is negative. Therefore, the eigenvector of $U(T)$ associated to v_2 can be chosen in the interior of the first quadrant, while the eigenvector associated to v_1 can be chosen in the interior of the second quadrant.

Now, note that the change of variable

$$(u, v) = e^{\lambda t}(x, y)$$

transforms (6.14) into

$$\begin{cases} x' = \alpha x + \beta y, \\ y' = \gamma x + \rho y. \end{cases}$$

Thus, for any $(u_0, v_0) \in \mathbb{R}^2$, the unique solution of (6.14) satisfying

$$\bigl(u(0), v(0)\bigr) = (u_0, v_0)$$

is given through

$$\big(u(t), v(t)\big) = e^{\lambda t}\, U(t)(u_0, v_0).$$

Therefore, the Poincaré map of (6.14) is given by

$$P_T := e^{\lambda T}\, U(T)$$

and, consequently, the multipliers of (6.14) are $e^{\lambda T} v_j$, $j \in \{1, 2\}$. Thus, the eigenvalue problem (6.14) possesses a T-periodic solution if, and only, if

$$\lambda \in \left\{ -\frac{1}{T}\mathrm{Log}\, v_1, \ -\frac{1}{T}\mathrm{Log}\, v_2 \right\}.$$

By the analysis already done, the unique value of λ providing us with a T-periodic solution having both components positive is the one given by (6.13), as the other eigenvalue provides us with a T-periodic solution whose components have contrary sign. This concludes the proof.
$\qquad\square$

Subsequently, the value λ_0 defined by (6.13) will be called the *principal eigenvalue* of (6.14). It should be noted that (6.1) is asymptotically stable if, and only if $\lambda_0 > 0$, whereas it is unstable if, and only if, $\lambda_0 < 0$. The following result provides us with an extremely useful criterion for ascertaining the stability of (6.1).

THEOREM 6.4. *The following properties are satisfied:*
(a) *System* (6.1) *is stable* ($v_2 < 1$) *if, and only if, there exists* $(\bar{u}, \bar{v}) \in C_T^1 \times C_T^1$ *with* $\bar{u} \gg 0$ *and* $\bar{v} \gg 0$ *such that*

$$\begin{cases} \bar{u}' \geqslant \alpha \bar{u} + \beta \bar{v}, \\ \bar{v}' \geqslant \gamma \bar{u} + \rho \bar{v}, \end{cases} \tag{6.15}$$

with some of these inequalities strict.
(b) *System* (6.1) *is unstable* ($v_2 > 1$) *if, and only if, there exits* $(\underline{u}, \underline{v}) \in C_T^1 \times C_T^1$ *with* $\underline{u} \gg 0$ *and* $\underline{v} \gg 0$ *such that*

$$\begin{cases} \underline{u}' \leqslant \alpha \underline{u} + \beta \underline{v}, \\ \underline{v}' \leqslant \gamma \underline{u} + \rho \underline{v}, \end{cases} \tag{6.16}$$

with some of these inequalities strict.

PROOF. First we will prove part (a). If $v_2 < 1$, then the principal eigenvalue of (6.14) is positive. Indeed,

$$\lambda_0 = -\frac{1}{T}\mathrm{Log}\, v_2 > 0$$

and, hence, any T-periodic strongly positive solution associated to λ_0 provides us with a strict supersolution of (6.1). This shows the necessity of the existence of the supersolution. To show the sufficiency of this condition, suppose there exists $(\bar{u}, \bar{v}) \in C_T^1 \times C_T^1$ with $\bar{u} \gg 0$ and $\bar{v} \gg 0$ satisfying (6.15) with some of the inequalities strict. Now, pick

$$(u_0, v_0) \in N\big[U(T) - v_2 I\big]$$

with $u_0 > 0$ and $v_0 > 0$, and let $(u(t), v(t))$ denote the unique solution of (6.3). Then,

$$\big(u(nT), v(nT)\big) = v_2^n(u_0, v_0), \quad n \geqslant 0.$$

Now, pick $R > 0$ satisfying

$$0 < u_0 < R\bar{u}(0), \qquad 0 < v_0 < R\bar{v}(0),$$

and consider the auxiliary function

$$x := R\bar{u} - u, \qquad y := R\bar{v} - v.$$

Then, by construction,

$$\begin{cases} x' \geqslant \alpha x + \beta y, \\ y' \geqslant \gamma x + \rho y, \\ x(0) > 0, \quad y(0) > 0, \end{cases} \tag{6.17}$$

and, hence, thanks to Proposition 6.2,

$$x(t) > 0 \quad \text{and} \quad y(t) > 0 \quad \text{for each } t \in \mathbb{R}_+.$$

Equivalently,

$$0 < u(t) < R\bar{u}(t) \quad \text{and} \quad 0 < v(t) < R\bar{v}(t) \quad \text{for each } t \in \mathbb{R}_+,$$

since $u_0 > 0$ and $v_0 > 0$. In particular, the sequence

$$\big(u(nT), v(nT)\big) = v_2^n(u_0, v_0), \quad n \geqslant 0,$$

is bounded above and, therefore, $v_2 \leqslant 1$. Suppose $v_2 = 1$. Then, $\lambda_0 = 0$ and, thanks to Theorem 6.3, (6.1) possesses a T-solution (p, q) with $p \gg 0$ and $q \gg 0$. On the other hand, setting

$$d := \bar{u}' - \alpha\bar{u} - \beta\bar{v}, \qquad \delta := \bar{v}' - \gamma\bar{u} - \rho\bar{v},$$

we have that $d \geqslant 0$, $\delta \geqslant 0$, $d + \delta > 0$, and

$$\begin{cases} \bar{u}' = \alpha\bar{u} + \beta\bar{v} + d, \\ \bar{v}' = \gamma\bar{u} + \rho\bar{v} + \delta. \end{cases}$$

Consequently, for each $\xi \in \mathbb{R}$, the pair

$$(u_\xi, v_\xi) := (\bar{u} - \xi p, \bar{v} - \xi q)$$

provides us with a T-periodic solution of

$$\begin{cases} u' = \alpha u + \beta v + d, \\ v' = \gamma u + \rho v + \delta. \end{cases} \tag{6.18}$$

Since $\bar{u} \gg 0$ and $\bar{v} \gg 0$, we have that $u_\xi \gg 0$ and $v_\xi \gg 0$ for sufficiently small $\xi > 0$. Moreover, since $p \gg 0$ and $q \gg 0$, we have that $u_\xi \ll 0$ and $v_\xi \ll 0$ for sufficiently large $\xi > 0$. Thus, there exists $\xi > 0$ such that $u_\xi \geq 0$, $v_\xi \geq 0$ and, for some $t_0 \in \mathbb{R}$, either $u_\xi(t_0) = 0$, or $v_\xi(t_0) = 0$. Without lost of generality we can assume that $u_\xi(t_0) = 0$. Then, for each $t \geq t_0$, we have that

$$u_\xi(t) = \int_{t_0}^{t} e^{\int_\tau^t \alpha} \big[\beta v_\xi(\tau) + d(\tau)\big]\, d\tau. \tag{6.19}$$

If $\beta v_\xi + d > 0$, then, we find from (6.19) that $u_\xi(t_0 + nT) > 0$ for each $n \geq 2$, which contradicts the periodicity of u_ξ, since $u_\xi(t_0) = 0$. Thus,

$$\beta v_\xi = d = 0, \tag{6.20}$$

and, thanks again to (6.19),

$$u_\xi = 0.$$

Thus, substituting into the v-equation of (6.18) shows that

$$v'_\xi = \rho v_\xi + \delta.$$

Necessarily,

$$v_\xi = \mathfrak{R}_{-\rho}(\delta) \gg 0,$$

since $\delta > 0$, and, therefore, $\beta v_\xi > 0$, which contradicts (6.20) and shows $v_2 < 1$. This concludes the proof of part (a).

Now, we will prove part (b). Suppose $v_2 > 1$. Then,

$$\lambda_0 = -\frac{1}{T}\mathrm{Log}\, v_2 < 0$$

and, hence, any T-periodic strongly positive solution of (6.14) associated to λ_0 provides us with a strict subsolution of (6.1). To show the converse, let $(\underline{u}, \underline{v}) \in \mathcal{C}_T^1 \times \mathcal{C}_T^1$ with $\underline{u} \gg 0$ and $\underline{v} \gg 0$ satisfying (6.16) with some of the inequalities strict. Pick

$$(u_0, v_0) \in N\big[U(T) - v_2 I\big]$$

with $u_0 > 0$ and $v_0 > 0$, and let $(u(t), v(t))$ denote the unique solution of (6.3). Then,

$$\big(u(nT), v(nT)\big) = v_2^n (u_0, v_0), \quad n \geqslant 0. \tag{6.21}$$

Now, for each $\varepsilon > 0$ we consider the auxiliary function

$$(U_\varepsilon, V_\varepsilon) := (u - \varepsilon \underline{u}, v - \varepsilon \underline{v}).$$

For any $\varepsilon > 0$, $(U_\varepsilon, V_\varepsilon)$ is a strict supersolution of (6.1). Moreover, for sufficiently small $\varepsilon > 0$ we have that

$$U_\varepsilon(0) = u_0 - \varepsilon \underline{u}(0) > 0 \quad \text{and} \quad V_\varepsilon(0) = v_0 - \varepsilon \underline{v}(0) > 0.$$

Choose one of these ε's. Then, thanks to Proposition 6.2, $U_\varepsilon(t) > 0$ and $V_\varepsilon(t) > 0$ for each $t \geqslant 0$. In particular, for each $n \geqslant 0$,

$$\varepsilon \underline{u}(0) = \varepsilon \underline{u}(nT) < u(nT) = v_2^n u_0$$

and, therefore, $v_2 \geqslant 1$.

Suppose $v_2 = 1$. Then, $\lambda_0 = 0$ and, thanks to Theorem 6.3, (6.1) possesses a T-solution (p, q) with $p \gg 0$ and $q \gg 0$. On the other hand, setting

$$d := \underline{u}' - \alpha \underline{u} - \beta \underline{v}, \qquad \delta := \underline{v}' - \gamma \underline{u} - \rho \underline{v},$$

we have that $d \leqslant 0$, $\delta \leqslant 0$, $d + \delta < 0$, and

$$\begin{cases} \underline{u}' = \alpha \underline{u} + \beta \underline{v} + d, \\ \underline{v}' = \gamma \underline{u} + \rho \underline{v} + \delta. \end{cases}$$

Consequently, for each $\xi \in \mathbb{R}$, the pair

$$(u_\xi, v_\xi) := (p - \xi \underline{u}, q - \xi \underline{v})$$

provides us with a T-periodic solution of

$$\begin{cases} u' = \alpha u + \beta v - d, \\ v' = \gamma u + \rho v - \delta. \end{cases} \tag{6.22}$$

Since $p \gg 0$ and $q \gg 0$, we have that $u_\xi \gg 0$ and $v_\xi \gg 0$ for sufficiently small $\xi > 0$. Moreover, since $\underline{u} \gg 0$ and $\underline{v} \gg 0$, we have that $u_\xi \ll 0$ and $v_\xi \ll 0$ for sufficiently large $\xi > 0$. Thus, there exists $\xi > 0$ such that $u_\xi \geqslant 0$, $v_\xi \geqslant 0$ and, for some $t_0 \in \mathbb{R}$, either $u_\xi(t_0) = 0$, or $v_\xi(t_0) = 0$. The result obtained in ending the proof of part (a) shows that this is impossible. It should be noted that $-d \geqslant 0$, $-\delta \geqslant 0$ and $-(\delta + d) > 0$. Therefore, $v_2 > 1$, which concludes the proof of the theorem. $\qquad \square$

6.2. *Structure of the set of coexistence states*

In this section, the coexistence states of (1.2) will be regarded as zeroes of the operator

$$\mathfrak{F}:\mathbb{R}^2 \times \mathcal{C}_T \times \mathcal{C}_T \to \mathcal{C}_T \times \mathcal{C}_T$$

defined by

$$\mathfrak{F}(\lambda, \mu, u, v) := \begin{pmatrix} u - \mathfrak{R}_M\big[(\lambda\ell + M)u - au^2 - buv\big] \\ v - \mathfrak{R}_M\big[(\mu m + M)v - dv^2 - cuv\big] \end{pmatrix} \tag{6.23}$$

where $M > 0$ is fixed. The following result will provide us with the sharp structure of the whole surface, in terms of the parameters (λ, μ), of the coexistence states of (1.2) linking the surfaces of semi-trivial states of (1.2) along their curves of neutral stability.

PROPOSITION 6.5. *Suppose* $(u_0, v_0) \in \mathcal{C}_T \times \mathcal{C}_T$ *satisfies* $u_0 \gg 0$, $v_0 \gg 0$, *and*

$$\mathfrak{F}(\lambda, \mu, u_0, v_0) = 0.$$

Then, each of the operators

$$D_{(\lambda,u,v)}\mathfrak{F}(\lambda, \mu, u_0, v_0):\mathbb{R} \times \mathcal{C}_T \times \mathcal{C}_T \to \mathcal{C}_T \times \mathcal{C}_T$$

and

$$D_{(\mu,u,v)}\mathfrak{F}(\lambda, \mu, u_0, v_0):\mathbb{R} \times \mathcal{C}_T \times \mathcal{C}_T \to \mathcal{C}_T \times \mathcal{C}_T$$

is surjective. In other words, $(0, 0)$ *is a regular value of* $\mathfrak{F}$ *with respect to* $\mathbb{R}^2 \times \mathrm{Int}\, P \times \mathrm{Int}\, P$.

PROOF. By differentiating, we have that

$$D_{(u,v)}\mathfrak{F}(\lambda, \mu, u_0, v_0) = I_{\mathcal{C}_T \times \mathcal{C}_T} - \mathfrak{R}_M \begin{pmatrix} \alpha_M & -bu_0 \\ -cv_0 & \beta_M \end{pmatrix}, \tag{6.24}$$

$$D_\lambda\mathfrak{F}(\lambda, \mu, u_0, v_0) = -\mathfrak{R}_M \begin{pmatrix} \ell u_0 \\ 0 \end{pmatrix}, \tag{6.25}$$

and

$$D_\mu\mathfrak{F}(\lambda, \mu, u_0, v_0) = -\mathfrak{R}_M \begin{pmatrix} 0 \\ m v_0 \end{pmatrix}, \tag{6.26}$$

where, for any $\xi \geq 0$, we are denoting

$$\alpha_\xi := \lambda\ell + \xi - 2au_0 - bv_0, \qquad \beta_\xi := \mu m + \xi - 2dv_0 - cu_0. \tag{6.27}$$

Subsequently, it will be said that (λ, μ, u_0, v_0) is a non-degenerate coexistence state of (1.2) if $D_{(u,v)}\mathfrak{F}(\lambda, \mu, u_0, v_0)$ is an isomorphism. In the contrary case, it will be said that (λ, μ, u_0, v_0) is degenerate.

Obviously, $D_{(\lambda,u,v)}\mathfrak{F}(\lambda, \mu, u_0, v_0)$ and $D_{(\mu,u,v)}\mathfrak{F}(\lambda, \mu, u_0, v_0)$ are surjective if (λ, μ, u_0, v_0) is non-degenerate. So, suppose (λ, μ, u_0, v_0) is degenerate. Then, since $D_{(u,v)}\mathfrak{F}(\lambda, \mu, u_0, v_0)$ is a compact perturbation of the identity, it is Fredholm of index zero and, hence,

$$N := \dim N\big[D_{(u,v)}\mathfrak{F}(\lambda, \mu, u_0, v_0)\big] = \operatorname{codim} R\big[D_{(u,v)}\mathfrak{F}(\lambda, \mu, u_0, v_0)\big] \geqslant 1.$$

Now, note that $N[D_{(u,v)}\mathfrak{F}(\lambda, \mu, u_0, v_0)]$ is the set of T-periodic solutions (u, v) of the linear system

$$\begin{cases} u' = \alpha_0 u - bu_0 v, \\ v' = -cv_0 u + \beta_0 v. \end{cases} \tag{6.28}$$

Since $b < 0$ and $c < 0$, we have that

$$-bu_0 > 0 \quad \text{and} \quad -cv_0 > 0.$$

Moreover, since (λ, μ, u_0, v_0) is a coexistence state of (1.2), relations (6.7) hold with $(u_2, v_2) = (u_0, v_0)$ and, hence,

$$\int_0^T \alpha_0 = -\int_0^T a(t)u_0(t)\,dt < 0 \quad \text{and} \quad \int_0^T \beta_0 = -\int_0^T d(t)v_0(t)\,dt < 0.$$

Therefore, (6.28) fits into the abstract setting of Section 6.1. Consequently, thanks to Theorems 6.1 and 6.3, the biggest multiplier of (6.28), denoted by v_2, equals one and any T-periodic solution of (6.28) must be a multiple of a fixed T-periodic solution (φ, ψ) such that $\varphi \gg 0$ and $\psi \gg 0$. In particular, $N = 1$. So, thanks to (6.25) and (6.26), to conclude the proof it suffices to show the following:

$$\begin{aligned} -\mathfrak{R}_M(\ell u_0, 0) &\notin R\big[D_{(u,v)}\mathfrak{F}(\lambda, \mu, u_0, v_0)\big], \\ -\mathfrak{R}_M(0, mv_0) &\notin R\big[D_{(u,v)}\mathfrak{F}(\lambda, \mu, u_0, v_0)\big]. \end{aligned} \tag{6.29}$$

On the contrary, suppose that, e.g.,

$$-\mathfrak{R}_M(\ell u_0, 0) \in R\big[D_{(u,v)}\mathfrak{F}(\lambda, \mu, u_0, v_0)\big].$$

Then, the system

$$\begin{cases} u' = \alpha_0 u - bu_0 v + \ell u_0, \\ v' = -cv_0 u + \beta_0 v, \end{cases} \tag{6.30}$$

possesses a T-periodic solution, say (u, v). Actually, for each $R > 0$, the pair

$$(u_R, v_R) := (u + R\varphi, v + R\psi)$$

provides us with a T-periodic solution of (6.30). Since $\ell u_0 > 0$, for sufficiently large $R > 0$, (u_R, v_R) provides us with a T-periodic strongly positive strict supersolution of (6.28) and, therefore, thanks to Theorem 6.4(a), $v_2 < 1$, which is a contradiction, since $v_2 = 1$. Similarly, the second relation of (6.29) holds. This concludes the proof. $\qquad\square$

The following result, which is an easy consequence from Proposition 6.5, ascertains the global structure of the components of coexistence states $\mathfrak{C}_\mu^\lambda$ and $\mathfrak{C}_\lambda^\mu$ of (1.2) constructed in Theorems 5.4 and 5.5.

THEOREM 6.6. *The following properties are satisfied*:
(a) *For each $\lambda > 0$, the component of coexistence states $\mathfrak{C}_\mu^\lambda$ is unbounded in $\mathbb{R} \times C_T \times C_T$ and it consists of a real analytic curve. More precisely, there exits a real analytic map $(\mu, u, v) : [0, \infty) \to \mathbb{R} \times \mathrm{Int}\, P \times \mathrm{Int}\, P$ such that*

$$\big(\mu(0), u(0), v(0)\big) = (\mu_0, \theta_{[\lambda\ell,a]}, 0)$$

and

$$\mathfrak{C}_\mu^\lambda = \big\{\big(\mu(s), u(s), v(s)\big) \colon s \in (0, \infty)\big\}.$$

(b) *For each $\mu > 0$, the component of coexistence states $\mathfrak{C}_\lambda^\mu$ is unbounded in $\mathbb{R} \times C_T \times C_T$ and it consists of a real analytic curve. More precisely, there exists a real analytic map $(\lambda, u, v) : [0, \infty) \to \mathbb{R} \times \mathrm{Int}\, P \times \mathrm{Int}\, P$ such that*

$$\big(\lambda(0), u(0), v(0)\big) = (\lambda_0, 0, \theta_{[\mu m, d]})$$

and

$$\mathfrak{C}_\lambda^\mu = \big\{\big(\lambda(s), u(s), v(s)\big) \colon s \in (0, \infty)\big\}.$$

(c) *The set of coexistence states of (1.2) $(\lambda, \mu, u, v) \in \mathbb{R}^2 \times \mathrm{Int}\, P \times \mathrm{Int}\, P$ such that $\lambda > 0$ and, for some $\mu \in \mathbb{R}$, $(\mu, u, v) \in \mathfrak{C}_\mu^\lambda$, or $\mu > 0$ and, for some $\lambda \in \mathbb{R}$, $(\lambda, u, v) \in \mathfrak{C}_\lambda^\mu$ is a real analytic surface linking the surfaces of semi-trivial states*

$$\big\{(\lambda, \mu, \theta_{[\lambda\ell,a]}, 0) \colon \lambda > 0,\ \mu \in \mathbb{R}\big\}$$

and

$$\big\{(\lambda, \mu, 0, \theta_{[\mu m, d]}) \colon \lambda \in \mathbb{R},\ \mu > 0\big\}$$

along their respective curves of neutral stability

$$\mu = \frac{1}{T} \int_0^T c(t)\theta_{[\lambda\ell,a]}(t)\, \mathrm{d}t \quad and \quad \lambda = \frac{1}{T} \int_0^T b(t)\theta_{[\mu m,d]}(t)\, \mathrm{d}t.$$

PROOF. Suppose (λ, μ, u_0, v_0) is a coexistence state of (1.2) with, e.g., $\lambda > 0$. Subsequently, we fix λ and regard to μ as the main path-following parameter. Suppose (λ, μ, u_0, v_0) is non-degenerate. Then, the implicit function theorem shows that (μ, u_0, v_0) lies in an analytic curve through it that can be locally parameterized by μ, or … by some length, or pseudo-length of arc of curve, as it is usual in numerical analysis of bifurcation problems. Now, suppose that (λ, μ, u_0, v_0) is degenerate and let $\omega_0 \in \mathbb{R} \times C_T \times C_T$ such that

$$N\big[D_{(\mu,u,v)}\mathfrak{F}(\lambda, \mu, u_0, v_0)\big] = \mathrm{span}\,[\omega_0].$$

Let Z be any closed supplement of ω_0 in $\mathbb{R} \times C_T \times C_T$, i.e.,

$$\mathbb{R} \times C_T \times C_T = \mathrm{span}\,[\omega_0] \oplus Z.$$

Then, any element $(\mu, u, v) \in \mathbb{R} \times C_T \times C_T$ admits a unique decomposition as

$$(\mu, u, v) = s\omega_0 + z, \quad (s, z) \in \mathbb{R} \times Z.$$

So, the problem of solving $\mathfrak{F}(\lambda, \mu, u, v) = 0$, with λ fixed, around (μ, u_0, v_0) is equivalent to the problem of solving

$$0 = \mathfrak{G}(s, z) := \mathfrak{F}\big(\lambda, (\mu, u_0, v_0) + s\omega_0 + z\big)$$

around $(s, z) = (0, 0)$. The map $\mathfrak{G} : \mathbb{R} \times Z \to C_T \times C_T$ is real analytic and it satisfies $\mathfrak{G}(0, 0) = 0$. Moreover, thanks to Proposition 6.5, the linearized operator

$$D_z\mathfrak{G}(0, 0) = D_{(\mu,u,v)}\mathfrak{F}(\mu, u_0, v_0)|_Z$$

is a linear isomorphism. Therefore, thanks to the implicit function, the set of coexistence states of (1.2) around (λ, μ, u_0, v_0) consists of an analytic curve. Actually, the curve can be parameterized by the projection of the coexistence states on ω_0.

Now, fix $\lambda > 0$. Since $b < 0$, for each $\mu > 0$ we have that

$$\lambda > 0 > \frac{1}{T} \int_0^T b(t)\theta_{[\mu m,d]}(t)\, \mathrm{d}t$$

and, therefore, thanks to Theorem 5.4, $\mathfrak{C}_\mu^\lambda$ is unbounded. Moreover, thanks to Theorem 5.6, there exists $L > 0$ and a real analytic map

$$(\mu, u, v) : [0, L] \to \mathbb{R} \times \mathrm{Int}\, P \times \mathrm{Int}\, P$$

such that

$$\big(\mu(0), u(0), v(0)\big) = (\mu_0, \theta_{[\lambda\ell, a]}, 0)$$

and, in a neighborhood of $(\mu_0, \theta_{[\lambda\ell, a]}, 0)$, $\mathfrak{C}^\lambda_\mu$ consists of

$$\big\{\big(\mu(s), u(s), v(s)\big): 0 < s \leqslant L\big\}.$$

By a global continuation argument based on the implicit function theorem, the *local curve* $(\mu(s), u(s), v(s))$, $s \in (0, L]$, admits a *maximal continuation* (non-prolonging) to a real analytic curve, $(\mu(s), u(s), v(s))$, $s \in (0, \infty)$, of coexistence states of (1.2). Actually, the parameter s can be taken as the length of arc of curve. Clearly,

$$\Gamma := \big\{\big(\mu(s), u(s), v(s)\big): s > 0\big\} \subset \mathfrak{C}^\lambda_\mu,$$

since $\mathfrak{C}^\lambda_\mu$ is connected. Moreover, the uniqueness obtained as an application of the implicit function theorem shows that there exists an open subset $\mathcal{O} \subset \mathbb{R} \times \mathcal{C}_T \times \mathcal{C}_T$ such that

$$\Gamma \subset \mathcal{O} \quad \text{and} \quad \mathfrak{F}^{-1}(0) \cap (\mathbb{R} \times \operatorname{Int} P \times \operatorname{Int} P) \cap \mathcal{O} = \Gamma.$$

Therefore, $\mathfrak{C}^\lambda_\mu = \Gamma$, since $\mathfrak{C}^\lambda_\mu$ is connected. Part (b) follows from the symmetry of the problem. part (c) is a two-parameter re-interpretation of parts (a) and (b) that can be easily obtained from the fact that $\mathfrak{F}$ is real analytic in all its arguments. This concludes the proof. $\qquad\qquad\Box$

For the autonomous counterpart of (1.2) with $bc = 1$, is very simple to check that, for each $\lambda > 0$, the curve

$$\big(\mu(s), u(s), v(s)\big) = (c\lambda, \lambda - bs, s), \quad s \geqslant 0,$$

provides us with the component $\mathfrak{C}^\lambda_\mu$ bifurcating from the semi-trivial solution $(\lambda, 0)$ at $\mu = c\lambda$. As this simple example shows, in general, condition $\mu' = 0$ cannot be avoided unless some additional assumptions are made, as, e.g., having a priori bounds for the coexistence states of the model at $\mu = \mu_0$. Note that in case $\mu' = 0$, one necessarily has $\mu(s) = \mu_0$ for each $s \geqslant 0$ and

$$\limsup_{s \uparrow \infty} \big\| \big(u(s), v(s)\big) \big\|_{\mathcal{C}_T \times \mathcal{C}_T} = \infty,$$

since $\mathfrak{C}^\lambda_\mu$ is unbounded. Actually, as an immediate consequence from the analysis carried out in the next section, in case $\mu' = 0$ one has that, for each $s \geqslant 0$, $\dot{u}(s) \gg 0$ and $\dot{v}(s) \gg 0$, where $\cdot := \frac{d}{ds}$. Therefore,

$$\lim_{s \uparrow \infty} \big\| \big(u(s), v(s)\big) \big\|_{\mathcal{C}_T \times \mathcal{C}_T} = \infty.$$

6.3. *Stability of the coexistence states along* $\mathfrak{C}_\mu^\lambda$ *and* $\mathfrak{C}_\lambda^\mu$

Fix $\lambda > 0$ and let $(\mu(s), u(s), v(s))$, $s \geqslant 0$, be the real analytic map constructed in Theorem 6.6, where s is assumed to be the length of arc of curve. Subsequently, we shall denote $\cdot := \frac{\mathrm{d}}{\mathrm{d}s}$. Since

$$\mathfrak{F}(\lambda, \mu(s), u(s), v(s)) = 0, \quad s \geqslant 0,$$

differentiating with respect to s gives

$$D_{(\mu,u,v)}\mathfrak{F}(\lambda, \mu(s), u(s), v(s))(\dot\mu(s), \dot u(s), \dot v(s)) = 0, \quad s > 0, \tag{6.31}$$

and, hence,

$$N[D_{(\mu,u,v)}\mathfrak{F}(\lambda, \mu(s), u(s), v(s))] = \mathrm{span}\left[(\dot\mu(s), \dot u(s), \dot v(s))\right], \quad s > 0.$$

Moreover, since s has been taken as the length of arc of curve,

$$\left\|(\dot\mu(s), \dot u(s), \dot v(s))\right\|_{\mathbb{R}\times C_T\times C_T} = 1, \quad s > 0. \tag{6.32}$$

On the other hand, since

$$(\mu(0), u(0), v(0)) = (\mu_0, \theta_{[\lambda\ell,a]}, 0),$$

it follows from Theorem 5.6 that

$$\dot v(s) = \varphi_v + \mathrm{O}(s) = e^{\int_0^{\cdot}(\mu_0 m - c\theta_{[\lambda\ell,a]})} + \mathrm{O}(s) \gg 0, \quad s \downarrow 0. \tag{6.33}$$

Now, set

$$(p_s, q_s) := (\dot u(s), \dot v(s)), \quad s > 0,$$

and note that (6.31) can be equivalently rewritten in the form

$$\begin{cases} (p_s)' = [\lambda\ell - 2au(s) - bv(s)]p_s - bu(s)q_s, \\ (q_s)' = -cv(s)p_s + [\mu(s)m - 2dv(s) - cu(s)]q_s + \dot\mu(s)mv(s). \end{cases} \tag{6.34}$$

Note that, since $-bu(s) > 0$, $-cv(s) > 0$,

$$\int_0^T [\lambda\ell - 2au(s) - bv(s)] = -2\int_0^T [au(s)] < 0, \tag{6.35}$$

and

$$\int_0^T [\mu(s)m - 2dv(s) - cu(s)] = -2\int_0^T [dv(s)] < 0, \tag{6.36}$$

the linear system (6.34) fits within the general setting of Section 6.1. In particular, thanks to Proposition 2.1, we find from the first equation of (6.34) that, for each $s > 0$ sufficiently small,

$$p_s = \Re_{-[\lambda\ell - 2au(s) - bv(s)]}\big(-bu(s)q_s\big) \gg 0.$$

Therefore, there exists $\varepsilon > 0$ such that

$$p_s \gg 0 \quad \text{and} \quad q_s \gg 0 \quad \text{for each } s \in [0, \varepsilon]. \tag{6.37}$$

Suppose $\mu = \mu_0$ for each $s \geqslant 0$. Then, (6.34) becomes into

$$\begin{cases} (p_s)' = \big[\lambda\ell - 2au(s) - bv(s)\big]p_s - bu(s)q_s, \\ (q_s)' = -cv(s)p_s + \big[\mu(s)m - 2dv(s) - cu(s)\big]q_s. \end{cases} \tag{6.38}$$

Thanks to (6.32), $(p_s, q_s) \neq (0, 0)$ and, hence, thanks to Theorem 6.1, for each $s \geqslant 0$, either $p_s \gg 0$ and $q_s \gg 0$, or $p_s \ll 0$ and $q_s \ll 0$. Thanks to (6.37), necessarily

$$p_s \gg 0 \quad \text{and} \quad q_s \gg 0 \quad \text{for each } s \in [0, \infty). \tag{6.39}$$

Indeed, since $s \to (p_s, q_s)$ is real analytic, if (6.39) is not satisfied, then, due to (6.37), there is some value of s, say $\tilde{s} > 0$, where $p_{\tilde{s}} > 0$, $q_{\tilde{s}} > 0$, and, e.g., $p_{\tilde{s}} \gg 0$ fails. Thanks to Theorem 6.1, this would imply $p_{\tilde{s}} \gg 0$ and $q_{\tilde{s}} \gg 0$, which is a contradiction. Thus, (6.39) is satisfied, as it was claimed in the last paragraph of Section 6.2, and, consequently, thanks to Theorem 6.3, the principal characteristic multiplier of $(\mu(s), u(s), v(s))$, which will be subsequently denoted by $v_2(s)$, satisfies $v_2(s) = 1$ for each $s \geqslant 0$. Therefore, if $\dot{\mu} = 0$, then $\dot{v}_2 = 0$ and, hence, all the coexistence states of $\mathcal{C}_\mu^\lambda$ are neutrally stable. When $\dot{\mu} \neq 0$, then the following result is satisfied.

THEOREM 6.7. *Suppose $\dot{\mu} \neq 0$ in $[0, \infty)$. Then, the set*

$$\mathcal{S} := \big\{s \in (0, \infty) : \dot{\mu}(s) = 0\big\} \tag{6.40}$$

is discrete, each $s \in \mathcal{S}$ is a zero of finite order of $\dot{\mu}$, and

$$\mathcal{S} = \big\{s \in (0, \infty) : v_2(s) = 1\big\}. \tag{6.41}$$

Moreover,

$$\dot{\mu}(s)\big(1 - v_2(s)\big) > 0, \quad s \in (0, \infty) \setminus \mathcal{S}, \tag{6.42}$$

and

$$p_s \gg 0 \quad \text{and} \quad q_s \gg 0 \quad \text{if} \quad \dot{\mu}(s) \geqslant 0. \tag{6.43}$$

PROOF. The fact that $\mathcal{S}$ is discrete and that each $s \in \mathcal{S}$ has finite order, as a zero of $\dot{\mu}$, are consequences from the fact that $\dot{\mu}$ is real analytic. Now, we shall prove (6.41). Pick $s \in \mathcal{S}$. Then, (p_s, q_s) satisfies (6.38) and, due to (6.32), $(p_s, q_s) \neq (0, 0)$. Thus, thanks to Theorem 6.1, $p_s \gg 0$ and $q_s \gg 0$, or $p_s \ll 0$ and $q_s \ll 0$. In any of these cases, it follows from Theorem 6.3 that $\lambda = 0$ is the *principal characteristic exponent* of $(\mu(s), u(s), v(s))$ and, hence, $\nu_2(s) = 1$. To prove the converse inclusion, let $s \in (0, \infty)$ such that $\nu_2(s) = 1$. Then, thanks to Theorem 6.3, there exists a T-periodic pair (P, Q) with $P \gg 0$ and $Q \gg 0$ such that

$$\begin{cases} P' = \left[\lambda \ell - 2au(s) - bv(s)\right]P - bu(s)Q, \\ Q' = -cv(s)P + \left[\mu(s)m - 2dv(s) - cu(s)\right]P. \end{cases} \tag{6.44}$$

Now, choose a sufficiently small $\varepsilon > 0$ so that the auxiliary pair

$$(U, V) := (P - \varepsilon p_s, Q - \varepsilon q_s),$$

satisfy $U \gg 0$ and $V \gg 0$. Thanks to (6.34) and (6.44) we have that

$$\begin{cases} U' = \left[\lambda \ell - 2au(s) - bv(s)\right]U - bu(s)V, \\ V' = -cv(s)U + \left[\mu(s)m - 2dv(s) - cu(s)\right]V - \varepsilon\dot{\mu}(s)mv(s). \end{cases}$$

Assume $\dot{\mu}(s) > 0$. Then, since $\varepsilon mv(s) > 0$, (U, V) is a strict subsolution of

$$\begin{cases} x' = \left[\lambda \ell - 2au(s) - bv(s)\right]x - bu(s)y, \\ y' = -cv(s)x + \left[\mu(s)m - 2dv(s) - cu(s)\right]y, \end{cases} \tag{6.45}$$

and, hence, thanks to Theorem 6.4(b), $\nu_2(s) > 1$, which is a contradiction. Assume $\dot{\mu}(s) < 0$. Then, (U, V) is a strict supersolution of (6.45) and, due to Theorem 6.4(a), $\nu_2(s) < 1$, which is a contradiction again. Consequently, $\dot{\mu}(s) = 0$ and, therefore, (6.41) holds.

Now, pick $\tilde{s} \in \mathcal{S}$. Thanks to (6.32), it follows from Theorem 6.1 that some of the following alternatives occurs:

1. $p_{\tilde{s}} \gg 0$ and $q_{\tilde{s}} \gg 0$.
2. $p_{\tilde{s}} \ll 0$ and $q_{\tilde{s}} \ll 0$.

Assume alternative 1 occurs. As $\tilde{s}$ is an isolated zero of $\dot{\mu}(s)$, there exists $\varepsilon > 0$ such that $p_s = \dot{u}(s) \gg 0$ and $q_s = \dot{v}(s) \gg 0$ for each $s \in (\tilde{s} - \varepsilon, \tilde{s} + \varepsilon)$, and

$$\dot{\mu}^{-1}(0) \cap \left(\tilde{s} - \varepsilon, \tilde{s} + \varepsilon\right) \cap (0, \infty) = \{\tilde{s}\}.$$

Thus, for each

$$s \in J := (0, \infty) \cap \left[\left(\tilde{s} - \varepsilon, \tilde{s} + \varepsilon\right) \setminus \{\tilde{s}\}\right],$$

either $\dot{\mu}(s) > 0$, or $\dot{\mu}(s) < 0$. Suppose $\dot{\mu}(s) > 0$. Then, (p_s, q_s) provides us with a strongly positive strict supersolution of (6.45) and it follows from Theorem 6.4(a) that $\nu_2(s) < 1$.

Hence, $\dot{\mu}(s)(1 - v_2(s)) > 0$. Now, suppose $\dot{\mu}(s) < 0$. Then, (p_s, q_s) provides us with a strongly positive strict subsolution of (6.45) and it follows from Theorem 6.4(b) that $v_2(s) > 1$. Hence, $\dot{\mu}(s)(1 - v_2(s)) > 0$. Therefore, condition (6.42) is satisfied for each $\tilde{s} \in \mathcal{S}$ satisfying alternative 1.

Now, suppose $\tilde{s} \in \mathcal{S}$ satisfies alternative 2. As $\tilde{s}$ is an isolated zero of $\dot{\mu}(s)$, there exists $\varepsilon > 0$ such that $p_s = \dot{u}(s) \ll 0$ and $q_s = \dot{v}(s) \ll 0$ for each $s \in (\tilde{s} - \varepsilon, \tilde{s} + \varepsilon)$, and

$$\dot{\mu}^{-1}(0) \cap \left(\tilde{s} - \varepsilon, \tilde{s} + \varepsilon\right) \cap (0, \infty) = \{\tilde{s}\}.$$

Thus, for each

$$s \in J := (0, \infty) \cap \left[\left(\tilde{s} - \varepsilon, \tilde{s} + \varepsilon\right) \setminus \{\tilde{s}\}\right],$$

either $\dot{\mu}(s) > 0$, or $\dot{\mu}(s) < 0$. Suppose $\dot{\mu}(s) > 0$. Then, $(-p_s, -q_s)$ provides us with a strongly positive strict subsolution of (6.45) and it follows from Theorem 6.4(b) that $v_2(s) > 1$. Hence, $\dot{\mu}(s)(1 - v_2(s)) < 0$. Similarly, if $\dot{\mu}(s) < 0$, then, $(-p_s, -q_s)$ provides us with a strongly positive strict supersolution of (6.45) and Theorem 6.4(a) implies $v_2(s) < 1$. Therefore, the following condition is satisfied

$$\dot{\mu}(s)\left(1 - v_2(s)\right) < 0, \quad s \in (0, \infty) \cap \left[\left(\tilde{s} - \varepsilon, \tilde{s} + \varepsilon\right) \setminus \{\tilde{s}\}\right], \tag{6.46}$$

for each $\tilde{s} \in \mathcal{S}$ satisfying alternative 2.

Now, suppose

$$\mathcal{S} = \{s_n : n \geqslant 1\}, \quad 0 < s_1 < s_n < s_{n+1}, \ n \geqslant 2;$$

the subsequent argument can be easily adapted to cover the case when $\mathcal{S}$ is finite, possibly empty. Thanks to (6.37), for sufficiently small $s > 0$, $p_s \gg 0$ and $q_s \gg 0$. Moreover, either $\dot{\mu}(s) > 0$ for each $s \in (0, s_1)$, or $\dot{\mu}(s) < 0$ for each $s \in (0, s_1)$.

Suppose $\dot{\mu}(s) > 0$ for each $s \in (0, s_1)$. Then, for sufficiently small $s > 0$, (p_s, q_s) provides us with a strongly positive strict supersolution of (6.45) and, hence, thanks to Theorem 6.4(a), $v_2(s) < 1$. Actually, this is the inter-exchange stability principle between the semi-trivial state $(\mu, \theta_{[\lambda\ell, a]}, 0)$ and the bifurcating coexistence state $(\mu(s), u(s), v(s))$ as μ crosses μ_0. Thanks to (6.41), necessarily $v_2(s) < 1$ for every $s \in (0, s_1)$ and, therefore,

$$\dot{\mu}(s)\left(1 - v_2(s)\right) > 0, \quad s \in (0, s_1). \tag{6.47}$$

Necessarily $p_{s_1} \gg 0$ and $q_{s_1} \gg 0$. Indeed, if $p_{s_1} \ll 0$ and $q_{s_1} \ll 0$, then (6.46) holds, which contradicts (6.47). Therefore,

$$\dot{\mu}(s)\left(1 - v_2(s)\right) > 0, \quad s \in (0, s_2) \setminus \{s_2\}. \tag{6.48}$$

Reiterating this argument, (6.42) holds readily.

Now, suppose $\dot{\mu}(s) < 0$ for each $s \in (0, s_1)$. Then, (p_s, q_s) provides us with a strongly positive strict subsolution of (6.45) and, hence, thanks to Theorem 6.4(b), $v_2(s) > 1$. As in the previous case, this provides us with the inter-exchange stability principle between the semi-trivial state $(\mu, \theta_{[\lambda\ell,a]}, 0)$ and the bifurcating coexistence state $(\mu(s), u(s), v(s))$. Thanks to (6.41), $v_2(s) > 1$ for every $s \in (0, s_1)$ and, therefore, (6.47) is as well satisfied. Arguing as above, the proof of (6.42) is concluded.

Note that, as a result from the previous analysis, we have that

$$p_{\tilde{s}} \gg 0 \quad \text{and} \quad q_{\tilde{s}} \gg 0 \quad \text{for each } \tilde{s} \in \mathcal{S} \cup \{0\}. \tag{6.49}$$

Let $s_1^*, s_2^* \in \mathcal{S} \cup \{0\}$ such that

$$\dot{\mu}(s) > 0, \quad s \in \left(s_1^*, s_2^*\right).$$

Thanks to (6.49), $p_s \gg 0$ and $q_s \gg 0$ for each $s > s_1^*$ sufficiently close to s_1^*. Thus, either $p_s \gg 0$ and $q_s \gg 0$ for each $s \in (s_1^*, s_2^*)$, or there exist $s_0 \in (s_1^*, s_2^*)$ and $t \in \mathbb{R}$ such that $p_{s_0} \geq 0$ and $q_{s_0} \geq 0$ and either $p_{s_0}(t) = 0$, or $q_{s_0}(t) = 0$. Since

$$-cv(s_0)p_{s_0} + \dot{\mu}(s_0)mv(s_0) > 0,$$

thanks to Proposition 2.1, it follows from the second equation of (6.34) that $q_{s_0} \gg 0$. Thus,

$$-bu(s_0)q_{s_0} > 0$$

and, thanks again to Proposition 2.1, the first equation of (6.34) gives $p_{s_0} \gg 0$, which is impossible. Therefore, for each $s \in (s_1^*, s_2^*)$, $p_s \gg 0$ and $q_s \gg 0$, which concludes the proof of (6.43). The proof is completed. $\qquad\square$

It should be noted that, since the dependence of the Poincaré map of (6.45) on s is analytic, the map $s \to v_2(s)$ is as well real analytic. Moreover,

$$v_2(0) = 1,$$

because this is the critical value of the parameter where the stability of the semi-trivial state changes. As far as to $\dot{\mu}(0)$ concerns, it might take any real value, according with the value of the integral of (5.53).

Subsequently, we shall assume

$$\dot{\mu} \neq 0$$

in order to use Theorem 6.7 for ascertaining the attractive character of all the coexistence states along the curve $\mathfrak{C}_\mu^\lambda$. In this case, setting

$$s_1 = \min \mathcal{S},$$

some of the following alternatives occurs. Either

$$\dot{\mu}(s) > 0, \quad s \in (0, s_1) \tag{6.50}$$

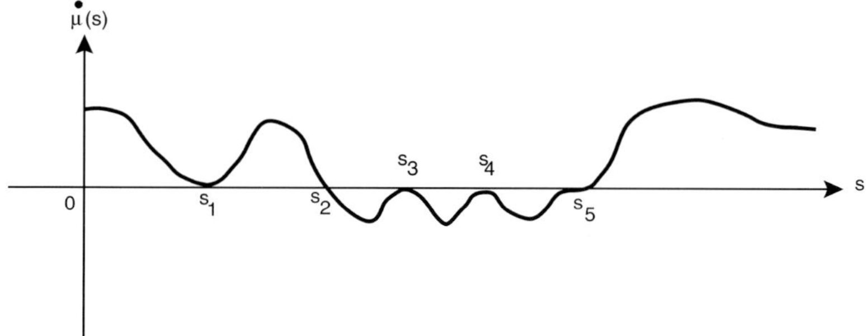

Fig. 11. An admissible graph of $\dot{\mu}(s)$ in case (6.50).

or

$$\dot{\mu}(s) < 0, \quad s \in (0, s_1). \tag{6.51}$$

If $S = \emptyset$, then either $\dot{\mu}(s) > 0$ for each $s > 0$, or $\dot{\mu}(s) < 0$ for each $s > 0$. The bifurcation of $\mathfrak{C}_\mu^\lambda$ from $(\mu_0, \theta_{[\lambda\ell,a]}, 0)$ is said to be *super-critical* if (6.50) occurs, while it is said to be *sub-critical* if (6.51) is satisfied.

The following concepts will be extremely useful in the forthcoming discussion. Given $\tilde{s} \in S$, we shall denote by $\mathrm{ord}\,(\tilde{s})$ the order of $\tilde{s}$ as a zero of $\dot{\mu}(s)$.

- A value $\tilde{s} \in S$ is said to be a *hysteresis point* of $\mathfrak{C}_\mu^\lambda$ if $\mathrm{ord}\,(\tilde{s})$ is even.
- A value $\tilde{s} \in S$ is said to be a *super-critical turning point* of $\mathfrak{C}_\mu^\lambda$ if $\mathrm{ord}\,(\tilde{s})$ is odd and $\mu(s) > \mu(\tilde{s})$ in a perforated neighborhood of $\tilde{s}$.
- A value $\tilde{s} \in S$ is said to be a *sub-critical turning point* of $\mathfrak{C}_\mu^\lambda$ if $\mathrm{ord}\,(\tilde{s})$ is odd and $\mu(s) < \mu(\tilde{s})$ in a perforated neighborhood of $\tilde{s}$.

Suppose (6.50) and, e.g., the graph of $\dot{\mu}(s)$ is of the type sketched in Fig. 11. Although far from necessary we shall assume that $S = \{s_1, s_2, s_3, s_4, s_5\}$. Hence, $\dot{\mu}(s) > 0$ for each $s > s_5$. For this special configuration, s_1, s_3 and s_4 are hysteresis points, while s_2 is a sub-critical turning point and s_5 is a super-critical turning point. Thanks to Theorem 6.7, the principal characteristic multiplier $\nu_2(s)$ of the coexistence state $(\mu(s), u(s), v(s))$ must adjust to the pattern shown in Fig. 12. Therefore, the coexistence state $(\mu(s), u(s), v(s))$ is asymptotically stable if $s \in (0, s_1) \cup (s_1, s_2) \cup (s_5, \infty)$, whereas it is unstable if $s \in (s_2, s_5) \setminus \{s_3, s_4\}$. In Fig. 13 we have represented the corresponding bifurcation diagram of coexistence states, where we are representing the value of $u(t)$ for a given time $t \in \mathbb{R}$ versus the parameter μ, not s. Except for $(\mu, 0, 0)$, continuous lines represent stable solutions, while dashed lines represent unstable solutions. The horizontal lines represent the curves of states $(\mu, 0, 0)$ and $(\mu, \theta_{[\lambda\ell,0]}, 0)$. The curve $\mathfrak{C}_\mu^\lambda$ bifurcates from $(\mu, \theta_{[\lambda\ell,a]}, 0)$ at

$$\mu = \mu_0 = \frac{1}{T} \int_0^T c(t)\theta_{[\lambda\ell,a]}(t)\,\mathrm{d}t < 0,$$

since $c < 0$. The wiggled thick line represent $\mathfrak{C}_\mu^\lambda$, whose bifurcation from the semi-trivial state is super-critical. At the sub-critical turning point $(\mu(s_2), u(s_2), v(s_2))$ the stability of

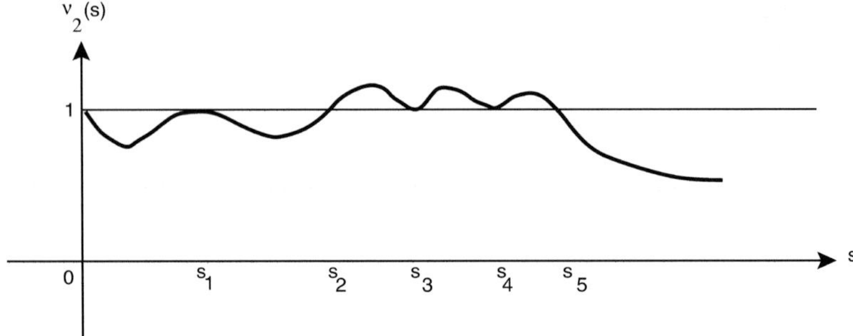

Fig. 12. The corresponding principal characteristic multiplier $\nu_2(s)$.

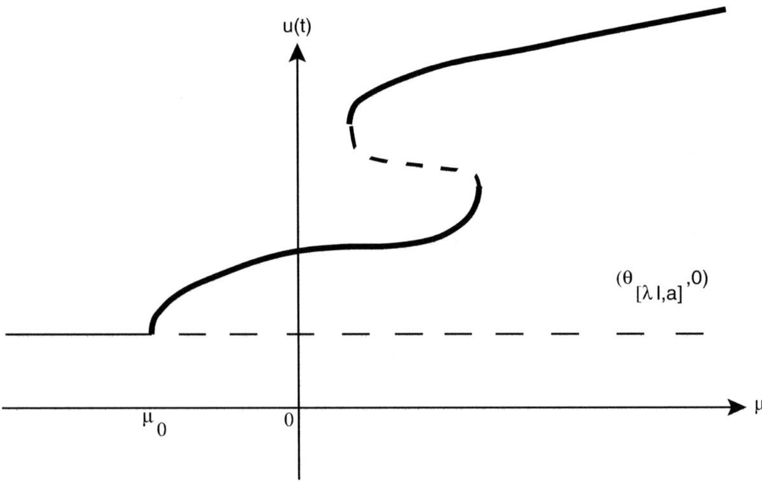

Fig. 13. The corresponding bifurcation diagram.

the solutions along the lower arc of curve of $\mathfrak{C}_\mu^\lambda$ is lost. Then, the solutions stay unstable until the second turning point, $(\mu(s_5), u(s_5), v(s_5))$, which is super-critical, is passed by. Once crossed the super-critical turning point, all remaining solutions for further values of s stay stable. It should be noted that, thanks to Theorem 6.7, $u(s)$ and $v(s)$ are strongly increasing along the pieces of the curve $\mathfrak{C}_\mu^\lambda$ where $\dot\mu(s) \geqslant 0$.

Now, suppose (6.51), instead of (6.50), and, e.g., the graph of $\dot\mu(s)$ adjust to the profile shown in Fig. 14. According to Theorem 6.7, the principal characteristic multiplier $\nu_2(s)$ of the coexistence state $(\mu(s), u(s), v(s))$ must adjust to the pattern shown in Fig. 15. In Fig. 16 we have represented an admissible bifurcation diagram corresponding to the graph of $\dot\mu(s)$ shown in Fig. 14. Now the bifurcation of $\mathfrak{C}_\mu^\lambda$ from $(\mu, \theta_{[\lambda\ell,a]}, 0)$ at μ_0 is sub-critical, the solutions along the lower curve being unstable until they reach the super-critical turning point $(\mu(s_2), u(s_2), v(s_2))$, where they become stable until reaching the second turning point, $(\mu(s_5), u(s_5), v(s_5))$, where they become unstable and stay unstable for any further value of s, $s > s_5$.

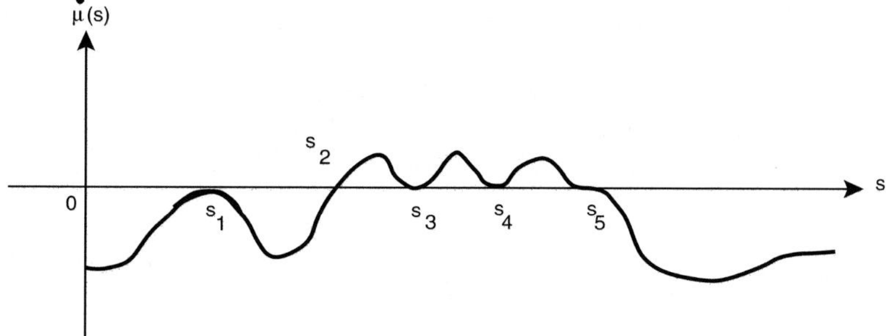

Fig. 14. An admissible graph for $\dot{\mu}(s)$ in case (6.51).

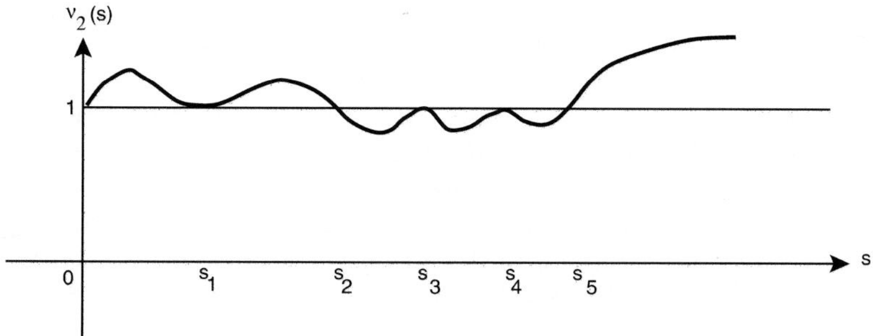

Fig. 15. The corresponding principal characteristic multiplier $v_2(s)$.

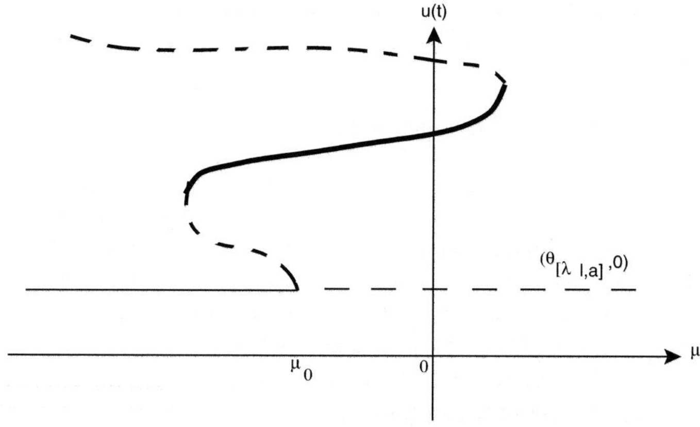

Fig. 16. The corresponding bifurcation diagram.

As in the previous example, as a result of Theorem 6.7, $s \mapsto (u(s), v(s))$ is strongly increasing along any arc of curve filled in by stable solutions, though some of the components might decrease along some pieces of the arcs of curve where the coexistence states are unstable.

Generically, when some additional parameter of the model is varied, each hysteresis point will either disappear, or it will generate an even number of turning points; half among them sub-critical and the other half super-critical. This might provide with a mechanism to generate models with an arbitrarily large number of coexistence states.

It should be noted that (1.2) might possess some additional coexistence state not lying in $\mathcal{C}_\mu^\lambda$. If this is the case, i.e., if (1.2) possesses a coexistence state $(\mu, u_0, v_0) \notin \mathcal{C}_\mu^\lambda$, then the global continuation argument used to ascertain the analytic structure of $\mathcal{C}_\mu^\lambda$ also shows that there is an analytic curve

$$(\mu, u, v) : (-\infty, \infty) \to \mathbb{R} \times \mathcal{C}_T \times \mathcal{C}_T$$

of coexistence states of (1.2) such that

$$\big(\mu(0), u(0), v(0)\big) = (\mu, u_0, v_0)$$

and $\dot{\mu} \neq 0$. Indeed, if $\dot{\mu} = 0$, then $\mu(s) = \mu$ for each s, and $s \mapsto (u(s), v(s))$ must be strongly monotone, and, as a result, it should bifurcate from $(\mu_0, \theta_{[\lambda\ell,a]}, 0)$, which is not possible, because $(\mu, u_0, v_0) \notin \mathcal{C}_\mu^\lambda$. Theorem 6.7 also applies along these curves, showing that $v_2(s) - 1$ must change of sign at any turning point along them. The corresponding curves should be either isolated bounded components, or unbounded components, but a sharper analysis of this and other related problems escapes from the general scope of this chapter.

By symmetry, all the results that we have obtained for $\mathcal{C}_\mu^\lambda$ are also valid for the components $\mathcal{C}_\lambda^\mu$ obtained by fixing $\mu > 0$ and using λ as the main bifurcation parameter.

6.4. *Low and strong symbiosis effects*

Throughout this section we will assume that

$$\ell \gg 0, \quad m \gg 0, \quad a \gg 0, \quad d \gg 0, \quad -b \gg 0, \quad -c \gg 0. \tag{6.52}$$

Then, it is said that (1.2) is lowly symbiotic if

$$\left(\frac{-b}{a}\right)_M \left(\frac{-c}{d}\right)_M < 1, \tag{6.53}$$

while it is said that (1.2) is highly symbiotic if

$$\left(\frac{-b}{a}\right)_L \left(\frac{-c}{d}\right)_L > 1. \tag{6.54}$$

Although in the intermediate situation cases where conditions (6.53) and (6.54) fail, the curves of coexistence states $\mathcal{C}_\mu^\lambda$ and $\mathcal{C}_\lambda^\mu$ might exhibit a bounded μ projection and a bounded λ projection, respectively, because of the lost of uniform priori bounds in $\mathcal{C}_T \times \mathcal{C}_T$, in the cases when some of these conditions is satisfied, the μ and λ-projections are unbounded. Actually, the following result is satisfied for the case of low symbiosis.

THEOREM 6.8. *Assume* (6.52), (6.53), *and* (1.2) *possesses a coexistence state, say* (λ, μ, u, v). *Then,*

$$
\begin{aligned}
&\lambda \geqslant -(-b/\ell)_M (m/d)_M \mu \quad &\text{if } \mu > 0, \\
&\mu \geqslant -(-c/m)_M (\ell/a)_M \lambda \quad &\text{if } \lambda > 0.
\end{aligned}
\tag{6.55}
$$

Moreover,

$$
\begin{aligned}
\|u\|_{C_T} &\leqslant \frac{(\ell/a)_M |\lambda| + (-b/a)_M (m/d)_M |\mu|}{1 - (-b/a)_M (-c/d)_M}, \\[2ex]
\|v\|_{C_T} &\leqslant \frac{(m/d)_M |\mu| + (-c/d)_M (\ell/a)_M |\lambda|}{1 - (-b/a)_M (-c/d)_M},
\end{aligned}
\tag{6.56}
$$

and, therefore, for each $\lambda > 0$ and $\mu > 0$ there exist

$$
\mu_1 = \mu_1(\lambda) \in (-\infty, \mu_0] \quad \text{and} \quad \lambda_1 = \lambda_1(\mu) \in (-\infty, \lambda_0]
$$

such that

$$
P_\mu \mathfrak{C}_\mu^\lambda \in \big\{ [\mu_1, \infty), (\mu_0, \infty) \big\} \quad \text{and} \quad P_\lambda \mathfrak{C}_\lambda^\mu \in \big\{ [\lambda_1, \infty), (\lambda_0, \infty) \big\},
$$

where P_γ stands for the γ-projection operator, $\gamma \in \{\lambda, \mu\}$.

PROOF. Pick $t_0, t_1 \in \mathbb{R}$ such that

$$
u(t_0) = \|u\|_{C_T} \quad \text{and} \quad v(t_1) = \|v\|_{C_T}.
$$

Then, $u'(t_0) = v'(t_1) = 0$ and, hence,

$$
\lambda \ell(t_0) - a(t_0)u(t_0) - b(t_0)v(t_0) = 0, \qquad \mu m(t_1) - d(t_1)v(t_1) - c(t_1)u(t_1) = 0.
$$

Thus,

$$
u(t_0) = \lambda \frac{\ell(t_0)}{a(t_0)} - \frac{b(t_0)}{a(t_0)} v(t_0), \qquad v(t_1) = \mu \frac{m(t_1)}{d(t_1)} - \frac{c(t_1)}{d(t_1)} u(t_1).
\tag{6.57}
$$

Now, since

$$
v(t_0) \leqslant v(t_1) = \mu \frac{m(t_1)}{d(t_1)} - \frac{c(t_1)}{d(t_1)} u(t_1) \leqslant \mu \frac{m(t_1)}{d(t_1)} - \frac{c(t_1)}{d(t_1)} u(t_0),
$$

the first identity of (6.57) implies

$$u(t_0) \leqslant \frac{\ell(t_0)}{a(t_0)}\lambda - \frac{b(t_0)}{a(t_0)}\frac{m(t_1)}{d(t_1)}\mu + \frac{b(t_0)}{a(t_0)}\frac{c(t_1)}{d(t_1)}u(t_0)$$

and, hence,

$$\left[1 - \frac{b(t_0)}{a(t_0)}\frac{c(t_1)}{d(t_1)}\right]u(t_0) \leqslant \frac{\ell(t_0)}{a(t_0)}\lambda - \frac{b(t_0)}{a(t_0)}\frac{m(t_1)}{d(t_1)}\mu. \tag{6.58}$$

Similarly,

$$\left[1 - \frac{b(t_0)}{a(t_0)}\frac{c(t_1)}{d(t_1)}\right]v(t_1) \leqslant \frac{m(t_1)}{d(t_1)}\mu - \frac{c(t_1)}{d(t_1)}\frac{\ell(t_0)}{a(t_0)}\lambda. \tag{6.59}$$

Moreover, thanks to (6.53),

$$0 < 1 - \left(\frac{-b}{a}\right)_M\left(\frac{-c}{d}\right)_M \leqslant 1 - \frac{b(t_0)}{a(t_0)}\frac{c(t_1)}{d(t_1)} \tag{6.60}$$

and, consequently, it follows from (6.58) and (6.59) that, necessarily,

$$\lambda \geqslant -\frac{-b(t_0)}{\ell(t_0)}\frac{m(t_1)}{d(t_1)}\mu \quad \text{and} \quad \mu \geqslant -\frac{-c(t_1)}{m(t_1)}\frac{\ell(t_0)}{a(t_0)}\lambda.$$

Therefore, (6.55) are necessary conditions for the existence of a coexistence state. Also, substituting (6.60) into (6.58) and (6.59), gives (6.56).

Now, fix $\lambda > 0$ and consider the analytic curve $\mathfrak{C}_\mu^\lambda$. Thanks to Theorem 6.6(a), $\mathfrak{C}_\mu^\lambda$ is unbounded in $\mathbb{R} \times C_T \times C_T$. Thus, thanks to the uniform a priori estimates (6.56), $P_\mu \mathfrak{C}_\mu^\lambda$ must be unbounded. Note that it is connected, since $\mathfrak{C}_\mu^\lambda$ is a curve. Therefore, the conclusion follows from the second necessary condition of (6.55). Similarly, one can obtain the corresponding assertion for $\mathfrak{C}_\lambda^\mu$. $\square$

As a consequence from Theorem 6.8, for each $\lambda > 0$, the curve $\mathfrak{C}_\mu^\lambda$ looks like the one shown in Fig. 13, whereas, for the case of strong symbiosis, the following result shows that $\mathfrak{C}_\mu^\lambda$ looks like shows Fig. 16.

THEOREM 6.9. *Assume* (6.52), (6.54),

$$(a/d)_L(d/a)_L > \left[(-b/a)_L(-c/d)_L\right]^{-1}, \tag{6.61}$$

and (1.2) *possesses a coexistence state, say* (λ, μ, u, v). *Then,*

$$\lambda \leqslant -(-b/a)_L(a/d)_L\mu, \qquad \mu \leqslant -(-c/d)_L(d/a)_L\lambda, \tag{6.62}$$

and

$$\int_0^T u \leqslant \frac{-T(\lambda + (-b/a)_L(a/d)_L\mu)}{a_L[(-b/a)_L(-c/d)_L(a/d)_L(d/a)_L - 1]},$$

$$\int_0^T v \leqslant \frac{-T(\mu + (-c/d)_L(d/a)_L\lambda)}{d_L[(-b/a)_L(-c/d)_L(a/d)_L(d/a)_L - 1]}.$$

(6.63)

Therefore, for each $\lambda > 0$ *and* $\mu > 0$ *there exist*

$$\mu_1 = \mu_1(\lambda) \in [\mu_0, 0) \quad \text{and} \quad \lambda_1 = \lambda_1(\mu) \in [\lambda_0, 0)$$

such that

$$P_\mu \mathcal{C}_\mu^\lambda \in \{(-\infty, \mu_1], (-\infty, \mu_0)\} \quad \text{and} \quad P_\lambda \mathcal{C}_\lambda^\mu \in \{(-\infty, \lambda_1], (-\infty, \lambda_0)\}.$$

PROOF. Dividing by u the u-equation, by v the v-equation, and integrating in $[0, T]$ gives

$$\lambda = \frac{1}{T} \int_0^T (au + bv), \qquad \mu = \frac{1}{T} \int_0^T (dv + cu). \tag{6.64}$$

Thus,

$$\int_0^T (au) = \lambda T - \int_0^T (bv) \geqslant \lambda T + (-b/a)_L \int_0^T (av)$$

$$\geqslant \lambda T + (-b/a)_L(a/d)_L \int_0^T (dv)$$

$$= T\lambda + (-b/a)_L(a/d)_L T\mu + (-b/a)_L(a/d)_L \int_0^T (-cu)$$

$$\geqslant T\lambda + (-b/a)_L(a/d)_L T\mu$$

$$+ (-b/a)_L(-c/d)_L(a/d)_L(d/a)_L \int_0^T (au)$$

and, hence,

$$\left[1 - (-b/a)_L(-c/d)_L(a/d)_L(d/a)_L\right] \int_0^T (au) \geqslant T\left[\lambda + (-b/a)_L(a/d)_L\mu\right].$$

Similarly,

$$\left[1 - (-b/a)_L(-c/d)_L(a/d)_L(d/a)_L\right] \int_0^T (dv) \geqslant T\left[\mu + (-c/d)_L(d/a)_L\lambda\right].$$

On the other hand, thanks to (6.61), we have that

$$(-b/a)_L(-c/d)_L(a/d)_L(d/a)_L > 1,$$

and, hence, (6.62) holds true. Therefore,

$$a_L \int_0^T u \leqslant \int_0^T (au) \leqslant -T \frac{\lambda + (-b/a)_L(a/d)_L\mu}{(-b/a)_L(-c/d)_L(a/d)_L(d/a)_L - 1},$$

$$d_L \int_0^T v \leqslant \int_0^T (dv) \leqslant -T \frac{\mu + (-c/d)_L(d/a)_L\lambda}{(-b/a)_L(-c/d)_L(a/d)_L(d/a)_L - 1}, \tag{6.65}$$

and, consequently, (6.63) is as well true.

Now, fix $\lambda > 0$ and consider $\mathfrak{C}_\mu^\lambda$. Thanks to Theorem 6.6(a), $\mathfrak{C}_\mu^\lambda$ is unbounded in $\mathbb{R} \times \mathcal{C}_T \times \mathcal{C}_T$. Moreover, due to second estimate of (6.62),

$$P_\mu \mathfrak{C}_\mu^\lambda \cap [0, \infty) = \emptyset.$$

On the other hand, thanks to (6.63), u and v possess uniform a priori bounds in $\mathcal{C}_T$ on compact subsets of (λ, μ). Indeed, it is rather clear that (6.63) shows that u_L and v_L are uniformly bounded. Now, pick $t_u, t_v \in \mathbb{R}$ such that

$$u_L = u(t_u), \qquad v_L = v(t_v).$$

Then,

$$u(t) = e^{\int_{t_u}^t (\lambda\ell - au - bv)} u(t_u), \qquad v(t) = e^{\int_{t_v}^t (\mu m - dv - cu)} v(t_v),$$

must be uniformly bounded, as well. This concludes the proof.   □

Even in the case when the model possesses a stable coexistence state, the solutions of (1.1) might blow up in finite time for sufficiently large initial data. In the case of high symbiosis, where according to Theorem 6.9 Problem (1.2) does not admit a coexistence state if $\lambda > 0$ and $\mu > 0$, all solutions of (1.1) will blow-up in finite time independently of the size of the initial populations. Actually, this is the reason why (1.2) cannot admit a coexistence state. We refrain of giving more details here in, though we send to the interested reader to [11] for further technical details concerning the elliptic counterpart of the model studied here in.

7. The competing species model ($b > 0$ and $c > 0$)

Undoubtedly, this is the case that has attracted more attention in the mathematical literature. Perhaps, because in a certainly pioneer chapter, P. de Mottoni and A. Schiaffino [29] showed that any solution of (1.1) converges in the large to a T-periodic component-wise non-negative solution of (1.2), which extraordinary facilitates the analysis of the dynamics of (1.1). Among the most important works related to this model one should include the most pioneering chapter of J.M. Cushing [7], where the average of the birth rates, λ and μ, were used by the first time as the main bifurcation parameters to show the existence of a continuum, not necessarily a curve, of coexistence states of (1.2) linking the two semi-trivial states of the model. Then, a huge industry grew from these pioneering chapters. Among the most important posterior works one should quote those of S. Ahmad [1], S. Ahmad and A.C. Lazer [2], C. Alvárez and A.C. Lazer [3], J.C. Eilbeck and J. López-Gómez [12], J.K. Hale and A. Somolinos [14], P. Hess and A.C. Lazer [15], J. López-Gómez [17], R. Ortega and A. Tineo [30], H. Smith [32,33], A. Tineo [34,35], and F. Zanolin [36], among many others that are omitted here; many of them authorized by the previous authors.

Rather naturally, in this section we will focus our attention into the results obtained in [17] and [12], where it was proved that the components $\mathfrak{C}^\lambda_\mu$ and $\mathfrak{C}^\mu_\lambda$ are real analytic curves consisting of coexistence states whose attractive character changes when a turning point is crossed. This theorem is a very sharp improvement of a previous result by R. Ortega and A. Tineo [30], were it was shown that either the model possess a segment of coexistence states, or it exhibits a finite number of coexistence states.

Although in the previous references some uniqueness results were given, the problem of the uniqueness remains, in its wide amplitude, entirely open, and it will be treated elsewhere.

7.1. *Periodic systems of quasi-cooperative type*

In this section we study the periodic linear system

$$\begin{cases} u' = \alpha u - \beta v, \\ v' = -\gamma u + \rho v, \end{cases} \tag{7.1}$$

where $\alpha, \beta, \gamma, \rho \in \mathcal{C}_T$ satisfy

$$\int_0^T \alpha < 0, \quad \int_0^T \rho < 0, \quad \beta > 0, \quad \gamma > 0, \tag{7.2}$$

and its associated Cauchy problem

$$\begin{cases} u' = \alpha u - \beta v, \\ v' = -\gamma u + \rho v, \\ u(0) = u_0, \quad v(0) = v_0. \end{cases} \tag{7.3}$$

As an immediate consequence from Theorem 6.1, it follows the next result.

THEOREM 7.1. *Suppose* (7.1) *possesses a solution* $(u, v) \in \mathcal{C}^1_T \times \mathcal{C}^1_T \setminus \{(0, 0)\}$. *Then, either* $u \gg 0$ *and* $v \ll 0$, *or else* $u \ll 0$ *and* $v \gg 0$. *Therefore, all the coexistence states* *of (1.2) must be ordered, in the sense that for any pair of coexistence states, $(u_1, v_1) \neq (u_2, v_2)$, either $u_1 \gg u_2$ and $v_1 \ll v_2$, or $u_1 \ll u_2$ and $v_1 \gg v_2$.*

PROOF. The change of variable

$$(x, y) = (u, -v) \tag{7.4}$$

transforms (7.1) into

$$\begin{cases} x' = \alpha x + \beta y, \\ y' = \gamma x + \rho y, \end{cases} \tag{7.5}$$

which is of the same type as (6.1). Therefore, the result is a corollary of Theorem 6.1. $\square$

Similarly, the following quasi-cooperative counterpart of Proposition 6.2 holds.

PROPOSITION 7.2. *Suppose $(\bar{u}, \bar{v}) \in C^1(\mathbb{R}_+) \times C^1(\mathbb{R}_+)$, $\mathbb{R}_+ := [0, \infty)$, satisfy*

$$\begin{cases} \bar{u}' \geq \alpha \bar{u} - \beta \bar{v}, \\ \bar{v}' \leq -\gamma \bar{u} + \rho \bar{v}, \\ \bar{u}(0) > 0, \quad \bar{v}(0) < 0. \end{cases} \tag{7.6}$$

Then, $\bar{u}(t) > 0$ and $\bar{v}(t) < 0$ for each $t \geq 0$.

If $u_0 - v_0 > 0$ and $u_0 v_0 = 0$, then there exists $t_0 > 0$ such that the unique solution (u, v) of (7.3) satisfies $u(t) > 0$ and $v(t) < 0$ for each $t \geq t_0$.

PROOF. The auxiliary pair

$$(\bar{x}, \bar{y}) := (\bar{u}, -\bar{v})$$

satisfies

$$\begin{cases} \bar{x}' \geq \alpha \bar{x} + \beta \bar{y}, \\ \bar{y}' \geq \gamma \bar{x} + \rho \bar{y}, \\ \bar{x}(0) > 0, \quad \bar{y}(0) > 0, \end{cases}$$

and, hence, thanks to Proposition 6.2, $\bar{x}(t) > 0$ and $\bar{y}(t) > 0$ for each $t \geq 0$, which concludes the proof of the first claim.

Now, suppose $u_0 - v_0 > 0$ and $u_0 v_0 = 0$, and let $(u(t), v(t))$ denote the unique solution of (7.3). Then,

$$(x, y) := (u, -v)$$

satisfies

$$\begin{cases} x' = \alpha x + \beta y, \\ y' = \gamma x + \rho y, \\ x(0) = u_0, \quad y(0) = -v_0, \end{cases}$$

and

$$x(0) + y(0) = u_0 - v_0 > 0, \qquad x(0)y(0) = 0.$$

Thus, it follows from Proposition 6.2 that there exists $t_0 > 0$ such that $x(t) > 0$ and $y(t) > 0$ for each $t \geq t_0$. Therefore, $u(t) > 0$ and $v(t) < 0$ for each $t \geq t_0$, which concludes the proof. $\qquad\square$

Also, the following counterpart of Theorem 6.3 holds. Subsequently, we shall denote by $\Phi(T)$ the Poincaré map of (7.1).

THEOREM 7.3. *The characteristic multipliers of* (7.1), v_1 *and* v_2, *are real and distinct, and satisfy* $0 < v_1 v_2 < 1$. *Moreover, if we order them so that* $0 < v_1 < v_2$, *then,* $0 < v_1 < 1$,

$$N\big[\Phi(T) - v_1 I\big] = \mathrm{span}\big[(x_0, y_0)\big] \quad \textit{with } x_0 > 0 \textit{ and } y_0 > 0,$$

and

$$N\big[\Phi(T) - v_2 I\big] = \mathrm{span}\big[(x_0, y_0)\big] \quad \textit{with } x_0 > 0 \textit{ and } y_0 < 0.$$

In particular, (7.1) *is asymptotically stable if* $v_2 < 1$, *neutrally stable if* $v_2 = 1$, *and unstable if* $v_2 > 1$. *Furthermore, the value*

$$\lambda_0 := -\frac{1}{T}\mathrm{Log}\, v_2 \tag{7.7}$$

provides us with the unique value of $\lambda \in \mathbb{R}$ *for which the eigenvalue problem*

$$\begin{cases} u' = \alpha u - \beta v + \lambda u, \\ v' = -\gamma u + \rho v + \lambda v \end{cases} \tag{7.8}$$

admits a solution $(u, v) \in C_T^1 \times C_T^1$ *such that* $u \gg 0$ *and* $v \ll 0$.

PROOF. As the change of variable (7.4) transforms (7.1) into (7.5), if

$$U(T) = \begin{pmatrix} u_{11} & u_{12} \\ u_{21} & u_{22} \end{pmatrix}$$

denotes the Poincaré map of (7.5), necessarily

$$\Phi(T) = \begin{pmatrix} u_{11} & -u_{12} \\ -u_{21} & u_{22} \end{pmatrix}. \tag{7.9}$$

Indeed, if we denote by $(u(t), v(t))$ the unique solution of (7.3) and $(x(t), y(t))$ stands for the unique solution of (7.5) satisfying $(x(0), y(0)) = (u_0, -v_0)$, then

$$\begin{pmatrix} u(T) \\ v(T) \end{pmatrix} = \begin{pmatrix} x(T) \\ -y(T) \end{pmatrix} = \begin{pmatrix} u_{11} & -u_{12} \\ -u_{21} & u_{22} \end{pmatrix} \begin{pmatrix} x(0) \\ -y(0) \end{pmatrix} = \Phi(T) \begin{pmatrix} u_0 \\ v_0 \end{pmatrix}.$$

Clearly, the eigenvalues of $U(T)$ and $\Phi(T)$ are the same. Moreover, thanks to Theorem 6.3,

$$N[U(T) - \nu_1 I] = \mathrm{span}\big[(\tilde{x}_0, \tilde{y}_0)\big] \quad \text{with } \tilde{x}_0 < 0 \text{ and } \tilde{y}_0 > 0,$$
$$N[U(T) - \nu_2 I] = \mathrm{span}\big[(\tilde{x}_0, \tilde{y}_0)\big] \quad \text{with } \tilde{x}_0 > 0 \text{ and } \tilde{y}_0 > 0,$$

and a direct calculation shows that, for each $j \in \{1, 2\}$, (x, y) is an eigenvector of $U(T)$ associated to the eigenvalue ν_j if, and only if, $(x, -y)$ is an eigenvector of $\Phi(T)$ associated to ν_j. Therefore,

$$N[\Phi(T) - \nu_1 I] = \mathrm{span}\big[(x_0, y_0)\big] \quad \text{with } x_0 > 0 \text{ and } y_0 > 0,$$

and

$$N[\Phi(T) - \nu_2 I] = \mathrm{span}\big[(x_0, y_0)\big] \quad \text{with } x_0 > 0 \text{ and } y_0 < 0,$$

which concludes the proof of the first block of statements.

Finally, note that the change of variable (7.4) transforms (7.8) into the equivalent system

$$\begin{cases} x' = \alpha x + \beta y + \lambda x, \\ y' = \gamma x + \rho y + \lambda y, \end{cases}$$

for which the last assertion of Theorem 6.3 is valid. This concludes the proof. $\square$

Finally, by performing the change of variable (7.4), it is easy to see that the following counterpart of Theorem 6.4 is satisfied.

THEOREM 7.4. *The following properties are satisfied:*

(a) *System (7.1) is stable ($\nu_2 < 1$) if, and only if, there exists $(\bar{u}, \bar{v}) \in C_T^1 \times C_T^1$ with $\bar{u} \gg 0$ and $\bar{v} \ll 0$ such that*

$$\begin{cases} \bar{u}' \geqslant \alpha \bar{u} - \beta \bar{v}, \\ \bar{v}' \leqslant -\gamma \bar{u} + \rho \bar{v}, \end{cases} \tag{7.10}$$

with some of these inequalities strict.

(b) *System (6.1) is unstable ($\nu_2 > 1$) if, and only if, there exits $(\underline{u}, \underline{v}) \in C_T^1 \times C_T^1$ with $\underline{u} \gg 0$ and $\underline{v} \ll 0$ such that*

$$\begin{cases} \underline{u}' \leqslant \alpha \underline{u} - \beta \underline{v}, \\ \underline{v}' \geqslant -\gamma \underline{u} + \rho \underline{v}, \end{cases} \tag{7.11}$$

with some of these inequalities strict.

7.2. *Structure of the components $\mathfrak{C}_\mu^\lambda$ and $\mathfrak{C}_\lambda^\mu$*

As in Section 6.2, the coexistence states of (1.2) will be regarded as zeroes of the operator $\mathfrak{F} : \mathbb{R}^2 \times C_T \times C_T \to C_T \times C_T$ defined by (6.23), where $M > 0$ is fixed. The following counterpart of Proposition 6.5 is satisfied.

PROPOSITION 7.5. *Suppose $(u_0, v_0) \in C_T \times C_T$ satisfies $u_0 \gg 0$, $v_0 \gg 0$, and*

$$\mathfrak{F}(\lambda, \mu, u_0, v_0) = 0.$$

Then, each of the operators

$$D_{(\lambda, u, v)} \mathfrak{F}(\lambda, \mu, u_0, v_0) : \mathbb{R} \times C_T \times C_T \to C_T \times C_T$$

and

$$D_{(\mu, u, v)} \mathfrak{F}(\lambda, \mu, u_0, v_0) : \mathbb{R} \times C_T \times C_T \to C_T \times C_T$$

is surjective. In other words, $(0, 0)$ is a regular value of $\mathfrak{F}$ with respect to $\mathbb{R}^2 \times \mathrm{Int}\, P \times \mathrm{Int}\, P$.

PROOF. As in the proof of Proposition 6.5, differentiating gives (6.24), (6.25) and (6.26). Similarly, $D_{(\lambda, u, v)} \mathfrak{F}(\lambda, \mu, u_0, v_0)$ and $D_{(\mu, u, v)} \mathfrak{F}(\lambda, \mu, u_0, v_0)$ are surjective if (λ, μ, u_0, v_0) is non-degenerate. So, suppose (λ, μ, u_0, v_0) is degenerate. Then, since $D_{(u, v)} \mathfrak{F}(\lambda, \mu, u_0, v_0)$ is a compact perturbation of the identity,

$$N := \dim N \big[D_{(u,v)} \mathfrak{F}(\lambda, \mu, u_0, v_0) \big] = \mathrm{codim}\, R \big[D_{(u,v)} \mathfrak{F}(\lambda, \mu, u_0, v_0) \big] \geqslant 1,$$

and $N[D_{(u,v)} \mathfrak{F}(\lambda, \mu, u_0, v_0)]$ is the set of T-periodic solutions (u, v) of the linear system

$$\begin{cases} u' = \alpha_0 u - b u_0 v, \\ v' = -c v_0 u + \beta_0 v. \end{cases} \tag{7.12}$$

Since $b > 0$ and $c > 0$, we have that

$$-b u_0 < 0 \quad \text{and} \quad -c v_0 < 0.$$

Moreover, since (λ, μ, u_0, v_0) is a coexistence state of (1.2),

$$\int_0^T \alpha_0 = -\int_0^T a(t)u_0(t)\,dt < 0 \quad \text{and} \quad \int_0^T \beta_0 = -\int_0^T d(t)v_0(t)\,dt < 0.$$

Therefore, (7.12) fits into the abstract setting of Section 7.1. Thus, thanks to Theorem 7.3, the biggest multiplier of (7.12), denoted by ν_2, equals one and any T-periodic solution of (7.12) must be a multiple of a fixed T-periodic solution (φ, ψ) such that $\varphi \gg 0$ and $\psi \ll 0$. In particular, $N = 1$. Now, due to (6.25) and (6.26), to conclude the proof it suffices to prove (6.29). On the contrary, suppose that, e.g.,

$$-\mathfrak{R}_M(\ell u_0, 0) \in R\big[D_{(u,v)}\mathfrak{F}(\lambda, \mu, u_0, v_0)\big].$$

Then, the system

$$\begin{cases} u' = \alpha_0 u - b u_0 v + \ell u_0, \\ v' = -c v_0 u + \beta_0 v, \end{cases} \tag{7.13}$$

possesses a T-periodic solution (u, v). Actually, for each $R > 0$, the pair

$$(u_R, v_R) := (u + R\varphi, v + R\psi)$$

provides us with a T-periodic solution of (7.13) such that, for sufficiently large $R > 0$,

$$u_R \gg 0 \quad \text{and} \quad v_R \ll 0.$$

Moreover, it follows from $\ell u_0 > 0$ that

$$\begin{cases} u_R' > \alpha_0 u_R - b u_0 v_R, \\ v_R' = -c v_0 u_R + \beta_0 v_R, \end{cases}$$

and, therefore, thanks to Theorem 7.4(a), $\nu_2 < 1$, which is a contradiction, since $\nu_2 = 1$. Similarly, the second relation of (6.29) holds. This concludes the proof. $\qquad\square$

The following result ascertains the global structure of the coexistence states of (1.2) for the competing species model. To reduce the complexity of the number of cases to be considered in our analysis we shall throughout assume that conditions (4.10) are satisfied, though most of the results are still valid in the absence of (4.10)—such restrictions are exclusively needed in order to get some non-existence results entailing the relative compactness of the components $\mathfrak{C}_\mu^\lambda$ and $\mathfrak{C}_\lambda^\mu$.

THEOREM 7.6. *Assume $b > 0$, $c > 0$, and (1.2) exhibits a coexistence state, say (λ, μ, u_0, v_0). Then, $\lambda > 0$, $\mu > 0$, and*

$$u_0 \ll \theta_{[\lambda \ell, a]}, \qquad v_0 \ll \theta_{[\mu m, d]}. \tag{7.14}$$

Moreover, the following properties are satisfied:

(a) *For each $\lambda > 0$ there exists $\mu_1(\lambda) > 0$ such that (1.2) does not admit a coexistence state if $\mu \geqslant \mu_1(\lambda)$. Moreover, the component of coexistence states $\mathfrak{C}^{\lambda}_{\mu}$ constructed in Theorem 5.4 is bounded in $\mathbb{R} \times C_T \times C_T$ and it consists of a real analytic curve. More precisely, there exits a real analytic map $(\mu, u, v) : [0, 1] \to \mathbb{R} \times \mathrm{Int}\, P \times \mathrm{Int}\, P$ such that*

$$\big(\mu(0), u(0), v(0)\big) = (\mu_0, \theta_{[\lambda\ell,a]}, 0),$$

$$\big(\mu(1), u(1), v(1)\big) = (\mu_*, 0, \theta_{[\mu_*m,d]}),$$

and

$$\mathfrak{C}^{\lambda}_{\mu} = \big\{ \big(\mu(s), u(s), v(s)\big) : s \in (0, 1)\big\},$$

where $\mu_* > 0$ is the unique value of μ satisfying (5.40).

(b) *For each $\mu > 0$ there exists $\lambda_1(\mu) > 0$ such that (1.2) does not admit a coexistence state if $\lambda \geqslant \lambda_1(\mu)$. Moreover, the component of coexistence states $\mathfrak{C}^{\mu}_{\lambda}$ constructed in Theorem 5.5 is bounded in $\mathbb{R} \times C_T \times C_T$ and it consists of a real analytic curve. More precisely, there exists a real analytic map $(\lambda, u, v) : [0, 1] \to \mathbb{R} \times \mathrm{Int}\, P \times \mathrm{Int}\, P$ such that*

$$\big(\lambda(0), u(0), v(0)\big) = (\lambda_0, 0, \theta_{[\mu m,d]}),$$

$$\big(\lambda(1), u(1), v(1)\big) = (\lambda_*, \theta_{[\lambda_*\ell,a]}, 0),$$

and

$$\mathfrak{C}^{\mu}_{\lambda} = \big\{ (\lambda(s), u(s), v(s)) : s \in (0, 1)\big\},$$

where $\lambda_* > 0$ is the unique value of λ satisfying (5.46).

(c) *The set of coexistence states of (1.2) $(\lambda, \mu, u, v) \in \mathbb{R}^2 \times \mathrm{Int}\, P \times \mathrm{Int}\, P$ such that, for some $\mu \in \mathbb{R}$, $(\mu, u, v) \in \mathfrak{C}^{\lambda}_{\mu}$, or, for some $\lambda \in \mathbb{R}$, $(\lambda, u, v) \in \mathfrak{C}^{\mu}_{\lambda}$, is a real analytic surface linking the surfaces of semi-trivial states*

$$\big\{ (\lambda, \mu, \theta_{[\lambda\ell,a]}, 0) : \lambda > 0, \ \mu \in \mathbb{R} \big\}$$

and

$$\big\{ (\lambda, \mu, 0, \theta_{[\mu m,d]}) : \lambda \in \mathbb{R}, \ \mu > 0 \big\}$$

along their respective curves of neutral stability

$$\mu = \frac{1}{T} \int_0^T c(t)\theta_{[\lambda\ell,a]}(t)\, \mathrm{d}t \quad and \quad \lambda = \frac{1}{T} \int_0^T b(t)\theta_{[\mu m,d]}(t)\, \mathrm{d}t.$$

Therefore, (1.2) possesses a coexistence state if both semi-trivial states are, simultaneously, linearly unstable, or linearly stable.

PROOF. The fact that the components $\mathfrak{C}^{\lambda}_{\mu}$ and $\mathfrak{C}^{\mu}_{\lambda}$ are real analytic curves follows from Proposition 7.5 adapting the argument of the proof of Theorem 6.6. Now, we shall show that both components are bounded. The remaining assertions of the theorem follow straight ahead from Theorems 5.4 and 5.5.

Suppose (λ, μ, u_0, v_0) is a coexistence state of (1.2). Then, dividing by u_0 the u-equation, by v_0 the v-equation, and integrating in $[0, T]$ gives

$$\lambda = \frac{1}{T} \int_0^T (au_0 + bv_0) > 0 \quad \text{and} \quad \mu = \frac{1}{T} \int_0^T (dv_0 + cu_0) > 0.$$

Thus, $\lambda > 0$ and $\mu > 0$ are necessary for the existence of a coexistence state. So, throughout the remaining of the proof we will assume that $\lambda > 0$ and $\mu > 0$.

Fix $\lambda > 0$. Then, $u_0 \gg 0$ is a positive strict subsolution of the problem

$$u' = \lambda \ell - au^2, \quad u \in \mathcal{C}^1_T,$$

since $bu_0 v_0 > 0$, and, hence, thanks to Proposition 2.6, $u_0 \ll \theta_{[\lambda \ell, a]}$. Similarly, $v_0 \ll \theta_{[\mu m, d]}$. Thus,

$$v_0 = \mu m v_0 - dv_0^2 - cu_0 v_0 > (\mu m - c\theta_{[\lambda \ell, a]})v_0 - dv_0^2.$$

Hence, if

$$\mu > \frac{1}{T} \int_0^T c(t)\theta_{[\lambda \ell, a]}(t)\, dt,$$

then, thanks again to Proposition 2.6, we have that

$$v_0 \gg \theta_{[\mu m - c\theta_{[\lambda \ell, a]}, d]} := \Theta_\mu,$$

and, so,

$$u_0 < (\lambda \ell - b\Theta_\mu)u_0 - au_0^2.$$

Consequently, dividing by u_0 and integrating in $[0, T]$ gives

$$\lambda > \frac{1}{T} \int_0^T b(t)\Theta_\mu(t)\, dt. \tag{7.15}$$

To show the existence of $\mu_1(\lambda)$ such that (1.2) cannot admit a coexistence state if $\mu \geq \mu_1$ it suffices to show that (7.15) cannot be satisfied for sufficiently large μ. To show this, note that the function Ψ_μ defined through $\Theta_\mu := \mu \Psi_\mu$ satisfies

$$\frac{1}{\mu}\Psi'_\mu = \left(m - \frac{c\theta_{[\lambda \ell, a]}}{\mu}\right)\Psi_\mu - d\Psi_\mu^2.$$

Adapting the argument of the proof of (2.20) it is apparent that $\lim_{\mu\uparrow\infty}\Psi_\mu = \frac{m}{d}$ uniformly in $\mathbb{R}$, since $\lim_{\mu\uparrow\infty}\frac{c\theta_{[\lambda\ell,a]}}{\mu} = 0$. Therefore,

$$\lim_{\mu\uparrow\infty}\Theta_\mu = \infty$$

uniformly in $[0, T]$, and, consequently, (7.15) fails to be true for sufficiently large μ. By symmetry, for any $\mu > 0$ there exists $\lambda_1(\mu) > 0$ such that (1.2) does not admit a coexistence state if $\lambda \geqslant \lambda_1(\mu)$.

The previous bounds show that $\mathfrak{C}_\mu^\lambda$ and $\mathfrak{C}_\lambda^\mu$ must be bounded for each $\lambda > 0$ and $\mu > 0$. The remaining assertions of the theorem follow readily from Theorems 5.4, 5.5 and 5.6. $\square$

For the autonomous counterpart of (1.2) with $bc = 1$, is easy to check that, for any $\lambda > 0$, the curve

$$\big(\mu(s), u(s), v(s)\big) = (c\lambda, \lambda - bs, s), \quad 0 \leqslant s \leqslant \frac{\lambda}{b},$$

provides us with the component $\mathfrak{C}_\mu^\lambda$ bifurcating from the semi-trivial solution $(\lambda, 0)$ at $\mu = c\lambda$. As this simple example shows, condition $\dot\mu = 0$ cannot be avoided in general, unless some additional assumptions are imposed; e.g., that the semi-trivial states cannot be simultaneously neutrally stable.

As in the symbiotic model, fixing λ, or μ, and due to Proposition 7.5, any coexistence state of (1.2), say (λ, μ, u_0, v_0), must lay into a real analytic curve of coexistence states of (1.2). Fix, e.g., λ and let Γ denote the trajectory of the μ–curve of coexistence states containing (λ, μ, u_0, v_0). As a consequence from the existence of uniform a priori bounds in $\mathcal{C}_T$ for the coexistence states of (1.2) (cf. Theorem 7.6), Γ must be bounded. If $\Gamma \cap \mathfrak{C}_\mu^\lambda \neq \emptyset$, then $\Gamma = \mathfrak{C}_\mu^\lambda$. Thus, if $\Gamma \neq \mathfrak{C}_\mu^\lambda$, necessarily $\Gamma \cap \mathfrak{C}_\mu^\lambda = \emptyset$, and, consequently, Γ must be an isolated component with respect of the surfaces of semi-trivial positive solutions of (1.2). Suppose such a component exists. Then, by varying an additional parameter of the model, e.g., the amplitude of b, or c, it is possible to show that when b, or c are sufficiently small, then (1.2) possesses a unique coexistence state, necessarily belonging to $\mathfrak{C}_\mu^\lambda$. Therefore, at some intermediate value of the *unfolding parameter*, the isolated component Γ must reduce to a single point. Thanks to Proposition 7.5 this is impossible. Therefore, the only coexistence states of (1.2) are those contained in the components $\mathfrak{C}_\mu^\lambda$. The technical details of this analysis will appear elsewhere.

As an easy consequence from Theorem 7.6, except in the case when both semi-trivial states are neutrally stable, (1.2) possesses, at most, a finite number of coexistence states and, according to Theorem 7.1, all of them must be ordered, which provides us with the main result of R. Ortega and A. Tineo [30].

In the case when both semi-trivial states are neutrally stable, either there is an analytic curve of coexistence states linking both states, or the model possesses a finite number of coexistence states; its cardinal depending on the values of the several parameter functions involved in the setting of the model.

7.3. *Stability of the coexistence states along $\mathfrak{C}_\mu^\lambda$ and $\mathfrak{C}_\lambda^\mu$*

Fix $\lambda > 0$ and let $(\mu(s), u(s), v(s))$, $s \geq 0$, be the real analytic map constructed in Theorem 7.6, where $s \in (0, L)$ is assumed to be the length of arc of curve. Subsequently, we shall denote $\cdot := \frac{d}{ds}$. Repeating the argument of Section 6.3, we have (6.31) and (6.32). Moreover, since

$$\big(\mu(0), u(0), v(0)\big) = (\mu_0, \theta_{[\lambda\ell,a]}, 0),$$

it follows from Theorem 5.6 that

$$q_s := \dot{v}(s) = \varphi_v + O(s) = e^{\int_0^s (\mu_0 m - c\theta_{[\lambda\ell,a]})} + O(s) \gg 0, \quad s \downarrow 0. \tag{7.16}$$

Similarly, setting

$$(p_s, q_s) := \big(\dot{u}(s), \dot{v}(s)\big), \quad s > 0,$$

(6.31) can be rewritten in the form

$$\begin{cases} (p_s)' = \big[\lambda\ell - 2au(s) - bv(s)\big]p_s - bu(s)q_s, \\ (q_s)' = -cv(s)p_s + \big[\mu(s)m - 2dv(s) - cu(s)\big]q_s + \dot{\mu}(s)mv(s). \end{cases} \tag{7.17}$$

In the competing species model, we have that $-bu(s) < 0$, $-cv(s) < 0$, (6.35) and (6.36). Thus, (7.17) fits within the general setting of Section 7.1. In particular, thanks to Proposition 2.1, we find from the first equation of (7.17) that, for each sufficiently small $s > 0$,

$$p_s = \mathfrak{R}_{-[\lambda\ell - 2au(s) - bv(s)]}\big(-bu(s)q_s\big) \ll 0.$$

Therefore, there exists $\varepsilon > 0$ such that

$$p_s \ll 0 \quad \text{and} \quad q_s \gg 0 \quad \text{for each } s \in [0, \varepsilon]. \tag{7.18}$$

Suppose $\mu(s) = \mu_0$ for each $s \in [0, L]$. Then, (7.17) becomes into (6.38) and, due to (6.32), $(p_s, q_s) \neq (0, 0)$. Thus, thanks to Theorem 7.1, for each $s \in [0, L]$, either $p_s \gg 0$ and $q_s \ll 0$, or $p_s \ll 0$ and $q_s \gg 0$. Thanks to (7.18), necessarily

$$p_s \ll 0 \quad \text{and} \quad q_s \gg 0 \quad \text{for each } s \in [0, L]. \tag{7.19}$$

Indeed, since $s \to (p_s, q_s)$ is analytic, if (7.19) is not satisfied, then, due to (6.37), there is some value of s, say $\tilde{s} \in (0, L]$ where $p_{\tilde{s}} < 0$, $q_{\tilde{s}} > 0$, and, e.g., $p_{\tilde{s}} \ll 0$ fails. Due to Theorem 7.1, this implies $p_{\tilde{s}} \ll 0$ and $q_{\tilde{s}} \gg 0$, which is a contradiction. Thus, (7.19) holds true. Consequently, thanks to Theorem 7.3, the principal characteristic multiplier of $(\mu(s), u(s), v(s))$, $v_2(s)$, equals 1 for each $s \in [0, L]$. Therefore, if $\dot{\mu} = 0$, then $\dot{v}_2 = 0$ and, hence, all the coexistence states of $\mathfrak{C}_\mu^\lambda$ are neutrally stable. When $\dot{\mu} \neq 0$, the following result is satisfied.

THEOREM 7.7. *Suppose $\dot{\mu} \neq 0$ in $[0, L]$. Then, the set*

$$\mathcal{S} := \left\{ s \in (0, L) \colon \dot{\mu}(s) = 0 \right\} \tag{7.20}$$

is finite (possibly empty), each $s \in \mathcal{S}$ is a zero of finite order of $\dot{\mu}$, and

$$\mathcal{S} = \left\{ s \in (0, L) \colon \nu_2(s) = 1 \right\}. \tag{7.21}$$

Moreover,

$$\dot{\mu}(s)\left(1 - \nu_2(s)\right) > 0, \quad s \in (0, L) \setminus \mathcal{S}, \tag{7.22}$$

and

$$p_s \ll 0 \quad \text{and} \quad q_s \gg 0 \quad \text{whenever} \quad \dot{\mu}(s) \geqslant 0. \tag{7.23}$$

PROOF. The fact that $\mathcal{S}$ is, at most, finite and that each $s \in \mathcal{S}$ has finite order—as a zero of $\dot{\mu}$—are consequences from the fact that $\dot{\mu}$ is real analytic. Now, we shall prove (7.21). Pick $s \in \mathcal{S}$. Then, (p_s, q_s) satisfies (6.38) and, due to (6.32), $(p_s, q_s) \neq (0, 0)$. Thus, thanks to Theorem 7.1, $p_s \ll 0$ and $q_s \gg 0$, or $p_s \gg 0$ and $q_s \ll 0$. In any of these cases, we find from Theorem 7.3 that $\lambda = 0$ is the *principal characteristic exponent* of $(\mu(s), u(s), v(s))$ and, hence, $\nu_2(s) = 1$. To prove the converse inclusion, let $s \in (0, L)$ such that $\nu_2(s) = 1$. Then, thanks to Theorem 7.3, there exists a T-periodic pair (P, Q) with $P \gg 0$ and $Q \ll 0$ such that

$$\begin{cases} P' = \left[\lambda \ell - 2au(s) - bv(s)\right]P - bu(s)Q, \\ Q' = -cv(s)P + \left[\mu(s)m - 2dv(s) - cu(s)\right]P. \end{cases} \tag{7.24}$$

Now, choose a sufficiently small $\varepsilon > 0$ so that the auxiliary pair

$$(U, V) := (P - \varepsilon p_s, Q - \varepsilon q_s),$$

satisfy $U \gg 0$ and $V \ll 0$. Thanks to (7.17) and (7.24) we have that

$$\begin{cases} U' = \left[\lambda \ell - 2au(s) - bv(s)\right]U - bu(s)V, \\ V' = -cv(s)U + \left[\mu(s)m - 2dv(s) - cu(s)\right]V - \varepsilon\dot{\mu}(s)mv(s). \end{cases}$$

Assume $\dot{\mu}(s) > 0$. Then, since $\varepsilon mv(s) > 0$, the pair (U, V) satisfies

$$\begin{cases} U' = \left[\lambda \ell - 2au(s) - bv(s)\right]U - bu(s)V, \\ V' < -cv(s)U + \left[\mu(s)m - 2dv(s) - cu(s)\right]V, \end{cases}$$

and, hence, thanks to Theorem 7.4(a), $v_2(s) < 1$, which is a contradiction. Assume $\dot{\mu}(s) < 0$. Then, (U, V) satisfies

$$\begin{cases} U' = \big[\lambda\ell - 2au(s) - bv(s)\big]U - bu(s)V, \\ V' > -cv(s)U + \big[\mu(s)m - 2dv(s) - cu(s)\big]V, \end{cases}$$

and, due to Theorem 7.4(b), $v_2(s) > 1$, which is a contradiction again. Consequently, $\dot{\mu}(s) = 0$ and, therefore, (7.21) holds true.

Now, pick $\tilde{s} \in \mathcal{S}$. Thanks to (6.32), it follows from Theorem 7.1 that some of the following alternatives occurs:

1. $p_{\tilde{s}} \gg 0$ and $q_{\tilde{s}} \ll 0$.
2. $p_{\tilde{s}} \ll 0$ and $q_{\tilde{s}} \gg 0$.

Assume alternative 1 occurs. As $\tilde{s}$ is an isolated zero of $\dot{\mu}(s)$, there exists $\varepsilon > 0$ such that $p_s = \dot{u}(s) \gg 0$ and $q_s = \dot{v}(s) \ll 0$ for each $s \in (\tilde{s} - \varepsilon, \tilde{s} + \varepsilon)$, and

$$\dot{\mu}^{-1}(0) \cap (\tilde{s} - \varepsilon, \tilde{s} + \varepsilon) \cap (0, L) = \{\tilde{s}\}.$$

Thus, for each

$$s \in J := (0, L) \cap \big[(\tilde{s} - \varepsilon, \tilde{s} + \varepsilon) \setminus \{\tilde{s}\}\big],$$

either $\dot{\mu}(s) > 0$, or $\dot{\mu}(s) < 0$. Suppose $\dot{\mu}(s) > 0$. Then,

$$\begin{cases} (p_s)' = \big[\lambda\ell - 2au(s) - bv(s)\big]p_s - bu(s)q_s, \\ (q_s)' > -cv(s)p_s + \big[\mu(s)m - 2dv(s) - cu(s)\big]q_s, \end{cases} \tag{7.25}$$

and it follows from Theorem 7.4(b) that $v_2(s) > 1$. Hence, $\dot{\mu}(s)(1 - v_2(s)) < 0$. Now, suppose $\dot{\mu}(s) < 0$. Then,

$$\begin{cases} (p_s)' = \big[\lambda\ell - 2au(s) - bv(s)\big]p_s - bu(s)q_s, \\ (q_s)' < -cv(s)p_s + \big[\mu(s)m - 2dv(s) - cu(s)\big]q_s, \end{cases} \tag{7.26}$$

and it follows from Theorem 7.4(a) that $v_2(s) < 1$. Hence, $\dot{\mu}(s)(1 - v_2(s)) < 0$. Therefore, the following condition is satisfied

$$\dot{\mu}(s)\big(1 - v_2(s)\big) < 0, \quad s \in (0, L) \cap \big[(\tilde{s} - \varepsilon, \tilde{s} + \varepsilon) \setminus \{\tilde{s}\}\big], \tag{7.27}$$

for each $\tilde{s} \in \mathcal{S}$ satisfying alternative 1.

Now, suppose $\tilde{s} \in \mathcal{S}$ satisfies alternative 2. As $\tilde{s}$ is an isolated zero of $\dot{\mu}(s)$, there exists $\varepsilon > 0$ such that $p_s = \dot{u}(s) \ll 0$ and $q_s = \dot{v}(s) \gg 0$ for each $s \in (\tilde{s} - \varepsilon, \tilde{s} + \varepsilon)$, and

$$\dot{\mu}^{-1}(0) \cap (\tilde{s} - \varepsilon, \tilde{s} + \varepsilon) \cap (0, L) = \{\tilde{s}\}.$$

Thus, for each

$$s \in J := (0, L) \cap \left[(\tilde{s} - \varepsilon, \tilde{s} + \varepsilon) \setminus \{\tilde{s}\} \right],$$

either $\dot{\mu}(s) > 0$, or $\dot{\mu}(s) < 0$. Suppose $\dot{\mu}(s) > 0$. Then,

$$\begin{cases} (-p_s)' = \left[\lambda \ell - 2au(s) - bv(s) \right](-p_s) - bu(s)(-q_s), \\ (-q_s)' < -cv(s)(-p_s) + \left[\mu(s)m - 2dv(s) - cu(s) \right](-q_s), \end{cases} \tag{7.28}$$

with $-p_s \gg 0$ and $-q_s \ll 0$, and, hence, due to Theorem 7.4(a), $v_2(s) < 1$. Thus, $\dot{\mu}(s)(1 - v_2(s)) > 0$. Similarly, if $\dot{\mu}(s) < 0$, then, due to Theorem 7.4(b), $v_2(s) < 1$. Therefore, condition (7.22) is satisfied for each $\tilde{s} \in \mathcal{S}$ satisfying alternative 2.

Now, suppose

$$\mathcal{S} = \{s_n : 1 \leqslant n \leqslant N\}, \quad 0 < s_n < s_{n+1}, \quad 1 \leqslant n \leqslant N - 1;$$

the subsequent argument is easily adapted to cover the case when $\mathcal{S}$ is empty. Thanks to (7.19), for sufficiently small $s > 0$, $p_s \ll 0$ and $q_s \gg 0$. Moreover, either $\dot{\mu}(s) > 0$ for each $s \in (0, s_1)$, or $\dot{\mu}(s) < 0$ for each $s \in (0, s_1)$.

Suppose $\dot{\mu}(s) > 0$ for each $s \in (0, s_1)$. Then, for sufficiently small $s > 0$, $(-p_s, -q_s)$ satisfies (7.28) and, hence, thanks to Theorem 7.4(a), $v_2(s) < 1$. Actually, this is the inter-exchange stability principle between the semi-trivial state $(\mu, \theta_{[\lambda\ell, a]}, 0)$ and the bifurcating coexistence state $(\mu(s), u(s), v(s))$ as μ crosses the bifurcation value μ_0. Thanks to (7.21), necessarily $v_2(s) < 1$ for each $s \in (0, s_1)$ and, therefore,

$$\dot{\mu}(s)\left(1 - v_2(s)\right) > 0, \quad s \in (0, s_1). \tag{7.29}$$

Necessarily $p_{s_1} \ll 0$ and $q_{s_1} \gg 0$. Indeed, if $p_{s_1} \gg 0$ and $q_{s_1} \ll 0$, then (7.27) holds, which contradicts (7.29). Therefore,

$$\dot{\mu}(s)\left(1 - v_2(s)\right) > 0, \quad s \in (0, s_2) \setminus \{s_2\}. \tag{7.30}$$

Reiterating this argument, (7.22) holds true.

Now, suppose $\dot{\mu}(s) < 0$ for each $s \in (0, s_1)$. Then, $(-p_s, -q_s)$ satisfies

$$\begin{cases} (-p_s)' = \left[\lambda \ell - 2au(s) - bv(s) \right](-p_s) - bu(s)(-q_s), \\ (-q_s)' > -cv(s)(-p_s) + \left[\mu(s)m - 2dv(s) - cu(s) \right](-q_s), \end{cases} \tag{7.31}$$

and, hence, thanks to Theorem 7.4(b), $v_2(s) > 1$. As in the previous case, this provides us with the inter-exchange stability principle between the semi-trivial state $(\mu, \theta_{[\lambda\ell, a]}, 0)$ and the bifurcating coexistence state $(\mu(s), u(s), v(s))$. Thanks to (7.21), $v_2(s) > 1$ for each $s \in (0, s_1)$ and, therefore, (7.22) is as well satisfied. Arguing as above, the proof of (7.22) is concluded.

Note that, as a result from the previous analysis, we have that

$$p_{\tilde{s}} \ll 0 \quad \text{and} \quad q_{\tilde{s}} \gg 0 \quad \text{for each } \tilde{s} \in \mathcal{S} \cup \{0\}. \tag{7.32}$$

Finally, let $s_1^*, s_2^* \in \mathcal{S} \cup \{0\}$ such that

$$\dot{\mu}(s) > 0, \quad s \in \left(s_1^*, s_2^*\right).$$

Thanks to (7.32), $p_s \ll 0$ and $q_s \gg 0$ for each $s > s_1^*$ sufficiently close to s_1^*. Thus, either $p_s \ll 0$ and $q_s \gg 0$ for each $s \in (s_1^*, s_2^*)$, or there exist $s_0 \in (s_1^*, s_2^*)$ and $t \in \mathbb{R}$ such that $p_{s_0} \leqslant 0$ and $q_{s_0} \geqslant 0$ and either $p_{s_0}(t) = 0$, or $q_{s_0}(t) = 0$. Since

$$-cv(s_0)p_{s_0} + \dot{\mu}(s_0)mv(s_0) > 0,$$

we find from Proposition 2.1 applied to the second equation of (7.17) that $q_{s_0} \gg 0$. Thus, $-bu(s_0)q_{s_0} < 0$ and, thanks again to Proposition 2.1, the first equation of (7.17) gives $p_{s_0} \ll 0$, which is impossible. Therefore, for each $s \in (s_1^*, s_2^*)$, $p_s \ll 0$ and $q_s \gg 0$, which concludes the proof of (7.23). The proof of the theorem is completed. $\qquad\square$

7.4. *Some bifurcation diagrams*

In order to study the curves $\mathfrak{C}_\lambda^\mu$ we did carried out some numerical investigations in [12], where we solved (1.2) using spectral collocation methods coupled with path-following techniques. To compute T-periodic solutions of (1.2), the most appropriate set of basis functions to use is trigonometric polynomials, and these were used to generate the bifurcation diagrams of Figs. 17 and 18. Fig. 17 shows the approximated components $\mathfrak{C}_\lambda^\mu$ for the following problems, with $T = 1$ and $a = d = 1$,

(a) $\mu = 3.32$, $m(t) = 1 + 0.7 \sin(2\pi t)$, $\ell(t) = 1 + 0.7 \cos(2\pi t)$, $b(t) = 0.9 + 0.7 \sin(2\pi t + \pi/4)$, $c(t) = 0.9 + 0.7 \cos(2\pi t + \pi/4)$.

(b) $\mu = 3.32$, $m(t) = 1 + 0.7 \sin(2\pi t)$, $\ell(t) = 1 + 0.7 \cos(2\pi t)$, $b(t) = 0.9 + 0.7 \sin(2\pi t)$, $c(t) = 0.9 + 0.7 \cos(2\pi t)$.

(c) $\mu = 5.00$, $\ell(t) = m(t) = 1 + 0.7 \sin(2\pi t)$, $b(t) = c(t) = 0.9 + 0.7 \sin(2\pi t)$.

(d) $\mu = 3.32$, $\ell(t) = m(t) = 1 - 0.7 \sin(2\pi t)$, $b(t) = c(t) = 1.1 + 0.7 \sin(2\pi t)$.

Given a $\mu > 0$, we have plotted the parameter λ along the horizontal axis, and along the vertical one we give the value $\frac{\|u\|_2}{2} + \|v\|_2$, where

$$\|f\|_2 = \sqrt{\sum_{j=1}^{\infty} c_j^2} \quad \text{if} \quad f = c_1 + \sum_{j=1}^{\infty} \left[c_{2j} \sin(2\pi j t) + c_{2j+1} \cos(2\pi j t) \right].$$

This particular choice of norm makes it easy to distinguish between the semi-trivial states $(\lambda, \mu, \theta_{[\lambda \ell, a]}, 0)$ and $(\lambda, \mu, 0, \theta_{[\mu m, d]})$, which would otherwise cross near the bifurcation points and confuse the picture. Stable solutions are represented by continuous lines, unstable ones by dashed lines. Several different cases are shown according to the nature of the

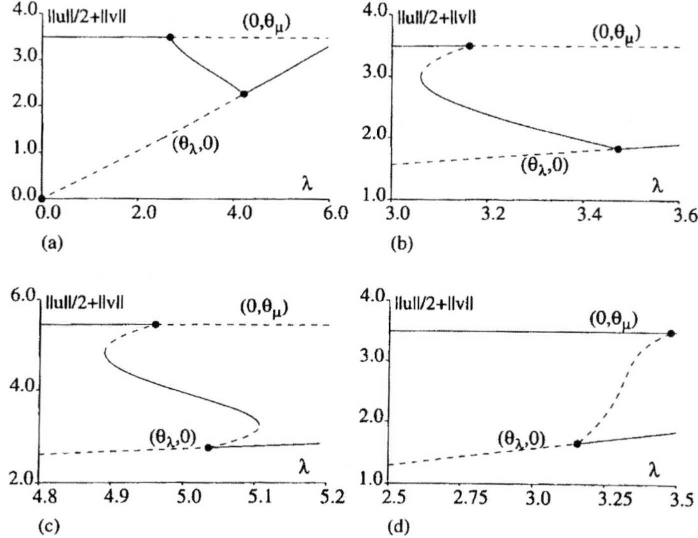

Fig. 17. Bifurcation diagrams for $\mu > 0$ constant.

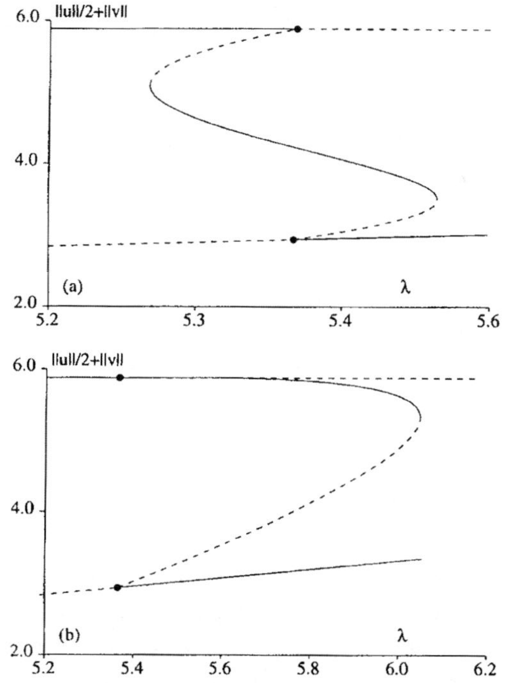

Fig. 18. Two examples with both semi-trivial states neutrally stable.

bifurcation to coexistence states from the semi-trivial solutions. In all cases, $\mathcal{C}_\lambda^\mu$ possesses at most a finite number of hysteresis and turning points and the stability of the coexistence states change according to Theorem 7.7. Fig. 18 shows some of the possible situations when both semi-trivial states are neutrally stable. In some cases, (1.2) possesses a coexistence states, while in others it does not. Fig. 18 shows the following two cases, with $T = 1$ and $a = d = 1$:

(a) $\mu = 5.365$, $m(t) = 1 + 0.7\sin(2\pi t)$, $\ell(t) = 1 + 0.7\cos(2\pi t)$, $b(t) = c(t) = 0.9 + 0.7\sin(2\pi t)$.

(b) $\mu = 5.365$, $m(t) = 1 + 0.7\sin(2\pi t)$, $\ell(t) = 1 + 0.7\cos(2\pi t)$, $b(t) = 0.9 + 0.7\sin(2\pi t)$, $c(t) = 0.9 + 0.7\cos(2\pi t)$.

Case (a) provides us with an example where both semi-trivial states are neutrally stable and (1.2) does not admit a coexistence state. It should be noted that, thanks to the discussion closing Section 7.3, (1.2) cannot admit a further coexistence state outside $\mathcal{C}_\lambda^\mu$. Case (b) provides us with an example where both semi-trivial states are neutrally stable and (1.2) possesses a unique coexistence state, which is stable.

7.5. *Dynamics of* (1.1)

Thanks to the theory developed by P. de Mottoni and A. Schiaffino [29], all solutions of (1.1) must approach a T-periodic solution of (1.2) as time passes by. Combining this information with the theory developed in this section, one can easily obtain the validity of the following assertions. Subsequently, we shall denote by $(u(t), v(t))$ the unique positive solution of (1.1).

1. If $\lambda \leqslant 0$ and $\mu \leqslant 0$, then $\lim_{t\uparrow\infty} (u(t), v(t)) = (0, 0)$.
2. If $\lambda > 0$ and $\mu \leqslant 0$, then

$$\lim_{t\uparrow\infty} \big(u(t), v(t)\big) = (\theta_{[\lambda\ell,a]}, 0). \tag{7.33}$$

3. If $\lambda \leqslant 0$ and $\mu > 0$, then

$$\lim_{t\uparrow\infty} \big(u(t), v(t)\big) = (0, \theta_{[\mu m,d]}). \tag{7.34}$$

4. If

$$\lambda > 0, \quad 0 < \mu < \frac{1}{T}\int_0^T c(t)\theta_{[\lambda\ell,a]}(t)\,\mathrm{d}t, \tag{7.35}$$

and (1.2) does not admit a coexistence state, then, (7.33) holds true.
5. If (7.35) is satisfied, and (1.2) has a coexistence state, then, either (7.33) holds true, or there exists a coexistence state (u_0, v_0) of (1.2) such that

$$\lim_{t\uparrow\infty} \big(u(t), v(t)\big) = (u_0, v_0). \tag{7.36}$$

Moreover, in this case the number of stable coexistence states of (1.2) equals the number of unstable coexistence states, if there is some.

6. If

$$\mu > 0, \quad 0 < \lambda < \frac{1}{T} \int_0^T b(t)\theta_{[\mu m,d]}(t)\,dt, \tag{7.37}$$

and (1.2) does not admit a coexistence state, then, (7.34) holds true.

7. If (7.37) is satisfied and (1.2) has a coexistence state, then, either (7.34) holds true, or there exists a coexistence state (u_0, v_0) of (1.2) satisfying (7.36). Moreover, in this case the number of stable coexistence states of (1.2) equals the number of unstable coexistence states, if there is some.

8. If

$$\lambda > \frac{1}{T} \int_0^T b(t)\theta_{[\mu m,d]}(t)\,dt, \quad \mu > \frac{1}{T} \int_0^T c(t)\theta_{[\lambda\ell,a]}(t)\,dt, \tag{7.38}$$

then (1.2) possesses a stable coexistence state. Moreover, if $n \geqslant 1$ stands for its number of stable coexistence states, then (1.2) possesses $n - 1$ unstable coexistence states and a finite number of neutrally stable coexistence states. Furthermore, there exists a coexistence state (u_0, v_0) for which (7.36) holds true.

9. If

$$0 < \lambda < \frac{1}{T} \int_0^T b(t)\theta_{[\mu m,d]}(t)\,dt, \quad 0 < \mu < \frac{1}{T} \int_0^T c(t)\theta_{[\lambda\ell,a]}(t)\,dt,$$

then (1.2) possesses an unstable coexistence state. Moreover, if $n \geqslant 1$ stands for its number of unstable coexistence states, then (1.2) possesses $n - 1$ stable coexistence states and a finite number of neutrally stable coexistence states. Furthermore, either there exists a coexistence state (u_0, v_0) for which (7.36) holds, or either (7.33), or (7.34), holds, according to the values of (x, y).

8. The species v preys on u ($b > 0$ and $c < 0$)

Now, the linearization of (1.2) around any coexistence state can be expressed in the form

$$\begin{cases} u' = \alpha u - \beta v, \\ v' = \gamma u + \rho v, \end{cases} \tag{8.1}$$

where $\alpha, \beta, \gamma, \rho \in \mathcal{C}_T$ satisfy

$$\int_0^T \alpha < 0, \quad \int_0^T \rho < 0, \quad \beta > 0, \quad \gamma > 0, \tag{8.2}$$

which is a non-cooperative periodic system. As a result, the nice monotonicity properties available for competing species and symbiotic models are lost, and, in general, the components of coexistence states $\mathfrak{C}_\mu^\lambda$ and $\mathfrak{C}_\lambda^\mu$ bifurcating from the semi-trivial states are far from being analytic curves, except in a neighborhood of their bifurcation points from the semi-trivial states. Nevertheless, in strong contrast with the classical cases of competition and symbiosis, in this model the existence of a coexistence state is characterized by the linear stability of the semi-trivial states.

8.1. *Characterizing the existence of coexistence states*

Subsequently, we suppose $m \gg 0$ and $d \gg 0$. Then,

$$\lim_{\mu \uparrow \infty} \frac{1}{T} \int_0^T b(t)\theta_{[\mu m,d]}(t)\,dt = \infty,$$

since

$$\lim_{\mu \uparrow \infty} \frac{\theta_{[\mu m,d]}}{\mu} = \frac{m}{d},$$

and, hence, for each $\lambda > 0$ there exists a unique value of μ, $\mu_* > 0$, such that

$$\lambda = \frac{1}{T} \int_0^T b(t)\theta_{[\mu_* m,d]}(t)\,dt. \tag{8.3}$$

The following result is satisfied.

THEOREM 8.1. *The following conditions are equivalent:*
  (a) *Problem (1.2) possesses a coexistence state.*
  (b) $\lambda > 0$ *and any semi-trivial state of (1.2) is linearly unstable.*
  (c) *The parameters (λ, μ) satisfy*

$$\lambda > \max\left\{0, \frac{1}{T}\int_0^T (b\theta_{[\mu m,d]})\right\}, \quad \mu > \frac{1}{T}\int_0^T (c\theta_{[\lambda\ell,a]}), \tag{8.4}$$

where, for each $\mu \in \mathbb{R}$, $\theta_{[\mu m,d]} \in C_T^1$ stands for the maximal non-negative solution of $u' = \lambda\ell u - au^2$. In other words, $\theta_{[\mu m,d]} \gg 0$ if $\mu > 0$, while $\theta_{[\mu m,d]} = 0$ if $\mu \leqslant 0$. Moreover, for each $\lambda > 0$, the component $\mathfrak{C}_\mu^\lambda$ of coexistence states constructed in Theorem 5.4 is bounded in $\mathbb{R} \times C_T \times C_T$ and it satisfies

$$(\lambda, \mu_0, \theta_{[\lambda\ell,a]}, 0), (\lambda, \mu_*, 0, \theta_{[\mu_* m,d]}) \in \bar{\mathfrak{C}}_\mu^\lambda, \tag{8.5}$$

where μ_* is the unique value of μ satisfying (8.3). Therefore, the μ-projection of $\mathfrak{C}_\mu^\lambda$ is the interval (μ_0, μ_*).

PROOF. Suppose (λ, μ, u_0, v_0) is a coexistence state of (1.2). Then, dividing by u_0 the u-equation and integrating in $[0, T]$ gives

$$\lambda = \frac{1}{T} \int_0^T (a u_0 + b v_0) > 0.$$

So, $\lambda > 0$ is necessary for the existence. Subsequently, we will assume that $\lambda > 0$. Then, thanks again to the u-equation,

$$u_0' = \lambda \ell u_0 - a u_0^2 - b u_0 v_0 < \lambda \ell u_0 - a u_0^2,$$

and, due to Proposition 2.6,

$$u_0 \ll \theta_{[\lambda \ell, a]}. \tag{8.6}$$

So, dividing by v_0 the v-equation, integrating in $[0, T]$, and using (8.6) shows that

$$\mu = \frac{1}{T} \int_0^T (c u_0 + d v_0) > \frac{1}{T} \int_0^T c(t) \theta_{[\lambda \ell, a]}(t)\, dt,$$

since $c < 0$. Thus, the second estimate of (8.4) is necessary for the existence.

Now, suppose $\mu > 0$. Then, $\theta_{[\mu m, d]} \gg 0$,

$$v_0' = \mu m v_0 - d v_0^2 - c u_0 v_0 > \mu m v_0 - d v_0^2,$$

and, it is apparent from Proposition 2.6 that

$$v_0 \gg \theta_{[\mu m, d]}. \tag{8.7}$$

Consequently,

$$\lambda = \frac{1}{T} \int_0^T (a u_0 + b v_0) > \frac{1}{T} \int_0^T b(t) \theta_{[\mu m, d]}(t)\, dt$$

and, therefore, (8.4) is necessary for the existence of a coexistence state. It should be noted that (8.4) provides us with the set of values of (λ, μ) with $\lambda > 0$ for which any semi-trivial state of (1.2) is linearly unstable, i.e., (b) and (c) are equivalent and we have just seen that (a) implies (c).

Now, fix $\lambda > 0$ and pick $(\mu, u_0, v_0) \in \mathfrak{C}_\mu^\lambda$. Then, (8.7) is satisfied. Moreover, at any time t_0 where $v_0(t_0) = (v_0)_M$, we have that $v_0'(t_0) = 0$ and, hence, it follows from the v-equation that

$$\mu m(t_0) - d(t_0) v_0(t_0) - c(t_0) u_0(t_0) = 0.$$

Thus, thanks to (8.6),

$$\|v_0\|_{C_T} = v_0(t_0) = \frac{m(t_0)}{d(t_0)}\mu + \frac{-c(t_0)}{d(t_0)}u(t_0)$$

$$\leqslant \left(\frac{m}{d}\right)_M |\mu| + \left(\frac{-c}{d}\right)_M \theta_{[\lambda\ell,a]},$$

and, consequently, $\mathfrak{C}^\lambda_\mu$ is bounded. Now, it is apparent that Theorem 5.4 concludes the proof. Indeed, the μ-projection of $\mathfrak{C}^\lambda_\mu$ must be the interval (μ_0, μ_*) and, therefore, (8.4) is sufficient for the existence of a coexistence state of (1.2). $\qquad\square$

8.2. *Dynamics of* (1.1)

Thanks to the theory developed in [25], the following results, whose proofs are omitted here, are satisfied. Subsequently, we shall denote by $(u(t), v(t))$ the unique positive solution of (1.1).

1. If $(0, 0)$ is linearly stable, i.e., if $\lambda \leqslant 0$ and $\mu \leqslant 0$, then $\lim_{t\uparrow\infty} (u(t), v(t)) = (0, 0)$.
2. If $\lambda > 0$ and $(\theta_{[\lambda\ell,a]}, 0)$ is linearly stable, i.e., $\mu \leqslant \frac{1}{T}\int_0^T c(t)\theta_{[\lambda\ell,a]}(t)\,dt$, then $\lim_{t\uparrow\infty} (u(t), v(t)) = (\theta_{[\lambda\ell,a]}, 0)$.
3. If $\mu > 0$ and $(0, \theta_{[\mu m,d]})$ is linearly stable, i.e., $\lambda \leqslant \frac{1}{T}\int_0^T b(t)\theta_{[\mu m,d]}(t)\,dt$, then $\lim_{t\uparrow\infty} (u(t), v(t)) = (0, \theta_{[\mu m,d]})$.

As far as to the problem of the uniqueness, multiplicity and stability of the coexistence states things are much more difficult to deal with than in the competing species and symbiotic prototype counterparts, since the linearization of (1.2) at any coexistence state has a *non-cooperative structure*—the off-diagonal entries of the coupling matrix of the system have contrary signs—and for such problems a general comparison principle within the spirit of Theorems 6.1 and 7.1 is far from being available. Therefore, the problem of ascertaining the signs of the Floquet exponents of a given coexistence state is extremely difficult to treat with.

Quite strikingly, although the one-dimensional predator–prey model with diffusion under homogeneous Dirichlet boundary conditions is known to admit at most a coexistence state [26,10,18], there are choices of the several coefficient functions in the setting of (1.2) for which the predator–prey model admits two $2T$-periodic coexistence states and one unstable T-periodic coexistence state [5,8,18]. Hence, the uniqueness result for the one-dimensional elliptic model does not extend in general to the periodic prototype models, however both prototypes, viewed as abstract non-cooperative systems, possess the same operator structure [18]. In [25] were obtained some persistence results in the presence of coexistence states, and in [5] it was shown that if the product of the interactions b and c is sufficiently small, then the model admits at most a coexistence state, though we refrain of giving more details here. Actually, this uniqueness principle holds true in the general case when b and c are arbitrary. The details will appear elsewhere.

References

[1] S. Ahmad, *On the non-autonomous Volterra–Lotka competition equations*, Proc. Amer. Math. Soc. **117** (1993), 199–204.

[2] S. Ahmad and A.C. Lazer, *Asymptotic behaviour of solutions of periodic competition diffusion systems*, Nonlinear Anal. **13** (1989), 263–284.

[3] C. Alvárez and A.C. Lazer, *An application of topological degree to the periodic competing species problem*, J. Austral. Math. Soc. Ser. B **28** (1986), 202–219.

[4] H. Amann, *Fixed point equations and nonlinear eigenvalue problems in ordered Banach spaces*, SIAM Rev. **18** (1976), 620–709.

[5] Z. Amine and R. Ortega, *A periodic prey-predator system*, J. Math. Anal. Appl. **185** (1994), 477–489.

[6] M.G. Crandall and P.H. Rabinowitz, *Bifurcation from simple eigenvalues*, J. Funct. Anal. **8** (1971), 321–340.

[7] J.M. Cushing, *Two species competition in a periodic environment*, J. Math. Biol. **10** (1980), 385–390.

[8] E.N. Dancer, *Turing instabilities for systems of two equations with periodic coefficients*, Differential Integral Equations **7** (1994), 1253–1264.

[9] E.N. Dancer, *Bifurcation from simple eigenvalues and eigenvalues of geometric multiplicity one*, Bull. London Math. Soc. **34** (2002), 533–538.

[10] E.N. Dancer, J. López-Gómez and R. Ortega, *On the spectrum of some linear noncooperative elliptic systems with radial symmetry*, Differential Integral Equations **8** (1995), 515–523.

[11] M. Delgado, J. López-Gómez and A. Suárez, *On the symbiotic Lotka–Volterra model with diffusion and transport effects*, J. Differential Equations **160** (2000), 175–262.

[12] J.C. Eilbeck and J. López-Gómez, *On the periodic Lotka–Volterra competition model*, J. Math. Anal. Appl. **210** (1997), 58–87.

[13] J.E. Furter, J.C. Eilbeck and J. López-Gómez, *Coexistence in the competition model with diffusion*, J. Differential Equations **107** (1994), 96–139.

[14] J.K. Hale and A. Somolinos, *Competition for fluctuating nutrient*, J. Math. Biol. **18** (1983), 255–280.

[15] P. Hess and A.C. Lazer, *On an abstract competition model and applications*, Nonlinear Anal. **16** (1991), 917–940.

[16] M.A. Krasnosel'skii, *Translations Along Trajectories of Differential Equations*, Math. Monogr. of the AMS, Amer. Math. Soc., Providence, RI (1968).

[17] J. López-Gómez, *Structure of positive solutions in the periodic Lotka–Volterra competition model when a parameter varies*, Nonlinear Anal. **27** (1996), 739–768.

[18] J. López-Gómez, *A bridge between Operator Theory and Mathematical Biology*, Fields Inst. Commun. **25** (2000), 383–397.

[19] J. López-Gómez, *Spectral Theory and Nonlinear Functional Analysis*, Pitman Res. Notes Math. Ser., Vol. 426, Chapman & Hall/CRC, Boca Raton, FL (2001).

[20] J. López-Gómez, *Classifying smooth supersolutions in a general class of elliptic boundary value problems*, Adv. Differential Equations **8** (2003), 1025–1042.

[21] J. López-Gómez and M. Molina-Meyer, *The maximum principle for cooperative weakly coupled elliptic systems and some applications*, Differential Integral Equations **7** (1994), 383–398.

[22] J. López-Gómez and M. Molina-Meyer, *Singular perturbations in Economy and Ecology. The effect of strategic symbiosis in random competitive environments*, Adv. Math. Sci. Appl. **14** (2004), 87–107.

[23] J. López-Gómez and M. Molina-Meyer, *Symbiosis in Competitive environments*, Res. Inst. Math. Sci. Kyoto **12** (2003), 118–141.

[24] J. López-Gómez and M. Molina-Meyer, *Bounded components of positive solutions of abstract fixed point equations: mushrooms, loops and isolas*, J. Differential Equations **209** (2005) 416–441.

[25] J. López-Gómez, R. Ortega and A. Tineo, *The periodic predator–prey Lotka–Volterra model*, Adv. Differential Equations **1** (1996), 403–423.

[26] J. López-Gómez and R. Pardo, *Existence and uniqueness for the predator-prey model with diffusion: The scalar case*, Differential Integral Equations **6** (1993), 1025–1031.

[27] J. Mawhin, *Les héritiers de Pierre-François Verhulst: une population dynamique*, Acad. Roy. Belg. Bull. Cl. Sci. (6) **13** (2003), 349–378.

[28] J. Mawhin, *The legacy of P.F. Verhulst and V. Volterra in population dynamics*, The First 60 Years of Nonlinear Analysis of Jean Mawhin, M. Delgado, J. López-Gómez, R. Ortega and A. Suárez, eds., World-Scientific (2004).

[29] P. de Mottoni and A. Schiaffino, *Competition systems with periodic coefficients: a geometric approach*, J. Math. Biol. **11** (1981), 319–335.

[30] R. Ortega and A. Tineo, *On the number of positive solutions for planar competing Lotka–Volterra systems*, J. Math. Anal. Appl. **193** (19945), 975–978.

[31] P.H. Rabinowitz, *Some global results for nonlinear eigenvalue problems*, J. Funct. Anal. **7** (1971), 487–513.

[32] H. Smith, *Periodic competitive differential equations and the discrete dynamics of competitive maps*, J. Differential Equations **64** (1986), 165–194.

[33] H. Smith, *Periodic solutions of periodic competitive and cooperative systems*, SIAM J. Math. Anal. **17** (1986), 1289–1238.

[34] A. Tineo, *Existence of global coexistence state for periodic competition diffusion systems*, Nonlinear Anal. **19** 81992), 335–344.

[35] A. Tineo, *An iterative scheme for the N-competing species problem*, J. Differential Equations (1995), 1–15.

[36] F. Zanolin, *Permanence and positive periodic solutions of Kolmogorov competing species models*, Results Math. **21** (1992), 224–250.

CHAPTER 6

Nonlocal Initial and Boundary Value Problems: A Survey

S.K. Ntouyas

Department of Mathematics, University of Ioannina, 451 10 Ioannina, Greece
E-mail: sntouyas@cc.uoi.gr

Contents

1. Introduction 463
   1.1. Examples of nonlocal conditions 463
2. Nonlocal initial value problems 465
   2.1. Existence and uniqueness 465
3. Nonlocal boundary value problems 469
   3.1. First-order three-point boundary value problems 470
   3.2. Second-order three-point boundary value problems 475
   3.3. Third-order three-point boundary value problems 484
   3.4. m-point boundary value problems reduced to three-point boundary value problems 489
   3.5. Nonresonance results 490
   3.6. Results at resonance 498
   3.7. m-point boundary value problems reduced to four-point boundary value problems 505
   3.8. The general case 513
   3.9. Positive solutions of some nonlocal boundary value problems 518
   3.10. Positive solutions of nonlocal boundary value problems with dependence on the first-order derivative 542
   3.11. Positive solutions of nonlocal problems for p-Laplacian 547
References 554

HANDBOOK OF DIFFERENTIAL EQUATIONS
Ordinary Differential Equations, volume 2
Edited by A. Cañada, P. Drábek and A. Fonda
© 2005 Elsevier B.V. All rights reserved

1. Introduction

The evolution of a physical system in time is described by an initial value problem, i.e. a differential equation (ordinary or partial) and an initial condition. In many cases it is better to have more initial information. The local condition is replaced then by a nonlocal condition, which gives better effect than the local initial condition, since the measurement given by a nonlocal condition is usually more precise than the only one measurement given by a local condition.

The study of initial value problems (IVP for short) with nonlocal conditions is of significance, since they have applications in problems in physics and other areas of applied mathematics. Conditions of this type can be applied in the theory of elasticity with better effect than the initial or Darboux conditions.

Consider the differential equation

$$x'(t) = f\big(t, x(t)\big), \quad x(t_0) = x_0, \quad t \geqslant 0, \ x(t) \in \mathbb{R}^n. \tag{1.1}$$

The fundamental assumption used when modeling a system using a differential equation is that the time rate at time t, given as $x'(t)$, depends only on the current status at time t, given as $f(t, x(t))$. Moreover, the initial condition is given in the form $x(t_0) = x_0$. In applications, this assumption and the initial condition should be improved so that we can model the situations more accurately and therefore derive better results.

One improvement of Eq. (1.1) is to assume that the time rate depends not only on the current status, but also on the status in the past; that is, the past history will contribute to the future development, or, there is a *time-delay* effect.

Another improvement of Eq. (1.1) is the so-called *differential equations with nonlocal conditions*. That is, we extend the initial condition (also called *local condition*)

$$x(t_0) = x_0$$

to the following *nonlocal condition*

$$x(t_0) + g\big(x(\cdot)\big) = x_0, \tag{1.2}$$

where $x(\cdot)$ denotes a solution (that is, $x(\cdot)$ is a function) and g is a mapping defined on some space consisting of certain functions. (Of course, g may be identically zero, in which case it reduces to the local condition or initial condition $x(t_0) = x_0$.) The advantage of using nonlocal conditions is that measurements at more places can be incorporated to get better models.

1.1. *Examples of nonlocal conditions*

Let $p \in \mathbb{N}$ and let $t_1, t_2, \ldots, t_p$ be given real numbers such that

$$t_0 < t_1 < t_2 < \cdots < t_p \leqslant t_0 + \alpha.$$

Then g can be defined by the formula

$$g(x) = \sum_{i=1}^{p} c_i x(t_i), \tag{1.3}$$

where $c_i, i = 1, 2, \ldots, p$ are given constants or by the relation

$$g(x) = \sum_{i=1}^{p} \frac{c_i}{\epsilon_i} \int_{t_i - \epsilon_i}^{t_i} x(t)\, dt, \tag{1.4}$$

where $\epsilon_i, i = 1, 2, \ldots, p$ are given positive constants such that: $t_{i-1} < t_i - \epsilon_i < t_i, i = 1, 2, \ldots, p$. Particularly, if $x_0 = 0, p = 1, t_1 = t_0 + \alpha$ then

$$u(t_0) = -c_1 x(t_0 + \alpha) \quad \text{and} \quad x(t_0) = \frac{-c_1}{\epsilon_1} \int_{t_0 + \alpha - \epsilon_1}^{t_0 + \alpha} x(t)\, dt$$

respectively.

The constants $c_i, i = 1, 2, \ldots, p$ in the nonlocal condition (1.3) can satisfy the inequalities $|c_i| > 1, i = 1, 2, \ldots, p$. It is remarkable that if $c_i \neq 0, i = 1, 2, \ldots, p$ the results can be applied to kinematics to determine the evolution $t \to x(t)$ of the location of a physical object for which we do not know the positions $x(0), x(t_1), \ldots, x(t_p)$, but we know that the nonlocal condition of (1.3) holds. Consequently, to describe some physical phenomena, the nonlocal condition can be more useful than the standard initial condition $x(0) = x_0$. From the nonlocal condition of (1.3) it is clear that when $c_i = 0, i = 1, 2, \ldots, p$ we have the classical initial condition.

Let us have one more example for partial differential equations. In the theory of diffusion and heat conduction one can encounter a mathematical model of the form:

$$Lx + c(w, t)x = f(w, t), \quad t \in \Omega, \ 0 < t < T,$$

$$x(w, t) = \phi(w, t), \quad w \in \partial\Omega, \ 0 < t < T,$$

$$x(w, 0) + \sum_{k=1}^{N} \beta_k(w)x(w, t_k) = \psi(w), \quad w \in \Omega \text{ and } t_k \in (0, T], \ k = 1, \ldots, N,$$

where Ω is a bounded region in $\mathbb{R}^n$ and L is a uniformly parabolic operator with continuous and bounded coefficients. It represents the diffusion phenomenon of a small amount of gas in a transparent tube. If there is very little gas at the initial time, the measurement $x(w, 0)$ of the amount of the gas in this instant may be less precise that the measurement $x(w, 0) + \sum_{k=1}^{N} \beta_k(x)x(w, t_k)$ of the sum of the amount of this gas.

2. Nonlocal initial value problems

Here, we consider the following initial value problem

$$x' = f(t, x), \quad t \in I := [t_0, T], \ x \in \mathbb{R}^n, \tag{2.1}$$

$$x(t_0) + g(x(\cdot)) = x_0, \tag{2.2}$$

where $x(\cdot)$ denotes a solution and g is a mapping acting on some space of functions defined on $[0, T]$.

Since $g(x(\cdot))$ in Eq. (2.2) is defined on the interval $[t_0, T]$ rather than a single point, the IVP (2.1)–(2.2) is called an *"initial value problem with nonlocal conditions"* or *"nonlocal initial value problem"*, and can be applied with better effect than just using the classical initial value problem with $x(t_0) = x_0$, because new measurements at more places are allowed, thus more information is available.

For the IVP (2.1)–(2.2), observe that if x is a solution, then

$$x(t) = x(t_0) + \int_{t_0}^{t} f(s, x(s)) \, \mathrm{d}s = \left[x_0 - g(x(\cdot)) \right] + \int_{t_0}^{t} f(s, x(s)) \, \mathrm{d}s. \tag{2.3}$$

On the other hand, assume that a function x on the interval $[t_0, T]$ satisfies (2.3). For this fixed function x, $g(x(\cdot))$ is a fixed element in $\mathbb{R}^n$, hence $\frac{\mathrm{d}}{\mathrm{d}t} g(x(\cdot)) = 0$. Therefore, if we take a derivative in t, then x is a solution of IVP (2.1)–(2.2). That is, we conclude that a continuous function x is a solution of the IVP (2.1)–(2.2) if and only if

$$x(t) = \left[x_0 - g(x(\cdot)) \right] + \int_{t_0}^{t} f(s, x(s)) \, \mathrm{d}s.$$

This leads to a mapping P on $C([t_0, T], \mathbb{R}^n)$ such that

$$(Px)(t) = \left[x_0 - g(x(\cdot)) \right] + \int_{t_0}^{t} f(s, x(s)) \, \mathrm{d}s.$$

2.1. Existence and uniqueness

Motivated by physical problems, Byszewski and Lakshmikantham [12] considered the following nonlocal IVP

$$x' = f(t, x), \quad t \in I := [t_0, b], \tag{2.4}$$

$$x(t_0) + g(t_1, t_2, \ldots, t_p, x(\cdot)) = x_0, \tag{2.5}$$

where $t_0 < t_1 < \cdots < t_p < b$, $p \in \mathbb{N}$, $x = (x_1, x_2, \ldots, x_n) \in \Omega$, $x_0 = (x_{10}, \ldots, x_{n0}) \in \Omega$, $f = (f_1, \ldots, f_n) \in C(I \times \Omega, E)$, $g = (g_1, \ldots, g_n) : I^p \times \Omega \to E$, $g(t_1, \ldots, t_p, \cdot) \in$

$C(\Omega, E)$ and $\Omega \subset E$ is a suitable subset. The symbol $g(t_1, \ldots, t_p, x(\cdot))$ is used in the sense that in the place of $\cdot$ we can substitute only elements of the set $\{t_1, \ldots, t_p\}$.

Let $E = E_1 \times E_2 \times \cdots \times E_n$, where E_i, $i = 1, 2, \ldots, n$ are Banach spaces with norm $\| \cdot \|$.

THEOREM 2.1. *Assume that:*
  (i) *E is a Banach space with norm $\| \cdot \|$, $x_0 \in E$, and $\Omega := \overline{B(x_0, r)} = \{y \colon \|y - x_0\| \leqslant r\} \subset E$.*
  (ii) *There exist constants $L_i > 0$, $i = 1, \ldots, n$ such that*

$$\left\| f_i(s, y) - f_i(s, \bar{y}) \right\| \leqslant L_i \|y - \bar{y}\|$$

  for (s, y), $(s, \bar{y}) \in I \times \Omega$, $i = 1, \ldots, n$.
  (iii) *There exist constants $K_i > 0$, $i = 1, \ldots, n$ such that*

$$\left\| g_i(t_1, \ldots, t_p, z) - g_i(t_1, \ldots, t_p, \bar{z}) \right\| \leqslant K_i \|z - \bar{z}\|, \quad z, \bar{z} \in \Omega, \ i = 1, \ldots, n.$$

  (iv) $(Md + N)\sqrt{n} \leqslant r, \qquad (Ld + K)\sqrt{n} < 1,$
  where

$$M_i = \sup_{(s,y) \in I \times \Omega} \left\| f_i(s, y) \right\|, \quad N_i = \sup_{z \in \Omega} \left\| g(t_1, \ldots, t_p, z) \right\|, \quad i = 1, \ldots, n,$$

  and

$$d = b - t_0, \quad L = \max_{i=1,\ldots,n} L_i, \quad K = \max_{i=1,\ldots,n} K_i,$$

$$M = \max_{i=1,\ldots,n} M_i, \quad N = \max_{i=1,\ldots,n} N_i.$$

Then there exists a unique solution on the nonlocal IVP (2.4)–(2.5).

PROOF. It is easy to see, that problem (2.4)–(2.5) is equivalent to the integral equation

$$x(t) = x_0 - g(t_1, \ldots, t_p, x(\cdot)) + \int_{t_0}^{t} f(s, x(s)) \, ds, \quad t \in I,$$

which is also equivalent to the following system of the integral equations

$$x_i(t) = x_{i0} - g_i(t_1, \ldots, t_p, x(\cdot)) + \int_{t_0}^{t} f_i(s, x(s)) \, ds, \quad t \in I, \ i = 1, \ldots, n.$$

We introduce the operator T given by

$$(Ty)(t) := \left((T_1 y)(t), \ldots, (T_n y)(t) \right), \quad t \in I,$$

where

$$(T_i y)(t) = x_{i0} - g_i\left(t_1, \ldots, t_p, x(\cdot)\right) + \int_{t_0}^{t} f_i\left(s, x(s)\right) ds, \quad t \in I, \; i = 1, \ldots, n.$$

Let $X := C(I, \Omega)$. We will show that the operator T maps X into X. We observe that

$$
\begin{aligned}
\|Ty - x_0\| &= \left(\sum_{i=1}^{n} \|T_i y - x_{i0}\|^2\right)^{1/2} \\
&= \left(\sum_{i=1}^{n} \left[\sup_{t \in I} \|(T_i y)(t) - x_{i0}\|\right]^2\right)^{1/2} \\
&\leqslant \left(\sum_{i=1}^{n} \left[\sup_{t \in I} \int_{t_0}^{t} \|f_i\left(s, y(s)\right)\| \, ds + \|g_i\left(t_1, \ldots, t_p, y(\cdot)\right)\|\right]^2\right)^{1/2} \\
&\leqslant \left(\sum_{i=1}^{n} \left[\sup_{t \in I}\left(|t - t_0| M_i + N_i\right)\right]^2\right)^{1/2} \\
&\leqslant (Md + N)\sqrt{n} \leqslant r,
\end{aligned}
$$

for $y \in X$. Therefore $T : X \to X$.

Next, we will show that T is a contraction on X. We have:

$$
\begin{aligned}
\|Ty - T\bar{y}\| &= \left(\sum_{i=1}^{n} \|T_i y - T_i \bar{y}\|^2\right)^{1/2} \\
&= \left(\sum_{i=1}^{n} \left[\sup_{t \in I} \|(T_i y)(t) - (T_i \bar{y})(t)\|\right]^2\right)^{1/2} \\
&\leqslant \left(\sum_{i=1}^{n} \left[\sup_{t \in I} \int_{t_0}^{t} \|f_i\left(s, y(s)\right) - f_i\left(s, \bar{y}(s)\right)\| \, ds \right.\right. \\
&\qquad\qquad \left.\left. + \|g_i\left(t_1, \ldots, t_p, y(\cdot)\right) - g_i\left(t_1, \ldots, t_p, \bar{y}(\cdot)\right)\|\right]^2\right)^{1/2} \\
&\leqslant \left(\sum_{i=1}^{n} \left[\sup_{t \in I}\left(|t - t_0| L_i \|y(s) - \bar{y}(s)\| + K_i \|y(\cdot) - \bar{y}(\cdot)\|\right)\right]^2\right)^{1/2} \\
&\leqslant (Ld + K)\sqrt{n}\|y - \bar{y}\|
\end{aligned}
$$

for arbitrary $y, \bar{y} \in X$ which proves that T is a contraction. Hence, by Banach fixed point theorem, in space X there is the only one fixed point of T and this point is the solution of the nonlocal problem (2.4)–(2.5). So, the proof of the theorem is complete. $\qquad\square$

Consider now a nonlocal IVP with a special form of the nonlocal initial condition,

$$x'(t) = f\big(t, x(t)\big), \quad t \in J := [0, T], \tag{2.6}$$

$$x(0) + \sum_{k=1}^{p} c_k x(t_k) = x_0, \tag{2.7}$$

where $0 < t_1 < \cdots < t_p \leqslant T$, $f : J \times E \to E$ is a given function and E is a Banach space with norm $\| \cdot \|$, $x_0 \in E$, $c_k \neq 0$ and $p \in \mathbb{N}$.

By X we denote the Banach space $C(J, E)$ with the standard norm $\| \cdot \|_X$.

DEFINITION 2.2. Assume that $\sum_{k=1}^{p} c_k \neq -1$. A function $x \in X$ is called a mild solution of (2.6)–(2.7) if satisfies the integral equation

$$x(t) = A\left(x_0 - \sum_{k=1}^{p} c_k \int_0^{t_k} f\big(s, x(s)\big)\, ds\right) + \int_0^t f\big(s, x(s)\big)\, ds,$$

where $A = (1 + \sum_{k=1}^{p} c_k)^{-1}$.

DEFINITION 2.3. A function $x : J \to E$ is said to be a classical solution of the nonlocal problem (2.6)–(2.7) if

    (i) x is continuous on J and continuously differentiable on J,

    (ii) $x'(t) = f(t, x(t))$ for $t \in J$, and

    (iii) $x(0) + \sum_{k=1}^{p} c_k x(t_k) = x_0$.

It is easy to see that if $f \in C(J \times E, E)$ and $\sum_{k=1}^{p} c_k \neq -1$, then x is the unique classical solution of the nonlocal problem (2.6)–(2.7) if and only if x is the unique mild solution of this problem.

THEOREM 2.4. *Assume that:*

    (1) $f : J \times E \to E$ *is continuous with respect to the first variable on J and there is* $L > 0$ *such that*

$$\big\| f(t, z) - f(t, \bar{z}) \big\| \leqslant L \sum_{i=1}^{2} \| z - \bar{z} \|$$

    for $t \in J, z, \bar{z} \in E$.

    (2) $\sum_{k=1}^{p} c_k \neq -1$.

    (3) $2LT(1 + |A \sum_{k=1}^{p} c_k|) < 1$.

Then the nonlocal IVP (2.6)–(2.7) has a unique classical solution.

PROOF. Introduce the operator P by the formula

$$(Px)(t) = A\left(x_0 - \sum_{k=1}^{p} c_k \int_0^{t_k} f\big(s, x(s)\big)\,ds\right)$$
$$+ \int_0^t f\big(s, x(s)\big)\,ds, \quad t \in J,\ x \in X.$$

It is easy to see that $P : X \to X$. We will prove that P is a contraction on X. We have

$$\big\|(Px)(t) - (P\bar{x})(t)\big\| = \left\|A\left(-\sum_{k=1}^{p} c_k \int_0^{t_k}\big[f\big(s, x(s)\big) - f\big(s, \bar{x}(s)\big)\big]\,ds\right)\right.$$
$$\left. + \int_0^t \big[f\big(s, x(s)\big) - f\big(s, \bar{x}(s)\big)\big]\,ds\right\|$$
$$\leqslant 2LT\left(1 + \left|A\sum_{k=1}^{p} c_k\right|\right)\|x - \bar{x}\|_X.$$

By (3) operator P is a contraction. Consequently by Banach fixed point theorem in space X there is only one fixed point of P and this fixed point is the unique classical solution of the nonlocal IVP (2.6)–(2.7). The proof of the theorem is complete. $\quad\square$

Theorem 2.4 is a special case of Theorem 2.4 of [11], where the functional differential equation $x'(t) = f(t, x(t), x(a(t)))$, $t \in J$ was considered. We point out that when we model some situations from physics and other applied sciences, we typically end up with partial differential equations. In recent years there has been increasing interest in studying evolution differential equations with nonlocal conditions. The interested reader is referred to [10] and [86] and the references cited therein.

3. Nonlocal boundary value problems

Let $f : [0, 1] \times \mathbb{R}^2 \to \mathbb{R}$ be a continuous function and let $e : [0, 1] \to \mathbb{R}$ be a function in $L^1[0, 1]$, $a_i \in \mathbb{R}$, $\xi_i \in (0, 1)$, $i = 1, 2, \ldots, m - 2$, $0 < \xi_1 < \xi_2 < \cdots < \xi_{m-2} < 1$. We consider the following second-order ordinary differential equation

$$x''(t) = f\big(t, x(t), x'(t)\big) + e(t), \quad t \in (0, 1) \tag{3.1}$$

subject to the following boundary value conditions:

$$x(0) = 0, \qquad x(1) = \sum_{i=1}^{m-2} a_i x(\xi_i). \tag{3.2}$$

It is well known, that if a function $x \in C^1[0, 1]$ satisfies the boundary conditions (3.2) and all α_i have the same sign then there exists an $\eta \in [\xi_1, \xi_{m-2}]$ such that

$$x(0) = 0, \qquad x(1) = \alpha x(\eta),$$

with $\alpha = \sum_{k=1}^{m-2} a_i$. Accordingly, the problem of the existence of a solution for the boundary value problem (BVP for short) (3.1)–(3.2) can be studied via the existence of a solution of the following three-point BVP

$$x(0) = 0, \qquad x(1) = \alpha x(\eta). \tag{3.3}$$

The above so-called "*nonlocal*" or "*multi-point*" or "*m-point*" BVP was initiated by Il'in and Moiseev in [43,44], motivated by the work of Bitsadze and Samarskii on nonlocal linear elliptic BVPs [7,8]. The nonlocal BVP for ordinary differential equations arise in a variety of different areas of applied mathematics and physics, and describe many phenomena in the applied mathematical sciences. For example, the vibrations of a guy wire of a uniform cross-section and composed on N parts of different densities can be set up as a multi-point BVP (see [83]); many problems in the theory of elastic stability can be handled by the method of multi-point problems (see [90]).

We will study nonlocal BVP for first, second or higher order ordinary differential equations.

3.1. *First-order three-point boundary value problems*

Given $a, b, c \in \mathbb{R}$ with $a < b < c$. Let $J := [a, c]$ and $M_{n \times n}$ the Banach space of all constant matrices of order n with the norm

$$\|B\| = \max_{1 \leqslant i, j \leqslant n} |b_{i,j}|.$$

Let $C(J), L^1(J)$, and $AC(J)$ the usual Banach spaces of all continuous functions, all Lebesque integrable functions and all absolutely continuous functions with the norms

$$\|x\|_\infty = \max\{|x(t)| : t \in J\}, \qquad \|x\|_{L^1} = \int_J |x(t)| \, dt,$$

$$\|x\|_{AC} = \|x\|_\infty + \|x'\|_{L^1},$$

respectively. For $\alpha = (\alpha_1, \ldots, \alpha_n) \in \mathbb{R}^n$, define

$$\|\alpha\| = \max\{|\alpha_i| : i = 1, \ldots, n\}.$$

Denote by X, Y and Z the Banach spaces $C[J, \mathbb{R}^n] = \{(x_1, \ldots, x_n) : x_i \in C(J)\}$, $L^1[J, \mathbb{R}^n] = \{(x_1, \ldots, x_n) : x_i \in L^1 C(J)\}$, and $AC[J, \mathbb{R}^n] = \{(x_1, \ldots, x_n) : x_i \in AC(J)\}$ with the norms $\|x\|_X = \max\{\|x_i\|_\infty : i = 2, \ldots, n\}$, $\|x\|_Y = \sum_{i=1}^n \|x_i\|_{L^1}$ and $\|x\|_Z = \max\{\|x_i\|_{AC} : i = 2, \ldots, n\}$ respectively.

DEFINITION 3.1. We say that $f : J \times \mathbb{R}^n \to \mathbb{R}^n$ is a Carathéodory function if
   (i) $f(\cdot, u)$ is Lebesgue measurable on J for each $u \in \mathbb{R}^n$,
   (ii) $f(t, \cdot)$ is continuous on $\mathbb{R}^n$ for a.e. $t \in J$,
   (iii) for each $r \in (0, \infty)$, there exists an $h_r \in L^1(J, \mathbb{R}^n)$ such that

$$\left| f_i(t, u) \right| \leqslant (h_r)_i(t), \quad \text{for a.e. } t \in J, \ \|u\| \leqslant r, \ i = 1, \ldots, n.$$

We study the problem of existence of solutions for the three-point BVP

$$x'(t) = f\big(t, x(t)\big), \quad \text{for a.e. } t \in J, \tag{3.4}$$

$$Mx(a) + Nx(b) + Rx(c) = \alpha, \tag{3.5}$$

where $f : [a, c] \times \mathbb{R}^n \to \mathbb{R}^n$ is a Carathéodory function, $M, N, R \in M_{n \times n}$, and $\alpha \in \mathbb{R}^n$ are given.

THEOREM 3.2. *Let $f : [a, c] \times \mathbb{R}^n \to \mathbb{R}^n$ be a Carathéodory function and $\alpha \in \mathbb{R}^n$. Assume that*
   (A1) *M, N, R are constant square matrices of order n such that*

$$\det(M + N + R) \neq 0.$$

   (A2) *There exist functions $p, r \in L^1(J)$ such that*

$$\left\| f(t, u) \right\| \leqslant p(t)\|u\| + r(t),$$

   for a.e. $t \in J$ and $u \in \mathbb{R}^n$.
Then the three-point BVP (3.4)–(3.5) has at least one solution in $C(J, \mathbb{R}^n)$ provided

$$\Gamma \|p\|_1 < 1,$$

where

$$\Gamma = \max\big\{ \left\| (M + N + R)^{-1}R \right\|, \left\| (M + N + R)^{-1}M \right\|,$$
$$\left\| (M + N + R)^{-1}(N + R) \right\|, \left\| (M + N + R)^{-1}(N + M) \right\| \big\}.$$

PROOF. Let $e(\cdot) \in L^1(J, \mathbb{R}^n)$ and $\alpha \in \mathbb{R}^n$, $x \in AC(J)$ be such that

$$x'(t) = e(t), \quad \text{for a.e. } t \in J,$$

$$Mx(a) + Nx(b) + Rx(c) = \alpha.$$

Then

$$x(t) = \int_a^t e(s)\,\mathrm{d}s + (M + N + R)^{-1}\left[\alpha - N \int_a^b e(s)\,\mathrm{d}s - R \int_a^c e(s)\,\mathrm{d}s\right].$$

If E is the identity matrix, for $a \leqslant t \leqslant b < c$, we have

$$\begin{aligned}
x(t) &= \int_a^t \left[E - (M+N+R)^{-1}N - (M+N+R)^{-1}R \right] e(s)\,ds \\
&\quad - \int_t^b \left[(M+N+R)^{-1}N + (M+N+R)^{-1}R \right] e(s)\,ds \\
&\quad - \int_b^c \left[(M+N+R)^{-1}R \right] e(s)\,ds \\
&= \int_a^t (M+N+R)^{-1}Me(s)\,ds \\
&\quad - \int_t^b (M+N+R)^{-1}(N+R)e(s)\,ds \\
&\quad - \int_b^c (M+N+R)^{-1}Re(s)\,ds,
\end{aligned}$$

and hence

$$\begin{aligned}
\|x(t)\| \leqslant \max\{ &\|(M+N+R)^{-1}M\|, \|(M+N+R)^{-1}(N+R)\|, \\
&\|(M+N+R)^{-1}R\| \} \|e\|_Y.
\end{aligned}$$

For $a \leqslant b \leqslant t \leqslant c$ we have

$$\begin{aligned}
x(t) &= \int_a^b \left[E - (M+N+R)^{-1}N - (M+N+R)^{-1}R \right] e(s)\,ds \\
&\quad - \int_b^t \left[-E + (M+N+R)^{-1}R \right] e(s)\,ds \\
&\quad - \int_t^c \left[(M+N+R)^{-1}R \right] e(s)\,ds \\
&= \int_a^b (M+N+R)^{-1}Me(s)\,ds \\
&\quad - \int_b^t (M+N+R)^{-1}(-M-N)e(s)\,ds \\
&\quad - \int_t^c (M+N+R)^{-1}Re(s)\,ds,
\end{aligned}$$

and hence

$$\begin{aligned}
\|x(t)\| \leqslant \max\{ &\|(M+N+R)^{-1}M\|, \|(M+N+R)^{-1}(M+N)\|, \\
&\|(M+N+R)^{-1}R\| \} \|e\|_Y.
\end{aligned}$$

By combining the above relations we get

$$|x(t)| \leqslant \Gamma \|e\|_1$$

where

$$\Gamma = \max\{\|(M+N+R)^{-1}R\|, \|(M+N+R)^{-1}M\|,$$
$$\|(M+N+R)^{-1}(N+R)\|, \|(M+N+R)^{-1}(N+M)\|\}.$$

It is obvious that, if $e = 0$, then the problem

$$x'(t) = 0, \quad \text{for a.e. } t \in J,$$
$$Mx(a) + Nx(b) + Rx(c) = \alpha,$$

has the unique solution $x(t) = w$, where

$$w = (M+M+R)^{-1}\alpha.$$

Moreover the problem

$$x'(t) = f(t, x), \quad \text{for a.e. } t \in J,$$
$$Mx(a) + Nx(b) + Rx(c) = \alpha,$$

has a solution $x = y_1 + w$ if and only if y_1 is a solution of

$$y'(t) = f(t, y + w), \quad \text{for a.e. } t \in J, \tag{3.6}$$
$$My(a) + Ny(b) + Ry(c) = 0. \tag{3.7}$$

Now we define $L: D(L) \subset X \to Y$ by

$$D(L) = \{y \in AC(J): My(a) + Ny(b) + Ry(c) = 0\},$$

and for $y \in D(L)$,

$$Ly = y'.$$

We also define a nonlinear mapping $G: X \to X$ by

$$Gy = f(t, y(t) + w), \quad t \in J.$$

We note that G is a bounded operator from X into Y. Next it is easy to see from (A1) that the linear mapping $L: D(L) \subset X \to Y$, is a one-to-one mapping. Let the linear mapping $K: Y \to X$ be defined for $e \in Y$ by

$$(Ke)(t) = \int_a^t e(s)\,ds + (M+M+R)^{-1}\left[\alpha - N\int_a^b e(s)\,ds - R\int_a^c e(s)\,ds\right].$$

Then for $e \in Y$, we have $Ke \in D(L)$ and $LKe = e$. For $y \in D(L)$, we have $KLy = y$. Furthermore, it follows easily using the Arzelá–Ascoli theorem that KG maps a bounded subset of X into a relatively compact subset of X. Hence, $KG : X \to X$ is a compact mapping.

We next note that $y \in C(J, \mathbb{R}^n)$ is a solution of BVP (3.6)–(3.7) if and only if y is a solution to operator equation

$$Ly - Gy,$$

which is equivalent to the equation

$$y = KGy.$$

We apply the Leray–Schauder continuation theorem (see, e.g., [82, Corollary IV.7]) to obtain existence of a solution to $y = KGy$ or equivalently to the BVP (3.6)–(3.7).

To do this, it suffices to verify that the set of all possible solutions of the family of equations

$$y'(t) = \lambda f(t, y + w), \quad \text{for a.e. } t \in J, \tag{3.8}$$

$$My(a) + Ny(b) + Ry(c) = 0, \tag{3.9}$$

is bounded in $X = C(J, \mathbb{R}^n)$ by a constant independent of $\lambda \in [0, 1]$.

Let $y(t)$ be a solution of (3.8)–(3.9) for some $\lambda \in [0, 1]$. It follows from our assumptions that $f(t, y(t) + w(t)) \in L^1(J, \mathbb{R}^n)$. We have

$$\|y\|_X \leqslant \lambda \Gamma \| f(t, y(t) + w) \|_Y$$

$$\leqslant \Gamma \big[\|p\|_{L^1} \|y + w\|_X + \|r\|_{L^1} \big]$$

$$\leqslant \Gamma \big[\|p\|_{L^1} \big(\|y\|_X + \|w\|_X \big) + \|r\|_{L^1} \big].$$

It follows from $\Gamma \|p\|_{L^1} < 1$ that there exists a constant c_0, independent of $\lambda \in [0, 1]$, such that

$$\|y\|_X \leqslant c_0.$$

This completes the proof of the theorem. $\qquad\qquad\qquad\qquad\qquad\qquad\square$

THEOREM 3.3. *Let $f : [a, c] \times \mathbb{R}^n \to \mathbb{R}^n$ be a Carathéodory function and $\alpha \in \mathbb{R}^n$. Assume that* (A1) *holds. Moreover we suppose that*
   (A3) *There exists function $p \in L^1(J)$ such that*

$$\big\| f(t, u) - f(t, v) \big\| \leqslant p(t) \|u - v\|$$

for a.e. $t \in J$ and all $u, v \in \mathbb{R}^n$.

Then the three-point BVP (3.4)–(3.5) has a unique solution in $C(J, \mathbb{R}^n)$ provided

$$\Gamma \|p\|_1 < 1.$$

PROOF. We note that (A3) implies

$$\|f(t, u)\| \leq \|f(t, 0)\| + p(t)\|u\|,$$

for a.e. $t \in J$ and all $u \in \mathbb{R}^n$. Hence the BVP (3.4)–(3.5) has at least one solution in $C(J, \mathbb{R}^n)$ by Theorem 3.2. Suppose that u_1, u_2 are two solutions of (3.4)–(3.5) in $C(J, \mathbb{R}^n)$. Setting $z = u_1 - u_2$, we have

$$z' = f\big(t, u_1(t)\big) - f\big(t, u_2(t)\big),$$
$$Mz(a) + Nz(b) + Rz(c) = 0.$$

Then we have

$$\|z\|_X \leq \lambda \Gamma \big\| f\big(t, u_1(t)\big) - f\big(t, u_2(t)\big) \big\|_Y$$
$$\leq \Gamma \big(\|p\|_{L^1} \|u_1 - u_2\|_X \big)$$
$$\leq \Gamma \|p\|_{L^1} \|z\|_X,$$

which implies that $z(t) = 0$, for $t \in J$ and hence $u_1 = u_2$. This completes the proof of the theorem. $\qquad\square$

Theorems 3.2 and 3.3 were taken from [74]. Existence and uniqueness of solutions of the BVP (3.4)–(3.5) were studied in [84] by using the successive over relaxation iteration and the Banach contraction mapping principle.

3.2. *Second-order three-point boundary value problems*

In this section we consider the following three-point BVP

$$x''(t) = f\big(t, x(t), x'(t)\big) + e(t), \quad t \in (0, 1) \tag{3.10}$$
$$x(0) = 0, \qquad x(1) = x(\eta), \tag{3.11}$$

where $f : [0, 1] \times \mathbb{R}^2 \to \mathbb{R}$, $e : [0, 1] \to \mathbb{R}$ are given functions and $\eta \in (0, 1)$.

We will study the existence and uniqueness of the BVP (3.10)–(3.11). We use the classical spaces $C[0, 1]$, $C^k[0, 1]$, $L^k[0, 1]$, and $L^\infty[0, 1]$ of continuous, k-times continuously differentiable, measurable real valued functions whose kth power of the absolute value is Lebesgue integrable on $[0, 1]$, or measurable functions that are essentially bounded on $[0, 1]$. We also use the Sobolev space $W^{2,k}(0, 1)$, $k = 1, 2$ defined by

$$W^{2,k}(0, 1) = \big\{ x : [0, 1] \to \mathbb{R} : \ x, x' \text{ are absolutely continuous}$$
$$\text{on } [0, 1] \text{ with } x \in L^k[0, 1] \big\}$$

with its usual norm. We denote the norm in $L^k[0, 1]$ by $\|\cdot\|_k$, and the norm in $L^\infty[0, 1]$ by $\|\cdot\|_\infty$.

The BVP (3.10)–(3.11) can be put in the form of an operator equation

$$Lx + Nx = w,$$

where $L : D(L) \subset X \to Y$ is a linear operator, $N : X \to Y$ is a nonlinear operator, and X, Y are suitable spaces in duality. Clearly the linear operator L in (3.10) is given by

$$Lx = -x''$$

where the boundary conditions (3.11) are used to define the domain, $D(L)$, of L.

We begin by giving the following definition.

DEFINITION 3.4. A function $f : [0, 1] \times \mathbb{R}^2 \to \mathbb{R}$ satisfies Carathéodory's conditions if:
  (i) for each $(x, y) \in \mathbb{R}^2$, the function $t \in [0, 1] \to f(t, x, y) \in \mathbb{R}$ is measurable on $[0, 1]$,
  (ii) for a.e. $t \in [0, 1]$, the function $(x, y) \in \mathbb{R}^2 \to f(t, x, y) \in \mathbb{R}$ is continuous on $\mathbb{R}^2$, and
  (iii) for each $r > 0$, there exists $g_r \in L^1[0, 1]$ such that $|f(t, x, y)| \leqslant g_r(t)$ for a.e. $t \in [0, 1]$ and $(x, y) \in \mathbb{R}^2$ with $\sqrt{x^2 + y^2} \leqslant r$.

In the next existence theorems, we obtain appropriate *a priori* bounds by applying degree theory. We use Wirtinger-type inequalities to obtain the necessary a priori bounds needed to apply the Leray–Schauder continuation theorem [82, Corollary IV.7].

THEOREM 3.5. *Let $f : [0, 1] \times \mathbb{R}^2 \to \mathbb{R}$ satisfy Carathéodory's conditions. Assume that there exist functions $p, q, r \in L^1[0, 1]$ such that*

$$\left| f(t, u, v) \right| \leqslant p(t)|u| + q(t)|v| + r(t), \tag{3.12}$$

for a.e. $t \in [0, 1]$ and all $(u, v) \in \mathbb{R}^2$. Then for every given function $e \in L^1[0, 1]$, the BVP (3.10)–(3.11), has at least one solution in $C^1[0, 1]$ provided

$$\|q\|_1 \leqslant 1,$$

and

$$\sqrt{\eta}\|p\|_{L^1[\eta,1]} \frac{1 - \|q\|_{L^1[0,1]}}{1 - \|q\|_{L^1[\eta,1]}} + \|p\|_{L^1[0,1]} + \|q\|_{L^1[0,1]} < 1.$$

PROOF. Let X be the Banach space $C^1[0, 1]$ and Y denote the Banach space $L^1(0, 1)$ with their usual norms. We define a linear mapping $L : D(L) \subset X \to Y$ by setting

$$D(L) = \left\{ x \in W^{2,1}(0, 1) : x(0) = 0,\ x(1) = x(\eta) \right\},$$

and for $x \in D(L)$,

$$Lx = -x''.$$

We also define a nonlinear mapping $N : X \to Y$ by setting

$$(Nx)(t) = f\big(t, x(t), x'(t)\big), \quad t \in [0, 1].$$

We note that N is a bounded mapping from X into Y. Next, it is easy to see that the linear mapping $L : D(L) \subset X \to Y$, is one-to-one mapping. Next, the linear mapping $K : Y \to X$, defined for $y \in Y$ by

$$(Ky)(t) = -\int_0^t (t - s)y(s)\,\mathrm{d}s + t\int_0^\eta y(s)\,\mathrm{d}s + \frac{t}{1 - \eta}\int_0^1 (1 - s)y(s)\,\mathrm{d}s$$

is such that for $y \in Y$, $Ky \in D(L)$ and $LKy = y$; and for $u \in D(L)$, $KLu = u$. Furthermore, it follows easily using the Arzelá–Ascoli theorem that KN maps a bounded subset of X into a relatively compact subset of X. Hence $KN : X \to X$ is a compact mapping.

We next note that $x \in C^1[0, 1]$ is a solution of the BVP (3.10)–(3.11) if and only if x is a solution to the operator equation

$$Lx = Nx + e.$$

Now, the operator equation $Lx = Nx + e$ is equivalent to the equation

$$x = KNx + Ke.$$

We apply the Leray–Schauder continuation theorem (see, e.g., [82, Corollary IV.7]) to obtain the existence of a solution for $x = KNx + Ke$ or equivalently to the BVP (3.10)–(3.11).

To do this, it suffices to verify that the set of all possible solutions of the family of equations

$$x''(t) = \lambda f\big(t, x(t), x'(t)\big) + \lambda e(t), \quad t \in (0, 1), \tag{3.13}$$

$$x(0) = 0, \qquad x(1) = x(\eta), \tag{3.14}$$

is, a priori, bounded in $C^1[0, 1]$ by a constant independent of $\lambda \in [0, 1]$.

We observe first that for $x \in W^{2,1}(0, 1)$ with $x(0) = 0, x(1) = x(\eta)$ there exists a $\zeta \in (0, 1), \eta < \zeta < 1$, such that $x'(\zeta) = 0$. It follows that

$$\|x\|_2 \leqslant \frac{2}{\pi}\|x'\|_2, \qquad \|x\|_\infty \leqslant \|x'\|_2,$$

$$\big|x(1)\big| = \big|x(\eta)\big| = \left|\int_0^\eta x'(t)\,\mathrm{d}t\right| \leqslant \sqrt{\eta}\|x'\|_2,$$

$$\left| x'(1) \right| = \left| \int_{\zeta}^{1} x''(t)\,dt \right| \leqslant \int_{\zeta}^{1} \left| x''(t) \right|\,dt.$$

We multiply Eq. (3.13) by x and integrate from 0 to 1 to get

$$0 = -\int_{0}^{1} x''(t)x(t)\,dt + \lambda \int_{0}^{1} f\big(t, x(t), x'(t)\big)x(t)\,dt - \lambda \int_{0}^{1} e(t)x(t)\,dt$$

$$\geqslant \|x'\|_2^2 - \left| x'(1) \right|\left| x(1) \right| - \int_{0}^{1} \left| f\big(t, x(t), x'(t)\big) \right|\left| x(t) \right|\,dt - \|e\|_1\|x\|_\infty$$

$$\geqslant \|x'\|_2^2 - \left| x'(1) \right|\left| x(1) \right|$$

$$\quad - \int_{0}^{1} \Big[p(t)\left| x(t) \right|^2 + q(t)\left| x(t) \right|\left| x'(t) \right| + r(t)\left| x(t) \right| \Big]\,dt - \|e\|_1\|x\|_\infty$$

$$\geqslant \|x'\|_2^2 - \sqrt{\eta}\|x'\|_2\|x''\|_{L^1[\eta,1]}$$

$$\quad - \Big[\|p\|_1\|x\|_2^2 + \|q\|_1\|x\|_\infty\|x'\|_\infty + \|r\|_1\|x\|_\infty \Big] - \|e\|_1\|x\|_\infty$$

$$\geqslant \|x'\|_2^2 - \sqrt{\eta}\|x'\|_2\|x''\|_{L^1[\eta,1]} - \Big[\|p\|_1\|x'\|_2^2 + \|q\|_1\|x'\|_2\|x''\|_1 \Big]$$

$$\quad - \big(\|r\|_1 + \|e\|_1 \big)\|x'\|_2.$$

From Eq. (3.13) we have

$$\|x''\|_1 \leqslant \|p\|_1\|x\|_\infty + \|q\|_1\|x'\|_\infty + \|r\|_1$$

$$\leqslant \|p\|_1\|x'\|_2 + \|q\|_1\|x''\|_1 + \|r\|_1,$$

so that

$$\|x''\|_1 \leqslant \frac{\|p\|_1\|x'\|_2}{1 - \|q\|_1} + \frac{\|r\|_1}{1 - \|q\|_1}.$$

Similarly we have

$$\|x''\|_{L^1[\eta,1]} \leqslant \frac{\|p\|_{L^1[\eta,1]}\|x'\|_2}{1 - \|q\|_{L^1[\eta,1]}} + \frac{\|r\|_1}{1 - \|q\|_{L^1[\eta,1]}}.$$

Consequently

$$\left\{ 1 - \|q\|_1 - \sqrt{\eta}\|p\|_{L^1[\eta,1]}\frac{1 - \|q\|_1}{1 - \|q\|_{L^1[\eta,1]}} - \|p\|_1 \right\}\|x'\|_2$$

$$\leqslant \|q\|_1\|r\|_1 + \sqrt{\eta}\|r\|_1\frac{1 - \|q\|_1}{1 - \|q\|_{L^1[\eta,1]}} + \big(\|r\|_1 + \|e\|_1 \big)\big(1 - \|q\|_1 \big).$$

Hence, there exists a constant C, independent of $\lambda \in [0, 1]$ such that $\|x'\|_2 \leqslant C$. It follows that there is a constant, still denoted by C, independent of $\lambda \in [0, 1]$ such that

$$\|x\|_{C^1[0,1]} \leqslant C.$$

This completes the proof of the theorem. $\qquad\square$

THEOREM 3.6. *Let* $f : [0, 1] \times \mathbb{R}^2 \to \mathbb{R}$ *satisfy Carathéodory's conditions. Assume that* $a \geqslant 0, b \geqslant 0, \alpha(t) \in L^1[0, 1]$ *are such that*

$$\left| f(t, u, v) \right| \leqslant a|u| + b|v| + \alpha(t),$$

for a.e. $t \in [0, 1]$ *and all* $(u, v) \in \mathbb{R}^2$. *Then for every given function* $e(t) \in L^1[0, 1]$, *the BVP* (3.10)–(3.11) *has at least one solution in* $C^1[0, 1]$ *provided*

$$\left(\frac{2}{\pi}a + b \right)\left(\frac{2}{\pi} + \sqrt{\eta(1 - \eta)} \right) < 1.$$

PROOF. As in the proof of Theorem 3.5 it suffices to verify that the set of all possible solutions of the family of equations (3.13)–(3.14) is a priori bounded in $C^1[0, 1]$ by a constant independent of $\lambda \in [0, 1]$.

Letting x to be a solution of (3.13)–(3.14) for some $\lambda \in [0, 1]$ we get, as in Theorem 3.5, that

$$0 = -\int_0^1 x''(t)x(t)\,dt + \lambda \int_0^1 f\big(t, x(t), x'(t)\big)x(t)\,dt - \lambda \int_0^1 e(t)x(t)\,dt$$

$$\geqslant \|x'\|_2^2 - |x'(1)||x(1)| - \int_0^1 \left[a|x|^2 + b|x||x'| + \alpha(t)|x|\right]dt - \|e\|_1\|x\|_\infty$$

$$\geqslant \|x'\|_2^2 - \sqrt{\eta}\|x'\|_2\|x''\|_{L^1[\eta,1]} - a\|x\|_2^2 - b\|x\|_2\|x'\|_2$$

$$\quad - \left(\|\alpha\|_1 + \|e\|_1\right)\|x\|_\infty$$

$$\geqslant \|x'\|_2^2 - \sqrt{\eta}\|x'\|_2\|x''\|_{L^1[\eta,1]} - \left(\frac{4}{\pi^2}a + \frac{2}{\pi}b\right)\|x'\|_2^2 - \left(\|\alpha\|_1 + \|e\|_1\right)\|x'\|_2.$$

Also

$$\|x''\|_{L^1[\eta,1]} \leqslant \left\| f\big(t, x(t), x'(t)\big) \right\|_{L^1[\eta,1]} + \|e\|_1$$

$$\leqslant a \int_\eta^1 |x|\,dx + b \int_\eta^1 |x'|\,dx + \|\alpha\|_1 + \|e\|_1$$

$$\leqslant \sqrt{1 - \eta}\left(\frac{2}{\pi}a + b\right)\|x'\|_2 + \|\alpha\|_1 + \|e\|_1.$$

480   *S.K. Ntouyas*

Consequently

$$\|x'\|_2 \leqslant \frac{(\|\alpha\|_1 + \|e\|_1)(1 + \sqrt{\eta})}{1 - ((2/\pi)a + b)(2/\pi + \sqrt{\eta(1 - \eta)})}.$$

It then follows as in proof of Theorem 3.5 that there exists a constant C independent of $\lambda \in [0, 1]$ such that $\|x\|_{C^1[0,1]} \leqslant C$, which proves the theorem. $\qquad\square$

We give now a uniqueness result for the BVP (3.10)–(3.11).

THEOREM 3.7. *Let $f : [0, 1] \times \mathbb{R}^2 \to \mathbb{R}$ satisfy Carathéodory's conditions. Assume that $a \geqslant 0$, $b \geqslant 0$, are such that*

$$\left| f(t, u_1, v_1) - f(t, u_2, v_2) \right| \leqslant a|u_1 - u_2| + b|v_1 - v_2|, \tag{3.15}$$

for a.e. $t \in [0, 1]$ and all $(u_i, v_i) \in \mathbb{R}^2, i = 1, 2$. Then for every given function $e(t) \in L^1[0, 1]$, the BVP (3.10)–(3.11) has unique solution in $C^1[0, 1]$ provided

$$\left(\frac{2}{\pi}a + b\right)\left(\frac{2}{\pi} + \sqrt{\eta(1 - \eta)}\right) < 1.$$

PROOF. We note that (3.15) implies that

$$\left| f(t, u, v) \right| < a|u| + b|v| + \left| f(t, 0, 0) \right|$$

for all $t \in [0, 1]$ and all $(u, v) \in \mathbb{R}^2$. Hence the BVP (3.10)–(3.11) has a solution by Theorem 3.6. Let now u_1, u_2 be two solutions of (3.10)–(3.11) in $C^1[0, 1]$. Setting $w = u_1 - u_2$ we have

$$w'' = f(t, u_1, u_1') - f(t, u_2, u_2'), \quad t \in (0, 1), \tag{3.16}$$

$$w(0) = 0, \qquad w(1) = w(\eta). \tag{3.17}$$

Multiplying Eq. (3.16) by w and integrating over $[0, 1]$ we get

$$0 = -\int_0^1 w'' w \, dt + \int_0^1 \left[f\big(t, u_1(t), u_1'(t)\big) - f\big(t, u_2(t), u_2'(t)\big) \right] w(t) \, dt$$

$$\geqslant \|w'\|_2^2 - \left| w(1) \right| \left| w'(1) \right|$$

$$- \int_0^1 \left| f\big(t, u_1(t), u_1'(t)\big) - f\big(t, u_2(t), u_2'(t)\big) \right| \left| w(t) \right| dt$$

$$\geqslant \|w'\|_2^2 - \sqrt{\eta} \|w'\|_2 \|w''\|_{L^1[\eta, 1]}$$

$$- \int_0^1 \left[a|u_1 - u_2|^2 + b|u_1' - u_2'||u_1 - u_2|_2 \right] dt$$

$$\geqslant \|w'\|_2^2 - \sqrt{\eta}\|w'\|_2\|w''\|_{L^1[\eta,1]} - \left(\frac{4}{\pi^2}a + \frac{2}{\pi}b\right)\|w'\|_2^2.$$

But

$$\|w''\|_{L^1[\eta,1]} = \left\| f\big(t, u_1(t), u_1'(t)\big) - f\big(t, u_2(t), u_2'(t)\big) \right\|_{L^1[\eta,1]}$$

$$\leqslant \sqrt{1-\eta}\left(\frac{2}{\pi}a + b\right)\|w'\|_2^2,$$

and consequently

$$\left(1 - \left(\sqrt{\eta(1-\eta)} + \frac{2}{\pi}\right)\left(\frac{2}{\pi}a + b\right)\right)\|w'\|_2^2 \leqslant 0,$$

so that $\|w'\|_2 = 0$. Since $w(0) = 0$ we obtain using $w(t) = \int_0^t w'(t)\,dt$ that $\|w\|_\infty \leqslant \|w'\|_2 = 0$. Hence $w = 0$ a.e. in $[0, 1]$ and thus $w(t) = 0$ for every $t \in [0, 1]$ because w is continuous. $\qquad\square$

Theorems 3.5–3.7 were proved by Gupta in [25]. By means of rather different approach based on an existence theorem for operator inclusions, Marano proved in [80], that if p, q, r are in $L^1[0, 1]$ and (3.12) holds, then the BVP (3.10)–(3.11) has at least one generalized solution $x \in W^{2,1}[0, 1]$ provided

$$\|p\|_1 + \|q\|_1 < 1,$$

which is an improvement of Theorem 3.5. For this result of Marano, Gupta in [26] gave a simple proof by using degree theory and Wirtinger-type inequalities. Marano in [80] proved also that if p, q, r are in $L^2[0, 1]$ and (3.12) holds, then the BVP (3.10)–(3.11) has at least one generalized solution $x \in W^{2,1}[0, 1]$ provided

$$\frac{\sqrt{\eta}+1}{\pi}\|p\|_2 + \|q\|_2 < 1.$$

In the case when p, q, r are in $L^2(0, 1)$, and (3.12) holds, Gupta in [32] proved that the BVP (3.10)–(3.11) has at least one generalized solution $x \in C^1[0, 1]$ provided

$$\frac{2}{\pi}\|p\|_2 + \|q\|_2 < 1.$$

The Leray–Schauder continuation theorem can be used without the use of Wirtinger-type inequalities and using instead Green's functions. The Green's functions for various BVPs are easily computed. Then one uses Green's functions to reduce the BVPs to integral equations. Thus finding a priori estimates can be realized by estimating the norms of integral operators, which can be solved by Green's functions and its derivatives. In this direction Zhang and Han [93] proved that if p, q, r are in $L^1[0, 1]$ and (3.12) holds, then the BVP (3.10)–(3.11) has at least one solution in $C^1[0, 1]$ provided

$$\max\left\{ \int_0^{\eta} \left(sp(s) + q(s)\right) ds + \int_{\eta}^1 \frac{1-s}{1-\eta}\left(sp(s) + q(s)\right) ds, \right.$$
$$\left. \int_{\eta}^1 \frac{s-\eta}{1-\eta}\left(sp(s) + q(s)\right) ds \right\} < 1.$$

In [9] Boucherif and Bouguima proved the existence of solutions of the BVP (3.10)–(3.11) by considering two types of nonlinearities, continuous and Carathéodory's nonlinearities. The methods used in the first case rely on the topological transversality theorem of Granas [14], and in the second case on the technique of cut-off functions. They motivated their results by the remark that if $f(t, u, v) = u + v + R(t, u, v)$ with $R(t, u, v) \leqslant r(t)$ for a.e. $t \in [0, 1]$ then f satisfies condition (3.12) i.e. $|f(t, u, v)| \leqslant p_0(t)|u| + q_0|v| + r(t)$ but $\|p_0\|_1 + \|q_0\|_1 = 2 > 1$.

THEOREM 3.8. *Let $f : [0, 1] \times \mathbb{R}^2 \to \mathbb{R}$ be a continuous function satisfying the following conditions*:
    (f1) *there exists $r > 0$ such that*

$$\int_0^t f\left(s, x(s), x'(s)\right)x(s)\, ds > 0 \quad \text{whenever} \quad \left|x(t)\right| > r.$$

    (f2) *$|f(t, u, v)| \leqslant \psi(|v|)$ for $(t, u) \in [0, 1] \times [-r, r]$, $v \in \mathbb{R}$ where $\psi : [0, \infty) \to (0, \infty)$ is such that $1/\psi$ is integrable over bounded intervals in $[0, \infty)$ and $\int_0^{+\infty} \frac{s}{\psi(s)}\, ds > 2r$.*

Then the BVP (3.10)–(3.11) has at least one solution.

PROOF. We begin by examining the linear problem

$$x'' = z, \qquad x(0) = 0, \qquad x(1) = x(\eta)$$

where $z \in C([0, 1], \mathbb{R})$. This problem has a unique solution $x \in C^2[0, 1]$, given by

$$x(t) = \int_0^t (t - s)z(s)\, ds + t \int_0^{\eta} z(s)\, ds + \frac{t}{1-\eta} \int_{\eta}^1 (1 - s)z(s)\, ds.$$

This relation defines a linear operator $K : C[0, 1] \to C^2[0, 1]$ such that $x = Kz$ is the unique solution. Moreover $L = K^{-1}$ is defined by

$$D(L) = \left\{x \in C^2[0, 1]\colon x(0) = 0, \ x(1) = x(\eta)\right\}, \qquad Lx = x''.$$

For $\lambda \in [0, 1]$ consider the one-parameter family of problems

$$x''(t) = \lambda f\left(t, x(t), x'(t)\right), \quad t \in (0, 1), \tag{3.18}$$

$$x(0) = 0, \qquad x(1) = x(\eta). \tag{3.19}$$

We shall prove that any solution x of (3.18)–(3.19) is a priori bounded, independently of λ.

Let x be a solution of (3.18)–(3.19). If $\lambda = 0$ then $x = 0$ and so $\|x\|_0 = 0 < r$. Assume $0 < \lambda \leqslant 1$. Then λf satisfies (f1). Suppose x takes on a positive maximum at $t_0 \in [0, 1]$. Then $x'(t_0) = 0$ and $x''(t_0) \leqslant 0$. Now, if $x(t_0) > r$ it follows from (f1) that

$$\int_0^{t_0} x''(s)x(s)\,\mathrm{d}s = x(t_0)x'(t_0) - \int_0^{t_0} x'(s)^2\,\mathrm{d}s > 0.$$

Since $x'(t_0) = 0$ we have a contradiction. Hence $x(t_0) \leqslant r$ and therefore $\|x\|_0 \leqslant r$. Note that $t_0 \neq 0$ since $x(0) = 0$. Also, the case $t_0 = \eta$ or $t_0 = 1$ leads to a contradiction.

Now we want to show that there exists a constant m_0, such that $|x'(t)| \leqslant m_0$ for all $t \in [0, 1]$. Suppose on the contrary that for all positive constants m, there exists $t_0 \in (0, 1]$ such that $x'(t_0) > m$ or $x'(t_0) < -m$. We will assume only the case $x'(t_0) > m$, since the other case can be handled in a similar way. Let $\tau \in (0, 1)$ be such that $x'(\tau) = 0$ and $x'(t) > 0$ for all $t \in (\tau, t_0]$. This implies that

$$x''(t) \leqslant \psi\bigl(x'(t)\bigr) \quad \text{for all } t \in [\tau, t_0].$$

Hence

$$\frac{x''(t)x'(t)}{\psi(x'(t))} \leqslant x'(t) \quad \text{for all } t \in [\tau, t_0].$$

Integrating this inequality from τ to t_0 we get

$$\int_0^{x'(t_0)} \frac{s}{\psi(s)}\,\mathrm{d}s \leqslant 2r.$$

It follows from the assumptions on ψ that there exists a constant $\tilde{m}$ such that $x'(t_0) \leqslant \tilde{m}$, and so we get a contradiction.

Since f is continuous, we have $|x''(t)| \leqslant m_1 := \sup\{|f(t, u, v)|\colon 0 \leqslant t \leqslant 1, \|u\|_0 \leqslant r, \|v\|_0 \leqslant m_0\}$. Hence $\|x''\|_0 \leqslant m_1$. Finally, let $M = \max(r, m_0, m_1)$. Then we get, from $\|x\| = \max(\|x\|_0, \|x'\|_0, \|x''\|_0)$ that $\|x\| \leqslant M$, and M is independent of λ.

We apply the topological transversality of Granas [14]. Let $U = \{x \in D(L)\colon \|x\| < M + 1\}$, and $j\colon D(L) \to C^1[0, 1]$ with $j(x) = x$, the natural embedding. The Nemitskii operator $F\colon C^1[0, 1] \to C[0, 1]$ is defined by $F(x)(t) := f(t, x(t), x'(t))$ for all $t \in [0, 1]$.

For $0 \leqslant \lambda \leqslant 1$ define a family of mappings

$$H_\lambda\colon \overline{U} \to D(L) \quad \text{by } H_\lambda(x) = \lambda K F j(x).$$

H_λ is a compact homotopy which has no fixed points on the boundary ∂U of U. Since the constant map H_0 is essential, the topological transversality theorem implies that H_1 is essential i.e. has a fixed point. This completes the proof. $\qquad\square$

In the case of Carathéodory's nonlinearities we state the following result whose proof is given in [9].

THEOREM 3.9. *Let $f : [0, 1] \times \mathbb{R}^2 \to \mathbb{R}$ be a function satisfying Carathéodory's conditions. Assume that*

(f) $|f(t, u, v)| \leqslant \omega(t, |u| + |v|)$ *for a.e. $t \in [0, 1]$ and all $(u, v) \in \mathbb{R}^2$ where $\omega : [0, 1] \times \mathbb{R}^+ \to \mathbb{R}^+$ is nondecreasing in its second argument and*

$$\lim_{\rho \to \infty} \frac{1}{\rho} \int_0^1 \omega(t, \rho)\, \mathrm{d}t = 0.$$

Then the BVP (3.10)–(3.11) has at least one solution.

3.3. *Third-order three-point boundary value problems*

In this section we consider third-order three-point BVPs. The existence and uniqueness of third-order BVPs deserve a good deal of attention, since they occur in a wide variety of applications. For example, a three-layer beam is formed by parallel layers of different materials. For an equally loaded beam of this type, has shown that the deflection ψ is governed by an ordinary third-order linear differential equation

$$\psi''' - K^2 \psi' + a = 0,$$

where K^2 and a are physical parameters depending on the elasticity of the layers. The condition of zero moment at the free ends implies the boundary conditions

$$\psi'(0) = \psi'(1) = 0,$$

and the symmetry yields the third boundary condition

$$\psi(1/2) = 0.$$

We will study existence and uniqueness of the following three-point BVP

$$x'''(t) = f\big(t, x(t), x'(t), x''(t)\big) + e(t), \quad t \in (0, 1), \tag{3.20}$$

$$x(0) = x(\eta) = x(1) = 0 \tag{3.21}$$

where $e : [0, 1] \to \mathbb{R}$ is a function in $L^1[0, 1]$, $\eta \in (0, 1)$, and $f : [0, 1] \times \mathbb{R}^3 \to \mathbb{R}$ is a function satisfying Carathéodory's conditions, that is

(i) for each $(x, y, z) \in \mathbb{R}^3$, the function $t \in [0, 1] \to f(t, x, y, z) \in \mathbb{R}$ is measurable on $[0, 1]$,

(ii) for a.e. $t \in [0, 1]$, the function $(x, y, z) \in \mathbb{R}^3 \to f(t, x, y, z) \in \mathbb{R}$ is continuous on $\mathbb{R}^3$, and

(iii) for each $r > 0$, there exists $g_r \in L^1[0, 1]$ such that $|f(t, x, y, z)| \leqslant g_r(t)$ for a.e. $t \in [0, 1]$ and $(x, y, z) \in \mathbb{R}^3$ with $\sqrt{x^2 + y^2 + z^2} \leqslant r$.

THEOREM 3.10. *Let* $f : [0, 1] \times \mathbb{R}^3 \to \mathbb{R}$ *satisfy Carathéodory's conditions. Assume that*
(i) *there exist* $a, b, c \in \mathbb{R}$ *and* $\alpha(t) \in L^1[0, 1]$ *such that*

$$f(t, x, y, z)y \geqslant ay^2 + b|y||x| + c|y||z| + \alpha(x)y,$$

for a.e. $t \in [0, 1]$ *and all* $(x, y, z) \in \mathbb{R}^3$;
(ii) *there exist functions* p, q, r *in* $L^2[0, 1]$ *and a function* $S(t) \in L^1[0, 1]$ *such that*

$$\left| f(t, x, y, z) \right| \leqslant p(t)|x| + q(t)|y| + r(t)|z| + S(t),$$

for a.e. $t \in [0, 1]$ *and all* $(x, y, z) \in \mathbb{R}^3$.
Let $\eta \in (0, 1)$ *be given. Then for any given function* $e \in L^1[0, 1]$, *the tree-point BVP* (3.20)–
(3.21) *has at least one solution in* $C^2[0, 1]$ *provided*

$$\frac{4}{\pi^2}|a| + \frac{4}{\pi^3}|b| + \frac{2}{\pi}|c| + \frac{2}{\pi^2}\|p\|_2 + \frac{2}{\pi}\|q\|_2 + \|r\|_2 < 1.$$

PROOF. Let X be the Banach space $C^2[0, 1]$ and Y denote the Banach space $L^1(0, 1)$ with their usual norms. We define a linear mapping $L : D(L) \subset X \to Y$ by setting

$$D(L) = \{x \in W^{3,1}(0, 1): \ x(0) = x(\eta) = x(1) = 0\},$$

and for $x \in D(L)$,

$$Lx = -x'''.$$

We also define a nonlinear mapping $N : X \to Y$ by setting

$$(Nx)(t) = f\big(t, x(t), x'(t), x''(t)\big), \quad t \in [0, 1].$$

We note that N is a bounded mapping from X into Y. Next, it is easy to see that the linear mapping $L : D(L) \subset X \to Y$, is one-to-one mapping. Next, the linear mapping $K : Y \to X$, defined for $y \in Y$ by

$$(Ky)(t) = \frac{1}{2}\int_0^t (t - s)^2 y(s)\,ds + Bt + Ct^2,$$

where

$$B = \frac{1}{2\eta(\eta - 1)}\int_0^\eta (t - s)^2 y(s)\,ds + \frac{\eta}{2(1 - \eta)}\int_0^1 (t - s)^2 y(s)\,ds,$$

$$C = \frac{1}{2\eta(1 - \eta)}\int_0^\eta (t - s)^2 y(s)\,ds - \frac{1}{2(1 - \eta)}\int_0^1 (t - s)^2 y(s)\,ds,$$

is such that for $y \in Y$, $Ky \in D(L)$ and $LKy = y$; and for $u \in D(L)$, $KLu = u$. Furthermore, it follows easily using the Arzelá–Ascoli theorem that KN maps a bounded subset of X into a relatively compact subset of X. Hence $KN : X \to X$ is a compact mapping.

We next note that $x \in C^1[0, 1]$ is a solution of the BVP (3.20)–(3.21) if and only if x is a solution to the operator equation

$$Lx = Nx + e.$$

Now, the operator equation $Lx = Nx + e$ is equivalent to the equation

$$x = KNx + Ke.$$

We apply the Leray–Schauder continuation theorem (see, e.g., [82, Corollary IV.7]) to obtain the existence of a solution for $x = KNx + Ke$ or equivalently to the BVP (3.20)–(3.21).

To do this, it suffices to verify that the set of all possible solutions of the family of equations

$$x'''(t) = \lambda f\big(t, x(t), x'(t), x''(t)\big) + \lambda e(t), \quad t \in (0, 1), \tag{3.22}$$

$$x(0) = x(\eta) = x(1) = 0, \tag{3.23}$$

is, a priori, bounded in $C^2[0, 1]$ by a constant independent of $\lambda \in [0, 1]$.

We observe that for $x \in W^{3,1}(0, 1)$ with $x(0) = x(\eta) = x(1) = 0$, there exist ξ_1, ξ_2, ξ_3 in $(0, 1)$ with $\xi_1 < \eta < \xi_2$, $x'(\xi_1) = x'(\xi_2) = 0$, $x''(\xi_3) = 0$. This gives

$$\|x\|_2 \leqslant \frac{1}{\pi}\|x'\|_2, \qquad \|x'\|_2 \leqslant \frac{2}{\pi}\|x''\|_2,$$

$$\big|x'(0)\big| = \left|\int_0^{\xi_1} x''(t)\,dt\right| \leqslant \int_0^{\eta} \big|x''(t)\big|\,dt,$$

$$\big|x'(1)\big| = \left|\int_{\xi_2}^{1} x''(t)\,dt\right| \leqslant \int_{\eta}^{1} \big|x''(t)\big|\,dt,$$

$$\big|x''(0)\big| = \left|\int_0^{\xi_3} x'''(t)\,dt\right| \leqslant \|x'''\|_1, \qquad \big|x''(1)\big| = \left|\int_{\xi_3}^{1} x'''(t)\,dt\right| \leqslant \|x'''\|_1.$$

We multiply Eq. (3.22) by x' and integrate from 0 to 1 to get

$$0 = -\int_0^1 x'''(t)x'(t)\,dt + \lambda \int_0^1 f\big(t, x(t), x'(t), x''(t)\big)x'(t)\,dt$$

$$- \lambda \int_0^1 e(t)x'(t)\,dt$$

$$\geq \|x''\|_2^2 - |x''(1)||x'(1)| - |x''(0)||x'(0)|$$

$$+ \lambda \int_0^1 \left\{ ax'^2(t) + b|x(t)||x'(t)| + c|x'(t)||x''(t)| + \alpha(t)x'(t) \right\} dt$$

$$- \lambda \|e\|_1 \|x'\|_\infty$$

$$\geq \|x''\|_2^2 - \left(|x'(0)| + |x'(1)| \right) \|x''\|_1 - |a| \|x'\|_2^2 - |b| \|x\|_2 \|x'\|_2$$

$$- |c| \|x'\|_2 \|x''\|_2 - \left(\|\alpha\|_1 + \|e\|_1 \right) \|x'\|_\infty$$

$$\geq \|x''\|_2^2 - \left(\int_0^\eta |x''(t)| \, dt + \int_\eta^1 |x''(t)| \, dt \right) \|x'''\|_1$$

$$- \left(\frac{4}{\pi^2} |a| + \frac{4}{\pi^3} |b| + \frac{2}{\pi} |c| \right) \|x''\|_2^2 - \left(\|\alpha\|_1 + \|e\|_1 \right) \|x''\|_2$$

$$\geq \|x''\|_2^2 - \|x''\|_1 \|x'''\|_1 - \left(\frac{4}{\pi^2} |a| + \frac{4}{\pi^3} |b| + \frac{2}{\pi} |c| \right) \|x''\|_2^2$$

$$- \left(\|\alpha\|_1 + \|e\|_1 \right) \|x''\|_2.$$

From Eq. (3.22) we have

$$\|x'''\|_1 \leq \|p\|_2 \|x\|_2 + \|q\|_2 \|x'\|_2 + \|r\|_2 \|x''\|_2 + \|S\|_1 + \|e\|_1$$

$$\leq \left(\frac{2}{\pi^2} \|p\|_2 + \frac{2}{\pi} \|q\|_2 + \|r\|_2 \right) \|x''\|_2 + \|S\|_1 + \|e\|_1.$$

Putting this in the previous inequality we find

$$\|x''\|_2 \leq \frac{\|S\|_1 + 2\|e\|_1 + \|\alpha\|_1}{1 - \left(\frac{4}{\pi^2} |a| + \frac{4}{\pi^3} |b| + \frac{2}{\pi} |c| + \frac{2}{\pi^2} \|p\|_2 + \frac{2}{\pi} \|q\|_2 + \|r\|_2 \right)}.$$

Hence, there exists a constant C, independent of $\lambda \in [0, 1]$ such that $\|x'''\|_1 \leq C$. It follows that there is a constant, still denoted by C, independent of $\lambda \in [0, 1]$ such that

$$\|x\|_{C^2[0,1]} \leq C.$$

This completes the proof of the theorem. $\qquad\square$

In Theorem 3.10 we can do without assumption (i), if we observe that

$$f(t, x, y, z)y \geq |f(t, x, y, z)||y|$$

$$\geq -|p(t)||y||x| - |q(t)||y|^2 - |r(t)||y||z| - |S(t)||y|.$$

488   S.K. Ntouyas

It is easy to see from the proof of Theorem 3.10 that the BVP (3.20)–(3.21) has at least one
solution in $C^2[0, 1]$ provided

$$\frac{4}{\pi^2}\|p\|_2 + \frac{4}{\pi}\|q\|_2 + 2\|r\|_2 < 1.$$

THEOREM 3.11. *Let* $f : [0, 1] \times \mathbb{R}^3 \to \mathbb{R}$ *satisfy Carathéodory's conditions. Suppose that*
   (iii) *there exist functions* $p, q, r \in L^2[0, 1]$ *such that*

$$\left| f(t, x_1, y_1, z_1) - f(t, x_2, y_2, z_2) \right|$$
$$\leqslant p(t)|x_1 - x_2| + q(t)|y_1 - y_2| + r(t)|z_1 - z_2|,$$

   for a.e. $t \in [0, 1]$ *and all* $(x_i, y_i, z_i) \in \mathbb{R}^3, i = 1, 2$.
Let $\eta \in (0, 1)$ *be given. Then for every* $e \in L^1[0, 1]$ *the BVP* (3.20)–(3.21) *has exactly one*
solution in $C^2[0, 1]$ *provided*

$$\frac{4}{\pi^2}\|p\|_2 + \frac{4}{\pi}\|q\|_2 + 2\|r\|_2 < 1.$$

PROOF. We note that (iii) implies that

$$\left| f(t, x, y, z) \right| \leqslant p(t)|x| + q(t)|y| + r(t)|z| + \left| f(t, 0, 0, 0) \right|$$

for a.e. $t \in [0, 1]$ and all $(x, y, z) \in \mathbb{R}^3$. Accordingly the BVP (3.20)–(3.21) has at least one
solution in $C^2[0, 1]$.

   Now, to prove the uniqueness, let x_1, x_2 are two solutions of the BVP (3.20)–(3.21).
Setting $u(t) = x_1(t) - x_2(t)$ we then get

$$u'''(t) = f\big(t, x_1(t), y_1(t), z_1(t)\big) - f\big(t, x_2(t), y_2(t), z_2(t)\big), \tag{3.24}$$

$$u(0) = u(\eta) = u(1) = 0. \tag{3.25}$$

We multiply Eq. (3.24) by u' and integrate over $[0, 1]$ to get

$$0 = -\int_0^1 u'''(t)u'(t)\,dt$$

$$+ \int_0^1 \left[f\big(t, x_1(t), y_1(t), z_1(t)\big) - f\big(t, x_2(t), y_2(t), z_2(t)\big) \right] u'(t)\,dt$$

$$\geqslant \|u''\|_2^2 - \|u''\|_2\|u'''\|_1 - \|p\|_2\|u\|_2\|u'\|_\infty - \|q\|_2\|u'\|_2\|u'\|_\infty$$

$$- \|r\|_2\|u'\|_\infty\|u''\|_2$$

$$\geqslant \|u''\|_2^2 - \|u''\|_2\big(\|p\|_2\|u\|_2 + \|q\|_2\|u'\|_2 + \|r\|_2\|u''\|_2\big)$$

$$- \left(\frac{2}{\pi^2}\|p\|_2 + \frac{2}{\pi}\|q\|_2 + \|r\|_2\right)\|u''\|_2^2$$

$$\geqslant \left(1 - \left[\frac{4}{\pi^2}\|p\|_2 + \frac{4}{\pi}\|q\|_2 + 2\|r\|_2\right]\right)\|u''\|_2^2,$$

hence $\|u''\|_2 = 0$. It now follows from the estimate $\|u\|_2 \leqslant (2/\pi^2)\|u''\|_2$ and the continuity of u that $u(t) = 0$ for every $t \in [0, 1]$, i.e. $x_1(t) = x_2(t)$ for every $t \in [0, 1]$. This completes the proof of the theorem. $\qquad\Box$

Theorems 3.10 and 3.11 were proved in [31]. For other results for third-order three-point BVPs the interested reader is referred to [1,24,81] and the references cited therein.

3.4. *m-point boundary value problems reduced to three-point boundary value problems*

Let $f:[0, 1] \times \mathbb{R}^2 \to \mathbb{R}$ be either a continuous or a Carathéodory's function and let $e:[0, 1] \to \mathbb{R}$ be a function in $L^1[0, 1]$, $a_i \in \mathbb{R}$, with all of the a_i's having the same sign, $\xi_i \in (0, 1)$, $i = 1, 2, \ldots, m - 2$, $0 < \xi_1 < \xi_2 < \cdots < \xi_{m-2} < 1$. We consider the following second-order ordinary differential equation

$$x''(t) = f\big(t, x(t), x'(t)\big) + e(t), \quad t \in (0, 1) \tag{3.26}$$

subject to one of the following boundary value conditions:

$$x(0) = 0, \qquad x(1) = \sum_{i=1}^{m-2} a_i x(\xi_i), \tag{3.27}$$

$$x'(0) = 0, \qquad x(1) = \sum_{i=1}^{m-2} a_i x(\xi_i), \tag{3.28}$$

$$x(0) = 0, \qquad x'(1) = \sum_{i=1}^{m-2} a_i x'(\xi_i). \tag{3.29}$$

It is well known, that the existence of a solution for these BVPs can be studied via the existence of a solution for Eq. (3.26) subject to one of the following three-point boundary value conditions:

$$x(0) = 0, \qquad x(1) = \alpha x(\eta), \tag{3.30}$$

$$x'(0) = 0, \qquad x(1) = \alpha x(\eta), \tag{3.31}$$

$$x(0) = 0, \qquad x'(1) = \alpha x'(\eta), \tag{3.32}$$

where $\alpha = \sum_{i=1}^{m-2} a_i$ and $\eta \in [\xi_1, \xi_{m-2}]$.

For certain boundary condition case such that the linear operator $Lx = x''$, defined in a suitable Banach space, is invertible, is the so-called *nonresonance* case. Otherwise, the so-called *resonance* case. In the next sections we will give existence results for both, non-resonance and resonance cases.

3.5. *Nonresonance results*

Using the Leray–Schauder continuation theorem (see [82]) we first give an existence result for the three-point BVP (3.26), (3.30) in the case when $\alpha \neq 1$.

THEOREM 3.12. *Let $f : [0, 1] \times \mathbb{R}^2 \to \mathbb{R}$ be a function satisfying Carathéodory's conditions. Assume that (3.12) holds. Also let $\alpha \in \mathbb{R}$, and $\eta \in (0, 1)$ be given. Then the BVP (3.26), (3.30) has at least one solution in $C^1[0, 1]$ provided*

$$\begin{cases} \|p\|_1 + \|q\|_1 < 1, & \text{if } \alpha \leqslant 1, \\ \|p\|_1 + \|q\|_1 < \dfrac{1 - \alpha\eta}{\alpha(1 - \eta)}, & \text{if } 1 < \alpha < 1/\eta. \end{cases}$$

PROOF. Let X be the Banach space $C^1[0, 1]$ and Y denote the Banach space $L^1(0, 1)$ with their usual norms. We denote a linear mapping $L : D(L) \subset X \to Y$ by setting

$$D(L) = \left\{ x \in W^{2,1}(0, 1) \colon x(0) = 0, x(1) = \alpha x(\eta) \right\},$$

and for $x \in D(L)$,

$$Lx = x''.$$

We also define a nonlinear mapping $N : X \to Y$ by setting

$$(Nx)(t) = f\big(t, x(t), x'(t)\big), \quad t \in [0, 1].$$

We note that N is a bounded mapping from X into Y. Next, it is easy to see that the linear mapping $L : D(L) \subset X \to Y$, is one-to-one mapping. We note that since $\alpha \leqslant 1, \alpha \neq 1/\eta$. Next, the linear mapping $K : Y \to X$, defined for $y \in Y$ by

$$(Ky)(t) = \int_0^t (t - s)y(s)\, \mathrm{d}s + \frac{\alpha t}{1 - \alpha\eta} \int_0^\eta (\eta - s)y(s)\, \mathrm{d}s$$

$$- \frac{t}{1 - \alpha\eta} \int_0^1 (1 - s)y(s)\, \mathrm{d}s$$

is such that for $y \in Y$, $Ky \in D(L)$ and $LKy = y$; and for $u \in D(L)$, $KLu = u$. Furthermore, it follows easily using the Arzelá–Ascoli theorem that KN maps a bounded subset of X into a relatively compact subset of X. Hence $KN : X \to X$ is a compact mapping.

We next note that $x \in C^1[0, 1]$ is a solution of the BVP (3.26), (3.30) if and only if x is a solution to the operator equation

$$Lx = Nx + e.$$

Now, the operator equation $Lx = Nx + e$ is equivalent to the equation

$$x = KNx + Ke.$$

We apply the Leray–Schauder continuation theorem (see, e.g., [82, Corollary IV.7]) to obtain the existence of a solution for $x = KNx + Ke$ or equivalently to the BVP (3.26), (3.30).

To do this, it suffices to verify that the set of all possible solutions of the family of equations

$$x''(t) = \lambda f\big(t, x(t), x'(t)\big) + \lambda e(t), \quad t \in (0, 1), \tag{3.33}$$

$$x(0) = 0, \qquad x(1) = \alpha x(\eta), \tag{3.34}$$

is, a priori, bounded in $C^1[0, 1]$ by a constant independent of $\lambda \in [0, 1]$.

We observe that if $x \in W^{2,1}(0, 1)$, with $x(0) = 0$, $x(1) = \alpha x(\eta)$, with $\alpha \leqslant 1$, there exists a $\zeta \in (0, 1)$ with $x'(\zeta) = 0$. Indeed, suppose that $\alpha \leqslant 1$ and $x'(t) > 0$ for every $t \in (0, 1)$. Then $x(t) > 0$ for every $t \in (0, 1)$ and $x(t)$ is strictly increasing on $[0, 1]$. We then get that $0 < x(1) - x(\eta) = (\alpha - 1)x(\eta) \leqslant 0$, a contradiction. Then we have

$$\|x\|_\infty \leqslant \|x'\|_\infty \leqslant \|x''\|_1. \tag{3.35}$$

Now let $x(t)$ be a solution of (3.33)–(3.34) for some $\lambda \in [0, 1]$, so that $x \in W^{2,1}(0, 1)$ with $x(0) = 0$, $x(1) = \alpha x(\eta)$. We then get from Eq. (3.33) that

$$\begin{aligned}
\|x''\|_1 &= \lambda \big\| f\big(t, x(t), x'(t)\big) + e(t) \big\|_1 \\
&\leqslant \|p\|_1 \|x\|_\infty + \|q\|_1 \|x'\|_\infty + \|r\|_1 + \|e\|_1 \\
&\leqslant \big(\|p\|_1 + \|q\|_1\big) \|x''\|_1 + \|r\|_1 + \|e\|_1.
\end{aligned}$$

It follows from the assumption that there is a constant c, independent of $\lambda \in [0, 1]$, such that

$$\|x''\|_1 \leqslant c.$$

It is now immediate from (3.35) that the set of solutions of the family of equations (3.33)–(3.34) is, a priori, bounded in $C^1[0, 1]$ by a constant independent of $\lambda \in [0, 1]$. This completes the proof of the theorem when $\alpha \leqslant 1$.

Let now $\alpha > 1$ and $\eta \in (0, 1)$ with $\alpha\eta < 1$. It suffices to verify that the set of all possible solutions of the family of equations (3.33)–(3.34) is, a priori, bounded in $C^1[0, 1]$ by a constant independent of $\lambda \in [0, 1]$.

For $x \in W^{2,1}(0, 1)$, with $x(0) = 0$, $x(1) = \alpha x(\eta)$, if $\alpha > 1$, there exists a $\zeta \in (\eta, 1)$ such that $(\alpha - 1)x(\eta) = x(1) - x(\eta) = (1 - \eta)x'(\zeta)$, or $x(\eta) = (1 - \eta)/(\alpha - 1)x'(\zeta)$.

Then we have

$$\|x\|_\infty \leqslant \|x'\|_\infty \leqslant \frac{\alpha(1 - \eta)}{1 - \alpha\eta}.$$

We then see from Eq. (3.33) that

$$\|x''\|_1 = \lambda \big\| f\big(t, x(t), x'(t)\big) + e(t) \big\|_1$$

$$\leqslant \|p\|_1 \|x\|_\infty + \|q\|_1 \|x'\|_\infty + \|r\|_1 + \|e\|_1$$

$$\leqslant \left(\|p\|_1 + \|q\|_1\right)\left(\frac{\alpha(1-\eta)}{1-\alpha\eta}\right)\|x''\|_1 + \|r\|_1 + \|e\|_1$$

for a solution x of the family of equations (3.33)–(3.34) for some $\lambda \in [0, 1]$. It is then immediate that the set of solutions of the family of equations (3.33)–(3.34) is, a priori, bounded in $C^1[0, 1]$ by a constant, independent of $\lambda \in [0, 1]$. This completes the proof of the theorem. $\qquad\square$

Now we study the m-point BVP (3.26)–(3.27) using the a priori estimates that we obtained for the three-point BVP (3.26), (3.30). This is because for every solution $x(t)$ of the BVP (3.26)–(3.27) there exists $\eta \in [\xi_1, \xi_{m-2}]$, depending on $x(t)$, such that $x(t)$ is also a solution of the BVP (3.26), (3.30) with $\alpha = \sum_{i=1}^{m-2} a_i \neq 1$.

THEOREM 3.13. *Let* $f : [0, 1] \times \mathbb{R}^2 \to \mathbb{R}$ *be a function satisfying Carathéodory's conditions. Assume that (3.12) holds. Also let* $\alpha = \sum_{i=1}^{m-2} a_i$, *and* $\eta \in (0, 1)$ *be given. Then the BVP (3.26)–(3.27) has at least one solution in* $C^1[0, 1]$ *provided*

$$\begin{cases} \|p\|_1 + \|q\|_1 < 1, & \text{if } \alpha \leqslant 1, \\[2mm] \|p\|_1 + \|q\|_1 < \dfrac{1 - \alpha\xi_{m-2}}{\alpha(1 - \xi_1)}, & \text{if } 1 < \alpha < 1/\xi_{m-2}. \end{cases}$$

PROOF. We will give only the proof for $\alpha = \sum_{i=1}^{m-2} a_i \leqslant 1$; the other case is similar. The proof is quite similar to that of Theorem 3.12 and uses the a priori bounds for the set of solutions of the family of equations (3.33)–(3.34).

Let X be the Banach space $C^1[0, 1]$ and Y denote the Banach space $L^1(0, 1)$ with their usual norms. We denote a linear mapping $L : D(L) \subset X \to Y$ by setting

$$D(L) = \left\{ x \in W^{2,1}(0, 1) : x(0) = 0, x(1) = \sum_{i=1}^{m-2} a_i x(\xi_i) \right\},$$

and for $x \in D(L)$,

$$Lx = x''.$$

We also define a nonlinear mapping $N : X \to Y$ by setting

$$(Nx)(t) = f\big(t, x(t), x'(t)\big), \quad t \in [0, 1].$$

We note that N is a bounded mapping from X into Y. Next, it is easy to see that the linear mapping $L : D(L) \subset X \to Y$, is one-to-one mapping. Next, the linear mapping $K : Y \to X$, defined for $y \in Y$ by

$$(Ky)(t) = \int_0^t (t - s)y(s)\,\mathrm{d}s + At$$

where A is given by

$$A\left(1 - \sum_{i=1}^{m-2} a_i \xi_i\right) = \sum_{i=1}^{m-2} a_i \int_0^{\xi_1} (\xi_i - s) y(s)\,ds - \int_0^1 (1-s) y(s)\,ds,$$

is such that for $y \in Y$, $Ky \in D(L)$ and $LKy = y$; and for $u \in D(L)$, $KLu = u$. Furthermore, it follows easily using the Arzelá–Ascoli theorem that KN maps bounded subsets of X into a relatively compact subsets of X. Hence $KN : X \to X$ is a compact mapping.

We, next, note that $x \in C^1[0, 1]$ is a solution of the BVP (3.26)–(3.27) if and only if x is a solution to the operator equation

$$Lx = Nx + e.$$

Now, the operator equation $Lx = Nx + e$ is equivalent to the equation

$$x = KNx + Ke.$$

We apply the Leray–Schauder continuation theorem to obtain the existence of a solution for $x = KNx + Ke$ or equivalently to the BVP (3.26)–(3.27).

To do this, it suffices to verify that the set of all possible solutions of the family of equations

$$x''(t) = \lambda f\big(t, x(t), x'(t)\big) + \lambda e(t), \quad t \in (0, 1), \tag{3.36}$$

$$x(0) = 0, \qquad x(1) = \sum_{i=1}^{m-2} a_i x(\xi_i), \tag{3.37}$$

is, a priori, bounded in $C^1[0, 1]$ by a constant independent of $\lambda \in [0, 1]$.

Let, now, $x(t)$ be a solution of (3.36)–(3.37) for some $\lambda \in [0, 1]$, so that $x \in W^{2,1}(0, 1)$ with $x(0) = 0$, $x(1) = \sum_{i=1}^{m-2} a_i x(\xi_i)$. Accordingly, there exists, an $\eta \in [\xi_1, \xi_{m-2}]$, depending on x, such that $x(t)$ is a solution of the three-point BVP

$$x''(t) = \lambda f\big(t, x(t), x'(t)\big) + \lambda e(t), \quad t \in (0, 1),$$

$$x(0) = 0, \qquad x(1) = \alpha x(\eta).$$

It then follows, as in the proof of Theorem 3.12 that there is a constant c, independent of $\lambda \in [0, 1]$, and $\eta \in [\xi_1, \xi_{m-2}]$, such that

$$\|x\|_\infty \leqslant \|x'\|_\infty \leqslant \|x''\|_1 \leqslant c.$$

Thus the set of solutions of the family of equations (3.36)–(3.37) is, a priori, bounded in $C^1[0, 1]$ by a constant, independent of $\lambda \in [0, 1]$.

This completes the proof of the theorem. $\square$

Theorems 3.12 and 3.13 were proved in [33]. In Theorem 3.12 the functions p, q, r in the condition (3.12) are in $L^1(0, 1)$. If p, q are in $L^\infty(0, 1)$ and r, e in $L^2(0, 1)$ then the BVP (3.26), (3.30) has at least one solution in $C^1[0, 1]$ provided

$$\begin{cases} \dfrac{2}{\pi}\left(\dfrac{2}{\pi}\|p\|_\infty + \|q\|_\infty\right) < 1, & \text{if } \alpha \leqslant 1, \\[3mm] \dfrac{2}{\pi}\|p\|_\infty + \|q\|_\infty < \dfrac{\sqrt{2}(1 - \alpha\eta)}{\alpha(1 - \eta)}, & \text{if } 1 < \alpha < 1/\eta. \end{cases}$$

In the case when $\alpha \geqslant 1$ and $\alpha\eta \neq 1$ we have a sharper condition for the solvability of BVP (3.26), (3.30). Thus, if tp, q, r are in $L^1[0, 1]$ and (3.12) holds, then the BVP (3.26), (3.30) has at least one solution in $C^1[0, 1]$ provided

$$\begin{cases} \|tp(t)\|_1 + \|q(t)\|_1 < \dfrac{1 - \alpha\eta}{1 - \eta}, & \text{if } \alpha\eta \leqslant 1, \\[3mm] \|tp(t)\|_1 + \|q(t)\|_1 < \dfrac{\alpha\eta - 1}{(\alpha - 1)\eta} & \text{if } \alpha\eta > 1. \end{cases}$$

The above conditions were proved in [33] and based on a priori estimates obtained by a different method than that was used in Theorem 3.12.

In the case when $\alpha \in \mathbb{R}$, $\alpha \leqslant 1$ Gupta and Trofimchuk in [38] gave constructive conditions for the solvability of the BVP (3.26), (3.30) using the spectral radius of a compact linear operator, and in [39] new solvability conditions based on a new type Wirtinger inequality. For other results see [37].

In the case when $\alpha \in \mathbb{R}$ with $\alpha\eta \neq 1$ a better condition for the solvability of the BVP (3.26), (3.30) is given by Gupta and Trofimchuk in [38]. They proved that, if $tp, q, t[tp + q], r$ are in $L^1[0, 1]$ and (3.12) holds, then the BVP (3.26), (3.30) has at least one solution in $C^1[0, 1]$ provided

$$\max\left\{ \int_0^\eta \left| \frac{1 - \alpha}{1 - \alpha\eta} - 1 \right| \left[sp(s) + q(s)\right] \mathrm{d}s \right.$$
$$+ \frac{1}{|1 - \alpha\eta|} \int_\eta^1 (1 - s)\left[sp(s) + q(s)\right] \mathrm{d}s,$$
$$\left. \int_\eta^1 \frac{|s - \alpha\eta|}{|1 - \alpha\eta|}\left[sp(s) + q(s)\right] \mathrm{d}s + \frac{|\alpha - 1|}{|1 - \alpha\eta|} \int_0^\eta s\left[sp(s) + q(s)\right] \mathrm{d}s \right\} < 1.$$

Moreover, Gupta in [30], when $\alpha \neq 1$ and $\alpha\eta \neq 1$, obtained new a priori estimates which are sharper than those used in the previous theorems. More precisely, he proved that if p, q, r are in $L^1(0, 1)$ and (3.12) holds, then the BVP (3.26), (3.30) has at least one solution in $C^1[0, 1]$ provided

$$M\|p\|_1 + \frac{1}{1 - \tau}\|q\|_1 < 1,$$

where M and τ are appropriate constants.

Finally, we note that in [40] Gupta and Trofimchuk studied examples of three-point BVP by making extensive use of computer algebra systems like Maple and MathCad.

Concerning the uniqueness of the solutions of the BVPs (3.26), (3.30) and (3.26)–(3.27) we have the following results.

THEOREM 3.14. *Let $f : [0, 1] \times \mathbb{R}^2 \to \mathbb{R}$ be a function satisfying Carathéodory's conditions. Assume that there exist nonnegative constants c, d such that*

$$\left| f(t, x_1, x_2) - f(t, y_1, y_2) \right| \leqslant c|x_1 - x_2| + d|y_1 - y_2|$$

for a.e. $t \in [0, 1]$ and all $(x_i, x_i) \in \mathbb{R}^2, i = 1, 2$. Also let $\alpha \in \mathbb{R}$, and $\eta \in (0, 1)$ be given. Then the BVP (3.26), (3.30) has exactly one solution in $C^1[0, 1]$ provided

$$\begin{cases} \dfrac{2}{\pi}\left(\dfrac{2}{\pi}c + d\right) < 1, & \text{if } \alpha \leqslant 1, \\[3mm] \dfrac{2}{\pi}c + d < \dfrac{\sqrt{2}(1 - \alpha\eta)}{\alpha(1 - \eta)}, & \text{if } 1 < \alpha < 1/\eta. \end{cases}$$

PROOF. The existence of a solution for the BVP (3.26), (3.30) follows from Theorem 3.12. Let now $x_1(t), x_2(t)$ be two solutions of the BVP (3.26), (3.30). We then get

$$x_1''(t) - x_2''(t) = f\left(t, x_1(t), x_1'(t)\right) - f\left(t, y_1(t), y_1'(t)\right), \quad 0 < t < 1,$$

$$(x_1 - x_2)(0) = 0, \qquad (x_1 - x_2)(1) = \alpha(x_1 - x_2)(\eta).$$

It follows that

$$\begin{aligned} \left\| x_1'' - x_2'' \right\|_2 &= \left\| f\left(t, x_1(t), x_1'(t)\right) - f\left(t, y_1(t), y_1'(t)\right) \right\|_2 \\[2mm] &\leqslant c\|x_1 - x_2\|_2 + \left\| x_1' - x_2' \right\|_2 \\[2mm] &\leqslant \frac{2}{\pi}\left(\frac{2}{\pi}c + d\right) \left\| x_1'' - x_2'' \right\|_2. \end{aligned}$$

Hence, we get that $\| x_1'' - x_2'' \|_2 = 0$. It now follows from the inequalities

$$\|x_1 - x_2\|_2 \leqslant \frac{2}{\pi}\left\| x_1' - x_2' \right\|_2 \leqslant \frac{4}{\pi^2}\left\| x_1'' - x_2'' \right\|_2 = 0,$$

that $x_1(t) = x_2(t)$ for a.e. $t \in [0, 1]$ and hence for every $t \in [0, 1]$ because of the continuity of $x_1(t)$ and $x_2(t)$ on $[0, 1]$. This completes the proof of the theorem in the case when $\alpha \leqslant 1$. The proof for the case $1 < \alpha < 1/\eta$ is similar. $\qquad\square$

THEOREM 3.15. *Let $f : [0, 1] \times \mathbb{R}^2 \to \mathbb{R}$ be as in Theorem 3.14. Then for any given $e(t)$ in $L^2(0, 1)$, the m-point BVP (3.26)–(3.27) has exactly one solution in $C^1[0, 1]$ provided*

$$\begin{cases} \|p\|_1 + \|q\|_1 < 1, & \text{if } \alpha \leqslant 1, \\ \|p\|_1 + \|q\|_1 < \dfrac{1 - \alpha \xi_{m-2}}{\alpha(1 - \xi_1)}, & \text{if } 1 < \alpha < 1/\eta. \end{cases}$$

We now give a theorem which allows f to have nonlinear growth. We do this by imposing a decomposition condition for f. We need the following simple lemma.

LEMMA 3.16. *Let $x \in C^1[0, 1]$ satisfy $x(0) = 0$, $x(1) = \alpha x(\eta)$, where $\eta \in (0, 1)$, and $\alpha > 1$, $\alpha \neq 1/\eta$, then there exist ζ and $C_0 \in (0, 1)$ such that $x'(\zeta) = C_0 x(1)$.*

PROOF. If $\alpha\eta > 1$, let $C_0 = 1/(\alpha\eta)$. There exists $\zeta \in (0, 1)$ with

$$x'(\zeta) = \frac{x(\eta) - x(0)}{\eta} = C_0 x(1).$$

If $\alpha\eta < 1$, let $C_0 = \frac{\alpha - 1}{\alpha(1 - \eta)}$. There exists $\zeta \in (\eta, 1)$ with

$$x'(\zeta) = \frac{x(1) - x(\eta)}{1 - \eta} = \frac{\alpha - 1}{\alpha(1 - \eta)} x(\eta) = C_0 x(1). \qquad \square$$

THEOREM 3.17. *Assume that $f : [0, 1] \times \mathbb{R}^2 \to \mathbb{R}$ is continuous and has the decomposition*

$$f(t, x, p) = g(t, x, p) + h(t, x, p)$$

such that
 (1) *$pg(t, x, p) \leqslant 0$ for all $(t, x, p) \in [0, 1] \times \mathbb{R}^2$;*
 (2) *$h(t, x, p)| \leqslant a(t)|x| + b(t)|p| + u(t)|x|^r + v(t)|p|^k + c(t)$ for all $(t, x, p) \in [0, 1] \times \mathbb{R}^2$ where a, b, u, v are in $L^1[0, 1]$ and $0 \leqslant r, k < 1$.*
Then, for $\alpha \neq 1/\eta$, there exists a solution $x \in C^1[0, 1]$ to BVP (3.26), (3.30) provided that

$$\begin{cases} \|a\|_1 + \|b\|_1 < \dfrac{1}{2}, & \text{if } \alpha \leqslant 1, \\[2mm] \|a\|_1 + \|b\|_1 < \dfrac{1}{2}\left(1 - \dfrac{(\alpha - 1)^2}{\alpha^2(1 - \eta)^2}\right), & \text{if } 1 < \alpha < \dfrac{1}{\eta}, \\[2mm] \|a\|_1 + \|b\|_1 < \dfrac{1}{2}\left(1 - \dfrac{1}{\alpha^2\eta^2}\right), & \text{if } \dfrac{1}{\eta} < \alpha. \end{cases}$$

PROOF. By the same argument as in the proof of Theorem 3.12, it suffices to show that all possible solutions of the following family of equations (3.33)–(3.34) is bounded in $C^1[0, 1]$ by a constant independent of $\lambda \in [0, 1]$.

Suppose that x is a solution of (3.33) and let ζ be as in Lemma 3.16 and write $C = \|c\|_1 + \|e\|_1$. Multiplying both sides of Eq. (3.33) with x' and integrating, we obtain

$$
\begin{aligned}
\frac{1}{2}x'^2(t) &\leqslant \frac{1}{2}x'^2(\zeta) + \|x'\|_\infty \big(\|a\|_1 \|x\|_\infty + \|b\|_1 \|x'\|_\infty \\
&\quad + \|u\|_1 \|x\|_\infty^r + \|v\|_1 \|x'\|_\infty^k + C \big) \\
&= \frac{1}{2}C_0^2 x'^2(1) + \|x'\|_\infty \big(\|a\|_1 \|x\|_\infty + \|b\|_1 \|x'\|_\infty \\
&\quad + \|u\|_1 \|x\|_\infty^r + \|v\|_1 \|x'\|_\infty^k + C \big).
\end{aligned}
$$

Since $\|x\|_\infty \leqslant \|x'\|_\infty$ and by assumption

$$
\|a\|_1 + \|b\|_1 < \frac{1}{2}\big(1 - C_0^2\big).
$$

If $\|x'\|_\infty \neq 0$, we have

$$
\left(\frac{1}{2}\big(1 - C_0^2\big) - \big(\|a\|_1 + \|b\|_1 \big) \right) \|x'\|_\infty \leqslant \|u\|_1 \|x\|_\infty^r + \|v\|_1 \|x'\|_\infty^k + C.
$$

This implies that $\|x'\|_\infty$ is bounded since $0 \leqslant r, k < 1$, that is, there is $M > 0$ such that $\|x\|_\infty \leqslant \|x'\|_\infty \leqslant M$. This completes the proof. $\qquad\square$

Theorem 3.17 is taken over from Feng and Webb [15] and is more general than Theorem 3.12, since the assumptions of Theorem 3.12 are special cases of Theorem 3.17, when $g(t, x, p) \equiv 0$, $u(t) \equiv 0$ and $v(t) \equiv 0$. But in Theorem 3.12 we need $\|a\|_1 + \|b\|_1 < 1$ and in Theorem 3.17 in these special cases we need $\|a\|_1 + \|b\|_1 < 1/2$. In [15] also interesting examples are given to compare their results with other existence results.

We now give existence results for the BVPs (3.26), (3.31) and (3.26), (3.32). By using the same argument as in Theorem 3.12 we can prove the following existence results (see [32,34]).

THEOREM 3.18. *Let* $f : [0, 1] \times \mathbb{R}^2 \to \mathbb{R}$ *be a function satisfying Carathéodory's conditions. Assume that* (3.12) *holds. Also let* $\alpha \in \mathbb{R}$, *and* $\eta \in (0, 1)$ *be given. Then the BVP* (3.26), (3.31) *has at least one solution in* $C^1[0, 1]$ *provided*

$$
\begin{cases}
\|p\|_1 + \|q\|_1 < 1, & \text{if } \alpha \leqslant 0, \\
\left(1 + \max\left\{ \dfrac{\alpha(1 - \eta)}{|\alpha - 1|}, \dfrac{1 - \eta}{|\alpha - 1|} \right\} \right) \|p\|_1 + \|q\|_1 < 1, & \text{if } 0 < \alpha \neq 1.
\end{cases}
$$

In Theorem 3.18 if p, q, r are in $L^2(0, 1)$ then the BVP (3.26), (3.31) has at least one solution in $C^1[0, 1]$ provided

$$
\begin{cases}
\dfrac{2}{\pi}\left(\dfrac{2}{\pi}\|p\|_2 + \|q\|_2\right) < 1, & \text{if } \alpha \leqslant 0, \\[3mm]
\left(\dfrac{\sqrt{2}}{\pi} + \max\left\{\dfrac{\alpha(1-\eta)}{|\alpha-1|}, \dfrac{1-\eta}{|\alpha-1|}\right\}\right)\|p\|_1 + \dfrac{2}{\pi}\|q\|_1 < 1, & \text{if } 0 < \alpha \neq 1.
\end{cases}
$$

THEOREM 3.19. *Let $f : [0, 1] \times \mathbb{R}^2 \to \mathbb{R}$ be a function satisfying Carathéodory's conditions. Assume that* (3.12) *holds. Also let $\alpha \in \mathbb{R}$, $\alpha \neq 1$ and $\eta \in (0, 1)$ be given. Then the BVP* (3.26), (3.32) *has at least one solution in $C^1[0, 1]$ provided*

$$
\left(\max\left\{\frac{1 + |\alpha - 1|}{|\alpha - 1|}, \frac{|\alpha| + |\alpha - 1|}{|\alpha - 1|}\right\}\right)(\|p\|_1 + \|q\|_1) < 1.
$$

As in Theorem 3.13 we prove existence results for the m-point BVPs (3.26), (3.28) and (3.29), (3.31), by using the a priori bounds obtained for the BVPs (3.26), (3.31) and (3.26), (3.32).

3.6. *Results at resonance*

In the following we shall give existence results for BVP (3.26), (3.30) when $\alpha\eta = 1$. In this case, the linear operator L is noninvertible and the Leray–Schauder continuation theorem cannot be used. We shall apply the continuation theorem of Mawhin [82]. For the convenience of the reader, we recall this theorem.

Let X and Z be Banach spaces and $L : D(L) \subset X \to Z$ be a linear operator which is Fredholm of index zero (that is $\operatorname{im}(L)$ (the image of L) is closed in Z, and $\ker(L)$ (the kernel of L) and $Z/\operatorname{im}(L)$ (the cokernel of L) are finite-dimensional with equal dimension). Let $P : X \to \ker(L)$ and $Q : Z \to Z_1$, where $X = \ker(L) \oplus X_1$ and $Z = \operatorname{im}(L) \oplus Z_1$, be continuous projections. Let L_1 denote L restricted to $D(L) \cap X_1$, an invertible operator into $\operatorname{im}(L)$, and write $K = L_1^{-1}$. Let Ω be a bounded, open subset of X such that $D(L) \cap \Omega \neq \emptyset$ and let $N : \overline{\Omega} \to Z$ be an L-compact mapping, that is, the maps $QN : \overline{\Omega} \to Z$ and $K(I - Q)N : \overline{\Omega} \to X$ are compact.

THEOREM 3.20. *Let L be Fredholm of index zero and let N be L-compact on $\overline{\Omega}$. Assume that the following conditions are satisfied.*
   (1) *$Lx + \lambda Nx \neq 0$ for each $(x, \lambda) \in [D(L)\backslash \ker L \cap \partial\Omega] \times (0, 1)$.*
   (2) *$Nx \notin \operatorname{im}(L)$ for each $x \in \ker(L) \cap \partial\Omega$.*
   (3) *$\deg(QN|_{\ker L}, \Omega \cap \ker(L), 0) \neq 0$, where $Q : Z \to Z$ is a continuous projection as above.*
Then the equation $Lx + Nx = 0$ has at least one solution in $D(L) \cap \overline{\Omega}$.

The following existence theorems hold.

THEOREM 3.21. *Let $f : [0, 1] \times \mathbb{R}^2 \to \mathbb{R}$ be a continuous function. Assume that:*
(1) *There exist functions p, q, r in $L^1[0, 1]$ such that*

$$\left| f(t, u, v) \right| \leqslant p(t)|u| + q(t)|v| + r(t), \quad \text{for } t \in [0, 1] \text{ and } (u, v) \in \mathbb{R}^2.$$

(2) *There exists $N > 0$ such that for $v \in \mathbb{R}$ with $|v| > N$, one has*

$$\left| f(t, u, v) \right| \geqslant -l|u| + n|v| - M \quad \text{for } t \in [0, 1], \ u \in \mathbb{R}$$

where $n > l \geqslant 0$, $M \geqslant 0$.
(3) *There exists $R > 0$ such that for $|v| > R$ one has either*

$$vf(t, vt, v) \leqslant 0 \quad \text{for } t \in [0, 1]$$

or else

$$vf(t, vt, v) \geqslant 0 \quad \text{for } t \in [0, 1].$$

Then, for every continuous function e, the BVP (3.26), (3.30) with $\alpha\eta = 1$ has at least one solution in $C^1[0, 1]$ provided that

$$2\left(\|p\|_1 + \|q\|_1\right) + \frac{l}{n} < 1.$$

We give the proof via several lemmas. We now let L be the linear operator from $D(L) \subset X \to Z = L^1(0, 1)$ defined by

$$D(L) = \left\{ x \in W^{2,1}(0, 1) \colon x(0) = 0, x(1) = x(\eta)/\eta \right\}$$

and for $x \in D(L)$, $Lx = x''$. Let $X_1 = \{x \in X \colon x(0) = 0\}$.

LEMMA 3.22. *Suppose L is as above. Then $L : D(L) \subset X \to Z$ is Fredholm of index zero. Furthermore, the linear operator $K : \mathrm{im}(L) \to D(L) \cap X_1$ defined by*

$$(Ky)(t) = \int_0^t \int_0^\tau y(s) \, ds \, d\tau, \quad y \in \mathrm{im}(L)$$

is such that $K = L_P^{-1}$, where $L_P = L|_{D(L) \cap X_1}$. Also $\|Ky\| \leqslant \|y\|_1$ for all $y \in \mathrm{im}(L)$.

PROOF. It is easy to see that $\ker(L) = \{ct \colon c \in \mathbb{R}\}$. We show that

$$\mathrm{im}(L) = \left\{ y \in L^1[0, 1] \colon \int_0^1 Y(t) \, dt = \int_0^1 Y(\eta t) \, dt, \text{ where } Y(t) = \int_0^t y(s) \, ds \right\}.$$

500   *S.K. Ntouyas*

In fact, for $y \in \mathrm{im}(L)$, there is $x \in D(L)$ with $y(t) = x''(t)$. Therefore

$$\int_0^1 Y(t)\,\mathrm{d}t = \int_0^1 \left[x'(t) - x'(0)\right]\mathrm{d}t = x(1) - x'(0),$$

and

$$\int_0^1 Y(\eta t)\,\mathrm{d}t = \int_0^1 \left[x'(\eta t) - x'(0)\right]\mathrm{d}t = x(\eta)/\eta - x'(0).$$

Since $x(1) = \frac{1}{\eta}x(\eta)$, we have $\int_0^1 Y(t)\,\mathrm{d}t = \int_0^1 Y(\eta t)\,\mathrm{d}t$. On the other hand, suppose $y \in L^1[0, 1]$ is such that $\int_0^1 Y(t)\,\mathrm{d}t = \int_0^1 Y(\eta t)\,\mathrm{d}t$. Let $x(t) = \int_0^t Y(s)\,\mathrm{d}s$, then $x \in D(L)$ and $x''(t) = y(t)$.

For $y \in L^1[0, 1]$, let

$$Qy = \frac{2}{1 - \eta} \int_0^1 \int_{\eta t}^t y(s)\,\mathrm{d}s\,\mathrm{d}t,$$

and let $y_1(t) = y(t) - Qy$. Then $Y_1(t) = \int_0^t y(s)\,\mathrm{d}s - (Qy)(t)$, and writing Qy in the form

$$Qy = \frac{2}{1 - \eta}\left\{\int_0^1 \int_0^t y(s)\,\mathrm{d}s\,\mathrm{d}t - \int_0^1 \int_0^{\eta t} y(s)\,\mathrm{d}s\,\mathrm{d}t\right\},$$

we see that

$$\int_0^1 \int_0^t y(s)\,\mathrm{d}s\,\mathrm{d}t - \frac{Qy}{2} = \int_0^1 \int_0^{\eta t} y(s)\,\mathrm{d}s\,\mathrm{d}t - \eta\frac{Qy}{2},$$

that is

$$\int_0^1 Y_1(t)\,\mathrm{d}t = \int_0^1 Y_1(\eta t)\,\mathrm{d}t.$$

This shows that $y_1 \in \mathrm{im}(L)$. Hence $Z = \mathrm{im}(L) + \mathbb{R}$. Since $\mathrm{im}(L) \cap \mathbb{R} = \{0\}$, we have $Z = \mathrm{im}(L) \oplus \mathbb{R}$ and therefore L is Fredholm of index zero.

Now we define a projection from X onto $\ker(L)$ by setting $(Px)(t) = x'(0)t$ and let $X_1 = \{x \in X, x'(0) = 0\}$. Then, for $x \in D(L) \cap X_1$, we have

$$(KL_Px)(t) = Kx''(t) = \int_0^t \int_0^\tau x''(s)\,\mathrm{d}s\,\mathrm{d}\tau = \int_0^t \left(x'(\tau) - x'(0)\right)\mathrm{d}\tau = x(t),$$

and for $y \in \mathrm{im}(L)$, we have

$$(L_PKy)(t) = \left(\int_0^t \int_0^\tau y(s)\,\mathrm{d}s\,\mathrm{d}\tau\right)'' = y(t).$$

This shows that $K = L_P^{-1}$. Also we have

$$\left\|(Ky)(t)\right\|_\infty \leqslant \int_0^1 \int_0^1 |y(s)| \, ds \, d\tau \leqslant \left\|y(t)\right\|_1,$$

and

$$(Ky)'(t) = \int_0^t y(s) \, ds, \quad \text{so} \quad \left\|(Ky)'(t)\right\|_\infty \leqslant \left\|y(t)\right\|_1,$$

and hence $\|Ky\| \leqslant \|y\|_1$ which completes the proof. $\qquad\Box$

LEMMA 3.23. *Let* $U_1 = \{x \in D(L) \setminus \ker(L), Lx + \lambda Nx = 0 \text{ for some } \lambda \in [0, 1]\}$. *Then* U_1 *is a bounded subset of* X.

PROOF. Suppose that $x \in U_1$, and $Lx = -\lambda Nx$, then $\lambda \neq 0$ and $QNx = 0$. Therefore

$$\int_0^1 \int_{\eta\tau}^\tau \left(f\big(t, x(t), x'(t)\big) + e(t) \right) dt \, d\tau = 0,$$

and hence there exists $\gamma \in (0, 1)$ such that

$$\left| f\big(\gamma, x(\gamma), x'(\gamma)\big) \right| = \left| e(\gamma) \right| \leqslant \left\| e(t) \right\|_\infty.$$

Also for $x \in D(L) \setminus \ker(L)$, by Lemma 3.22 and condition (1)

$$\left\| (I - P)x \right\| = \left\| KL(I - P)x \right\| \leqslant \left\| L(I - P)x \right\|_1 = \left\| L(x) \right\|_1$$
$$\leqslant \left\| N(x) \right\|_1 \leqslant \|p\|_1 \|x\|_\infty + \|q\|_1 \|x'\|_\infty + \|r\|_1 + \|e\|_1.$$

If for some $t_0 \in [0, 1]$, $|x'(t_0)| \leqslant N$ then we have

$$\left| x'(0) \right| = \left| x'(t_0) - \int_0^{t_0} x''(t) \, dt \right| \leqslant N + \|x''\|_1.$$

Otherwise, if $|x'(t)| > N$ for all t, from condition (2) we obtain

$$\left| x'(\gamma) \right| \leqslant \frac{\|e\|_\infty + M}{n} + \frac{l}{n} \|x\|_\infty,$$

so that

$$\left| x'(0) \right| = \left| x'(\gamma) - \int_0^\gamma x''(t) \, dt \right| \leqslant \frac{\|e\|_\infty + M}{n} + \frac{l}{n} \|x\|_\infty + \|x''\|_1.$$

Therefore we have in either case

$$\|P(x)\| = |x'(0)| \leqslant \max\left\{\frac{\|e\|_\infty + M}{n}, N\right\} + \frac{l}{n}\|x\|_\infty + \|x''\|_1.$$

Writing $x(t) = \int_0^t x'(s)\,ds$, we obtain

$$\|x\|_\infty \leqslant \|x'\|_1 \leqslant \|x'\|_\infty.$$

Therefore

$$\|x'\|_\infty \leqslant \|x\| \leqslant \|(I - P)(x)\| + \|Px\|$$

$$\leqslant \left(\|p\|_1 + \|q\|_1 + \frac{l}{n}\right)\|x'\|_\infty + \|x''\|_1 + C,$$

where $C = \|r\|_1 + \max\{\|e\|_\infty + M/n, N\}$. Let $C_1 = 1 - [\|p\|_1 + \|q\|_1 + \frac{l}{n}]$, so that $C_1 > 0$.
Then we can write

$$\|x'\|_\infty \leqslant \frac{1}{C_1}\|x''\|_1 + \frac{C}{C_1},$$

and then

$$\|x''\|_1 = \|L(x)\|_1 \leqslant \|Nx\|_1 \leqslant \|p\|_1\|x\|_\infty + \|q\|_1\|x'\|_\infty + \|r\|_1 + \|e\|_1$$

$$\leqslant (\|p\|_1 + \|q\|_1)\|x'\|_\infty + \|r\|_1 + \|e\|_1$$

$$\leqslant \frac{\|p\|_1 + \|q\|_1}{C_1}\|x''\|_1 + C_2,$$

where $C_2 = \frac{\|p\|_1 + \|q\|_1}{C_1} + \|r\|_1 + \|e\|_1$. If $C_3 = \frac{\|p\|_1 + \|q\|_1}{C_1}$ then $C_3 < 1$ and hence

$$\|x\|_\infty \leqslant \|x'\|_\infty \leqslant \frac{C_2}{C_1(1 - C_3)} + \frac{C}{C_1},$$

which proves that U_1 is bounded. $\square$

LEMMA 3.24. *The set $U_2 := \{x \in \ker(L): N(x) \in \mathrm{im}(L)\}$ is bounded.*

PROOF. Suppose that $x \in U_2$ so that $x(t) = ct$, where c is a constant. Since $QNx = 0$, we
have

$$\int_0^1 \int_{\eta\tau}^\tau f(t, ct, c)\,dt\,d\tau = -\int_0^1 \int_{\eta\tau}^\tau e(t)\,dt\,d\tau.$$

Therefore there exists $\xi \in (0, 1)$ such that $|f(\xi, c\xi, c)| \leqslant |e(\xi)| \leqslant \|e\|_\infty$. It follows that

$$|c| \leqslant \max\left\{N, \frac{M + \|e\|_\infty}{n - l}\right\}.$$

For, if $|c| > N$, then by condition (2) we obtain $\|e\|_\infty \geqslant -l|c\xi| + n|c| - M$ and hence $|c| \leqslant (M + \|e\|_\infty)/(n - l)$. Thus U_2 is bounded. $\qquad\square$

LEMMA 3.25. *If in condition* (3) *we assume that there exists* $R > 0$ *such that for all* $|v| > R$, $vf(t, vt, v) \leqslant 0$, *for* $t \in [0, 1]$, *then the set*

$$U_3 := \left\{x \in \ker(L): \ H(x, \lambda) = \lambda J x + (1 - \lambda) Q N x = 0, \ \lambda \in [0, 1]\right\},$$

is bounded, where $J : \ker(L) \to Z_0$, *is the linear isomorphism given by* $J(ct) = c$.

PROOF. Assume that $x_n(t) - c_n t \in U_3$ and $\|c_n t\| = |c_n| \to \infty$ as $n \to \infty$. Then there exists $\lambda_n \in [0, 1]$ such that

$$\lambda_n c_n + (1 - \lambda_n)(QN)(c_n t) = 0.$$

$\{\lambda_n\}$ has a convergent subsequence, for simplicity we write $\lambda_n \to \lambda_0$. We show that $\lambda_0 \neq 1$. Indeed, otherwise we have

$$\lambda_n = -(1 - \lambda_n)\frac{(QN)c_n t}{c_n},$$

and as previously we find that

$$(1 - \lambda_n)\frac{\|(QN)c_n t\|}{|c_n|} \to 0 \quad (n \to \infty),$$

contradicting $\lambda_n \to 1$. Hence for n large enough, $1 - \lambda_n \neq 0$, and therefore

$$\frac{\lambda_n}{1 - \lambda_n} c_n = Q\big(f(t, c_n t, c_n) + e(t)\big),$$

and so

$$\frac{\lambda_n}{1 - \lambda_n} = \frac{2}{1 - \eta} \int_0^1 \int_{\eta\tau}^\tau \frac{f(s, c_n s, c_n)}{c_n} \, ds \, d\tau + \frac{2}{c_n(1 - \eta)} \int_0^1 \int_{\eta\tau}^\tau e(s) \, ds \, d\tau.$$

Since $|c_n| \to \infty$, we may assume that $|c_n| > \max\{N, R\}$. Then for large enough n we have

$$\left|\frac{f(s, c_n s, c_n)}{c_n}\right| \geqslant n - l - \frac{M}{|c_n|} \geqslant \frac{n - l}{2}.$$

By our assumption $c_n f(t, c_n t, c_n) \leqslant 0$ this yield $f(t, c_n t, c_n)/c_n \leqslant -(n-l)/2$. Hence, by Fatou's lemma,

$$\limsup_{n\to\infty}\left\{\int_0^1\int_{\eta\tau}^\tau \frac{f(s, c_n s, c_n)}{c_n}\,ds\,d\tau + \frac{1}{c_n}\int_0^1\int_{\eta\tau}^\tau e(s)\,ds\,d\tau\right\}$$

$$\leqslant \limsup_{n\to\infty}\int_0^1\int_{\eta\tau}^\tau \frac{f(s, c_n s, c_n)}{c_n}\,ds\,d\tau$$

$$\leqslant \int_0^1\int_{\eta\tau}^\tau \limsup_{n\to\infty} \frac{f(s, c_n s, c_n)}{c_n}\,ds\,d\tau$$

$$\leqslant -\frac{(n-l)(1-\eta)}{4}.$$

This is a contradiction with $\lambda_n/(1-\lambda_n) \geqslant 0$. Thus U_3 is bounded. $\qquad\square$

PROOF OF THEOREM 3.21. Firstly, by Arzelá–Ascoli theorem, it can be shown that the linear operator $K : \mathrm{im}(L) \to D(L) \cap X_1$ in Lemma 3.22 is compact operator, so N is L-compact. Let Ω be a bounded open set containing $\bigcup_{i=1}^3 U_i$. Then by the above lemmas the conditions of Theorem 3.20 are satisfied and therefore $Lx + Nx = 0$ has at least one solution in $D(L) \cap \overline{\Omega}$ so that the BVP (3.26), (3.30) with $\alpha\eta = 1$ has at least one solution. $\qquad\square$

THEOREM 3.26. *Assume that $f : [0, 1] \times \mathbb{R}^2 \to \mathbb{R}$ is continuous and has the decomposition*

$$f(t, x, p) = g(t, x, p) + h(t, x, p).$$

Assume that

(1) *There exists $M_1 > 0$ such that for $x \in D(L) := \{x \in W^{2,1}(0, 1): x(0) = 0,\ x(1) = \frac{1}{\eta}x(\eta)\}$, if $|x'(t)| > M_1$ for all $t \in [0, 1]$, then*

$$\int_0^1\int_{\eta s}^s \big(f\big(t, x(t), x'(t)\big) + e(t)\big)\,dt\,ds \neq 0;$$

(2) *There exists $M_2 > 0$, such that for all $v \in \mathbb{R}$ with $|v| > M_2$ one has either*

$$v\big(f(t, vt, v) + e(t)\big) \geqslant 0 \quad \text{for } t \in [0, 1],$$

or

$$v\big(f(t, vt, v) + e(t)\big) \leqslant 0 \quad \text{for } t \in [0, 1];$$

(3) *$pg(t, x, p) \leqslant 0$ for all $(t, x, p) \in [0, 1] \times [-M, M] \times \mathbb{R}$;*

(4) $|h(t, x, p)| \leq a(t)|x| + b(t)|p| + u(t)|x|^r + v(t)|p|^k + c(t)$ *for* $(t, x, p) \in [0, 1] \times [-M, M] \times \mathbb{R}$ *where* a, b, u, v *are in* $L^1[0, 1]$ *and* $0 \leq r, k < 1$.

Then, the BVP (3.26), (3.30) *with* $\alpha\eta = 1$ *has at least one solution in* $C^1[0, 1]$ *provided that*

$$\|a\|_1 + \|b\|_1 < \frac{1}{2}.$$

Theorem 3.21 is taken over from [16]. For the proof of Theorem 3.26 we refer to [15]. Other results are given by Gupta in [27,34]. In [16] is also proved the following uniqueness result.

THEOREM 3.27. *Suppose that the conditions* (1) *and* (2) *in Theorem 3.21 are replaced by the following conditions respectively:*
   (1) *There exist functions* p, q *in* $L^1[0, 1]$ *such that*

$$\left|f(t, u_1, v_1) - f(t, u_2, v_2)\right| \leq p(t)|u_1 - u_2| + q(t)|v_1 - v_2|$$

   for $t \in [0, 1]$ *and* $(u_1, v_1), (u_2, v_2) \in \mathbb{R}^2$.
   (2) *There exists* $n > l \geq 0$ *such that*

$$\left|f(t, u_1, v_1) - f(t, u_2, v_2)\right| \geq -l|u_1 - u_2| + n|v_1 - v_2|$$

   for $t \in [0, 1]$, $(u_1, v_1), (u_2, v_2) \in \mathbb{R}^2$.
Then the BVP (3.26), (3.30) *with* $\alpha\eta = 1$ *has exactly one solution in* $C^1[0, 1]$ *provided that*

$$2\left(\|p\|_1 + \|q\|_1\right) + \frac{l}{n} < 1.$$

3.7. *m-point boundary value problems reduced to four-point boundary value problems*

Let $f : [0, 1] \times \mathbb{R}^2 \to \mathbb{R}$ be either a continuous or a Carathéodory's function and $e : [0, 1] \to \mathbb{R}$ be a function in $L^1[0, 1]$, $c_i, a_j \in \mathbb{R}$, with all of the c_i's, and all of a_j's, having the same sign, $\xi_i, \tau_j \in (0, 1)$, $i = 1, 2, \ldots, m - 2$, $j = 1, 2, \ldots, n - 2$, $0 < \xi_1 < \xi_2 < \cdots < \xi_{m-2} < 1$, $0 < \tau_1 < \tau_2 < \cdots < \tau_{n-2} < 1$.

Consider the following second-order ordinary differential equation

$$x''(t) = f\left(t, x(t), x'(t)\right) + e(t), \quad t \in (0, 1) \tag{3.38}$$

subject to one of the following boundary conditions

$$x(0) = \sum_{i=1}^{m-2} c_i x'(\xi_i), \qquad x(1) = \sum_{j=1}^{n-2} a_j x(\tau_j), \tag{3.39}$$

$$x(0) = \sum_{i=1}^{m-2} c_i x'(\xi_i), \qquad x'(1) = \sum_{j=1}^{n-2} a_i x'(\tau_j), \tag{3.40}$$

$$x(0) = \sum_{i=1}^{m-2} c_i x'(\xi_i), \qquad x(1) = \sum_{j=1}^{n-2} a_i x(\tau_j), \tag{3.41}$$

$$x(0) = \sum_{i=1}^{m-2} c_i x(\xi_i), \qquad x'(1) = \sum_{j=1}^{n-2} a_i x'(\tau_j). \tag{3.42}$$

It is well known that if a function $x \in C^1$ satisfies one of the boundary conditions (3.39)–(3.42) and c_i, a_j, $i = 1, 2, \ldots, m-2$, $j = 1, 2, \ldots, n-2$ are as above, then there exist $\zeta \in [\xi_1, \xi_{m-2}]$, $\eta \in [\tau_1, \tau_{n-2}]$ such that

$$x(0) = \gamma x'(\zeta), \qquad x(1) = \alpha x(\eta), \tag{3.43}$$

$$x(0) = \gamma x'(\zeta), \qquad x'(1) = \alpha x'(\eta), \tag{3.44}$$

$$x(0) = \gamma x(\zeta), \qquad x(1) = \alpha x(\eta), \tag{3.45}$$

$$x(0) = \gamma x(\zeta), \qquad x'(1) = \alpha x'(\eta), \tag{3.46}$$

respectively with $\gamma = \sum_{i=1}^{m-2} c_i$, $\alpha = \sum_{j=1}^{n-2} a_j$.

As in the case of m-point BVPs which are reduced to three-point BVPs, we shall prove an existence result the BVP (3.38)–(3.39), proving first an existence result for the BVP (3.38), (3.43), and using the a priori bounds obtained for this problem for the BVP (3.38)–(3.39). In the next lemma we obtain the needed a priori bounds, which are independent of ζ and η, for the BVP (3.38), (3.43).

LEMMA 3.28. *Let $\zeta, \eta \in (0, 1)$ is given and $x(t) \in W^{2,1}(0, 1)$ be such that $x(0) = \gamma x'(\zeta)$, $x(1) = \alpha x(\eta)$. Then*

$$\|x\|_\infty \leqslant A \|x'\|_\infty, \qquad \|x'\|_\infty \leqslant B \|x''\|_1$$

where

$$A = \begin{cases} 1, & \text{if } \alpha \leqslant 0, \\ L, & \text{if } \alpha > 0,\, \alpha \neq 1, \\ 1 + |\gamma|, & \text{if } \alpha = 1 \end{cases}$$

and

$$B = \begin{cases} 1, & \text{if } \alpha \leqslant 0,\, \gamma = 0, \\ \dfrac{1}{1-Q}, & \text{if } \alpha \leqslant 0,\, \gamma \neq 0, \\ \dfrac{1}{1-S}, & \text{if } \alpha > 0,\, \alpha \neq 1, \\ 1, & \text{if } \alpha = 1, \end{cases}$$

where for $\alpha > 0$, $\alpha \neq 1$,

$$
\begin{cases}
M = \min\left\{\alpha, \dfrac{1}{\alpha}\right\} < 1, \\[3mm]
L = \min\left\{\dfrac{1}{1 - M}, 1 + \dfrac{1 - \eta}{|1 - \alpha|}, 1 + \dfrac{|\alpha|(1 - \eta)}{|1 - \alpha|}, 1 + |\gamma|\right\}, \\[3mm]
S = \min\left\{\dfrac{|1 - \alpha|}{1 - \eta}L, \dfrac{|1 - \alpha|}{\alpha(1 - \eta)}L, \dfrac{1}{|\gamma|}L\right\}
\end{cases}
$$

for $\alpha < 0$,

$$
Q = \min\left\{\frac{1 - \alpha}{1 - \eta}, \frac{1 - \alpha}{|\alpha|(1 - \eta)}, \frac{1}{|\gamma|}\right\}
$$

and for $\alpha = 0$, $Q = \frac{1}{|\gamma|}$ provided $Q < 1$ and $S < 1$.

PROOF. We consider the following cases:

Case 1: $\alpha \leqslant 0$. In this case $x(1)x(\eta) \leqslant 0$ and accordingly there exists a $\theta \in [\eta, 1]$ such that $x(\theta) = 0$. Hence it follows that $\|x\|_\infty \leqslant \|x'\|_\infty$. Also if $\gamma = 0$, we have from $x(0) = 0$ and $x(\theta) = 0$ that there exists a $z \in (0, \theta)$ such that $x'(z) = 0$. Accordingly, we get that $\|x'\|_\infty \leqslant \|x''\|_1$. Suppose, now, that $\alpha < 0$ and $\gamma \neq 0$. Next we see from Mean Value Theorem there exists $\omega \in (\eta, 1)$ such that

$$
(\alpha - 1)x(\eta) = x(1) - x(\eta) = (1 - \eta)x'(\omega)
$$

and hence

$$
x(\eta) = \frac{1 - \eta}{\alpha - 1}x'(\omega).
$$

Also, since $x(1) = \alpha x(\eta)$ we get

$$
x(1) = \frac{\alpha(1 - \eta)}{\alpha - 1}x'(\omega).
$$

From the relations

$$
x'(t) = x'(\omega) + \int_\omega^t x''(s)\,ds = \frac{\alpha - 1}{1 - \eta}x(\eta) + \int_\omega^t x''(s)\,ds,
$$

$$
x'(t) = x'(\omega) + \int_\omega^t x''(s)\,ds = \frac{\alpha - 1}{\alpha(1 - \eta)}x(1) + \int_\omega^t x''(s)\,ds
$$

and

$$
x'(t) = x'(\zeta) + \int_0^t x''(s)\,ds = \frac{1}{\gamma}x(0) + \int_0^t x''(s)\,ds
$$

then follows that

$$\|x'\|_\infty \leqslant \frac{1}{1-Q}\|x''\|_1,$$

where $Q = \min\{\frac{1-\alpha}{1-\eta}, \frac{1-\alpha}{|\alpha|(1-\eta)}, \frac{1}{|\gamma|}\}$ if $Q < 1$. Finally, for $\alpha = 0$, $\gamma \neq 0$ it is easy to see that $Q = \frac{1}{|\gamma|}$ since we require that $Q < 1$ and $\frac{1}{1-\eta} > 1$.

Case 2: $\alpha > 0$, $\alpha \neq 1$. We first consider the relations

$$x(t) = x(1) + \int_1^t x'(s)\,ds = \alpha x(\eta) + \int_1^t x'(s)\,ds$$

and

$$x(t) = x(\eta) + \int_\eta^t x'(s)\,ds = \frac{1}{\alpha}x(1) + \int_\eta^t x'(s)\,ds.$$

Since, now, $M = \min\{\alpha, \frac{1}{\alpha}\} < 1$, we get from the above relations that

$$\|x\|_\infty \leqslant \frac{1}{1-M}\|x'\|_\infty.$$

Next, we get the relations

$$x(t) = x(1) + \int_1^t x'(s)\,ds = \frac{\alpha(1-\eta)}{\alpha-1}x'(\omega) + \int_1^t x'(s)\,ds$$

and

$$x(t) = x(\eta) + \int_\eta^t x'(s)\,ds = \frac{1-\eta}{\alpha-1}x'(\omega) + \int_\eta^t x'(s)\,ds.$$

From these relations and

$$x(t) = x(0) + \int_0^t x'(s)\,ds = \gamma x'(\zeta) + \int_0^t x'(s)\,ds$$

it is immediate that

$$\|x\|_\infty \leqslant L\|x'\|_\infty,$$

where $L = \min\{\frac{1}{1-M}, 1 + \frac{1-\eta}{|\alpha-1|}, 1 + \frac{|\alpha|(1-\eta)}{|\alpha-1|}, 1 + |\gamma|\}$. Further, we have

$$\|x'\|_\infty \leqslant \frac{1}{1-S}\|x''\|_1,$$

where $S = \min\{\frac{|1-\alpha|}{1-\eta}L, \frac{|1-\alpha|}{\alpha(1-\eta)}L, \frac{1}{|\gamma|}L\}$ if $S < 1$.

Case 3: $\alpha = 1$. Since $x(1) = x(\eta)$ there exists an $\omega \in (\eta, 1)$ with $x'(\omega) = 0$. It is then immediate that $\|x'\|_\infty \leqslant \|x''\|_1$. Also since $x(t) = x(0) + \int_0^t x'(s)\,\mathrm{d}s = \gamma x'(\zeta) + \int_0^t x'(s)\,\mathrm{d}s$, it is immediate that $\|x\|_\infty \leqslant (1 + |\gamma|)\|x'\|_\infty$.

This completes the proof of the lemma. $\qquad\square$

THEOREM 3.29. *Let $f : [0, 1] \times \mathbb{R}^2 \to \mathbb{R}$ be a function satisfying Carathéodory's conditions. Assume that (3.12) holds. Also let $\eta \in (0, 1)$ be given and $\alpha, \gamma \in \mathbb{R}$ with $1 + \gamma \neq \alpha(\gamma + \eta)$. Moreover we assume that $Q < 1$ and $S < 1$.*

Then the BVP (3.38), (3.43) has at least one solution in $C^1[0, 1]$ provided

$$\begin{cases} \|p\|_1 + \|q\|_1 < 1, & \alpha \leqslant 0, \ \gamma = 0, \\ \|p\|_1 + \|q\|_1 < 1 - Q, & \alpha \leqslant 0, \ \gamma \neq 0, \\ L\|p\|_1 + \|q\|_1 < 1 - S, & \alpha > 0, \ \alpha \neq 1, \\ \left(1 + |\gamma|\right)\|p\|_1 + \|q\|_1 < 1, & \alpha = 1. \end{cases}$$

PROOF. Let X be the Banach space $C^1[0, 1]$ and Y denote the Banach space $L^1(0, 1)$ with their usual norms. We denote a linear mapping $L : D(L) \subset X \to Y$ by setting

$$D(L) = \left\{x \in W^{2,1}(0, 1): \ x(0) = \gamma x'(\zeta), \ x(1) = \alpha x(\eta)\right\},$$

and for $x \in D(L)$,

$$Lx = x''.$$

We also define a nonlinear mapping $N : X \to Y$ by setting

$$(Nx)(t) = f\left(t, x(t), x'(t)\right), \quad t \in [0, 1].$$

We note that N is a bounded mapping from X into Y. Next, it is easy to see that the linear mapping $L : D(L) \subset X \to Y$, is one-to-one mapping. Next, the linear mapping $K : Y \to X$, defined for $y \in Y$ by

$$(Ky)(t) = \int_0^t (t - s)y(s)\,\mathrm{d}s + \gamma \int_0^\zeta y(s)\,\mathrm{d}s$$

$$+ \frac{\gamma + t}{1 + \gamma - \alpha(\gamma + \eta)}\left[\alpha \int_0^\eta (\eta - s)y(s)\,\mathrm{d}s\right.$$

$$\left. - \int_0^1 (1 - s)y(s)\,\mathrm{d}s + \gamma(\alpha - 1)\int_0^\zeta y(s)\,\mathrm{d}s\right], \quad t \in [0, 1]$$

is such that for $y \in Y$, $Ky \in D(L)$ and $LKy = y$; and for $u \in D(L)$, $KLu = u$. Furthermore, it follows easily using the Arzelá–Ascoli theorem that KN maps bounded subsets of X into a relatively compact subsets of X. Hence $KN : X \to X$ is a compact mapping.

We, next, note that $x \in C^1[0, 1]$ is a solution of the BVP (3.38), (3.43) if and only if x is a solution to the operator equation

$$Lx = Nx + e.$$

Now, the operator equation $Lx = Nx + e$ is equivalent to the equation

$$x = KNx + Ke.$$

We apply the Leray–Schauder continuation theorem to obtain the existence of a solution for $x = KNx + Ke$ or equivalently to the BVP (3.38), (3.43).

To do this, it suffices to verify that the set of all possible solutions of the family of equations

$$x''(t) = \lambda f\big(t, x(t), x'(t)\big) + \lambda e(t), \quad t \in (0, 1), \tag{3.47}$$

$$x(0) = \gamma x'(0), \qquad x(1) = \alpha x(\eta) \tag{3.48}$$

is, a priori, bounded in $C^1[0, 1]$ by a constant independent of $\lambda \in [0, 1]$.

(I) Assume that $\alpha \leqslant 0$, $\gamma = 0$. From Lemma 3.28 we have

$$\|x\|_\infty \leqslant \|x'\|_\infty \leqslant \|x''\|_1.$$

Let, now, $x(t)$ be a solution of (3.47)–(3.48) for some $\lambda \in [0, 1]$, so that $x \in W^{2,1}(0, 1)$ with $x(0) = \gamma x'(\zeta)$, $x(1) = \alpha x(\eta)$. We then get

$$\begin{aligned}
\|x''\|_1 &= \lambda \big\| f\big(t, x(t), x'(t)\big) + e(t) \big\|_1 \\
&\leqslant \|p\|_1 \|x\|_\infty + \|q\|_1 \|x'\|_\infty + \|r\|_1 + \|e\|_1 \\
&\leqslant \big(\|p\|_1 + \|q\|_1\big) \|x''\|_1 + \|r\|_1 + \|e\|_1.
\end{aligned}$$

It follows from our assumption that there is a constant c, independent of $\lambda \in [0, 1]$, such that

$$\|x''\|_1 \leqslant c.$$

It is now immediate that the set of solutions of the family of equations (3.47)–(3.48) is, a priori, bounded in $C^1[0, 1]$ by a constant independent of $\lambda \in [0, 1]$.

(II) Assume that $\alpha \leqslant 0$, $\gamma \neq 0$. Then we have, by Lemma 3.28 that

$$\|x\|_\infty \leqslant \|x'\|_\infty, \qquad \|x'\|_\infty \leqslant \frac{1}{1 - Q} \|x''\|_1.$$

We then get

$$\|x''\|_1 = \lambda \big\| f\big(t, x(t), x'(t)\big) + e(t) \big\|_1$$

$$\leqslant \|p\|_1 \|x\|_\infty + \|q\|_1 \|x'\|_\infty + \|r\|_1 + \|e\|_1$$

$$\leqslant \left[\|p\|_1 + \|q\|_1\right] \frac{1}{1-Q} \|x''\|_1 + \|r\|_1 + \|e\|_1.$$

We proceed as in case (I).

The process for the other cases is similar to the previous cases and we omit the details. This completes the proof of the theorem. $\qquad\square$

We study now the multi-point BVP (3.38)–(3.39) using the a priori estimates that can be obtained for a four-point BVP (3.38), (3.43). This is because for every solution $x(t)$ of the BVP (3.38)–(3.39), there exist $\eta \in [\xi_1, \xi_{m-2}]$, $\zeta \in [\tau_1, \tau_{n-2}]$, depending on, $x(t)$, such that $x(t)$ is also a solution of the BVP (3.38), (3.43) with $\gamma = \sum_{i=1}^{m-2} c_i$ and $\alpha = \sum_{j=1}^{n-2} a_j$. The proof is quite similar to the proof of Theorem 3.29 and uses the a priori estimates obtained in the proof of Lemma 3.28 for the set of solutions of the family of equations (3.38), (3.43).

THEOREM 3.30. *Let $f : [0, 1] \times \mathbb{R}^2 \to \mathbb{R}$ be a function satisfying Carathéodory's conditions. Assume that (3.12) holds. Let $c_i, a_j \in \mathbb{R}$, with all of the c_i's (respectively, a_j's), having the same sign, $\xi_i, \tau_j \in (0, 1)$, $i = 1, 2, \ldots, m-2$, $j = 1, 2, \ldots, n-2$, $0 < \xi_1 < \xi_2 < \cdots < \xi_{m-2} < 1$, $0 < \tau_1 < \tau_2 < \cdots < \tau_{n-2} < 1$ be given. Suppose that*

$$1 + \left(\sum_{i=1}^{m-2} c_i\right)\left(1 - \sum_{j=1}^{n-2} a_j\right) - \sum_{j=1}^{n-2} a_j \tau_j \neq 0.$$

Let $\gamma = \sum_{i=1}^{m-2} c_i$ and $\alpha = \sum_{j=1}^{n-2} a_j$. Moreover we assume that $Q' < 1$, and $S' < 1$, where $M = \min\{\alpha, \frac{1}{\alpha}\} < 1$,

$$L' = \min\left\{\frac{1}{1-M}, 1 + \frac{1-\tau_1}{|1-\alpha|}, 1 + \frac{|\alpha|(1-\tau_1)}{|1-\alpha|}, 1 + |\gamma|\right\},$$

$$S' = \min\left\{\frac{|1-\alpha|}{1-\tau_{n-2}}L, \frac{|1-\alpha|}{\alpha(1-\tau_{n-2})}L, \frac{1}{|\gamma|}L\right\},$$

$$Q' = \min\left\{\frac{1-\alpha}{1-\tau_{n-2}}, \frac{1-\alpha}{|\alpha|(1-\tau_{n-2})}, \frac{1}{|\gamma|}\right\}.$$

Then the BVP (3.38)–(3.39) has at least one solution in $C^1[0, 1]$ provided

$$\begin{cases} \|p\|_1 + \|q\|_1 < 1, & \alpha \leqslant 0, \gamma = 0, \\ \|p\|_1 + \|q\|_1 < 1 - Q', & \alpha \leqslant 0, \gamma \neq 0, \\ L'\|p\|_1 + \|q\|_1 < 1 - S', & \alpha > 0, \alpha \neq 1, \\ (1 + |\gamma|)\|p\|_1 + \|q\|_1 < 1, & \alpha = 1. \end{cases}$$

PROOF. Let X be the Banach space $C^1[0, 1]$ and Y denote the Banach space $L^1(0, 1)$ with their usual norms. We denote a linear mapping $L : D(L) \subset X \to Y$ by setting

$$D(L) = \left\{ x \in W^{2,1}(0, 1) \colon x(0) = \sum_{i=1}^{m-2} c_i x'(\xi_i),\ x(1) = \sum_{j=1}^{n-2} a_j x(\tau_j) \right\},$$

and for $x \in D(L)$,

$$Lx = x''.$$

We also define a nonlinear mapping $N : X \to Y$ by setting

$$(Nx)(t) = f\big(t, x(t), x'(t)\big), \quad t \in [0, 1].$$

We note that N, is a bounded mapping from X into Y. Next, it is easy to see that the linear mapping $L : D(L) \subset X \to Y$, is one-to-one mapping. Next, the linear mapping $K : Y \to X$, defined for $y \in Y$ by

$$(Ky)(t) = \int_0^t (t - s)y(s)\,ds + ct + k, \quad t \in [0, 1],$$

where c and k are given by

$$\left[1 + \left(\sum_{i=1}^{m-2} c_i \right) \left(1 - \sum_{j=1}^{n-2} a_j \right) - \sum_{j=1}^{n-2} a_j \tau_j \right] c$$

$$= \left(\sum_{j=1}^{n-2} a_j - 1 \right) \left(\sum_{i=1}^{m-2} c_i \int_0^{\xi_i} y(s)\,ds \right)$$

$$+ \sum_{j=1}^{n-2} a_j \int_0^{\tau_j} (\tau_j - s)y(s)\,ds - \int_0^1 (1 - s)y(s)\,ds$$

and

$$\left[1 + \left(\sum_{i=1}^{m-2} c_i \right) \left(1 - \sum_{j=1}^{n-2} a_j \right) - \sum_{j=1}^{n-2} a_j \tau_j \right] k = \sum_{i=1}^{m-2} c_i \sum_{j=1}^{n-2} a_j \int_0^{\tau_j} (\tau_j - s)y(s)\,ds$$

$$- \sum_{i=1}^{m-2} c_i \int_0^1 (1 - s)y(s)\,ds + \left(1 - \sum_{j=1}^{n-2} a_j \tau_j \right) \sum_{i=1}^{m-2} c_i \int_0^{\xi_i} y(s)\,ds,$$

is such that for $y \in Y$, $Ky \in D(L)$ and $LKy = y$; and for $u \in D(L)$, $KLu = u$. Furthermore, it follows easily using the Arzelá–Ascoli theorem that KN maps bounded subsets of X into a relatively compact subset of X. Hence $KN : X \to X$ is a compact mapping.

We, next, note that $x \in C^1[0, 1]$ is a solution of the BVP (3.38)–(3.39) if and only if x is a solution to the operator equation

$$Lx = Nx + e.$$

Now, the operator equation $Lx = Nx + e$ is equivalent to the equation

$$x = KNx + Ke.$$

We apply the Leray–Schauder continuation theorem to obtain the existence of a solution for $x = KNx + Ke$ or equivalently to the BVP (3.38)–(3.39).

To do this, it suffices to verify that the set of all possible solutions of the family of equations

$$x''(t) = \lambda f\big(t, x(t), x'(t)\big) + \lambda e(t), \quad t \in (0, 1), \tag{3.49}$$

$$x(0) = \sum_{i=1}^{m-2} c_i x'(\xi_i), \qquad x(1) = \sum_{j=1}^{n-2} a_j x(\tau_j) \tag{3.50}$$

is, a priori, bounded in $C^1[0, 1]$ by a constant independent of $\lambda \in [0, 1]$.

Let, now, $x(t)$ be a solution of (3.49)–(3.50) for some $\lambda \in [0, 1]$, so that $x \in W^{2,1}(0, 1)$ with $x(0) = \sum_{i=1}^{m-2} c_i x'(\xi_i)$, $x(1) = \sum_{j=1}^{n-2} a_j x(\tau_j)$. Accordingly, there exist $\zeta \in [\xi_1, \xi_{m-2}]$ and $\eta \in [\tau_1, \tau_{n-2}]$ depending on $x(t)$, such that $x(t)$ is a solution of the four point BVP

$$x''(t) = \lambda f\big(t, x(t), x'(t)\big) + \lambda e(t), \quad t \in (0, 1),$$

$$x(0) = \gamma x'(\zeta), \qquad x(1) = \alpha x(\eta).$$

It then follows, as in the proof of Theorem 3.30 that there is a constant c, independent of $\lambda \in [0, 1]$, and $\eta \in [\xi_1, \xi_{m-2}]$, $\zeta \in [\tau_1, \tau_{n-2}]$ such that

$$\|x\|_\infty \leqslant c_1 \|x'\|_\infty \leqslant c_2 \|x''\|_1 \leqslant c,$$

where c_1, c_2 are constants independent of λ, η, ζ. Thus the set of solutions of the family of equations is, a priori, bounded in $C^1[0, 1]$ by a constant, independent of $\lambda \in [0, 1]$. This completes the proof of the theorem. $\qquad\square$

Theorems 3.29 and 3.30 are taken from [35]. Similar results for the BVP (3.38), (3.40) are proved in [35] and for the BVPs (3.38), (3.41) and (3.38), (3.42) in [36].

3.8. *The general case*

Let $f : [0, 1] \times \mathbb{R}^2 \to \mathbb{R}$ be either a continuous or a Carathéodory's function and $e(t) \in L^1[0, 1]$. Let $\xi_i, \eta_j, c_i, a_j, i = 1, 2, \ldots, m - 2, j = 1, 2, \ldots, n - 2, 0 < \xi_1 < \xi_2 < \cdots <$

$\xi_{m-2} < 1, 0 < \eta_1 < \eta_2 < \cdots < \eta_{n-2} < 1$ be given. Consider the following second-order ordinary differential equation:

$$x''(t) = f\big(t, x(t), x'(t)\big) + e(t), \quad t \in (0, 1), \tag{3.51}$$

subject to one of the following boundary value conditions:

$$x(0) = \sum_{i=1}^{m-2} c_i x(\xi_i), \qquad x(1) = \sum_{j=1}^{n-2} a_j x(\eta_j), \tag{3.52}$$

$$x(0) = \sum_{i=1}^{m-2} c_i x(\xi_i), \qquad x'(1) = \sum_{j=1}^{n-2} a_j x(\eta_j), \tag{3.53}$$

$$x'(0) = \sum_{i=1}^{m-2} c_i x(\xi_i), \qquad x(1) = \sum_{j=1}^{n-2} a_j x(\eta_j), \tag{3.54}$$

$$x(0) = \sum_{i=1}^{m-2} c_i x'(\xi_i), \qquad x(1) = \sum_{j=1}^{n-2} a_j x'(\eta_j), \tag{3.55}$$

$$x(0) = \sum_{i=1}^{m-2} c_i x'(\xi_i), \qquad x'(1) = \sum_{j=1}^{n-2} a_j x(\eta_j), \tag{3.56}$$

$$x(0) = \sum_{i=1}^{m-2} c_i x(\xi_i), \qquad x'(1) = \sum_{j=1}^{n-2} a_j x'(\eta_j), \tag{3.57}$$

$$x(0) = \sum_{i=1}^{m-2} c_i x'(\xi_i), \qquad x'(1) = \sum_{j=1}^{n-2} a_j x'(\eta_j). \tag{3.58}$$

When all the c_i's have the same sign and also all the a_j's have the same sign, it is known that existence of a solution for (3.51) with boundary conditions (3.52)–(3.58) can be obtained via existence subject to the respective four-point boundary conditions

$$x(0) = \gamma x(\xi), \qquad x(1) = \alpha x(\eta), \tag{3.59}$$

$$x(0) = \gamma x(\xi), \qquad x'(1) = \alpha x(\eta), \tag{3.60}$$

$$x'(0) = \gamma x'(\xi), \qquad x(1) = \alpha x(\eta), \tag{3.61}$$

$$x'(0) = \gamma x'(\xi), \qquad x'(1) = \alpha x'(\eta), \tag{3.62}$$

$$x(0) = \gamma x'(\xi), \qquad x(1) = \alpha x(\eta), \tag{3.63}$$

$$x(0) = \gamma x'(\xi), \qquad x'(1) = \alpha x'(\eta), \tag{3.64}$$

$$x(0) = \gamma x(\xi), \qquad x'(1) = \alpha x'(\eta), \tag{3.65}$$

where $\xi \in [\xi_1, \xi_{m-2}]$, $\eta \in [\eta_1, \eta_{n-2}]$, $\gamma = \sum_{i=1}^{m-2} c_i$, $\alpha = \sum_{j=1}^{n-2} a_j$.

When all the c_i's have the same sign and also all the a_j's have no same sign, the existence of a solution for (3.51) with boundary value conditions (3.52)–(3.58) can be obtained via existence subject to the following boundary conditions

$$x(0) = \gamma x(\xi), \qquad x(1) = \sum_{j=1}^{n-2} a_j x(\eta_j), \tag{3.66}$$

$$x(0) = \gamma x(\xi), \qquad x'(1) = \sum_{j=1}^{n-2} a_j x'(\eta_j), \tag{3.67}$$

$$x'(0) = \gamma x(\xi), \qquad x(1) = \sum_{j=1}^{n-2} a_j x(\eta_j), \tag{3.68}$$

$$x'(0) = \gamma x'(\xi), \qquad x'(1) = \sum_{j=1}^{n-2} a_j x'(\eta_j), \tag{3.69}$$

$$x(0) = \gamma x'(\xi), \qquad x(1) = \sum_{j=1}^{n-2} a_j x(\eta_j), \tag{3.70}$$

$$x(0) = \gamma x'(\xi), \qquad x'(1) = \sum_{j=1}^{n-2} a_j x'(\eta_j), \tag{3.71}$$

$$x(0) = \gamma x(\xi), \qquad x'(1) = \sum_{j=1}^{n-2} a_j x'(\eta_j), \tag{3.72}$$

where $\xi \in [\xi_1, \xi_{m-2}]$, $\gamma = \sum_{i=1}^{m-2} c_i$.

When all the c_i's do not have the same sign and also all the a_j's have the same sign, the existence of a solution for (3.51) with boundary value conditions (3.52)–(3.58) can be obtained via existence subject to the following boundary conditions

$$x(0) = \sum_{i=1}^{m-2} c_i x(\xi_i), \qquad x(1) = \alpha x(\eta), \tag{3.73}$$

$$x(0) = \sum_{i=1}^{m-2} c_i x(\xi_i), \qquad x'(1) = \alpha x'(\eta), \tag{3.74}$$

$$x'(0) = \sum_{i=1}^{m-2} c_i x'(\xi_i), \qquad x(1) = \alpha x(\eta), \tag{3.75}$$

$$x'(0) = \sum_{i=1}^{m-2} c_i x'(\xi_i), \qquad x'(1) = \alpha x'(\eta), \tag{3.76}$$

$$x(0) = \sum_{i=1}^{m-2} c_i x'(\xi_i), \qquad x(1) = \alpha x(\eta), \tag{3.77}$$

$$x(0) = \sum_{i=1}^{m-2} c_i x'(\xi_i), \qquad x'(1) = \alpha x'(\eta), \tag{3.78}$$

$$x(0) = \sum_{i=1}^{m-2} c_i x(\xi_i), \qquad x'(1) = \alpha x'(\eta), \tag{3.79}$$

where $\eta \in [\eta_1, \eta_{n-2}], \alpha = \sum_{j=1}^{n-2} a_j$.

For certain boundary condition case such that the linear operator $Lx = x''$, defined in a suitable Banach space, is invertible, is the so-called nonresonance case. Otherwise, the so-called resonance case.

Existence results were given for the BVPs:

- (3.51)–(3.52) in [29] for the nonresonance case and in [59] for resonance case,
- (3.51), (3.54) and (3.51), (3.55) in [59] for resonance case,
- (3.51), (3.57) in [29] for nonresonance case,
- (3.51), (3.58) in [28] for nonresonance case,
- (3.51), (3.59), (3.51), (3.61), (3.51), (3.62) and (3.51), (3.65) in [61] for resonance case,
- (3.51), (3.66), (3.51), (3.67), (3.51), (3.68) and (3.51), (3.69) in [60] for resonance case,
- (3.51), (3.73), (3.51), (3.74), (3.51), (3.75) and (3.51), (3.76) in [62] for resonance case.

For the BVP (3.38), (3.52) we report an existence result for the nonresonance case from [29].

THEOREM 3.31. *Let $f : [0, 1] \times \mathbb{R}^2 \to \mathbb{R}$ be a function satisfying Carathéodory's conditions. Assume that (3.12) holds. Let $c_i, a_j \in \mathbb{R}$, with all of the c_i's (respectively, a_j's), not having the same sign, $\xi_i, \eta_j \in (0, 1)$, $i = 1, 2, \ldots, m - 2$, $j = 1, 2, \ldots, n - 2$, $0 < \xi_1 < \xi_2 < \cdots < \xi_{m-2} < 1$, $0 < \eta_1 < \eta_2 < \cdots < \eta_{n-2} < 1$ be given. Suppose that*

$$\left(\sum_{i=1}^{m-2} c_i \xi_i\right)\left(1 - \sum_{j=1}^{n-2} a_j\right) \neq \left(1 - \sum_{j=1}^{n-2} c_i\right)\left(\sum_{j=1}^{n-2} a_j \eta_j - 1\right).$$

Then the BVP (3.51)–(3.52) has at least one solution in $C^1[0, 1]$ provided

$$Q\big(M\|p\|_1 + \|q\|_1\big) < 1,$$

where

$$M = \min\left\{ \frac{1}{\left|\sum_{i=1}^{m-2} c_i\right|} \left(\sum_{i=1}^{m-2} |c_i|\lambda_i + \frac{\sum_{i=1}^{m-2} |c_i\xi_i|}{\left|1 - \sum_{i=1}^{m-2} c_i\right|} \right), \right.$$

$$\frac{1}{\left|\sum_{j=1}^{n-2} a_j\right|} \left(\sum_{j=1}^{n-2} |a_i|\mu_j + \frac{\sum_{j=1}^{n-2} |c_j(1 - \eta_j)|}{\left|1 - \sum_{j=1}^{n-2} a_j\right|} \right),$$

$$\left. 1 + \frac{\sum_{i=1}^{m-2} |c_i\xi_i|}{\left|1 - \sum_{i=1}^{m-2} c_i\right|}, 1 + \frac{\sum_{j=1}^{n-2} |c_j(1 - \eta_j)|}{\left|1 - \sum_{j=1}^{n-2} a_j\right|} \right\},$$

$\lambda_i = \max\{\xi_i, 1 - \xi_i\}, i = 1, \ldots, m - 2, \mu_j = \max\{\eta_j, 1 - \eta_j\}, j = 1, \ldots, n - 2$ *and*

$$Q = \min\left\{ \frac{\sum_{i=1}^{m-2} |c_i\xi_i|}{\left|\sum_{i=1}^{m-2} c_i\xi_i\right| - M\left|1 - \sum_{i=1}^{m-2} c_i\right|}, \right.$$

$$\left. \frac{\sum_{j=1}^{n-2} |a_j(1 - \eta_j)|}{\left|\sum_{i=j}^{n-2} a_j(1 - \eta_j)\right| - M\left|\sum_{j=1}^{n-2} a_j - 1\right|} \right\}.$$

For resonance case Liu in [59] gave existence results for the BVP (3.51)–(3.52) in the following cases:

(1) $\sum_{i=1}^{m-2} c_i\xi_i = 0, \gamma = 1, \alpha = 1, \sum_{i=1}^{m-2} c_i\xi_i^2 \neq 0, \sum_{j=1}^{n-2} a_j\eta_j^2 \neq 1.$

(2) $\sum_{i=1}^{m-2} c_i\xi_i \neq 0, \gamma = 1, \alpha = 1, \sum_{i=1}^{n-2} a_j\eta_j = 1, \sum_{j=1}^{n-2} a_j\eta_j^2 \neq 1.$

(3) $\sum_{i=1}^{m-2} c_i\xi_i = 0, \gamma \neq 1, \alpha = 1, \sum_{i=1}^{n-2} a_j\eta_j = 1, \sum_{j=1}^{n-2} a_j\eta_j^2 \neq 1.$

(4) $\sum_{i=1}^{m-2} c_i\xi_i = 0, \gamma = 1, \alpha \neq 1, \sum_{i=1}^{n-2} a_j\eta_j = 1, \sum_{i=1}^{m-2} c_i\xi_i^2 \neq 0.$

We give an existence result at resonance, for the BVP (3.51)–(3.52) in the case (1), i.e. when

$$\sum_{i=1}^{m-2} c_i\xi_i = 0, \qquad \gamma = 1, \qquad \alpha = 1, \qquad \sum_{i=1}^{m-2} c_i\xi_i^2 \neq 0, \qquad \sum_{j=1}^{n-2} a_j\eta_j^2 \neq 1.$$

THEOREM 3.32. *Let $f : [0, 1] \times \mathbb{R}^2 \to \mathbb{R}$ be a continuous function. Assume that either there exists $m_1 \in \{1, 2, \ldots, m - 3\}$ such that $c_i < 0$ $(1 \leqslant i \leqslant m_1)$, $c_i > 0$ $(m_1 + 1 \leqslant i \leqslant m - 2)$, or there exists $n_1 \in \{1, 2, \ldots, n - 3\}$ such that $a_j > 0$ $(1 \leqslant j \leqslant n_1)$, $a_j < 0$ $(n_1 + 1 \leqslant j \leqslant n - 2)$, furthermore*

(H1) *There exist functions a, b, c, r in $L^1[0, 1]$, and constant $\theta \in [0, 1)$ such that for all $(x, y) \in \mathbb{R}^2, t \in [0, 1]$ either*

$$\left|f(t, x, y)\right| \leqslant a(t)|x| + b(t)|y| + c(t)|y|^\theta + r(t)$$

or else

$$\left|f(t, x, y)\right| \leqslant a(t)|x| + b(t)|y| + c(t)|x|^\theta + r(t).$$

(H2) *There exists constant $M > 0$ such that, for $x \in \text{dom}(L)$, if $|x(t)| > M$ for all $t \in [0, 1]$, then*

$$\sum_{i=1}^{m-2} c_i \int_0^{\xi_i} \int_0^s \left[f\left(\tau, x(\tau), x'(\tau)\right) + e(\tau) \right] d\tau \, ds \neq 0.$$

(H3) *There exists constant $M^* > 0$ such that for any $d \in \mathbb{R}$, if $|d| > M^*$, then either*

$$d \sum_{i=1}^{m-2} c_i \int_0^{\xi_i} \int_0^s \left[f(\tau, d, 0) + e(\tau) \right] d\tau \, ds < 0$$

or else

$$d \sum_{i=1}^{m-2} c_i \int_0^{\xi_i} \int_0^s \left[f(\tau, d, 0) + e(\tau) \right] d\tau \, ds > 0.$$

Then, for every $e \in L^1[0, 1]$, the BVP (3.51)–(3.52) in the case (1), i.e. when $\sum_{i=1}^{m-2} c_i \xi_i = 0$, $\gamma = 1$, $\alpha = 1$, $\sum_{i=1}^{m-2} c_i \xi_i^2 \neq 0$, $\sum_{j=1}^{n-2} a_j \eta_j^2 \neq 1$, has at least one solution in $C^1[0, 1]$ provided

$$\|a\|_1 + \|b\|_1 < \frac{1}{\Delta_1 + 1}, \quad \text{where } \Delta_1 = a + \frac{\sum_{i=j}^{n-2} |a_j|(1 - \eta_j)}{\left| 1 - \sum_{j=1}^{n-2} a_j \eta_j \right|}.$$

3.9. *Positive solutions of some nonlocal boundary value problems*

There is much attention focused on question of positive solutions of BVPs for ordinary differential equations. Much of interest is due to the applicability of certain Krasnosel'skii fixed point theorems or the Leggett–Williams multiple fixed point theorem, or a synthesis of both to obtain positive solutions or multiple positive solutions which lie in a cone. Here we present some of the results on positive solutions of some nonlocal BVPs.

Consider the differential equation

$$x'' + a(t)f(x) = 0, \quad t \in (0, 1) \tag{3.80}$$

with the boundary conditions

$$x(0) = 0, \qquad \alpha x(\eta) = x(1), \tag{3.81}$$

where $0 < \eta < 1$. Our purpose here is to give some existence results for positive solutions to (3.80)–(3.81), assuming that $\alpha\eta < 1$ and f is either superlinear or sublinear.

Set

$$f_0 = \lim_{x \to 0+} \frac{f(x)}{x}, \qquad f_\infty = \lim_{x \to \infty} \frac{f(x)}{x}.$$

Then $f_0 = 0$ and $f_\infty = \infty$ correspond to the superlinear case, and $f_0 = \infty$ and $f_\infty = 0$ correspond to the sublinear case. By the *positive solution* of the BVP (3.80)–(3.81) we understand a function $x(t)$ which is positive on $0 < t < 1$ and satisfies the differential equation (3.80) and the boundary conditions (3.81).

The key tool in our approach is the following Krasnosel'skii's fixed point theorem in a cone [54] (see also [21]).

THEOREM 3.33. *Let E be a Banach space, and let $K \subset E$ be a cone. Assume Ω_1, Ω_2 are open bounded subsets of E with $0 \in \Omega_1, \overline{\Omega}_1 \subset \Omega_2$, and let*

$$A : K \cap \left(\overline{\Omega}_2 \setminus \Omega_1 \right) \longrightarrow K$$

be a completely continuous operator such that
   (i) $\|Au\| \leqslant \|u\|, u \in K \cap \partial\Omega_1$, *and* $\|Au\| \geqslant \|u\|, u \in K \cap \partial\Omega_2$; *or*
   (ii) $\|Au\| \geqslant \|u\|, u \in K \cap \partial\Omega_1$, *and* $\|Au\| \leqslant \|u\|, u \in K \cap \partial\Omega_2$.
Then A has a fixed point in $K \cap (\overline{\Omega}_2 \setminus \Omega_1)$.

We give first some preliminary lemmas.

LEMMA 3.34. *Let $\alpha\eta \neq 1$ then for $y \in C[0, 1]$, the problem*

$$x''(t) + y(t) = 0, \quad t \in (0, 1), \tag{3.82}$$

$$x(0) = 0, \qquad \alpha x(\eta) = x(1) \tag{3.83}$$

has a unique solution

$$x(t) = -\int_0^t (t - s)y(s)\,ds - \frac{\alpha t}{1 - \alpha\eta} \int_0^\eta (\eta - s)y(s)\,ds$$

$$+ \frac{t}{1 - \alpha\eta} \int_0^1 (1 - s)y(s)\,ds.$$

LEMMA 3.35. *Let $0 < \alpha < \frac{1}{\eta}$. If $y \in C[0, 1]$ and $y \geqslant 0$, for $t \in (0, 1)$ then the unique solution x of the problem (3.82)–(3.83) satisfies $x(t) \geqslant 0, t \in [0, 1]$.*

PROOF. From the fact that $x''(t) = -y(t) \leqslant 0$, we know that the graph of $x(t)$ is concave down on $(0, 1)$. So, if $x(1) \geqslant 0$, then the concavity of x and the boundary condition $x(0) = 0$ imply that $x \geqslant 0$ for $t \in [0, 1]$. If $x(1) < 0$, then we have that $x(\eta) < 0$ and $x(1) = \alpha x(\eta) > \frac{1}{\eta}x(\eta)$ which contradicts to the concavity of x. $\qquad\square$

LEMMA 3.36. *Let $\alpha\eta > 1$. If $y \in C[0, 1]$ and $y \geqslant 0$, for $t \in (0, 1)$ then (3.82)–(3.83) has no positive solution.*

PROOF. Assume that (3.82)–(3.83) has a positive solution x. If $x(1) > 0$, then $x(\eta) > 0$ and $\frac{x(1)}{1} = \frac{\alpha x(\eta)}{1} > \frac{x(\eta)}{\eta}$ which contradicts to the concavity of x. If $x(1) = 0$ and $x(\tau) > 0$ for some $\tau \in (0, 1)$, then $x(\eta) - x(1) = 0$, $\tau = \eta$. If $\tau \in (0, \eta)$, then $x(\tau) > x(\eta) = x(1)$, which contradicts to the concavity of x again. $\qquad\square$

LEMMA 3.37. *Let $0 < \alpha < \frac{1}{\eta}$. If $y \in C[0, 1]$ and $y \geqslant 0$, for $t \in (0, 1)$ then the unique solution x of the problem (3.82)–(3.83) satisfies*

$$\inf_{t \in [\eta, 1]} x(t) \geqslant \gamma \|x\|,$$

where $\gamma = \min\{\alpha\eta, \frac{\alpha(1-\eta)}{1-\alpha\eta}, \eta\}$.

PROOF. We divide the proof into two steps.

Step 1. We deal with the case $0 < \alpha < 1$. In this case, by Lemma 3.35, we know that $x(\eta) \geqslant x(1)$. Set

$$x(\bar{t}) = \|x\|.$$

If $\bar{t} \leqslant \eta < 1$, then $\min_{t \in [\eta, 1]} x(t) = x(1)$ and

$$x(\bar{t}) \leqslant x(1) + \frac{x(1) - x(\eta)}{1 - \eta}(0 - 1) = x(1)\frac{1 - \alpha\eta}{\alpha(1 - \eta)}.$$

Then

$$\min_{t \in [\eta, 1]} x(t) \geqslant \frac{\alpha(1 - \eta)}{1 - \alpha\eta}.$$

If $\eta < \bar{t} < 1$, then $\min_{t \in [\eta, 1]} x(t) = x(1)$. From the concavity of x, we know that $\frac{x(\eta)}{\eta} \geqslant \frac{x(\bar{t})}{\bar{t}}$ which combined with boundary condition $\alpha x(\eta) = x(1)$, gives $\frac{x(1)}{\alpha\eta} \geqslant \frac{x(\bar{t})}{\bar{t}} \geqslant x(\bar{t}) = \|x\|$. Therefore

$$\min_{t \in [\eta, 1]} x(t) \geqslant \alpha\eta \|x\|.$$

Step 2. We deal with the case $1 < \alpha < \frac{1}{\eta}$. In this case, we have $x(\eta) \leqslant x(1)$. Set

$$x(\bar{t}) = \|x\|.$$

Then we can choose $\bar{t}$ such that $\eta \leqslant \bar{t} \leqslant 1$. This contradicts to the concavity of x. From $x(\eta) \leqslant x(1)$ and the concavity of x, we know that $\min_{t \in [\eta, 1]} x(t) = x(\eta)$. Using the concavity of x and Lemma 3.35, we have $\frac{x(\eta)}{\eta} \geqslant \frac{x(\bar{t})}{\bar{t}}$ which implies that

$$\min_{t \in [\eta, 1]} x(t) \geqslant \eta \|x\|.$$

This completes the proof. $\qquad\square$

The main result here is the following:

THEOREM 3.38. *Assume that*
   (A1) $f \in C([0, \infty), [0, \infty))$;
   (A2) $a \in C([0, 1], [0, \infty))$ *and there exists* $x_0 \in [\eta, 1]$ *such that* $a(x_0) > 0$.
Then the BVP (3.80)–(3.81) *has at least one positive solution in the case*
   (i) $f_0 = 0$ *and* $f_\infty = \infty$ *(superlinear) or*
   (ii) $f_0 = \infty$ *and* $f_\infty = 0$ *(sublinear).*

PROOF. (i) Suppose that $f_0 = 0$ and $f_\infty = \infty$. We wish to show the existence of a positive solution of (3.80)–(3.81). Now (3.80)–(3.81) has a solution $x = x(t)$ if and only if x solves the operator equation

$$x(t) = -\int_0^t (t - s)a(s)f\big(x(s)\big)\, ds - \frac{\alpha t}{1 - \alpha\eta} \int_0^\eta (\eta - s)a(s)f\big(x(s)\big)\, ds$$

$$+ \frac{t}{1 - \alpha\eta} \int_0^1 (1 - s)a(s)f\big(x(s)\big)\, ds$$

$$:= Ax(t).$$

Denote

$$K = \left\{ x\colon x \in C[0, 1], x \geqslant 0, \min_{\eta \leqslant t \leqslant 1} x(t) \geqslant \gamma \|x\| \right\}.$$

It is obvious that K is a cone in $C[0, 1]$. Moreover, by Lemma 3.37, $AK \subset K$. It is also easy to check that $A\colon K \to K$ is completely continuous.

Now $f_0 = 0$, we may choose $H_1 > 0$ so that $f(x) \leqslant \epsilon x$, for $0 < x < H_1$, where $\epsilon > 0$ satisfies

$$\frac{\epsilon}{1 - \alpha\eta} \int_0^1 (1 - s)a(s)\, ds \leqslant 1.$$

Thus, if $x \in K$ and $\|x\| = H_1$ we get

$$Ax(t) \leqslant \frac{t}{1 - \alpha\eta} \int_0^1 (1 - s)a(s)f\big(x(s)\big)\, ds$$

$$\leqslant \frac{t}{1-\alpha\eta} \int_0^1 (1-s)a(s)\epsilon x(s)\,\mathrm{d}s$$

$$\leqslant \frac{\epsilon}{1-\alpha\eta} \int_0^1 (1-s)a(s)\,\mathrm{d}s\,\|x\|$$

$$\leqslant \frac{\epsilon}{1-\alpha\eta} \int_0^1 (1-s)a(s)\,\mathrm{d}s\,H_1.$$

Now if we let

$$\Omega_1 = \big\{x \in C[0,1]\colon \|x\| < H_1\big\},$$

then $\|Ax\| \leqslant \|x\|$, for $x \in K \cap \partial\Omega_1$.

Further since $f_\infty = \infty$, there exists $\widehat{H}_2 > 0$ such that $f(x) \geqslant \rho x$, for $x \geqslant \widehat{H}_2$, where ρ is chosen so that

$$\rho \frac{\eta\gamma}{1-\alpha\eta} \int_\eta^1 (1-s)a(s)\,\mathrm{d}s > 1.$$

Let $H_2 = \max\{2H_1, \frac{\widehat{H}_2}{\gamma}\}$ and $\Omega_2 = \{x \in C[0,1]\colon \|x\| < H_2\}$, then $x \in K$ and $\|x\| = H_2$ implies

$$\min_{\eta \leqslant t \leqslant 1} x(t) \geqslant \gamma\|x\| \geqslant \widehat{H}_2,$$

and so

$$Ax(\eta) = -\int_0^\eta (\eta-s)a(s)f\big(x(s)\big)\,\mathrm{d}s - \frac{\alpha\eta}{1-\alpha\eta} \int_0^\eta (\eta-s)a(s)f\big(x(s)\big)\,\mathrm{d}s$$

$$+ \frac{\eta}{1-\alpha\eta} \int_0^1 (1-s)a(s)f\big(x(s)\big)\,\mathrm{d}s$$

$$= -\frac{1}{1-\alpha\eta} \int_0^\eta (\eta-s)a(s)f\big(x(s)\big)\,\mathrm{d}s$$

$$+ \frac{\eta}{1-\alpha\eta} \int_0^1 (1-s)a(s)f\big(x(s)\big)\,\mathrm{d}s$$

$$= \frac{\eta}{1-\alpha\eta} \int_\eta^1 a(s)f\big(x(s)\big)\,\mathrm{d}s + \frac{1}{1-\alpha\eta} \int_0^\eta sa(s)f\big(x(s)\big)\,\mathrm{d}s$$

$$- \frac{\eta}{1-\alpha\eta} \int_0^1 sa(s)f\big(x(s)\big)\,\mathrm{d}s$$

$$\geqslant \frac{\eta}{1-\alpha\eta} \int_\eta^1 a(s)f\big(x(s)\big)\,\mathrm{d}s$$

$$-\frac{\eta}{1-\alpha\eta}\int_{\eta}^{1}sa(s)f\big(x(s)\big)\,ds \quad \text{(by } \eta < 1\text{)}$$

$$=\frac{\eta}{1-\alpha\eta}\int_{\eta}^{1}(1-s)a(s)f\big(x(s)\big)\,ds.$$

Hence, for $x \in K \cap \partial\Omega_2$,

$$\|Ax\| \geqslant \rho\frac{\eta\gamma}{1-\alpha\eta}\int_{\eta}^{1}(1-s)a(s)\,ds\,\|x\| \geqslant \|x\|.$$

Therefore, by the first part of Theorem 3.33, it follows that A has a fixed point in $K \cap (\overline{\Omega}_2 \setminus \Omega_1)$, such that $H_1 \leqslant \|x\| \leqslant H_2$. This completes the superlinear part of the theorem.

(ii) Suppose next that $f_0 = \infty$ and $f_\infty = 0$. We first choose $H_3 > 0$ such that $f(x) \geqslant Mx$ for $0 < x < H_3$, where

$$M\gamma\left(\frac{\eta}{1-\alpha\eta}\right)\int_{\eta}^{1}(1-s)a(s)\,ds \geqslant 1.$$

Then

$$Ax(\eta) = -\int_{0}^{\eta}(\eta-s)a(s)f\big(x(s)\big)\,ds - \frac{\alpha\eta}{1-\alpha\eta}\int_{0}^{\eta}(\eta-s)a(s)f\big(x(s)\big)\,ds$$

$$+\frac{\eta}{1-\alpha\eta}\int_{0}^{1}(1-s)a(s)f\big(x(s)\big)\,ds$$

$$\geqslant \frac{\eta}{1-\alpha\eta}\int_{\eta}^{1}(1-s)a(s)f\big(x(s)\big)\,ds$$

$$\geqslant \frac{\eta}{1-\alpha\eta}\int_{\eta}^{1}(1-s)a(s)Mx(s)\,ds$$

$$\geqslant \frac{\eta}{1-\alpha\eta}\int_{\eta}^{1}(1-s)a(s)M\gamma\,ds\,\|x\|$$

$$\geqslant H_3.$$

Thus, we may let $\Omega_3 = \{x \in C[0,1]: \|x\| < H_3\}$ so that

$$\|Ax\| \geqslant \|x\|, \quad x \in K \cap \partial\Omega_3.$$

Now, since $f_\infty = 0$, there exist $\widehat{H}_4 > 0$ so that $f(x) \leqslant \lambda x$ for $x \geqslant \widehat{H}_4$, where $\lambda > 0$ satisfies

$$\frac{\lambda}{1-\alpha\eta}\int_{0}^{1}(1-s)a(s)\,ds \leqslant 1.$$

524   *S.K. Ntouyas*

We consider two cases:

Case 1. Suppose f is bounded, say $f(x) \leqslant N$ for all $x \in [0, \infty)$. In this case choose

$$H_4 = \max\left\{2H_3, \frac{N}{1-\alpha\eta} \int_0^1 (1-s)a(s)\,ds\right\}$$

so that for $x \in K$ with $\|x\| = H_4$ we have

$$\begin{aligned}
Ax(t) &= -\int_0^t (t-s)a(s)f\big(x(s)\big)\,ds - \frac{\alpha t}{1-\alpha\eta} \int_0^\eta (\eta-s)a(s)f\big(x(s)\big)\,ds \\
&\quad + \frac{t}{1-\alpha\eta} \int_0^1 (1-s)a(s)f\big(x(s)\big)\,ds \\
&\leqslant \frac{t}{1-\alpha\eta} \int_0^1 (1-s)a(s)f\big(x(s)\big)\,ds \\
&\leqslant \frac{1}{1-\alpha\eta} \int_0^1 (1-s)a(s)N\,ds \\
&\leqslant H_4
\end{aligned}$$

and therefore $\|Ax\| \leqslant \|x\|$.

Case 2. If f is unbounded, then we know from (A1) that there is H_4: $H_4 > \max\{2H_3, \frac{1}{\gamma}\widehat{H}_4\}$ such that

$$f(x) \leqslant f(H_4) \quad \text{for } 0 < x \leqslant H_4.$$

Then for $x \in K$ and $\|x\| = H_4$ we have

$$\begin{aligned}
Ax(t) &= -\int_0^t (t-s)a(s)f\big(x(s)\big)\,ds - \frac{\alpha t}{1-\alpha\eta} \int_0^\eta (\eta-s)a(s)f\big(x(s)\big)\,ds \\
&\quad + \frac{t}{1-\alpha\eta} \int_0^1 (1-s)a(s)f\big(x(s)\big)\,ds \\
&\leqslant \frac{t}{1-\alpha\eta} \int_0^1 (1-s)a(s)f(H_4)\,ds \\
&\leqslant \frac{1}{1-\alpha\eta} \int_0^1 (1-s)a(s)\lambda H_4\,ds \\
&\leqslant H_4.
\end{aligned}$$

Therefore, in either case we may put

$$\Omega_4 = \big\{x \in C[0, 1]: \|x\| < H_4\big\},$$

and for $x \in K \cap \partial \Omega_4$ we may have $\|Ax\| \leqslant \|x\|$. By the second part of Theorem 3.33, if follows that the BVP (3.80)–(3.81) has a positive solution. The proof of the theorem is completed. $\qquad\square$

The previous results are taken from [69]. Theorem 3.38 was proved either for sub- or superlinear case. Therefore the following questions are natural:
- *Whether or not we can obtain similar conclusion, if $f_0 = f_\infty = 0$ or $f_0 = f_\infty = \infty$;*
- *Whether or not we can obtain similar conclusion, if $f_0, f_\infty \notin \{0, \infty\}$.*

Motivated by the results in [69], Liu in [57] established some simple criteria for the existence of positive solutions of the BVP (3.80)–(3.81), which gives a positive answer to the questions stated above. The key tool in his approach is the following fixed point index theorem [21].

THEOREM 3.39. *Let E be Banach space and $K \subset E$ be a cone in E. Let $r > 0$, and define $\Omega_r = \{x \in K: \|x\| < r\}$. Assume $A: \overline{\Omega}_r \to K$ is a completely continuous operator such that $Ax \neq x$ for $x \in \partial \Omega_r$.*
  (i) *If $\|Ax\| \leqslant \|x\|$ for $x \in \partial \Omega_r$, then $i(A, \Omega_r, K) = 1$.*
  (ii) *If $\|Ax\| \geqslant \|x\|$ for $x \in \partial \Omega_r$, then $i(A, \Omega_r, K) = 0$.*

In what follows, for the sake of convenience, set

$$\Lambda_1 = (1 - \alpha \eta) \left(\int_0^1 (1 - s) a(s) \, ds \right)^{-1},$$

$$\Lambda_2 = (1 - \alpha \eta) \left(\eta \gamma \int_\eta^1 (1 - s) a(s) \, ds \right)^{-1}.$$

THEOREM 3.40. *Assume that the following assumptions are satisfied:*
  (H1) *$f_0 = f_\infty = \infty$.*
  (H2) *There exist constants $\rho_1 > 0$ and $M_1 \in (0, \Lambda_1)$ such that $f(u) \leqslant M_1 \rho_1$, $u \in [0, \rho_1]$.*
Then the BVP (3.80)–(3.81) has at least two positive solutions x_1 and x_2 such that

$$0 < \|x_1\| < \rho_1 < \|x_2\|.$$

PROOF. At first, in view of $f_0 = \infty$, then for any $M_* \in (\Lambda_2, \infty)$, there exist $\rho_* \in (0, \rho_1)$ such that

$$f(x) \geqslant M_* x, \quad 0 \leqslant x \leqslant \rho_*.$$

Set $\Omega_{\rho_*} = \{x \in K: \|x\| < \rho_*\}$. Since $x \in \partial \Omega_{\rho_*} \subset K$, we have $\min_{\eta \leqslant t \leqslant 1} x(t) \geqslant \gamma \|x\|$. Thus, for any $x \in \partial \Omega_{\rho_*}$, we have

$$Ax(\eta) = - \int_0^\eta (\eta - s) a(s) f(x(s)) \, ds - \frac{\alpha \eta}{1 - \alpha \eta} \int_0^\eta (\eta - s) a(s) f(x(s)) \, ds$$

$$+ \frac{\eta}{1 - \alpha\eta} \int_0^1 (1 - s)a(s)f\big(x(s)\big)\,ds$$

$$= \frac{\eta}{1 - \alpha\eta} \int_\eta^1 a(s)f\big(x(s)\big)\,ds + \frac{1 - \eta}{1 - \alpha\eta} \int_0^\eta sa(s)f\big(x(s)\big)\,ds$$

$$- \frac{\eta}{1 - \alpha\eta} \int_\eta^1 sa(s)f\big(x(s)\big)\,ds$$

$$\geqslant \frac{\eta}{1 - \alpha\eta} \int_\eta^1 (1 - s)a(s)f\big(x(s)\big)\,ds$$

$$\geqslant \frac{\eta}{1 - \alpha\eta} \int_\eta^1 (1 - s)a(s)M_* x(s)\,ds$$

$$\geqslant \frac{\gamma\eta M_*}{1 - \alpha\eta} \int_\eta^1 (1 - s)a(s)\|x\|\,ds$$

$$> \|x\|,$$

which yields

$$\|Ax\| > \|x\|, \quad \text{for } y \in \partial\Omega_{\rho_*}.$$

Hence, Theorem 3.39 implies $i(A, \Omega_{\rho_*}, K) = 0$.

Next, since $f_\infty = \infty$, then for any $M^* \in (\Lambda_2, \infty)$, then there exist $\rho^* > \rho_1$ such that

$$f(x) \geqslant M^* x, \quad 0 \leqslant x \leqslant \gamma\rho^*.$$

Set $\Omega_{\rho^*} = \{x \in K : \|x\| < \rho^*\}$ for $x \in \partial\Omega_{\rho^*}$. Since $x \in K$, we have $\min_{\eta \leqslant t \leqslant 1} x(t) \geqslant \gamma\|x\| = \gamma\rho^*$. Thus, for any $x \in \partial\Omega_{\rho^*}$, we have

$$Ax(\eta) = -\int_0^\eta (\eta - s)a(s)f\big(x(s)\big)\,ds - \frac{\alpha\eta}{1 - \alpha\eta} \int_0^\eta (\eta - s)a(s)f\big(x(s)\big)\,ds$$

$$+ \frac{\eta}{1 - \alpha\eta} \int_0^1 (1 - s)a(s)f\big(x(s)\big)\,ds$$

$$\geqslant \frac{\eta}{1 - \alpha\eta} \int_\eta^1 (1 - s)a(s)f\big(x(s)\big)\,ds$$

$$\geqslant \frac{\eta}{1 - \alpha\eta} \int_\eta^1 (1 - s)a(s)M^* x(s)\,ds$$

$$\geqslant \frac{\gamma\eta M^*}{1 - \alpha\eta} \int_\eta^1 (1 - s)a(s)\|x\|\,ds$$

$$> \|x\|,$$

which implies

$$\|Ax\| > \|x\|, \quad \text{for } y \in \partial\Omega_{\rho^*}.$$

Hence, Theorem 3.39 yields $i(A, \Omega_{\rho_*}, K) = 0$.

Finally, set $\Omega_{\rho_1} = \{x \in K : \|x\| < \rho_1\}$. For any $x \in \partial\Omega_{\rho_1}$ we obtain

$$\begin{aligned}
Ax(t) &\leqslant \frac{t}{1 - \alpha\eta} \int_0^1 (1 - s)a(s)f\big(x(s)\big)\,ds \\
&\leqslant \frac{M_1\rho_1}{1 - \alpha\eta} \int_0^1 (1 - s)a(s)\,ds \\
&< \rho_1 = \|x\|,
\end{aligned}$$

which yields

$$\|Ax\| < \|x\|, \quad \text{for } y \in \partial\Omega_{\rho_1}.$$

Thus, Theorem 3.39 again shows that $i(A, \Omega_{\rho_1}, K) = 1$.

Hence, since $\rho_* < \rho_1 < \rho^*$ it follows from the additivity of the fixed point index that

$$i\big(A, \Omega_{\rho_1} \setminus \overline{\Omega}_{\rho_*}, K\big) = 1, \qquad i\big(A, \Omega_{\rho^*} \setminus \overline{\Omega}_{\rho_1}, K\big) = -1.$$

Thus, A has a fixed point x_1 in $\Omega_{\rho_1} \setminus \overline{\Omega}_{\rho_*}$, and a fixed point x_2 in $\Omega_{\rho^*} \setminus \overline{\Omega}_{\rho_1}$. Both are positive solutions, $0 < \|x_1\| < \rho_1 < \|x_2\|$, and the proof of the theorem is complete. $\square$

THEOREM 3.41. *Assume that the following assumptions are satisfied:*
  (H3) $f_0 = f_\infty = 0$.
  (H4) *There exist constants* $\rho_2 > 0$ *and* $M_2 \in (\Lambda_2, \infty)$ *such that* $f(u) \geqslant M_2\rho_2$, $u \in$
     $[\gamma\rho_2, \rho_2]$.
Then the BVP (3.80)–(3.81) *has at least two positive solutions* x_1 *and* x_2 *such that*

$$0 < \|x_1\| < \rho_2 < \|x_2\|.$$

PROOF. First, since $f_0 = 0$, for any $\varepsilon \in (0, \Lambda_1)$, there exists $\rho_* \in (0, \rho_2)$ such that

$$f(x) \leqslant \varepsilon x, \quad \text{for } x \in [0, \rho_*].$$

Setting $\Omega_{\rho_*} = \{x \in K : \|x\| < \rho_*\}$ for any $x \in \partial\Omega_{\rho_*}$, we get

$$\begin{aligned}
Ax(t) &\leqslant \frac{t}{1 - \alpha\eta} \int_0^1 (1 - s)a(s)f\big(x(s)\big)\,ds \\
&\leqslant \frac{\varepsilon\rho_*}{1 - \alpha\eta} \int_0^1 (1 - s)a(s)\,ds < \rho_* = \|x\|,
\end{aligned}$$

which yields

$$\|Ax\| < \|x\|, \quad \text{for } x \in \partial\Omega_{\rho_*}.$$

Thus Theorem 3.39 implies $i(A, \Omega_{\rho_*}, K) = 1$.

Second, in view $f_\infty = 0$, then for any $\varepsilon \in (0, \Lambda_1)$, there exists $\rho_0 > \rho_2$ such that

$$f(x) \leqslant \varepsilon x, \quad \text{for } x \in [\rho_0, \infty),$$

and we consider two cases.

Case (i). Suppose that f is unbounded; then from $f \in C([0, \infty), [0, \infty))$, we know that there is $\rho^* > \rho_0$ such that $f(x) \leqslant f(\rho^*)$, for $x \in [0, \rho^*]$. Since $\rho^* > \rho_0$, one has $f(x) \leqslant f(\rho^*) \leqslant \varepsilon\rho^*$, for $x \in [0, \rho^*]$. For $x \in K$, $\|x\| = \rho^*$ we obtain

$$Ax(t) \leqslant \frac{t}{1 - \alpha\eta} \int_0^1 (1 - s)a(s)f\big(x(s)\big)\,ds$$

$$\leqslant \frac{\varepsilon\rho_*}{1 - \alpha\eta} \int_0^1 (1 - s)a(s)\,ds < \rho_* = \|x\|.$$

Case (ii). Suppose that f is bounded, say $f(x) \leqslant L$. Taking $\rho^* \geqslant \max\{L/\varepsilon, \rho_2\}$, for $x \in K$, $\|x\| = \rho^*$ one has

$$Ax(t) \leqslant \frac{t}{1 - \alpha\eta} \int_0^1 (1 - s)a(s)f\big(x(s)\big)\,ds$$

$$\leqslant \frac{L}{1 - \alpha\eta} \int_0^1 (1 - s)a(s)\,ds < \rho_* = \|x\|.$$

Hence, in either case, we always may set $\Omega_{\rho^*} = \{x \in K: \|x\| < \rho^*\}$ such that

$$\|Ax\| < \|x\|, \quad \text{for } x \in \partial\Omega_{\rho_*}.$$

Thus Theorem 3.39 implies $i(A, \Omega_{\rho_*}, K) = 1$.

Finally set $\Omega_{\rho_2} = \{x \in K: \|x\| < \rho_2\}$, for $x \in \partial\Omega_{\rho_2}$, since $x \in K$, $\min_{\eta \leqslant t \leqslant 1} x(t) \geqslant \gamma\|x\| = \gamma\rho_2$, and hence we can get

$$Ax(\eta) \geqslant \frac{\eta}{1 - \alpha\eta} \int_\eta^1 (1 - s)a(s)f\big(x(s)\big)\,ds$$

$$\geqslant \frac{\gamma\eta M_2}{1 - \alpha\eta}\rho_2 \int_\eta^1 (1 - s)a(s)\,ds > \rho_2 = \|x\|,$$

which yields

$$\|Ax\| > \|x\|, \quad \text{for } x \in \partial\Omega_{\rho_2}.$$

Thus Theorem 3.39 implies $i(A, \Omega_{\rho_2}, K) = 0$.

Hence, since $\rho_* < \rho_2 < \rho^*$ it follows from the additivity of the fixed point index that

$$i\left(A, \Omega_{\rho_2} \setminus \overline{\Omega}_{\rho_*}, K\right) = -1, \qquad i\left(A, \Omega_{\rho^*} \setminus \overline{\Omega}_{\rho_2}, K\right) = 1.$$

Thus, A has a fixed point x_1 in $\Omega_{\rho_2} \setminus \overline{\Omega}_{\rho_*}$, and a fixed point x_2 in $\Omega_{\rho^*} \setminus \overline{\Omega}_{\rho_2}$. Both are positive solutions, $0 < \|x_1\| < \rho_2 < \|x_2\|$, and the proof of the theorem is complete. $\qquad\square$

For the case when $f_0, f_\infty \notin \{0, \infty\}$ we have the following

THEOREM 3.42. *Suppose that* (H2) *and* (H4) *hold and that* $\rho_1 \neq \rho_2$. *Then the BVP* (3.80)–(3.81) *has at least one positive solution* x *satisfying* $\rho_1 < \|x\| < \rho_2$ *or* $\rho_2 < \|x\| < \rho_1$.

Theorems 3.40–3.42 were taken from [57]. A method to study the existence of positive solutions of some three-point BVPs is to write the BVP as an equivalent Hammerstein integral equation of the form

$$x(t) = \int_0^1 k(t, s) a(s) f\left(x(s)\right) ds \equiv Tx(t),$$

and seek fixed points of T in the cone of positive functions in the space $C[0, 1]$. We use the fixed point index for compact maps, which is based on Leray–Schauder degree theory, and use a well-known nonzero fixed point theorem in order to prove that T has a positive fixed point.

We apply this method for the BVP (3.80)–(3.81) using the cone defined by

$$K = \left\{x \in C[0, 1]\colon x \geqslant 0, \ \min\{x(t)\colon a \leqslant t \leqslant b\} \geqslant c\|x\|\right\},$$

and the following kernel (Green's function) in the Hammerstein integral operator

$$k(t, s) = \frac{1}{1 - \alpha\eta} t(1 - s) - \begin{cases} \dfrac{\alpha t}{1 - \alpha\eta}(\eta - s), & s \leqslant \eta \\ 0, & s > \eta \end{cases} - \begin{cases} t - s, & s \leqslant t \\ 0, & s > t. \end{cases}$$

Let $a, b \in (0, 1]$ and suppose that $\int_a^b a(t) \, dt > 0$. We are able to find upper and lower bounds for $k(t, s)$ with s fixed, of the same type. We have the same freedom in choosing the numbers a and b. Let us find the upper and lower bounds for the BVP (3.80)–(3.81). We have to exhibit $\Phi(s)$, a subinterval $[a, b] \subset [0, 1]$ and a constant $c \in (0, 1]$ such that

$$\begin{aligned} k(t, s) &\leqslant \Phi(s), && \text{for every } t, s \in [0, 1], \\ k(t, s) &\geqslant c\Phi(s), && \text{for every } s \in [0, 1], t \in [a, b]. \end{aligned}$$

3.9.1. *Upper bounds* We shall show that we may take $\Phi(s) = \max\{1, \alpha\}\frac{s(1-s)}{1-\alpha\eta}$. (In this case $\alpha\eta < 1$ but it is possible to have $\alpha > 1$.)

Case 1: $s > \eta$. $t < s$ is simple so consider $t \geqslant s$. Then $k(t,s) = s + \frac{(\alpha\eta - s)t}{1 - \alpha\eta}$. If also $s \geqslant \alpha\eta$ then it is a decreasing function of t so the maximum occurs when $t = s$ and

$$k(t,s) \leqslant \frac{s - s\alpha\eta + s(\alpha\eta - s)}{1 - \alpha\eta} = \frac{s(1-s)}{1-\alpha\eta} \leqslant \Phi(s).$$

If $s < \alpha\eta$ (which can happen only if $\alpha > 1$) the maximum occurs at $t = 1$ and then

$$k(t,s) \leqslant \frac{\alpha\eta(1-s)}{1-\alpha\eta} < \alpha \frac{s(1-s)}{1-\alpha\eta} = \Phi(s).$$

Case 2: $s \leqslant \eta$. For $t \leqslant s$ we clearly have $k(t,s) \leqslant \frac{s(1-s)}{1-\alpha\eta} \leqslant \Phi(s)$. So consider $t \geqslant s$. Then $k(t,s) = s + \frac{st(\alpha - 1)}{1 - \alpha\eta}$. For $\alpha > 1$ the maximum occurs when $t = 1$ and then

$$k(t,s) \leqslant \frac{\alpha s(1-\eta)}{1-\alpha\eta} \leqslant \alpha \frac{s(1-s)}{1-\alpha\eta} = \Phi(s),$$

using $s \leqslant \eta$. For $\alpha \leqslant 1$ the maximum occurs when $t = s$ and then

$$k(t,s) = \frac{\alpha s(s-\eta) + s(1-s)}{1-\alpha\eta} \leqslant \frac{s(1-s)}{1-\alpha\eta} = \Phi(s).$$

3.9.2. *Lower bounds*   We will show that we may take arbitrary $a > 0$ and $b \leqslant 1$.

Case 1: $s > \eta$. For $t < s$, $k(t,s) = \frac{t(1-s)}{1-\alpha\eta} \geqslant a \frac{s(1-s)}{1-\alpha\eta} \geqslant a\eta\Phi(s)$. For $t \geqslant s$, $k(t,s) = s - \frac{t(s-\alpha\eta)}{1-\alpha\eta}$. If $s > \alpha\eta$ then the minimum occurs when $t = 1$ so

$$k(t,s) \geqslant \frac{\alpha\eta(1-s)}{1-\alpha\eta} \geqslant \alpha\eta \frac{s(1-s)}{1-\alpha\eta} = \begin{cases} \alpha\eta\Phi(s), & \alpha \leqslant 1, \\ \eta\Phi(s), & \alpha > 1. \end{cases}$$

If $s \leqslant \alpha\eta$ (only possible if $\alpha > 1$) the minimum is at $t = s$ and

$$k(t,s) \geqslant \frac{s(1-s)}{1-\alpha\eta} \geqslant \eta\Phi(s).$$

Case 2: $s \leqslant \eta$. First suppose $a \leqslant t \leqslant s$. (This case cannot occur if $\alpha \geqslant \eta$.) Then $k(t,s) = \frac{t[1-\alpha\eta+(\alpha-1)s]}{1-\alpha\eta}$. For $\alpha < 1$ we have $[1 - \alpha\eta + (\alpha - 1)s] \geqslant 1 - \alpha\eta + (\alpha - 1)\eta = 1 - \eta$ so $k(t,s)$ is increasing in t and has a minimum when $t = a$. Thus

$$k(t,s) \geqslant \frac{a}{1-\alpha\eta}[1 - \eta] \geqslant 4a(1 - \eta)\Phi(s),$$

since $s(1 - s) \leqslant 1/4$. Note that $4a(1 - \eta) < 1$ since $\alpha < \eta$.

For $\alpha > 1$, $k(t,s)$ is clearly increasing so we have

$$k(t,s) \geqslant \frac{a}{1-\alpha\eta}\big[1 - \alpha\eta + (\alpha - 1)s\big] \geqslant \frac{a}{1-\alpha\eta}[1 - \alpha\eta] \geqslant \frac{4(1-\alpha\eta)a}{\alpha}\Phi(s).$$

Note that $\frac{4(1-\alpha\eta)a}{\alpha} = \frac{4a\alpha\eta(1-\alpha\eta)}{\alpha^2\eta} < \frac{1}{\alpha^2} < 1$ since $\alpha < \eta$.

Now suppose that $t \geqslant s$. Then $k(t,s) = s + \frac{t(\alpha-1)s}{1-\alpha\eta}$. For $\alpha \leqslant 1$ the minimum occurs when $t = 1$ and then

$$k(t,s) \geqslant \frac{\alpha(1-\eta)s}{1-\alpha\eta} \geqslant \alpha(1-\eta)\Phi(s).$$

For $\alpha > 1$ the minimum occurs either at $t = s$ or $t = a$ but in both cases

$$k(t,s) \geqslant s \geqslant \frac{1-\alpha\eta}{\alpha}\Phi(s) \geqslant \eta(1-\alpha\eta)\Phi(s).$$

The conclusion is that we may take

$$c = \begin{cases} \min\{a, \alpha\eta, 4a(1-\eta), \alpha(1-\eta)\}, & \alpha < 1, \\ \min\{a\eta, 4a(1-\alpha\eta), \eta(1-\alpha\eta)\}, & \alpha \geqslant 1. \end{cases}$$

Hence we have:

THEOREM 3.43. *Let $a, b \in (0, 1]$ and suppose that $\int_a^b a(t)\,dt > 0$. Let c as defined above and*

$$m = \left(\max_{0\leqslant t\leqslant 1} \int_0^1 k(t,s)a(s)\,ds \right)^{-1}, \qquad M = \left(\min_{0\leqslant t\leqslant 1} \int_a^b k(t,s)a(s)\,ds \right)^{-1}.$$

Then for $0 < \alpha\eta < 1$ the BVP (3.80)–(3.81) has at least one positive solution if either

$$\text{(h1)} \quad 0 \leqslant \limsup_{x\to 0} \frac{f(x)}{x} < m \quad \text{and} \quad M < \liminf_{x\to\infty} \frac{f(x)}{x} \leqslant \infty,$$

or

$$\text{(h2)} \quad 0 \leqslant \limsup_{x\to\infty} \frac{f(x)}{x} < m \quad \text{and} \quad M < \liminf_{x\to 0} \frac{f(x)}{x} \leqslant \infty,$$

and has two positive solutions if there is $\rho > 0$ such that either

$$\text{(E1)} \quad \begin{cases} 0 \leqslant \limsup_{x\to 0} \dfrac{f(x)}{x} < m, \\[2mm] \min\left\{ \dfrac{f(x)}{\rho} : x \in [c\rho, \rho] \right\} \geqslant cM, \ x \neq Tx \ \ \text{for } x \in \partial\Omega_\rho, \quad \text{and} \\[2mm] 0 \leqslant \limsup_{x\to\infty} \dfrac{f(x)}{x} < m, \end{cases}$$

or

$$(E2) \quad \begin{cases} M < \limsup_{x \to 0} \dfrac{f(x)}{x} \leqslant \infty, \\[2mm] \max\left\{\dfrac{f(x)}{\rho} : x \in [0, \rho]\right\} \leqslant m, \ x \neq Tx \quad \text{for } x \in \partial K_\rho, \quad \text{and} \\[2mm] M < \limsup_{x \to \infty} \dfrac{f(x)}{x} \leqslant \infty, \end{cases}$$

where

$$\Omega_\rho = \left\{x \in K : c\|x\| \leqslant \min_{a \leqslant t \leqslant b} x(t) < c\rho\right\}, \qquad K_r = \left\{x \in K : \|x\| < \rho\right\}$$

and K is a closed convex set in a Banach space.

Theorem 3.43 is taken from Webb [92]. Similar results are given by Infante [45] who proved existence of eigenvalues of some nonlocal BVP, including nonlocal boundary conditions other than (3.81), as

(a) $x'(0) = 0$, $\alpha x'(\eta) = x(1)$, $0 < \eta < 1$,
(b) $x(0) = 0$, $\alpha x'(\eta) = x(1)$, $0 < \eta < 1$,
(c) $x'(0) = 0$, $\alpha x(\eta) = x(1)$, $0 < \eta < 1$.

Infante and Webb [47] obtained more general results for existence of positive solutions of some m-point BVPs in the case when all the parameters occurring in the boundary conditions are not positive and the nonlinear term allow more general behaviour than being either sub- or superlinear. Two four-points BVPs are studied in details in [47].

Also interesting results were given for the following three-point BVP

$$x'' + a(t)f(x) = 0, \quad t \in (0, 1), \tag{3.84}$$

$$x'(0) = 0, \qquad \alpha x(\eta) = x(1), \tag{3.85}$$

where $0 < \eta < 1$, $0 < \alpha < 1$ by Liu in [56]. By applying Krasnosel'skii's fixed point theorem he proved existence

- of single positive solution under $f_0 = 0$, $f_\infty = \infty$ or $f_0 = \infty$, $f_\infty = 0$,
- of two positive solution under $f_0 = f_\infty = \infty$ or $f_0 = f_\infty = 0$,
- of positive solutions under $f_0, f_\infty \notin \{0, \infty\}$.

In [56] some interesting examples were given to demonstrate the results.

A more general three-point BVP was studied by Ma and Wang. In [79] they studied the existence of positive solutions of the following BVP

$$x''(t) + a(t)x'(t) + b(t)x(t) + h(t)f(x) = 0, \quad t \in (0, 1), \tag{3.86}$$

$$x(0) = 0, \qquad x(1) = \alpha x(\eta), \tag{3.87}$$

where $f \in C([0, 1], [0, \infty))$, $h \in C([0, 1], [0, \infty))$ and there exists $x_0 \in [0, 1]$ such that $h(x_0) > 0$, and $a \in C[0, 1]$, $b \in C([0, 1], (-\infty, 0))$. Under these assumptions they proved the existence of at least one positive solution of the BVP (3.86)–(3.87) either in superlinear or sublinear case, by applying the Krasnosel'skii's fixed point theorem in cones.

If in the BVP (3.80)–(3.81) the boundary condition (3.81) is nonhomogeneous, i.e.

$$x(0) = 0, \qquad x(1) - \alpha x(\eta) = b, \tag{3.88}$$

Ma in [78], by using Schauder's fixed point theorem, proved in the superlinear case that, if (A1) and (A2) of Theorem 3.38 hold, then there exists a positive number b^* such that (3.80), (3.88) has at least one positive solution for b: $0 < b < b^*$ and no solution for $b > b^*$. This result was improved and complemented by Zhang and Wang in [95], where they studied the BVP (3.80), (3.88) under the following conditions:

(H1) $\alpha \in (0, 1/\eta)$, $\eta \in (0, 1)$;

(H2) $a(t)$ is a nonnegative measurable function defined on $(0, 1)$ and

$$0 \leqslant \int_0^\eta sa(s)\,ds < +\infty, \qquad 0 < \int_\eta^1 (1-s)a(s)\,ds < +\infty;$$

(H3) $f : [0, \infty) \to [0, \infty)$ is continuous;

(H3*) $f : [0, \infty) \to [0, \infty)$ is locally Lipschitz continuous;

(H4) $\limsup_{u \to +\infty} \frac{f(u)}{u} < \delta$;

(H5) $\liminf_{u \to +\infty} \frac{f(u)}{u} > M$;

(H6) $\lim_{u \to 0+} \frac{f(u)}{u} < \delta$,

where δ, M are suitable defined constants, and proved that:

- the BVP (3.80), (3.88) has a positive solution for all $b > 0$ if (H1)–(H4) hold;
- there exists a positive number b^* such that (3.80), (3.88) has at least one positive solution for b: $0 < b < b^*$ and no solution for $b > b^*$ if (H1)–(H3), (H5) and (H6) hold;
- there exists a positive number b^* such that (3.80), (3.88) has at least two solution for b: $0 < b < b^*$, at least one for $b = b^*$, none for $b > b^*$ if (H1), (H2), (H3*), (H5) and (H6) hold.

We remark that (H2) allows $a(t)$ to be singular at $t = 0$ and/or $t = 1$, and (H5) and (H6) allow but not require the nonlinearity $f(x)$ to be superlinear at zero and infinity.

Now, we consider the following three-point nonlinear second-order BVP

$$x''(t) + \lambda a(t) f(x(t)) = 0, \quad t \in (0, 1), \tag{3.89}$$

$$x(0) = 0, \qquad \alpha x(\eta) = x(1), \tag{3.90}$$

where $0 < \eta < 1$. For this problem, an open interval of eigenvalues is determined, which in return, imply the existence of a positive solution of (3.89)–(3.90) by appealing to Krasnosel'skii's fixed point theorem.

For the sake of simplicity, we let

$$A = (1 - \alpha\eta)^{-1} \int_0^1 (1 - \eta)a(s)\,ds, \qquad B = \eta(1 - \alpha\eta)^{-1} \int_0^1 (1 - \eta)a(s)\,ds.$$

THEOREM 3.44. *Assume that*

  (A1) $f \in C([0, \infty), [0, \infty))$;

  (A2) $a \in C([0, 1], [0, \infty))$ *and does not vanish identically on any subinterval.*

  (A3) $\lim_{x \to 0+} \frac{f(x)}{x} = l$ *with* $0 < l < \infty$.

  (A4) $\lim_{x \to \infty} \frac{f(x)}{x} = L$ *with* $0 < L < \infty$.

Then, for each λ *satisfying either*

  (i) $\frac{1}{\gamma BL} < \lambda < \frac{1}{Al}$

or

  (ii) $\frac{1}{\gamma Bl} < \lambda < \frac{1}{AL}$

the BVP (3.89)–(3.90) *has at least one positive solution.*

PROOF. Let λ given as in (i), and choose $\epsilon > 0$ such that

$$\frac{1}{\gamma B(L - \epsilon)} \leqslant \lambda \leqslant \frac{1}{A(l + \epsilon)}.$$

Consider also the cone K and the operator A defined in the proof of Theorem 3.38. By (A3), there exists $H_1 > 0$ such that $f(x) \leqslant (l + \epsilon)x$, for $0 < x \leqslant H_1$. So, choosing $x \in K$ with $\|x\| = H_1$, we have

$$
\begin{aligned}
Ax(t) &= \lambda \frac{t}{1 - \alpha\eta} \int_0^1 (1 - s)a(s)f\big(x(s)\big)\,ds \\
&\leqslant \lambda \frac{t}{1 - \alpha\eta} \int_0^1 (1 - s)a(s)(l + \epsilon)x(s)\,ds \\
&\leqslant \lambda \frac{1}{1 - \alpha\eta} \int_0^1 (1 - s)a(s)(l + \epsilon)\|x\|\,ds \\
&\leqslant \lambda \frac{1}{1 - \alpha\eta} \int_0^1 (1 - s)a(s)(l + \epsilon)H_1\,ds \\
&\leqslant \lambda A(l + \epsilon)\|x\| \leqslant \|x\|.
\end{aligned}
$$

Consequently, $\|Ax\| \leqslant \|x\|$. So, if we set

$$\Omega_1 = \big\{x \in C[0, 1]: \|x\| < H_1\big\},$$

then $\|Ax\| \leqslant \|x\|$, for $x \in K \cap \partial\Omega_1$.

Next we construct the set Ω_2. Considering (A4) there exists $\overline{H}_2$ such that $f(x) \geqslant (L - \epsilon)x$, for $x \geqslant \overline{H}_2$. Let $H_2 = \max\{2H_1, \frac{\widehat{H}_2}{\gamma}\}$ and $\Omega_2 = \{x \in C[0, 1]: \|x\| < H_2\}$, then $x \in K$ and $\|x\| = H_2$ implies

$$\min_{\eta \leqslant t \leqslant 1} x(t) \geqslant \gamma\|x\| \geqslant \widehat{H}_2,$$

and so

$$Ax(\eta) = \lambda \frac{\eta}{1 - \alpha\eta} \int_{\eta}^{1} (1 - s)a(s)f\big(x(s)\big)\,ds$$

$$\geqslant \lambda \frac{\eta}{1 - \alpha\eta} \int_{\eta}^{1} (1 - s)a(s)(L - \epsilon)x(s)\,ds$$

$$\geqslant \lambda \frac{\eta}{1 - \alpha\eta} \int_{\eta}^{1} (1 - s)a(s)(L - \epsilon)\gamma\,\|x\|\,ds$$

$$= \lambda \frac{\gamma\eta}{1 - \alpha\eta} \int_{\eta}^{1} (1 - s)a(s)(L - \epsilon)H_2\,ds$$

$$\geqslant \lambda B\gamma(L - \epsilon)\|x\|$$

$$= \|x\|.$$

Hence, for $x \in K \cap \partial\Omega_2$,

$$\|Ax\| \geqslant \|x\|.$$

Therefore, by the first part of Theorem 3.33, it follows that A has a fixed point in $K \cap (\overline{\Omega}_2 \setminus \Omega_1)$, completing the proof in case (i).

Consider now the case (ii). Let λ given as in (ii) and choose $\epsilon > 0$ such that

$$\frac{1}{\gamma B(l - \epsilon)} \leqslant \lambda \leqslant \frac{1}{A(L + \epsilon)}.$$

We omit the rest of the proof, since it is similar to that of Theorem 3.38. $\qquad\square$

Theorem 3.44 is taken from Raffoul [89]. He proved also that:

- if (A1), (A2), (A4) hold and $\lim_{x \to 0+} \frac{f(x)}{x} = \infty$ then the BVP (3.89)–(3.90) has at least one positive solution for each λ satisfying $0 < \lambda < \frac{1}{AL}$;
- if (A1), (A2), (A3) hold and $\lim_{x \to \infty} \frac{f(x)}{x} = \infty$ then the BVP (3.89)–(3.90) has at least one positive solution for each λ satisfying $0 < \lambda < \frac{1}{Al}$.

Other existence and multiplicities of positive solutions for the BVP (3.89)–(3.90) were proved by Ma in [70] by using fixed point index theory and the method of upper and lower solutions. He proved in [70] the following result.

THEOREM 3.45. *Assume that*:
(A1) *λ is a positive parameter; $\eta \in (0, 1)$ and $\alpha\eta < 1$.*
(A2) *$a : [0, 1] \to [0, \infty)$ is continuous and does not vanish identically on any subset of positive measure.*
(A3) *$\lim_{x \to \infty} \frac{f(x)}{x} = \infty$.*

Then there exists a positive λ^* *such that the BVP* (3.89)–(3.90) *has at least two positive solutions for* $0 < \lambda < \lambda^*$, *at least one positive solution for* $\lambda = \lambda^*$, *and no positive solutions for* $\lambda > \lambda^*$.

Interesting results on the existence of multiple positive solutions for three-point BVP can be obtained by using the Leggett–Williams fixed point theorem. We will give such a result proving the existence of triple positive solutions of the following BVP

$$x'' + f(t, x) = 0, \quad t \in (0, 1), \tag{3.91}$$

$$x(0) = 0, \qquad x(1) = \alpha x(\eta). \tag{3.92}$$

In this approach, we do not need the assumption that f is sublinear or superlinear, which was required in Theorem 3.38.

For convenience of the reader, we present here the necessary definitions from cone theory in Banach spaces.

DEFINITION 3.46. Let E be a Banach space over $\mathbb{R}$. A nonempty, closed set $P \subset E$ is said to be a cone provided that
(a) $\alpha u + \beta v \in P$ for all $u, v \in P$ and all $\alpha, \beta \geqslant 0$ and
(b) $u, -u \in P$ implies $u = 0$.

If $P \subset E$ is a cone, we denote the order induced by P on E by $\leqslant$. For $u, v \in P$, we write $u \leqslant v$ if and only if $v - u \in P$.

DEFINITION 3.47. The map ψ is said to be a nonnegative continuous concave functional on P provided $\psi : P \to [0, \infty)$ is continuous and

$$\psi\big(\lambda x + (1 - \lambda)y\big) \geqslant \lambda\psi(x) + (1 - \lambda)\psi(y)$$

for all $x, y \in P$ and $\lambda \in [0, 1]$.

Let $0 < a < b$ be given and let ψ be a nonnegative continuous concave functional on the cone P. Define the convex sets P_r and $P(\psi, a, b)$ by

$$P_r = \big\{y \in P\colon \|y\| < r\big\}$$

and

$$P(\psi, a, b) = \big\{y \in P\colon a \leqslant \psi(y), \text{ and } \|y\| \leqslant b\big\}.$$

Our consideration is based on the following fixed point theorem given by Leggett and Williams in 1979 [55] (see also Guo and Lakshmikantham [21]).

THEOREM 3.48 (Legget–Williams fixed-point theorem). *Let* $T : \overline{P}_c \to \overline{P}_c$ *be a completely continuous operator and let* ψ *be a nonnegative continuous concave functional* *on P such that $\psi(y) \leqslant \|y\|$ for $y \in \overline{P}_c$. Suppose that there exist $0 < a < b < d \leqslant c$ such that*

(C1) *$\{y \in P(\psi, b, d): \psi(y) > b\} \neq \emptyset$ and $\psi(Ty) > b$ for all $y \in P(\psi, b, d)$;*

(C2) *$\|Ty\| < a$ for all $\|y\| \leqslant a$, and*

(C3) *$\psi(Ty) > b$ for $y \in P(\psi, b, c)$ with $\|Ty\| > d$.*

Then T has at least three fixed points y_1, y_2, and y_3 such that $\|y_1\| < a$, $b < \psi(y_2)$ and $\|y_3\| > a$ with $\psi(y_3) < b$.

THEOREM 3.49. *Assume that $\eta \in (0, 1)$, $\alpha > 0$ and $\alpha\eta < 1$. Assume that:*

(A1) *$f : [0, 1] \times [0, \infty) \to [0, \infty)$ is continuous and $f(t, \cdot)$ does not vanish identically on any subset of $[0, 1]$ with positive measure.*

Suppose also that there exist constants $0 < a < b < b/\gamma \leqslant c$ such that

(D1) *$f(t, x) < ma$, for $0 \leqslant t \leqslant 1, 0 \leqslant x \leqslant a$,*

(D2) *$f(t, x) \geqslant b/\delta$, for $\eta \leqslant t \leqslant 1, b \leqslant x \leqslant b/\gamma$,*

(D3) *$f(t, x) \leqslant mc$, for $0 \leqslant t \leqslant 1, 0 \leqslant x \leqslant c$,*

where

$$m = \left(\frac{2 - \alpha\eta + \alpha\eta^2}{2(1 - \alpha\eta)}\right)^{-1}, \qquad \delta = \min\left\{\frac{\alpha\eta(1 - \eta)^2}{2(1 - \alpha\eta)}, \frac{\eta(1 - \eta)^2}{2(1 - \alpha\eta)}\right\},$$

and γ is defined in Lemma 3.37. Then the BVP (3.91)–(3.92) has at least three positive solutions x_1, x_2 and x_3 satisfying $\|x_1\| < a$, $b < \psi(x_2)$ and $\|x_3\| > a$ with $\psi(x_3) < b$.

PROOF. Let $E = C[0, 1]$ be endowed with the maximum norm, $\|x\| = \max_{0 \leqslant t \leqslant 1} |x(t)|$, and the ordering $x \leqslant y$ if $x(t) \leqslant y(t)$ for all $t \in [0, 1]$. From the fact $x''(t) = -f(t, x) \leqslant 0$, we know that $x(t)$ is concave on $[0, 1]$. So, define the cone $P \subset E$ by

$$P = \{x \in E : x \text{ is concave and nonnegative valued on } [0, 1]\}.$$

Finally, let the nonnegative concave functional $\psi : P \to [0, \infty)$ be defined by

$$\psi(x) = \min_{\eta \leqslant t \leqslant 1} x(t), \quad x \in P.$$

We notice that for each $x \in P$, $\psi(x) \leqslant \|x\|$. Define an operator $T : P \to E$ by

$$Tx(t) = -\int_0^t (t - s) f(s, x(s)) \, ds - \frac{\alpha t}{1 - \alpha\eta} \int_0^\eta (\eta - s) f(s, x(s)) \, ds$$

$$+ \frac{t}{1 - \alpha\eta} \int_0^1 (1 - s) f(s, x(s)) \, ds.$$

We note that, if $x \in P$ then $Tx(t) \geqslant 0, 0 \leqslant t \leqslant 1$. In view of $(Tx)''(t) = -f(t, x(t)) \leqslant 0, 0 \leqslant t \leqslant 1$, we see that $Tx \in P$; that is to say that $T : P \to P$. Also T is completely continuous.

We now show that all the conditions of Theorem 3.48 are satisfied. Now if $x \in \overline{P}_c$, then $\|x\| \leqslant c$ and (D3) implies $f(t, x(t)) \leqslant mc$, $0 \leqslant t \leqslant 1$. Consequently,

$$
\begin{aligned}
\|Tx\| &= \max_{0 \leqslant t \leqslant 1} -\int_0^t (t-s) f(s, x(s)) \, \mathrm{d}s - \frac{\alpha t}{1-\alpha \eta} \int_0^\eta (\eta - s) f(s, x(s)) \, \mathrm{d}s \\
&\quad + \frac{t}{1-\alpha \eta} \int_0^1 (1-s) f(s, x(s)) \, \mathrm{d}s \\
&\leqslant \max_{0 \leqslant t \leqslant 1} \int_0^t (t-s) f(s, x(s)) \, \mathrm{d}s + \frac{\alpha t}{1-\alpha \eta} \int_0^\eta (\eta - s) f(s, x(s)) \, \mathrm{d}s \\
&\quad + \frac{t}{1-\alpha \eta} \int_0^1 (1-s) f(s, x(s)) \, \mathrm{d}s \\
&\leqslant \max_{0 \leqslant t \leqslant 1} \left(\int_0^t (t-s) \, \mathrm{d}s + \frac{\alpha t}{1-\alpha \eta} \int_0^\eta (\eta - s) \, \mathrm{d}s \right. \\
&\quad \left. + \frac{t}{1-\alpha \eta} \int_0^1 (1-s) \, \mathrm{d}s \right) \cdot mc \\
&= \frac{2 - \alpha \eta + \alpha \eta^2}{2(1 - \alpha \eta)} \cdot mc \\
&= c.
\end{aligned}
$$

Hence, $T : \overline{P}_c \to \overline{P}_c$. In the same way, if $x \in \overline{P}_c$, then assumption (D1) yields $f(t, x(t)) < ma$, $0 \leqslant t \leqslant 1$. As in the argument above, we can obtain that $T : \overline{P}_a \to P_a$. Therefore condition (C2) of Theorem 3.48 is satisfied.

To check condition (C1) of Theorem 3.48, we choose $x(t) = b/\gamma$, $0 \leqslant t \leqslant 1$. It is easy to see that $x(t) = b/\gamma \in P(\psi, b, b/\gamma)$ and $\psi(x) = \psi(b/\gamma) > b$, and so $\{x \in P(\psi, b, b/\gamma): \psi(x) > b\} \neq \emptyset$. Hence, if $x \in P(\psi, b, b/\gamma)$, then $b \leqslant x(t) \leqslant b/\gamma$, $\eta \leqslant t \leqslant 1$. From assumption (D2), we have $f(t, x(t)) \geqslant b/\delta$, $\eta \leqslant t \leqslant 1$, and by the definition of ψ and the cone P, we have to distinguish two cases, (i) $\psi(Tx) = Tx(\eta)$ and (ii) $\psi(Tx) = Tx(1)$.

In case (i) we have

$$
\begin{aligned}
\psi(Tx) &= Tx(\eta) \\
&= -\int_0^\eta (\eta - s) f(s, x(s)) \, \mathrm{d}s - \frac{\alpha \eta}{1-\alpha \eta} \int_0^\eta (\eta - s) f(s, x(s)) \, \mathrm{d}s \\
&\quad + \frac{\eta}{1-\alpha \eta} \int_0^1 (1-s) f(s, x(s)) \, \mathrm{d}s \\
&\leqslant -\frac{1}{1-\alpha \eta} \int_0^\eta (\eta - s) f(s, x(s)) \, \mathrm{d}s + \frac{\eta}{1-\alpha \eta} \int_0^1 (1-s) f(s, x(s)) \, \mathrm{d}s \\
&= -\frac{\eta}{1-\alpha \eta} \int_0^\eta f(s, x(s)) \, \mathrm{d}s + \frac{1}{1-\alpha \eta} \int_0^\eta s f(s, x(s)) \, \mathrm{d}s
\end{aligned}
$$

$$+ \frac{\eta}{1 - \alpha\eta} \int_0^1 f(s, x(s)) \, ds - \frac{\eta}{1 - \alpha\eta} \int_0^1 s f(s, x(s)) \, ds$$

$$= \frac{\eta}{1 - \alpha\eta} \int_\eta^1 f(s, x(s)) \, ds + \frac{1}{1 - \alpha\eta} \int_0^\eta s f(s, x(s)) \, ds$$

$$- \frac{\eta}{1 - \alpha\eta} \int_0^1 s f(s, x(s)) \, ds$$

$$> \frac{\eta}{1 - \alpha\eta} \int_\eta^1 f(s, x(s)) \, ds - \frac{\eta}{1 - \alpha\eta} \int_\eta^1 s f(s, x(s)) \, ds$$

$$\geqslant \frac{\eta}{1 - \alpha\eta} \int_\eta^1 (1 - s) \, ds \cdot \frac{b}{\delta}$$

$$= \frac{\eta(1 - \eta)^2}{2(1 - \alpha\eta)} \cdot \frac{b}{\delta}$$

$$\geqslant b.$$

In case (ii), we have

$$\psi(Tx) = Tx(1)$$

$$= - \int_0^1 (1 - s) f(s, x(s)) \, ds - \frac{\alpha}{1 - \alpha\eta} \int_0^\eta (\eta - s) f(s, x(s)) \, ds$$

$$+ \frac{1}{1 - \alpha\eta} \int_0^1 (1 - s) f(s, x(s)) \, ds$$

$$= \frac{\alpha\eta}{1 - \alpha\eta} \int_0^1 (1 - s) f(s, x(s)) \, ds - \frac{\alpha}{1 - \alpha\eta} \int_0^\eta (\eta - s) f(s, x(s)) \, ds$$

$$= \frac{\alpha\eta}{1 - \alpha\eta} \int_0^1 f(s, x(s)) \, ds - \frac{\alpha\eta}{1 - \alpha\eta} \int_0^1 s f(s, x(s)) \, ds$$

$$- \frac{\alpha\eta}{1 - \alpha\eta} \int_0^\eta f(s, x(s)) \, ds + \frac{\alpha}{1 - \alpha\eta} \int_0^\eta s f(s, x(s)) \, ds$$

$$= \frac{\alpha\eta}{1 - \alpha\eta} \int_\eta^1 f(s, x(s)) \, ds - \frac{\alpha\eta}{1 - \alpha\eta} \int_0^1 s f(s, x(s)) \, ds$$

$$+ \frac{\alpha}{1 - \alpha\eta} \int_0^\eta s f(s, x(s)) \, ds$$

$$> \frac{\alpha\eta}{1 - \alpha\eta} \int_\eta^1 f(s, x(s)) \, ds - \frac{\alpha\eta}{1 - \alpha\eta} \int_\eta^1 s f(s, x(s)) \, ds$$

$$= \frac{\alpha\eta}{1 - \alpha\eta} \int_\eta^1 (1 - s) f(s, x(s)) \, ds$$

$$\geqslant \frac{\alpha\eta(1-\eta)^2}{2(1-\alpha\eta)} \cdot \frac{b}{\delta}$$

$$\geqslant \delta \cdot \frac{b}{\delta} = b,$$

i.e.

$$\psi(Tx) > b, \quad \forall x \in P(\psi, b, b/\gamma).$$

This shows that condition (C1) of Theorem 3.48 is satisfied. We finally show that (C3) of Theorem 3.48 also holds. Suppose that $x \in P(\psi, b, c)$ with $\|Tx\| > b/\gamma$. Then we have

$$\psi(Tx) = \min_{\eta \leqslant t \leqslant 1} Tx(t) \geqslant \gamma \cdot \|Tx\| > \gamma \cdot b/\gamma = b.$$

So, condition (C3) of Theorem 3.48 is satisfied. Therefore an application of Theorem 3.48 completes the proof. $\qquad\qquad\qquad\qquad\qquad\qquad\qquad\qquad\qquad\qquad\qquad\qquad\qquad\Box$

Theorem 3.49 was taken from [42]. For the BVP (3.91)–(3.92) Ma in [76] studied multiplicity results at resonance, i.e. when $\alpha\eta = 1$, by developing the methods of lower and upper solutions when the lower and upper solutions are well ordered as well as opposite ordered, by the connectivity properties of the solution set of parameterized families of compact vector fields.

Let us discuss some extensions of the above results.

Consider the following m-point BVP

$$x'' + a(t)f(x) = 0, \quad t \in (0, 1), \tag{3.93}$$

$$x(0) = 0, \qquad x(1) = \sum_{i=1}^{m-2} \alpha_i x(\xi_i) \tag{3.94}$$

where $\alpha_i \geqslant 0$ for $i = 1, 2, \ldots, m - 3$ and $\alpha_{m-2} > 0$, $0 < \xi_1 < \xi_2 < \cdots < \xi_{m-2} < 1$, $\sum_{i=1}^{m-2} \alpha_i \xi_i < 1$. By using the Krasnosel'skii's fixed point theorem in cones Ma in [72] proved the existence of positive solution of the BVP (3.93)–(3.94) under the assumptions that $a \in C([0, 1], [0, \infty))$, and there exists $x_0 \in [\xi_{m-2}, 1]$ such that $a(x_0) > 0$ and $\sum_{i=1}^{m-2} \alpha_i \xi_i < 1$, either in superlinear or sublinear case. The steps of the proof are parallel to that of the proof of Theorem 3.38.

A natural generalization of the boundary condition (3.94) is

$$x(0) = 0, \qquad \int_a^b h(t)x(t)\,dt = x(1), \tag{3.95}$$

where $[a, b] \subset (0, 1)$, $h \in C([a, b], [0, \infty))$, $\int_a^b th(t)\,dt \neq 1$ and $b\int_a^b h(t)\,dt < 1$. The existence of positive solutions for the BVP (3.80), (3.95) is studied by Ma [73] when f is either superlinear or sublinear.

Another result for the following m-point BVP

$$x'' + a(t)f(x) = 0, \quad t \in (0, 1), \tag{3.96}$$

$$x'(0) = \sum_{i=1}^{m-2} \beta_i x'(\xi_i), \qquad x(1) = \sum_{i=1}^{m-2} \alpha_i x(\xi_i) \tag{3.97}$$

where $\alpha_i, \beta_i \geqslant 0$ and $\sum_{i=1}^{m-2} \alpha_i < 1$, $\sum_{i=1}^{m-2} \beta_i < 1$, was given by Ma and Castaneda in [71] also by using the Krasnosel'skii's fixed point theorem in cones and either in superlinear or sublinear case. These results were generalized and improved by Liu [58] by applying topological degree methods.

The result for the BVP (3.80), (3.88) was generalized by Guo et al. in [23], where they studied the following BVP (3.80), (3.98) where (3.98) stands for the boundary condition

$$x(0) = 0, \qquad x(1) - \sum_{i=1}^{m-2} \alpha_i x(\xi_i) = b, \tag{3.98}$$

where $b, \alpha_i > 0$ $(i = 1, 2, \ldots, m - 2)$. In this problem $a(t)$ is allowed to be singular at $t = 0, 1$.

The m-point BVP

$$x'' + f(t, x) = 0, \quad t \in (0, 1), \tag{3.99}$$

$$x(0) = 0, \qquad x(1) = \sum_{i=1}^{m-2} \alpha_i x(\xi_i) \tag{3.100}$$

where $\alpha_i > 0$ for $i = 1, 2, \ldots, m - 2$, $0 < \xi_1 < \xi_2 < \cdots < \xi_{m-2} < 1$, $\sum_{i=1}^{m-2} \alpha_i \xi_i < 1$ and $f : [0, 1] \times [0, \infty) \to [0, \infty)$ a continuous function, was studied by Liu et al. in [64] by using the Legget–Williams fixed point theorem and Green's functions.

Existence and multiplicity results for positive solutions for the m-point BVP

$$\bigl(p(t)x'\bigr)' - q(t)x + f(t, x) = 0, \quad 0 < t < 1,$$

$$ax(0) - bp(0)x'(0) = \sum_{i=1}^{m-2} \alpha_i x(\xi_i),$$

$$cx(1) - dp(1)x'(1) = \sum_{i=1}^{m-2} \beta_i x(\xi_i),$$

where $p, q \in C([0, 1], (0, \infty))$, $a, b, c, d \in [0, \infty)$ are given by Ma in [77]. See also [75] for results for superlinear semipositone m-point BVPs.

Three-point BVP of higher-order ordinary differential equations was studied by Liu and Ge in [65] where they consider BVP consisting of the equation

$$x^{(n)} + \lambda a(t) f(x(t)) = 0, \quad t \in (0, 1)$$

with one of the following boundary value conditions:

$$x(0) = \alpha x(\eta), \qquad x(1) = \beta x(\eta),$$
$$x^{(i)}(0) = 0, \quad i = 1, 2, \ldots, n - 2,$$

and

$$x^{(n-2)}(0) = \alpha x^{(n-2)}(\eta), \qquad x^{(n-2)}(1) = \beta x^{(n-2)}(\eta),$$
$$x^{(i)}(0) = 0, \quad i = 1, 2, \ldots, n - 3,$$

where $\eta \in (0, 1)$, $\alpha \geqslant 0$, $\beta \geqslant 0$, and $a : (0, 1) \to \mathbb{R}$ may change sign, $f(0) > 0$, and $\lambda > 0$ is a parameter.

For other recent results on positive solutions for nonlocal BVP the interested reader is referred to [13,18,46,48,51,64,67,68] and the references cited therein.

3.10. *Positive solutions of nonlocal boundary value problems with dependence on the first-order derivative*

All the results in the previous section were proved under the assumption that the first derivative x' is not involved explicitly in the nonlinear term. In this section we are concerned with the existence of positive solutions for the second-order three-point BVP

$$x''(t) + f(t, x(t), x'(t)), \quad 0 < t < 1, \tag{3.101}$$
$$x(0) = 0, \qquad x(1) = \alpha x(\eta), \tag{3.102}$$

where $f : [0, 1] \times [0, \infty) \times \mathbb{R} \to [0, \infty)$ is continuous, $\alpha > 0$, $0 < \eta < 1$ and $1 - \alpha \eta > 0$.

To show the existence of positive solutions to BVP (3.101)–(3.102), we use an extension of Krasnosel'skii's fixed point theorem in cones, proved in [19]. To state this fixed point theorem some notations are necessary.

Let X be a Banach space and $K \subset X$ a cone. Suppose $\alpha, \beta : X \to \mathbb{R}^+$ are two convex functionals satisfying

$$\alpha(\lambda x) = |\lambda| \alpha(x), \quad \beta(\lambda x) = |\lambda| \beta(x), \quad \lambda \in \mathbb{R},$$

and

$$\|x\| \leqslant M \max\{\alpha(x), \beta(x)\} \quad \text{for } x \in X \quad \text{and}$$
$$\alpha(x) \leqslant \alpha(y) \qquad\qquad\qquad \text{for } x, y \in K, \ x \leqslant y,$$

where $M > 0$ is a constant.

THEOREM 3.50. *Let* $r_2 > r_1 > 0$, $L > 0$ *be constants and*

$$\Omega_i = \{x \in X : \alpha(x) < r_i, \ \beta(x) < L\}, \quad i = 1, 2,$$

two bounded open sets in X. *Set*

$$D_i = \{x \in X : \alpha(x) = r_i\}.$$

Assume $T : K \to K$ *is a completely continuous operator satisfying*
   (a) $\alpha(Tu) < r_1, u \in D_1 \cap K$; $\alpha(Tu) > r_2, u \in D_2 \cap K$;
   (b) $\beta(Tu) < L, u \in K$;
   (c) *there is a* $p \in (\Omega \cap K) \setminus \{0\}$ *such that* $\alpha(p) \neq 0$ *and* $\alpha(x + \lambda p) \geqslant \alpha(x)$ *for all* $x \in K$
      and $\lambda \geqslant 0$.
Then T *has at least one fixed point in* $(\Omega_2 \setminus \overline{\Omega}_1) \cap K$.

From Lemmas 3.34 and 3.37 we know that the unique solution of the BVP

$$x'' + y(t) = 0, \quad 0 < t < 1, \qquad x(0) = 0, \qquad x(1) = \alpha x(\eta)$$

is given by

$$x(t) = -\int_0^t (t - s) y(s)\, ds - \frac{\alpha t}{1 - \alpha \eta} \int_0^\eta (\eta - s) y(s)\, ds$$

$$+ \frac{t}{1 - \alpha \eta} \int_0^1 (1 - s) y(s)\, ds$$

and satisfies

$$\min_{t \in [\eta, 1]} x(t) \geqslant \gamma \|x\|,$$

where $\gamma = \min\{\alpha \eta, \frac{\alpha(1 - \eta)}{1 - \alpha \eta}, \eta\}$.
   Moreover it is easy to see that, if $1 - \alpha \eta \neq 0$, the Green function for the BVP

$$-x'' = 0, \quad 0 < t < 1, \qquad x(0) = 0, \qquad x(1) = \alpha x(\eta)$$

is given by

$$
G(t,s) = \begin{cases}
\dfrac{s[(1-t)-\alpha(\eta-t)]}{1-\alpha\eta}, & s \leqslant t,\ s \leqslant \eta, \\[2mm]
\dfrac{s(1-t)+\alpha\eta(t-s)}{1-\alpha\eta}, & \eta \leqslant s \leqslant t, \\[2mm]
\dfrac{t[(1-s)-\alpha(\eta-s)]}{1-\alpha\eta}, & t \leqslant s \leqslant \eta, \\[2mm]
\dfrac{t(1-s)}{1-\alpha\eta}, & t \leqslant s,\ s \geqslant \eta.
\end{cases}
$$

Let $X = C^1([0,1], \mathbb{R})$ with $\|x\| = \max_{0\leqslant t\leqslant 1}[x^2(t) + (x'(t))^2]^{1/2}$, and $K = \{x \in X\colon x(t) \geqslant 0,\ x$ is concave on $[0,1]\}$. Define functionals $\alpha(x) = \max_{0\leqslant t\leqslant 1}|x(t)|$ and $\beta(x) = \max_{0\leqslant t\leqslant 1}|x'(t)|$ for each $x \in X$. Then $\|x\| \leqslant \sqrt{2}\max\{\alpha(x), \beta(x)\}$ and

$$\alpha(\lambda x) = |\lambda|\alpha(x), \quad \beta(\lambda x) = |\lambda|\beta(x), \quad x \in X,\ \lambda \in \mathbb{R},$$

$$\alpha(x) \leqslant \alpha(y), \quad \text{for } x, y \in K,\ x \leqslant y.$$

We set

$$M = \max_{0\leqslant t\leqslant 1}\int_0^1 G(t,s)\,\mathrm{d}s, \qquad m = \max_{0\leqslant t\leqslant 1}\int_\eta^1 G(t,s)\,\mathrm{d}s,$$

$$Q = \frac{3-\alpha\eta+\alpha\eta^2}{2(1-\alpha\eta)}.$$

THEOREM 3.51. *Let $f:[0,1] \times [0,\infty) \times \mathbb{R} \to [0,\infty)$ be a continuous function, $\alpha > 0$, $0 < \eta < 1$ and $1 - \alpha\eta > 0$. Suppose that there are $L > b > \gamma b > c > 0$ such that $f(t,u,v)$ satisfies the growth conditions:*
   (1) $f(t,u,v) < c/M$ for $(t,u,v) \in [0,1] \times [0,c] \times [-L,L]$;
   (2) $f(t,u,v) \geqslant b/m$ for $(t,u,v) \in [0,1] \times [\gamma b, b] \times [-L,L]$;
   (3) $f(t,u,v) < L/Q$ for $(t,u,v) \in [0,1] \times [0,b] \times [-L,L]$.
Then the BVP (3.101)–(3.102) has at least one positive solution $y(t)$ satisfying

$$c < \alpha(x) < b, \qquad |y'(t)| < L.$$

PROOF. Take

$$\Omega_1 = \left\{x \in X\colon |x(t)| < c,\ |x'(t)| < L\right\},$$

$$\Omega_2 = \left\{x \in X\colon |x(t)| < b,\ |x'(t)| < L\right\}$$

two bounded open sets in X, and

$$D_1 = \left\{x \in X\colon \alpha(x) = c\right\}, \qquad D_2 = \left\{x \in X\colon \alpha(x) = b\right\}.$$

Let

$$f^*(t, u, v) = \begin{cases} f(t, u, v), & (t, u, v) \in [0, 1] \times [0, b] \times (-\infty, \infty), \\ f(t, b, v), & (t, u, v) \in [0, 1] \times (b, \infty) \times (-\infty, \infty), \end{cases}$$

and

$$f_1(t, u, v) = \begin{cases} f^*(t, u, v), & (t, u, v) \in [0, 1] \times [0, \infty) \times [-L, L], \\ f^*(t, u, -L), & (t, u, v) \in [0, 1] \times [0, \infty) \times (-\infty, -L], \\ f^*(t, u, L), & (t, u, v) \in [0, 1] \times [0, \infty) \times [L, \infty). \end{cases}$$

Then $f_1 \in C([0, 1] \times [0, \infty) \times \mathbb{R}, \mathbb{R}^+)$. Define

$$(Tx)(t) = \int_0^1 G(t, s) f_1(s, x(s), x'(s)) \, ds.$$

Obviously, $T : K \to K$ is completely continuous, and there is a $p \in (\Omega_2 \cap K) \setminus \{0\}$ such that $\alpha(x + \lambda p) \geq \alpha(x)$ for all $x \in K$ and $\lambda \geq 0$. For $x \in D_1 \cap K$, $\alpha(x) = c$. From (1), we get

$$\alpha(Tx) = \max_{t \in [0,1]} \left| \int_0^1 G(t, s) f_1(s, x(s), x'(s)) \, ds \right|$$

$$< \max_{t \in [0,1]} \int_0^1 G(t, s) \frac{c}{M} \, ds = \frac{c}{M} \max_{t \in [0,1]} \int_0^1 G(t, s) \, ds = c.$$

Whereas for $x \in D_2 \cap K$, $\alpha(x) = b$. We have $x(t) \geq \gamma \alpha(x) = \gamma b$ for $t \in [\eta, 1]$. So, from (2), we get

$$\alpha(Tx) = \max_{t \in [0,1]} \left| \int_0^1 G(t, s) f_1(s, x(s), x'(s)) \, ds \right|$$

$$> \max_{t \in [0,1]} \left| \int_\eta^1 G(t, s) f_1(s, x(s), x'(s)) \, ds \right|$$

$$> \max_{t \in [0,1]} \int_\eta^1 G(t, s) \frac{b}{m} \, ds = \frac{b}{m} \max_{t \in [0,1]} \int_\eta^1 G(t, s) \, ds = b.$$

For $x \in K$, from (3), we get

$$\beta(Tx) = \max_{t \in [0,1]} \left| - \int_0^t f_1(s, x(s), x'(s)) \, ds \right.$$

$$+ \frac{1}{1 - \alpha\eta} \int_0^1 (1 - s) f_1(s, x(s), x'(s)) \, ds$$

$$\left. - \frac{\alpha}{1 - \alpha\eta} \int_0^\eta (\eta - s) f_1(s, x(s), x'(s)) \, ds \right|$$

$$< \left[1 + \frac{1}{1-\alpha\eta} \int_0^1 (1-s)\,ds + \frac{\alpha}{1-\alpha\eta} \int_0^\eta (\eta-s)\,ds \right] \frac{L}{Q}$$

$$= \frac{3 - \alpha\eta + \alpha\eta^2}{2(1-\alpha\eta)} \frac{L}{Q} = L.$$

Theorem 3.50 implies there is $y \in (\Omega_2 \setminus \overline{\Omega}_1) \cap K$ such that $y = Ty$. So, y is a positive solution for BVP (3.101)–(3.102) satisfying

$$c < \alpha(x) < b, \qquad |y'(t)| < L.$$

Thus, the proof of the theorem is complete. $\qquad\qquad\qquad\qquad\qquad\qquad\qquad\qquad$ $\square$

Theorem 3.51 was proved in [19], where one more result was proved in the case when $0 < \alpha \leqslant 1$ and $\eta \in (0, 1)$, under the assumptions (1), (2) and the following one

    (4) $f(t, u, v) < L^2/(2b)$ for $(t, u, v) \in [0, 1] \times [0, b] \times [-L, L]$.

Consider now a three-point BVP

$$\big(p(t)x'(t)\big)' = f\big(t, x(t), x'(t)\big), \quad t \in [0, 1], \tag{3.103}$$

$$x'(0) = 0, \qquad x(1) = x(\eta) \tag{3.104}$$

where $f : [0, 1] \times \mathbb{R} \times \mathbb{R} \to \mathbb{R}$ is continuous, $p : [0, 1] \to \mathbb{R}$ is a positive continuous differentiable function, $\min_{t \in [0,1]} p(t) := p_1 > 0$ and $\eta \in (0, 1)$. For this problem the linear operator $Lx(t) = (p(t)x'(t))'$ is not invertible, so the three-point BVP (3.103)–(3.104) is a resonance problem. By using a theorem of a fixed point index for A-proper semilinear operators Bai and Fang [6] proved the following theorem.

THEOREM 3.52. *Suppose*

    (1) $f : [0, 1] \times \mathbb{R} \times \mathbb{R} \to \mathbb{R}$ *is continuous and there exist constants $a > 0$, $b > 0$, and $0 < c < p_1^2/(2\|p\|_\infty)$ such that*

$$|f(t, x, y)| \leqslant a + b|x| + c|y|, \quad \forall t \in [0, 1], \ x, y \in \mathbb{R}.$$

    (2) *There exists $M > 0$ such that*

$$f(t, x, 0) > 0, \quad \forall t \in [0, 1], \ x > M.$$

    (3) $f(t, x, y) \geqslant 0$, $\forall t \in [0, 1]$, $y \in \mathbb{R}$ *and there exist constants $0 < \sigma < b$ and $\varepsilon > 0$ such that*

$$f(t, x, y) \geqslant \sigma x, \quad \forall t \in [0, 1], \ 0 \leqslant x \leqslant \varepsilon, \ y \in \mathbb{R}.$$

Then the BVP (3.103)–(3.104) has at least one positive solution.

A more general nonlocal BVP was studied by Palamides [87]. He consider the following BVP

$$x''(t) = f\bigl(t, x(t), x'(t)\bigr), \quad t \in (0, 1),$$

$$\alpha x(0) - \beta x'(0) = 0, \qquad x(1) = \sum_{i=0}^{m-2} \alpha_i x(\xi_i)$$

with $\alpha \geqslant 0$, $\beta > 0$. Existence of positive solution is given, under superlinear and/or sublinear growth rate in f. A different method is employed based on analysis of the corresponding vector field on the (x, x')-face plane and Kneser's property of the solutions funnel.

Finally positive solutions for $2n$th order nonlocal BVP was studied by Guo et al. in [22] and [20] where they consider BVPs consisting of the equation

$$x^{(2n)}(t) = f\bigl(t, x(t), x''(t), \dots, x^{2(n-1)}(t)\bigr), \quad 0 \leqslant t \leqslant 1,$$

and one of the following boundary conditions:

$$x^{(2i)}(0) = 0, \quad x^{(2i)}(1) = \sum_{j=1}^{m-2} k_{ij} x^{(2i)}(\xi_j), \quad 0 \leqslant i \leqslant n - 1,$$

and

$$x^{(2i)}(0) - \beta_i x^{(2i+1)}(0) = 0, \quad x^{(2i)}(1) = \sum_{j=1}^{m-2} k_{ij} x^{(2i)}(\xi_j), \quad 0 \leqslant i \leqslant n - 1.$$

They use the Legget–Williams fixed point theorem and a fixed point theorem in double cones respectively.

For other recent results we refer to [49,50,52,53].

3.11. *Positive solutions of nonlocal problems for p-Laplacian*

The purpose of this section is to establish the existence of positive solutions to the following three-point boundary value problem for p-Laplacian

$$\bigl(g(u')\bigr)' + a(t)f(u) = 0, \quad \text{for } 0 < t < 1, \tag{3.105}$$

$$u(0) = 0, \quad \text{and} \quad u(v) = u(1), \tag{3.106}$$

where $g(v) = |v|^{p-2}v$, with $p > 1$, and $v \in (0, 1)$.

We will prove the existence of at least three positive pseudo-symmetric solutions of the BVP (3.105)–(3.106) where we now define what we mean by a pseudo-symmetric function.

DEFINITION 3.53. For $v \in (0, 1)$ a function $u \in C[0, 1]$ is said to be pseudo-symmetric if u is symmetric over the interval $[v, 1]$. That is, for $t \in [v, 1]$ we have $u(t) = u(1 - (t - v))$.

In this setting we are able to verify that for all x in our cone $\sigma_u = \frac{1+v}{2}$.

In our approach we will use a new multiple fixed point theorem, now called the five functionals fixed point theorem, obtained by Avery [2], which generalized the Leggett–Williams fixed point theorem [55] in terms of functionals rather than norms. We state first the five functional fixed point theorem.

Let γ, β, θ be nonnegative, continuous, convex functionals on P and α, ψ be nonnegative, continuous, concave functionals on P. Then, for nonnegative real numbers h, a, b, d and c, we define the convex sets,

$$P(\gamma, c) = \{x \in P \colon \gamma(x) < c\},$$

$$P(\gamma, \alpha, a, c) = \{x \in P \colon a \leqslant \alpha(x), \ \gamma(x) \leqslant c\},$$

$$Q(\gamma, \beta, d, c) = \{x \in P \colon \beta(x) \leqslant d, \ \gamma(x) \leqslant c\},$$

$$P(\gamma, \theta, \alpha, a, b, c) = \{x \in P \colon a \leqslant \alpha(x), \ \theta(x) \leqslant b, \ \gamma(x) \leqslant c\},$$

and

$$Q(\gamma, \beta, \psi, h, d, c) = \{x \in P \colon h \leqslant \psi(x), \ \beta(x) \leqslant d, \ \gamma(x) \leqslant c\}.$$

THEOREM 3.54. *Let P be a cone in a real Banach space E. Suppose there exist positive numbers c and M, nonnegative, continuous, concave functionals α and ψ on P, and nonnegative, continuous, convex functionals $\gamma, \beta,$ and θ on P, with*

$$\alpha(x) \leqslant \beta(x) \quad and \quad \|x\| \leqslant M\gamma(x)$$

for all $x \in \overline{P(\gamma, c)}$. Suppose

$$A : \overline{P(\gamma, c)} \to \overline{P(\gamma, c)}$$

is completely continuous and there exist nonnegative numbers h, a, k, b, with $0 < a < b$ such that:
    (i) *$\{x \in P(\gamma, \theta, \alpha, b, k, c) \colon \alpha(x) > b\} \neq \emptyset$ and $\alpha(Ax) > b$ for $x \in P(\gamma, \theta, \alpha, b, k, c)$;*
    (ii) *$\{x \in Q(\gamma, \beta, \psi, h, a, c) \colon \beta(x) < a\} \neq \emptyset$ and $\beta(Ax) < a$ for $x \in Q(\gamma, \beta, \psi, h, a, c)$;*
    (iii) *$\alpha(Ax) > b$ for $x \in P(\gamma, \alpha, b, c)$ with $\theta(Ax) > k$;*
    (iv) *$\beta(Ax) < a$ for $x \in Q(\gamma, \beta, a, c)$ with $\psi(Ax) < h$.*
Then A has at least three fixed points $x_1, x_2, x_3 \in \overline{P(\gamma, c)}$ such that,

$$\beta(x_1) < a, \qquad b < \alpha(x_2),$$

and

$$a < \beta(x_3) \quad with \quad \alpha(x_3) < b.$$

Assume $f:[0,\infty) \to [0,\infty)$ is continuous, and let $a:[0,1] \to [0,\infty)$ be continuous. Let $\nu \in (0,1)$, and let $E = C[0,1]$ be the Banach space with the sup-norm, $\|u\| = \sup\{|u(x)|: 0 \leqslant x \leqslant 1\}$, and define the cone $P \subset E$ by

$$P = \{u \in E: u(0) = 0, \ u \text{ is concave, and } u \text{ is symmetric on } [\nu, 1]\}.$$

Hereafter, suppose $\nu \in (0,1)$, with $\int_0^\nu a(r)\,dr > 0$, and we will choose a $\delta \in (0, \nu)$, such that $\int_\delta^\nu a(t)\,dt > 0$, and the constants M^* and m^*, defined by

$$M^* = \int_0^\nu G\left(\int_s^\sigma a(r)\,dr\right) ds,$$

and

$$m^* = \int_0^\delta G\left(\int_s^\sigma a(r)\,dr\right) ds,$$

satisfy the inequality $\nu < \frac{m^*}{M^*}$. Trivially, we have $\delta < \frac{m^*}{M^*}$. We let

$$m = \int_0^\delta G\left(\int_\delta^\nu a(r)\,dr\right) ds = \delta G\left(\int_\delta^\nu a(r)\,dr\right)$$

and

$$M = \int_0^\sigma G\left(\int_s^\sigma a(r)\,dr\right) ds,$$

as well as,

$$h_1 = \int_\delta^\nu G\left(\int_s^\sigma a(r)\,dr\right) ds,$$

$$h_2 = \int_0^\delta G\left(\int_s^\delta a(r)\,dr\right) ds,$$

and

$$h_3 = \int_0^\delta G\left(\int_\delta^\sigma a(r)\,dr\right) ds.$$

Define the nonnegative, continuous, concave functionals α, ψ, and the nonnegative, continuous, convex functionals β, θ, γ on the cone P by:

$$\gamma(x) = \theta(x) := \max_{t\in[0,1]} x(t) = x(\sigma),$$

$$\beta(x) := \max_{t\in[\delta,\nu]} x(t) = x(\nu),$$

$$\alpha(x) = \psi(x) := \min_{t \in [\delta, v]} x(t) = x(v).$$

In our main result, we will make use of the following lemma. The lemma is easily proved using the concavity and the pseudo-symmetry of all $u \in P$.

LEMMA 3.55. *Let $x \in P$. Then*
  (D1) $u(\delta) \geqslant \delta u(1) = \delta u(v)$, *and*
  (D2) $\sigma u(v) \geqslant v u(\sigma) = v\|u\|$.

We are now ready to apply the five functionals fixed point theorem to an operator A to give sufficient conditions for the existence of at least three positive pseudo-symmetric solutions to (3.105)–(3.106).

THEOREM 3.56. *Assume that $v \in (0, 1)$, $a : [0, 1] \to [0, \infty)$ is a pseudo-symmetric continuous function, $\delta \in (0, v)$ such that*

$$\int_{\delta}^{v} a(t)\,dt > 0 \quad \text{with } v < \frac{m^*}{M^*},$$

and $f : [0, \infty) \to [0, \infty)$ is continuous. Let $0 < a < b < c$, with $ch_2 < aM$, and suppose that f satisfies the following conditions:
  (i) $f(x) > g(\frac{b}{m})$ *for all $b \leqslant x \leqslant \frac{b}{\delta}$,*
  (ii) $f(x) < g(\frac{Ma - ch_2}{M(h_1 + h_3)})$ *for all $a\delta \leqslant x \leqslant \frac{a\sigma}{v}$, and*
  (iii) $f(x) \leqslant g(\frac{c}{M})$ *for all $0 \leqslant x \leqslant c$.*
Then the three-point boundary value problem (3.105)–(3.106) has at least three positive pseudo-symmetric solutions u_1, u_2, u_3 such that

$$\max_{t \in [\delta, v]} u_1(t) < a < \max_{t \in [\delta, v]} u_2(t) \quad \text{and} \quad \min_{t \in [\delta, v]} u_2(t) < b < \min_{t \in [\delta, v]} u_3(t).$$

PROOF. Define the completely continuous operator A on P by

$$Au(t) = w(t) = \begin{cases} \displaystyle\int_0^t G\left(\int_s^\sigma a(r) f\big(u(r)\big)\,dr\right) ds, & 0 \leqslant t \leqslant \sigma, \\[3mm] \displaystyle w(v) + \int_t^1 G\left(\int_\sigma^s a(r) f\big(u(r)\big)\,dr\right) ds, & \sigma \leqslant t \leqslant 1, \end{cases}$$

where

$$\sigma = \frac{v + 1}{2}.$$

We first note that for $u \in P$ we have $Au(t) \geqslant 0$, $Au(0) = 0$, and applying the Fundamental Theorem of Calculus we have that Au is concave. Furthermore, for $t \in [v, 1]$

$$Au(t) = Au\big(1 - (t - v)\big).$$

Consequently, $Au \in P$, that is, $A : P \to P$. Moreover, since

$$\left(g\big((Au)'\big)\right)'(t) = -a(t) f\big(u(t)\big) \leqslant 0 \quad \text{for all } 0 < t < 1,$$

we have that all fixed points of A are solutions of (3.105)–(3.106). Thus we set out to verify that the operator A satisfies the five functionals fixed point theorem which will prove the existence of three fixed points of A which satisfy the conclusion of the theorem.

If $u \in \overline{P(\gamma, c)}$, then $\gamma(u) = \max_{t \in [0,1]} u(t) \leqslant c$. Thus

$$\gamma(Au) = Au(\sigma)$$

$$= \int_0^\sigma G\left(\int_s^\sigma a(r) f\big(u(r)\big) \, dr\right) ds$$

$$\leqslant \int_0^\sigma G\left(\int_s^\sigma a(r) g\left(\frac{c}{M}\right) dr\right) ds$$

$$= \left(\frac{c}{M}\right) \int_0^\sigma G\left(\int_s^\sigma a(r) \, dr\right) ds$$

$$= c.$$

Hence,

$$A : \overline{P(\gamma, c)} \to \overline{P(\gamma, c)}.$$

Next, let $N = \frac{m^* + \delta M^*}{2}$. Thus $m^* > N > \delta M^*$, and if we define

$$u_P(t) = \begin{cases} \left(\dfrac{b}{N}\right) \displaystyle\int_0^t G\left(\int_s^\sigma a(r) \, dr\right) ds, & 0 \leqslant t \leqslant \sigma, \\[1.5em] u_P(v) + \left(\dfrac{b}{N}\right) \displaystyle\int_t^1 G\left(\int_\sigma^s a(r) \, dr\right) ds, & \sigma \leqslant t \leqslant 1, \end{cases}$$

and

$$u_Q(t) = \begin{cases} \left(\dfrac{a\delta}{N}\right) \displaystyle\int_0^t G\left(\int_s^\sigma a(r) \, dr\right) ds, & 0 \leqslant t \leqslant \sigma, \\[1.5em] u_Q(v) + \left(\dfrac{a\delta}{N}\right) \displaystyle\int_t^1 G\left(\int_\sigma^s a(r) \, dr\right) ds, & \sigma \leqslant t \leqslant 1, \end{cases}$$

then, clearly $u_P, u_Q \in P$. Furthermore,

$$\alpha(u_P) = u_P(\delta) = \left(\frac{b}{N}\right) \int_0^\delta G\left(\int_s^\sigma a(r) \, dr\right) ds = \frac{b m^*}{N} > b,$$

and

$$\theta(u_P) = u_P(v) = \left(\frac{b}{N}\right) \int_0^v G\left(\int_s^\sigma a(r) \, dr\right) ds = \frac{b M^*}{N} < \frac{b}{\delta},$$

as well as,

$$\psi(u_Q) = u_Q(\delta) = \left(\frac{a\delta}{N}\right) \int_0^\delta G\left(\int_s^\sigma a(r)\,\mathrm{d}r\right)\mathrm{d}s = \frac{a\delta m^*}{N} > a\delta,$$

and

$$\beta(u_Q) = u_Q(v) = \left(\frac{a\delta}{N}\right) \int_0^v G\left(\int_s^\sigma a(r)\,\mathrm{d}r\right)\mathrm{d}s = \frac{aM^*\delta}{N} < a.$$

Therefore,

$$u_P \in \left\{ u \in P\left(\gamma, \theta, \alpha, b, \frac{b}{\delta}, c\right) : \alpha(u) > b \right\},$$

and

$$u_Q \in \left\{ u \in Q(\gamma, \beta, \psi, a\delta, a, c) : \beta(u) < a \right\},$$

hence, these sets are nonempty.

If $u \in P(\gamma, \theta, \alpha, b, \frac{b}{\delta}, c)$, then $b \leqslant u(t) \leqslant \frac{b}{\delta}$, for all $t \in [\delta, v]$, and thus by condition (i) of this theorem,

$$\alpha(Au) = Au(\delta)$$
$$= \int_0^\delta G\left(\int_s^\sigma a(r)f\big(u(r)\big)\,\mathrm{d}r\right)\mathrm{d}s$$
$$\geqslant \int_0^\delta G\left(\int_\delta^v a(r)f\big(u(r)\big)\,\mathrm{d}r\right)\mathrm{d}s$$
$$> \int_0^\delta G\left(\int_\delta^v a(r)g\left(\frac{b}{m}\right)\,\mathrm{d}r\right)\mathrm{d}s$$
$$= \left(\frac{b}{m}\right)\int_0^\delta G\left(\int_\delta^v a(r)\,\mathrm{d}r\right)\mathrm{d}s$$
$$= b.$$

Hence, condition (i) of the five functionals fixed point theorem is satisfied.

If $u \in P(\gamma, \alpha, b, c)$ with $\theta(Au) > \frac{b}{\delta}$, then by Lemma 3.55(D1), we have

$$\alpha(Au) = Au(\delta) \geqslant \delta Au(1) = \delta Au(v) = \delta\theta(Au) > b.$$

Thus, condition (iii) of the five functionals fixed point theorem is satisfied.

If $u \in Q(\gamma, \beta, \psi, a\delta, a, c)$, then $a\delta \leqslant u(t) \leqslant a$, for all $t \in [\delta, \nu]$, and thus by Lemma 3.55(D2), $a\delta \leqslant u(t) \leqslant \frac{a\sigma}{\nu}$, for all $t \in [\delta, \sigma]$. Thus by condition (ii) of this theorem,

$$\beta(Au) = Au(\nu)$$

$$= \int_0^\nu G\left(\int_s^\sigma a(r) f(u(r)) \, dr\right) ds$$

$$= \int_0^\delta G\left(\int_s^\sigma a(r) f(u(r)) \, dr\right) ds + \int_\delta^\nu G\left(\int_s^\sigma a(r) f(u(r)) \, dr\right) ds$$

$$\leqslant \int_0^\delta G\left(\int_s^\delta a(r) f(u(r)) \, dr\right) ds + \int_0^\delta G\left(\int_\delta^\sigma a(r) f(u(r)) \, dr\right) ds$$

$$\quad + \int_\delta^\nu G\left(\int_s^\sigma a(r) f(u(r)) \, dr\right) ds$$

$$< \left(\frac{c}{M}\right) \int_0^\delta G\left(\int_s^\delta a(r) \, dr\right) ds$$

$$\quad + \left(\frac{Ma - ch_2}{M(h_1 + h_3)}\right) \int_0^\delta G\left(\int_\delta^\sigma a(r) \, dr\right) ds$$

$$\quad + \left(\frac{Ma - ch_2}{M(h_1 + h_3)}\right) \int_\delta^\nu G\left(\int_s^\sigma a(r) \, dr\right) ds$$

$$= \frac{ch_2}{M} + \frac{(Ma - ch_2)h_3}{M(h_1 + h_3)} + \frac{(Ma - ch_2)h_1}{M(h_1 + h_3)}$$

$$= a.$$

Hence, condition (ii) of the five functionals fixed point theorem is satisfied.

If $u \in Q(\gamma, \beta, a, c)$, with $\psi(Au) < a\delta$, then by Lemma 3.55, we have

$$\beta(Au) = Au(\nu) \leqslant \frac{Au(\delta)}{\delta} = \frac{\psi(Au)}{\delta} < a.$$

Consequently, condition (iv) of the five functionals fixed point theorem is also satisfied. Therefore, the hypotheses of the five functionals fixed point theorem 3.54 are satisfied, and there exist at least three positive pseudo-symmetric solutions $u_1, u_2, u_3 \in \overline{P(\gamma, c)}$ for the three-point boundary value problem (3.105)–(3.106) such that,

$$\beta(u_1) < a < \beta(u_2) \quad \text{and} \quad \alpha(u_2) < b < \alpha(u_3). \qquad \square$$

Theorem 3.56 was taken from [3]. Existence of positive solutions for the BVP (3.105)–(3.106) was proved also in [91] and [41], relying on Krasnosel'skii's and Legget–Williams fixed point theorems respectively. BVP with p-Laplacian and more general boundary conditions was studied in [66], where the boundary conditions $u(0) - B_0(u'(\eta)) = 0$, $u'(1) = 0$

and $u'(0) = 0$, $u(1) + B_1(u'(\eta)) = 0$ are used and in [4,5] where the m-point boundary conditions $u(0) = 0$, $u(1) = \sum_{i=1}^{m-2} a_i u(\xi_i)$ and $u'(0) = \sum_{i=1}^{m-2} b_i u'(\xi_i)$, $u(1) = \sum_{i=1}^{m-2} a_i u(\xi_i)$ are used. Finally multi-point BVP for p-Laplacian at resonance was proved by Ni and Ge in [85] and Garcia-Huidobro et al. in [17].

References

[1] A. Aftabizadeh, C. Gupta and J.-M. Xu, *Existence and uniqueness theorems for three-point boundary value problems*, SIAM J. Math. Anal. **20** (1989), 716–726.

[2] R. Avery, *A generalization of the Legget–Willaimas fixed point theorem*, Math. Res. Hot-Line **2** (1998), 9–14.

[3] R. Avery and J. Henderson, *Existence of three positive pseudo-symmetric solutions for a one-dimensional p-Laplacian*, J. Math. Anal. Appl. **277** (2003), 395–404.

[4] C. Bai and J. Fang, *Existence of multiple positive solutions for nonlinear m-point boundary value problems*, J. Math. Anal. Appl. **281** (2003), 76–85.

[5] C. Bai and J. Fang, *Existence of multiple positive solutions for nonlinear m-point boundary-value problems*, Appl. Math. Comput. **140** (2003), 297–305.

[6] C. Bai and J. Fang, *Existence of positive solutions for three-point boundary value problems at resonance*, J. Math. Anal. Appl. **291** (2004), 538–549.

[7] A. Bitsadze, *On the theory of nonlocal boundary value problems*, Russian Acad. Sci. Dokl. Math. **30** (1984), 8–10.

[8] A. Bitsadze and A. Samarskii, *On some simple generalizations of linear elliptic boundary problems*, Russian Acad. Sci. Dokl. Math. **10** (1969), 398–400.

[9] A. Boucherif and S. Bouguima, *Nonlinear second order ordinary differential equations with nonlocal boundary conditions*, Comm. Appl. Nonlinear Anal. **5** (1998), 73–85.

[10] L. Byszewski, *Theorems about existence and uniqueness of solutions of a semilinear evolution nonlocal Cauchy problem*, J. Math. Anal. Appl. **162** (1991), 494–505.

[11] L. Byszewski, *Existence and uniqueness of a classical solution to a functional-differential abstract nonlocal Cauchy problem*, J. Appl. Math. Stochastic Anal. **12** (1999), 91–97.

[12] L. Byszewski and V. Lakshmikantham, *Theorem about the existence and uniqueness of a solution of a nonlocal abstract Cauchy problem in a Banach space*, Appl. Anal. **40** (1990), 11–19.

[13] B. Calvert and C. Gupta, *Multiple solutions for a super-linear three-point boundary value problem*, Nonlinear Anal. **50** (2002), 115–128.

[14] J. Dugundji and A. Granas, *Fixed Point Theory*, Vol. I, Monographie Matematyczne, PNW, Warsawa (1982).

[15] W. Feng and J.R.L. Webb, *Solvability of three-point boundary value problems at resonance*, Proceedings of the Second World Congress of Nonlinear Analysts, Part 6 (Athens, 1996), Nonlinear Anal. **30** (1997), 3227–3238.

[16] W. Feng and J.R.L. Webb, *Solvability of m-point boundary value problems with nonlinear growth*, J. Math. Anal. Appl. **212** (1997), 467–480.

[17] M. Garcia-Huidobro, C.P. Gupta and R. Manasevich, *An m-point boundary value problem of Neumann type for a p-Laplacian like operator*, Nonlinear Anal. **56** (2004), 1071–1089.

[18] J. Graef, C. Qian and B. Yang, *A three-point boundary value problem for nonlinear fourth order differential equations*, J. Math. Anal. Appl. **287** (2003), 217–233.

[19] Y. Guo and W. Ge, *Positive solutions for three-point boundary value problems with dependence on the first order derivative*, J. Math. Anal. Appl. **290** (2004), 291–301.

[20] Y. Guo, W. Ge and Y. Gao, *Twin positive solutions for higher order m-point boundary value problems with sign changing nonlinearities*, Appl. Math. Comput. **146** (2003), 299–311.

[21] D. Guo and V. Lakshmikantham, *Nonlinear Problems in Abstract Cones*, Springer-Verlag, New York (1985).

[22] Y. Guo, X. Liu and J. Qiu, *Three positive solutions for higher order m-point boundary value problems*, J. Math. Anal. Appl. **289** (2004), 545–553.

[23] Y. Guo, W. Shan and W. Ge, *Positive solutions for second-order m-point boundary value problems*, J. Comput. Appl. Math. **151** (2003), 415–424.

[24] C. Gupta, *On a third-order three-point boundary value problem at resonance*, Differential Integral Equations **2** (1989), 1–12.

[25] C. Gupta, *Solvability of a three-point nonlinear boundary value problem for second order ordinary differential equations*, J. Math. Anal. Appl. **168** (1992), 540–551.

[26] C. Gupta, *A note on a second order three-point boundary value problem*, J. Math. Anal. Appl. **186** (1994), 277–281.

[27] C. Gupta, *A second order m-point boundary value problem at resonance*, Nonlinear Anal. **24** (1995), 1483–1489.

[28] C. Gupta, *Solvability of a generalized multi-point boundary value problem of mixed type for second order ordinary differential equations*, Proceedings of Dynamic Systems and Applications, Vol. 2 (Atlanta, GA, 1995), 215–222, Dynamic Publishers, Atlanta, GA (1996).

[29] C. Gupta, *A generalized multi-point boundary value problem for second order ordinary differential equations*, Appl. Math. Comput. **89** (1998), 133–146.

[30] C. Gupta, *A new a priori estimate for multi-point boundary-value problems*, Proceedings of the 16th Conference on Applied Mathematics (Edmond, OK, 2001), Electron. J. Differ. Equ. Conf. **7** (2001), 47–59 (electronic).

[31] C. Gupta and V. Lakshmikantham, *Existence and uniqueness theorems for a third-order three-point boundary value problem*, Nonlinear Anal. **16** (1991), 949–957.

[32] C. Gupta, S.K. Ntouyas and P.Ch. Tsamatos, *Existence results for m-point boundary value problems*, Differential Equations Dynam. Systems **2** (1994), 289–298.

[33] C. Gupta, S.K. Ntouyas and P.Ch. Tsamatos, *On an m-point boundary-value problem for second-order ordinary differential equations*, Nonlinear Anal. **23** (1994), 1427–1436.

[34] C. Gupta, S.K. Ntouyas and P.Ch. Tsamatos, *Solvability of an m-point boundary value problem for second order ordinary differential equations*, J. Math. Anal. Appl. **189** (1995), 575–584.

[35] C. Gupta, S.K. Ntouyas and P.Ch. Tsamatos, *On the solvability of some multi-point boundary value problems*, Appl. Math. **41** (1996), 1–17.

[36] C. Gupta, S.K. Ntouyas and P.Ch. Tsamatos, *Existence results for multi-point boundary value problems for second order ordinary differential equations*, Bull. Greek Math. Soc. **43** (2000), 105–123.

[37] C. Gupta and S. Trofimchuk, *Solvability of a multi-point boundary value problem and related a priori estimates*, Geoffrey J. Butler Memorial Conference in Differential Equations and Mathematical Biology (Edmonton, AB, 1996), Canad. Appl. Math. Quart. **6** (1998), 45–60.

[38] C. Gupta and S. Trofimchuk, *Existence of a solution of a three-point boundary value problem and the spectral radius of a related linear operator*, Nonlinear Anal. **34** (1998), 489–507.

[39] C. Gupta and S. Trofimchuk, *A Wirtinger type inequality and a three-point boundary value problem*, Dynam. Systems Appl. **8** (1999), 127–132.

[40] C. Gupta and S. Trofimchuk, *An interesting example for a three-point boundary value problem*, Bull. Belg. Math. Soc. Simon Stevin **7** (2000), 291–302.

[41] X. He and W. Ge, *A remark on some three-point boundary value problems for the one-dimensional p-Laplacian*, Z. Angew. Math. Mech. **82** (2002), 728–731.

[42] X. He and W. Ge, *Triple solutions for second-order three-point boundary value problems*, J. Math. Anal. Appl. **268** (2002), 256–265.

[43] V. Il'in and E. Moiseev, *Nonlocal boundary value problem of the first kind for a Sturm–Liouville operator in its differential and finite difference aspects*, Differential Equations **23** (1987), 803–810.

[44] V. Il'in and E. Moiseev, *Nonlocal boundary value problem of the second kind for a Sturm–Liouville operator*, Differential Equations **23** (1987), 979–987.

[45] G. Infante, *Eigenvalues of some nonlocal boundary-value problems*, Proc. Edinburgh Math. Soc. (2) **46** (2003), 75–86.

[46] G. Infante and J.R.L. Webb, *Nonzero solutions of Hammerstein integral equations with discontinuous kernels*, J. Math. Anal. Appl. **272** (2002), 30–42.

[47] G. Infante and J.R.L. Webb, *Positive solutions of some nonlocal boundary value problems*, Abstr. Appl. Anal. **18** (2003), 1047–1060.

[48] G.L. Karakostas and P.Ch. Tsamatos, *Positive solutions for a nonlocal boundary-value problem with increasing response*, Electron. J. Differential Equations **73** (2000), 8 pp. (electronic).

[49] G.L. Karakostas and P.Ch. Tsamatos, *Positive solutions of a boundary-value problem for second order ordinary differential equations*, Electron. J. Differential Equations **49** (2000), 9 pp. (electronic).

[50] G.L. Karakostas and P.Ch. Tsamatos, *On a nonlocal boundary value problem at resonance*, J. Math. Anal. Appl. **259** (2001), 209–218.

[51] G.L. Karakostas and P.Ch. Tsamatos, *Multiple positive solutions for a nonlocal boundary-value problem with response function quiet at zero*, Electron. J. Differential Equations **13** (2001), 10 pp. (electronic).

[52] G.L. Karakostas and P.Ch. Tsamatos, *Sufficient conditions for the existence of nonnegative solutions of a nonlocal boundary value problem*, Appl. Math. Lett. **15** (2002), 401–407.

[53] G.L. Karakostas and P.Ch. Tsamatos, *Nonlocal boundary vector value problems for ordinary differential systems of higher order*, Nonlinear Anal. **51** (2002), 1421–1427.

[54] M. Krasnosel'skii, *Positive Solutions of Operator Equations*, Noordhoff, Groningen (1964).

[55] R. Legget and L. Williams, *Multiple positive fixed points of nonlinear operators on ordered Banach spaces*, Indiana Univ. Math. J. **28** (1979), 673–688.

[56] B. Liu, *Positive solutions of a nonlinear three-point boundary value problem*, Appl. Math. Comput. **132** (2002), 11–28.

[57] B. Liu, *Positive solutions of a nonlinear three-point boundary value problem*, Comput. Math. Appl. **44** (2002), 201–211.

[58] B. Liu, *Positive solutions of a nonlinear three-point boundary value problem*, Appl. Math. Comput. **132** (2002), 11–28.

[59] B. Liu, *Solvability of multi-point boundary value problem at resonance IV*, Appl. Math. Comput. **143** (2003), 275–299.

[60] B. Liu, *Solvability of multi-point boundary value problem at resonance II*, Appl. Math. Comput. **136** (2003), 353–377.

[61] B. Liu and J. Yu, *Solvability of multi-point boundary value problems at resonance I*, Indian J. Pure Appl. Math. **33** (2002), 475–494.

[62] B. Liu and J. Yu, *Solvability of multi-point boundary value problem at resonance III*, Appl. Math. Comput. **129** (2002), 119–143.

[63] X. Liu, *Nontrivial solutions of singular m-point boundary value problems*, J. Math. Anal. Appl. **284** (2003), 576–590.

[64] X. Liu, J. Qiu and Y. Guo, *Three positive solutions for second-order m-point boundary value problems*, Appl. Math. Comput. **156** (2004), 733–742.

[65] Y. Liu and W. Ge, *Positive solutions for $(n-1, 1)$ three-point boundary value problems with coefficient that changes sign*, J. Math. Anal. Appl. **282** (2003), 816–825.

[66] Y. Liu and W. Ge, *Multiple positive solutions to a three-point boundary value problem with p-Laplacian*, J. Math. Anal. Appl. **277** (2003), 293–302.

[67] R. Ma, *Existence theorems for a second order m-point boundary value problem*, J. Math. Anal. Appl. **211** (1997), 545–555.

[68] R. Ma, *Existence theorems for a second order three-point boundary value problem*, J. Math. Anal. Appl. **212** (1997), 430–442.

[69] R. Ma, *Positive solutions of a nonlinear three-point boundary value problem*, Electron. J. Differential Equations **34** (1999), 8 pp. (electronic).

[70] R. Ma, *Multiplicity of positive solutions for second-order three-point boundary value problems*, Comput. Math. Appl. **40** (2000), 193–204.

[71] R. Ma, *Existence of solutions of a nonlinear m-point boundary value problems*, J. Math. Anal. Appl. **256** (2001), 556–567.

[72] R. Ma, *Positive solutions of a nonlinear m-point boundary value problem*, Comput. Math. Appl. **42** (2001), 755–765.

[73] R. Ma, *Positive solutions for second order functional differential equations*, Dynam. Systems Appl. **10** (2001), 215–223.

[74] R. Ma, *Existence and uniqueness of solutions to first-order three-point boundary value problems*, Appl. Math. Lett. **15** (2002), 211–216.

[75] R. Ma, *Existence of positive solutions for superlinear semipositone m-point boundary value problems*, Proc. Edinburgh Math. Soc. **46** (2003), 279–292.

[76] R. Ma, *Multiplicity results for a three-point boundary value problem at resonance*, Nonlinear Anal. **53** (2003), 777–789.

[77] R. Ma, *Multiple positive solutions for nonlinear m-point boundary value problems*, Appl. Math. Comput. **148** (2004), 249–262.

[78] R. Ma and N. Castaneda, *Positive solutions for second-order three-point boundary value problems*, Appl. Math. Lett. **14** (2001), 1–5.

[79] R. Ma and H. Wang, *Positive solutions of nonlinear three-point boundary-value problems*, J. Math. Anal. Appl. **279** (2003), 216–227.

[80] S. Marano, *A remark on a second order three-point boundary value problem*, J. Math. Anal. Appl. **183** (1994), 518–522.

[81] S. Marano, *A remark on a third-order three-point boundary value problem*, Bull. Austral. Math. Soc. **49** (1994), 1–5.

[82] J. Mawhin, *Topological Degree Methods in Nonlinear Boundary Value Problems*, NSF-CBMS Reg. Conf. Ser. Math., Vol. 40, Amer. Math. Soc., Providence, RI (1979).

[83] M. Moshinsky, *Sobre los problemas de condiciones a la frontiera en una dimension de caracteristicas discontinuas*, Bol. Soc. Mat. Mexicana **7** (1950), 1–25.

[84] K. Murty and S. Sivasundaram, *Existence and uniqueness of solutions to three-point boundary value problems associated with nonlinear first order systems of differential equations*, J. Math. Anal. Appl. **173** (1993), 158–164.

[85] X. Ni and W. Ge, *Multi-point boundary value problems for the p-Laplacian at resonance*, Electron. J. Differential Equations **112** (2003), 7 pp. (electronic).

[86] S.K. Ntouyas and P.Ch. Tsamatos, *Global existence for semilinear evolution equations with nonlocal conditions*, J. Math. Anal. Appl. **210** (1997), 679–687.

[87] P.K. Palamides, *Positive and monotone solutions of an m-point boundary value problem*, Electron. J. Differential Equations **18** (2002), 1–16 pp. (electronic).

[88] B. Przeradski and R. Stanczy, *Solvability of a multi-point boundary value problem at resonance*, J. Math. Anal. Appl. **264** (2001), 253–261.

[89] Y. Raffoul, *Positive solutions of three-point nonlinear second order boundary value problem*, Electron. J. Qual. Theory Differ. Equ. **15** (2002), 11 pp. (electronic).

[90] S. Timoshenko, *Theory of Elastic Stability*, McGraw-Hill, New York (1961).

[91] J. Wang and D. Zheng, *On the existence of positive solutions to a three-point boundary value problem for the one-dimensional p-Laplacian*, Z. Angew. Math. Mech. **77** (1997), 477–479.

[92] J.R. Webb, *Positive solutions of some three-point boundary value problems via fixed point index theory*, Nonlinear Anal. **47** (2001), 4319–4322,

[93] M. Zhang and Y. Han, *On the applications of Leray–Schauder continuation theorem to boundary value problems of semilinear differential equations*, Ann. Differential Equations **13** (1997), 189–207.

[94] Z. Zhang and J. Wang, *The upper and lower solution method for a class of singular nonlinear second order three-point boundary value problems*, J. Comput. Appl. Math. **147** (2002), 41–52.

[95] Z. Zhang and J. Wang, *Positive solutions to a second order three-point boundary value problem*, J. Math. Anal. Appl. **285** (2003), 237–249.

Author Index

Roman numbers refer to pages on which the author (or his/her work) is mentioned. Italic numbers refer to reference pages. Numbers between brackets are the reference numbers. No distinction is made between first and co-author(s).

Abbondandolo, A. 80, 106, 121, *142* [1]
Adams, R.A. 98, 126, 129, *142* [2]
Aftabizadeh, A. 489, *554* [1]
Agrachev, A.A. 3, 16, 18, 57, 67, *74* [1]; *74* [2]
Ahmad, S. 439, *459* [1]; *459* [2]
Albrecht, F. 297, *348* [1]
Alexandroff, A. 315, *348* [2]
Alikakos, N. 314, 321, 332, *348* [3]
Alvárez, C. 439, *459* [3]
Amann, H. 104, 106, 121, *142* [3]; *142* [4]; 275, 314, 315, 320, 332, 338, 340, 345, *348* [4]; *349* [5]; *349* [6]; *349* [7]; *349* [8]; *349* [9]; *349* [10]; *349* [11]; 398, *459* [4]
Ambrosetti, A. 80, *142* [5]
Amine, Z. 458, *459* [5]
Angenent, S. *349* [12]; *349* [27]
Arioli, G. 90, 129, 138, *142* [6]
Arnold, V.I. 5, 6, 8, 19, 27, *74* [3]
Aubin, J.P. 154, 188, 228, 229, 233, 234, *235* [1]; *235* [2]; *235* [3]; *235* [4]
Avery, R. 548, 553, *554* [2]; *554* [3]

Bahri, A. 91, *142* [7]; *142* [8]
Bai, C. 546, 554, *554* [4]; *554* [5]; *554* [6]
Barbu, V. 3, 11, 38, 47, 67, 70, *74* [4]; *74* [5]; *74* [6]; 230, *235* [5]
Bartsch, T. 88–90, 97, 107, 125, *142* [9]; *142* [10]; *142* [11]; *143* [12]; *143* [13]; *143* [14]; *143* [15]; *143* [16]
Batty, C. *349* [13]
Bebernes, W. 227, *235* [6]
Bellman, R. 3, *74* [7]; 223, *235* [7]; *235* [8]; *235* [9]
Benaïm, M. 296, 302, *349* [14]; *349* [15]; *349* [16]; *349* [17]
Benci, V. 80, 88, 97, 106, 121, *143* [17]; *143* [18]; *143* [19]

Berestycki, H. 91, 115, 121, *142* [7]; *142* [8]; *143* [20]
Berliocchi, H. 190, *235* [10]
Berman, A. 282, 284, 286–289, *349* [18]; *349* [19]
Birkhoff, G. 315, *349* [20]
Bitsadze, A. 470, *554* [7]; *554* [8]
Bliss, G.A. 4, *74* [8]
Bolotin, S.V. 139, *143* [21]
Boltyanskii, V.G. 29, *75* [30]; 223, *235* [11]; *235* [12]; *237* [83]
Bony, J.M. 153, *235* [13]
Bothe, D. 233, 234, *235* [14]; *235* [15]
Boucherif, A. 482, 484, *554* [9]
Bouguima, S. 482, 484, *554* [9]
Bouligand, H. 159, 160, *235* [16]
Bressan, A. 233, *235* [17]
Brezis, H. 11, 13, *74* [9]; 153, 154, 156, 157, *235* [18]; *235* [19]
Browder, F. 156, 157, *235* [19]
Brunovsky, P. 296, 314, *349* [21]; *349* [22]
Byszewski, L. 465, 469, *554* [10]; *554* [11]; *554* [12]

Calvert, B. 542, *554* [13]
Carathéodory, C. 223, *235* [29]
Cârjă, O. 154, 158, 176, 178, 190, 203, 210, 213, 225, 227, 228, 231, 233, 234, *235* [20]; *235* [21]; *235* [22]; *235* [23]; *235* [24]; *235* [25]; *235* [26]; *235* [27]; *235* [28]
Castaneda, N. 533, *557* [78]
Cellina, A. 154, 188, 228, 229, 233, 234, *235* [3]
Cerami, G. 96, *143* [22]
Cesari, L. 223, *236* [30]
Chang, K.C. 81, 82, 106, 110, 121, 124, *143* [23]
Chen, M. 314, *349* [24]
Chen, X. 314, *349* [24]
Chen, X.-Y. 314, *349* [23]

Chiş-Şter, I. 232, *236* [31]
Cholewa, J. 332, 345, *349* [25]
Chow, S.N. 331, *349* [26]
Clarke, F.H. 3, 14, 15, 32, 50, 73, *74* [10]; *74* [11];
    74 [12]; *74* [13]; *74* [14]; 163, 175, 223, 226,
    236 [32]; *236* [33]; *236* [34]; *236* [35]
Clements, Ph. *349* [27]
Coddington, E.A. 16, *74* [15]; *349* [28]
Cohen, D.S. 314, *352* [95]
Colombo, F. 340, *349* [29]
Conley, C. 124, *143* [24]; 293, *349* [30]; *349* [31]
Coppel, W. *349* [32]
Cornet, B. 166, *236* [36]
Cosner, C. 345, *349* [33]
Coti Zelati, V. 80, 126, 135, 136, 138–140,
    142 [5]; *143* [25]; *143* [26]
Courant, R. 7, 8, *74* [16]; 223, *236* [37]; 314,
    349 [34]
Crandall, M.G. 68, 70, *74* [17]; *75* [18]; *75* [19];
    75 [20]; 151, 152, 154, 233, *236* [38]; *236* [39];
    371, 404, 407, *459* [6]
Cushing, J.M. 392, 439, *459* [7]

Da Prato, G. 70, *74* [5]
Dafermos, C.M. 327, *350* [35]
Dancer, E.N. 316, 318, 320–322, 332, *350* [36];
    350 [37]; *350* [38]; 392, 397, 458, *459* [8];
    459 [9]; *459* [10]
Dancer, N. 107, 109, *143* [27]
de Leenheer, P. 302, *350* [41]
de Mottoni, P. 314, 329–331, *350* [42]; 438, 454,
    460 [29]
de Pagter, B. *349* [27]
DeAngelis, D.L. 242, *350* [39]
Degiovanni, M. 121, *143* [28]
Deimling, K. *350* [40]
Delgado, M. 438, *459* [11]
Ding, Y.H. 97, 129, 132, 135, 136, 138, 140,
    143 [12]; *143* [13]; *143* [29]; *143* [30]; *143* [31]
Dlotko, T. 332, 345, *349* [25]
Dugundji, J. 482, 483, *554* [14]
Dunford, N. 201, 234, *236* [40]

Eden, A. *350* [43]
Eilbeck, J.C. 363, 384, 439, 452, *459* [12];
    459 [13]
Ekeland, I. 15, 50, *74* [13]; *75* [21]; 80, 115, 118,
    126, 135, 139, 140, *143* [25]; *143* [32];
    143 [33]; *143* [34]; *143* [35]
Elsner, L. 286, *350* [44]
Evans, L.C. 68, 70, *74* [17]

Fadell, E.R. 88, 109, 121, *143* [36]

Fang, J. 546, 554, *554* [4]; *554* [5]; *554* [6]
Federer, H. 161, *236* [41]
Feferman, S. 156, *236* [42]
Felmer, P. 118, 124, *143* [37]; *143* [38]; *144* [39]
Feng, W. 497, 505, *554* [15]; *554* [16]
Fife, P. *350* [49]; *350* [50]
Filippov, A.F. 213, *236* [43]
Fleckinger-Pellé, J. 327, *350* [45]; *350* [46]
Fleming, W.H. 3, 73, *75* [22]
Foias, C. *350* [43]
Fomin, S.V. 4, 8, *75* [24]
Fonda, A. 124, *144* [40]
Fournier, G. 92, 94, 124, *144* [42]
Frankowska, H. 190, 228, 234, *235* [4]; *236* [44];
    236 [45]
Freedman, H.I. 320, 313, *356* [235]; *357* [236]
Friedman, A. *350* [47]; *350* [48]
Frobenius, G. 315, *350* [51]
Fryszkowski, A. 233, *236* [46]
Fukaya, K. 125, *144* [41]
Fukuhara, M. 219, 220, *236* [47]
Furter, J.E. 384, *459* [13]

Gamkrelidze, R.V. 16, 29, *74* [1]; *75* [23]; *75* [30];
    223, *237* [83]
Gao, Y. 547, *554* [20]
Garcia-Huidobro, M. 554, *554* [17]
Gatzke, H. 297, *348* [1]
Gautier, S. 228, *236* [48]
Ge, W. 540–542, 546, 547, 553, 554, *554* [19];
    554 [20]; *555* [23]; *555* [41]; *555* [42];
    556 [65]; *556* [66]; *557* [85]
Gelfand, I.M. 4, 8, *75* [24]
Ginzburg, V.L. 110, 113, *144* [43]; *144* [44];
    144 [45]
Girardi, M. 115, 136, 138, *143* [30]; *144* [46]
Girsanov, I.V. 161, *236* [49]
Glas, D. 254, *350* [55]
Gonzales, R. 223, *236* [50]
Goursat, E. 223, *236* [51]
Graef, J. 542, *554* [18]
Granas, A. 482, 483, *554* [14]
Greiner, G. *350* [52]
Grossberg, S. 297, *350* [53]; *350* [54]
Guckenheimer, J. 80, 138, 142, *144* [47]
Guo, D. 519, 525, 536, *554* [21]
Guo, Y.X. 121, *144* [48]; 541, 542, 546, 547,
    554 [19]; *554* [20]; *554* [22]; *555* [23]; *556* [64]
Gupta, C.P. 481, 489, 494, 495, 497, 505, 513,
    516, 542, 554, *554* [1]; *554* [13]; *554* [17];
    555 [24]; *555* [25]; *555* [26]; *555* [27];
    555 [28]; *555* [29]; *555* [30]; *555* [31];
    555 [32]; *555* [33]; *555* [34]; *555* [35];
    555 [36]; *555* [37]; *555* [38]; *555* [39]; *555* [40]

Gürel, B.Z. 113, *144* [44]

Haddad, A. 297, *348* [1]
Haddad, G. 227, *236* [52]
Hadeler, K. 254, *350* [55]
Hájek, O. 223, *236* [53]
Hale, J.K. 301–303, 305, 307, 309, 313, 314, 329,
    331, 332, 344, *349* [24]; *349* [26]; *350* [56];
    350 [57]; *350* [58]; *350* [59]; *350* [60];
    350 [61]; 439, *459* [14]
Hall, G.R. 80, *145* [75]
Han, Y. 482, *557* [93]
Hartman, P. 151, 152, 220, *236* [54]; *236* [55]
He, X. 540, 553, *555* [41]; *555* [42]
Heijmans, H. *349* [27]
Henderson, J. 553, *554* [3]
Henry, D. 332–334, *350* [62]
Herman, M. 113, *144* [49]
Hess, P. 314, 316, 318, 320, 321, 329, 332,
    348 [3]; *350* [38]; *350* [63]; *350* [64]; *350* [65];
    439, *459* [15]
Hilbert, D. 7, 8, *74* [16]; 223, *236* [37]; 314,
    349 [34]
Hirsch, M.W. 242, 247, 249, 250, 254–256, 258,
    264, 265, 269, 271, 272, 275, 281, 291, 292,
    296, 297, 299, 300, 302, 317, 319, *349* [16];
    349 [17]; *351* [66]; *351* [67]; *351* [68];
    351 [69]; *351* [70]; *351* [71]; *351* [72];
    351 [73]; *351* [74]; *351* [75]; *351* [76];
    351 [77]; *351* [78]; *351* [79]; *351* [80]
Hobson, E.W. 175, *236* [56]
Hofbauer, J. 301, 302, *351* [81]; *351* [82]
Hofer, H. 80, 113, 115, 125, *144* [50]; *144* [51];
    144 [52]
Holmes, P. 80, 138, 142, *144* [47]
Hopf, H. 315, *348* [2]
Howard, P. 156, *236* [57]
Hsu, S.-B. 302, 320, 332, *351* [83]; *351* [84]

Il'in, V. 470, *555* [43]; *555* [44]
Infante, G. 532, 542, *555* [45]; *555* [46]; *555* [47]
Ishii, H. 70, *75* [18]; *75* [19]; 223, *236* [58]
Isidori, A. 3, *75* [25]
Izydorek, M. 121, 124, *144* [53]; *144* [54]

Jeanjean, L. 133, *144* [55]
Jentzsch, R. 315, *351* [85]
Jiang, J. 265, 296, 314, 319, 332, *351* [86];
    351 [87]; *351* [88]; *351* [89]; *351* [90];
    352 [120]; *352* [121]; *356* [229]; *356* [230];
    356 [231]
Jurdjevic, V. 3, 67, *75* [26]

Kamke, E. 223, *237* [59]; 241, 285, *351* [91]
Karakostas, G.L. 542, 547, *556* [48]; *556* [49];
    556 [50]; *556* [51]; *556* [52]; *556* [53]
Kato, T. 274, 275, *351* [92]
Keller, H.B. 314, *351* [93]; *352* [94]; *352* [95]
Kenmochi, N. 181, *237* [60]
Kerman, E. 110, *144* [45]
Kerscher, W. *352* [96]
Kneser, H. 219, 221, *237* [61]
Kobayashi, J. 213, *237* [62]
Kolmogorov, A. 297, *352* [97]
Krasnosel'skii, M.A. 88, *144* [56]; 314, 323, 324,
    352 [98]; *352* [99]; *352* [100]; *352* [101]; 363,
    411, *459* [16]; 519, *556* [54]
Krause, U. 315, 323, 324, 327, *352* [102];
    352 [103]
Krein, M.G. 274, 315, *352* [104]
Krisztin, T. 303, 313, *352* [105]; *352* [106];
    352 [107]; *352* [108]
Kryszewski, W. 95, 106, 121, 136, *144* [57];
    144 [58]
Kucia, A. 190, *237* [63]
Kunisch, K. 305, *352* [109]
Kunze, H. 289, *352* [110]; *352* [111]; *352* [112]
Kuznetsov, Y.A. 328, *352* [113]

Ladyzhenskaya, L. 314, *352* [100]; *352* [114]
Lakshmikantham, V. 465, 489, 519, 525, 536,
    554 [12]; *554* [21]; *555* [31]
LaSalle, J.P. 316, *352* [116]
Lasry, J.M. 115, 121, *143* [20]; *143* [35]; 190,
    235 [10]
Lassoued, L. 115, *143* [34]
Lazer, A.C. 332, *350* [64]; 439, *459* [2]; *459* [3];
    459 [15]
Ledyaev, Yu.S. 175, 226, 228, *236* [33]; *236* [34];
    237 [64]
Legget, R. 536, 548, *556* [55]
Lemmens, B. 315, *352* [117]
Lemmert, R. 284, *352* [118]
Leung, A.W. *352* [119]
Levinson, N. 16, *74* [15]; *349* [28]
Li, G. 96, *144* [59]
Li, M. 301, *352* [115]
Li, S.J. 83, 86, 91, 106, 118, 121, *144* [60];
    144 [61]; *144* [62]
Li, W. 301, *357* [237]
Li, Y. 54, *75* [36]
Li, Y.Q. 92, *144* [63]
Liang, X. 314, 332, *352* [120]; *352* [121]
Lieb, E.H. 129, *144* [64]
Lions, J.-L. 3, *75* [27]
Lions, P.-L. 68, 70, *74* [17]; *75* [18]; *75* [19];
    75 [20]; 95, 129, *144* [65]; 233, *236* [39]

Liu, B. 516, 517, 525, 529, 532, 541, *556* [56];
  556 [57]; *556* [58]; *556* [59]; *556* [60];
  556 [61]; *556* [62]
Liu, C.G. 115, *144* [67]
Liu, J.Q. 83, 91, 106, 121, 124, *144* [60]; *144* [61];
  144 [66]
Liu, X. 541, 542, 547, *554* [22]; *556* [63]; *556* [64]
Liu, Y. 542, 553, *556* [65]; *556* [66]
Loewy, R. *353* [122]
Long, Y. 80, 106, 115, 118, 121, *144* [67];
  145 [68]; *145* [69]; *145* [70]; *145* [71]
López-Gómez, J. 363, 384, 390, 392, 394–397,
  404, 411, 438, 439, 452, 458, *459* [10];
  459 [11]; *459* [12]; *459* [13]; *459* [17];
  459 [18]; *459* [19]; *459* [20]; *459* [21];
  459 [22]; *459* [23]; *459* [24]; *459* [25]; *459* [26]
Loss, M. 129, *144* [64]
Lui, R. 315, *353* [123]
Lunardi, A. 332, 334, 339–342, 345, *353* [124]
Lupo, D. 92, 94, 124, *144* [42]
Lupulescu, V. 154, *237* [65]
Lyapunov, A.M. 106, *145* [72]

Ma, R. 475, 525, 532, 533, 535, 540–542,
  556 [67]; *556* [68]; *556* [69]; *556* [70];
  556 [71]; *556* [72]; *556* [73]; *556* [74];
  557 [75]; *557* [76]; *557* [77]; *557* [78]; *557* [79]
Manasevich, R. 554, *554* [17]
Mancini, G. 115, 121, *143* [20]
Marano, S. 481, 489, *557* [80]; *557* [81]
Marsden, J.E. 6, 19, *75* [28]
Martin, R.H. 305, 332, 335–338, 343, *353* [125];
  353 [126]; *353* [127]; *353* [128]; *353* [129];
  353 [130]
Martin, R.H., Jr. 153, 154, *237* [66]
Matano, H. 247, 314, 320, 321, 332, *348* [3];
  349 [23]; *353* [131]; *353* [132]; *353* [133];
  353 [134]; *353* [135]
Matzeu, M. 115, *144* [46]
Mawhin, J. 80, 89, 124, *144* [40]; *145* [73]; 363,
  459 [27]; *460* [28]; 474, 476, 477, 486, 490,
  491, 498, *557* [82]
May, R. 315, *353* [136]
McDuff, D. 80, *145* [74]
McShane, E.J. 293, *353* [137]
Meyer, K.R. 80, *145* [75]
Michael, E. 212, 229, *237* [67]; *237* [68]
Mierczyński, J. 265, 280, 296, 327, 328,
  353 [138]; *353* [139]; *353* [140]; *353* [141];
  353 [142]; *353* [143]
Miller, R.K. 320, *357* [236]
Mincheva, M. 345, *353* [144]; *353* [145]
Mischaikow, K. *353* [146]

Mishchenko, E.F. 29, *75* [30]; 223, *237* [83]
Moiseev, E. 470, *555* [43]; *555* [44]
Molina-Meyer, M. 363, 390, *459* [21]; *459* [22];
  459 [23]; *459* [24]
Monteiro Marques, M.D.P. 190, 227, 228, 231,
  235 [21]; *235* [22]; *235* [23]
Mora, X. 332, 340, *353* [147]
Mordukhovich, B.S. 153, *237* [69]
Moreau, J. 13, *75* [29]
Moser, J. 106, 107, *145* [76]
Moshinsky, M. 470, *557* [83]
Muldowney, J. 301, *352* [115]
Müller, M. 241, 285, *353* [148]
Murty, K. 475, *557* [84]

Nagel, R. *352* [96]
Nagumo, M. 151, 166, *237* [70]
Necula, M. 154, 163, 178, 231, 233, *235* [24];
  237 [65]; *237* [71]; *237* [72]; *237* [73]; *237* [74]
Neumann, M. 282, 284, 286–289, *349* [18]
Ni, X. 554, *557* [85]
Nicolaenko, B. *350* [43]
Ntouyas, S.K. 469, 481, 494, 497, 505, 513,
  555 [32]; *555* [33]; *555* [34]; *555* [35];
  555 [36]; *557* [86]
Nussbaum, R.D. 315, 327, *352* [102]; *352* [117];
  354 [149]; *354* [150]; *354* [151]

Ohta, Y. 305, *354* [152]
Olian Fannio, L. 121, *143* [28]
Ono, K. 125, *144* [41]
Ortega, R. 302, *354* [153]; 439, 447, 458, *459* [5];
  459 [10]; *459* [25]; *460* [30]
Oster, G. 315, *353* [136]
Oxtoby, J. 257, *354* [154]

Palamides, P.K. 547, *557* [87]
Palmer, K.J. 138, 140, *145* [77]
Pao, C.V. 341, 343, *354* [155]
Pardo, R. 458, *459* [26]
Pavel, N.H. 230–232, 234, *235* [5]; *237* [75];
  237 [76]; *237* [77]; *237* [78]; *237* [79]
Pazy, A. 333, *354* [156]
Peano, G. 149, *237* [80]
Peixoto, M. 246, *354* [157]
Perron, O. 223, *237* [81]; 315, *354* [158]
Pituk, M. 313, *354* [159]
Plaskacz, S. 190, 227, 228, *236* [45]; *237* [82]
Plemmons, R. 287–289, *349* [19]
Poláčik, P. 280, 281, 314, 320, 329, 332, 343, 345,
  349 [22]; *350* [65]; *354* [160]; *354* [161];
  354 [162]; *354* [163]; *354* [164]; *354* [165]
Pontryagin, L.S. 29, *75* [30]; 223, *237* [83]
Post, W.M. 242, *350* [39]

Precupanu, Th. 11, 47, *74* [6]
Protter, M.H. 346, *354* [166]
Przeradski, B. *557* [88]

Qian, C. 542, *554* [18]
Qiu, J. 541, 542, 547, *554* [22]; *556* [64]
Quincampoix, M. 161, *237* [84]

Rabinowitz, P.H. 79, 80, 82, 87, 88, 99, 100, 109,
    111, 118, 121, 136, 138, 139, *143* [19];
    143 [21]; *143* [26]; *143* [36]; *145* [78];
    145 [79]; *145* [80]; *145* [81]; 363, 371, 392,
    395–397, 404, 407, *459* [6]; *460* [31]
Raffoul, Y. 535, *557* [89]
Ramos, M. 92, 94, 124, *144* [42]
Ranft, P. 323, 324, 327, *352* [103]
Ratiu, T.S. 6, 19, *75* [28]
Raugel, G. *354* [167]
Redheffer, R. 153, 154, *237* [85]
Rescigno, A. 297, *354* [168]
Richardson, I. 297, *354* [168]
Rishel, R.W. 3, 73, *75* [22]
Robinson, C. *349* [13]; *354* [169]
Robinson, J. *354* [170]
Rockafellar, R.T. 11, 14, 47, *75* [31]; *75* [32]
Roxin, E. 174, *237* [86]
Rubin, J.E. 156, *236* [57]
Rudin, W. 306, *354* [171]
Ruelle, D. *354* [172]
Ruf, B. 115, 121, *143* [20]
Rutman, M.A. 274, 315, *352* [104]; *354* [173]
Rybicki, S. 107, 109, *143* [27]
Rzeżuchowski, T. 190, 228, *236* [45]

Sachkov, Y.L. 3, 16, 18, 57, 67, *74* [2]
Sacker, R. 332, *354* [174]
Salamon, D. 80, *145* [74]
Samarskii, A. 470, *554* [8]
Sánchez, L. 302, *354* [153]
Sandholm, W. 302, *351* [81]
Sandstede, B. 314, *349* [22]
Saperstone, S.H. 246, *354* [175]
Sattinger, D. 314, *354* [176]
Schappacher, W. 305, *352* [109]
Schiaffino, A. 314, 329–331, *350* [42]; 438, 454,
    460 [29]
Schneider, H. 282, 284, 286, 301, *353* [122];
    354 [177]
Schuur, I.D. 227, *235* [6]
Schwartz, J.T. 201, 234, *236* [40]
Scorza Dragoni, G. 190, *237* [87]; *237* [88]
Secchi, S. 135, *145* [82]

Selgrade, J. 297, 315, *354* [178]; *355* [179];
    355 [180]; *355* [181]
Sell, G. 332, *354* [174]
Séré, E. 126, 135, 139, 140, 142, *143* [25];
    145 [83]
Severi, F. 159, 222, *237* [89]; *238* [90]
Shan, W. 541, *555* [23]
Shi, S.Z. 231, 232, *238* [91]
Siegel, D. 289, 345, *352* [110]; *352* [111];
    352 [112]; *353* [145]
Sivasundaram, S. 475, *557* [84]
Sleeman, B.D. 332, *357* [246]
Slemrod, M. 327, *350* [35]
Smale, S. 294, 297, *351* [80]; *355* [182]
Smith, H.L. 242, 247, 249, 250, 255, 266, 267,
    269, 271, 272, 281, 293, 296, 297, 299, 301,
    302, 304, 305, 311–315, 318, 320, 327–329,
    332, 337, 343, 345, 346, *350* [41]; *351* [83];
    353 [129]; *353* [130]; *353* [146]; *355* [183];
    355 [184]; *355* [185]; *355* [186]; *355* [187];
    355 [188]; *355* [189]; *355* [190]; *355* [191];
    355 [192]; *355* [193]; *355* [194]; *355* [195];
    355 [196]; *355* [197]; *355* [198]; *355* [199];
    355 [200]; *355* [201]; *355* [202]; *355* [203];
    357 [248]; 439, *460* [32]; *460* [33]
Smith, R.A. 302, *355* [204]
Smoller, J. 345, *355* [205]
So, J.W.-H. 301, *351* [82]
Somolinos, A.S. 314, 329, 332, *350* [60]; 439,
    459 [14]
Sontag, E.D. 3, *75* [33]
Spanier, E.H. 83, *145* [84]
Staicu, V. 233, *235* [17]
Stanczy, R. *557* [88]
Stern, R.J. 175, 226, *236* [33]; *236* [34]; 282,
    284–289, *349* [18]; *355* [206]
Stewart, H. *355* [207]
Struwe, M. 87, 113, *145* [85]; *145* [86]
Stuart, C.A. 126, 131, 135, *145* [82]; *145* [87]
Suárez, A. 438, *459* [11]
Szlenk, W. 142, *145* [88]
Szulkin, A. 86, 90, 92, 95, 96, 106, 109, 115, 118,
    121, 124, 129, 135, 136, 138, *142* [6]; *144* [57];
    144 [58]; *144* [59]; *144* [62]; *145* [89];
    145 [90]; *145* [91]; *145* [92]; *145* [93]; *145* [94]

Takáč, P. 274, 281, 315, 316, 318, 321, 322, 327,
    328, 332, *350* [45]; *350* [46]; *356* [208];
    356 [209]; *356* [210]; *356* [211]; *356* [212];
    356 [213]; *356* [214]; *356* [215]; *356* [216]
Takahashi, T. 181, *237* [60]
Tallos, P. 228, *238* [92]
Tanaka, K. 135, *145* [95]

Tang, M. *350* [50]

Temam, R. *350* [43]

Tereščák, I. 329, *354* [164]; *354* [165]; *356* [217]

Thieme, H.R. 247, 250, 255, 266, 267, 269, 271, 272, 281, 304, 313, 315, 320, 332, *353* [146]; *355* [197]; *355* [198]; *355* [199]; *355* [200]; *355* [201]; *356* [218]

Thompson, A.C. 324, *356* [219]

Tikhomirov, V.M. 4, *75* [34]

Timoshenko, S. 470, *557* [90]

Tineo, A. 296, *356* [220]; 439, 447, 458, 459 [25]; *460* [30]; *460* [34]; *460* [35]

Travis, C.C. 242, *350* [39]

Treiman, J.S. 166, *238* [93]

Trofimchuk, S. 494, 495, *555* [37]; *555* [38]; *555* [39]; *555* [40]

Tsamatos, P.Ch. 469, 481, 494, 497, 505, 513, 542, 547, *555* [32]; *555* [33]; *555* [34]; *555* [35]; *555* [36]; *556* [48]; *556* [49]; *556* [50]; *556* [51]; *556* [52]; *556* [53]; *557* [86]

Uhl, R. 282, *356* [221]

Ursescu, C. 151, 152, 154, 158, 161, 162, 166, 175, 176, 210, 213, 222, 225, 227, 228, 234, *235* [25]; *235* [26]; *237* [77]; *238* [94]; *238* [95]; *238* [96]; *238* [97]; *238* [98]; *238* [99]; *238* [100]; *238* [101]

Uryson, P.S. 314, *356* [222]

van den Driessche, P. 296, 301, *356* [223]

van Duijn, C. *349* [27]

Verduyn Lunel, S.M. 302, 303, 305, 307, 309, 315, *350* [61]; *352* [117]

Vespri, V. 340, *349* [29]

Vidyasagar, M. 282, 284, 286, 301, *354* [177]

Vinter, R.B. 73, *74* [14]; 223, *236* [35]

Viterbo, C. 113, 115, *145* [96]; *146* [97]

Volkmann, P. 282, 284, *352* [118]; *356* [224]

Vrabie, I.I. 21, *75* [35]; 151, 152, 154, 171, 173, 174, 177, 178, 228, 230–234, *235* [24]; *235* [27]; *235* [28]; *237* [74]; *237* [78]; *237* [79]; *238* [102]; *238* [103]; *238* [104]; *238* [105]; *238* [106]

Vulikh, B. 338, *356* [225]

Walcher, S. 282, *356* [226]

Walter, W. 282, 289, 343, *356* [227]; *356* [228]

Walther, H.-O. 303, 313, *352* [105]

Waltman, P. 249, 302, 320, 332, 343, *351* [83]; *351* [84]; *355* [202]; *355* [203]

Wang, H.Z. 54, *75* [36]; 532, *557* [79]

Wang, J. 533, 553, *557* [91]; *557* [94]; *557* [95]

Wang, Y. 296, 314, 332, *356* [229]; *356* [230]; *356* [231]

Wang, Z.-Q. 125, *143* [14]; *143* [15]; *144* [39]

Wax, N. 297, *348* [1]

Ważewski, T. 227, *238* [107]

Webb, J.R.L. 497, 505, 532, 542, *554* [15]; *554* [16]; *555* [46]; *555* [47]; *557* [92]

Weinberger, H.F. 315, 346, *354* [166]; *356* [232]

Weinstein, A. 106, 107, 111, *146* [98]; *146* [99]

Wiggins, S. 138, 140, *146* [100]

Wilder, R. 322, *356* [233]

Willem, M. 80–82, 89, 90, 92, 94, 99, 124, 127, 129, 132, 135, 140, *143* [16]; *143* [31]; *144* [42]; *145* [73]; *146* [101]; *146* [102]

Williams, L. 536, 548, *556* [55]

Wolenski, P.R. 226, *236* [34]

Wolkowicz, H. 285, *355* [206]

Wu, J. 303, 313, 320, *352* [105]; *352* [106]; *352* [107]; *352* [108]; *356* [234]; *356* [235]; *357* [236]

Wysocki, K. 115, *144* [50]

Xiao, D. 301, *357* [237]

Xu, J.-M. 489, *554* [1]

Xu, X. 118, 135, 138, *145* [70]; *146* [103]

Yang, B. 542, *554* [18]

Yorke, J.A. 151–154, 217, 220, *238* [108]; *238* [109]

Yoshizawa, T. 216, *238* [110]

Yosida, S. 47, *75* [37]

Yu, J. 516, *556* [61]; *556* [62]

Yu, S. 265, 319, *351* [90]

Zabczyk, J. 3, 67, *75* [38]

Zabreiko, P.P. 314, *352* [101]

Zanolin, F. 332, *357* [238]; 439, *460* [36]

Zaremba, S.C. 227, *238* [111]

Zeeman, E.C. 296, 301, *357* [239]; *357* [240]; *357* [241]; *357* [242]

Zeeman, M.L. 296, 301, *356* [223]; *357* [240]; *357* [241]; *357* [242]; *357* [243]

Zehnder, E. 80, 104, 106, 113, 115, 121, 124, 125, *142* [3]; *142* [4]; *143* [24]; *144* [50]; *144* [51]; *144* [52]

Zeidler, E. 274, *357* [244]

Zhang, M. 482, *557* [93]

Zhang, Z. 533, *557* [94]; *557* [95]

Zhao, X.-Q. 314, 332, *357* [245]; *357* [246]

Zheng, D. 553, *557* [91]

Zhu, C. 115, *144* [67]; *145* [71]

Zhu, H.-R. 302, *357* [247]; *357* [248]

Ziehe, M. 315, *355* [181]

Zou, W. 121, 135, *145* [93]; *145* [94]

Subject Index

τ-topology, 95

accessible point, 244
adapted space, 335
admissible pair, 9
Arnold conjecture, 125
asymptotically order stable, 262
attainable set, 55
attract, 245
attractor, 245
axiom
– of choice, 156
– of countable choice, 156
– of dependent choice, 156

Banach lattice, 338
basin, 258
Bernoulli shift, 140
Bochner integral, 21
Bolza problem, 9, 30, 40, 56
Brezis–Browder ordering principle, 157

C_0-semigroup, 229
Carathéodory function, 471
Cartan formula, 24
Cerami sequence, 96
chain recurrent, 293
chronological exponential
– left, 22
– right, 19
classical solution, 468
closed loop system, 62
closed order interval, 244
cocycle identities, 333
coexistence state, 361, 401
compact map, 335
comparison property
– Carathéodory, 198
– of a function, 180
– of a multifunction, 212
competing species, 379, 438
competitive map, 329
completely continuous map, 266

component of coexistence states, 404
component of positive solutions, 394
concave functional, 536
conditionally completely continuous map, 266
cone, 536
– conjugate, 163
– contingent, 160
– in the sense of Bony, 162
– normal, 11, 15
– proximal normal, 161
– tangent, 15
conjugate mapping, 141
conjugate point, 8, 67
Conley–Zehnder index, 106
contingent derivative, 227
contraction, 325
control, 9
– bang-bang, 9, 47
– closed loop, 62
– constraints, 9
– feedback, 62
– open loop, 62
– optimal, 9, 55, 62
convergent, 246, 317
cooperative map, 329
cost functional, 9
critical orbit, 88
crossing number, 393
curve of change of stability, 376–378
curve of neutral stability, 376, 377

deformation lemma, 81
derivation, 18
derivative
– contingent, 227
– Dini right lower, 175
– right directional
– – of the norm, 162
diameter, 258
directional derivative, 11, 14
domain, 149
doubly accessible point, 257
dual cone, 281

eigenvalue, 393
Ekeland variational principle, 15
entire orbit, 320
equation
– Bellman, 63
– dynamic programming, 63
– eikonal, 72
– Euler–Lagrange, 5
– generalized Bellman, 224
– Hamilton, 6, 27
– Hamilton–Jacobi, 7, 63
– semilinear differential, 230
equilibrium, 245, 316
equivariant mapping, 87
escape time, 334
eventually strongly monotone, 247
eventually strongly monotone map, 316
extremal, 5

falling interval, 249, 317
five functional fixed point theorem, 548
fixed point, 316
– index, 393
– index theorem, 525
– set, 87
Floquet multipliers, 375, 377, 415
flow, 17, 245
– Hamiltonian, 27
function
– Carathéodory comparison, 197
– comparison, 154
– conjugate, 12
– convex, 10
– effective domain, 10
– epigraph, 10
– integrable, 21
– locally Lipschitz, 14
– lower semicontinuous, 10
– of Carathéodory type, 190
– positively sublinear, 173
– proper, 10
– regulated, 229
– Severi differentiable, 222
– strongly measurable, 21
– synthesis, 63
– weakly measurable, 21
functional ordering, 245
fundamental matrix, 414

generalized gradient, 14
genus, 86
geometrically distinct homoclinic solutions, 128
geometrically distinct periodic solutions, 100
global attractor, 246

global process, 333
group representation, 87

Hamiltonian, 6, 27
Hamiltonian lift, 28
high intensity competition, 384
homoclinic solutions, 126
– geometrically distinct, 128
hyperbolic point solution, 126
hysteresis point, 431

index, 88
infimum, 244
infinitesimal generator, 229
invariant, 245
– forward, 152
– local, 152
– locally right, 152
– mapping, 87
– right, 152
– set, 87
invariantly connected set, 316
isometric group representation, 87

k-bump solution, 138
K-competitive, 289
K-cooperative, 285
K-irreducible matrix, 288
K-positive matrix, 284
Krasnoselskii's fixed point theorem, 519
Krein–Rutman operator, 274

Lagrange problem, 9, 56
Lebesgue point, 34
Legendre transform, 6, 12
Legget–Williams fixed point theorem, 536
lemma of Cârjă–Ursescu, 177, 210
limit genus, 92
limit index, 92
Liouville formula, 414
Lipschitz system, 292
local linking, 83
local semiflow, 251
locally closed, 335
locally closed set, 86
locally Hölder map, 333
locally Lipschitz map, 333
locally monotone map, 290
locally strongly monotone map, 290
logistic equation, 364
Lotka–Volterra periodic system, 362
low intensity competition, 384
lower boundary, 244

$\mathcal{M}$-maximal element, 157
m-point BVP, 470
mapping
– positively sublinear, 208
Maslov index, 106
maximal element, 244
Mayer problem, 9, 56
mild solution, 334, 468
minimal element, 244
minimal set, 269
monotone map, 316
monotone system, 282
multi-point BVP, 470
multibump solution, 138
multifunction, 164
– l.s.c. at a point, 200
– l.s.c. on a set, 200
– u.s.c. at a point, 200
– u.s.c. on a set, 200

necessary conditions, 5, 29
– Jacobi, 8
– Legendre, 6
– Weierstrass–Erdmann, 6
negative semiorbit, 289
non-cooperative periodic system, 456
nonexpansive map, 325
nonlocal BVP, 470
nonlocal condition, 463
nonlocal initial value problem, 465
nonresonance, 489
nonwandering set, 319
normality constant, 259
normally ordered space, 259

omega limit set, 246
open order interval, 244
operator pencil, 393
optimal control problem, 9
optimal pair, 9
optimal state, 9
orbit, 87, 245, 333
– of a set, 266
order bounded, 244
order compact, 322
order neighborhood, 261
order norm, 261
order related, 244
order stable, 262
order topology, 261
ordered Banach space, 245
ordered space, 243
ordered subspace, 244

p-convex, 245
Palais–Smale condition, 81
Palais–Smale sequence, 81
period, 245
periodic, 245, 316
periodic solutions
– geometrically distinct, 100
Poincaré map, 377, 414, 417, 441
point dissipative, 329
Poisson bracket, 27
polyhedral, 281
positive operator, 316
positive semiorbit, 289
positive solution, 519
positively invariant, 245, 333
positivity properties, 364
predator–prey model, 386, 455
preorder, 176
– closed, 176
principal characteristic exponent, 428, 449
principal eigenvalue, 417
problem
– abnormal, 30
– normal, 30
– of local invariance, 152
process, 333
projection
– subordinated to $\mathbb{V}$, 158
proposition
– Roxin, 174
proximal neighborhood, 158
(PS)-attractor, 96
$(PS)_c$-condition, 81
$(PS)^*$-condition, 91
$(PS)^*$-sequence, 91
pseudo-gradient vector field, 82
pseudo-symmetric function, 547

quasi-cooperative system, 440
quasiconvergent, 246, 317
quasiderivative map, 273
quasimonotone condition, 282
quasimonotone condition (QMD), 304

reachable set, 55
reaction–diffusion systems, 339
recurrence, 293
regularization, 13
relation
– preorder, 156
relative Morse index, 105, 106
representation
– diffeomorphisms, 18

– points, 18
– tangent vectors, 18
– vector fields, 18
reproducing cone, 322
resolvent operator, 366
resonance, 489
rising interval, 249, 317

sectorial, 333
semi-trivial positive state, 361
semiconjugate mapping, 141
semiflow, 245
semiorbit, 251
set
– reachable, 224
– target, 224
Severi differential
– at x, 222
– at x in the direction u, 222
singular value of an operator pencil, 393
solution
– ε-approximate, 168, 206
– Carathéodory, 190
– contingent, 227, 233
– flow, 245
– global, 172
– mild, or C^0, 229
– noncontinuable, 172, 207
– of a differential inclusion, 202
– of inequality, 224
– process, 335
– right funnel, 219
– semigroup, 229
– to a differential equation, 149
spectral radius, 274
stable fixed point, 320
stable from above, 258
stable from below, 258
standard cone, 245
state of the system, 9
strictly monotone, 246
strictly monotone map, 316
strong approximation, 265
strong maximum principle, 393
strong ordering, 244
strongly accessible, 265
strongly increasing, 370
strongly monotone, 246
strongly negative, 366
strongly order-preserving, 246
strongly order-preserving (SOP) map, 316
strongly positive, 366
strongly positive operator, 316
strongly sublinear map, 324

sub-critical turning point, 431
subdifferential, 11
subgradient, 11
sublinear map, 323
subset
– Carathéodory invariant, 197
– Carathéodory locally invariant, 197
– closed relative to $\mathbb{D}$, 151
– flow invariant, 153
– invariant, 178, 210
– locally closed, 151, 158
– locally invariant, 178, 210
– right viable with respect to f, 149
– right viable with respect to F, 202
– viable, 166
subsolution, 367, 368
super-critical turning point, 431
superlinear indefinite model, 388
supersolution, 367, 368
supremum, 244
symbiotic species, 387, 411
synthesis problem, 63
system
– strongly decreasing, 226
– weakly decreasing, 226

t-cross-section of $F_{\tau,\xi}$, 219
theorem
– Bony, 153
– Brezis–Browder, 157
– Hukuhara, 220
– Kneser, 221
– Nagumo, 151, 166
– Pavel, 231
– Peano, 149
– Redheffer, 154
– Scorza Dragoni, 190
– Shi, 232
– Vitali, 234
– Yorke, 217, 220
– Yoshizawa, 216
topological degree, 393
topologically equivalent, 292
totally competitive, 294
totally ordered arc, 255
trajectory, 245, 333
transversality conditions, 7, 30, 61
trivial state, 361

uniform global attractor, 321
unilateral bifurcation theorem, 397
unordered, 244
unstable from above, 268

unstable from below, 269
upper boundary, 244

variation of constants formula, 24
variation of parameters method, 24
vector
– metric normal to $\mathbb{K}$ at ξ, 153, 161
– tangent in the sense of
– – Bony, 162
– – Bouligand–Severi, 159
– – Clarke, 163

– – Federer, 161
vector field
– complete, 17
– Hamiltonian, 27
– nonautonomous, 16
vector ordering, 245
very strongly order preserving (VSOP) process, 336
viability, 149

Yosida approximation, 13